Haematology and Blood Transfusion
Hämatologie und Bluttransfusion 32

Modern Trends in Human Leukemia VIII

New Results in Clinical and Biological Research Including Pediatric Oncology

Organized on behalf of the Deutsche Gesellschaft für Hämatologie und Onkologie, Hamburg, June 18, 1988 and Wilsede, June 19–23, 1988

Wilsede Joint Meeting on Pediatric Oncology V
Hamburg, June 17, 1988

Edited by
Rolf Neth, Robert C. Gallo, Melvyn F. Greaves,
Gerhard Gaedicke, Sven Gohla,
Klaus Mannweiler, and Jörg Ritter

With 251 Figures and 105 Tables

Springer-Verlag
Berlin Heidelberg New York
London Paris Tokyo Hong Kong

Dr. Rolf Neth, Universitäts-Krankenhaus Eppendorf,
Medizinische Klinik, Abteilung für klinische Chemie,
Martinistraße 52, 2000 Hamburg 20, FRG

Dr. Robert C. Gallo, National Cancer Institute, Laboratory
of Tumor Cell Biology, Bethesda, MD 20892, USA

Dr. Melvyn F. Greaves, Leukemia Research Fund Centre, Institute of
Cancer Research, Chester Beatty Laboratories, Fulham Road,
London SW3 6JB, UK

Dr. Gerhard Gaedicke, Universitätskinderklinik und Poliklinik Ulm,
Abteilung Pädiatrie II, Universität Ulm, Prittwitzstr. 43, 7900 Ulm, FRG

Dr. Sven Gohla, Beiersdorf-AG, Labor für Hautimmunologie,
Bioverträglichkeit-Immunologie, Paul-Gerson-Unna-Forschungszentrum,
Unnastr. 48, 2000 Hamburg 20, FRG

Dr. Klaus Mannweiler, Heinrich-Pette-Institut für Experimentelle
Virologie und Immunologie, Martinistr. 52, 2000 Hamburg 20, FRG

Dr. Jörg Ritter, Kinderklinik und Poliklinik der Westfälischen
Wilhelms-Universität, Abteilung Hämatologie und Onkologie,
Albert-Schweitzer-Str. 33, 4400 Münster, FRG

ISBN-13:978-3-540-50967-7 e-ISBN-13:978-3-642-74621-5
DOI: 10.1007/978-3-642-74621-5

2127/3335-543210 – Printed on acid-free paper

Scientific Organisation of the Sessions

Clinical Aspects
Mertelsmann, Roland, Universität Mainz
Waldmann, Thomas, National Cancer Institute, Bethesda, Maryland

Cell Biology
Dexter, Mike, Paterson Laboratories, Christie Hospital, Manchester
Moore, Malcolm, Memorial Sloan-Kettering Cancer Center, New York

Immunology
Mitchison, N. Avrion, Dept. of Zoology, Tumour Immun. Unit, London
Rajewsky, Klaus, Inst. f. Genetik d. Univ. Köln, Köln

Molecular Genetics
Greaves, Melvyn, Leukemia Research Fund Centre, London
Melchers, Fritz, Basel Institute for Immunology, Basel

Early Life and Evolution
Degens, Egon T., Universität Hamburg, Geologisch-Paläont. Institut und Museum, Hamburg
Gajdusek, Carleton, National Inst. of Health, Bethesda, Maryland

Wilsede Joint Meeting in Pediatric Hematology and Oncology V
Gaedicke, Gerhard, Univ.-Kinderklinik, Ulm
Ritter, Jörg, Westph. Wilhelms-Universität, Münster

Local Organisation
Mannweiler, Klaus, Heinrich-Pette-Institut, Hamburg
Neth, Rolf, II. Med. Universitätsklinik, Hamburg
Schubert, Johannes, St.-Joseph-Hospital, Bremerhaven

Contents

Cell Biology

Immunology

Molecular Genetics

Early Life and Evolution

Participants of the Meeting

Australia

Alexander, Warren, Molecular Biology Unit, The Walter and Eliza Hall Inst. PO. Royal Melbourne Hospital, Victoria 3050

Gearing, David, Cancer Research Unit, The Walter and Eliza Hall Inst. PO. Royal Melbourne Hospital, Victoria 3050

Belgium

Burny, Arsene, Laboratoire de Chemie Biologique, Rue des Chevaux, 67, 1640 Rhode-St.-Genese

Van den Broeke, Anne, Dept. De Biologie Molekulaire Université Libre de Bruxelles, Rue des Chevaux, 67, 1640 Rhode-Saint-Genese

Canada

Minden, Mark, Ontario Cancer Institute, 500 Sherbourne St., Toronto, Ontario M4X 1K9

Tellier, Raymond, Royal Victoria Hospital, Dev. of Hematology, McGill Univ. and Dept. of Genetics 687, Avenue des Pins Quest, Montreal, Quebec H3A 1A1

Czechoslovakia

Petrakova, Alena, Research Inst. of Child Devel. Fac. of Pediatrics, Charles University V uvalu 84, 150 18 Prague 5

Denmark

Killmann, S. A., Rigs Hospitalet, Kopenhagen

Fed. Rep. of Germany

Anders, Fritz, Genetisches Institut der Justus-Liebig-Universität, Heinrich-Buff-Ring 58–62, 6300 Gießen

Atzpodien, Jens, Med. Hochschule Hannover, Abt. Hämatologie u. Onkologie, Postfach 61 01 80, 3000 Hannover 6

Bier, Verena, Kinderklinik der Universität Heidelberg, Im Neuenheimer Feld 150, 6900 Heidelberg

Boos, Joachim, im Heidgarten, 4400 Münster

Brandenburg, Irene, Abt. f. Humangenetik der Christian Albrecht Universität, Schwanenweg 24, 2300 Kiel

Burdach, Universitäts Kinderklinik, Moorenstr. 5, 4000 Düsseldorf

Bührer, Christoph, Kinderklinik der Medizinischen Hochschule, 3000 Hannover 61

Debatin, Klaus-Michael, Sektion Onkologie/Immunologie Universitäts-Kinderklinik, Im Neuenheimer Feld 150, 6900 Heidelberg

Debus, Otfried, Adolfstr. 33, 2300 Kiel 1

Degens, E. T., Universität Hamburg, Geologisch-Paläont. Institut und Museum, Bundesstr. 55, 2000 Hamburg 13

Dittmann, Klaus, Klinik. Großhadern/Univ. München Medizinische Klinik III, Klinische Hämatologie, Landwehrstr. 61, 8000 München 2

Faller, Gerhard, Pathologisches Institut Universität Würzburg, Luitpoldkrankenhaus, 8700 Würzburg

Fiedler, Walter, II Medizinische Klinik, Abt. Onkologie u. Hämatologie, Univ. Klinik Eppendorf, Martinistr. 52, 2000 Hamburg 20

Gaedicke, Gerhard, Universitäts-Kinderklinik, Abt. Pädiatrie II, Prittwitzstr. 43, 7900 Ulm

Gattermann, Norbert, Universitätsklinik Düsseldorf, Moorenstr. 5, 4000 Düsseldorf 1

Gelderblom, Hans, Robert-Koch-Institut, Nordufer 20, 1000 Berlin 65

Gohla, Sven, Labor für Hautimmunologie, Bioverträglichkeit/Immunologie, P-G-U-Forschungszentrum, Beiersdorf AG, Unnastr. 48, 2000 Hamburg 20

Grund, Christian, Robert-Koch-Institut, Wilhelmshavener Str. 14, 1000 Berlin 21

Hölzer, D., Zentrum d. Inneren Medizin, Universität Frankfurt, Theodor-Stern-Kai 7, 6000 Frankfurt/a. M. 70

Halberstadt, Antje, Am Zollstock 24, 6300 Gießen

Harbott, Jochen, Klinikum d. Universität Gießen, Abt. Allgemeine Pädiatrie, Hämatologie und Onkologie, Feulgenstr. 12, 6300 Gießen

Haubeck, Hans-Dieter, Abt. Klinische Chemie und Pathobiochemie der Med. Fakultät der RWTH, Pauwelsstr., 5100 Aachen

Hehlmann, R., Medizinische Poliklinik Universität München, Pettenkoferstr. 8a, 8000 München 2

Heisig, Volker, Universität Hamburg, IMMI, Martinistr. 52, 2000 Hamburg 20

Herrmann, F., Klinikum d. Johannes-Guttenberg Univ., Abt. für Hämatologie, Langenbeckstr. 1, 6500 Mainz

Jahn, Gabriele, Universität Hamburg, IMMI, Martinistr. 52, 2000 Hamburg 20

Janssen, J. W. G., Klinikum der Universität Ulm, Zentrum für Kinderheilkunde, Postfach 38 80, 7900 Ulm

Jähde, Eckhard, Baumsatzstr. 64, 7401 Pliezhausen

Kabisch, Universitätsklinikum Eppendorf, Martinistr. 52, 2000 Hamburg 20

Kalthoff, Holger, Abt. f. klin. Immunologie, Med. Universität Eppendorf, Martinistr. 52, 2000 Hamburg 20

Kawai, Makoto, Heinrich-Pette-Institut, Universitätskrankenhaus Eppendorf (UKE), Martinisstr. 52, 2000 Hamburg 20

Klausmann, Martine, Klinikum der Philipps Univ. Marburg/Zent. Innere Med., Med. Klinik u. Poliklinik, Baldingerstr., 3550 Marburg

Koch, Ina, Heinrich-Pette-Institut, Martinistr. 52, 2000 Hamburg 20

Löffler, Helmut, II. Medizinische Univ. Klinik und Poliklinik, Metzstr., 2300 Kiel

Lampert, Fritz, Kinder-Poliklinik, Feulgenstr. 12, 6300 Gießen

Landfester, Cornelia, Albert-Ludwigs-Universität, Dreikönigstr. 26, 7800 Freiburg

Leib-Mösch, Christine, Ges. f. Strahlen- und Umweltforschung, Abt. Molekulare Zellpathologie, Ingolstädter Landstr. 1, 8042 Neuherberg

Lindemann, Albrecht, Joh. Gutenberg-Uni. Klinikum, I. Med. Klinik u. Poliklinik, Langenbeckstr. 1, 6500 Mainz

Ludwig, Wolf-Dieter, Freie Universität Berlin, Universitätsklinikum Steglitz, Abt. für Innere Medizin, Hindenburgdamm 30, 1000 Berlin 45

Mannweiler, Klaus, Heinrich-Pette-Institut, Martinistr. 52, 2000 Hamburg 20

Mannweiler, Sebastian, Isestr. 23, 2000 Hamburg 13

Mertelsmann, R., Abt. für Hämatologie, Kln. Johannes Gutenberg-Univ., Langenbeckstr. 1, 6500 Mainz 1
Munk, Klaus, Deutsches Krebsforschungszentrum, Im Neuenheimer Feld 280, 6900 Heidelberg
Odenkirchen, Andreas, Samwerstr. 2, 2300 Kiel
Opalka, Bertram, Universitätsklinikum Essen, Inst. für Molekularbiologie, Tumorforschung, Hufelandstr. 55, 4300 Essen
Ostertag, Wolfram, Heinrich Pette Institut für Exp. Virologie u. Immunologie, 2000 Hamburg
Ottmann, Oliver, Klinikum der J. W. Goethe-Univ., Zentrum der Inneren Medizin, Theodor-Stern-Kai 7, 6000 Frankfurt/M. 70
Raffael, Alexander, Becton Dickson GmbH, Postfach 10 16 29, 6900 Heidelberg
Rajewski, Klaus, Institut für Genetik der Universität zu Köln, Weyertal 121, 5000 Köln 41
Ritter, J., Westphälische Wilhelms-Univ., Kinderklinik und Poliklinik, Abt. für pädiatrische Hämatologie und Onkologie, Albert-Schweitzer-Str. 33, 4400 Münster
Schedel, Roswitha, Universitäts Kinderklinik, Pädiatrie II, 7900 Ulm
Schmetzer, Helga, Forschungs Lab. A. Med. Klinik III, Klinikum Großhadern, 8000 München 70
Schmidt, Reinhold E., Abt. Immunologie Zentrum Innere Medizin, Med. Hochschule Hannover, 3000 Hannover 61
Schubert, Johannes, St. Joseph Hospital, Wiener Str. 1, 2850 Bremerhaven
Stoll, Matthias, Abt. Immunologie, Medizinische Hochschule Hannover, Konstanty-Gutschow-Str. 8, 3000 Hannover 61
Teichmann, Johannes, Freie Universität Berlin, Medizin. Klinik u. Poliklinik, Universitätsklinikum Steglitz, Hindenburgdamm 30, 1000 Berlin 45
Voltz, Raymond, Heiglhofstr. 64/410, 8000 München 70
Wörmann, Bernhard, Breul 4, 4400 Münster
Welte, Karl, Kinderklinik der Medizinischen Hochschule Hannover, Konstanty-Gutschow-Str. 8, 3000 Hannover 61
Wilhelm, Martin, Med. Poliklinik der Universität Würzburg, Klinikstr. 8, 8700 Würzburg
Wolf, Jürgen, Institut für Virusforschung, Deut. Krebsforschungszentrum, Im Neuenheimer Feld 280, 6900 Heidelberg
Zechel, Christina, Ebertalstr. 24, 6293 Lühnberg 2
Zeller, Wolf, UKE, II. Med. Klinik, Abt. Onkologie u. Hämatologie, Martinistr. 52, 2000 Hamburg 210
Zur Hausen, Harald, Deutsches Krebsforschungszentrum, Im Neuenheimer Feld 280, 6900 Heidelberg 1
Zylka, Vera, Die Welt, Bad Godesberger Allee 99, 5300 Bonn 2
Zühlsdorf, Michael, Westfälische Wilhelms-Univers. Münster, Klnk. u. Poliklnk Kinderheilknd., Albert-Schweitzer-Str. 33, 4400 Münster

German Dem. Rep.
Elstner, Elena, Universitäts-Klinik für Innere Medizin, Charité, Schumannstr. 20/21, 1040 Berlin
Helbig, Werner, Klinik für Innere Medizin der KMU Leipzig, Abt. Hämatologie/Onkologie, Johannisallee 32, 7010 Leipzig
Ihle, Rainer, I. Med. Klinik, Charité, Humboldt-Universität, Schumannstr. 20, 1040 Berlin
Wechter, Marlies, Charité, 1. Med. Klinik, 1040 Berlin
Zintl, Felix, Universitäts-Kinderklinik, Kochstr. 2, 6900 Jena

Israel
Ben-Neriah, Yinon, The Lautenberg Cent. f. General & Tumor Immunology/Hebrew University, Hadassah Medical School, P.O. Box 11 72, 91010 Jerusalem
Haran-Ghera, Nechama, Dept. of Chemical Immunology, The Weizmann Inst. of Science, P.O. Box 26, 76100 Rehovot
Peled, Alpha, Dept. of Chemical Immunology, The Weizmann Inst. of Science, P.O. Box 26, 76100 Rehovot
Sachs, Leo, The Weizmann Institute of Science, Dept. of Genetics, 76100 Rehovot

Italy
Chieco-Bianchi, Luigi, Istituto di Oncologia, Centro Studio Leucemie, Via Gattamelata 64, 35128 Padova
Saglio, Guiseppe, Dipt. Di Scienze Biomediche E Oncologia Umana, Sezione Clinica Via Genova. 3, 10126 Torino

Japan
Minowada, Jun, Fujiisaki Cell Center Hayashibara Biochemical Lab., 675-1 Fujisaki Okayama, 702 Japan
Mizutani, Shuki, Division of Virology, National Children's Medical, Research Center 3-35-31 Taishido, Setagaya-ku, 154 Tokyo

Netherlands
van der Feltz, Machtheld, The Dr. Daniel Hoed Cancer Center, P.O. Box 5201, 3008 AE Rotterdam

Peoples Rep. of China
Gou-Wei, Rong, Shanghai Second Medical Univ., 280 South Chongoing Road, Shanghai
Meng-er, Huang, Shanghei Second Medical Univ., 280 South Chongoing Road, Shanghai

Poland
Deptala, Andrzey, Inst. of Haematology, Chocomska 5, 00-951 Warsaw
Holowiecki, J., Dept. of Hematology, Silesian Medical Academy, 8 Reymonta, 40-029 Katowice
Kazanowska, Bernarda, Dept. of Haematology, Medical Academy of Wroclaw, Wroclaw
Konopka, Lech, Institute of Haematology, Chocomska 5, 00-951 Warsaw
Pietras, Wojciech, Dept. of Haematology Medical Academy of Wroclaw, Wroclaw
Rozmyslowicz, Tomasz, Inst. of Haematology, Dept. of Internal Medicine, Chocimska Str. 5, 00-957 Warsaw

Switzerland
Melchers, Fritz, Basel Inst. for Immunbiology, Postfach, 4005 Basel
Moroni, Christoph, Institut für Medizinische Mikrobiologie der Universität Basel, Petersplatz 10, 4003 Basel

United Kingdom
Bains, Margaret, LRF Preleukaemia Unit. Dept. of Haematology, University Hospital of Wales, Heath Park, Cardiff CF4 4XN
Bearpark, Andrew, LRF Centre at the Inst. of Cancer Research, Fulham Road, London SW3 6JB

Boehm, T., Laboratory of Molecular Biol. Hills Road, Cambridge CB 2 QH
Campana, Dario, Dept. of Immunology, Royal Free Hospital, School of Medicine, London NW3 2QG
Chessells, Judith Mary, Dept. of Haematology/Oncology, The Hospital for Sick Children, Great Ormond Street, London WC1N 3JH
Cotter, F. E., Imperial Cancer Research Fund, Dept. of Medical Oncology, St. Bartholomew's Hospital, London EC 1
Dexter, T. Michael, Patterson Laboratories, Christie Hospital and Holt Radium Institute, Manchester M20 9BX
Fisher, Amanda, HTIG, ICRF, Courtauld Building, 91 Riding House Str., London W1P 8BT
Galvani, David, Dept. of Haematology, Duncan Building, Royal Liverpool Hospital, Prescot Street/PO. Box 147, Liverpool L69 3BX
Gledhill, Sarah, LRF Virus Center at the Univ. of Glasgow, Dept. of Veterinary Pathology, Bearsden Road, Glasgow G61 1QH
Greaves, Melvyn F., Leukemia Research Fund Center, Institute of Cancer Research, Fulham Road, London SW3 6JB
Janossy, G., Dept. of Immunology, Royal Free Hospital, School of Medicine, London NW3 2Q9
Jarrett, Ruth, University of Glasgow, Bearsden Road, Bearsden, Glasgow G61 1QH
Just, Ursula, Paterson Inst. for Cancer Res., Christie Hospital & Holt Radium Institute Wilmslow Road, Manchester M20 9BX
Kölble, U., Dept. of Haematology, John Radcliffe Hospital, Oxford OX3 9DU, Haedington
Lister, T. A., Imperial Cancer Research Fund, Dept. of Med. Oncology, St. Bartholomew's Hospital, London EC1
Mason, Donald, Dept. of Haematology, John Radcliffe Hospital, Oxford OX3 9DU
McKinney, Patricia, LRF Centre f. Clinical Epidemiology at the University of Leeds, 17 Springfield Mount, Leeds LS2 9NG
Merkenschlager, Matthias, ICRF Human Tumor Immunology, Univ. College/Middlesex School, The Courtauld Inst. of Biochem., 91 Riding House Street, London W1P 8BT
Mitchell, Christopher, Dept. of Haematol. and Oncol., The Hospital for Sick Children, Great Ormond Street, London WC1N 3 JH
Mitchison, Avrion, Dept. of Biology (Medawar Blg.), University College London, Gower Street, London WC1E 6BT
Neil, James, The Beatson Inst. Garscube Estate, Switchback Road, Bearsden, Glasgow G61 1BD
Padua, Rose Ann, LRF Preleukaemia Unit, Dept. of Haematology, Univ. of Wales/Coll. of Medic., Heath Park, Cardiff CF4 4NX
Pezzella, Francesco, Dept. of Haematology, John Radcliffe Hospital, Headington, Oxford OX3 9DU
Schiro, Raffaella, Paterson Inst. for Cancer Res. Lab. Hexp. Haematology, Christie Hospital, Wilmslow Road, Manchester M20 9BX
Shan, Li C., Dept. of Haematology, St. George's Hospital, London
Stauss, Hans, University College London, Dept. of Biology, Gower Street, London WC1E-6BT
Wright, Eric, MCR Radiobiology Unit Chilton, Didcot, Oxon OX11 0RD

USSR

Afanasyev, Boris, N. N. Petrov Inst. of Oncology, Dept. Bone Marrow Transplan., Pesochny-2, 188646 Leningrad

Chertkov, J. L., Central Inst. of Hematology & Blood Transfusion, Novozykovsky pr. 4, 125167 Moscow
Friedenstein, Alexander, Gamaleya Institute for Epidemiology and Microbiology, Academy of Medical Sciences, Gamaleya Str. 18, D-98 Moscow
Frolova, Elena, USSR Academy of Sciences Shemyakin Inst. of Bioorganic, U1. Miklukho-Maklaya, 16/10, 117871 GSP Moscow V-437
Knyazev, Peter, N. N. Petrov Institute of Oncology, Pesochny-2, 188646 Leningrad

USA
Böhnlein, Ernst, Duke University HHMI, P.O. Box 3037, Durham, NC 27710
Buchow, Hartmut, Dept. of Health & Human Service, National Institute of Health, National Cancer Institute, Bethesda, Maryland 20892
Cooper, Max, University of Alabama at Birmingham/Univ. Station, 263 Tumor Inst., Birmingham AL 35213
Crist, William, St. Jude Children's Research Hospital, 332, North Lauderdale P.O. Box 318, Memphis-Tennessee 38101-0318
Dean, Michael, National Cancer Institute, Frederick Cancer Research Fac., P.O. Box B, Maryland 21701
Duesberg, Peter, University of California, Dept. of Molecular Biology, 229 Stanley Hall, Berkeley, CA 94720
Eichberg, Jörg, South West Foundation Biomedical Research, P.O. Box 28 11 47, St. Antonio, TX 78284
Ensoli, Barbara, Laboratory of Tumor Cell Biology NIH-NCI, Bethesda, MD 20892
Ernst, Timothy, Dana-Faber Cancer Inst., 44 Binney Street, Boston, MA 02115
Fisher, Robert, Lab. of Molecular Oncology, National Cancer Inst., NCI-FCRF, Bldg. 469, Rm. 203, Frederick, MD 21701-1013
Gajdusek, Carleton, National Inst. of Health, National Cancer Inst. Bethesda, Maryland 20892
Griesinger, Frank, Dept. of Lab. Medicine/Pathol., Box 86 UMHC, Harvard Street/ East River Road, Minneapolis, MN 55455-0315
Houghton, Peter, St. Jude Children's Research Hospital, 332 North Lauderdale P.O. Box 318, Memphis-Tennessee 38101-0318
Kern, Francis, Breast Cancer Section Medicine Branch, Building 10, Room 12N 226, NCI, NIH, Bethesda, MD 20892
Kirsten, Werner, The University of Chicago, Dept. of Pathology, 5841 S. Maryland Ave., Chicago, IL 60637
Koeppen, Hartmut, The University of Chicago, Dept. of Pathology, Committee on Immunology, 5841 S Maryland Ave./Box 414, Chicago, IL 60637
Kramer, Gisela, Dept. Chemistry, University of Texas, Austin, TX 78712
Leibson, Paul, Mayo Clinic, Dept. of Immunology, Rochester, MN 55902
Lenardo, Michael, Whitehead Institute for Biomedical Research, 9 Cambridge Center, Cambridge, MA 02142
Look, Thomas, St. Jude Children's Research Hospital, 332 North Lauderdale P.O. Box 318, Memphis-Tennessee 38101-0318
McCredie, Kenneth B., Univ. of Texas, Syst. Cancer Cen. Anderson Hosp. + Tumor Inst., Texas Medical Center, 6723 Bertner Avenue, Houston, Texas 77030
McKeithan, Timothy, Dept. of Medicine, University of Chicago, Box 420 5841 S. Maryland Ave., Chicago, IL 60637
McNagny, Kelly, The University of Alabama, School of Medicine, Comprehensive Cancer Center, Birmingham/Alabama 35294
Melchner, H. von, Centre for Cancer Research, Massachusett Inst. of Technol., 40 Ames Street E 17-517, Cambridge, M.A. 02139

Meyer, William, St. Jude Children's Research Hospital, 332 North Lauderdale P.O. Box 318, Memphis-Tennessee 38101-0318

Mirro, Joseph, St. Jude Children's Research Hospital, 332, North Lauderdale P.O. Box 318, Memphis-Tennessee 38101-0318

Moore, Malcolm, Memorial Sloan-Kettering Cancer Center, 1275 York Avenue, New York, N.Y. 10021

Ohno, Susumu, Beckman Research Institute of the City of Hope, 1450 East Durarte Road, Duarte, CA 91010

Pinkel, Donald, 2501 Addison Road, Houston, TX 77030

Rapp, Ulf, National Cancer Inst. Lab. of Viral Carcinogenesis, Viral Pathology Section Bldg. 560, Room 21-75, Frederick, MD 21701-1013

Rettenmier, Carl, St. Jude Children's Research Hospital, 332 North Lauderdale P.O. Box 318, Memphis-Tennessee 38101-0318

Rowley, Donald, Dept. Medicine, Section of Hematology/Oncology, Box 420 5841 South Maryland Ave., Chicago, IL 60637

Rowley, Janet, Dept. Medicine, Section of Haematology/Oncology, Box 420 5841 South Maryland Ave., Boston, MA 022115

Ruprecht, Ruth, Dana-Farber Cancer Institute, 44 Binney Street, Boston, MA 022115

Sadaie, Reza, Laboratory of Tumor Cell Biol. National Cancer Institute, Bld. 37, Room 6A09, Bethesda, Maryland 20892

Schreiber, Hans, La Rabiba Child. Hospital and Research Center, Dept. of Patho., East 65th Street at Lake Michigan, Chicago, Illinois 60649

Storb, R., Dept. of Microbiology and Immunology, University of Washington, Seattle, Washington 98195

Terhorst, Cox, Dana-Farber Cancer Institute, 44 Binney Street, Boston, MA 02115

Tsujimoto, Yoshihide, The Wistar Institute, 3601 Spruce St. Philadelphia, PA 19104

Waldmann, Thomas A., National Inst. of Health, National Cancer Inst., Metabolism Branch, Bethesda, Maryland 20892

Warren, David, Memorial Sloan-Kettering, Cancer Center, 145 Bosten Post RD, Rye, N.Y. 10580

Yohn, David, Intern. Ass. of Comparative Res. on Leukemia + Related Diseases, 410 West 12th Avenue, Columbia, Ohio 43210

In Memoriam Dr. Mildred Scheel

The Wilsede Meeting is also supported by the Wilsede Fellowship Programme of the Dr. Mildred Scheel Stiftung, which is part of the Deutsche Krebshilfe. Since that Foundation was established by Dr. Mildred Scheel, it is appropriate that we should reflect and comment on the great contribution which she made to cancer prevention, treatment and research.

Who was Mildred Scheel? What were her ideas and what did she achieve with her Foundation?

Mildred Scheel was born in Cologne in 1932, daughter of a physician and radiologist. She studied medicine and specialized in radiology. Later, she married Mr. Walter Scheel before he was appointed Minister for Foreign Affairs. When Mr. Scheel subsequently became President of the Federal Republic of Germany, she became the "First Lady" of this country. No doubt this helped her to fulfil her noble ambition to contribute to the fight against cancer. As a consequence of this, she founded the Deutsche Krebshilfe in 1974. From that time on, all her efforts were directed towards encouraging people to contribute money for this crucial purpose. She developed many significant ideas for organizing cancer prevention, early diagnosis and treatment that was applicable on a large scale. She initiated the establishment of the first five cancer centers in this country. Once they were functioning successfully, she was able to convince the Government to assume full responsibility for maintaining them. She then prepared to launch new undertakings. It became apparent to people that she had unique qualities that enabled her to initiate new ideas for fighting cancer, and this added significantly to her personal success. She also supported in particular the treatment of childhood cancer in many hospitals, and initiated the psychosocial after-care of patients and their families. In addition, she aided individuals who were economically affected by having cancer.

The Dr. Mildred Scheel Stiftung was established to promote and support cancer research. It supports a great number of research projects in many institutes and provides a fellowship programme for sicentists to work and study at institutions abroad. Included in that programme is the Wilsede Fellowship Programme. The Dr. Mildred Scheel Stiftung is now an important body in the Federal Republic of Germany for the granting of fellowships. Many of Mildred Scheel's initiatives were not broadly accepted at first, but through her continued energy they are now accepted as common practices in the oncological field in this country.

When she had a particular goal in sight, no obstacles could prevent her from reaching it. Yet, for all her tenacity, Mildred Scheel was a warm, loving and sensitive person who had special understanding for cancer patients, together with a human touch. She was always very hard-working and enthusiastic, and stimulating for all of us. None of those who, like myself, had worked with her in the Foundation for over 10 years can remember her ever missing a meeting of the board or the scientific councils of the Deutsche Krebshilfe or the Dr. Mildred Scheel Stiftung, until the last few weeks of her life. During those meetings she listened carefully to the experts, although sometimes she came to her own conclusions when she was convinced that a particular step forward had to be made. She never lost her enthusiasm for helping others, even when she realized what would be the consequence of her own illness. She always seemed to be positive in her attitude and could always stimulate others with her spirit and her personality. She could have done so much more in the future and she is sadly missed by all of us. We all will always remember her with great devotion.

The Mildred Scheel Memorial Lectures are our tribute. The first lecture will be held by F. Anders, a classic geneticist and molecular biologist, who has made important contributions to the understanding of cancer in the field of molecular genetics.

Klaus Munk
Heidelberg

The Mildred Scheel 1988 Memorial Lecture

A Biologist's View of Human Cancer*

F. Anders

This biennial lecture reflects the generosity of Dr. Mildred Scheel, whose life was dedicated to the fight against cancer. I met Mildred Scheel personally on the occasion of several conferences on human cancer and remember her with gratitude. It is an honor to have been invited to present the 1988 lecture.

The ultimate purpose of all who study cancer biology falls within the general goal of the efforts of Dr. Scheel: to analyze the biological factors that are involved in tumor development for the purpose of preventing cancer. At times the analytical work of many scientists of Mildred Scheel's generation appeared to meet certain opposition when they have seen printed in large letters "cancer is not inherited" and "genes that determine cancer do not exist." Such statements came from well-meaning people intent on calming the fears of families that have had cancer in their ancestry.

We all are involved in the fight against cancer, the physicians, epidemiologists, biochemists, immunologists, virologists; everybody in his place. I am a zoologist, trained as a geneticist who views human beings as products of nature with all their potentials, limitations, and inadequacies arising from their animal background.

* The work of the author's research group was supported by the Justus-Liebig-Universität Giessen, by the Deutsche Forschungsgemeinschaft, by the Bundesminister für Forschung und Technologie, and by the Umweltbundesamt

A. Oncogenes in Phylogeny

Neoplasia is not limited to human beings, or to mammals, but develops in all taxonomic groups of recent Eumetazoa and even in multicellular plants. Neoplasia was also found in Jurassic Sauria and in other fossils including humans. Neoplasia, therefore, was not created by human civilization, but is inherent in the multicellular organization of life [1]. It is, therefore, not surprising that the genes coding for human cancer are distributed throughout the animal kingdom (Fig. 1, [2–10]).

The most venerable oncogene seems to be the *ras* oncogene, which probably has evolved together with the heterotrophic organization of the early **Eucaryotes.** This supposition does not exclude the idea that certain sequences of *ras* (and other oncogenes) might have been evolved before the heterotrophs in the history of life. Actually *ras* is distributed as a normal genomic constituent from yeast [11], where one obviously cannot recognize a cancerous state, through all groups of the animal kingdom studied up to humans and is possibly involved in the development of human tumors such as bladder carcinoma, melanoma, neuroblastoma, fibrosarcoma, lung sarcoma, lung carcinoma, and acute myeloid leukemia (for review see [12, 13]). Its early appearance in the history of life suggests fundamental functions for our life. Its product is a GTP-binding protein which probably activates phospholipase C that generates the internal promoter diazylglycerol for kinase C, thus signalizing cell proliferation [14–16].

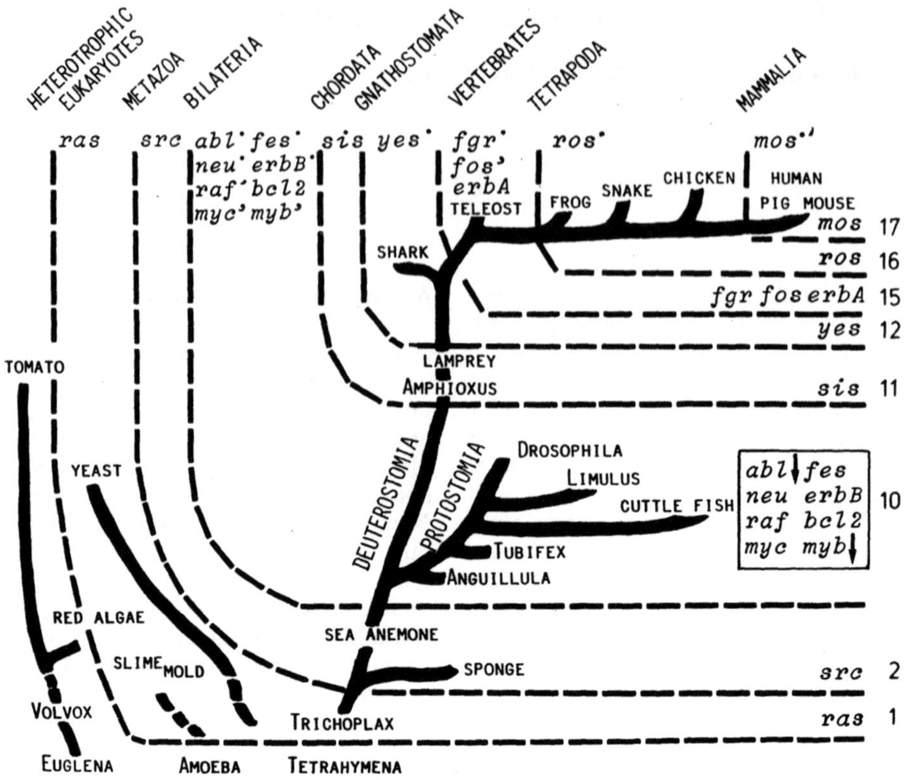

Fig. 1. Attempted outline of the evolution of oncogene systems in the animal kingdom (compiled from [2–10]). See text

As one moves up the evolutionary scale to the multicellular organization of the living beings, i.e., to the **Metazoa,** the *src* oncogene appears in the parazoic sponges and is, thereafter, traceable through the Eumetazoa up to humans [2, 17, 18]. We have not identified cancer in sponges, but *src* was found highly active in the sponges which, because of the autonomy of their cells, can be considered to grow as independently as tumors. In Coelenterata such as sea anemone both *src* activity and abnormal growth comparable to teratomas of higher species have been observed. High activity of *src,* measured as activity of its product, the pp60$^{c\text{-}src}$ kinase, was detected in the nervous cell systems of all groups of animals tested. Its activity is also high in animal and human melanoma [19, 20], the cells of which are probably all derived from the neural crest cell-system. The *src* oncogene is possibly, like *ras,* involved in the transmission of proliferation signals which, on this evolutionary level, possibly include the phosphoinositide phosphoinositol turnover [15]. It serves probably in intercellular communication for coordination of growth and function of the Metazoa, perhaps through gap junctions.

As we go up to the **Bilateria** the Metazoa branch out to the **Protostomia** and **Deuterostomia.** This period must have been evolutionarily very active and successful. A large variety of taxonomic groups containing a large packet of oncogenes has been evolved. In addition to *ras* and *src*, the following have been identified: (a) *abl, fes, neu, erb*B, which belong to the *src* family and exhibit tyrosine kinase function, (b) *myc* and *myb*, which are assumed to fulfill regulatory functions of gene expression in the nucleus, (d) *raf*, coding for a serine/threonine kinase, and (e) *bcl*2, isolated from human B-cell lymphoma. Since the viral oncogenes which mostly have been used as probes originate from higher vertebrates (i.e., Deuterostomia), one can conclude that the respective cellular genes must have been already present in the last common ancestor of both Protostomia and Deuterostomia. The clear hybridization signals always found with *abl* and *myb* lead to the presumption that they evolved still earlier in the history of life as can be shown by present data (see arrows in Fig. 1). Nothing is known about the tumorigenic function of these oncogenes in the tumors observed in invertebrates. Little is known about these functions in human tumors [12]. *abl, myb, fes, bcl*2 present in *Drosophila, Limulus,* etc., organisms which have no blood in the sense of the blood of mammals, are possibly involved in human hematopoietic malignancies; but no convincing data from human biopsy specimens or fresh cells from a variety of human leukemias und lymphomas are available showing that these early oncogenes are crucial in human neoplasia [12].

The appearance of the *sis* oncogene, which codes for the platelet-derived growth factor (PDGF) in the **Chordata,** represented by *Amphioxus* and lamprey in the outline of the phylogenetic tree, might be critical for the evolution of the closed blood circulation apparatus that exposes the blood to pressure. Up to the teleosts this oncogene is represented by only one copy. Later on, moving from lower Tetrapoda to Mammalia, a second *sis* copy occurs. In humans PDGF is coded by two distinct but related genes, namely the PDGF-A gene and the PDGF-B gene, the latter one being known as human c-*sis,* which is less homologous to the teleost c-*sis* than the PDGF-A gene [6]. Although human c-*sis* is apparently inactive in most human cells, it is supposed that both PDGF A and B (and their receptors) are involved in general regulatory processes, cell proliferation, and tumor formation [12].

The *yes* oncogene occurs in the animal kingdom together with the appearance of the **Gnathostomata,** which are represented in our studies by sharks. This gene is a member of the *src* family which is highly homologous to *src* itself. This poses the question of gene duplication in evolution. Another example, the single *sis* copy of the teleosts that corresponds to the human PDGF A became duplicated (probably), as mentioned above. One could extend this question asking whether the large *src* family including the already mentioned *abl, fes, neu, erb*B, *yes* and the not yet mentioned *fgr, ros,* and *mos* could have been evolved by gene duplication. The idea that oncogene families might have been evolved by gene duplication contributes to the general concept of evolution by gene duplication proposed by Ohno [21] almost 20 years ago.

At the evolutionary level of **Vertebrates,** *fgr,* a member of the *src* family, *fos,* a nember of the *myc*/*myb* family, and *erb*A, a partial homolog of the receptors of thyroid hormone, estrogen, progesterone, glucocorticoid hormone of humans, and the human X-factor, appear together in the teleosts. Since *erb*A of the fish shows strong homologies to the viral gene, one could assume that it has evolved earlier in the history of life than the present data indicate. It seems not to be involved in neoplastic transformation but in tumor promotion, perhaps supporting *erb*B, which appears to be involved in transformation [22].

It is notable that, based on our earlier genetic and histogenetic experiments, not only have gene patterns favorable to neoplasia been observed in teleost species but also genes which limit the action of these genes to certain cell types [23]. This is an important point to consider in human neoplasia [3]. It appears that nature's way of keeping the oncogenes from their transforming capacity as soon as they became too dangerous for the increasing complexity of life has been to establish a new category of genes, namely the oncogene-specific regulatory genes [24], today sometimes called anti-oncogenes or oncostatic genes.

Finally, *ros*, a member of the *src* family, possibly involved in cell proliferation and tumor promotion through the internal promoter diazyglycerol [14–16], appears to be specific to the **Tetrapoda,** and *mos*, related to the *src* family and also to *raf*, appears to be specific to **Mammalia** [4, 5, 25]. Nothing is known, at least to my knowledge, about the specificity of these genes to the organization of the Tetrapoda and Mammalia, respectively. *mos* is probably involved in human acute myelogeneous leukemia [12].

In conclusion it appears that, in parallel with the advancement of the animal kingdom, particular oncogenes were subject to their own evolution and that, furthermore, the systems of the oncogenes corresponding to this advancement increased in number, several of them probably by gene duplication. From yeast to mammals we found an increase from 1 to 17 (see Fig. 1, right). This increase might reflect the increase of complexity required for advancement in the animal evolution but might in addition reflect an increase of sensitivity to any endogenous and exogenous impairment of the systems. Therefore, our phylogenetic view might reflect some rough observations on the tumor incidence in the animal kingdom which so far have never been studied seriously. Although both oncogenes and cancer have been observed in all systematic categories of the Eumetazoa, it appears that mammals are more afflicted with cancer than any other group of animals.

B. Low and High Susceptibility to Neoplasia

Neoplasia occurs infrequently in the natural populations of Eumetazoa, and induction of cancer by initiating carcinogens and tumor promoters is difficult to achieve [26]. This phenomenon was studied in detail in the Central American teleost genus *Xiphophorus* [26–29] and in East Asiatic mice [30]. Natural selection in Mendelian populations will not favor one population or race and discriminate against the other but will always work against susceptibility to cancer in all populations and races. However, certain nontaxonomically defined groups of animals are highly susceptible to spontaneously developing, carcinogen-initiated, and promoter-stimulated neoplasms (Table 1). These groups consist mainly of animals of hybrid origin, such as naturally occurring or experimentally produced interspecific, interracial, and interpopulational hybrids as well as laboratory and domesticated animals which actually are also hybrids, i.e., homozygous combinations of chromosomes of different populational or racial provenance. These animals share their high susceptibility to neoplasia with humans [26, 31].

While we do not have data on the relationship between hybridization and cancer in human beings comparable to the data on animals, it is interesting to speculate whether the many facts on tumor incidence in humans that do not agree with the concept of the primacy of environmental factors in carcinogenesis can be explained by interpopulational and interracial matings in our ancestry. Certainly interpopulational and interracial mating may have occurred at any time in any place. Because of the high and increasing mobility of modern humans as compared with other species, one should expect high heterogeneity. Various estimates based on enzyme variation showed that heterogeneity in humans is comparable to that of domestic animals such as cats, but is about six times higher than that of wild macaques, about ten times higher than that observed in the large wild mammals such as elk, moose, polar bear, and elephant seal, and about twice as great as that of most feral rodents studied so far [32–34]. Based on these

Table 1. Animals that exhibit a high tumor incidence (for references see [26, 31])

Species	Tumor
Insects	
Drosophila laboratory stocks	Various neoplasms
Solenobia hybrids	Various neoplasms
Teleosts	
Xiphophorus hybrids	Various neoplasms
Girardinus laboratory stocks	Promoter-induced melanoma
Ornamental guppy strains	Carcinogen-induced hepatoma
Orange medeka	Hepatoma
Domesticated trout	Aflatoxin-induced hepatoma
Salvelinus hybrids	Fibrosarcoma
Domestic carp	Neuroepithelioma
Ornamental hybrid carp	Ovarian neoplasia
Lake Ontario hybrid carp	Pollution-conditioned gonadal tumors
Goldfish	Erythrophoroma
Amphibia	
Bufo calamita and *B. viridis* hybrids	Chordomas
Birds	
Musk duck and mallard hybrids	Gonadal tumors
Peacock and guinea fowl hybrids	Gonadal tumors
Improved breeds of fowl	Leukosis
Mammals	
Mus musculus and *M. bactrianus* hybrids	Various neoplasms
Laboratory mice strains	Various neoplasms
Hybrids of mice strains	Increased incidence of various neoplasms
BALB/c and NZB hybrids	Plasma cell tumors (50%)
Blue ribbon mice	Mammary tumors (100%)
Sprague-Dawley and Long Evans hybrids	Increased mammary tumor incidence
Domestic dogs	Various neoplasms
Boxers	Very high tumor incidence
Domestic cats	Various neoplasms
Sinclair swine	Melanoma
Lipizzaner horses	Melanoma

data and on the assumption that tumor incidence in general is related to interpopulational and interracial matings, one could explain why humans have a high incidence of neoplasia comparable to that of the domestic animals.

Furthermore, there are some data on chromosomal heteromorphisms in human populations that might be useful for estimates of heterogeneity within and among different populations. According to such estimates it appears that, for instance, Japanese populations exhibit a low degree of Q- and C-band chromosome heteromorphisms, whereas Americans have a much higher degree of this heteromorphism, with blacks having more prominent heteromorphisms than whites [35, 36]. One is tempted to assume that this heteromorphism refects the differences in the degree of heterogeneity among the Japanese and white and black United States populations. In this context it is notable that the ratio of prostatic cancer in Japanese, United States whites, and United States blacks is reported as 1:10:30 and that the black citizens in San Francisco have double the risk of developing neoplasia as compared with their Japanese fellow citizens [37, 38]. We cannot explain these facts by environmental factors or racial differences. The high susceptibility to neoplasia in domestic or hybrid animals, respectively, could show us how to approach the problem. Of course, it is very difficult to study the heterogeneity of a recent human population of a city or country in terms of biological measures. However, new methods such as the determination of restriction fragment length polymorphisms available today could be helpful in revealing the possible relationship between genetic heterogeneity and tumor incidence in modern human populations.

C. Cancer in *Xiphophorus* as a Model for Cancer in Humans

Human biology is unique, but is not so unique in its fundamentals as to make studies on animal models irrelevant for an explanation of human diseases including cancer. Although mice and rats are the classical laboratory animals used in experimental cancer research, several genera of small teleost fish serve increasingly as models in new cancer research programs [39]. One of these genera is *Xiphophorus* (Fig. 2; for portraits of different phenotypes see [2, 3, 22, 23, 29, 31]), the animal model from Central America that we have used in our laboratories for 30 years [24, 40]. Neoplasia appears to develop only very exceptionally in the wild populations of xiphophorine fish. In spite of the fact that thousands of individuals of many wild populations that are isolated from each other have been collected by several investigators and myself, no tumor has been detected. In the progeny of the wild populations that have been inbred in the laboratory for about 80–100 generations, no tumor has occurred spontaneously and almost

Fig. 2. Female and male of the "spotted dorsal" platyfish, *Xiphophorus maculatus*, from Rio Jamapa (Mexico)

Table 2. Oncogenes in *Xiphophorus*

Oncogene	In *Xiphophorus*	Probe from
erb A	+	Avian erythroblastosis virus
erb B	+	Avian erythroblastosis virus
sis	+	Simian sarcoma virus
myc	+	Avian myelocytomatosis virus
myc	+	Human
N-*myc*	(+)	Human neuroblastoma
myb	+	Avian myeloblastosis virus
myb	(+)	Human
fos	+	FBJ osteosarcoma virus
fos	?	Human
Ha-*ras*	+	Harvey murine sarcoma virus
Ki-*ras*	+	Kirsten murine sarcoma virus
N-*ras*	(+)	Human promyelotic leukemia
abl	+	Abelson murine leukemia virus
yes	+	Yamaguichi-73 sarcoma virus
fms	+	McDonough feline sarcoma virus
fgr	+	Gardner-Rasheed feline sarcoma virus
src	+	Rous sarcoma virus
raf/mil	+	Murine sarcoma virus
neu	+	Human neuroblastoma
fes/fps	+	Gardner-Arnstein Virus
fes/fps	?	Human
bcl 2	+	Burkitt's lymphoma
Not found		
ros		UR-II sarcoma virus
mos		Moloney murine sarcoma virus
mos		Human

Twenty-six probes were used (gifts from R.C. Gallo, K. Toyoshima, M. Cleary, R. Müller)

no tumor could be induced even with the strongest mutagens-carcinogens such as X-rays and *N*-methyl-*N*-nitrosourea (MNU). This fact requires special clarification since most of the oncogenes that are known to transform the cells and to drive the tumors are present in the fish (Table 2). If, however, interpopulational and interspecific crossings are performed, depending on the genotype, the progeny spontaneously or following treatment with initiating carcinogens (X-rays, MNU, ethylnitrosourea, diethylnitrosamine, 2-amino-3-methylimidazo-(4,5-f)quinoline, etc.) and/or tumor promoters (12-O-tetradecanoylphorbol-13-acetate = TPA, 5-azacytidine, phenobarbital, cyclamate, testosterone, nortestosterone, methyltestosterone, trenbolone, ethinylestradiol, cAMP, biphenyl, butylhydroxytoluene, deoxycholic acid, thioacetamine, bis(2-ethylhexyl)-phthalate, betel nut extract, etc.) develops neoplasia (data in [41]). Neoplasms originate from all neurogenic, epithelial, and mesenchymal tissues (Table 3). The suitability of the model is, except for research on mammalian-specific tumors such as breast cancer, lung cancer, etc., beyond question and its efficiency is more economic and time-saving than that of the laboratory mammals. Agents that induce neoplasia in certain high-risk genotypes of the fish hybrids, might, in principle, also affect certain high-risk human individuals.

Table 3. Neoplasms in xiphophorine hybrid fish induced by physical and chemical agents (i) or spontaneously developed (s)

Neurogenic		Epithelial		Mesenchymal	
Pigment cell system		Surrounding epithelium		Connective tissues	
Benign melanoma	i, s	Epidermal papilloma	i	Intestinal fibroma	i
Malignant melanoma	i, s	Carcinoma	i	Fibrosarcoma	i
Pterinophoroma	i, s	Squamous cell carcinoma	i		
		Epithelioma	i	Muscles	
Nervous cell system				Rhabdomyoma	i
Neurilemmoma	i	Glands	i	Rhabdomyosarcoma	i
Ganglioneuroma	i	Thyroid adenocarcinoma	i, s	Leiomyosarcoma of mesentery	i
Retinoblastoma	i	Pancreatic adenocarcinoma	i		
Neuroblastoma	i, s				
		Organs		Hematopoietic tissues	
		Liver cell carcinoma	i	Reticulosarcoma	i, s
		Kidney adenocarcinoma	i, s	Lymphosarcoma	
		Gallbladder carcinoma	i		

The neoplasms were determined by K. Frese, Institut für Veterinär-Pathologie, Universität Giessen, and by M. Schwab, S. Abdo, G. Kollinger, Genetisches Institut, Universität Giessen, according to Mawdesley-Thomas [42], and were classified essentially according to Weiss [43]

D. Classification of Tumor Etiology in *Xiphophorus* and Humans

The neoplasms of *Xiphophorus* can be classified as:

1. *Mating conditioned:* accessory oncogenes are introduced into, and/or regulatory genes for the oncogenes are eliminated from, the germ line by replacement of chromosomes carrying the respective genes or lacking them, and vice versa.
2. *Mendelian inherited:* regulatory genes for oncogenes are impaired, lost, or dislocated in the germ line by mutation.
3. *Mutagen-carcinogen conditioned:* regulatory genes for oncogenes are impaired, lost, or dislocated in a somatic cell by mutation.
4. *Nutrient and endocrine conditioned:* resting stem cells are pushed to differentiate by tumor promoters (the genetic preconditions according to a, b, and c are fulfilled by earlier events).
5. *Virus conditioned:* accessory oncogenes are introduced (so far not convincingly shown in the fish).

The same classification can be applied to human cancer comprising a small group of (a) "familial"; (b) "hereditary" neoplasms in which genetic factors are supposed to be involved, e.g., retinoblastoma, meningioma, melanoma; (c) a large group of "carcinogen-dependent" neoplasms, e.g., lung cancer; (d) a large group of "endocrine-dependent" and "digestion-related" neoplasms, e.g., breast, prostatic, colon cancer; and, finally, (e) a group of viral-conditioned neoplasms, e.g., leukemia, genital tumors.

In *Xiphophorus* derived from a wild population neoplasia develops in general only if different protocols for the induction of tumors are combined by the experimenter, for instance, (a) the elimination of regulatory genes by selective matings, (b) the induction of germ line mutations, and (c) the induction of somatic mutations, etc. The particular events that alone do not lead to neoplasia, summate, and appear as a multistep process that goes beyond the generations and, finally, reaches the last step that leads to neoplasia in a certain individual. The experimenter must detect the sequence of the different steps, and it is easy to see that the last step that completes the multistep process determines the etiological type of neoplasia. This was shown for *Xiphophorus* but might be helpful to explain the different types of tumor etiology in humans in which both the ancestry of an individual and the individual itself are involved.

In the following paragraphs we shall try to approach the biological basis of spontaneously developing, carcinogen-mutagen induced, and promoter-dependent neoplasms.

E. Tumors Appearing and Disappearing in the Succeeding Generations

Human tumors such as a certain colon cancer that afflicts individuals 15–20 years sooner than generally may appear "spontaneously" in a family in one generation and may disappear in the succeeding generation. This is demonstrated by means of a cartoon (Fig. 3, upper part) adapted from Lynch and his colleagues [50]. We cannot explain this phenomenon. The *Xiphophorus* model (Fig. 3, lower part) provided the opportunity to study a similar appearance through the fish generations.

Crossings of a spotted platyfish (A) with a nonspotted swordtail (B) result in F_1 hybrids (C) that develop enhanced spot expression and sometimes benign melanoma instead of the spots. Backcrossings of the F_1 hybrids with the swordtail as the recurrent parent result in BC_1 offspring (D, E, F), 50% of which exhibit neither spots nor melanomas (F) while 25% develop benign melanoma (D) and 25% develop malignant melanoma (E). Further backcrossings of the fish (not shown in Fig. 3) carrying benign melanoma with the swordtail result in a BC_2 that exhibits the same segregation as the BC_1.

As opposed to the crossing procedure that gave rise to the melanoma, backcrossings of the melanoma-bearing hybrids (E), with the platyfish as the recurrent parent (A), result in an alleviation of the melanoma in the offspring (C*), which in the following BC generation grow into healthy fish (A*). In conclusion, malignant melanoma of the BC animal (E) originates from the spots of the preceding platyfish generations (A) and is reduced to spots again in succeeding generations (A*).

The formal parallelism in the occurrence of neoplasia in the human family and in the experimental model is striking. In our search for causes of human cancer there might be some value in realizing the types of factors that can be passed from the fish parents to the fish offspring to influence the occurrence of cancer. The experiment with the model suggests that certain human cancers may be expected to occur in individuals because of a combination of factors from both parents that by themselves did not cause cancer in either parent. More data are required in order to compare more stringently human familiar cancer with mating-conditioned neoplasia in the model.

F. Oncogene Expression in the Tumors

The appearance of tumors in both human and model brings about the question for the oncogenes expressed in human and xiphophorine neoplasms. Data available for melanoma indicate an elevated expression of both the human and the xiphophorine *src*, *erb*B, *sis*, *ras*, and *myc* ([2, 6, 7, 18–20, 40, 44, 45] personal communication, U. Rodeck). Measurements concerning the significance of the xiphophorine *src* oncogene (*x-src*) for the development of melanoma and other kinds of neoplasia in the fish (Table 4) showed that the activity of its product, the $pp60^{x\text{-}src}$ kinase, may be elevated in the tumors up to 50 times over that of the controls [46]. Furthermore, the phosphoinositide phosphoinositol turnover, which is supposed to be linked to the *x-src* activity [14–16], was found up to more than ten times elevated over that of the controls (Table 5). This finding is im-

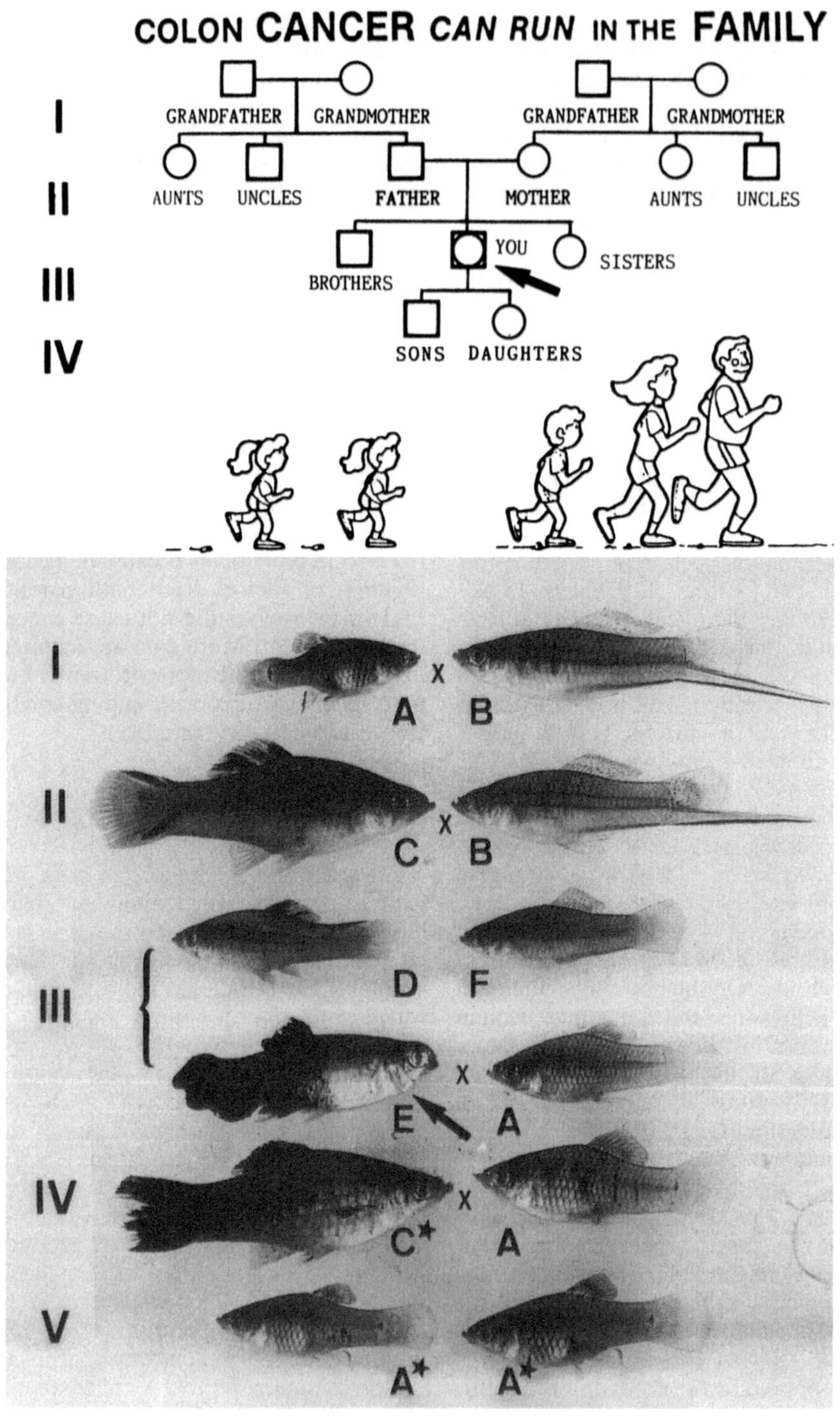

Fig. 3. Appearance and disappearance of neoplasia in succeeding generations (cartoon adapted from [50]). See text

Table 4. Elevation of pp60$^{x\text{-}src}$ kinase activity in tumors and brain of *Xiphophorus* hybrids. (Data from [46])

Tumor	Etiology	Character	Factor of elevation	
			Tumor	Brain
	Carcinogen-induced			
Melanomas	X-ray, adult	Invasive, malignant	5 [a]	No
Squamous cell carcinoma	X-ray, adult	Invasive	2 [a]	No
Epithelioma	X-ray, adult	Benign	2 [a]	No
Fibrosarcoma	ENU, adult	Invasive, malignant	13 [c]	No
Fibrosarcoma	MNU, adult	Malignant	10 [c]	1.4
Fibrosarcoma	MNU, adult	Invasive, malignant	10 [c]	NT
Fibrosarcoma	MNU, adult	Invasive, highly malignant	50 [c]	No
Retinoblastoma	MNU, adult	Progressive growth	3 [b]	No
Melanoma	MNU, embryo	Invasive	8 [a]	No
Melanoma	MNU, embryo	Invasive	10 [a]	No
Rhabdomyosarcoma	MNU, embryo		6 [c]	NT
Rhabdomyosarcoma	MNU, adult	High malignant, invasive	50 [c]	No
	Promoter-induced			
Mesenchymal tumor	MNU + testosterone	Exophytic, slow growing	7 [c]	NT
Melanoma, amelanotic	Testosterone	Highly malignant	30 [a]	No
	Hereditary			
Melanoma ($n=15$)	Spontaneous	Benign	2–3 [a]	1.5–2
Melanoma ($n=28$)	Spontaneous	Malignant	4–8 [a]	2, 3
	Unknown			
Rhabdomyosarcoma	Spontaneous	Invasive	20 [c]	NT

For comparison nontumorous tissues were used: [a] skin; [b] eye; [c] muscle

Table 5. [^{3}H]-Inositol incorporated into phosphoinositides of xiphophorine melanoma (cpm/10 mg neoplasm). (Data adapted from [47–49])

Melanoma	PtdIns	PtdInsP	PtdInsP$_2$
Benign	9000	1050	800
Malignant, "spontaneous"	29000	1200	2100
Malignant, "induced"	17000	600	2300
Extremely malignant, "spontaneous"	34000	2500	4300
Extremely malignant, inherited	30000	3300	700
Brain (control)	3000	500	300

PtdIns, phosphatidylinositol; PtdInsP, phosphatidylinositol-4-phosphate; PtdInsP$_2$, phosphatidylinositol-4,5-diphosphate

portant because the turnover may serve as a measure for the activation of phospholipase C, which generates the internal promoter diacylglycerol.

A tremendous amount of work on oncogene expression and its possible secondary processes in the tumors and in tumor-derived cell lines of experimental mammals and of humans [12] has been performed in the expectation of finding a particular tumor type-specific initial gene and the initial event of the formation of a particular neoplasm. While we were never able to identify what one could term a "liver cancer gene" or a "melanoma gene," others have thought they did. Our own studies on the *Xiphophorus* model showed only a relationship of a number of regulatory genes of a number of tissue-specific developmental genes which in total we called "tumor gene-complex" (*Tu* complex); but we interpreted this as an association rather than a true genetic entity, and we assigned the different kinds of neoplasms such as those listed in Tables 3 and 4 to the same *Tu* complex. The nature of the causality of neoplasia remained unclear.

G. An Approach to the Study of the Genetic and Molecular Basis of Neoplasia

The genes underlying neoplasia in *Xiphophorus* were most successfully studied in the generations developing the "spontaneously occurring" mating-conditioned tumors, and it appears to be in the nature of things that those laboratories working presently on the small group of familial and hereditary human tumors approached the fundamentals of neoplasia at least as closely as those working on the large groups of carcinogen- and promoter-dependent tumors.

Our approach in the model is described by means of Fig. 4, which refers to the same fish as indicated in Fig. 3 by the same capital letters (for the mutants see later). Based on breakpoint data the genes responsible for melanoma inheritance are located terminally in one Giemsa band of the X chromosome [51] and represent a complex consisting of (a) the pterinophore locus (*Ptr*) which is responsible for pterinophore differentiation, (b) the compartment-specific dorsal fin locus (*Df*, impaired to *Df'*) which restricts both pterinophore and macromelanophore differentiation to the dorsal part of the body, (c) the region in which a viral *erb*B-related oncogene (*erb*B*, an oncogene related to the receptor of the human epidermal growth factor, EGF, x-*egfr*) is located, (d) the melanophore locus (*Mel*), which appears to be under control of *Df* and *erb*B*, and (e) the arbitrarily symbolized "tumor gene" (*Tu*), which appears as a Mendelian factor but might possibly be composed of both *erb*B* and *Mel* [22, 52]. Oncogenes in addition to the xiphophorine *erb*B* (x-*erb*B*) could not be detected in the X chromosome. Based on our present knowledge, the respective region of the X chromosome of the platyfish, the "*Tu* complex," can be roughly mapped as follows (commas represent breaking points observed):

X . . ., Ptr, Df, erbB, Mel-Tu*

At least about 20 linked genes are involved in the regulation of the *Tu* complex, but there are also several nonlinked regulatory genes, e.g., the *Diff* gene, which, if present in the homozygous state, restrains the transformed pigment cells from proliferation by terminal differentiation [53].

The swordtail (B) has neither evolved a comparable *Tu* complex nor the linked and nonlinked regulatory genes.

Since platyfish and swordtails have a rather high number of chromosomes ($n = 48$) and since clear-cut chromosomal conditions concerning their origin were required, the experimental animals, besides the purebreds (A, B), were taken from the F_1 (C), which contains one platyfish and one swordtail genome, and from high backcross generations comprising BC_8 up to BC_{22} (F, E), the genome of which virtually consists of swordtail chromosomes except for the *Tu* complex containing X chromosome selected from the platyfish by the crossings. The phenotypic overexpression of the *Tu* complex thus depends mainly on the crossing-conditioned replacement of platyfish autosomes carrying regulatory genes such as the differentiation gene *Diff*, by swordtail autosomes lacking such genes.

More information about the *Tu* complex comes from studies on the restriction length polymorphism of the onco-

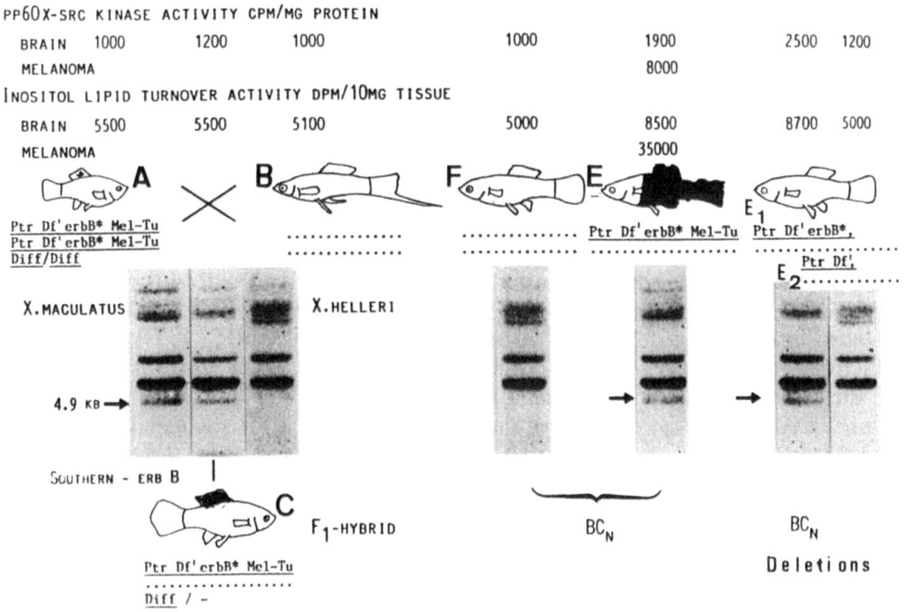

Fig. 4. Appearance of mating-conditioned development of melanoma after crossings of *X. maculatus* × *X. helleri* (platyfish × swordtail; A × B) and backcrossings of the F_1 hybrid (C) with *X. helleri*. F and E represent the backcross generation (BC_n). E_1 and E_2 represent deletions. The fish indicated by the *capital letters* correspond to those indicated in Fig. 3 by the same letters. Note that the 4.9-kb *Eco*R1 Southern fragment is inherited along with the tumor gene-complex. *Ptr*, pterinophore locus; *Df'*, impaired dorsal fin-specific regulatory gene; *erb*B*, xiphophorine copy of an oncogene related to the viral *erb*B; *Mel-Tu*, melanophore locus containing the potential for tumor formation. *Diff*, a nonlinked differentiation gene; ——, chromosomes of *X. maculatus*; . . ., chromosomes of *X. helleri*. See text

genes derived from platyfish and from swordtail. Some of the xiphophorine oncogenes (x-*onc*s) listed in Table 2 show restriction fragment length polymorphism (RFLP) the patterns of which have been differently evolved in the wild fish of different provenance [6–8, 22, 40]. For instance, the patterns of the lengths of the restriction fragments of x-*sis* are specific to each of the different species, but show no RFLP within each of the species; actually these species show a monomorphism of the restriction fragment lengths of x-*sis*. In contrast, the patterns of the lengths of restriction fragments of x-*erb*A and x-*erb*B are species nonspecific, but are specific to the different races and populations of the species. The lengths of certain fragments of x-*erb*B are even different in females and males of the same population.

We used the RFLP phenomenon as an indicator for the Mendelian inheritance of the x-*onc*s through the purebred and hybrid generations. If a certain oncogene fragment is independently inherited from the inheritance of spot or melanoma formation, then one can conclude that the respective oncogene is not "critical" for the first step of melanoma formation. This is not to say that such an oncogene is not involved in melanoma formation at all; as already mentioned, x-*src*, x-*sis*, x-*ras*, x-*myc* are expressed in the melanomas and are certainly involved in tumor growth or tumor progression, but they are not involved in the first step leading to melanoma because they are contributed by the swordtail to the hybrid genome whereas the appearance of the spots and the melanomas is contributed by the X chromosome of the platyfish. Furthermore, since 47 chromosomes of the malignant melanoma bearing backcross hybrids are contributed by the swordtail and only 1, namely the *Tu* complex carrying X chromosome, is contributed by the platyfish, one can assume that most of the oncogenes in the genome of the tumorous backcross animals are contributed by the swordtail genome. Actually, the only x-*onc* detected so far on the platyfish chromosome carrying the *Tu* complex is the x-*erb*B*. This oncogene is represented in Fig. 4 by a 4.9-kb *Eco*R1 Southern restriction fragment which is inherited along with spot and/or melanoma development (A, C, E) and is lacking in the melanoma-free swordtail (B) and the melanoma-free BC hybrid (F). The other *Eco*R1 fragments that also indicate *erb*B sequences could not be assigned to the X-chromosomal locus where the inheritance of the melanomas comes from.

Additional information about the correlation between the inheritance of melanoma formation and the inheritance of the x-*erb*B*-representing 4.9-kb Southern fragment comes from two mutants of the type E BC hybrids. Both types (Fig. 4, E_1 and E_2) have lost the locus *Mel-Tu*, i.e., the capability to develop melanoma, but only one type (E_2) has also lost x-*erb*B* as is shown by the lack of the 4.9-kb fragment. This result indicates that (a) x-*erb*B* is located between *Df* and *Mel-Tu* and (b) information crucial for melanoma formation depends on *Mel-Tu*, which codes for the differentiation of certain pigment cells. This is, however, not to say that there are no links in the chain of events leading to the very beginning of melanoma formation that precede the function of *Mel-Tu*.

As was already mentioned, pp60$^{x\text{-}src}$ kinase activity and inositol lipid turnover activity was found enormously elevated in the melanomas. This is true for all kinds of tumors so far studied and for all types of tumor etiology (Tables 4, 5). Unexpectedly, these activities were also found elevated in the healthy tissues of the fish carrying mating-conditioned and Mendelian-inherited melanomas. Figure 4 (upper part) shows the rounded data measured in the brain of the melanomatous BC hybrids type E in comparison to those of types A, B, C, F. The results suggest that the genes controlling pp60$^{x\text{-}src}$ and the inositol lipid turnover are expressed not only in the melanoma tissues but also in the healthy tissues of the tumorous individuals, independently of whether they are involved in neoplasia or not [46, 49]. Possibly this phenomenon corresponds to the often-occurring multiple tumors in combinations such as melanoma, neuroblastoma, rhabdomyosarcoma, and retinoblastoma in the BC segregants, sometimes even in a particular animal.

Multiple tumors and cancer family syndromes have been reported also in humans [54]. The working group of Lampert [55], for instance, studied a family which, despite a healthy ancestry, developed neuroblastoma, ganglioneuroma, and other neurogenic tumors running through two generations. Lynch and his colleagues [56] reported the pedigree of a family afflicted with cancer on breast, urinary bladder, brain, colon, cervix, endometrium, pancreas, prostate, skin, stomach, and uterus. We cannot explain this phenomenon, but the model shows us the possibility of an approach to the study of some of its molecular and biochemical fundamentals.

It appears that the measurements of pp60$^{x\text{-}src}$ kinase activity and inositol incorporation into phosphoinositides in the brain of the deletion mutants of the fish which are incapable of developing melanoma (Fig. 4, right) open new possibilities for intervention in key signals critical to the endogenous induction of

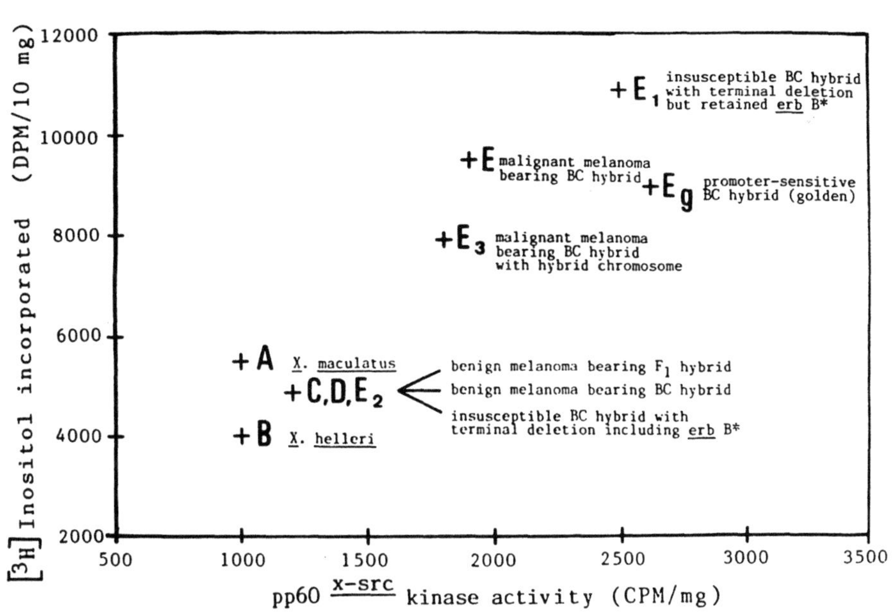

Fig. 5. Incorporation of [^{3}H] inositol in phosphatidylinositol in brain extracts, plotted against activity of pp60$^{x\text{-}src}$. *Capital letters* correspond to the fish indicated by the same letters in Figs. 3 and 4; E_3 ist not shown. E_g corresponds to the promoter-sensitive fish shown in Fig. 9 (*on right*). Note the high correlation of both parameters. Data from [49]. See text

neoplastic transformation in the animal model and possibly in humans. Both pp60$^{x\text{-}src}$ kinase activity and inositol lipid turnover activity are highly elevated in the brain of those insusceptible deletion animals that have lost the *Mel-Tu* locus but have retained the x-*erb*B* oncogene (Fig. 4, E_1). In contrast, the deletion animals having lost the x-*erb*B* together with the *Mel-Tu* locus (E_2) exhibit no elevation. This result suggests that the molecular and biochemical machinery supposedly involved in melanoma formation may be running for genetic reasons, without forming melanoma. Our results, moreover, suggest that there may be a particular type of activation of x-*src* and the inositol phospholipid system that is a marker for predisposition to cancer and could be used for the determination of pro-neoplasia conditions in cancer risk studies. Support for this suggestion comes from the excellent correlation existing between pp60$^{x\text{-}src}$ kinase activity and the [^{3}H]inositol incorporation into phosphatidylinositol (Fig. 5).

One more suggestion arises if one compares the different results obtained with the E_1 and E_2 BC hybrids. Because of the backcross procedure applied to the animals most of the genes involved in melanoma formation are contributed to the hybrids by the swordtail genome. In the deletion hybrid E_2 lacking x-*erb*B* they appear to rest in low activity, indicating that, in order to become involved in melanoma formation, they require a signal for the change from a resting to activated state. The results obtained with the deletion hybrid E_1 show that this signal is transmitted from that region of the *Tu* complex containing platyfish chromosome where x-*erb*B* is located and where the inheritance of the melanoma is determined.

In conclusion, based on the possibility of distinguishing between genes originating from platyfish and swordtail in the genome of certain hybrids, we found that development and growth of melanoma is mainly run by a set of genes that requires a signal for its activation which, due to the onset of the crossing experiments with the mutants, is transmitted from an x-*erb*B*-containing chromosome locus. This locus, however, is probably deregulated by the crossing-conditioned replacement of platyfish chromosomes carrying regulatory genes for the *Tu* complex (i.e., probably x-*erb*B*) by swordtail chromosomes lacking them.

The 4.9-kb *Eco*R1 restriction fragment was cloned, subcloned, and sequenced. It contains exon c and d of the kinase domain and shows high homology to the respective sequences of the human epidermal growth factor receptor (H-EGFR) gene and to the viral *erb*B (for complete data see [22, 40]). Hybridization of this xiphophorine fragment against genomic xiphophorine DNA revealed the presence of highly homologous sequences located on the Y-chromosome (6.7 kb; see later), on the Z-chromosome, and on an autosome present in all individuals. Another species, *Xiphophorus variatus,* which was studied for comparison, also exhibited an homologous fragment which is inherited along with tumor susceptibility. Each of the x-*erb*B* copies corresponding to these homologous fragments from different chromosomes is also part of a *Tu* complex [40]. Hybrids carrying these *Tu* complexes, however, require treatment with carcinogens as a precondition for melanoma development.

H. Carcinogen-Dependent Neoplasia

The remainder of my review of human cancer is devoted to the large groups of mutagen-carcinogen conditioned (somatic mutation conditioned) and nutrient and endocrine conditioned (promoter conditioned) neoplasms. Both types of etiology comprise probably more than 90% of all tumors. A large body of consistent and contradictory observations on their causation are available.

Lung tumors of humans probably offer the most convincing observations on the involvement of exogeneously induced somatic mutations in the initiation of the tumor. They appear not to be influenced by many environmental factors, and there is no evidence that hormonal or nutritional factors are involved in their causation. The simple interpretation of the induction of a somatic mutation by a physical or chemical carcinogen, however, does not explain the different susceptibility of the different individuals that are exposed to the carcinogen. There must exist hereditary factors that enable most of the individuals to escape lung cancer while others become victims. We cannot explain this observation.

Recently Newman and her colleagues [57] reported on breast cancer in an extended family (Fig. 6). A complex segregation analysis indicated that susceptibility to breast cancer in the family can be explained by autosomal inheritance of a defective regulatory gene while the appearance of the tumor requires a somatic mutation in a target cell. This example shows that steps toward breast cancer had already occurred unnoticed in the preceding generation; the somatic mutation represents only the last step that completes the chain of events leading to cancer.

The *Xiphophorus* model provided more details for the study of the complex situation in the somatic mutation-dependent tumors. In mutagenesis studies [52] we detected nontumorous hybrid genotypes which, following treatment with directly acting carcinogens (X-rays, MNU), develop after a latent period of 8–12 months foci of transformed pigment cells that grow out to compact melanomas (Fig. 7). The smallest cell clones to which these melanomas could be traced consisted of eight cells indicating that there were three cell divisions between a somatic mutation event and the occurrence of the transformed pigment cells [23]. The incidence of these tumors depends on the dosage of the treatment, and may reach up to 100%.

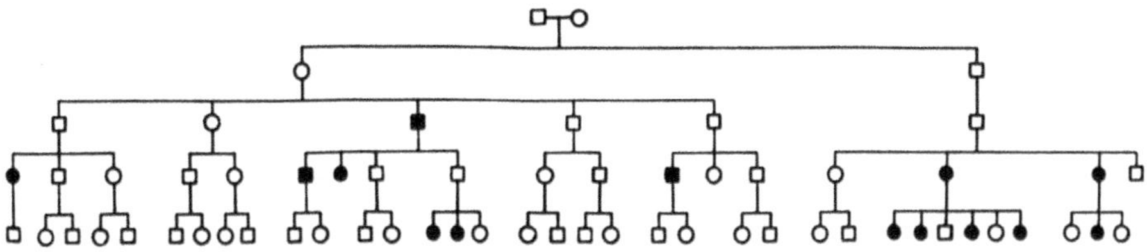

Fig. 6. Pedigree of a family at high risk of breast cancer, adapted from [57]. See text

Fig. 7. Mutagen-carcinogen-sensitive fish developing MNU-induced melanoma. Note the closely circumscribed growth reminiscent of the somatic mutation-conditioned unicellular origin of the tumor

These observations led to the assumption that the *Tu* complex of the treated hybrids is under control of only one regulatory gene which, following treatment, is impaired in a particular pigment cell. Assuming the total of the pigment cell precursors that are competent for neoplastic transformation is 10^6 (this is the average number in the pigment cell system of young fish), and the induced mutation rate is 10^{-6}, then the tumor incidence is 1 (on average 100% of the treated animals will develop one tumor). If, however, the *Tu* complex is under the control of two regulatory genes, the rate of simultaneous mutations of both of these regulatory genes in 1 cell is 10^{-12}, and the tumor incidence is 10^{-6}. This calculation shows that it is difficult to succeed in inducing somatic mutation-conditioned neoplasms if the *Tu* complex is controlled by more than one regulatory gene. This calculation also suggests that the insusceptibility of the animals of the purebred wild populations is based on a polygenic system of regulatory genes directed against cancer.

Support for the assumption that the *Tu* complex of these animals is controlled by only one regulatory gene comes from germinal mutation-conditioned melanoma which occurred in the same genotype. As a consequence of the inheritance of the mutation through the germ line, the *Tu* complex becomes active in the developing progeny as soon as the pigment cell precursors become competent for neoplastic transformation. This process starts in the embryo and continues in all areas of the developing fish where the pigment cell precursors become competent, thus building a lethal "whole body melanoma," which reflects the genuine effect of the *Tu* complex on the pigment cell system. It should be emphasized that the tumorous growths that appear on germinal inherited melanoma (and other hereditary neoplasms), i.e., both the mating-conditioned and the germ line mutation-conditioned melanomas, are not due to the occurrence of somatic mutations during development, because, in contrast to the somatic mutation-conditioned tumors, the transformed cells always occur simultaneously in large areas of the body and show permanent transformation and relapse after complete removal.

To study the molecular and biochemical background of the somatic mutation-conditioned melanomas we modified the experiment that led to mating-conditioned spontaneously occurring melanomas (see Fig. 8 and compare with Fig. 4). The *Tu* complex containing platyfish chromosome was replaced by another which, instead of the mutated dorsal fin-

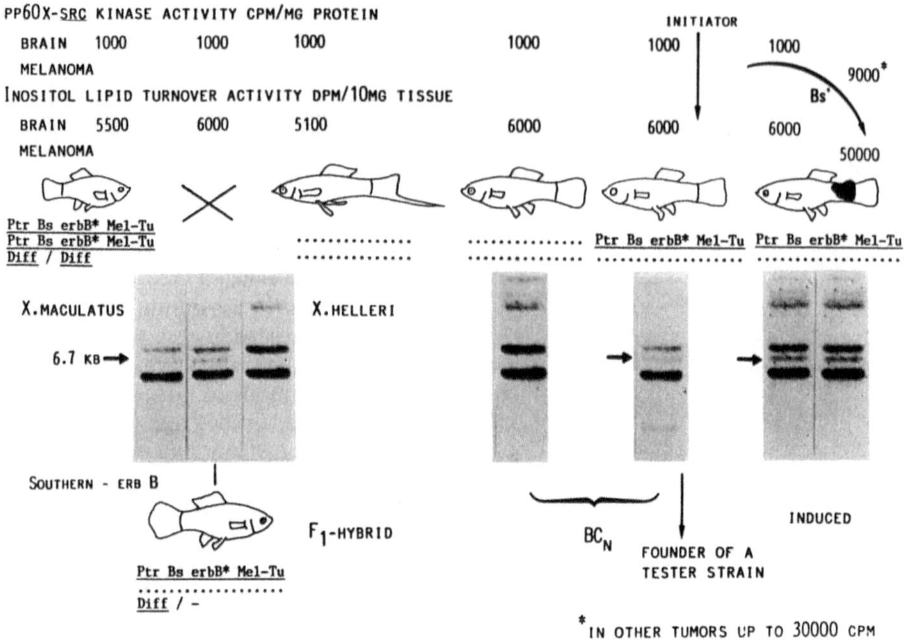

Fig. 8. Crossing procedure for the production of mutagen-carcinogen-sensitive backcross hybrids. Differences to the scheme shown in Fig. 3 are the replacement of the mutated *Df'* by the nonmutated body side-specific regulatory gene *Bs* that suppresses melanoma formation, and the replacement of the 4.9-kb *Eco*R1 fragment indicating *x-erb*B* by a 6.7-kb fragment indicating the same *x-erb*B* gene. See text

specific regulatory gene *Df'*, contains the nonmutated body side-specific regulatory gene *Bs;* in addition, the x-*erb*B* oncogene represented by the 4.9-kb fragment was replaced by a translocated Y-chromosomal copy that is represented by a 6.7-kb fragment. The other genetic conditions are the same as those described in Fig. 4. Melanoma development is suppressed by *Bs* in all purebred and hybrid animals carrying the *Tu* complex. All BC hybrids carrying the *Tu* complex including x-*erb*B* (they can be recognized by their pterinophore-specific reddish coloration coded by *Ptr*) are susceptible to melanoma (and other neoplasms) and may develop melanoma after treatment with physical or chemical carcinogens. Susceptibility to neoplasia or sensitivity to carcinogens, respectively, is inherited in a Mendelian fashion, but the tumors are, as a consequence of a somatic mutation of *Bs* to *Bs'*, nonhereditary and show no relapse after complete removal.

In contrast to the mating-conditioned spontaneous melanoma developing BC hybrids the carcinogen-sensitive BC hybrids show no elevation of $pp60^{x\text{-}src}$ activity as well as no elevation of inositol lipid turnover in the brain. Elevations of these functions are only detected in the neoplasm.

I. Nutrient- and Endocrine-Conditioned Neoplasia

Evidence for nutritional as well as endogenous and exogenous hormonal influences on human cancer has been accumulating over the past 20 years [58]. The agents exerting these influences, often called "promoters" or "cocarcinogens," are by no means mutagenic carcinogens, i.e., "initiators," but appear as agents affecting the course of differentation and the rate of proliferation of cells that have already undergone the genetic key event underlying neoplasia irrespective of whether they are tumor precursor cells or definite tumor cells; the changes in cell differentiation and cell proliferation appear as the last step in the chain of events resulting in cancer.

Many data on this subject come from epidemiological studies [59, 60]. It has been found that breast and colon cancer, which represent a high percentage of total neoplasias in humans, are highly correlated to animal fat intake in a large number of countries, and it has been proposed that low animal fat intake is responsible for a low incidence of these neoplasms, while high animal fat intake is responsible for a high. incidence. The order of countries begins (low fat intake, low tumor rate) with Thailand, the Philippines, Japan, Taiwan, continues to Czechoslovakia, Austria, France, Switzerland, Poland, the Netherlands, and Finland, and ends with the United States, Canada, Denmark, and New Zealand (high fat intake, high tumor rate). A more critical view, however, indicates that the tumor incidence of the Dutch is twice as high as that of the Finns, though both have the same fat intake. The same is true, if we compare the Swiss (high tumor incidence) with the Poles (low tumor incidence, but same fat intake). The Danes have an extremely high animal fat intake and an extremely high incidence of breast cancer. If one compares, however, the population of Copenhagen with that of the rural Denmark one finds that fat intake in Copenhagen is much lower than in rural Denmark while urban Danes have a higher tumor incidence than rural Danes.

This is not to say that fat intake will have no influence on the incidence of breast and colon cancer; however, our critical view of the data makes clear that fat intake alone cannot explain the differences in tumor incidence in different countries. There could be genetic factors involved in such a way that countries showing a high tumor incidence not only have a high fat intake but also contain a high percentage of individuals that are highly sensitive to the tumor-promoting effect of the fat. These genetic factors may also be related to an effect on normal body growth as has been reported in mouse studies [61, 62]. Thus, these genes might interact with a multitude of other nutritional factors, such as simple caloric intake, quantity and quality of protein ingested, as well as drugs that influence the general condition of an individual.

Our own studies concentrated first on the construction of strains of *Xiphophorus* that are highly sensitive to tumor promoters. Figure 9 shows the development of such a strain based upon the same genotypes and crossing procedures as were used for the production of BC hybrids that develop melanoma spontaneously (see Fig. 4). The only difference is that the genome of the animals contains a homozygous autosomal gene, "golden" (*g*/*g*), by which pigment cell differentiation is delayed in the stage of stem melanoblasts. Thus, the BC hybrids corresponding to those developing malignant melanoma spontaneously are incapable of developing a neoplasm. Chemical agents, such as cyclic AMP, corticotropin, a large variety of steroid hormones including testosterone, trenbolone [41], as well as general environmental changes, such as decrease in temperature and increase in salinity of the water in the tank, promote after a latent period of only 4 weeks (latent period of the carcinogen-dependent melanomas is 8–12 months; see preceding paragraph) almost simultaneously the differentiation of large amounts of the noncompetent cells to the competent ones, which subsequently give rise to the melanoma exactly at that place at the body of the fish where they are expected to grow according to the basic crossing experiment (compare Figs. 4, 9).

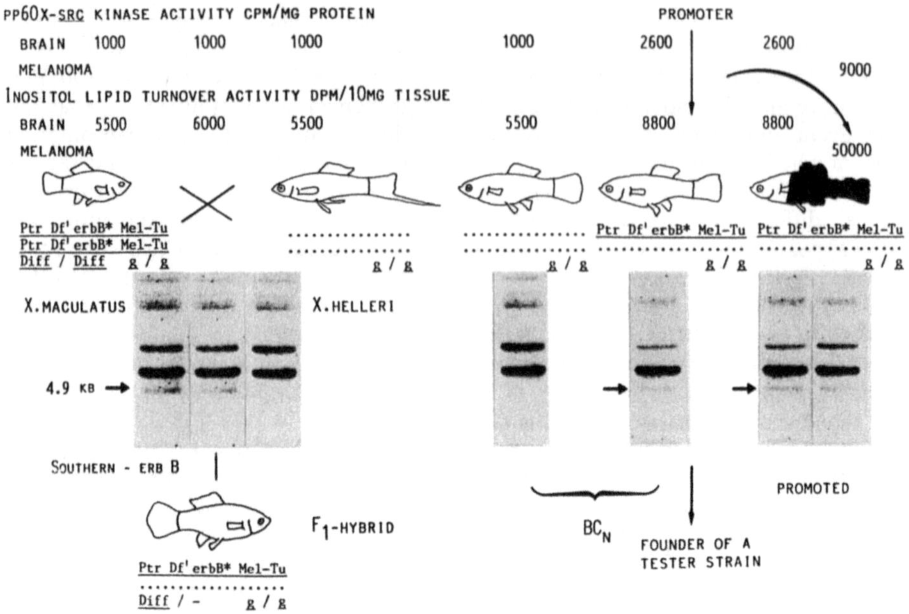

Fig. 9. Crossing procedure for the production of promoter-sensitive backcross hybrids. The only difference to the scheme shown in Fig. 3 is the presence of the homozygous gene "golden", *g*/*g*, by which pigment cell differentiation is delayed. See text

The very short latent period and the very fast growth of the occurring melanomas as compared with that of the carcinogen-induced tumors is remarkable, but is in line with the enhanced pp60$^{x\text{-}src}$ kinase activity and the enhanced phosphoinositide turnover found in the healthy tissues. It appears that, corresponding to the deletion mutant E_1 (see Fig. 4), the molecular and biochemical machinery leading to neoplasia is running in the susceptible but still tumor-free fish and becomes immediately effective as the competent cells become available for promotion of cell differentiation.

The latter results, again, indicate that both the enhanced activity of the xiphophorine *src* oncogene and the enhanced phosphoinositide turnover are intimately linked with the inheritance of x-*erb*B*, which is presumably involved in the key signal preceding melanoma formation in *Xiphophorus*. They furthermore show again that it should be possible to screen for sensitivity and insensitivity to tumor promoters.

J. Future Goals

In this lecture I have tried to explain some observations on human cancer from the view of a biologist working with a fish model. Of course, what I have presented is not altogether new. Nevertheless, what can we learn from the fish? First of all we should make informed decisions to control the chemical and physical carcinogens and promoters we receive today from our polluted environment. However, we should keep in mind that cancer not only depends on the agents but also on the genes that have been part of our evolution since life began. These genes have experienced mutation, duplication, selection, and genetic drift, and are controlled by oncostatic genes that keep a tight rein on them. To learn how these regulatory genes keep the oncogenes in check should be an challenging but fulfilling task in the future of cancer research.

Acknowledgment. Thanks are due to Professor Steward Sell, Houston (Texas) and to Professor Avril Woodhead, Brookhaven Natl. Lab. (New York), for discussions and critical reading of the manuscript.

References

1. Kaiser HE (1981) Neoplasms – comparative pathology of growth in animals, plants, and man. Williams and Wilkins, Baltimore
2. Anders F (1983) The biology of an oncogene, based upon studies on neoplasia. In: Neth R, Gallo RC, Greaves MF, Moore MAS, Winkler K (eds) Modern trends in human leukemia V. Springer, Berlin Heidelberg New York, pp 186–206 (Haematology and blood transfusion, vol 28.)
3. Anders F, Schartl M, Barnekow A, Schmidt CR, Lüke W, Jaenel-Dess G, Anders A (1985) The genes that carcinogens act upon. In: Neth R, Gallo CR, Greaves MF, Janka G (eds) Modern trends in human leukemia VI. Springer, Berlin Heidelberg New York, pp 228–252 (Haematology and blood transfusion, vol 29.)
4. Burk O (1987) Konservierung von homologen Sequenzen der kernprotein-kodierten Onkogene *v-fos, v-myc* und *v-myb* im Tierreich. Thesis, University of Giessen
5. Kaiser P (1988) Restriktionsanalytische Untersuchungen der Konservierung von homologen Sequenzen zu den Onkogenen *v-erb* A, *v-erb* B and *c-neu* im Tierreich. Thesis, University of Giessen
6. Schleenbecker U (1988) Untersuchungen zur Struktur und Funktion von homologen Sequenzen des Wachstumsfaktors und des Wachstumsfaktorrezeptors bei *Xiphophorus* (Teleostei: Poeciliidae). Thesis, University of Giessen
7. Zechel C (1988) Untersuchungen zur Struktur und Funktion *v-erb* A- und *v-erb* B-homologer Sequenzen im *Xiphophorus*-Tumorsystem. Thesis, University of Giessen
8. Schmidt D (1988) Restriktionsanalytische Untersuchungen an genomischer DNA von *Xiphophorus*. Thesis, University of Giessen
9. Gröger H (1987) Isolierung und Charakterisierung von *c-myc*-spezifischen Klonen aus einer lambda EMBL 4 Genbank von *Xiphophorus maculatus (DrLi/Sr)*. Thesis, University of Giessen
10. Pfütz M (1988) Sequenzierung *c-erb* A-spezifischer Sequenzen aus dem Genom von *Xiphophorus*. Thesis, University of Giessen
11. DeFeo-Jones D, Scolnik E, Koller R, Dhar R (1983): *ras*-Related gene sequences identified and isolated from *Saccharomyces cerevisiae*. Nature 306:707–709
12. Pimentel E (1986) Oncogenes. CRC, Boca Raton
13. Reddy EP, Skalka AM, Curran T (1988) The oncogene handbook. Elsevier, New York
14. Berridge MJ (1987) Insositol triphosphate and diacylglycerol: Two interacting second messengers. Annu Rev Biochem 56:159–193
15. Berridge MJ (1987) Inositol lipids and cell proliferation. Biochim Biophys Acta 907:33–45
16. Berridge MJ, Irvine RF (1984) Inositol triphosphate, a novel second messenger in cellular signal transduction. Nature 312:315–321
17. Schartl M, Barnekow A (1982) The expression in eukaryotes of a tyrosine kinase which is reactive with pp $60^{v\text{-}srr}$ antibodies. Differentiation 23:109–113
18. Barnekow A, Schartl M (1983) Cellular *src* gene product detected in the freshwater sponge *Spongilla lacrustis*. Mol Cell Biol 4:1179–1181
19. Albino AP, Strange RL, Oliff AL, Furth ME, Old LJ (1984) Transforming *ras* genes from human melanoma: a manifestation of tumor heterogeneity. Nature 308:69–72
20. Barnekow A, Paul E, Schartl M (1987) Expression of the c-*src* protooncogene in human skin tumors. Cancer Res 47:235–240
21. Ohno S (1970) Evolution by gene duplication. Springer, Berlin Heidelberg New York
22. Zechel C, Schleenbecker U, Anders A, Anders F (1988) v-*erb* B related sequences in *Xiphophorus* that map to melanoma determining Mendelian loci and overexpress in a melanoma cell line. Oncogene 3:605–617
23. Anders A, Anders F (1978) Etiology of neoplasia as studied in the platyfish-

swordtail system. Biochim Biophys Acta 516: 61–95
24. Anders F (1967) Tumor formation in platyfish – swordtail hybrids as a problem of gene regulation. Experientia 23: 1–10
25. Vielkind J, Dippel E (1984) Oncogene-related sequences in xiphophorine fish prone to hereditary melanoma formation. Can J Genet Cytol 26: 607–614
26. Anders F, Schartl M, Scholl E (1981) Evaluation of environmental and hereditary factors in carcinogenesis, based on studies in *Xiphophorus*. In: Dawe CD, Harshbarger JC, Kondo S, Sugimura T, Takayama S (eds) Phyletic approaches to cancer. Japan Scientific Societies, Tokyo, pp 289–309
27. Gordon M (1947) Speciation in fishes. Adv Genet 1: 95–132
28. Kallman KD (1975) The platyfish, *Xiphorus maculatus*. Handb Genet 4: 81–132
29. Anders F, Schartl M, Barnekow A, Anders A (1984) Xiphophorus as a in vivo model for studies on normal and defective control of oncogenes. Adv Cancer Res 42: 191–275
30. Moriwaki K, Shiroishi T, Miyashita N, Kondo N, Imai H (1980) Intersubspecies hybrid of mice as a tool for studying genetic governing tumor development. Gann Monogr 25: 165–176
31. Anders F (1981) Erb- und Umweltfaktoren im Ursachengefüge des neoplastischen Wachstums nach Studien an *Xiphophorus*. Verhandl Ges Dtsch Naturforscher Ärzte 22: 106–119
32. O'Brien S (1980) The extent and character of biochemical genetic variation in the domestic cat. J Hered 71: 2–8
33. Fuerst PA, Chakraborty R, Nei M (1977) Statistical studies on protein polymorphism in natural populations I. Distribution of single locus heterozygosity. Genetics 86: 455–483
34. Schull WJ (1979) Genetic structure of human populations. J Toxicol Environ Health 5: 17–25
35. Lubs HA, Kimberling WJ, Hecht F, Patil SR, Brown J, Gerald P, Summit RL (1977) Racial differences in the frequency of Q and G chromosomal heteromorphisms. Nature 268: 631–632
36. Yamada K, Hasegawa T (1978) Types and frequencies of Q-variant chromosomes in a Japanese population. Hum Genet 44: 89–98
37. Higginson J (1969) Present trends in cancer epidemiology. Proc 8th Canadian Cancer Conf. Pergamon, Toronto, pp 40–75
38. Higginson J (1988) Changing concepts in cancer prevention: limitations and implications for future research in environmental carcinogenesis. Cancer Res 48: 1383–1389
39. Hoover K (1984) Use of small fish species in carcinogenicity testing. NCI Monogr 65
40. Zechel C, Schleenbecker U, Anders A, Anders F (1988) Regulation of gene expression in the Gordon-Kosswig melanoma system. In: Schröder H (ed) Genetics and mutagenesis of fish. (in press)
41. Montesano R, Barth H, Vainio H, Wilborn J, Yamasaki H (1986) Long-term and short-term assays for carcinogenesis. IARC scientific publications no 83. IARC, Lyon, pp 103–126
42. Mawdesley-Thomas LE (1975) Neoplasia in fish. In: Tibelin WE, Migaki G (eds) The pathology of fishes. University of Wisconsin Press, pp 805–870
43. Dahme E, Weiss E (1988) Grundriß der speziellen pathologischen Anatomie der Haustiere, 4th edn. Enke, Stuttgart
44. Mäueler W (1988) Untersuchungen zur tumorspezifischen Genexpression bei *Xiphophorus* (Teleostei: Poeciliidae): 1. Enzyme des Intermediärstoffwechsels, 2. Expression von Proto-Onkogenen. Thesis, University of Giessen
45. Krekeler G (1988) Isolierung und Charakterisierung *v-myb*-homologer Sequenzen aus einer Genbank von *Xiphophorus* (Pisces: Poeciliidae). Thesis, University of Giessen
46. Schartl M, Schmidt CR, Anders A, Barnekow A (1985) Elevated expression of the cellular *src* gene of differing etiologies in *Xiphophorus*. Int J Cancer 36: 199–207
47. Gronau T (1987) Untersuchungen zur Organisation, Aktivität und Wirkung des zellulären Onkogens *c-src* im *Xiphophorus*-Tumorsystem. Thesis, University of Giessen
48. Pröfrock A (1988) Untersuchungen zum Phosphatidylinosit-Turnover an ausgewählten *Xiphophorus*-Genotypen. Thesis, University of Giessen
49. Smith AD, Gronau T, Pröfrock A, Zechel C, Bird IM, Lane PA, Barnekow A, Anders A, Anders F (1988) EGF receptor gene, inositol lipid turnover and c-*src* activity in key processes preceding melanoma in *Xiphophorus*. In: Lynch HT,

Fusaro RM (eds) Hereditary Malignant Melanoma. CRC, Boca Raton (in press)
50. Lynch HT, Lynch JF, Fusaro RM (1985) Clinical importance of familial cancer. In: Müller H, Weber W, Kuttapa T (eds) Familial cancer. Karger, Basel, pp 6–12
51. Ahuja MR, Lepper K, Anders F (1979) Sex chromosome aberrations involving loss and translocation of tumor-inducing loci in *Xiphophorus*. Experientia 35: 28–29
52. Anders A, Anders F, Klinke K (1973) Regulation of gene expression in the Gordon-Kosswig melanoma system. In: Schröder H (ed) Genetics and mutagenesis of fish. Springer, Berlin Heidelberg New York, pp 33–63
53. Anders A, Dess G, Nishimura S, Kersten H (1985) A molecular approach to the study of malignancy and benignancy in melanoma of *Xiphophorus*. In: Bagnara J, Klaus SN, Paul E, Schartl M (eds) Pigment cell 1985. University of Tokyo Press, pp 315–324
54. Müller HJ, Weber W, Kuttapa T (1985) Familial cancer. Karger, Basel
55. Rudolph B, Harbott J, Lampert F (1988) Fragile sites and neuroblastoma: Fragile site at one p 13.1 and other points on lymphocyte chromosomes from patients and family members. Cancer Genet Cytogenet 31: 83–94
56. Lynch HT, Lynch JF (1985) Genetics and colorectal cancer. In: Müller HJ, Weber W, Kuttapa T (eds) Familial cancer. Karger, Basel
57. Newman B, Austin MA, Lee M, King M-C (1988) Inheritance of human breast cancer: Evidence for autosomal dominant transmission in high-risk families. Proc Natl Acad Sci USA 85: 3044–3048
58. Cohen LA (1987) Diet and cancer. Sci Am 257 (5): 42–48
59. Hecker E, Fusenig NE, Kunz W, Marks F, Thielmann HW (1982) Cocarcinogenesis and biological effects of tumor promoters. Raven, New York
60. Carroll KK (1975) Experimental evidence of dietary factors and hormone-dependent cancers. Cancer Res 35: 3374–3383
61. Wynder EL (1975) The epidemiology of large bowel cancer. Cancer Res 35: 3388–3394
62. Heston WE (1957) Effects of genes located on chromosomes III, V, VII, IX, and XIV on the occurrence of pulmonary tumors in the mouse. Proc Intern Genetics Symposia, 1956. Cytologica [Suppl]: pp 219–224

Wilsede Scholarship Holders

Fed. Rep. of Germany

Dr. Atzpodien, Jens, Med. Hochschule Hannover Abt. Hämatologie u. Onkologie, Postfach 61 01 80, 3000 Hannover 6

Dr. Bier, Verena, Kinderklinik der Universität Heidelberg, Im Neuenheimer Feld 150, 6900 Heidelberg

Dr. Boos, Joachim, im Heidgarten, 4400 Münster

Dr. med. Brandenburg, Irene, Abt. f. Humangenetik der Christian Albrecht Universität, Schwanenweg 24, 2300 Kiel

Dr. Bührer, Christoph, Kinderklinik der Medizinischen Hochschule, 3000 Hannover 61

Debus, Otfried, Adolfstr. 33, 2300 Kiel 1

Dittmann, Klaus, Klinik. Großhadern/Univ. München Medizinische Klinik III, Klinische Hämatologie, Landwehrstr. 61, 8000 München 2

Faller, Gerhard, Pathologisches Institut Universität Würzburg, Luitpoldkrankenhaus, 8700 Würzburg

Dr. Fiedler, Walter, II Medizinische Klinik Abt. Onkologie u. Hämatologie, Univ. Klinik Eppendorf, Martinistr. 52, 2000 Hamburg 20

Dr. Gohla, Sven, Medizinische Klinik Abt. für Klinische Chemie, Univ. Krankenhaus Eppendorf, Martinistr. 52, 2000 Hamburg 20

Grund, Christian, Robert-Koch-Institut, Wilhelmshavener Str. 14, 1000 Berlin 21

Halberstadt, Antje, Am Zollstock 24, 6300 Gießen

Dr. Janssen, J. W. G., Klinikum der Universität Ulm, Zentrum für Kinderheilkunde, Postfach 38 80, 7900 Ulm

Dr. Jähde, Eckhard, Baumsatzstr. 64, 7401 Pliezhausen

Dr. Kalthoff, Holger, Abt. f. klin. Immunologie, Med. Universität Eppendorf, Martinistr. 52, 2000 Hamburg 20

Dr. Klausmann, Martine, Klinikum der Philipps Univ. Marburg/Zent. Innere Med., Med. Klinik u. Poliklinik, Baldingerstr., 3550 Marburg

Landfester, Cornelia, Albert-Ludwigs-Universität, Dreikönigstr. 26, 7800 Freiburg

Dr. Leib-Mösch, Christine, Ges. f. Strahlen- und Umweltforschung, Abt. Molekulare Zellpathologie, Ingolstädter Landstr. 1, 8042 Neuherberg

Lindemann, Albrecht, Joh. Gutenberg-Uni. Klinikum, I. Med. Klinik u. Poliklinik, Langenbeckstr. 1, 6500 Mainz

Odenkirchen, Andreas, Samwerstr. 2, 2300 Kiel

Dr. Opalka, Bertram, Universitätsklinikum Essen, Inst. für Molekularbiologie, Tumorforschung, Hufelandstr. 55, 4300 Essen

Schedel, Roswitha, Universitäts Kinderklinik, Abteilung Pädiatrie II, 7900 Ulm

Dr. Stoll, Matthias, Abt. Immunologie, Medizinische Hochschule Hannover, Konstanty-Guschow-Str. 8, 3000 Hannover 61

Dr. Teichmann, Johannes, Freie Universität Berlin, Medizin. Klinik u. Poliklinik, Universitätsklinikum Steglitz, Hindenburgdamm 30, 1000 Berlin 45

Voltz, Raymond, Heiglhofstr. 64/410, 8000 München 70
Dr. Wörmann, Bernhard, Breul 4, 4400 Münster
Dr. Wilhelm, Martin, Medizinische Poliklinik der Universität Würzburg, Klinikstr. 8, 8700 Würzburg
Dr. Wolf, Jürgen, Institut für Virusforschung, Deutsches Krebsforschungszentrum, Im Neuenheimer Feld 280, 6900 Heidelberg
Zechel, Christina, Ebertalstr. 24, 6293 Lühnberg 2
Dr. Zühlsdorf, Michael, Westfälische Wilhelms-Univers. Münster, Universitätskinderklinik und Poliklinik, Albert-Schweitzer-Str. 33, 4400 Münster

United Kingdom

Dr. Bains, Margaret, LRF Preleukaemia Unit Dept. of Haematology, University Hospital of Wales Heath Park, Cardiff CF4 4XN
Bearpark, Andrew, LRF Centre at the Inst. of Cancer Research, Fulham Road, London SW3 6JB
Dr. Chessells, Judith Mary, Dept. of Haematology/Oncology, The Hospital for Sick Children, Great Ormond Street, London WC1N 3JH
Dr. Galvani, David, Dept. of Haematology, Duncan Building, Royal Liverpool Hospital, Prescot Street/PO. Box 147, Liverpool L69 3BX
Gledhill, Sarah, LRF Virus Center at the Univ. of Glasgow, Dept. of Veterinary Pathology, Bearsden Road, Glasgow G61 1QH
Dr. Kölble, U., Dept. of Haematology, John Radcliffe Hospital, Oxford OX3 9DU, Haedington
Ph. D. McKinney, Patricia, LRF Centre f. Clinical Epidemiology at the University of Leeds, 17 Springfield Mount, Leeds LS2 9NG
Dr. Merkenschlager, Matthias, ICRF Human Tumor Immunology, Univ. College/ Middlesex School, The Courtauld Inst. of Biochem., 91 Riding House Street, London W1P 8BT
Dr. Mitchell, Christopher, Dept. of Haematology and Oncology, The Hospital for Sick Children, Great Ormond Street, London WC1N 3 JH
Dr. Pezzella, Francesco, Dept. of Haematology, John Radcliffe Hospital, Headington, Oxford OX3 9DU
Dr. Wright, Eric, MCR Radiobiology Unit Chilton, Didcot, Oxon OX11 0RD

USA

Dr. Buchow, Hartmut, Dept. of Health & Human Service, National Institute of Health, National Cancer Institute, Bethesda, Maryland 20892
Dr. Griesinger, Frank, Dept. of Lab. Medicine/Pathol., Box 86 UMHC, Harvard Street/East River Road, Minneapolis, MN 55455-0315
Dr. Koeppen, Hartmut, The University of Chicago, Dept. of Pathology, Committee on Immunology, 5841 S Maryland Ave./Box 414, Chicago, IL 60637
Dr. Melchner, H. von, Center for Cancer Research, Massachusett Inst. of Technol., 40 Ames Street E 17-517, Cambridge, M.A. 02139
Dr. Warren, David, Memorial Sloan-Kettering, Cancer Center, 145 Bosten Post RD, Rye, N.Y. 10580

Preface

You see things, and say why?
But I dream things that never were,
and I say, why not?

George Bernhard Shaw

Far ahead of his time, June 1st, 1909, Alexander Maximov communicated in a lecture, given in the Charité in Berlin, the fundamental knowledge, that there exists a lymphoid hemopoetic stem cell.

Alexander Friedenstein explained that during the following years, Maximov also showed that the idea of interaction between hemopoetic cells and their stroma to be one of the most significant experiences.

Monoclonal antibodies, recombinant DNA technics and the improvement of tissue culture models are the major developments to improve our possibilities to clarify growth and differentiation functions of hemopoetic cells.

During the last two decades it was shown that soluble products, released from T cells, were not only involved in inducing B cells to produce specific immunoglobulin secretion after antigen stimulation. Furthermore, lymphokines together with other cytokines regulate the growth and differentiation of hemopoetic cells.

As I have learned from Dick Gershon, our knowledge of the cellular basis for immunoregulation has come a long way since 450 B.C. Thucydides comments on the possible role of immune response in controlling the Black Death. Dick Gershon speculated that no scientific interest for these interesting observations was put forth at that time. Perhaps the problems, the Athenians were having with the Spartans, converted money from basis research into the military budget.

He also found documentations in western literature that scientific progress not only in our time could be made by industry rather than by academicians.

In Turkey, parents pretreated their girls with scabrous material obtained from subjects suffering from active smallpox to protect their daughters' beauty against pock marks.

But the first known inoculations have been invented in China during the Sung Age 1000 after Christ and here we have to deal with a different story from what Richard K. Gershon has learned in Turkey.

During the government of Jun Tsung (1023–1063) a Tibetan monk from the Omei mountains at the Tibetan border inoculated the son of an old Minister, who had lost already his other children. For this inoculation he used scabrous material obtained from cases with active smallpox. The material he kept in summer for 15 to 20 days and in winter for 40–50 days to get attenuated material. For this successful treatment, the Minister wanted to give to the monk a generous gift. The monk, however, refused any gift and instead he requested the Minister to serve his people faithfully. Then the monk returned to his mountains and continued to be a hermit there as he was before.

The inoculation, however, became quickly more and more popular in China. Out of the judging most probably, Turkish mothers have learned to protect the beauty of their daughters. Lady Wortly Montagu (1690–1762) got to know 1716 in Constantinople this treatment against smallpox, which was practised there already quite frequently. On March 18, 1718 she successfully inoculated her own son against strong opposition in England.

Fotos Regina Völz

She could publish her results only in a newspaper, the flying post. Later on, after Jenner and others, vaccination became a perfectly recommended protection against infection.

Av Mitchison made a summary of the modern trends in this field as discussed during the 8th Wilsede Meeting. I have to thank him for this and all the praticipants who made the 8th Wilsede Meeting again a high light.

Now I want to conclude my personal preface with a statement "I hate chaos" but especially in Wilsede I learned, "chaos loves me". Susumo Ohno taught me that there has been increasing realisation that there is order in chaos as well. So far, I don't understand the universal metric properties of non-linear transformations of Feigenbaum (1985 J. Stat. Physics 21: 669–706) and I don't succeed in understanding the secret laws which organise the chaos, like the weather, with the help of the fractal geometry of Mandelbrot. However, I have heard about a some hundred years old statement of Novalis about the sensible chaos.

We will prove in the 9th Wilsede Meeting 1990 whether there are comprehensive realities between chaos and biological aspects of human life and look on Egon Degens "Cosmic Composed" following his "Blueprints of Early Life".

Personal and scientific discussions around the Emmenhof

Fotos Regina Völz

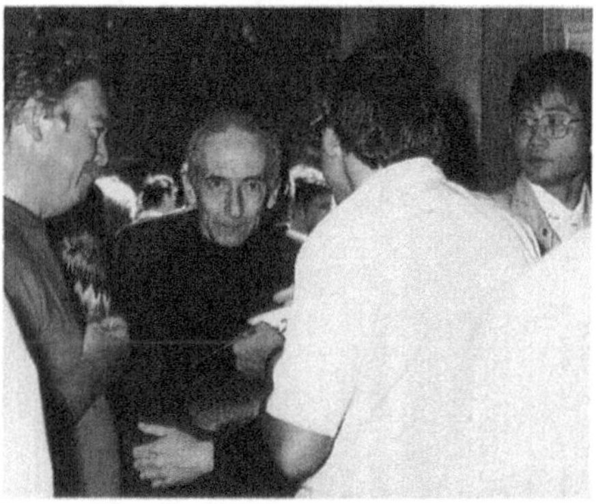

Acknowledgements

Bundesministerium für Jugend, Familie, Frauen und Gesundheit
Deutsche Forschungsgemeinschaft
Deutsche Krebshilfe
Erich und Gertrud Roggenbuck-Stiftung zur Krebshilfe, Hamburg
Freie und Hansestadt Hamburg
Hamburger Landesverband für Krebsbekämpfung und Krebsforschung e.V., Hamburg
Leukemia Research Fund, Great Britain
Leukemia Society of America
Elisabeth Jannsen-Stiftung

For their generous hospitality we thank the Stiftung F.V.S. zu Hamburg, Verein Naturschutzpark e.V., Hamburg, the Amerikahaus in Hamburg and the Freie und Hansestadt Hamburg.

I would like to thank Dr. Jürgen Wieczorek, Ms. Anne Clauss and Mr. J. W. Schmidt of Springer-Verlag for their assistance in the production of the book. On behalf of the authors and editors: Rolf Neth.

VIII Wilsede Symposium on Current Trends in Leukaemia Research Introduction: Differentiation-Oncogenes

N.A. Mitchison

Each of the successive Wilsede Symposia has provided a fine vantage point from which to survey current progress in leukaemia research. Because of their regularity, their consistently high standard of presentation, their broad international coverage, and their informal but highly critical atmosphere they have come to be accepted as providing authoritative statements of the achievements and research agenda of the day. This is the eighth of these biennial symposia, and it conforms to the same high standards.

From this vantage point in 1988 the most striking feature is the enormous strength and breadth of molecular genetics. This starts from the now universally accepted assumption that leukaemia, in common with other forms of neoplasia, develops as a result of mutational changes in the genome of a cell. These mutational changes occur serially and embrace a fairly wide range of options, that evidently include rearrangements, deletions, and loss of DNA as well as single-base substitution. Our background thinking on this subject is deeply coloured by knowledge of the cumulative increase of cancer incidence with age, cooperation between cytoplasmic and nuclear oncogenes, and the special susceptibility to DNA transformation of partially transformed cell lines. This view now dictates the main agenda of leukaemia research, which is to elucidate the mechanisms through which these mutational changes alter the behaviour of cells. An understanding of these mechanisms, it is confidently believed, will enable us to control and eventually eradicate leukaemia and other forms of cancer.

Not only does genetics dictate the agenda, but also in the form of recombinant DNA technology provides an immensely powerful set of tools. With these tools one felt at this symposium that almost any task can now be accomplished: sites of mutation can be located with total precision, functions identified by site-specific mutagenesis, and the interactions of a gene or its control elements with the rest of the cell can be analysed by transfection. Things move with great speed because DNA is after all just DNA: each protein confronts us with a new set of problems, but the problems posed by a gene and the methods for solving them by recombinant DNA technology are transferable. For instance this symposium includes reports on the first fruits in human leukaemia research of Cetus' new method of generating multiple copies of a short length of DNA in between oligonucleotide end-markers. The method is used by J. Rowley to study chromosome breakpoints, and by R.-A. Padua to identify *Ras* mutations.

Within the landscape defined by genetics a significant shift in emphasis is taking place, from dominant to recessive oncogenes. This is an abbreviated and somewhat misleading way of characterizing an important shift in the direction of research. The crucial point is not so much how many copies of a gene are needed for manifestation in the phenotype of a cell, but how the gene works. In general, genes that code for growth fac-

Imperial Cancer Research Fund, Tumour Immunology Unit, University College London, Department of Biology, Medawar Building, Gower Street, London, WC1E 6BT, UK tors, their receptors, or the cascade of messages that they trigger in the cell produce cancer through *activation*; while those that code for mechanisms of differentiation do so through *inactivation*. And, in general, since both copies of a diploid gene will need to be inactivated to prevent differentiation, the latter will behave as recessives. The new emphasis then is on genes that control differentiation, and their inactivation as differentiation-oncogenes in cancer. This emphasis goes back to the pioneering work of H. Harris in somatic genetics, where cancer cells upon fusion with normal cells generally display a normal phenotype. New experimental results to the same effect were presented in a poster by J. Wolf et al. on fusions between malignant and non-malignant B cells, and similar results in the papilloma virus system were discussed by H. zur Hausen. Over the last 2 decades this line of research became bogged down in sterile controversy about the generality of the result, and exactly what the famous Minz-Ihlmensee "suppression of malignancy by differentiation" experiments really mean. Now, thanks to molecular genetics, a way forward is open.

Leading on from the original ideas derived from work on somatic hybrids, three lines of approach to the recessive oncogene problem can be distinguished in this symposium. One is via formal genetics, and represents development of the concepts first formulated to explain the familial inheritance of retinoblastoma, Wilm's tumour, and coeliac polyposis. Another is via development, where our increasing understanding of molecular mechanisms in cell biology helps to identify situations in which recessive oncogenes able to inhibit normal differentiation might operate. And a third relates to the major growth factors that are normally associated with dominant oncogenes: research on these molecules and their receptors is beginning to identify control mechanisms that regulate their activity, and that may themselves be disrupted by recessive oncogenes.

The first of these lines of approach is represented by the contributions of J. Rowley, F. Anders, and M. Dean. In her outstanding Frederick Stohlmann Lecture, Rowley surveys the role that chromosome studies have played in identifying dominant oncogenes, and goes on to mention her current interest in monosomy of human chromosome 5 as indicative of recessive oncogene activity. As this chromosome also carries genes for growth factors and their receptors, it is possible – perhaps even likely – that closely linked recessive oncogenes may regulate the expression of these potentially dangerous molecules (at least that is what I understand her to have told me in conversation). Ander's vast effort in the genetics of congenital melanoma in fish (extended also to the genetics of carcinogen susceptibility) has revealed much about the control of dominant oncogenes by the rest of the genome. My guess is that in the future this branch of genetics will need to focus on these presumably recessive control elements, and that something like the mouse recombinant inbred lines will be needed for that task – a formidable undertaking. Dean describes an ongoing study of the long arm of human chromosome 7, often missing in myelodisplastic syndrome with all that that implies for the operation of recessive oncogenes.

The second line of approach, through development, is evident in the papers of M. Moore, T. Waldmann, N. Haran-Ghera, A. Friedenstein, T. M. Dexter, D. Mason, K. Rajewsky, and F. Melchers. Analysis of the interactions between haemopoietic cells and their surrounding stromal cells makes steady progress and the molecules involved in this binding are becoming clearer: this is an area that Friedenstein pioneered, and where Dexter is moving ahead with his studies of solid-phase-bound IL-3. Cell-bound and matrix-bound growth/differentiation functions are here to stay. Perhaps the best-characterized differentiation factor, and certainly the one where a potential for recessive oncogene activity is most evident in its title, is D. Gearing's leukaemic inhibitory factor.

The contribution of M. Lenardo and M. Greaves take us deep into the molecular mechanisms of transcriptional control that underline differentiation.

Finally there are the studies that sketch in the way that dominant oncogenes either respond to developmental control, or escape: those of W. Ostertag, W. Alexander, C. Moroni, and T. Ernst. The last of these provides novel and interesting evidence that enhanced levels of CSF production in leukaemic cells, and the autocrine stimulation that ensues from this, may reflect increased mRNA stability rather than increased transcription: a post-transcriptional modification, and therefore yet one more candidate site for the operation of recessive oncogenes.

Frederick Stohlmann Jr. Memorial Lecture

Wilsede, June 21, 1978

Memorial Tribute to Dr. Frederick Stohlman Presented by William C. Moloney

Gallo, Robert C.: Cellular and Virological Studies Directed to the Pathogenesis of the Human Myelogenous Leukemias

Pinkel, Donald: Treatment of Childhood Acute Lymphocytic Leukemia

Wilsede, June 18, 1980

Klein, George: The Relative Role of Viral Transformation and Specific Cytogenetic Changes in the Development of Murine and Human Lymphomas

Kaplan, Henry S.: On the Biology and Immunology of Hodgkin's Disease

Wilsede, June 21, 1982

Greaves, Melvyn, F.: Immunobiology of Lymphoid Malignancy

Thomas, E. Donnall: Bone Marrow Transplantation in Leukemia

Wilsede, June 17, 1984

Klein, Jan: Introduction for N. Avrion Mitchison

Mitchison, N. Avrion: Repertoire purging by medium-concentration self-macromolecules is the major factor determining helper and suppressor repertoire differences

June 18, 1984

Gallo, Robert C.: Introduction for Peter H. Duesberg

Duesberg, Peter H.: Are Activated Proto-*Onc* Genes Cancer Genes?

Wilsede, June 22, 1986

Vogt, Peter K.: Introduction for Harald zur Hausen

Harald zur Hausen: Viruses in Human Tumors

Moore, Malcolm A. S.: Introduction for Donald Metcalf

Metcalf, Donald: Hemopoietic Growth Factors and Oncogenes in Myeloid Leukemia Development

Wilsede, June 19, 1988

Frederick Stohlman Jr. Memorial Lecture

Rowley, J. D.: Molecular Analysis of Rearrangements in Philadelphia (Phl) Chromosome-Positive Leukemia

Moore, Malcolm A. S.: Interactions Between Hematopoietic Growth Factors: The Clinical Role of Combination Biotherapy

Molecular Analysis of Rearrangements in Philadelphia (Ph^1) Chromosome-Positive Leukemia

J.D. Rowley

A. Introduction

The close association of specific chromosome abnormalities with particular types of human cancer has been established by a number of investigators during the past decade [1–6]. A few of the genes involved in consistent chromosome rearrangements, notably translocations, have already been identified, and it is likely that the identity of most of the genes affected by these aberrations will be determined within the next decade. Moreover, for several of the rearrangements, some of the changes in gene structure and function have been defined. Therefore, some general principles that may be applicable to many chromosome rearrangements in human malignant disease are beginning to emerge. Chronic myeloid leukemia (CML) provides one of the clearest examples of our progress in first identifying a recurring chromosome abnormality and then cloning the genes involved in the abnormality. The analysis of these genes and their alteration as a result of the chromosome change is the subject of this lecture.

B. Cytogenetic and Clinical Features of Chronic Myeloid Leukemia

Chronic myeloid leukemia is important because it was the first human cancer in which a consistent chromosome abnormality was identified. The abnormality is the Philadelphia or Ph^1 chromosome [7], which was shown with banding to involve No. 22 (22q−). The correct chromosome defect was shown to be a translocation involving Nos. 9 and 22; this was the first consistent translocation specifically associated with any human or animal disease (Fig. 1) [8]. The reciprocal nature of the translocation was established only recently, when the Abelson protooncogene, *ABL,* normally on No. 9, was identified on the Ph^1 chromosome [9]. Other studies with fluorescent markers or chromosome polymorphisms have shown that, in a particular patient, the same No. 9 and No. 22 are involved in each cell. The Ph^1 chromosome is present in granulocytic, erythroid, and megakaryocytic cells, in some B cells, and probably in a few T cells. The karyotypes of many Ph^1+ patients with CML have been examined with banding techniques by a number of investigators; in a review of 1129 Ph^1+ patients, the 9;22 translocation was identified in 1036 (92%) [4]. Variant translocations have been discovered, however, in addition to the typical t(9;22). Until very recently, these were thought to be of two kinds; one appeared to be a simple translocation involving No. 22 and some chromosome other than No. 9 (about 4%), and the other was a complex translocation involving three or more different chromosomes, two of which were No. 9 and No. 22 (about 4%). Recent data clearly demonstrate that No. 9 is affected in the simple as well as the complex translocations, and that its involvement had been overlooked [10]. Virtually all chromosomes have been involved in these variant translocations, but No. 17 is affected

The Division of Biological Sciences and the Pritzker School of Medicine, Section of Hematology/Oncology, Box 420, 5841 S. Maryland Avenue, Chicago/Illinois 60637

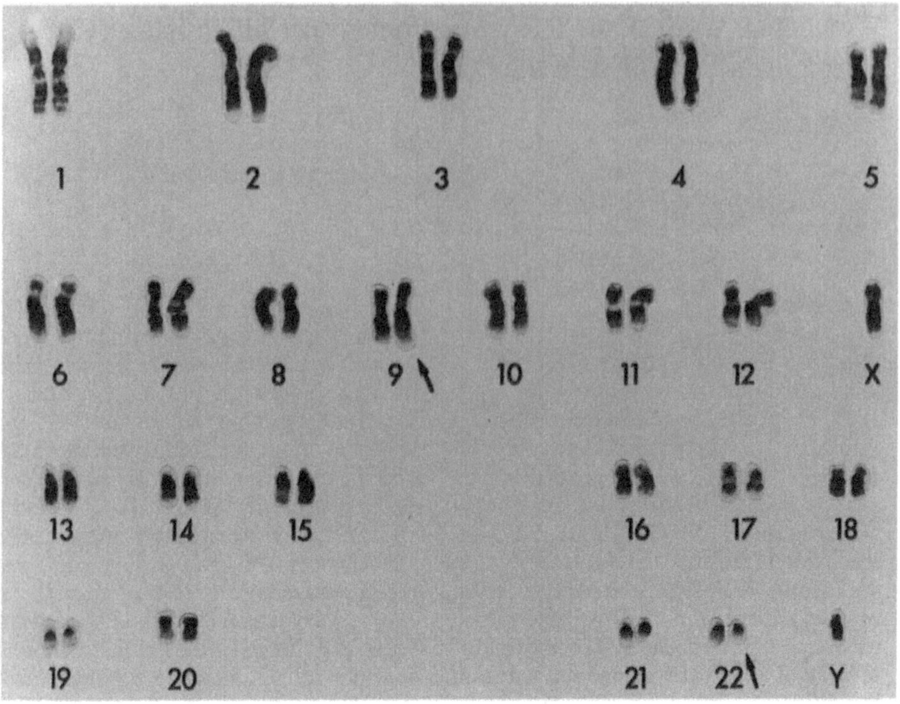

Fig. 1. Trypsin-Giemsa-stained karyotype of a metaphase cell from a bone marrow aspirate obtained from an untreated male with CML illustrating the t(9;22) (q34;q11). The Philadelphia chromosome (Ph^1) is the chromosome on the right in pair 22 (↑). The material missing from the long arm of this chromosome (22q−) is translocated to the long arm of chromosome 9 (9q+) (↑), and is the additional pale band that is not present on the normal chromosome 9

more often than are other chromosomes. The genetic consequences of the standard t(9;22) or the complex translocation involving at least three chromosomes is to move the *ABL* protooncogene on No. 9 next to a gene on No. 22, called *BCR,* whose function is currently unknown (Fig. 2).

Chronic myeloid leukemia usually terminates in an acute leukemia in which the blast cells have either lymphoid or myeloid morphology. In the acute phase, about 10%–20% appear to retain the 46, Ph^1+ cell line unchanged, whereas most patients show additional chromosome abnormalities resulting in cells with modal chromosome numbers of 47 to 50 [4]. Different abnormal chromosomes occur singly or in combination in a distinctly nonrandom pattern. In patients who have only a single new chromosome change, this most commonly involves a second Ph^1, an isochromosome for the long arm of No. 17 [i(17q)], or a +8, in descending order of frequency. Chromosome loss occurs only rarely; that most often seen is −7, which occurs in 3% of patients.

Early cases of acute leukemia in which the Ph^1 chromosome was present were classified as CML presenting in blast transformation; at present, patients who have no prior history suggestive of CML are classified as Ph^1+ acute leukemia. In fact, some of the patients with Ph^1+ ALL have a different breakpoint in the *BCR* gene on No. 22. In blast crisis, some blasts have intracytoplasmic IgM, which is characteristic of pre-B cells, and these cells have an immunoglobulin gene rearrangement [11].

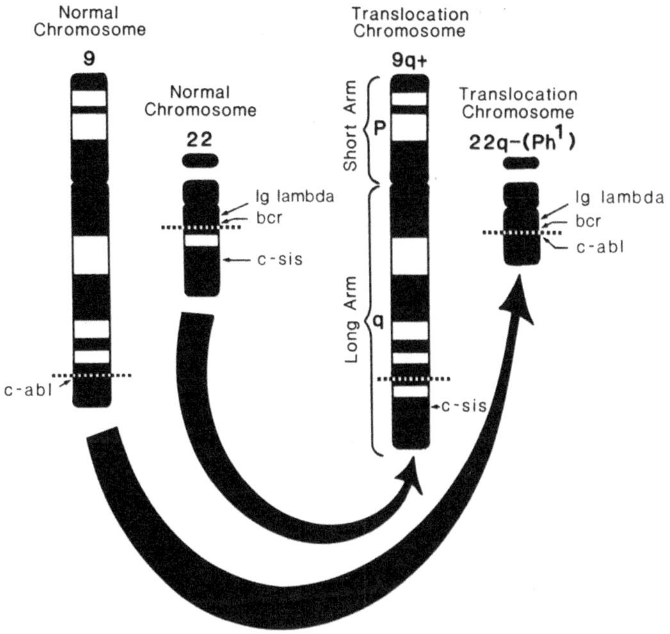

Fig. 2. Schematic drawing of chromosome No. 9 and No. 22 illustrating the chromosome translocation that produces the 9q+ and 22q− (Ph[1]) chromosomes. One protooncogene, *ABL*, is moved to No. 22 adjacent to a gene of unknown function called *BCR*; the break in No. 22 is distal to the IG lambda locus which is not involved in the translocation. The *SIS* protooncogene is moved to the 9q+ chromosome. It is located at some distance from the breakpoint on No. 22 and there is no evidence that it is altered as the result of the translocation

Marrow cells from some patients appear to lack a Ph[1] chromosome. The majority of these patients had a normal karyotype. Somewhat surprisingly, the survival of these patients was substantially shorter than those whose cells were Ph[1]+ [12]. Our recent review of the histology of 25 Ph[1]− patients showed that most of them did not have CML but they had some type of myelodysplasia, most commonly chronic myelomonocytic leukemia or refractory anemia with excess blasts [13]. However, the situation has become more complex because it has been shown recently that some patients with clinically typical CML who lack a Ph[1] chromosome cytogenetically have evidence of the insertion of *ABL* sequences into the *BCR* gene [14, 15]. Thus, it can be proposed that the sine qua non of CML is the juxtaposition of *BCR* and *ABL*.

C. Molecular Analysis of the 9;22 Translocation

Investigators are now in the process of unraveling the mystery of the Ph[1] translocation in CML and ALL. In the t(9;22) in CML and ALL, the Abelson protooncogene (*ABL*) is translocated to the Ph[1] chromosome [9]. The *ABL* gene was first identified because of its homology to the viral oncogene that had been isolated from a mouse pre-B-cell leukemia. The breakpoint junction in CML was cloned and the site on the Ph[1] was called *bcr*, for breakpoint cluster region, [16] since the majority of breaks cluster in a small 5.8-kilobase (kb) region. The gene in which this cluster is located has also been cloned. It is a very large gene greater than 100 kb and it is presently also called *BCR*, which leads to a great deal of confusion. In this lecture, *bcr* is used to denote the CML breakpoint region and *BCR* to identify the whole gene. In contrast to *bcr*, the breaks in *ABL* on No. 9 occur over an incredible distance of more than 200 kb. We have used pulse-

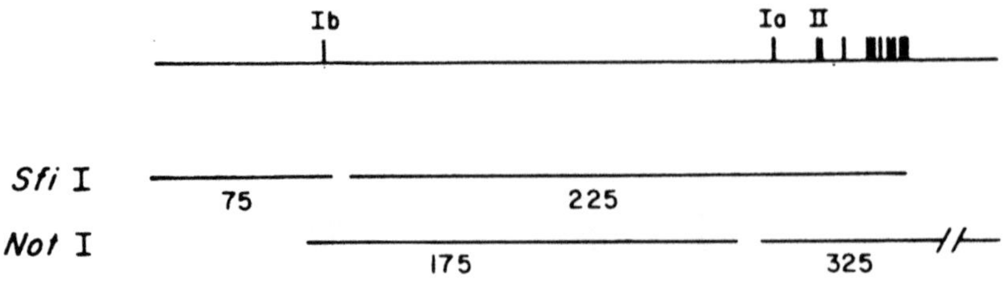

Fig. 3. Map of the *ABL* gene showing the position of the two alternative exons Ib and Ia relative to exon II. Exon Ib is the most 5′ exon, whereas Ia is less than 20 kb from exon II which is the common splice acceptor site. The *vertical bars* above the *horizontal line* represent the more 3′ exons which are homologous to the v-*abl* sequences. *Sfi* I and *Not* I are the enzymes used to determine the relative positions of exons Ia, Ib, and II. (Figure adapted from [17])

field gel electrophoresis (PFGE) to great advantage in the study of the *ABL* protooncogene. Southern blotting with standard gel electrophoresis leads to separation of DNA fragments in the size range of 2 to about 25 kb. Since the *ABL* gene is larger than 200 kb, mapping it in 10- to 20-kb pieces is a formidable task. In contrast, by using PFGE one can separate fragments more than 1000 kb in size, and this technique is also very effective in the 100- to 600-kb range. A normal chromosome band contains roughly 5000–10 000 kb and, thus, several very large, overlapping fragments could contain a single band. Using many probes for *ABL* provided by various investigators, Drs. Westbrook and Rubin have constructed a map of the normal *ABL* gene [17]. This is a very complex gene that normally uses one of two alternative beginnings, exon Ia or Ib. During transcription, either of these can be spliced at the same point on the remainder of the gene, which is called the common splice acceptor site or exon II (Fig. 3). One of their first discoveries was that the type Ib exon mapped more than 200 kb upstream from exon II. As a result, a very large segment of the RNA transcript is removed or spliced out to form the mature mRNA. This is a remarkable feat, not identified before in biological systems. The breakpoints in the chromosomes of various CML patients and cell lines occur in many locations upstream (5′) of exon II. However, the same size (8.5 kb) mRNA is found in all CML patients; this occurs because the *BCR* exons are spliced to *ABL* exon II, resulting in a chimeric mRNA which is translated into a chimeric protein ($p210^{BCR\text{-}ABL}$) [18, 19].

With regard to Ph^1-positive ALL, it has always been an enigma why the typical Ph^1 translocation is seen in ALL and in fact is the most common translocation in adults with ALL [20]. One relatively trivial explanation would be that the patients really had CML in lymphoid blast crisis with an undiagnosed chronic phase, and this may occur in some patients. However, analysis of DNA from some Ph^1-positive ALL cells indicates that the breakpoint in No. 22 is outside the *bcr* region. In one study, the majority of adult patients (13 of 17) appeared to have the same *bcr* rearrangement that is seen in CML whereas it has not been found in any of 7 children, who presumably had a more 5′ breakpoint in the *BCR* gene [21] (Table 1). Data from our

Table 1. Ph^1-positive leukemia

Diagnosis	Number of patients	Number with *bcr* rearrangements [a]
CML	135	133
ALL – Adults	32	17
ALL – Children	8	0

[a] *bcr* rearrangement in CML breakpoint cluster region

laboratory as well as others indicate that the breakpoints on No. 22 are greater than 50 kb proximal to the CML break but that they still are within the *BCR* gene [22]. The breakpoints on No. 9 are similar to those in CML. Several investigators have shown that these Ph^1+ ALL patients have an abnormal size chimeric BCR-ABL mRNA (7.0–7.4 kb) and ABL protein ($p185^{BCR\text{-}ABL}$) [23, 24].

These discoveries and the development of DNA probes that can detect rearrangements in the *BCR* and *ABL* genes have been applied very rapidly for use in diagnosis and monitoring of patients thought to have CML or Ph^1+ ALL. The results of the diagnostic use of the *bcr* probe are summarized in Table 1. Equally important is the ability to check for the recurrence of a Ph^1+ clone in CML patients who have undergone bone marrow transplantation or in Ph^1+ ALL patients in remission. These screening procedures have become even more sensitive with the use of the polymerase chain reaction to detect the *bcr-ABL* junction in leukemic cells.

We have recently studied seven patients with Ph^1+ ALL to determine whether the translocation breakpoints all occur within the *BCR* gene [25]. With PFGE we could show that every patient had a rearrangement within the *BCR* gene either in the 5′ portion of *BCR* in the first intron (five patients) or in *bcr* (two patients). Moreover *ABL* was fused with *BCR* in each patient. Of the seven patients, two were children, one of whom, age 12 years, had a *bcr* rearrangement. Further studies with additional patients will allow more precise correlations of the clinical features of the leukemias with the molecular abnormalities that underlie them.

In the future, we will understand the role of the BCR and ABL proteins in normal cells and that of the two different chimeric BCR-ABL proteins in CML and in ALL. Thus, the genetic analysis of what appeared to be a simple chromosome change, namely the 9;22 translocation, has revealed unexpected complexity. I am sure that, in the future, an understanding of the altered function of the ABL protein will be central to the development of more specific and more effective forms of therapy.

D. Biological Significance of Chromosomal Rearrangements

One of the most surprising revelations in the recent past has involved the cellular oncogenes and their chromosome location (Fig. 4). Much of the excitement derives from the observation that many protooncogenes are located in the bands that are involved in consistent translocations [3, 6]. There is a remarkable specificity of certain chromosome rearrangements for particular subtypes of tumors especially leukemia or lymphoma. The mechanism or mechanisms by which this specificity is achieved are unknown; however, a number of investigators have shown that certain proteins required for promotion of gene expression are synthesized in a very cell-type-specific manner [26]. These proteins are only present in the appropriate cell type and therefore the particular gene is activated only in that cell type. The chromosome rearrangements affecting *MYC* in B-cell [27, 28] and T-cell [29, 30] tumors strongly support the interpretation that the specificity resides in the gene that is uniquely active in the particular cell type. Thus the immunoglobulin genes are highly regulated in B cells and they can therefore serve as the switch or activator mechanism for *MYC* in B cells; on the other hand, the alpha chain of the T-cell receptor (*TCRA*) is an active gene in T cells with a strong enhancer/promotor and it clearly is an activator for *MYC* in T cells. A reasonable paradigm is that translocations bring together, in an inappropriate manner, a growth factor or growth factor receptor gene (the protooncogene in the examples defined to date) adjacent to an active cell-specific gene.

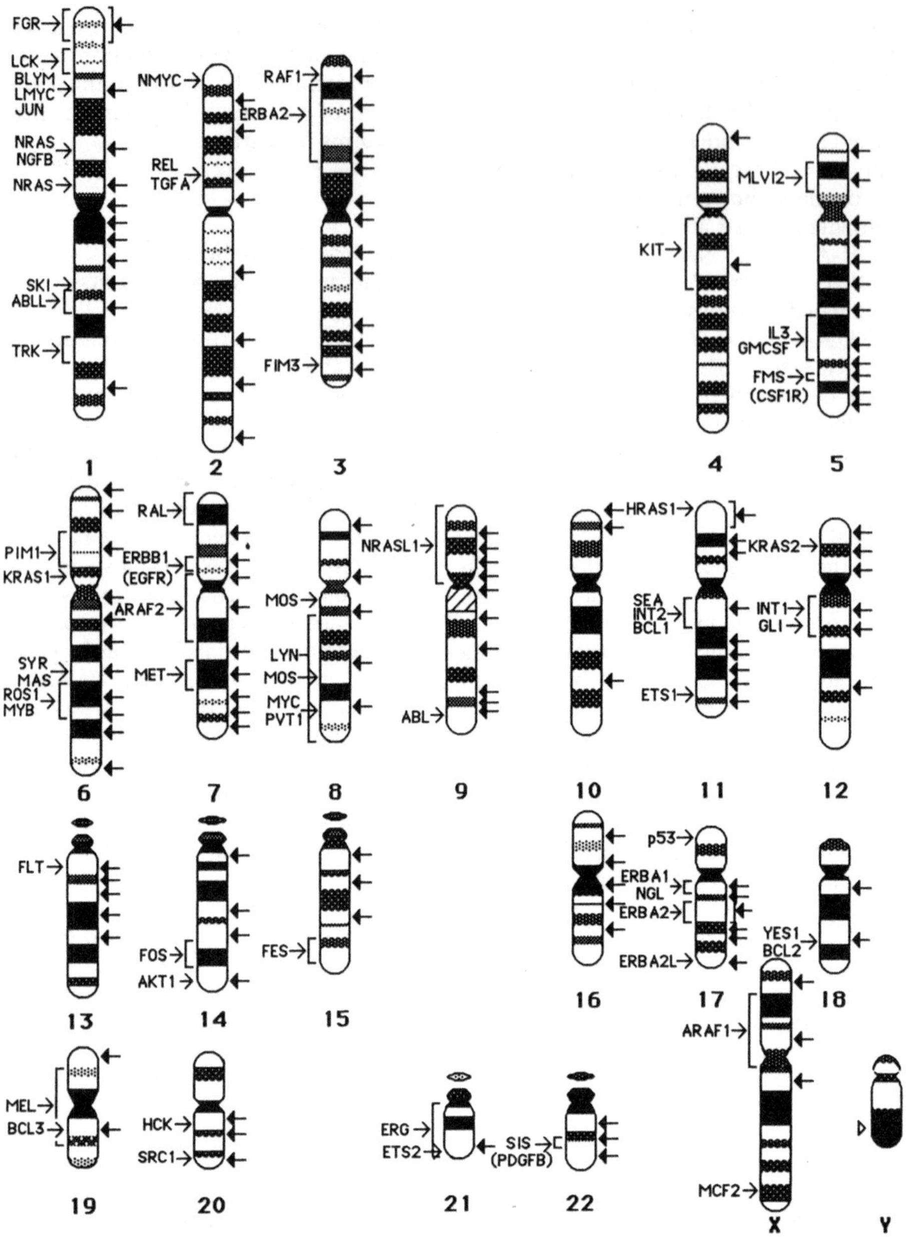

Fig. 4. Map of the chromosome location of protooncogenes or of genes with transforming properties and the breakpoints observed in recurring chromosome abnormalities in human leukemia, lymphoma, and solid tumors. The protooncogenes and their locations are placed to the left of the appropriate chromosome band (*arrow*) or region (*indicated by a bracket*). The breakpoints in recurring translocations, inversions, deletions, etc., are indicated with an *arrow to the right* of the affected chromosome band. The locations of the cancer specific breakpoints are based on the Human Gene Mapping 9 report [5]

It should be emphasized that many of the protooncogenes were identified in viruses that cause tumors. However, these genes have not been conserved through evolution from yeast and *Drosophila* to the chicken, mouse, and man to cause cancer! Where we have any insight into the function of these genes in normal cells, they are growth factors or growth factor receptors. It is not unexpected that the genes which a virus might coopt if it developed into a tumor-producing virus would be genes that control proliferation, genes which under viral regulation would function abnormally with regard to cell growth. Further support for the concept that oncogenes are growth factors gone wrong is provided by studies at the Hall Institute in Melbourne. There, investigators inserted the cloned gene for granulocyte-macrophage colony-stimulating factor into a viral vector, transfected mouse myeloid cells with this gene, and then injected the cells into mice which developed leukemia [31]. The term "oncogene" is too short and easy for it to be discarded, but it really refers to respectable genes for growth factors or their receptors.

The analysis of various tumors for alterations in protooncogenes has revealed that a number are abnormal as a result of translocations, amplification, or mutations [32]. In some situations the relationship of the change in the protooncogene to the multistage process of malignant transformation is unclear [33]. Such ambiguity is not a problem with chromosome translocations; the evidence is overwhelming that the t(8;14) in Burkitt's lymphoma and the t(9;22) in CML are an integral component of the cascade of events leading to the transformation of a normal to a malignant cell. The everincreasing number of translocations reviewed in this chapter provide a potential gold mine for identifying new genes that are unequivocally related to the malignant phenotype of the affected cell. The challenge is to isolate these translocation breakpoint junctions, to identify the genes that are located at these breakpoints, and then to determine the change in gene function that occurs as a consequence of the translocation. The ultimate measure of success, however, will be in the application of these new insights in the development of new, more effective treatments for cancer. In the future, each particular subtype of tumor will be treated in a uniquely defined way that is most appropriate for the specific genetic defect present in that tumor. This should lead to a new era of cancer therapy that is both more effective and less toxic.

References

1. Mitelman F (1988) Catalog of chromosome aberrations in cancer: Liss, New York
2. Heim S, Mitelman F (1987) Cancer cytogenetics. Liss, New York
3. Rowley JD (1988) Chromosome abnormalities in leukemia. J Clin Oncol 6:194–202
4. Rowley JD, Testa JR (1983) Chromosome abnormalities in malignant hematologic diseases. In: Advances in cancer research, Academic, New York, pp 103–148
5. Bloomfield CD, Trent JM, Van den Berghe H (1987) Report of the committee on structural chromosome changes in neoplasia (HGM9). Cytogenet Cell Genet 46:344–366
6. Yunis JJ (1983) The chromosomal basis of human neoplasia. Science 221:227–236
7. Nowell PC, Hungerford DA (1960) A minute chromosome in human granulocytic leukemia. Science 132:1497
8. Rowley JD (1973) A new consistent chromosomal abnormality in chronic myelogenous leukemia. Nature 243:290–293
9. De Klein A, Van Kessel AG, Grosveld G et al. (1982) A cellular oncogene is translocated to the Philadelphia chromosome in chronic myelocytic leukemia. Nature 300:765–767
10. De Klein A, Hagemeijer A (1984) Cytogenetic and molecular analysis of the Ph[1] translocation in chronic myeloid leukemia. Cancer Surv 3:515–529
11. Bakhshi A, Minowada J, Arnold A et al. (1983) Lymphoid blast crises of chronic myelogenous leukemia represent stages in the development of B-cell precursors. N Engl J Med 309:826–831

12. Whang-Peng J, Canellos GP, Carbone PP et al. (1968) Clinical implications of cytogenetic variants in chronic myelocytic leukemia (CML). Blood 32:755–766
13. Pugh WC, Pearson M, Vardiman JW et al. (1985) Philadelphia chromosome-negative chronic myelogenous leukaemia: a morphologic reassessment. Br J Haematol 60:457–467
14. Morris CM, Reeve AE, Fitzgerald PH et al. (1986) Genomic diversity correlates with clinical variation in Ph^1-negative chronic myeloid leukemia. Nature 320: 281–283
15. Bartram CR (1988) Molecular genetic analyses of chronic myelocytic leukemia. In: Huhn D, Hellriegel KP, Niederle N (eds) Chronic myelocytic leukemia and interferon. Springer, Berlin Heidelberg New York
16. Groffen J, Stevenson JR, Heisterkamp N et al. (1984) Philadelphia chromosomal breakpoints are clustered within a limited region, *bcr,* on chromosome 22. Cell 36: 93–99
17. Westbrook CA, Rubin CM, Carrino JJ et al. (1988) Long-range mapping of the Philadelphia chromosome by pulsed-field gel electrophoresis. Blood 79:697–702
18. Konopka JB, Watanabe SM, Witte ON (1984) An alteration of the human c-*abl* protein in K562 leukemia cells unmasks associate tyrosine kinase activity. Cell 37:1035–1042
19. Shtivelman E, Lifshitz B, Gale RP et al. (1985) Fused transcript of *abl* and *bcr* genes in chronic myelogenous leukaemia. Nature 315:550–554
20. Third international workshop on chromosomes in leukemia (1982). Cancer Genet Cytogenet 4:95–142
21. De Klein A, Hagemeijer A, Bartram CR et al. (1986) Rearrangement and translocation of the c-*abl* oncogene in Philadelphia positive acute lymphoblastic leukemia. Blood 68:1369–1375
22. Rubin CM, Carrino JJ, Dickler MN et al. (1988) Heterogeneity of genomic fusion of *BCR* and *ABL* in Philadelphia chromosome-positive acute lymphoblastic leukemia. Proc Natl Acad Sci USA 85:2795–2799
23. Clark SS, McLaughlin J, Christ WM et al. (1987) Unique forms of the *abl* tyrosine kinase distinguish Ph^1-positive CML from PH^1-positive ALL. Science 235: 85–88
24. Chan LC, Karhi KK, Rayter SI et al. (1987) A novel *abl* protein expressed in Philadelphia chromosome positive acute lymphoblastic leukaemia. Nature 325: 635–637
25. Hooberman A, Carrino JJ, Leibowitz D et al. (1989) Unexpected heterogeneity of BcR-ABL fusion mRNA detected by polymerase chain reaction in Philadelphia chromosome acute lymphoblastic leukemia. Proc Natl Acad Sci USA 86:4259–4263
26. Nomiyama H, Fromental C, Xiao JH et al. (1987) Cell-specific activity of the constituent elements of the simian virus 40 enhancer. Proc Natl Acad Sci USA 84:7881–7885
27. Leder P, Battey J, Lenoir G et al. (1983) Translocations among antibody genes in human cancer. Science 222:765–771
28. Croce CM, Isobe M, Palumbo A et al. (1985) Gene for α-chain of human T-cell receptor: location on chromosome 14 region involved in T-cell neoplasms. Science 227:1044–1047
29. Shima EA, Le Beau MM, McKeithan TW et al. (1986) Gene encoding the α-chain of the T-cell receptor is moved immediately downstream of c-*myc* in a chromosomal 8;14 translocation in a cell line from a human T-cell leukemia. Proc Natl Acad Sci USA 83:3439–3443
30. Mathieu-Mahul D, Caubet JF, Bernheim A et al. (1985) Molecular cloning of a DNA fragment from human chromosome 14 (14q11) involved in T cell malignancies. EMBO J 4:3427–3433
31. Lang RA, Metcalf D, Gough NM et al. (1985) Expression of a hemopoietic growth factor cDNA in a factor-dependent cell line results in autonomous growth and tumorigenicity. Cell 43:531–542
32. Bishop JM (1987) The molecular genetics of cancer. Science 235:305–311
33. Duesberg PH (1987) Retroviruses as carcinogens and pathogens: expectations and reality. Cancer Res 47:1199–1220

Interactions Between Hematopoietic Growth Factors: The Clinical Role of Combination Biotherapy *

M.A.S. Moore

A. Introduction

The Frederick Stohlman Memorial Lectures have, over the years, reflected the progression of leukemia research in areas initially thought to be as diverse as retrovirology/oncogenes, chemotherapy/immunotherapy, and hematopoietic growth factors. At the Seventh Wilsede Meeting we heard how convergent these areas were. The polypeptide growth factors which affect lymphohematopoietic cell proliferation and differentiation have expanded to at least 11 distinct gene products with pleiotropic and overlapping functions (G-, GM-, M-CSF, erythropoietin, and interleukins 1–7 [1]. The early characterization of hematopoietic growth factors was based on their ability to stimulate the clonal proliferation of human bone marrow progenitor cells. In addition, these factors were shown to activate a variety of functions of mature lymphoid and hematopoietic cells.

The purification, characterization, and gene cloning of human G-CSF followed shortly after the equivalent characterization of human GM-CSF [2, 3]. With the availability of recombinant G-CSF in quantities sufficient for in vivo evaluation, its action on hematopoietic function was assessed in normal murine and primate systems, and following chemotherapy or radiation-induced myelosuppression [1, 4, 5]. Efficacy was rapidly demonstrated and opened the way to clinical trials in human bone marrow suppression or failure systems [6–8]. Without reviewing this area in detail, I wish to touch on various aspects that illustrate the utility of G-CSF in various pathophysiological situations.

The James Ewing Laboratory of Developmental Hematopoiesis, Memorial Sloan-Kettering Cancer Center, New York, NY 10021, USA
* Supported by grants CA 20194, CA 32156, and CA 31780 from the National Cancer Institute; American Cancer Society Grant CH-3K, and the Gar Reichman Foundation.

B. Hematopoietic Role of G-CSF

I. G-CSF Treatment of Mice Receiving Cyclophosphamide, Myleran, or 5-Fluorouracil

In order to investigate the potential of G-CSF in preventing episodes of neutropenia following high-dose chemotherapy with cyclophosphamide (CY), C3H/HeJ mice were subject to weekly injections of 200 mg/kg CY intraperitoneally followed by G-CSF therapy (1.75 g $\times$ 2 daily) beginning 2 h after CY treatment and finished 48 h before the second cycle of CY. As shown in Fig. 1, this protocol prevented the subsequent nadirs of neutrophil counts between 4 and 5 days after CY treatment, and in all ten cycles of CY therapy G-CSF abrogated the neutrophil nadirs. The mice receiving CY alone had a substantial mortality evident by the eighth cycle of treatment. In eight cycles of CY treatment over the course of 70 days the G-CSF-treated animals were neutropenic (1000 ANC/mm^3) for only 3 days, whereas the untreated animals were neutropenic for 24 days. The time to recovery of absolute neutrophil counts (ANCs) to control levels was also substantially affected, with the non-G-CSF-treated groups requiring an additional 16 days for recovery. Neutropenia at levels

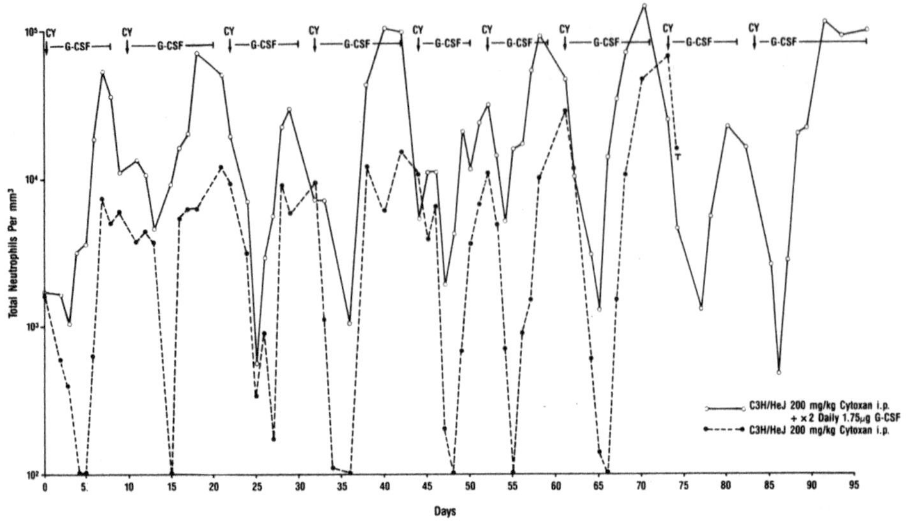

Fig. 1. Absolute neutrophil counts in the peripheral blood of C3H/HeJ mice treated with repeated dosed of 200 mg/kg cyclophosphamide (CY) administered at 7- to 9-day intervals over 3 months. Control mice (*broken line*) received saline alone; experimental mice (*continuous line*) received 1.75 mg rhG-CSF twice daily intraperitoneally beginning 2 h following CY and terminating 48 h prior to the next cycle of CY. Three mice per group. Note the log scale for ANC/mm^3

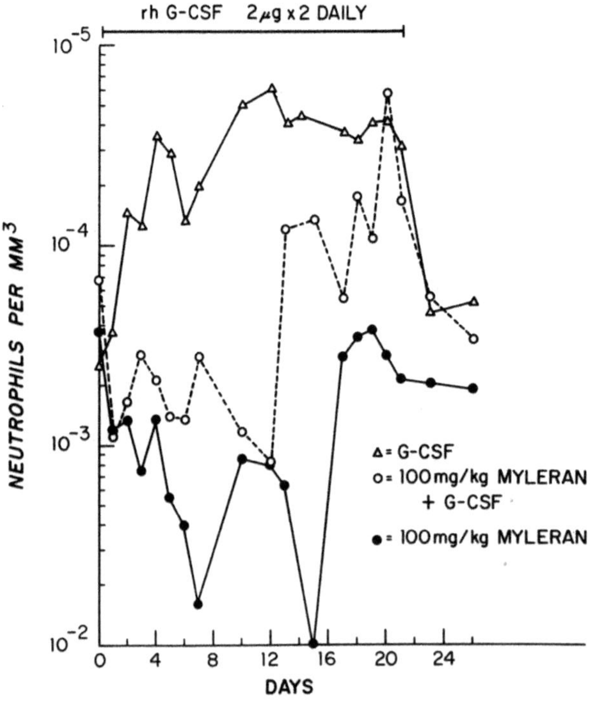

Fig. 2. Absolute neutrophil counts in the peripheral blood of C3H/HeJ mice administered 100 mg/kg myleran orally by intubation with or without twice daily i.p. injections of rh-G-CSF at 2 μg/mouse for 21 days. Control animals received corn oil alone with G-CSF. Four mice per group. Note the log scale

of 500 ANC/mm^3 indicated an even more significant difference, with such low levels observed on only 1 day out of 70 in the G-CSF + CY-treated mice, in contrast to 15 days in the CY-treated group.

This cyclophosphamide model allowed us to investigate the timing of initiation of G-CSF therapy following a single injection of a high dose of CY. Administration of G-CSF coincident with CY treatment resulted in an improvement in recovery of neutrophil counts, but this was substantially less than observed when G-CSF was administered 2 h after CY therapy. Administration of G-CSF 24 and 48 h after CY was less effective than after 2 h, indicating that treatment early in the course of chemotherapy is likely to be more effective.

Myleran (MY), administered orally (100 mg/kg), was also associated with a significant reduction in neutrophil counts and in bone marrow progenitor populations. G-CSF therapy initiated within 6 h of MY treatment led to an accelerated recovery of neutrophils so that by 1 week the neutrophil counts had returned to normal values, whereas the MY-treated controls remained profoundly neutropenic (ANC 200/mm^3) (Fig. 2). By 14 days MY + G-CSF-treated animals exhibited a neutrophil leukocytosis of between 10^4 and 2×10^4/mm^3 at times when the MY control animals had 100 neutrophils/mm^3.

Treatment with 5-fluorouracil (5-FU 150 mg/kg) led to profound myelosuppression in three different strains of mice tested. In C3H/HeJ mice, recovery of neutrophil counts to control values was delayed for 12–14 days. Administration of G-CSF accelerated recovery of neutrophil counts to normal values within 8 days [5].

II. Clinical Trials of G-CSF Following Chemotherapy

The use of rh-G-CSF in clinical trials has demonstrated efficacy in promoting regeneration of functioning neutrophil granulocytes. Our first clinical trial involved the administration of G-CSF in a phase I/II study of patients receiving M-VAC (methotrexate, vinblastine, doxarubicin, and cisplatin) in 22 patients with transitional cell carcinoma of the uroepithelium [6]. In all patients a specific dose-dependent increase in ANC was seen to levels of up to 12-fold above normal at the higher levels of G-CSF administration. At these high doses (10–60 g/kg) a tenfold increase in monocytes, but not in other hematopoietic lineages, was evident. Treatment after chemotherapy significantly reduced the number of days on which antibiotics were used. This resulted in a significant increase in the number of patients qualified to receive planned chemotherapy on day 14 of the treatment cycle (100% vs. 29%). In addition, the incidence of mucositis significantly decreased. Investigators at other centers have confirmed these observations. In patients with advanced small cell lung cancer receiving high-dose chemotherapy, repeated every 3 weeks, G-CSF was given to each patient for 14 days on alternate cycles of chemotherapy. This resulted in the reduction of the period of absolute neutropenia with return to normal or supranormal levels of ANC within 2 weeks [8]. This therapy also drastically reduced episodes of severe infections observed during these cycles of chemotherapy. In advanced malignancies treated with melphalan, neutropenia was significantly reduced even at dose levels of 1 or 3 µg/kg G-CSF. Preliminary clinical studies have also demonstrated that G-CSF hastens granulocyte recovery in Hodgkin's disease after high-dose chemotherapy and autologous bone marrow transplantation [9]. In all of these studies no significant dose-limiting toxicities have been observed to date.

III. In Vivo Studies of G-CSF in Congenital and Idiopathic Neutropenia

Cyclic neutropenia is an inherited disease of man and gray collie dogs, characterized by regular oscillations of the number of peripheral blood cells and of bone marrow progenitor cells. Cycling of serum or urine CSF levels has also been reported. However, induction or cure of cyclic neutropenia by bone marrow transplantation has suggested that this disorder is a disease of pluripotential hematopoietic stem cells. It should be noted that the hematopoietic stem cell itself can generate cells (macrophages, T cells) capable of hematopoietic growth factor production and thus it is not possible to exclude intrinsic growth factor production defects in the pathophysiology of cyclic neutropenia. In the dog model we have demonstrated that daily administration of rh-G-CSF (5 μg/kg × 2 daily for 30 days) caused an immediate (within 12 h) and persistent leukocytosis (>40 000 WBC) in both cyclic and normal dogs due to a ten fold increase in the numbers of circulating neutrophils and monocytes [10]. This therapy eliminated two predicted neutropenic episodes and suppressed the cycling of CFU-GM in the bone marrow. Comparable therapy with rh-GM-CSF induces a monocytosis and neutrophilia in normal dogs but did not eliminate the recurrent neutropenia in cyclic dogs. Analysis of the serum levels of CSF indicated a striking periodicity with peak levels of G-CSF coinciding with the peaks of monocytes and the nadirs of neutrophils. The continuing cycling of CSF levels was also observed even when animals showed a major neutrophil leukocytosis under exogenous G-CSF therapy.

A disorder of neutrophil regulation is also seen in chronic idiopathic neutropenia in which the peripheral neutrophil count is reduced to 2000 cells/mm^3 for prolonged periods, with maturation arrest of neutrophil granulocyte precursors in the bone marrow. Other hematopoietic cell counts are usually normal and antineutrophil antibodies are absent. Clinically, these patients experience recurrent episodes of life-threatening infections, ulcers of the mucus membrane, and periodontal disease. We have had the opportunity of studying a number of patients with this disorder utilizing primary clonogenic assay and in vitro suspension cultures of patient bone marrow cells [7]. The incidence of CFU-GM observed in primary bone marrow culture was always in the high range of normal with respect to each species of CSF (G-, GM-, M-, and IL-3), indicating that the neutropenia was not attributed to a shortage of appropriate myeloid progenitors. Furthermore, morphological studies showed normal neutrophil maturation within the colonies developing in the presence of G-, GM-CSF, and IL-3, with eosinophil maturation and macrophage development a significant feature with GM-CSF stimulation. In suspension cultures G-CSF was particularly effective in generating mature segmented neutrophils which were absent in input bone marrow, and did not develop spontaneously in over 3 weeks of culture. With the addition of G-CSF, myelocytes expanded in the 1st week of culture. By the 2nd and 3rd week high levels of production of segmented, functionally normal neutrophils were found. Based upon these in vitro studies, patients were treated with subcutaneous G-CSF on a daily continuous basis. In the first patient studied, the neutrophil count rose rapidly, and by 20 days the patient's ANC was >1000 cells/mm^3 and plateaued in the range of 2000–3000/mm^3 with evidence of a 40-day cycle (observed even when the dose of G-CSF was increased). The neutrophils were functionally normal and the patient has remained on this continuous G-CSF treatment without any notable toxic side effects [7].

A more severe form of neutropenia is found in patients with Kostmann's syndrome (congenital agranulocytosis). In this disorder, marked by severe neutropenia and maturation arrest at the myelocyte level, five patients received G-CSF therapy, and the neutrophils increased from 0%–1% to 10%–72% with clinical resolution of preexisting infections [11].

C. Hematopoietic Role of Interleukin-1

Interleukin-1 (IL-1) was first identified as an endogenous pyrogen, produced by macrophages following activation by endotoxin. Subsequent studies implicated IL-1 as a mediator of a variety of inflammatory phenomena involving production of cascades of cytokines and cyclooxagenase products [12]. Various lymphokines are released by the interaction of IL-1 with T cells, and IL-1 induces expression of G-, GM-, and M-CSF by a direct action on stromal cells such as fibroblasts and endothelial cells, both in vitro and in vivo [13–15]. Recent studies have revealed a more direct role for IL-1 in hematopoiesis, involving an interaction with the early hematopoietic stem cell [5, 16–18].

I. Interleukin-1 Identity with Hemopoietin 1/Synergistic Activity

Early studies involving mice treated with 5-fluorouracil (5-FU) suggested the necessity of a synergistic interaction between two factors, one being a direct hematopoietic colony stimulus such as M-CSF or IL-3, and the other, which lacked direct colony-stimulating activity, was shown to synergize with the preceding species of CSFs [19]. Synergistic activity was identified in conditioned media or extracts of various human tissues and was also identified and purified to homogeneity from the human bladder cancer cell line, 5637 [20]. This latter cell line had also been used to identify, purify, and clone G- and GM-CSF [2, 3, 21]. The constitutive expression of hemopoietin 1 (H-1), which was shown to be a 17 000-kd protein [20], led us to attempt to purify, sequence, and clone this molecule, using similar strategies to those that we had employed to clone the G-CSF gene from 5637 cells. The identity of IL-1 and H-1/synergistic activity was established on the basis of the following factors:

1. The bladder cancer cells evidently produced high levels of IL-1 active in the thymocyte comitogenesis assay.
2. The synergistic activity and IL-1 could be copurified using various protein purification procedures.
3. Abundant levels of IL-1-alpha and -beta mRNA could be detected in the 5637 cells – comparable to levels in activated macrophages.
4. Fractions of in RNA from 5637 expressed in oocytes resulted in intracellular production of hematopoietic synergistic activity and thymocyte comitogenesis factor production.
5. The synergistic activity and thymocyte comitogenesis activity were completely neutralized by monoclonal and polyclonal antibodies to IL-1.
6. rhIL-1 alpha and beta at 0.1–10 U/ml synergized with G-, GM- M-CSF, and IL-3 in stimulating high proliferative potential (HPP)-CFU in clonogenic assays of 5-FU-treated murine bone marrow [1, 5, 16–18].

IL-1 has a direct effect upon hematopoietic stem cells in addition to its ability to elicit production of various CSF species by accessory cell populations within hematopoietic tissues. In our original studies it was not possible to conclude that IL-1 was acting directly on early stem cells since accessory cell populations were not depleted from the target bone marrow cell population. In more recent studies we have established a linear dose-response relationship between the numbers of HPP-CFU, and the number of bone marrow cells plated, with a highly significant correlation ($r = 0.97$) indicative of a single-hit phenomenon.

II. Action of IL-1 in Short-term Marrow Suspension Culture (Delta Assay)

The rationale behind the delta assay is to demonstrate the ability of hematopoietic growth factors to promote the survival, recruitment, or expansion of stem cells and/or progenitor cells in relatively shortterm suspension culture systems. As originally developed, we utilized bone marrow from mice that had been treated with 5-FU for 24 h and then subjected to a 4- to 7-day suspension culture in the

Table 1. Interleukin-1 and CSF-induced amplification of CFU-GM in 7-day suspension cultures of 2.5×10^5 murine bone marrow cells obtained 24 h post 5-fluorouracil treatment

Stimulus in suspension phase	Delta (CFU-GM output/input)			
	"Readout" stimulus in 2° clonogenic assay			
	G+IL-1	GM+IL-1	IL-3+IL-1	CSF-1+IL-1
Medium	0	0	0	0
rh-IL-1 alpha	190	53	39	61
rh-G-CSF	0	5	2	5
IL-1+G-CSF	255	91	41	41
mCSF-1	0	2	1	2
IL-1+CSF-1	175	47	38	80
mGM-CSF	40	17	15	34
IL-1+GM-CSF	280	58	49	54
mIL-3	110	17	18	33
IL-1+IL-3	510	178	120	115

Femoral bone marrow cells taken from B6D2F1 mice 24 h after a single i.v. injection of 5-fluorouracil (150 mg/kg) were incubated at 2.5×10^5 cells/ml in Iscove's modified Dulbecco's medium with 20% fetal calf serum in 24-well cluster plates containing 100 units rh-IL-1 alpha, or 2000 u/ml rh-G-CSF; 1000 u/ml. m-GM-CSF (purified from murine post-endotoxin lung CM) 200 u m-IL-3 (purified from WEHI-3 cell line CM), or 1000 u CSF-1 (purified from L-cell CM). After 7 days of incubation, cells were recovered and assayed for CFU-GM in agarose cultures stimulated by the various CSF species alone or in combination with IL-1. The delta value (CFU-GM output/input) was calculated on a recovery from triplicate clonal assays from triplicate suspension cultures

presence of IL-1 alone, CSFs alone, or combinations of IL-1 with various CSF species [5, 16]. At the end of the suspension culture phase, total cellularity and morphology was determined, and cells were cloned in semisolid culture, again in the presence of IL-1 alone, CSFs alone, or combinations of IL-1 and CSFs. Table 1 shows that IL-1 caused an expansion of the numbers of CFU-GM recovered after 7 days of culture. Neither M-CSF nor G-CSF alone supported survival or expansion of CFU-GM; however, the combination of IL-1 and CSFs demonstrated additive or synergistic effects on the expansion of these progenitors. GM-CSF or IL-3 alone caused some expansion of progenitor populations, but again the combination with IL-1 evidenced synergism.

Human systems based on in vitro purging with 4-hydroperoxcyclophosphamide (4HC) and positive selection by "panning" with MY10 monoclonal antibody for $CD34^+$ cells followed by 7-day suspension culture also demonstrated synergistic interactions between IL-1 and various CSF species. The most dramatic effect was observed with the combination of IL-1 and IL-3, where an up to 85-fold increase in progenitor cells was noted following 7-day suspension culture (Table 2). In contrast to the murine system, synergism between IL-1 and CSF-1 was not evident.

III. In Vivo Interaction Between IL-1 and G-CSF in Mice Treated with 5-FU

In mice trated with 5-FU, G-CSF administration restores neutrophil counts to normal values some 5–6 days earlier than in mice not receiving the factor, but a period of profound neutropenia is still observed [5]. Administration of IL-1 alone, giving postchemotherapy for 4–10 days twice daily at doses of 0.2 μg/mouse per day, reduced the severity of

Table 2. Interleukin-1 and CSF-induced amplification of CFU-GM in 7-day suspension cultures of 2.5×10^5 4-HC-treated, $CD34^+$ human bone marrow cells

Stimulus in suspension phase	Delta (CFU-GM output/input)			
	"Readout" stimulus in 2° clonogenic assay			
	rh-G-CSF	rh-GM-CSF	rh-IL-3	rh-CSF-1
Medium	0	0	0	0
rh-IL-1 alpha	10	6	2	2
rh-G-CSF	8	9	2	2
IL-1 + G-CSF	8	9	2	1
rh-GM-CSF	8	9	2	2
IL-1 + GM-CSF	33	22	6	10
rh IL-3	5	30	17	2
IL-1 + IL-3	15	85	15	1

Bone marrow cells, obtained with informed consent from normal volunteers, were separated over Ficoll-Hypaque, subjected to plastic adherence, and incubated for 30 min with 100 μM 4-hydroperoxycyclophosphamide (4-HC). Cells were then treated with antimyl monoclonal antibody (anti-HPCA-1 Beckon Dickinson, Mountain View Ca.), on ice for 45 min, washed, and incubated for 1 h at 4 °C on bacteriological-grade plastic petri dishes previously coated with goat anti-mouse IgG. Adherent $CD34^+$ cells were harvested by vigorous pipetting. 2.5×10^5 harvested cells/well were incubated in Iscove's modified Dulbecco's medium (IMDM) plus 20% fetal calf serum in 24-well cluster plates containing test stimuli. In suspension phase, stimuli were 10 ng/ml rh-G-CSF, rh-GM-CSF, rh-IL-3 (Amgen), rh CSF-1 (Cetus), and rh-IL-1 alpha (Roche). 4-HC purged, $CD34^+$ marrow cells were plated at 2×10^4 cells/ml in semisolid agarose culture in the presence of 10 ng/ml GM-CSF, G-CSF, CSF-1, or IL-3 alone or in combination with IL-1 both pre- and postsuspension culture. Colony formation was assessed after 12 days. The delta value (CFU-GM output/input) was calculated on recovery from triplicate clonal assays from duplicate suspension cultures

the neutrophil nadir, and accelerated the recovery of the neutrophil count to an extent greater than observed with G-CSF alone. The combination of G-CSF and IL-1 administered after 5-FU therapy also resulted in accelerated hematopoietic reconstitution, although the results were additive rather than synergistic [5]. Analysis of total hematopoietic cell reconstitution in %-FU-treated mice also showed the efficacy of IL-1 therapy, or IL-1 plus G-CSF therapy, in accelerating total recovery of erythroid as well as granulocytic elements in the marrow, spleen, and blood.

The potential of combination biotherapy as an effective means of accelerating hemopoietic cell differentiation raises the issue of whether premature exhaustion of the stem cell and progenitor cell population may occur. Measurement of colony-forming units (CFU-s, CFU-c, BFU-e, CFU-GEMM, CFU-Meg) on an incidence basis proved to be misleading because of the redistribution of hematopoietic precursors in regenerating murine tissues. By quantitation of total numbers of these cell populations in marrow, spleen, blood, and other tissues, we demonstrated that there was an absolute increase in the recovery of all of these populations in 5-FU-treated mice exposed to both IL-1 alone and IL-1 plus G-CSF. The results indicate that the mechanism of action of these cytokines, both alone or in combination, is not simply mediated by accelerated differentiation, but involves an absolute expansion of the most primitive stem cell populations that can be measured, e.g., by the day 12 murine spleen colony (CFU-s) assay, or the high proliferative potential in vitro colony assay requiring IL-1 plus CSF as the read-out stimuli [1, 5, 16]. This is an important observation for extrapolation of these studies to clinical situations because it lays to rest some of the concerns that combination biotherapy may accelerate exhaustion of the stem cell compartment in patients receiving myeloablative therapy.

IV. Interleukin-1 and CSFs as Radioprotective Agents

Interleukin-1 has been shown to mediate a radioprotective action when given to mice 20 h before what would otherwise be a lethal dose of irradiation [12, 15]. This radioprotective effect cannot be duplicated by preadministration of various CSF species. Using BALB/c mice which are particularly susceptible to irradiation, we observed that 0.2 µg IL-1 given 20 h before 850 rads total body irradiation completely prevented what would otherwise be 100% mortality by 14 days, confirming earlier reports. However, we were able to show that administration of IL-1 or G-CSF after irradiation was also significantly radioprotective. While the mechanism of IL-1 radioprotection remains controversial, our down observations suggest that one mechanism may involve accelerated recruitment of primitive G_0 stem cells into the cell cycle with expression of CSF receptors. The administration of 750 rads total body irradiation to BALB/c mice, which resulted in 80% mortality within 14 days, provided a model in which we could test the interactions of IL-1 and G-CSF in reconstitution of hematopoiesis and in radioprotection. Recovery of neutrophils to normal levels in irradiated mice was delayed for up to 3 weeks; G-CSF administration begun immediately after irradiation, and carried out for 3 weeks, accelerated recovery of neutrophils and total hematopoietic cells, so that normal levels were observed by 14 days [22]. IL-1 administration as a single dose prior to irradiation resulted in accelerated recovery of hematopoiesis. When administered daily postirradiation for 4 days, IL-1 reduced the nadir of blood neutrophils and produced a degree of recovery of hematopoietic function comparable to that seen with G-CSF administration (Fig. 3). A significant synergistic interaction was noted when G-CSF was combined with IL-1 postirradiation,

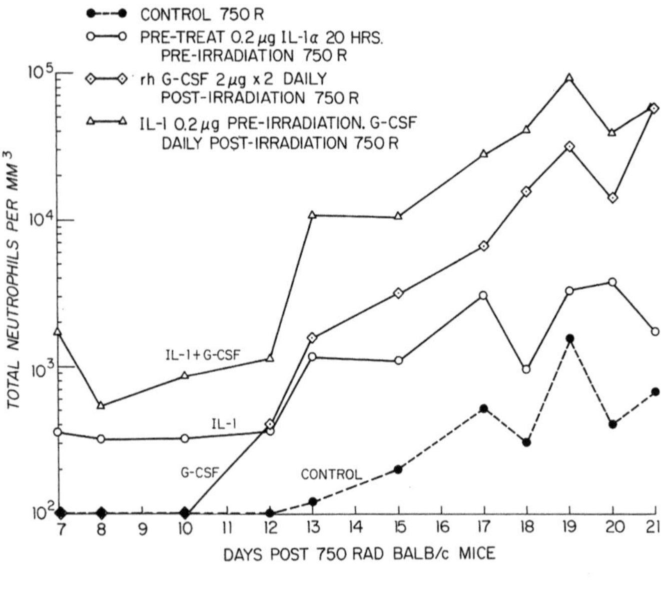

Fig. 3. Absolute neutrophil counts in the peripheral blood of Balb/c mice exposed to 750 rad total body irradiation and either pretreated with rhIL-1 alpha (0.2 µg/mouse i.p. 20 h before irradiation) or treated with 2 µg rh-G-CSF beginning 2 h postirradiation twice daily for 21 days, or the combination of IL-1 pre- and G-CSF post-treatment. Five mice per group. Control mice (*broken line*) received saline only

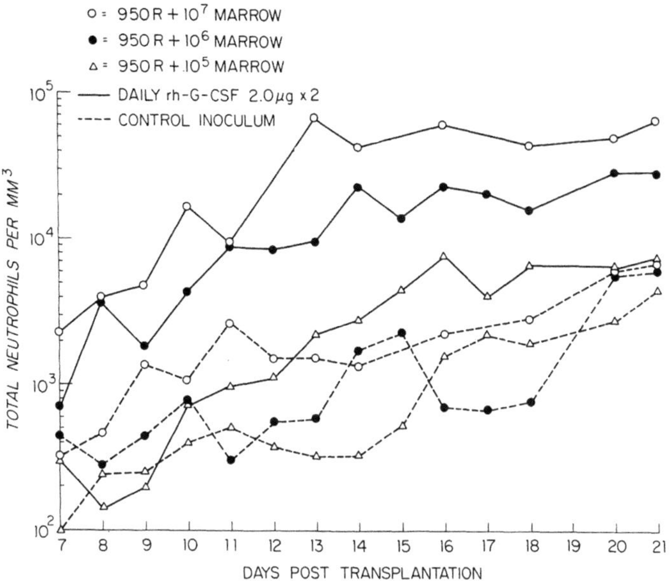

Fig. 4. Absolute neutrophil counts in the peripheral blood of B6D2F1 mice subject to 950 rads total irradiation and injected with varying numbers of syngeneic bone marrow cells, with or without subsequent twice daily administration of 2 µg rhG-CSF i.p. Five mice were used per group

with up to a tenfold greater increase in blood neutrophils, hematopoietic cells, and both stem and progenitor cells populations in both spleen and bone marrow.

V. Interactions Between IL-1 and G-CSF in Autologous Bone Marrow Transplantation

Transplantation of autologous bone marrow into lethally irradiated (950 rads) C3H/HeJ and B6D2F1 mice was undertaken using three different dose levels of donor marrow cells, 10^5, 10^6, and 10^7 cells/mouse. The LD_{50} in such transplanted animals is observed at dose levels of $1.0-2.5 \times 10^6$ marrow cells/mouse. It is generally accepted that the delayed recovery in hematopoietic cells in the bone marrow and spleen and in peripheral neutrophil counts was attributed to limitations in the number of donor stem cells as measured by the CFU-s assay. As can be seen in Fig. 4, the peripheral neutrophil counts reflected a marrow cell dose-dependent difference in a return to normal values. While this observation (which was also reflected in bone marrow and spleen hematopoietic cell reconstitution) might lead to the assumption that stem cell availability is the sole criterion for hematopoietic reconstitution, the administration of exogenous G-CSF suggested additional variables in the transplant equation. In all instances following autologous bone marrow transplantation at the different dose levels, exogenous G-CSF administration enhanced the recovery of neutrophils in the peripheral blood by three- to tenfold. This in turn was accompanied by accelerated regeneration of bone marrow and spleen hematopoietic progenitor cells and of pluripotential stem cells as measured by the CFU-s assay.

Studies in autologous bone marrow transplantation in cynomolgus monkeys confirmed the ability of G-CSF to accelerate bone marrow regeneration and recovery of peripheral blood neutrophil counts. These studies indicated that higher doses of G-CSF were required to enhance the recovery of hematopoietic parameters to normal, suggesting that the role of G-CSF involved recruitment of earlier stem cells by some indirect mechanism involving cytokine cascades or, alternatively, reflecting the relative paucity of G-CSF receptors on primitive hematopoietic cells. Based upon our in vitro studies indicating that IL-1 upregulated the receptors for G-CSF and other CSF species on primitive hematopoietic stem cells that were resistant to irradiation and chemotherapy, we cultured bone marrow from mice treated with 5-FU for 24 h with IL-1 prior to autologous transplantation in lethally irradiated mice. Following transplantation, recipients received IL-1 alone for 4 days, G-CSF alone for 21 days, or IL-1 plus G-CSF, and hematopoietic reconstitution was assessed. Preliminary studies indicated that in vitro pretreatment of bone marrow enhanced bone marrow reconstitution and accelerated recovery of peripheral neutrophil counts in mice that subsequently received IL-1, G-CSF, or a combination of IL-1 plus G-CSF in the post-transplant period. As measured by reconstitution of total stem cells and progenitor cells, hematopoietic recovery, and peripheral neutrophil counts, the most efficacious combination involved pretreatment of bone marrow in vitro with IL-1, followed by post-transplant treatment with a combination of IL-1 and G-CSF.

D. Conclusion

Numerous clinical trials are now underway using G-CSF and GM-CSF in iatrogenic myelosuppressive situations associated with cancer chemotherapy and irradiation therapy, and in conjunction with autologous and allogeneic bone marrow transplantation. Efficacy is also under study in congenital disorders of neutrophil production, and in myelodysplastic syndromes and myeloid leukemias. The initial promise of the CSFs and interleukins suggests that they may provide a major new therapeutic modality – but only if we can develop a deeper insight into their modes of action and interaction. The choice of the type of factor to be administered will be influenced by the particulars of the pathology being treated, making it meaningless to generalize as to whether one factor is "better" than another. It is already clear that the timing, dose, and route of administration of hematopoietic growth factors are important variables and efficacy will have to be balanced against potential adverse side effects (which fortunately have not emerged as serious limitations in the case of G-CSF and GM-CSF trials). In addition, preclinical studies point to the value of combination biotherapy using two or more factors administered at the same time, or sequentially. Finally, there is a need to recognize that proliferative processes are self-limiting and physiological mechanisms probably exist to counteract the action of hematopoietic growth factors. Unraveling these issues will undoubtedly occupy the agenda of many future Wilsede Meetings.

References

1. Moore MAS (1988) The use of hematopoietic growth and differentiation factors for bone marrow stimulation. In: De Vita VT, Hellman S, Rosenberg SA (eds) Important advances in oncology 1988. Lippincott, Philadelphia, pp 31–54
2. Welte K, Platzer E, Lu L, Gabrilove JL, Levi E, Mertelsmann R, Moore MAS (1985) Purification and biochemical characterization of human pluripotent hematopoietic colony-stimulating factor. Proc Natl Acad Sci USA 82:1526–1530
3. Souza L, Boone TC, Gabrilove JL, Lai PH, Zsebo KM, Murdock DC, Chazin VR, Bruszewski J, Lu H, Chen KK, Barendt J, Platzer E, Moore MAS, Mertelsmann R, Welte K (1986) Recombinant human granulocyte colony-stimulating factor; effects on normal and leukemic cells. Science 232:61–65
4. Welte K, Bonilla MA, Gillio AP, Boone TC, Potter GK, Gabrilove JL, Moore MAS, O'Reilly J, Souza LM (1987) Recombinant human granulocyte colony-stimulating factor: effects on hematopoiesis in normal and cyclophosphamide-treated primates. J Exp Med 1645:941–948
5. Moore MAS, Warren DJ (1987) Interleukin-1 and G-CSF synergism: in vivo

stimulation of stem cell recovery and hematopoietic regeneration following 5-fluorouracil treatment in mice. Proc Natl Acad Sci USA 84:7134-7138
6. Gabrilove JL, Jakobowski A, Scher H, Sternberg C, Wong G, Grous J, Yagoda A, Fain K, Moore MAS, Clarkson B, Oettgen HF, Alton K, Welte K, Souza L (1988) Effect of granulocyte colony-stimulating factor on neutropenia and associated morbidity due to chemotherapy for transitional-cell carcinoma of the urothelium. N Engl J Med 318:1414-1422
7. Jakubowski AA, Souza L, Kelly F, Fain K, Budman D, Clarkson B, Bonilla MA, Moore MAS, Gabrilove JL (1989) Effects of human granulocyte colony stimulating factor on a patient with idiopathic neutropenia. New Engl J Med 320:38-42
8. Hernandez Bronchud MH, Scargge JH, Thatcher N, Crowther D, Souza LM, Alton NK, Testa NG, Dexter TM (1987) Phase I/II study of recombinant human granulocyte colony-stimulating factor in patients receiving intensive chemotherapy for small cell lung cancer. Br J Cancer 56:809-813
9. Taylor K, Spitzer G, Jagannath S, Dicke K, Vincent M, Souza L (1988) rhG-CSF hastens granulocyte recovery in Hodgkin's disease after high-dose chemotherapy and autologous bone marrow transplant. Exp Hematol 16:413 (abstr)
10. Lothrop CD Jr, Warren DJ, Souza LM, Jones JB, Moore MAS (1988) Correction of canine cyclic hematopoiesis with recombinant human granulocyte colony stimulating factor. Blood 72:1324-1328
11. Bonilla MA, Gillio AP, Ruggiero M, Kernan NA, Brochstein JA, Fumagalli L, Bordignon C, Vincent M, Welte K, Souza LM, O'Reilly RJ (1988) Correction of neutropenia in patients with congenital agranulocytosis with recombinant human granulocyte colony stimulating factor in vivo. Exp Hematol 16:520 (abstr)
12. Oppenheim J, Kovacs E, Matsushima K, Durum SK (1986) There is more than one interleukin-1. Immunol Today 7:45-56
13. Lovhaug D, Pelus LM, Nordie EM, Boyum A, Moore MAS (1986) Monocyte-conditioned medium and interleukin 1 induce granulocyte macrophage colony stimulating factor production in the adherent cell layer of murine bone marrow cultures. Exp Hematol 14:1037-1042
14. Bagby GC, Dinarello CA, Wallace P, Wagner C, Hefeneider S, McCall E (1986) Interleukin 1 stimulates granulocyte macrophage colony stimulating activity release by vascular endothelial cells. J Clin Invest 78:1316-1320
15. Neta R, Oppenheim JJ (1988) Why should internists be interested in interleukin 1? Ann Intern Med 109:1-3
16. Moore MAS, Warren DJ, Souza L (1987) Synergistic interaction between interleukin-1 and CSFs in hematopoiesis. In: Gale RP, Golde DW (eds) Recent advances in leukemia and lymphoma: UCLA symposium on molecular and cellular biology. Liss, New York, pp 445-458
17. Mochizuki DY, Eisenman JR, Conlon PJ, Larsen AD, Tushinski RJ (1987) Interleukin 1 regulated hematopoietic activity, a role previously ascribed to hemopoietin 1. Proc Natl Acad Sci USA 84:5627
18. Warren DJ, Moore MAS (1988) Synergism among interleukin 1, interleukin 3, and interleukin 5 in the production of eosinophils from primitive hemopoietic stem cells. J Immunol 140:94-99
19. Stanley ER, Bartocci A, Patinkin D, Rosendaal M, Bradley TR (1986) Regulation of very primitive multipotent hemopoietic cells by hemopoietin-1. Cell 45:667-674
20. Jubinsky PI, Stanley ER (1985) Purification of hemopoietin-1: a multilineage hemopoietic growth factor. Proc Natl Acad Sci USA 82:2764-2767
21. Gabrilove J, Welte K, Harris P, Platzer E, Lu L, Levi E, Mertelsmann R, Moore MAS (1986) Pluripoietin alpha: a second human hematopoietic colony-stimulating factor produced by the human bladder carcinoma cell line 5637. Proc Natl Acad Sci USA 83:2478-2482
22. Moore MAS, Welte K, Gabrilove JL, Souza LM (1986) In vivo action of recombinant human G-CSF on chemotherapy or radiation myelosuppressed mice. Blood 68 [Suppl 1]:173a (abstr)

Clinical Aspects

Introduction of Donald Pinkel

F. Lampert

Organizers, colleagues, ladies and gentlemen, and above all, dear Don, everything is evolution, and every new field in medicine has its beginnings and its pioneers. The story of pediatric oncology is a short but fascinating one, now covering 40 years, a time span considered holy since biblical times. In Germany, that is, in West Germany or the Federal Republic of Germany, the real and effective pediatric oncology has an even shorter history, and its beginning is closely connected with Dr. Pinkel's name. It is thus a great pleasure and an honor to introduce Donald Pinkel.

I could now present to you Don Pinkel's scientific career, or when and where he was born – I think it must have been 1926, somewhere in the USA – or mention the many awards he received, the last, if I remember correctly, being the most prestigious and substantial Charles Kettering prize from the General Motors Foundation in 1986, but I will confine myself to some personal memories and what I have learned from him.

The first time I met Don Pinkel was on the last day of August in 1960, at a barbecue at Dr. Oleg Selawry's residence on Grand Island, near the Niagara falls, Buffalo, N.Y. I had just completed my internship in the Pacific Northwest, and, on the way back visited the famous Roswell Park Memorial Institute. Don Pinkel was chief of pediatrics there and had also been invited by Oleg Selawry. I will never forget what Don told me during that hot and humid Midwest evening about teaching medical students: "With a patient, the most important thing is, to observe, to examine, and to state the physical findings absolutely correctly, and then to do a proper evaluation. As to the pathogenesis of a disease, you can read about this everywhere in a book, but the patient is unique!"

The second time I spoke to Don was on the telephone, calling from Munich to Memphis. It was on a Tuesday, the 18th of May, 1971. The mother of a freshly diagnosed 6-year-old boy with acute lymphoblastic leukemia (ALL) urged me to call Dr. Pinkel. She had just read in the German news magazine *Der Spiegel* about an almost 20% 5-year-cure rate for ALL achieved by Dr. Pinkel's group at St. Jude's Hospital, published in the Journal of the American Medical Association at the end of April. I had my doubts, but on the phone he told me something about "total therapy," including radiation of the central nervous system, about aggressive multiagent therapy up to the biological tolerance of the patient, and about 300 or more active patients with ALL currently under treatment at St. Jude's; I was curious to see with my own eyes. Together with Dr. Gregor Heinze, our local radiotherapist, we arrived at St. Jude's on Friday, May 28, 1971, and were able to study all the treatment modalities in the utmost detail. Back home, our patient Markus received the first dose of cranial irradiation on June 8, 1971 – the first one in Germany; by the way, he is now a healthy university student – and was put on Pinkel therapy, as well called this ALL management all over Germany. This Pinkel therapy consisted of induction, CNS irradiation, and continuation therapy lasting 2–3 years. This treatment protocol was easy to understand and to apply. Within several months allmost all institutions in Germany were treating children with leukemia according to Pinkel. This was our first nationwide trial in ALL: 659 patients were registered, and about 240, or 36%, were cured. The lesson I learned from Don in pediatric oncology was that treatment should be not only specific and effective but also simple, safe, and cheap, so that everyone, everywhere can administer it. This so-called Pinkel therapy was the right beginning in our country, which is federally organized with so many small kings, kingdoms and opinions. We all gained experience with the side effects of toxic chemotherapy and with complications of the disease, and, above all, we were now able to cure children. We were ideally prepared to take on an even greater challenge, the prolongation and intensification of the induction period, developed by Hansjörg Riehm in our country.

Many more times thereafter I had the opportunity to come to Memphis, and many members of the St. Jude's family came to Germany in the years to follow. I was happy to arrange for a most remarkable lecture on treating children with leukemia that Don gave at Children's University Hospital in Munich, on September 8, 1971.

The third and most unforgettable lesson I owe to Don Pinkel, however, evolved during a dinner with live lobsters he invited me to Memphis. According to my diary, it was Wednesday, April 12, 1972. It was after the tenth anniversary of St. Jude's Children's Research Hospital, which now had become a world leader in the research and treatment of childhood cancer. Don told me about the initial difficulties this institution had had; it started as a dream, as an idea in the minds of Danny Thomas and Don. Going to Memphis at that time, he said, was like going into undeveloped territory. But setbacks are also challenges. He told me about those German Jewish professors who had been his teachers, who, after persecution and emigration were glad and humble to be in medicine again, even with a low income and no honors.

"Humiliation makes better men!" This is what you said, Don, and what always helped me to overcome frustrations and failures. And you, yourself, set examples, not minding working in clinics, even as a director, or making night calls. But I should not dwell on humiliation, but instead should close in your particular case with the beginning of the hillbilly song: "It is hard to be humble, if you are perfect in every way . . ."

It is a great thing that you are here, Don. Thank you for coming.

Species-Specific Therapy of Acute Lymphoid Leukemia

D. Pinkel

Forty years ago, Farber and associates described temporary remissions of acute leukemia in children produced by folic acid antagonists [13]. This ignited the hope that this most frequent and always fatal childhood cancer might be curable by drugs. Twenty years ago, Aur and associates completed accession of patients to total therapy study V, the first treatment protocol to result in 50% cure of acute lymphoid leukemia (ALL) [3]. Their results stand 20 years later (Fig. 1), and have been reproduced throughout the world in many thousands of children [6]. More important, recent national vital statistics of the United States and the United Kingdom indicate a 50% reduction in childhood leukemia mortality [4, 29]. Further, the cured children generally enjoy a normal life-style without need for medication.

In the past 20 years, efforts have been directed at improving the cure rate of ALL while simplifying curative treatment, reducing its side effects, and improving its availability and accessibility. In a Stohlman Lecture at Wilsede 10 years ago the following statement was made [32]:

- The most significant opportunity for improving the treatment of acute lymphoid leukemia in the past five years has been its biological and clinical classification by immunological cell surface markers. This allows species identification of the leukemia cells, the first step toward developing specific cytocidal or cytostatic therapy.

The University of Texas M.D. Anderson Cancer Center, Dept. of Pediatrics, Houston, Texas 77030, USA

The purpose of this communication is to review progress in immunophenotype-specific therapy of ALL, to discuss some alternate methods of guiding treatment, and to introduce the notion of genotype-specific chemotherapy of ALL.

A. Immunophenotype-Specific Therapy of ALL

I. Historical Perspective

When the first effective drugs were used to treat acute leukemia it became apparent that some cases were more responsive than others [12]. Methotrexate, prednisone, or mercaptopurine were most likely to produce remissions in children with ALL. Adults with ALL were less likely to experience remission. Both children and adults with acute nonlymphoid

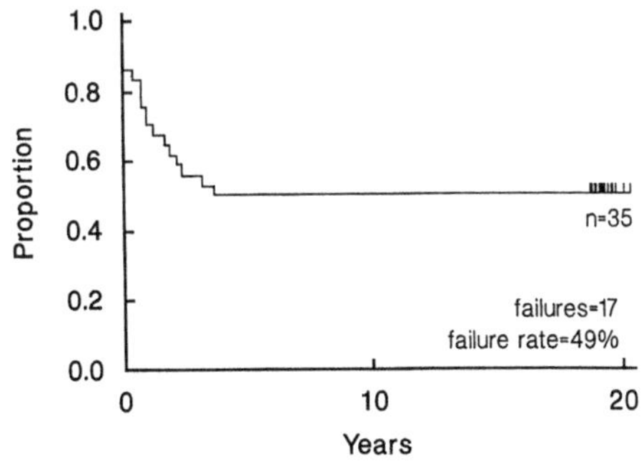

Fig. 1. Event-free survival (EFS) of 35 consecutive children with acute lymphoid leukemia admitted to St. Jude Children's Research Hospital from December 1967 to June 1968. Approximately one-half remain continuously free of leukemia for 20 years and off therapy for 18 years. Update of [3], kindly provided by Gaston Rivera

leukemia (ANLL) had few remissions with these agents. Some hematologists concluded that chemotherapy was of little use in adult acute leukemia and was perhaps better withheld in ANLL, in children as well as adults.

With the introduction of daunorubicin and cytarabine in the 1960s it became apparent that these drugs were highly active in the majority of patients with ANLL, especially when combined [18]. On the other hand, their value in childhood ALL was not so apparent. The concept of species-specific therapy was thus evolved and it became customary to utilize prednisone, vincristine, methotrexate, and mercaptopurine as the primary drugs for ALL, and daunorubicin and cytarabine as the mainstay of treatment of ANLL.

II. Species-Specific Therapy of T-Cell ALL

When T-cell ALL was first defined it was noted that children with this disease had short remissions and high mortality compared with children who had non-T ALL [43]. These observations were generally confirmed by others. However, in mice it was demonstrated that cyclophosphamide and cytarabine were more effective in AKR leukemia, a T-cell line, and Sullivan et al. suggested that cytarabine was specifically effective in human T-cell lymphoma/leukemia [42, 47]. A comparative study in children with ALL in remission demonstrated that the cure rate of T-cell ALL approached that of non-T ALL when the T-cell patients received cyclophosphamide and cytarabine in addition to methotrexate and mercaptopurine [26]. On the other hand, the cyclophosphamide and cytarabine provided no curative benefit, only additional toxicity, to children with non-T ALL receiving methotrexate and mercaptopurine. Thus, it became clear that immunophenotype of ALL was important in selecting and scheduling curative drug therapy.

The importance of immunophenotype-specific chemotherapy of T-cell lymphoma/leukemia was confirmed in a recent Pediatric Oncology Group study [1]. With a treatment plan that emphasizes the use of cytarabine, cyclophosphamide, Adriamycin, and teniposide, and excludes systemic methotrexate, actuarial event-free survival for 94 children with T-cell ALL is 71% at 18 months. Since most relapses of T-cell ALL occur within 18 months this is a meaningful figure.

III. Species-Specific Therapy of B-Cell ALL

When B-cell ALL was defined its rapidly fatal course despite chemotherapy was noted and confirmed [15]. However, two reports indicate that distinctive treatment plans emphasizing the use of cyclophosphamide, the most active agent in childhood B-cell lymphoma/ALL, and a concentrated, relatively brief multiple-drug program, result in a 40% cure rate [14, 30]. A Pediatric Oncology Group study appears to be confirming these observations (Bowman, personal communication).

IV. Species-Specific Therapy of Non-T Non-B ALL

The question rises whether species-specific therapy of subclasses of non-T non-B ALL might be appropriate. As with T-cell ALL and B-cell ALL, the first suggestion of the need for specific therapy is the appearance of an association between immunophenotype and prognosis on a given treatment regimen. Just as T-cell ALL and B-cell ALL demonstrated short remissions and very high mortality in early treatment programs, two immunophenotypic species of non-T non-B ALL have had less favorable courses in more recent studies. First is the "null" or pre-B lymphoid/monocytoid species associated with age less than 1 year, low CALLA antigen, chromosomal translocations involving chromosome 11, band

Table 1. Species-specific therapy, non-T, non-B ALL, treatment plan

Remission induction – 6 weeks	
Days 1–28:	prednisone, vincristine, asparaginase, triple intrathecal therapy
Days 29–42:	mercaptopurine, triple intrathecal therapy
Continuation therapy – 2–2½ years	
Methotrexate weekly, mercaptopurine daily	
Periodic triple intrathecal therapy	
Periodic pulses of prednisone, vincristine, asparaginase	
Periodic intensive therapy – during continuation chemotherapy	
Early pre-B	Intermediate dosage methotrexate
Early pre-B/monocytoid	Intermediate dosage etoposide + cytarabine
Early pre-B/T cell	Intermediate dosage methotrexate Intermediate dosage cyclophosphamide + cytarabine
Pre-B	Intermediate dosage methotrexate + cytarabine Intermediate dosage cyclophosphamide

The systemically administered mercaptopurine, methotrexate, cytarabine, cyclophosphamide, and etoposide are given in maximum tolerated dosage, using clinical status, absolute phagocyte count, and mean corpuscular volume as guides

q 23, presence of myeloid antigens, and monocytoid characteristics by electron microscopy and cell culture [23]. Second is pre-B ALL, which demonstrates cytoplasmic immunoglobulin and is sometimes associated with a t(1;19) chromosomal translocation [35]. A species of T-cell ALL that demonstrates CALLA antigen is reported to have a cure rate between that of T-cell ALL and common ALL on traditional therapy [9].

At UT MD Anderson Cancer Center a pilot protocol was designed and initiated for children newly diagnosed with non-T non-B ALL that provides different periodic consolidation therapy for four different species: common (early pre-B CALLA+), null (early pre-B lymphoid/monocytoid), early pre-B CALLA+ and thymic antigen+, and pre-B (Table 1). Each of the four regimens utilizes periodic consolidation drugs and drug schedules that are currently believed to be most effective for these specific subclasses, while retaining a core of conventional continuation therapy with daily mercaptopurine, weekly methotrexate, pulses of prednisone, vincristine and asparaginase, and periodic triple-intrathecal therapy. Early results suggest the feasibility of this pilot protocol. Of 26 consecutive children registered in the past 18 months, 24 developed complete remission. None have experienced relapse yet.

In summary, immunophenotype-specific selection and scheduling of chemotherapy has proven to be important for increasing the cure rate of T-cell and B-cell ALL. It may also be applicable to upgrading the curability of null ALL and pre-B ALL as well. Almost as important, immunophenotype-specific therapy allows one to exclude nonessential antineoplastic drugs from the combination chemotherapy regimens of ALL, thus avoiding unnecessary immediate and long-term toxic hazards. The prime example is hyperdiploid common ALL, which is highly curable with methotrexate and mercaptopurine continuation chemotherapy [6, 49]. There is no evidence that addition of anthracyclines or alkylating agents improves its cure rate [5]. Therefore, there is no reason to expose these highly vulnerable pre-school children to the risks of anthracycline cardiomyopathy or cyclophosphamide-induced bladder carcinoma [27, 31].

B. Selection and Scheduling Chemotherapy by "Prognostic Factors"

It was recognized decades ago that initial white blood cell count was predictive of response to leukemia chemotherapy [51]. Subsequently, other factors were identified and the term "high risk for treatment failure" was coined for patients with ALL who had such features [2]. It was suggested that more extensive remission induction chemotherapy be administered to such patients. Since then, terms such as "standard risk," "low risk," and "high risk" have become popular to define prognostic categories of patients with ALL and to select and schedule their chemotherapy [46]. In general, patients with "high-risk" ALL are given more drugs in higher dosage, particularly such agents as anthracyclines, alkylating compounds, and epipodophyllotoxins. Patients with "low-risk" ALL are given fewer drugs in lesser dosage, primarily corticosteroid, vinca alkaloid, and antimetabolites. In some treatment programs the decision to use cranial irradiation is based on "risk group" [46].

The problem with using prognostic factors to select therapy is that they are artifacts of data analysis and treatment [33, 34]. More aggressive and rapidly proliferating ALL tends to relapse early; less aggressive and slowly proliferating ALL tends to relapse late. When complete remission duration is used as the criterion for assessing prognostic factors undue weight is given to features associated with remission duration rather than to the true measure of efficacy of therapy, *cure*, as represented by the plateau of continuous complete remission. This problem with the use of prognostic factors could be corrected by using cure rate instead of remission duration to calculate prognostic variants.

However, the more important issue is treatment artifact. All leukemias are fatal when untreated. Survival and cure depend on the administration of appropriate drugs in appropriate schedules. For example, when T-cell ALL was treated with conventional non-T ALL chemotherapy it had a rapidly fatal course in most patients [26]. Features associated with T-cell ALL such as thymic mass, male sex, high white cell count, and older age were calculated to be "high-risk" or "bad-prognosis" factors. With appropriate chemotherapy of T-cell ALL these "risk factors" largely disappear.

In conclusion, there is no evidence that one type of ALL is inherently more lethal than another. All are equally lethal. Cure of ALL is solely a matter of developing and selecting the appropriate drug regimens for each specific type of ALL. The use of prognostic factors to guide leukemia therapy should be abandoned because it is based on artifacts and can give rise to erroneous conclusions.

C. All-Inclusive Multiple-Drug Chemotherapy for All ALL

Another method of selecting therapy for ALL is to avoid selection, but to give all patients all active antineoplastic drugs without regard to immunophenotypic species [37]. This approach carries multiple problems.

Unlike antibiotics, most antineoplastic drugs have overlapping short-term side effects. Administration of one drug usually interferes with the dosage of the other. If minimally effective or noneffective drugs are included in a combination, the dosage of the more effective drugs generally must be reduced. If numerous drugs with overlapping toxicities are utilized it is possible that the most effective drug or drugs may be given at minimally effective dosages and their benefit compromised or lost. Exposure to suboptimal dosage of drugs is an important mechanism of developing resistant cell lines in vitro and could be a mechanism in vivo.

In some all-inclusive multiple-drug regimens, drugs or drug combinations are alternated in order to minimize reduction of drug dosages [37]. The problem with this technique is that the leukemia, in effect, may be untreated or minimally treated during those intervals when drugs of minimal or no efficacy for that particular leukemia are being given. One might postulate the possibility of resurgence of leukemia cell proliferation during such periods of minimally effective or noneffective therapy.

A theoretical objection to the use of multiple drugs is the possibility of antagonistic interactions that might subtract from the efficacy of a given drug [21]. Little is known about subtractive drug interactions in human cancer chemotherapy. One would assume that the risk of such interactions would increase geometrically with linear increase in the number of drugs administered.

A major concern of cancer chemotherapy in children is the prospect of serious long-term sequelae. As noted previously, of special concern are the anthracyclines and the alkylating agents. In one study of children surviving ALL, 55% of those who had received doxorubicin demonstrated abnormal left ventricular function and/or afterload by echocardiography [27]. Cyclophosphamide not only produces sterility but carries a 10% risk of urinary bladder carcinoma 12 years later [31]. To exemplify this concern, it is known that children with hyperdiploid common ALL have a 70% or greater cure rate without alkylating agents or anthracyclines [6, 49]. The only comparative studies reported have failed to demonstrate that these agents contribute to the cure of common ALL in first remission [5]. For these reasons they should be avoided in children with hyperdiploid common ALL who are newly diagnosed or in first remission. The same can be said for any drug with demonstrated serious sequelae that has failed comparative testing for its value in contributing to the cure of a specific type of ALL.

A final objection to the all-inclusive multiple-drug chemotherapy approach is its excessive complexity and cost. This tends to limit the availability and accessibility of curative leukemia therapy to more privileged patients and more privileged nations. The objective of leukemia therapy is to reduce national and world leukemia mortality, not only that of well-financed medical centers.

D. Genotype-Specific Therapy of ALL

I. Acute Leukemias Are Genetic Disorders of Hematopoietic Cells

The most important advance in leukemia therapy in the past 10 years is the renewed realization that leukemias are genetic disorders of hematopoiesis [34, 38, 41]. Their abnormal morphology, immunophenotype, growth, and function are all reflections of their genetic abnormalities. This opens a pathway of drug therapy specific to their genetic properties, aimed at converting their genetic advantages to liabilities.

The evidence that acute leukemias are genetic disorders is convincing [34]. The risk of leukemia is increased in certain constitutional genetic disorders such as Down's, Fanconi's, and Bloom's syndromes and in persons exposed to mutagens such as ionizing irradiation. The morphology of leukemia cells tends to be disorderly and asynchronous, reflecting disordered genetic expression. Chromosome morphology is disturbed in most acute leukemias [41]. Nonrandom chromosome abnormalities are associated with specific types of acute leukemia, such as the t(1;19) translocation in pre-B ALL, the t(8;14) in B-cell ALL, and the t(15;17) in acute promyeloid leukemia [7, 35, 38].

Immunophenotypic and molecular genetic disorders are also prevalent in acute leukemias [20, 34, 45]. Some ALLs express surface antigens characteristic of B-cell and T-cell lineage simultaneously. Early pre-B-(common) ALL often demonstrates rearrangement of genes encoding the T-cell receptor while T-cell ALL may show gene rearrangement for immunoglobulins. It is now obvious that ALLs do not have true B-lymphocyte or T-lymphocyte lineage. Their genetic and phenotypic immunological markers are merely further reflections of their underlying genetic disorders. ALL is a genetic, not an immunological, disease.

The most recent evidence that acute leukemias are genetic disorders is the discovery of overexpression of certain oncogenes in some cases, for example, c-*myc* in B-cell ALL and c-*sis* in acute megakaryocytic leukemia [7, 48].

II. Chemotherapy May Cure Acute Leukemia by Genetic Mechanisms

Although chemotherapy appears to induce remissions of acute leukemia by direct cytolytic effects, it is possible to speculate that cures result from genetic alteration during chemotherapy [34]. Curative drugs such as methotrexate, cytarabine, cyclophosphamide, daunorubicin, and etoposide alter DNA structure as well as synthesis, while drugs without direct effect on DNA such as prednisone, vincristine, and asparaginase do not appear to be curative.

Secondly, curative chemotherapy eliminates genetically disturbed hematopoiesis but spares the capacity for genetically normal hematopoiesis [34]. The best example is the lymphoblastic and lymphocytic hyperplasia noted in the bone marrow of children with ALL after cessation of chemotherapy. Sometimes the frequency of CALLA+lymphoblasts in these children is sufficient to cause confusion with relapse.

Finally, the curative capacity of chemotherapy is strongly related to the genotype of the leukemia [34, 41]. For example, methotrexate and mercaptopurine is a highly curative drug combination in hyperdiploid common ALL, but not in common ALL with a t(9;22) translocation [45, 49]. Daunorubicin and cytarabine is more often curative in acute myeloid leukemia (AML) with a t(8;21) translocation than in AML without this translocation [41]. It is possible that leukemia chemotherapy, when it is curative, is more specific in affecting the genetic mechanism or genetic survival of leukemia strains than we have recognized.

III. Rationale for Genotype-Specific Therapy of ALL

The basis for attempting to target chemotherapy of ALL to its genotypic characteristics is severalfold. First is the convincing evidence that acute leukemias are genetic disorders of hematopoietic cells [34]. Their morphology, immunological markers, growth rate, and other phenotypic properties are reflections of their specific genetic disorders.

Secondly, genetic properties are the most significant variables in curability by a *given therapeutic regimen* [6, 49]. This indicates that therapeutic regimens should be varied in accordance with the genetic properties of the leukemias in order to achieve optimal cure rates. For example, common ALL with a t(9;22) translocation needs to be treated differently than common ALL with hyperdiploidy in order that the t(9;22) variety becomes as curable as the hyperdiploid type.

Thirdly, the current practices of altering chemotherapy regimens in accordance with morphology (ALL vs. ANLL), immunophenotype (T cell vs. B cell), and aggressiveness (white blood cell count) in fact do recognize genotypic properties because all these features reflect the genetic disorders. It would appear more rational to aim treatment directly at the genetic disorders that underly these features as we learn to define these disorders more precisely.

Finally, as noted above, there is reason to speculate that chemotherapy produces remissions by direct cytotoxicity but cures by genetic alteration.

IV. Relationships Between Genotype and Drug Efficacy in ALL

The relationships between the known genotypes of acute lymphoid leukemias and what appear to be the most effective drugs and drug combinations for curing them are summarized in Table 2. The data are yet fragmentary, only the beginning of an approach at targeting drug therapy to the genetic disorders of the

Table 2. Genotype and drug curability, acute lymphoid leukemia

Phenotype	Chromosomal rearrangements	Involved genes	Curative drugs	References
Common	Hyperdiploidy	?	Methotrexate + mercaptopurine	[6, 49]
	t(9;22) (q34;q11)	c-*abl*, *bcr*	?	[6, 39, 45]
Pre-B	t(1;19) (q23;p13)	Insulin receptor α	Methotrexate + cytarabine (?)	[50]
T cell	t(10;14) (q24;q11)	*tcr* α, *TdT* (?)	Cytarabine + cyclophosphamide	[8, 10, 11]
	t(11;14) (p13;q11)	*tcr* α, *WT* (?)		[16, 25, 26]
	t(8;14) (q24;q11)	*tcr* α, c-*myc*		
	t(1;14) (p32;q11)	*tcr* α		
	inv (14) (q11;q32)	*tcr* α, *Ig* μ		
	t(1;7) (p32;q32) t(2;7) (p21;q36) t(6;7) (p21;q36)	*tcr* β		[36]
B cell	t(8;14) (q24;q32)	*Ig* μ, c-*myc*	Cyclophosphamide	[7, 14, 30]
	t(8;22) (q24;q11)	*Ig* λ, c-*myc*		
	t(2;8) (p11;q24)	*Ig* κ, c-*myc*		
Null	t(4;11) (q21;q23)	IP-10 c-*ets*-1 (?)	Epipodophyllotoxins	[17, 19, 23, 28]
	t(1;2;11) (p36;p13;q21)	c-*fgr* (?) c-*src*-2 (?) c-*ets*-1 (?)		[40]
	t(11;19) (q23;p13) t(1;11) (p32;q23) t(10;11) (p15;q23)	c-*ets*-1 (?)		[22]

Many of the molecular genetic and drug data are unconfirmed or speculative

leukemias rather than to the phenotypic features that reflect the genetic disorders. As breakpoints of chromosomal translocations are defined in molecular terms and it becomes possible to classify leukemias as specific molecular genetic disorders it is to be expected that leukemias without apparent chromosomal rearrangements will be shown to have rearrangements of genes similar to those that do have the chromosomal changes. This has already been described in adult-type chronic myeloid leukemia where cases without the typical t(9;22) translocation have the same *bcr-abl* rearrangement that occurs in those with the translocation [24, 44]. As the acute leukemias become better defined in molecular genetic terms it seems plausible that genotype-specific therapy will become more apparent and feasible.

E. Summary

In the past 10 years immunophenotyping of ALL has been demonstrated to be useful for selecting and scheduling chemotherapy. Different drug regimens are now used for T-cell and B-cell ALL than for non-T non-B ALL with the result that survival and cure of T-cell and B-cell ALL have been considerably improved. The use of different drug regimens for different immunophenotypic varieties of non-T non-B ALL is being tested.

"Prognostic factors" of ALL are artifacts of data analysis and treatment and should no longer be used for guiding treatment. The administration of all-inclusive multiple-drug therapy to all patients with ALL regardless of species should also be abandoned. Minimally effective drugs can interfere with dosage and continuity of more effective drugs, and can result in side effects and sequelae that increase the mortality and morbidity of treatment.

Since acute leukemias are genetic disorders of hematopoiesis the future direction of leukemic therapy is toward genetic targeting.

References

1. Amylon M, Murphy S, Pullen J et al. (1988) Treatment of lymphoid malignancies according to immune phenotype: Preliminary results in T-cell disease (Abstr). Proc Am Soc Clin Oncol 7:225
2. Aur RJA, Simone JV, Pratt CB et al. (1971) Successful remission induction in children with acute lymphocytic leukemia at high risk for treatment failure. Cancer 27:1332–1336
3. Aur RJA, Simone JV, Hustu HO et al. (1971) Central nervous system therapy and combination chemotherapy of childhood lymphocytic leukemia. Blood 37:272–281
4. Birch JM, Marsden HB, Jones PH et al. (1988) Improvements in survival from childhood cancer: results of a population based survey over 30 years. Br Med J 296:1372–1376
5. Camitta BM, Pinkel D, Thatcher G et al. (1980) Failure of early intensive chemotherapy to improve prognosis in childhood acute lymphocytic leukemia. Med Pediatr Oncol 8:383–389
6. Crist WM, Furman W, Strother D et al. (1987) Acute lymphocytic leukemia in childhood: Immunologic marker, cytogenetic, and molecular studies. South Med J 80:841–847
7. Croce CM (1986) Chromosome translocations and human cancer. Cancer Res 46:6019–6023
8. Denny CT, Hollis GF, Hecht F et al. (1986) Common mechanism of chromosome inversion in B- and T-cell tumors: Relevance to lymphoid development. Science 234:197–200
9. Dowell BL, Borowitz MJ, Boyett JM et al. (1987) Immunologic and clinicopathologic features of common acute lymphoblastic leukemia antigen-positive childhood T-cell leukemia. Cancer 59:2020–2026
10. Dube ID, Raimondi SC, Pi D et al. (1986) A new translocation, t(10;14) (q24;q11), in T cell neoplasia. Blood 67:1181–1184
11. Erikson J, Finger L, Sun L et al. (1986) Deregulation of c-myc by translocation of the α-locus of the T-cell receptor in T-cell leukemias. Science 232:884–886
12. Farber S, Toch R, Sears EM, Pinkel D (1956) Advances in chemotherapy of cancer in man. Adv Cancer Res 4:1–71
13. Farber S, Diamond LK, Mercer RD et al. (1948) Temporary remissions in acute leukemia in children produced by folic acid antagonist, 4-aminopteroyl-glutamic acid (aminopterin). N Engl J Med 238:787–793
14. Feickert HJ, Göbel U, Ludwig W et al. (1987) Childhood acute lymphoblastic leukemia of B-cell type: Trials ALL-BFM 81 and ALL-BFM 83 (Abstr). Proc Am Soc Clin Oncol 6:149
15. Flandrin G, Brouet JC, Daniel MT et al. (1975) Acute leukemia with Burkitt's tumor cells: A study of six cases with special reference to lymphocyte surface markers. Blood 45:183–188
16. Finger LR, Harvey RC, Moore RC et al. (1986) A common mechanism of chromosomal translocation in T- and B-cell neoplasia. Science 234:982–985
17. Frankel LS, Ochs J, Shuster J et al. (1987) Pilot protocol improves remissions for infant leukemia and provides detailed laboratory characterization (Abstr). Proc Am Soc Clin Oncol 6:161
18. Gale RP (1979) Advances in the treatment of acute myelogenous leukemia. N Engl J Med 300:1189–1199
19. Goyns MH, Hann IM, Stewart J et al. (1987) The c-ets-1 proto-oncogene is rearranged in some cases of acute lymphoblastic leukaemia. Br J Cancer 56:611–613
20. Hurwitz CA, Loken MR, Graham ML, et al. (1988) Asynchronous antigen expression in B lineage acute lymphoblastic leukemia. Blood 72:299–307
21. Jolivet J, Cole D, Holcenberg JS et al. (1984) L-asparaginase (L-ASP) antagonism of methotrexate (MTX) cytotoxicity: An alternative explanation (Abstr). Pro-

ceedings of the American Association for Cancer Research 25:309
22. Kaneko Y, Maseki N, Takasaki N et al. (1986) Clinical and hematologic characteristics in acute leukemia with 11q23 translocations. Blood 67:484–491
23. Katz F, Malcolm S, Gibbons B et al (1988) Cellular and molecular studies on infant null acute lymphoblastic leukemia. Blood 71:1438–1447
24. Kurzrock R, Blick MB, Talpaz M et al. (1986) Rearrangement in the breakpoint cluster region and the clinical course in Philadelphia-negative chronic myelogenous leukemia. Ann Intern Med 105:673–679
25. Lampert F, Harbott J, Ritterbach J et al. (1988) T-cell acute childhood lymphoblastic leukemia with chromosome 14q11 anomaly: a morphologic, immunologic, and cytogenetic analysis of 10 patients. Blut 56:117–123
26. Lauer SJ, Pinkel D, Buchanan GR et al. (1987) Cytosine arabinoside/cyclophosphamide pulses during continuation therapy for childhood acute lymphoblastic leukemia. Cancer 60:2366–2371
27. Lipshultz SE, Colan SD, Sanders SP et al. (1987) Late cardiac effects of doxorubicin in childhood acute lymphoblastic leukemia (ALL) (Abstr). Proceedings of the American Society of Hematology, 234a
28. Luster AD, Jhanwar SC, Chaganti RSK et al. (1987) Interferon-inducible gene maps to a chromosomal band associated with a (4;11) translocation in acute leukemia cells. Proc Natl Acad Sci, USA 84:2868–2871
29. Miller RW, McKay FW (1984) Decline in US childhood cancer mortality 1950 through 1980. JAMA 251:1567–1570
30. Patte C, Philip T, Rodary C et al. (1986) Improved survival rate in children with Stage III and IV B cell non-Hodgkin's lymphoma and leukemia using multi-agent chemotherapy: Results of a study of 114 children from the French Pediatric Oncology Society. J Clin Oncol 4:1219–1226
31. Pedersen-Bjergaard J, Ersboll J, Hansen VL et al. (1988) Carcinoma of the urinary bladder after treatment with cyclophosphamide for non-Hodgkin's lymphoma. N Engl J Med 318:1028–1032
32. Pinkel D (1979) Treatment of childhood acute lymphocytic leukemia. Modern Trends in Human Leukemia III. R Neth, RC Gallo, P-H Hofschneider and K Mannweiler (eds). pp 25–33. New York
33. Pinkel D (1985) Current issues in the management of children with acute lymphocytic leukaemia. Postgrad Med J 61:93–102
34. Pinkel D (1987) Curing children of leukemia. Cancer 59:1683–1691
35. Pui C-H, Williams DL, Kalwinsky DK et al. (1986) Cytogenetic features and serum lactic dehydrogenase level predict a poor treatment outcome for children with pre-B-cell leukemia. Blood 67:1688–1692
36. Raimondi SC, Pui CH, Behm FG et al. (1987) 7q32-q36 Translocations in childhood T cell leukemia: Cytogenetic evidence for involvement of the T cell receptor β-chain gene. Blood 69:131–134
37. Rivera GK, Mauer AM (1987) Controversies in the management of childhood acute lymphoblastic leukemia: treatment intensification, CNS leukemia, and prognostic factors. Semin Hematol 24:12–26
38. Rowley JD (1979) Chromosome abnormalities in leukemia. Modern Trends in Human Leukemia III. R Neth, RC Gallo, P-H Hofschneider and K Mannweiller (eds.). pp 43–52. New York
39. Rubin CM, Carrino JJ, Dickler MN et al. (1988) Heterogeneity of genomic fusion of BCR and ABL in Philadelphia chromosome-positive acute lymphoblastic leukemia. Proc Natl Acad Sci USA 85:2795–2799
40. Sansone R, Strigini P (1988) Infantile leukemia with a new chromosomal rearrangement involving 11q. Cancer Genet Cytogenet 32:293–294
41. Sandberg AA (1986) The chromosomes in human leukemia. Semin Hematol 23:201–217
42. Schabel FM Jr, Skipper HE, Trader MW et al. (1974) Combination chemotherapy for spontaneous AKR lymphoma. Cancer Chemotherapy Reports 4:53–70
43. Sen L, Borella L (1975) Clinical importance of lymphoblasts with T markers in childhood acute leukemia. N Engl J Med 292:828–832
44. Stam K, Heisterkamp N, Grosveld G et al. (1985) Evidence of a new chimeric bcr/c-abl mRNA in patients with chronic myelocytic leukemia and the Philadelphia chromosome. N Engl J Med 313:1429–1433
45. Stass SA, Mirro J Jr (1986) Lineage heterogeneity in acute leukaemia: Acute mixed-lineage leukaemia and lineage switch. Clin Haematol 15:811–827

46. Steinherz PG, Gaynon P, Miller DR et al. (1986) Improved disease-free survival of children with acute lymphoblastic leukemia at high risk for early relapse with the New York regimen – A new intensive therapy protocol: A report from the Childrens Cancer Study Group. J Clin Oncol 4:744–752
47. Sullivan MP, Ramirez I (1982) Contribution of cytosar to T-antigen positive lymphoid disease control in children given 2nd generation LSA_2L_2 therapy. Proc Am Assoc Cancer Res 23:114
48. Sunami S, Fuse A, Simizu B et al. (1987) The c-sis gene expression in cells from a patient with acute megakaryoblastic leukemia and Down's Syndrome. Blood 70:368–371
49. Williams DL, Tsiatis A, Brodeur GM et al. (1982) Prognostic importance of chromosome number in 136 untreated children with acute lymphoblastic leukemia. Blood 60:864–871
50. Yang-Feng TL, Francke U, Ullrich A (1985) Gene for human insulin receptor: Localization to site on chromosome 19 involved in pre-B-cell leukemia. Science 228:728–730
51. Zuelzer WW, Flatz G (1960) Acute childhood leukemia: A ten-year study. Am J Dis Child 100:886–907

Current Issues and Future Directions in Marrow Transplantation *

R. Storb

A. Introduction

The use of allogeneic marrow transplantation as treatment for patients with various hematological diseases has increased in recent years [1–6]. A survey by the International Bone Marrow Transplantation Registry [7] estimated the number of transplants carried out through the year 1987 to be in the order of 20000, more than 10000 of these during the years of 1985 through 1987. Marrow transplantation has been employed in most cases (≥80%) for therapy of malignant hematologic diseases. Roughly 10% of all transplants have been for the treatment of patients with acquired or inherited marrow dysfunction (aplastic anemia), and 5%–6% have been for treatment of congenital defects of the hematopoietic and immune systems (thalassemia major, severe combined immuno-deficiency disease, and other inborn errors).

Since 1970, when marrow transplantation was restricted to patients with advanced hematologic malignancies and disease-free survival was in the order of 15%, remarkable advances have been made [8]. Recent studies in patients with acute nonlymphoblastic leukemia (ANL) in first chemotherapy-induced remission have shown actuarial survival to be superior in patients undergoing marrow transplantation (50%) compared with those given chemotherapy (20%) with a follow-up period of up to 10 years. Patients with acute lymphoblastic leukemia (ALL) given grafts in second or subsequent remission have shown disease-free survival of approximately 35%, whereas patients undergoing chemotherapy all died of recurrent disease within 3½ years of the initiation of therapy. Fifty to 60% of patients with chronic myelocytic leukemia (CML) transplanted while in chronic phase have obtained disease-free survival whereas none can be cured with chemotherapy alone. In patients with aplastic anemia treated with marrow grafting, survival has improved to 60%–80% compared with 40%–50% for patients treated with immunosuppression by antithymocyte globulin and only 20% for patients who receive only supportive therapy. Marrow grafting has produced 70% disease-free survival in patients with thalassemia major [9] and approximately 50%–60% survival in patients grafted for severe combined immunodeficiency disease and other inborn errors [10].

Despite impressive improvements, major problems and complications in marrow transplantation remain [1–6, 8, 10–23]. These are listed in Table 1. In patients grafted for leukemia 17%–75% of treatment failures are attributable to relapse, whereas graft rejection has resulted in the death of 5%–12% of patients grafted for aplastic anemia. Significant acute graft-versus-host disease (GVHD) with a case fatality rate of approximately 50% is seen in 18%–45% of all patients and it is responsible for 10%–25% of

---

The Fred Hutchinson Cancer Research Center and the University of Washington School of Medicine 1124 Columbia St., Seattle, WA, 98104, USA

* Supported in part by grant HL 36444 from the National Heart, Lung and Blood Institute, and by grants CA 18029, CA 18221, CA 31787, CA 18105 and CA 15704 awarded by the National Cancer Institute, National Institutes of Health, DHHS

Table 1. Incidence of transplant-related complications and long-term survival after HLA-identical marrow grafting. (Data reviewed in [1–6, 8, 10–23])

Disease[a]	CML			ALL				ANL			AA
Disease phase[b]	1st CP	AP	BC	1st CR	2nd CR	3rd CR	2nd+ Rel	1st CR	2nd+ CR	1st Rel	Severe
Five-year disease-free survival (%)	58	30	20	54	35	30	18	50	25	34	60–80
Relapse (%)	17	45	70	35	45	58	75	22	45	31	–
Grade II–IV acute GVHD (%)[c]		45			35				40–45		18–35
Chronic GVHD (%)		35			25				25–35		30–45
Interstitial pneumonia (%)[d]		22			15				15–35		5–15
VOD (%)[e]		25			7				28		<1
Bacterial + fungal infections											
During first 3 months											
Before engraftment (%)					20						15
After engraftment (%)					12						12
After first 3 months (%)					20						20
Graft failure (%)					<1						5–17
Secondary malignancies (%)					5						1

[a] CML, chronic myelocytic leukemia; ALL, acute lymphoblastic leukemia; ANL, acute nonlymphoblastic leukemia; AA, aplastic anemia. Patients with AA are usually conditioned with cyclophosphamide alone or in combination with total lymphoid or thoracoabdominal irradiation; patients with leukemia are usually given cyclophosphamide and total body irradiation

[b] CP, chronic phase; AP, accelerated phase; BC, blast crisis; CR, complete remission; Rel, relapse

[c] GVHD, graft-versus-host disease

[d] Includes both idiopathic and cytomegalovirus interstitial pneumonias

[e] VOD, veno-occlusive disease of the liver. VOD is rare in patients conditioned only with cyclophosphamide

treatment failures. Conditioning regimen related toxicity and bacterial or fungal infection during the early period of neutropenia result in 5%–10% of deaths. Fatal interstitial pneumonias are often associated with acute GVHD or may be the result of drug and radiation toxicity. Methods of improving the results of marrow transplantation are needed.

B. Graft-Versus-Host Disease Prophylaxis and Graft Failure

Reliable, nontoxic methods of preventing GVHD are needed (reviewed in [15, 22]). GVHD prophylaxis has customarily involved postgrafting immunosuppression. In many patients, immunosuppressive therapy can be discontinued by 3–6 months after transplantation when a stable state of graft-host tolerance has been achieved. Omission of immunosuppression in most patients has caused an unacceptably high incidence of acute GVHD and transplant-related death. Controlled randomized trials have shown methotrexate and cyclosporine to be comparable in their ability to prevent GVHD (reviewed in [24]), and a combination of the two drugs is significantly better than either alone in preventing GVHD [25, 26]. However, chronic GVHD remains a problem [27].

Graft-versus-host disease prevention has been attempted by immunological and mechanical removal of T cells from the donor marrow (reviewed in [6, 28, 29]), resulting in a reduction in the number of infused T cells by 1 to 3 logs. In this manner, most differentiated immune cells causing GVHD are eliminated and the immune system is returned to an early prenatal state. New stem cell-derived T cells accept the host's antigenic environment as "self" and become tolerant to it. T-cell depletion has worked well in rodent models but has been less successful in large random-bred animals where increased graft failure has been noted. Graft failure seems to be caused by host immune cells which survived the conditioning program and whose continued survival is assured through the absence of GVHD. Nearly all clinical studies have shown a significant reduction in acute GVHD in patients given T-cell-depleted marrow grafts (Table 2), providing convincing evidence for a favorable effect of T-cell depletion on GVHD. However, this was achieved at the price of substantial increases in graft rejection and leukemic relapse (Tables 2, 3): when T-cell-depleted marrow was used the overall incidence of graft rejection increased from 1% to 12% and from 5% to 32% in HLA-identical and in HLA-nonidentical recipients, respectively. Additionally, relapse rates in patients with leukemia increased significantly, most impressively in patients grafted for CML in chronic phase (Table 3), and graft failure rates increased in patients with aplastic anemia grafted with T-cell-depleted marrow.

Since graft rejection and leukemic relapse almost always result in death, improved survival has not been seen in patients given T-depleted marrow transplants. Nevertheless, the significant decrease in the incidence of acute GVHD suggests that T-cell depletion would be useful if the risks of graft rejection and relapse can be lessened. To achieve this, two different methods can be envisioned. One includes improvement of pretransplant-conditioning programs better to eradicate immune cells of host type as

Table 2. Effect of T-cell depletion on incidence of GVHD and graft failure in patients transplanted for leukemia. (Data from the International Bone Marrow Transplant Registry [28])

Marrow source	T-depletion	Acute GVHD	Graft failure
HLA-identical donor	Yes	11%	12%
	No	45%	1%
HLA-nonidentical donor	Yes	31%	32%
	No	75%	5%

Table 3. Leukemic relapse and T-cell depletion. (Data from the International Bone Marrow Transplant Registry [28])

Disease[a]	# Patients	Incidence of relapse	
		Untreated marrow	T-depleted marrow
CML–CP	309	8%	43%
ANL–1st CR	538	18%	45%
ALL–1st CR	205	25%	35%
ALL–2nd CR	179	55%	75%

[a] CML–CP, chronic myelocytic leukemia in chronic phase; ANL–CR, acute nonlymphoblastic leukemia in complete remission; ALL–CR, acute lymphoblastic leukemia in complete remission

well as malignant cells. As discussed below, this aim may be possible through better use of currently available chemoradiation therapy and through innovative approaches using antibody isotope conjugates in addition to chemoradiation therapy. The other method is based on the possibility that T cells causing GVHD are distinct from those which enhance engraftment and cause the graft-versus-leukemia effect. A better understanding of the precise role of lymphocytes in mediating these diverse immune functions might result in the development of strategies to eliminate GVHD without impairing engraftment or the graft-versus-leukemia effect.

C. Marrow Graft Rejection in Patients with Aplastic Anemia

A common problem in patients given HLA-identical marrow grafts for the treatment of severe aplastic anemia after conditioning with high-dose cyclophosphamide has been graft failure [3, 5, 21, 30]. In the early 1970s this problem was seen in 30%–60% of patients. Two factors were associated with rejection: positive in vitro tests of cell-mediated immunity, indicating reaction of host lymphocytes against antigens on donor cells before transplantation; and, secondly, a low number of transplanted marrow cells ($<3 \times 10^8$ cell/kg). As supported by studies in experimental animals, immunity of recipient against donor is thought to be the result of transfusion-induced sensitization. Canine studies have indicated that dendritic mononuclear cells in transfused blood products lead to sensitization of the recipient against minor antigens of the donor which may not be suppressed by the immunosuppressive conditioning programs [31]. Transplants carried out in patients who have not received preceding transfusions rarely result in graft failure: 80% of untransfused patients are alive with functioning grafts. This suggests that immunological mechanisms involved in graft failure are, for the most part, induced by previous blood transfusions.

Many regimens, mainly involving more intensive immunosuppression, are being used to avoid graft rejection in multiply-transfused patients. All programs include cyclophosphamide, but other features vary, such as the use of total body irradiation (TBI), total lymphoid irradiation, total nodal irradiation, and thoracoabdominal irradiation. The Seattle team has administered viable donor buffy coat cells along with the marrow infusion, since the donor's peripheral blood is a potential source of hemopoietic stem cells and/or lymphoid cells capable of abrogating rejection. Most transplant centers are now reporting that rejection rates have decreased and survival has increased in multiply-transfused patients with survivals between 60% and 70%.

Risks are associated with most of the conditioning programs. Buffy coat cells may lead to an increase in chronic GVHD. Irradiation carries the potential for future cancer. Because of these risks as well as the persistent possibility of rejection, emphasis should be placed on preventing rather than overcoming sensitization caused by blood transfusions. This is best done by performing transplantation before administering transfusions. In case transfusions are required, buffy coat-poor red blood cells and platelets should be used. Recent data in the canine model have shown that sensitization can be prevented if blood transfusion products are exposed to ultraviolet light irradiation [32].

D. Relapse in Patients Transplanted for Hematological Malignancies

Cyclophosphamide and TBI have been the most commonly used conditioning agents for patients with leukemia [1–4, 8, 11–13]. In the attempt to reduce the leukemic recurrence rate, numerous therapeutic reagents such as etoposide, high-dose cytosine arabinoside, piperazinedione, BCNU, and others have been used in addition to or instead of cyclophosphamide. Fractionated TBI has slowly replaced single-dose TBI over the past decade since a prospective comparison of the two schedules showed fractionated TBI to be better tolerated and to result in fewer long-term complications without any apparent increase in postgrafting relapse rates [33]. Hyperfractionated TBI followed by cyclophosphamide has been used in patients with ALL in second or subsequent remission by the Sloan-Kettering team with apparently superior results [11]. A combination of busulfan and cyclophosphamide has been used without TBI by the Johns-Hopkins team and they reported very low leukemic recurrence rates in patients with ANL in first remission, while relapse rates in patients with more advanced ANL appeared to be similar to those seen after cyclophosphamide/TBI regimens [14]. This appears to contrast with results reported by the Ohio State team, which suggest that relapse rates are low not only in patients with ANL in first remission but also in patients with advanced ANL and ALL, even with reduced doses of busulfan and cyclophosphamide [34]. It appears, however, that the limits of nonhemopoietic toxicity have been reached and no substantial improvements in relapse rates and survival can be expected using systemic chemotherapy and TBI.

In principle, the most efficient means of eradicating cancer would be to use agents which interact specifically with malignant cells. The method approaching this ideal most closely is the use of monoclonal antibodies directed against tumor-associated antigens. It is known that monoclonal antibodies injected in vivo can concentrate on tumor cells; however, the antitumor effect is limited, partly due to the fact that some tumor cells lack target antigens, and partly because some cells, though coated by antibody, may not be killed by it. Attempts are being made to link antibodies to toxins such as the ricin-A chain for more effective tumor cell kill. Also in progress are studies attaching monoclonal antibodies to short-lived radioactive isotopes which deposit most of their energy within a 1- to 2-mm radius. With these isotopes, cells expressing the target antigens as well as neighboring cells which may be antigen negative will be killed. In the case of hematologic malignancies, subsequent marrow "rescue" would be needed since this approach would ablate normal marrow cells. Initial experiments in a canine model of marrow transplantation have shown appropriate antibody isotope conjugates to localize preferentially in the marrow and spleen and also, to a lesser extent, in lymph nodes [35, 36], with the amount of isotope in the marrow achieving a ratio of 5:1 or better as compared with other organs. The marrow aplasia caused by radiolabeled antibodies can be reversed by infusion of cryopreserved autologous marrow at a time when very little radioactivity is left, about 8 days later. Canine studies are underway exploring the efficacy of various combinations of chemotherapy, TBI, and radiolabeled antibodies in conditioning dogs for T-cell-depleted marrow grafts. It is anticipated that refinements of this approach, particularly the use of high-energy beta-emitting isotopes with short linear energy transfer, will lead to less toxic and more efficient conditioning programs which will not only provide better elimination of malignant cells but will also ameliorate the problem of graft failure.

Radiolabeled antibodies might be useful in transplantation for nonmalignant as well as for malignant hematological diseases, by allowing engraftment to take place while eliminating busulfan in patients with thalassemia major or reducing the dose of cyclophosphamide in patients with aplastic anemia.

E. Prophylaxis and Therapy of Interstitial Pneumonia

Interstitial pneumonias are among the most serious complications arising during the first 3–4 months after transplantation (reviewed in [23, 37, 38]). Pneumonias are less frequent in patients grafted for aplastic anemia following cyclophosphamide than in patients with leukemia whose conditioning regimen included TBI or busulfan. *Pneumocystis carinii* infection, formerly the cause of about 10% of all interstitial pneumonias, is now being prevented by prophylactic trimethoprim sulfamethoxazole. Idiopathic interstitial pneumonia has been seen in approximately 13% of patients given single-dose TBI, but the incidence has declined to 3% with the use of fractionated TBI.

By far the most critical infection is cytomegalovirus (CMV). Evidence of CMV activation is seen in about 75% of all patients with positive CMV antibody titers before transplant. While often asymptomatic and manifested only by viral excretion in the urine or by increasing antibody titers, CMV activation can develop into a serious complication in the form of CMV pneumonia, which has a case fatality rate of approximately 85%. Patients who are CMV seronegative before transplant can be protected from infection by the use of CMV-sero-negative blood products during and after transplant. If possible, only CMV-negative blood products should be given to any CMV-negative patient who is a potential transplant candidate. Immunoprophylaxis using CMV immunoglobulin has been controversial, and there is currently no proven therapy for established CMV infection. The use of an acyclovir derivative, dihydroxymethyl-ethoxymethylguanine, has not been effective in treating CMV pneumonia although it has significantly reduced the amount of virus in the lung tissues, and it may prove to be beneficial when given along with CMV immunoglobulin in treating established CMV pneumonia. Also, it may be useful in prophylactic trials.

It is possible that the use of certain recombinant human hematopoietic growth factors, such as IL-1, IL-3, G-CSF, and GM-CSF, might shorten the period of granulocytopenia or thrombocytopenia after grafting, thus reducing the incidence of early infection and resulting in a modest improvement of survival.

F. Conclusions

In the early 1970s marrow transplants were only administered to patients who had advanced acute leukemia, severe aplastic anemia, or severe combined immunodeficiency diseases. Since then, the technique has been shown to be beneficial and even curative for patients with many different hematological conditions. In younger patients, marrow grafting is now the treatment of choice for aplastic anemia, immunodeficiency disease, certain genetic disorders of hemopoiesis, any leukemia which has relapsed at least once, ANL in first remission, and CML. For patients who have thalassemia major, CML in chronic phase, or ANL in first remission, the risk of early death from transplant-related complications must be weighed against the benefit of long-term cure.

Although impressive advances in transplantation have taken place, major problems persist. These include recurrence of leukemia, graft failure in patients given T-depleted or HLA-nonidentical grafts, acute and chronic GVHD, infections associated with prolonged immunodeficiency, and late-occurring complications resulting from the conditioning programs. Major improvements in the area of more effective and less toxic conditioning regimens are needed. In this regard, the use of monoclonal antibodies linked to short-lived radioactive isotopes with short linear energy transfer seems promising. It is expected that more effective conditioning programs will decrease the incidences of leukemic recurrence and graft failure. Better conditioning regimens should permit a broader application of T-cell depletion to prevent acute and chronic GVHD, thus extending marrow grafting to include more HLA-nonidentical and unrelated patients. The use of recombinant hemopoietic growth factors may prove to reduce the risk of early infections, but the problem of CMV infection in seropositive recipients will remain until effective antiviral drugs are identified.

References

1. Gratwohl A, Hermans J, Barrett AJ, Ernst P, Frassoni F, Gahrton G, Granena A, Kolb HJ, Marmont A, Prentice HG, Speck B, Vernant JP, Zwaan FJ (1988) Allogeneic bone marrow transplantation for leukaemia in Europe: report from the working party on leukaemia, European Group for Bone Marrow Transplantation. Lancet I: 1379–1382
2. Ringden O, Zwaan F, Hermans J, Gratwohl A (1987) for the Leukemia Working Party of the European Group for Bone Marrow Transplantation. European experience of bone marrow transplantation for leukemia. Transplant Proc 19: 2600–2604
3. Champlin R (1987) for the Advisory Committee of the International Bone Marrow Transplant Registry. Bone marrow transplantation for acute leukemia: a preliminary report from the International Bone Marrow Transplant Registry. Transplant Proc 19: 2626–2628
4. Gluckman E (1987) Current status of bone marrow transplantation for severe aplastic anemia: a preliminary report from the International Bone Marrow Transplant Registry. Transplant Proc 19: 2597–2599
5. Storb R, Doney K, Thomas ED, Anasetti C, Appelbaum F, Beatty P, Bensinger W, Buckner CD, Clift R, Fefer A, Hansen J, Hill R, Martin P, McGuffin R, Sanders J, Singer J, Stewart P, Sullivan K, Whitherspoon R (1988) Allogeneic and syngeneic marrow transplantation for aplastic anemia: overview of Seattle results. In: Baum SJ, Santos GW, Takaku F (eds) Experimental hematology today – 1987. Recent advances and future directions in bone marrow transplantation. Springer, Berlin Heidelberg New York, pp 119–124
6. Storb R (1987) Critical issues in bone marrow transplantation. Transplant Proc 19: 2774–2781
7. Bortin MM (1988) Key results from recent analyses: a report from the International Bone Marrow Transplant Registry. Proceedings of the 17th annual meeting of the International Society for Experimental Hematology, Houston Tex, August 21–25. Exp Hematol 16: 414, Abstr #7
8. Thomas ED, Storb R, Clift RA, Fefer A, Johnson FL, Neiman PE, Lerner KG, Glucksberg H, Buckner CD (1975) Bone-marrow transplantation. N Engl J Med 292: 832–843, 895–902
9. Lucarelli G, Galimberti M, Polchi P, Giardini C, Politi P, Baronciani D, Angelucci E, Manenti F, Delfini C, Aureli G, Muretto P (1987) Marrow transplantation in patients with advanced thalassemia. N Engl J Med 316: 1050–1055
10. O'Reilly RJ (1983) Allogeneic bone marrow transplantation: current status and future directions. Blood 62: 941–964
11. Brochstein JA, Kernan NA, Groshen S, Cirrincione C, Shank B, Emanuel D, Laver J, O'Reilly RJ (1987) Allogeneic bone marrow transplantation after hyperfractionated total-body irradiation and cyclophosphamide in children with acute leukemia. N Engl J Med 317: 1618–1624
12. Goldman JM, Apperley JF, Jones L, Marcus R, Goolden AWG, Batchelor R, Hale G, Waldmann H, Reid CD, Hows J, Gordon-Smith E, Catovsky D, Galton DAG (1986) Bone marrow transplantation for patients with chronic myeloid leukemia. N Engl J Med 314: 202–207
13. Dinsmore R, Kirkpatrick D, Flomenberg N, Gulati S, Kapoor N, Shank B, Reid A, Groshen S, O'Reilly RJ (1983) Allogeneic bone marrow transplantation for patients with acute lymphoblastic leukemia. Blood 62: 381–388
14. Santos GW, Tutschka PJ, Brookmeyer R, Saral R, Beschorner WE, Bias WB, Braine HG, Burns WH, Elfenbein GJ, Kaizer H, Mellits D, Sensenbrenner LL, Stuart RK, Yeager AM (1983) Marrow transplantation for acute nonlymphocytic leukemia after treatment with busulfan and cyclophosphamide. N Engl J Med 309: 1347–1353
15. Gale RP, Bortin MM, Van Bekkum DW, Biggs JC, Dicke KA, Gluckman E, Good RA, Hoffmann RG, Kay HEM, Kersey JH, Marmont A, Masaoka T, Rimm AA, Van Rood JJ, Zwaan FE (1987). Risk factors for acute graft-versus-host disease. Br J Haematol 67: 397–406
16. Thomas ED, Clift RA, Fefer A, Appelbaum FR, Beatty PG, Bensinger WI,

Buckner CD, Cheever MA, Deeg HJ, Doney K, Flournoy N, Greenberg P, Hansen JA, Martin P, McGuffin R, Ramberg R, Sanders JE, Singer J, Stewart P, Storb R, Sullivan K, Weiden PL, Witherspoon R (1986) Marrow transplantation for the treatment of chronic myelogenous leukemia. Ann Intern Med 104:155–163
17. Thomas ED, Buckner CD, Clift RA, Fefer A, Johnson FL, Neiman PE, Sale GE, Sanders JE, Singer JW, Shulman H, Storb R, Weiden PL (1979) Marrow transplantation for acute nonlymphoblastic leukemia in first remission. N Engl J Med 301:597–599
18. Thomas ED, Sanders JE, Flournoy N, Johnson FL, Buckner CD, Clift RA, Fefer A, Goodell BW, Storb R, Weiden P (1979) Marrow transplantation for patients with acute lymphoblastic leukemia in remission. Blood 54:468–476
19. Appelbaum FR, Dahlberg S, Thomas ED, Buckner CD, Cheever MA, Clift RA, Crowley J, Deeg HJ, Fefer A, Greenberg P, Kadin M, Smith W, Stewart P, Sullivan KM, Storb R, Weiden P (1984) Bone marrow transplantation or chemotherapy after remission induction for adults with acute nonlymphoblastic leukemia: a prospective comparison. Ann Intern Med 101:581–588
20. Clift RA, Buckner CD, Thomas ED, Kopecky KJ, Appelbaum FR, Tallman M, Storb R, Sanders J, Sullivan K, Banaji M, Beatty P, Bensinger W, Cheever M, Deeg J, Doney K, Fefer A, Greenberg P, Hansen JA, Hackman R, Hill R, Martin P, Meyers J, McGuffin R, Neiman P, Sale G, Shulman H, Singer J, Stewart P, Weiden P, Witherspoon R (1987) The treatment of acute nonlymphoblastic leukemia by allogeneic marrow transplantation. Bone Marrow Transplant 2:243–258
21. Storb R, Thomas ED, Buckner CD, Appelbaum FR, Clift RA, Deeg HJ, Doney K, Hansen JA, Prentice RL, Sanders JE, Stewart P, Sullivan KM, Witherspoon RP (1984) Marrow transplantation for aplastic anemia. Semin Hematol 21:27–35
22. Storb R, Thomas ED (1985) Graft-vs-host disease in dog and man: The Seattle experience. In: Moller G (ed) Immunological reviews, No. 88. Copenhagen, Munksgaard, 215–238
23. Meyers JD (1988) Prevention and treatment of cytomegalovirus infection after marrow transplantation. Bone Marrow Transplant 3:95–104
24. Storb R, Deeg HJ, Fisher LD, Appelbaum F, Buckner CD, Bensinger W, Clift R, Doney K, Irle C, McGuffin R, Martin P, Sanders J, Schoch G, Singer J, Stewart P, Sullivan K, Witherspoon R, Thomas ED (1988) Cyclosporine versus methotrexate for graft-versus-host disease prevention in patients given marrow grafts for leukemia: long-term follow-up of three controlled trials. Blood 71:293–298
25. Storb R, Deeg HJ, Whitehead J, Appelbaum F, Beatty P, Bensinger W, Buckner CD, Clift R, Doney K, Farewell V, Hansen J, Hill R, Lum L, Martin P, McGuffin R, Sanders J, Stewart P, Sullivan K, Witherspoon R, Yee G, Thomas ED (1986) Methotrexate and cyclosporine compared with cyclosporine alone for prophylaxis of acute graft versus host disease after marrow transplantation for leukemia. N Engl J Med 314:729–735
26. Storb R, Deeg HJ, Farewell V, Doney K, Appelbaum F, Beatty P, Bensinger W, Buckner CD, Clift R, Hansen J, Hill R, Longton G, Lum L, Martin P, McGuffin R, Sanders J, Singer J, Stewart P, Sullivan K, Witherspoon R, Thomas ED (1986) Marrow transplantation for severe aplastic anemia: methotrexate alone compared with a combination of methotrexate and cyclosporine for prevention of acute graft-versus-host disease. Blood 68:119–125
27. Sullivan KM, Witherspoon R, Storb R, Appelbaum F, Beatty P, Bensinger W, Bigelow C, Buckner CD, Cheever M, Clift R, Doney K, Fefer A, Greenberg P, Hansen J, Martin P, Matthews D, McDonald G, Meyers J, Petersen FB, Sanders J, Shulman H, Singer J, Stewart P, Thomas ED (1988) Chronic graft-versus-host disease: pathogenesis, diagnosis, treatment and prognostic factors. In: Baum SJ, Santos GW, Takaku F (eds) Experimental hematology today – 1987. Recent advances and future directions in bone marrow transplantation. Springer, Berlin Heidelberg New York, pp 150–157
28. Butturini A, Franceschini F, Gale RP (1988) Critical analysis of T-cell depletion in man. In: Martelli MF, Grignani F, Reisner Y (eds) T-cell depletion in allogeneic bone marrow transplantation. Ares-Serono symposia, Rome, pp 1–13
29. Goldman JM, Gale RP, Horowitz MM, Biggs JC, Champlin RE, Gluckman E, Hoffmann RG, Jacobsen SJ, Marmont AM, McGlave PB, Messner HA, Rimm AA, Rozman C, Speck B, Tura S, Weiner

RS, Bortin MM (1988) Bone marrow transplantation for chronic myelogenous leukemia in chronic phase: increased risk for relapse associated with T-cell depletion. Ann Intern Med 108:806-814

30. Gordon-Smith EC (1987) Recent advances and future trends in bone marrow transplantation for severe aplastic anemia. In: Baum SJ, Santos GW, Takaku F (eds) Experimental hematology today - 1987. Recent advances and future directions in bone marrow transplantation. Springer, Berlin Heidelberg New York, pp 125-129
31. Deeg HJ, Aprile J, Storb R, Graham T, Hackman R, Appelbaum F, Schuening F (1988) Functional dendritic cells are required for transfusion-induced sensitization in canine marrow graft recipients. Blood 71:1138-1140
32. Deeg HJ, Aprile J, Graham TC, Appelbaum FR, Storb R (1986) Ultraviolet irradiation of blood prevents transfusion-induced sensitization and marrow graft rejection in dogs. Concise report. Blood 67:537-539
33. Thomas ED, Clift RA, Hersman J, Sanders JE, Stewart P, Buckner CD, Fefer A, McGuffin R, Smith JW, Storb R (1982) Marrow transplantation for acute nonlymphoblastic leukemia in first remission using fractionated or single-dose irradiation. Int J Radiat Oncol Biol Phys 8:817-821
34. Tutschka PJ, Copelan EA, Klein JP (1987) Bone marrow transplantation for leukemia following a new busulfan and cyclophosphamide regimen. Blood 70:1382-1388
35. Appelbaum FR, Badger C, Deeg HJ, Nelp WB, Storb R (1987) Use of iodine-131-labeled anti-immune response-associated monoclonal antibody as a preparative regimen prior to bone marrow transplantation: initial dosimetry. NCI Monogr 3:67-71
36. Appelbaum FR, Brown PA, Graham TC, Sandmaier BM, Schuening FW, Storb R (1988) Characterization of malignant lymphoma in dogs and use as a model for the development of treatment strategies. In: Baum SJ, Santos GW, Takaku F (eds) Experimental hematology today - 1987. Recent advances and future directions in bone marrow transplantation. Springer, Berlin Heidelberg New York, pp 31-35
37. Meyers JD, Flournoy N, Thomas ED (1982) Nonbacterial pneumonia after allogeneic marrow transplantation: a review of ten years' experience. Rev Infect Dis 4:1119-1132
38. Winston DJ, Ho WG, Champlin RE, Gale RP (1984) Infectious complications of bone marrow transplantation. Exp Hematol 12:205-215

The Use of Long-term Bone Marrow Cultures (LTBMC) to Detect Hematotoxic Side Effects of Purging Methods

E. Schulze, W. Helbig, and U. Hofmann

A. Introduction

The development of suitable methods for the purging of malignant bone marrow contaminating cells in patients with acute leukemia may offer a better chance of success for autologous bone marrow transplantation. In addition to investigating the wanted effect, i.e., damage to leukemic cells, it is important to investigate the tolerance of normal hematopoietic stem cells within these manipulations in order to guarantee the grafting of the purged transplants.

Because VP 16-213 is discussed as a potent agent for eliminating tumor cells in vitro [1, 2], we incubated bone marrow with this drug.

Our aims were to determine what doses of VP 16-213 are tolerated by normal hematopoietic stem cells, and whether there is a difference between the behavior of GM-CFC and LTBMC stem cells after drug incubation.

B. Methods

I. Drug Incubation

Bone marrow cells (2×10^7/ml) were incubated for 2 h at 37 °C with different doses of VP 16-213 (50, 75, 100, 125 μ*M*/l), washed twice and cultivated thereafter in the GM-CFC and LTBMC assay.

Department of Internal Medicine, Division of Hematology and Oncology, Karl-Marx-University, 7010 Leipzig, German Democratic Republic

II. GM-CFC Assay

GM-CFC were assayed as described elsewhere [3]. The stimulator used was human umbilical cord conditioned medium. Colonies (≥ 50 cells) were counted after 10 days of incubation.

III. LTBMC Assay

LTBMCs were set up according to a modification of the method of Gartner and Kaplan [4]. Briefly, nucleated bone marrow cells (2×10^6/ml) were suspended in IMDM supplemented with 12.5% horse serum, 12.5% fetal calf serum, 10^{-6} *M*/l hydrocortisone sodium succinate, 10^{-4} *M*/l mercaptoethanol, 5×10^{-7} *M*/l sodium selenite, 2×10^{-6} *M*/l L-glutamine, and antibiotics. The cells were cultivated for 3–5 days at 37° C and thereafter until day 21 at 33° C. The cultures were fed weekly.

Two-stage LTBMCs were established on a 2- to 4-week old preirradiated (15 Gy) adherent layer of normal bone marrow. After 3 weeks the cultures were stopped; the adherent (after trypsinization) and nonadherent cells were united and assayed for GM-CFC.

C. Results

At initiation of all LTBMCs an aliquot of the sample was routinely tested for GM-CFC. The effects of VP 16-213 incubation on GM-CFC are shown in Fig. 1. It is obvious that all doses tested had a strong cytotoxic effect. Considering the mean values of recovery, the cytotoxic effect was more pronounced in bone

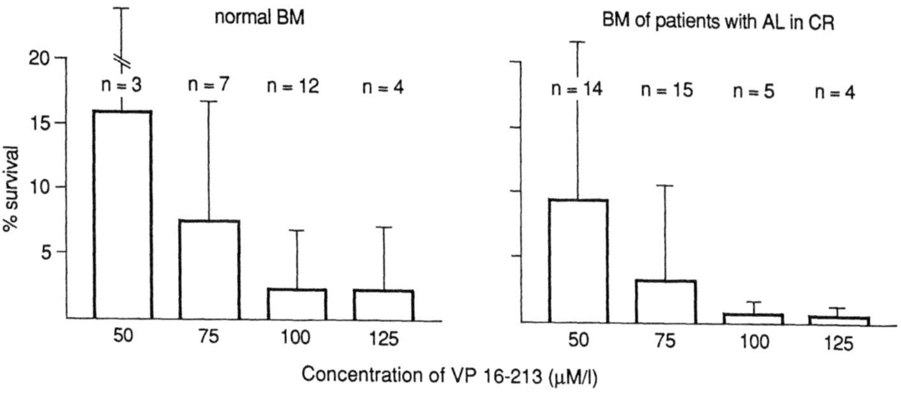

Fig. 1. Recovery at day-0 of GM-CFC after 2-h incubation with VP 16-213

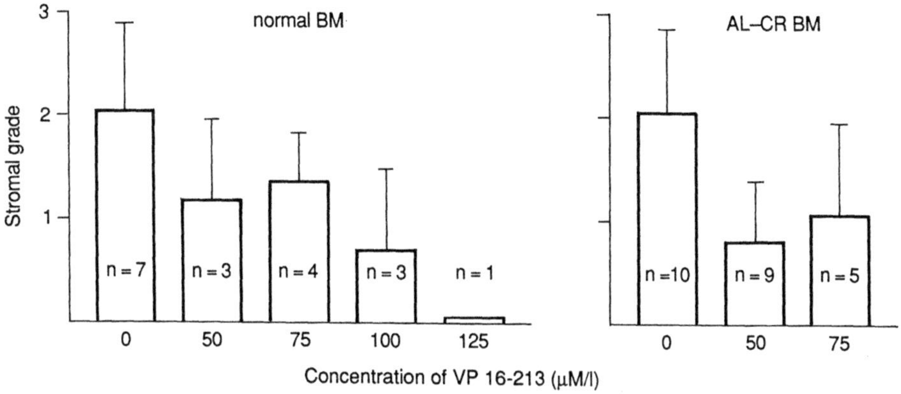

Fig. 2. Degree of adherent layer establishment in 3-week-old LTBMCs of VP 16-213-preincubated bone marrow compared with control cultures. Grade 1, adherent layer only patchy; grade 2, large adherent connected areas; grade 3, surface totally covered

Table 1. Recovery of GM-CFC after one-stage LTBMC of VP 16-213-treated normal bone marrow (% of control)

No. of experiment	Concentration of VP 16 (μ*M*/l)	GM-CFC day 0	GM-CFC after LTBMC
1	100	4.0	30.9
	125	0	1.5
2	50	7.0	14.4
	75	14.0	2.8
3	75	13.5	29.3
	100	0	23.2
4	50	0	6.0
5	100	0	0.1

Table 2. Recovery of GM-CFC after one-stage LTBMC of VP 16-213-treated bone marrow of patients with acute leukemia in complete remission (% of control)

No. of experiment	Concentration of VP 16 (μ*M*/l)	GM-CFC day 0	GM-CFC after LTBMC
1	50	0	10.6
	75	0	7.6
2	50	8.0	0
	75	1.3	0
3	50	17.4	1.0
	75	4.0	0
4	50	12.6	1.0
5	50	0	15.2
6	50	0.3	4.0
7	50	0	0
8	50	1.5	1.0
9	75	0.3	44.9
10	75	6.6	11.3
11	50	0	9.0
	75	0	27.3
12	75	0	0

Table 3. Recovery of GM-CFC after two-stage LTBMC compared with GM-CFC after one-stage LTBMC of VP 16-213-treated bone marrow of patients with acute leukemia in complete remission (% of control)

No. of experiment	Concentration of VP16 (μ*M*/l)	GM-CFC day 0	CM-CFC after 1-stage LTBMC	GM-CFC after 2-stage LTBMC
1	50	0	0	33.8
2	50	1.5	1.0	19.3
3	50	5.1	n.d.	43.1
	75	0.3	44.9	56.2
4	75	6.6	11.3	37.0
5	75	28.1	n.d.	4.8
6	50	0	9.0	18.0
	75	0	27.3	36.0
7	75	0	0	34.0

n.d., Not done

marrow of patients with acute leukemia in complete remission than in normal bone marrow. Whereas the critical dose of VP 16-213 (mean recovery $<5\%$) was 100 μ*M*/l in normal bone marrow, that of complete remission bone marrow was 75 μ*M*/l. However, the differences are not statistically significant.

VP 16-213-treated normal bone marrow showed in all but in one case a better recovery when cultured in one-stage LTBMC for 3 weeks and thereafter assayed for GM-CFC (Table 1). Bone marrow of patients with acute leukemia in complete remission showed an inconsistently different behavior in one stage

LTBMC (Table 2). Six patients had a higher and six a lower recovery compared with GM-CFC on day 0.

It was obvious that the VP 16-213 treatment caused a poorer and delayed establishment of the adherent layer in one-stage LTBMC (Fig. 2). In order to determine whether this might lead to an additional effect on GM-CFC recovery after LTBMC we compared the recovery of one- and two-stage LTBMC. The results are shown in Table 3. With the exception of one experiment, all bone marrow samples showed a distinctly higher recovery in two-stage LTBMC compared with day 0 GM-CFC and also with one-stage LTBMC. It must be pointed out that preirradiated cultures seeded with medium only did not give rise to any hematopoietic growth.

D. Discussion

VP 16-213 is known as a cell-cycle-dependent agent affecting cells in the S and G-2 phases [5, 6]. It shows a strong effect on GM-CFC, a population with a high number of proliferating cells. The possibly higher sensitivity of bone marrow from patients with acute leukemia in complete remission, shown by the lower than normal mean GM-CFC recovery, could be caused by a higher number of proliferating GM-CFC after chemotherapy. The higher recovery of GM-CFC after 3 weeks in one-stage LTBMC of normal bone marrow could indicate less damage to earlier stem cells, containing a lower number of cycling cells. These results agree with those of Ciobanu et al. [1] and Kushner et al. [7], who have also found a higher recovery of post-LTBMC GM-CFC after VP 16-213 incubation.

However, the behavior of bone marrow of patients with acute leukemia in complete remission, the real target of purging procedures, was very inconsistent in one-stage LTBMC after treatment with VP 16-213. This may reflect different answers to the hematopoietic stress of chemotherapy, i.e., a different activation of the early stem cell pool. Otherwise, it was obvious that the establishment of the adherent layer on one-stage LTBMCs was also delayed by drug treatment. Because the maintenance and survival of stem cells in LTBMC depends on an intact adherent layer, representing the hematopoietic microenvironment [8, 9] the possibility cannot be excluded that we measured a resultant of stem cell and stromal effects in our one-stage LTBMC system. To overcome this problem we used the two-stage LTBMC, where the adherent layer is preformed. The result was a much better recovery in this system, so we may assume that our first results with one-stage LTBMC indeed reflected both stem cell and stromal damage.

E. Summary

With a view to the establishment of purging methods it is necessary to investigate complete-remission bone marrow as the real target of purging, rather than bone marrow from healthy donors.

The results of LTBMC are superior to those of GM-CFC where the hematopoietic recovery of bone marrow is concerned. One-stage LTBMC after bone marrow manipulations may reflect mixed hematopoietic/stromal effects. The use of two-stage LTBMC allows the evaluation of the stem cell recovery without the possible influence of a damaged microenvironment.

Acknowledgment. We are grateful for the excellent technical assistance of C. Günther.

References

1. Ciobanu N, Paietta E, Andreef M, Papenhausen P, Wiernik P (1986) Etoposide as an in vitro purging agent for the treatment of acute leukemias and lymphomas in conjunction with autologous bone marrow transplantation. Exp Hematol 14: 626–635
2. Tamayo E, Hervé P (1988) Preclinical studies of the combination of mafosfamide (As-

ta-Z 7654) and etoposide (VP 16-213) for purging leukemic autologous marrow. Exp Hematol 16:97–101
3. Metcalf, D (1984) Clonal culture of hemopoietic cells: techniques and applications. Elsevier, New York
4. Gartner S, Kaplan HS (1980) Long-term culture of human bone marrow cells. Proc Natl Acad Sci USA 77:4756–4758
5. O'Dwyer PJ, Leyland-Jones B, Alonso MT, Marsoni S, Wittes RE (1985) Etoposide (VP 16-213): current status of an active anticancer drug. N Engl J Med 312:692–700
6. Abromovich M, Bowman WP, Ochs J, Rivera G (1985) Etoposide treatment of refractory acute lymphoblastic leukemia. J Clin Oncol 3:789–792
7. Kushner BH, Kwon J-H, Gualti SC, Castro-Malaspina H (1987) Preclinical assessment of purging with VP 16-213: key role for long-term marrow cultures. Blood 69:65–71
8. Schofield R (1978) The relationship between spleen colony-forming cell and the hemopoietic stem cell. Blood Cells 4:7–25
9. Singer JW (1985) The human hemopoietic microenvironment. In: Hoffbrand AV (ed) Recent advances of hematology. Churchill Livingstone, Edinburgh, p 1–24

Clinicopathological Features and Prognostic Implications of Immunophenotypic Subgroups in Childhood ALL: Experience of the BFM-ALL Study 83*

W.-D. Ludwig[1], H. Seibt-Jung[1], J.V. Teichmann[1], B. Komischke[1], A. Gatzke[1], G. Gassner[1], E. Odenwald[2], J. Hofmann[2], and H. Riehm[2]

A. Introduction

The application of immunological marker studies to acute lymphoblastic leukemias (ALL) has established a solid basis for precise diagnosis and classification [1] and, in combination with enzymatic, cytogenetic, and molecular analyses [2–4], has helped to unravel a great deal of the biological heterogeneity of childhood ALL.

Up to now, investigations examining the impact of the immunophenotype on treatment outcome have mostly reported results based upon conventional marker studies and have indicated a worse prognosis for children with pre-B, B-, and T-lineage ALL [5–8]. Due to the paucity of controlled prospective studies on clinical and prognostic implications of immunophenotypes, however, doubts have arisen regarding the value of the immunophenotype as an independent prognostic parameter in ALL [9]. Furthermore, the improvement of intensive therapy has affected the prognostic importance of most clinical and biological features in childhood ALL [10].

Therefore, the main objective of immunological marker studies in the therapy study ALL-BFM 83 was to determine prospectively the incidence, the clinical and hematological features, and the prognostic significance of immunophenotypic subgroups defined by monoclonal antibodies (MoAbs) in childhood ALL.

B. Materials and Methods

I. Patients

From October 1983 to September 1986, 709 previously untreated children with ALL were stratified for risk-adapted multidrug chemotherapy in the ALL-BFM 83 multicenter trial [11]. Complete immunophenotypic determinations were performed in 607 (85.6%) of these patients.

II. Immunophenotyping

Leukemic blasts from bone marrow and/or peripheral blood samples were isolated by Ficoll-Hypaque density gradient centrifugation. The following were performed as described elsewhere [12, 13]: a standard indirect immunofluorescence assay for detection of cell-surface antigens and conventional marker studies, including determination of surface immunoglobulin, sheep erythrocyte rosettes, and intranuclear terminal deoxynucleotidyl transferase. A marker was considered positive if reactive with ≥20% of leukemic blast cells.

III. Monoclonal Antibodies

The following MoAbs from cluster of differentiation (CD) groups defined by the International Workshops on Leukocyte Differentiation Antigens were used for immunophenotyping: (a) B-lineage-

* Supported in part by the Deutsche Krebshilfe and DAL/GPO

[1] Department of Hematology/Oncology, Klinikum Steglitz, Free University of Berlin, Federal Republic of Germany
[2] Department of Pediatrics, Hannover Medical School, Hannover, Federal Republic of Germany

associated antigens: HD37 (CD19) (B. Dörken, Heidelberg), B1 (CD20) (Coulter Clone, Hialeah, FL), VIB-C5 (CD24) (W. Knapp, Vienna); (b) T-lineage-associated antigens: Na1/34 (CD1) (Sera-Lab), OKT11 (CD2), OKT3 (CD3), OKT4 (CD4), OKT8 (CD8) (Orthodiagnostic Systems, Raritan, NJ), Leu-9 (CD7) (Becton Dickinson, Sunnyvale, CA); (c) myeloid-lineage-associated antigens: VIM-2 (CDw65), VIM-D5 (CD15) (W. Knapp, Vienna), My9 (CD33) (Coulter Clone); (d) non-lineage-restricted antigens: OKIa1 (not clustered) (Orthodiagnostic Systems), J5 (CD10) (Coulter Clone).

IV. Statistical Analysis

All *P*-values resulted from two-sided tests. The method of Kaplan and Meier [14] was used to construct the life-tables plotted in Figs. 1, 2, and 3. Comparisons of event-free survival (EFS) were calculated by the log-rank test [15]. Multivariate analyses were performed in a forward stepwise fashion, using the Cox regression model to analyze the importance of prognostic factors in influencing the duration of continuous complete remission [16].

C. Results

I. Treatment Outcome

Patients were classified according to their phenotypic profile into the following subgroups: null ALL, common ALL, B-ALL, pre-T-ALL, T-ALL. Three patients were unclassifiable by immunophenotypic analysis. The vast majority of children with ALL achieved complete remission (CR) (Table 1). The common ALL group had a significantly longer EFS after a median 27-month follow-up than children with pre-T/T-ALL and B-ALL (Fig. 1). The patients with null ALL, though only a small series in this study, appeared to have an intermediate remission duration (Fig. 1). Further subclassification of the common ALL group revealed significant differences between the $CD20^-$ and the $CD20^+$ patients, indicating that EFS was as poor for children with $CD20^+$ common ALL as for those with pre-T/T-ALL (Fig. 2).

Immunophenotypic subgroups in T-lineage ALL (pre-T ALL vs. early vs. cortical vs. mature T-ALL) (Fig. 3) as well as $CD10^+$ vs. $CD10^-$pre-T/T-ALL patients (data not shown) disclosed no significant differences with regard to EFS.

Table 1. Definition and distribution of immunophenotypic subgroups and their response to induction therapy among children with ALL in the ALL-BFM study 83

Immunological diagnosis	Immunophenotype	No. of patients (%)	Percent CR[a]
Null ALL	TdT^+, HLA-DR^+, $CD19^{+/(-)}$, $CD24^{+/(-)}$, $CD10^-$, $CD20^-$ $cIgM^{-/(+)}$, sIg^-, T-AG^-	21 (3.5)	100
Common ALL[b]	TdT^+, HLA-DR^+, $CD19^+$, $CD24^+$, $CD10^+$, $CD20^{-/+}$, $cIgM^{-/+}$, sIg^-, T-AG^-	481 (79.6)	99.2
B-ALL	TdT^-, HLA-DR^+, $CD19^+$, $CD24^+$, $CD20^+$, $CD10^{+/-}$, $cIgM^-$, sIg^+, T-AG^-	11 (1.8)	90.9
Pre-T ALL	TdT^+, HLA-$DR^{-/(+)}$, $CD7^+$, $CD5^{+/(-)}$, $CD2^-$, $CD1/3/4/8^-$, $CD10^{-/+}$, B-AG^-	18 (3.0)	94.4
T-ALL	TdT^+, HLA-DR^-, $CD7^+$, $CD5^+$, $CD2^+$, $CD1/3/4/8^{+/-}$, $CD10^{-/+}$, B-AG^-	73 (12.1)	95.9
Total		604 (100.0)	98.3

[a] CR, Complete remission

[b] $CD20^-$ $n=304$ (63.2% of cALL), $CD20^+$ $n=124$ (25.8%), CD20 not determined $n=53$ (11%)

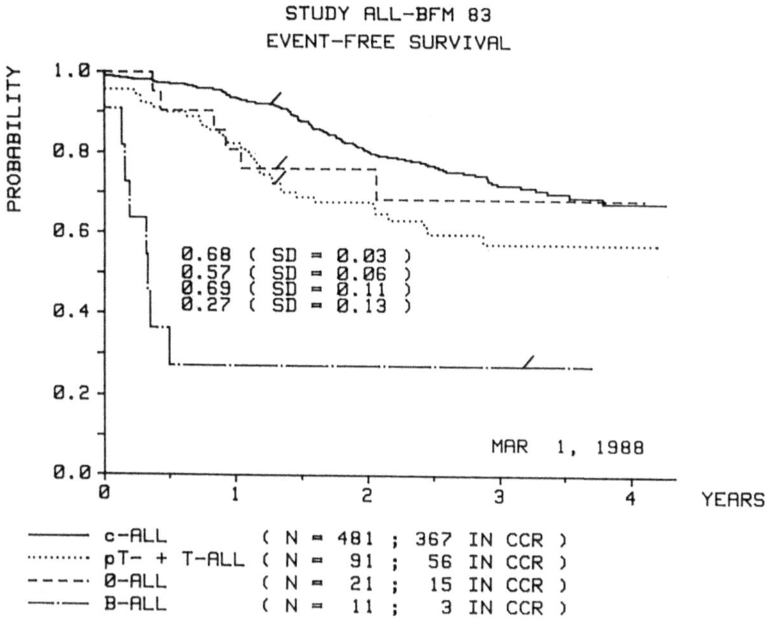

Fig. 1. Probability of event-free survival for 604 children with acute lymphoblastic leukemia (*ALL*) according to immunophenotyping subgroups. Median follow-up time 27 months. Slashes indicate last patient entering the group, as in Fig. 2 and Fig. 3. *P*-values: c-ALL vs. pT-/T-ALL <0.001; cALL vs. O-ALL 0.5; cALL vs. B-ALL <0.0001; pT-/T-All vs. O-ALL 0.48; pT-/T-ALL vs. B-ALL <0.001; O-ALL vs. B-ALL 0.001. *CCR*, Continuous complete remission

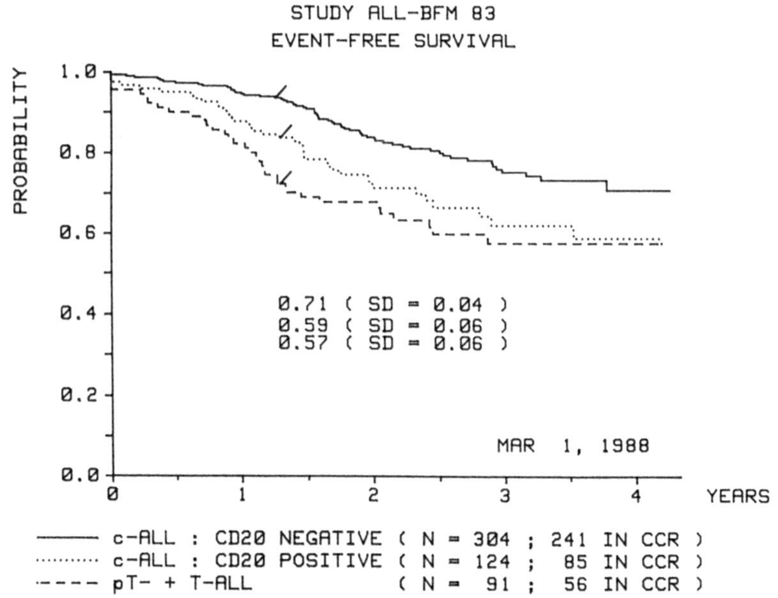

Fig. 2. Probability of event-free survival for patients with $CD20^-$ cALL, $CD20^+$ cALL, and pT-/T-ALL. *P*-values: $CD20^-$ cALL *vs.* $CD20^+$ cALL 0.004; $CD20^-$ cALL vs. pT-/T-ALL <0.0001; $CD20^+$ cALL vs. pT-/T-ALL 0.31. *CCR*, Continous complete remission

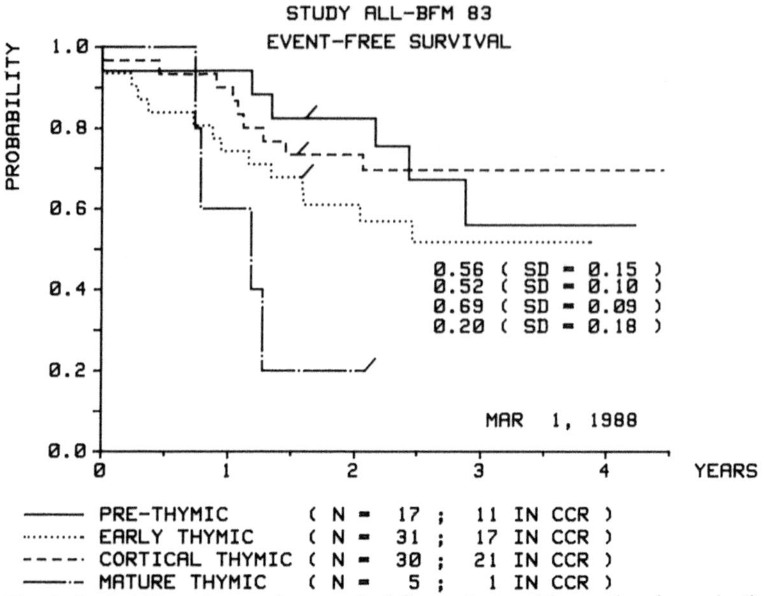

Fig. 3. Probability of event-free survival for patients with pre-thymic, early thymic, cortical thymic, and mature thymic T-ALL. *P*-values not significant. *CCR*, Continuous complete remission

Table 2. Clinical and hematological features of immunophenotypic subgroups at presentation

Feature	Units (%)	Null ($n=21$)	Common ($n=481$)	B ($n=11$)	Pre-T/T ($n=91$)
WBC ($\times 10^9$/l)	≥ 50	57.1	11.9	0	65
Age (years)	< 1	28.5	1	0	0
	$1-<10$	28.6	81.9	72.7	61.5
	≥ 10	42.9	17.1	27.3	38.5
Sex (% male)		47.6	53.8	90.9	70.3
Platelets ($\times 10^9$/l)	<100	57.1	74.2	45.5	70.3
Hemoglobin (g/dl)	<8.0	57.1	62.6	18.2	17.5
Splenomegaly[a]		57.1	35.6	27.3	69.2
Hepatomegaly[a]		57.1	52.6	18.2	71.4
Mediastinal mass (% present)		0	1.0	0	51.6
Lympadenopathy (% present)		33.3	36.6	54.5	73.6
CNS disease		9.5	1.5	45.5	9.9

[a] ≥ 4 cm below the costal margin

II. Clinicopathological Features

The clinical and hematological features of immunophenotypic subgroups at presentation are depicted in Table 2. Table 3 shows that there were pronounced clinical and hematological differences between common ALL and pre-T/T-ALL, whereas CD20⁻ and CD20⁺ common ALL did not differ significantly in their clinicopathological features. T-lineage immunophenotypic subgroups were sim-

Table 3. Comparison of clinical and hematological features at presentation within major immunological subgroups

Characteristics analyzed	Phenotype				Significance level [a]	
	$CD20^-$ cALL ($n=304$)	$CD20^+$ cALL ($n=124$)	pre-T ($n=18$)	T-ALL ($n=73$)	$CD20^-$ cALL v $CD20^+$ cALL	cALL v pre-T/T-ALL
Age (median, years)	4.5	3.9	9.4	8.0	NS [b]	$P<.001$
WBC (median, $\times 10^9/l$)	9.9	11.8	65.0	94.4	NS	$P<.001$
Risk group [c] (n)						
SR	195	73	5	12	NS	$P<.001$
MR	98	50	7	35	NS	$P<.01$
HR	11	1	6	26	NS	$P<.001$
Sex (% male)	54.9	53.2	66.6	71.2	NS	$P=.004$
Mediastinal mass (% present)	–	–	50.0	52.1	NS	$P<.001$
Hepatomegaly [d] (%)	54.6	50.8	72.2	71.2	NS	$P=.001$
Splenomegaly [d] (%)	39.1	37.1	72.2	68.5	NS	$P<.001$
CNS disease (%)	1.3	1.6	5.5	11	NS	$P<.001$
PAS (% positive)	70.1	60.5	28.8	31.5	$P=.07$	$P<.001$
Acid phosphatase (% positive)	5.6	16.9	83.3	93.2	$P<.001$	$P<.001$

[a] Pre-T and T-ALL are similar in all characteristics analyzed
[b] NS, Not significant
[c] Total tumor load at diagnosis estimated according to risk factor; SR, standard risk; MR, medium risk; HR, high risk
[d] ≥ 4 cm below costal margin

ilar in their clinical and hematological features, whereas children with $CD10^+$ pre-T/T-ALL were slightly older, had lower leukocyte counts, and presented with thymic mass more often than the $CD10^-$ patients.

D. Discussion

In the light of the progress made in the immunophenotyping of ALL, several studies have attempted to identify immunological subtypes with differing prognoses, the long-term goal being to individualize therapy according to the leukemic immunophenotype [17].

Identification of immunophenotypic features with potential prognostic significance in the large common-ALL group is rather difficult due to the relatively low failure rate for these patients. Recently, however, the prognosis in precursor B-cell-lineage ALL has been correlated with cytoplasmic μ chain expression [5] and quantitative levels of CD10 expression [18]. Since cytoplasmic Ig was not generally investigated in this study, we selected the CD20 antigen for further subclassification of the common-ALL group and observed that the duration of EFS was shorter to a statistically significant degree for patients with $CD20^+$ common ALL than for those in the $CD20^-$ common-ALL subgroup. This difference could not be explained by unequal distribution of the two well-established clinical prognostic factors, age and initial leukocyte count, nor could it be ascribed to other significant differences of clinical characteristics among these subgroups, e.g., incidence of extramedullary involvement or initial CNS disease. Within the common-ALL group, Cox regression analysis revealed independent prognostic value for only three factors, i.e., WBC, hemoglobin level, and expression of the CD20 antigen. These data suggest that common-ALL subgroups of potential prognostic significance can be defined by monoclonal antibodies and that the prognosis in precursor B-cell-lineage ALL is related to the degree of maturity of the malignant cells. Reasons for the poorer treatment outcome of the $CD20^+$ common-ALL patients are uncertain, and additional studies on the biological characteristics of this subgroup are necessary for clarification.

T-cell neoplasms have been categorized according to stages of normal differentiation into pre-T, early, cortical or common, and mature thymocyte subtypes [19]. The potential clinical relevance of subset designation, however, has not yet been demonstrated among patients with T-cell-lineage ALL. In the ALL-BFM 83 study, children with pre-T/T immunophenotype did not differ significantly in their response to induction therapy from other immunophenotypical subgroups, but they had a significantly shorter duration of EFS than children with common ALL. The poorer treatment outcome of T-lineage ALL, however, was mainly related to the association with unfavorable clinical features, and the pre-T/T-ALL phenotype did not retain independent prognostic significance in the multivariate model. Immunophenotypic subgroups of T-lineage ALL (i.e., pre-T vs. early vs. cortical vs. mature T-ALL) did not differ significantly with regard to clinical features, response to induction therapy, and EFS. Interestingly, four of five children in the small mature-T-cell subgroup relapsed within 16 months after diagnosis. The prognostic impact of this subgroup, however, has to be evaluated in larger series of patients. Furthermore, it should be emphasized that eight patients with T-lineage ALL did not fit into the T-cell-differentiation stages, indicating that any phenotypic categorization of T-lineage ALL is likely to be an oversimplification and does not reflect the real extent of heterogeneity of T-cell ontogeny. In contrast to a recent report from the Pediatric Oncology Group [20], the expression of CD10 within T-lineage ALL was not prognostically important in the ALL-BFM 83 study, but slight differences with regard to clinical features (age, WBC, mediastinal mass) were observed among $CD10^+$ and $CD10^-$ pre-T/T-ALL patients.

In conclusion, our data confirm the reported incidence of immunophenotypic subgroups and the clinical usefulness of monoclonal-antibody phenotyping in childhood ALL. The expression of the CD20 antigen could be identified as an independent prognostic factor in patients with precursor B-cell-lineage ALL and may be important for risk assignment in future treatment planning. The poorer treatment outcome of T-lineage ALL can be explained largely by the association with unfavorable clinical factors. In contrast to results in adult ALL [21], immunophenotypic subgroups in childhood T-lineage ALL (i.e., pre-T vs. T-ALL) did not differ significantly with regard to clinicopathological features and clinical outcome. Further studies of immunological features in combination with characterization by lineage-associated molecular probes are needed to evaluate the clinical significance of subset designation within T-lineage ALL.

Acknowledgements. The authors are grateful to Drs. B. Dörken and W. Knapp for supplying their monoclonal antibodies. We also thank J. Weirowski for helpful suggestions in preparing this manuscript.

References

1. Foon KA, Todd RF (1986) Immunologic classification of leukemia and lymphoma. Blood 68:1–31
2. Hoffbrand AV, Drexler HG, Ganeshaguru K, Piga A, Wickremasinghe RG (1986)

Biochemical aspects of acute leukaemia. In: Gale RP, Hoffbrand AV (eds) Acute leukaemia. Saunders, London, pp 669–694
3. Whang-Peng J, Knutsen T (1987) The role of cytogenetics in the characterization of acute leukemia. Acute lymphoblastic leukemia and acute myeloblastic leukemia. In: Stass SA (ed) The acute leukemias. Biologic, diagnostic, and therapeutic determinants. Dekker, New York, pp 153–201
4. Graninger WB, Wright JJ, Felix CA, Korsmeyer SJ (1987) Characterization of acute leukemia using lineage-associated molecular probes. In: Stass SJ (ed) The acute leukemias. Biologic, diagnostic, and therapeutic determinants. Dekker, New York, pp 299–325
5. Crist W, Boyett J, Roper M, Pullen J, Metzgar R, Van Eys J, Ragab A, Starling K, Vietti T, Cooper M (1984) Pre-B cell leukemia responds poorly to treatment: a Pediatric Oncology Group study. Blood 63:407–414
6. Wolff, LJ, Richardson ST, Neiburger JB, Neiburger RG, Irwin DS, Baehner RL (1976) Poor prognosis of children with acute lymphocytic leukemia and increased B-cell markers. J Pediatr 89:956–958
7. Sen L, Borella L (1975) Clinical importance of lymphoblasts with T markers in childhood acute leukemia. N Engl J Med 292:828–832
8. Sallan SE, Ritz J, Pesando J, Gelber R, O'Brien C, Hitchcock S, Coral F, Schlossman SF (1980) Cell surface antigens: prognostic implications in childhood acute lymphoblastic leukemia. Blood 55:395–402
9. Greaves MF, Janossy G, Peto J, Kay H (1981) Immunologically defined subclasses of acute lymphoblastic leukaemia in children: their relationship to presentation features and prognosis. Br J Haematol 48:179–197
10. Rivera G, Mauer AM (1987) Controversies in the management of childhood acute lymphoblastic leukemia: treatment intensification, CNS leukemia, and prognostic factors. Semin Hematol 24:12–26
11. Riehm H, Reiter A, Schrappe M, Berthold F, Dopfer R, Gerein V, Ludwig R, Ritter J, Stollmann B, Henze G (1987) Die Corticosteroid-abhängige Dezimierung der Leukämiezellzahl im Blut als Prognosefaktor bei der akuten lymphoblastischen Leukämie im Kindesalter (Therapiestudie ALL-BFM 83). Klin Pädiatr 199:151–160
12. Ludwig WD, Bartram CR, Ritter J, Raghavachar A, Hiddemann W, Heil G, Harbott J, Seibt-Jung H, Teichmann JV, Riehm H (1988) Ambiguous phenotypes and genotypes in 16 children with acute leukemia as characterized by multiparameter analysis. Blood 71:1518–1528
13. Hiddemann W, Ludwig WD, Herrmann F, Harbott J, Beck JD, Lampert F, Odenwald E, Riehm H (1986) New techniques in the diagnosis and pretherapeutic characterization of acute leukemias in children: analyses by flow cytometry, immunology and cytogenetics in the BFM studies. In: Riehm H (ed) Malignant neoplasms in childhood and adolescence. Karger, Basel, pp 106–127
14. Kaplan EL, Meier P (1958) Nonparametric estimation from incomplete observations. J Am Stat Assoc 53:457–481
15. Mantel N (1966) Evaluation of survival data and two new rank order statistics arising in its consideration. Cancer Chemother Rep 50:163–170
16. Cox DR (1972) Regression models and life tables. J R Stat Soc B 34:187–220
17. Pinkel D (1987) Curing children of leukemia. Cancer 59:1683–1691
18. Look AT, Melvin SL, Brown K, Dockter ME, Roberson PK, Murphy SB (1984) Quantitative variation of the common acute lymphoblastic leukemia antigen (gp100) on leukemic marrow blasts. J Clin Invest 73:1617–1628
19. Reinherz EL, Kung PC, Goldstein G, Levey RH, Schlossman SF (1980) Discrete stages of human intrathymic differentiation: analysis of normal thymocytes and leukemic lymphoblasts of T-cell lineage. Proc Natl Acad Sci USA 77:1588–1592
20. Dowell BL, Borowitz MJ, Boyett JM, Pullen J, Crist WM, Quddus FF, Russell EC, Falletta JM, Metzgar RS (1987) Immunologic and clinicopathologic features of common acute lymphoblastic leukemia antigen-positive childhood T-cell leukemia. Cancer 59:2020–2026
21. Thiel E, Kranz BR, Raghavachar A, Bartram CR, Löffler H, Messerer D, Ganser A, Ludwig WD, Büchner T, Hoelzer D (1989) Prethymic phenotype and genotype of pre-T (CD7+/ER−)-cell leukemia and its clinical significance within adult acute lymphoblastic leukemia. Blood 73:1247–1258

Current Approaches to Therapy for Childhood Lymphoblastic Leukemia: St. Jude Studies XI (1984–1988) and XII (1988)*

G. K. Rivera [1], D. K. Kalwinsky [1], J. Rodman [2], L. Kun [3], J. Mirro [1], Ching-Hon Pui [1], M. Abromowitch [1], J. Ochs [1], W. Furman [1], V. M. Santana [1], O. H. Hustu [3], J. Sandlund [1], and W. M. Crist [1]

A. Introduction

During the past decade, evidence from clinical trials for children with acute lymphoid leukemia (ALL) indicates that improved end results can be obtained by use of intensified early therapy [1–5]. Such improvement has been especially apparent for children judged to be at higher risk of relapse based on prognostic factor analysis. Risk assessment based upon the clinical features of patients at diagnosis (e.g., age, WBC count, and race) and the biologic features of their leukemic blast cells (e.g., immunophenotype and karyotype) has made it possible to modify therapy in rational ways. Additionally, alternative methods of irradiation for the treatment of preclinical CNS leukemia have been shown to be highly effective for children with non-B-cell ALL. Newer approaches to effective CNS prophylaxis include use of so-called, triple intrathecal (IT) chemotherapy [methotrexate (MTX), hydrocortisone, cytosine arabinoside (ara-C)] given early and throughout one or more years of treatment [6].

In the St. Jude Total XI study (1984–1988), intensive early treatment with seven agents is being used for remission induction; this phase of therapy is followed by consolidation with high-dose MTX. During continuation treatment, we are evaluating the use of pairs of effective antileukemic drugs alternated either weekly or every 6 weeks throughout the entire duration of therapy (120 weeks). The risk of relapse is assessed by a new system based on prognostic factor analysis of data obtained in the Total X study [7]. Age, WBC count, race, DNA index, and chromosome translocation were most predictive of relapse in our Total X Study [7] and, therefore, are being used to determine risk status in the Total XI study. A summary of the efficacy and toxicity of early therapy in the Total XI trial, and plans for our next study (Total XII), form the basis of this report.

B. Patients and Methods

Between March 1984 and May 1988, we treated 332 consecutive children with non-B-cell ALL according to the Total XI protocol. The diagnosis of ALL is based on morphologic evaluation of Wright-stained smears of bone marrow and negative marrow myeloperoxidase preparations (<3% positive blasts). CNS disease is defined as greater than five mononuclear cells/mm^3 on a CSF cell count and blasts seen on a cytospin preparation. Complete remission is defined as <5% marrow blasts and signs of recovery of normal hematopoiesis. Detailed immonologic marker studies including surface immunoglobulin (SIg) evaluation are performed on all cases to rule out B-cell ALL and to define other major

---

[1] Department of Hematology/Oncology*, [2] Pharmaceutical Division, and [3] Department of Radiation Oncology, St. Jude Children's Research Hospital; and the Departments of Pediatrics and Radiation Oncology, University of Tennessee, Memphis, College of Medicine, Memphis, Tennessee 38105, USA

* Supported by National Cancer Institute grants CA-21765 and CA-20180, and by the American Lebanese Syrian Associated Charities (ALSAC).

Table 1. Criteria for risk assignment in Total Therapy Study XI

Factor	Adverse feature	Estimated relative risk[a]
DNA index[b]	≤ 1.15	4.7
Race	Black	2.5
Chromosomal translocation	Present	2.2
Leukocyte count	$\geq 25 \times 10^9$/liter	1.9
Age	≤ 2 or ≥ 10 years	1.8

[a] Probability of relapse, at any time, in a child with a "worse" feature compared with that in a patient having the complementary feature
[b] Ratio of the modal DNA content of G_0/G_1-phase tumor versus normal cells

immunophenotypes (T, pre-B and common). All surface Ig-negative cases are included herein and detailed results of the immunologic studies will be summarized in a separate report. Flow cytometry to establish the ratio of the DNA content of leukemic versus normal cells (DNA index) and cytogenetic studies to determine the leukemic cell karyotype are successful in over 95% of cases. Results of these studies will be described separately.

The clinical and biologic features that determine risk assignment are shown in Table 1, together with the estimated relative risk of treatment failure for each factor, as determined by Cox regression model analysis of data from 431 patients treated from 1979 to 1983 in our preceding Total X study. We recently published two reports summarizing the results of Total X, including details of the risk model noted above and end results at a median follow-up time of 4 years [1, 7]. The five factors noted in Table 1 were most predictive of outcome in Total X and therefore were used for risk assignment in Total XI. The presence of any two of these features resulted in a worse-risk assignment; a WBC count of over 100×10^9/liter was sufficient by itself to confer a worse risk classification.

The details of therapy including drug schedule, dosage, and duration of administration are shown in Figs. 1 and 2. In brief, seven highly effective agents are used for induction treatment, given over 8 weeks; CNS prophylaxis with IT chemotherapy is started early in this phase of treatment. High-dose MTX (2 g/m^2 given intravenously over 2 h with 6 h of hydration and urinary alkalinization) is begun on day 43 and is repeated 1 week later. Leucovorin rescue is given at a dose of 30 mg/m^2 at 32, 44, and 48 h, followed by 5 mg/m^2 at 56 and 68 h. These doses are intended to achieve total reduced folate serum concentrations above MTX concentrations at 44 h and equimolar levels until MTX concentrations are less than 0.05 µmol/liter.

Three regimens were employed as continuation therapy. One-third of the better-risk patients receive conventional continuation treatment 1 (6-MP, MTX, vincristine, and prednisone) and two-thirds receive treatment 2, which is experimental and consists of drug pairs rotated weekly as shown in Fig. 2. Worse-risk patients are randomized unequally between experimental treatment 2 (two-thirds) and experimental treatment 3 (one-third). Relatively non-cross-resistant effective drug pairs with different toxicities are used and rotated either weekly (treatment 2) or every 6 weeks (treatment 3); this allows us to evaluate the effect of rapid versus slower rotation of drug pairs. Both treatments use the same drug pairs administered at identical doses. Further, this intensive continuation therapy is continued for the entire length of treatment rather than only during early therapy. Continuation therapy given is based on the patient's absolute phagocyte count (absolute number of

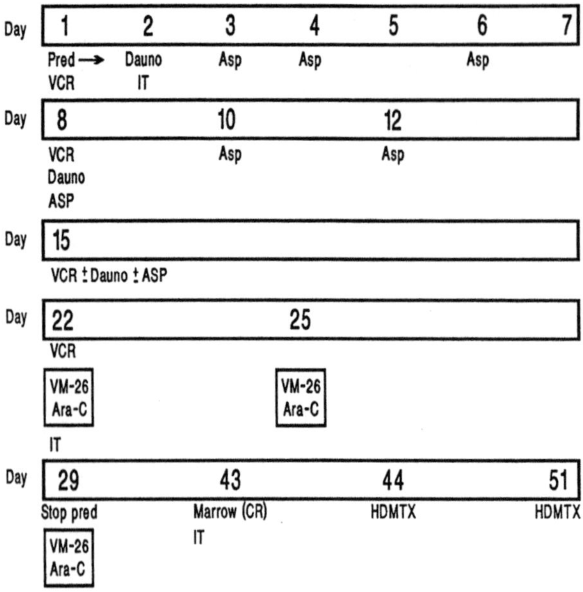

Fig. 1. Schema of early therapy, St. Jude Total Therapy Study XI. The induction phase continues until day 43; the remainder of therapy is considered a consolidation phase. *Pred,* prednisone (40 mg/m^2/day); *VCR*, vincristine (1.5 mg/m^2/week); *Dauno*, daunomycin (25 mg/m^2/week); *Asp*, L-asparaginase (10 000 U/m^2/dose i.m.); *VM-26*, teniposide (200 mg/m^2); *ara-C*, cytera-bine (300 mg/m^2/dose), *IT*, intrathecal methotrexate (12 mg), hydrocortisone (24 mg), and ara-C (36 mg) (dosages of IT chemotherapy adjusted for children less than 3 years of age). *HDMTX*, high-dose methotrexate (2 g/m^2/dose) with leucovorin rescue

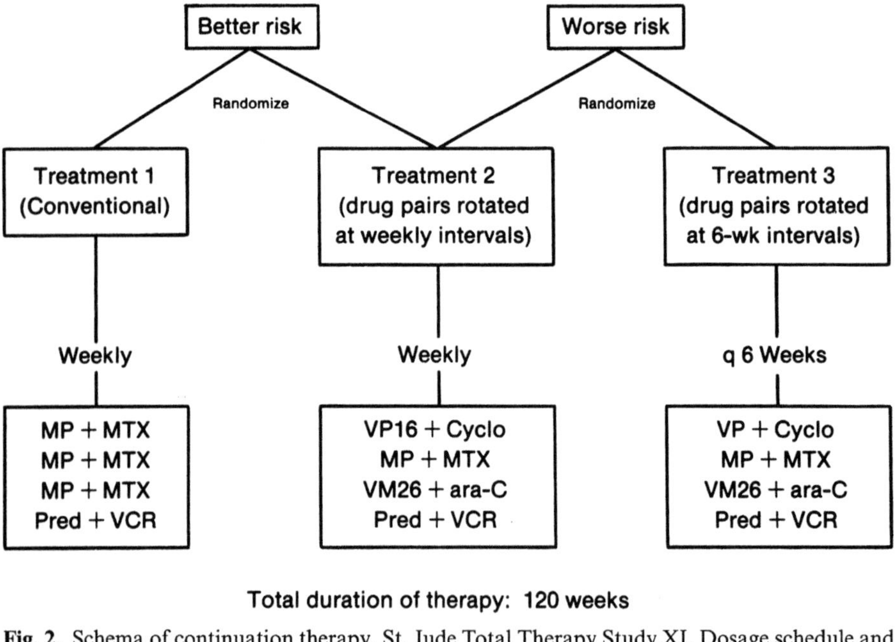

Fig. 2. Schema of continuation therapy, St. Jude Total Therapy Study XI. Dosage schedule and routes of administration; VP-16 (300 mg/m^2 i.v., once a week); cyclophosphamide (*CTX* 300 mg/m^2 i.v., once a week), Mercaptopurine (MP) (75 mg/m^2 p.o. daily $\times 7$), MTX (40 mg/m^2 i.m., once a week), VM-26 (150 mg/m^2 i.v., once a week), ara-C (300 mg/m^2 i.v., once a week), prednisone (40 mg/m^2 p.o. daily $\times 7$), VCR (1.5 mg/m^2 i.v., once a week)

early myeloid forms and segmented neutrophils plus monocytes). Full doses of chemotherapy are given if the absolute phagocyte count level exceeds 0.5×10^9/liter; otherwise doses are skipped and not delayed.

Central nervous system prophylaxis is continued for 1 year based on available data from studies of the Pediatric Oncology Group (ALinC 11, 12, 13), which suggest that such therapy is effective, especially for better-risk patients [6, 8]. Patients with CNS leukemia at diagnosis or those at worse risk also receive delayed cranial irradiation (2400 or 1800 cGy, respectively) after 1 year of remission. The results are sequentially analyzed and appropriate stopping rules are incorporated to avoid continuing the study unnecessarily.

C. Results

I. Patient Accrual and Risk Distribution

As of 1 May 1988, 335 patients had been enrolled in the Total XI study. Only three of these children were not evaluable (one refused therapy, two had incorrect diagnoses). Of 332 evaluable children, 107 (32%) were considered standard risk and 225 (68%) worse risk. Thirteen percent of the patients are black and 3.0% are infants (≤ 1 year of age). Twenty-five percent are over 10 years of age, but under 18 years. Seventeen percent have a WBC count greater than 100×10^9/liter. The proportion of worse-risk patients is much higher than we had anticipated largely due to an increased number of patients with chromosomal translocations, as compared with results in the Total X study. This circumstance resulted primarily from improvements in our cytogenetic techniques during the Total X study which led to better detection of subtle translocations in Total XI.

Ninety-seven percent (321/332) of our patients had successful immunologic marker studies and DNA indices performed on leukemic blast cells. Banded karyotypes were available for over 90% of the patients.

The toxicity encountered during the induction phase of therapy is summarized in Table 2. It can be seen that the major problems encountered were infectious in nature and developed during periods of neutropenia. A relatively increased rate of seizures noted early in the

Table 2. Toxic effects during induction therapy

No. of patients	335
No. evaluable	332
% receiving all planned treatment	85
Median No. (range) of inpatient days	10 (0–57)
% Patients with:	
VCR-induced toxicity	42
Asp-induced toxicity	22
VM-26-induced toxicity	4
Fever	75
Infections	51
Seizures	5
Early death	2
Median no. (range) of days when:	
APC $<0.5 \times 10^9$/liter	23 (0–56)
APC $<0.1 \times 10^9$/liter	8 (0–50)
Median No. of days to recover	
APC $>0.5 \times 10^9$/liter	40

VCR, vincristine; Asp, L-asparaginase; VM-26, teniposide; APC, absolute phagocyte count course of Total XI was largely corrected by changing the schedule of intrathecal therapy during induction treatment.

II. Early Results of Therapy

The complete remission (CR) rate for all patients is 96% (318/332). Of the 318 patients attaining CR, 106 (99%) of 107 better-risk patients and 212 (94%)/225 worse-risk patients achieved CR. Of the 14 patients who did not attain initial remission (early failures), 6 had primary drug resistance and 8 died of chemotherapy-related toxicity. Since we revised the induction phase of chemotherapy, the tolerance to treatment has improved and, in fact, one-third of children receive the early phase of therapy as outpatients. Ninety-seven percent of the planned doses of induction treatment have been given and 85% of patients received all planned therapy (Table 2). Ninety-six percent of children have received the two courses of high-dose MTX consolidation in the "day hospital" setting.

Three hundred and eight patients have been randomized to receive continuation therapy as follows: better risk, treatment 1 ($n=37$); better risk, treatment 2 ($n=61$); worse risk, treatment 2 ($n=139$); worse risk, treatment 3 ($n=71$). Three patients were not randomized and received other therapies after CR induction. The tolerance to continuation treatment for each of the randomized treatment regimens has been excellent and most treatment is delivered in the outpatient setting. It is too early to report the durations of remission on the three treatment regimens of Study XI (the median follow-up time is only 2 years). However, event-free survival (EFS) is estimated to be significantly better than the result obtained in Total Therapy Study X.

D. Comments

These results indicate that a high remission induction rate (96%) can be obtained without undue toxicity in children with non-B ALL using a seven-drug induction regimen that features early administration of VM-26 and ara-C (at days 22, 25, and 26) and high-dose MTX (at days 44 and 51). The rationale for this treatment is that very early use of multiple effective agents may eradicate larger numbers of leukemic cells before the emergence of drug resistance, yielding improved end results with acceptable toxicity. Another notable feature of the Total XI treatment is use of intensification therapy, not only early but sustained throughout 120 weeks of continuation treatment. It is also of interest that CNS prophylaxis for one-third of patients who have had a lower risk of relapse does not include irradiation and that cranial irradiation for worse-risk patients or those with CNS disease at diagnosis is delayed for 1 year, permitting better tolerance to systemic therapy and sparing patients who relapse early in the bone marrow from the toxic effects of irradiation. The isolated CNS relapse rate in Study XI is currently $<2\%$, and all continuation therapy has been administered in an outpatient setting. The toxicity has been modest in all phases of treatment. We are encouraged by the comparatively low relapse rate in this study; at the time of the most recent analysis, duration of event-free survival was superior to that seen in the preceding clinical trial, although follow-up times are too short for Study XI for statistically meaningful evaluation.

From experience in Total XI, combined with that in studies of patients with ALL in relapse, we believe that the epidophyllotoxins VM-26 and VP-16 warrant further evaluation in clinical trials for children with newly diagnosed ALL. Information demonstrating that high-dose MTX may be more effective if systemic exposure is optimized [7, 9] indicates that such therapy may be especially effective for patients with better-risk ALL. Fixed dosing of each of these three agents results in a wide variation in systemic exposure among patients and

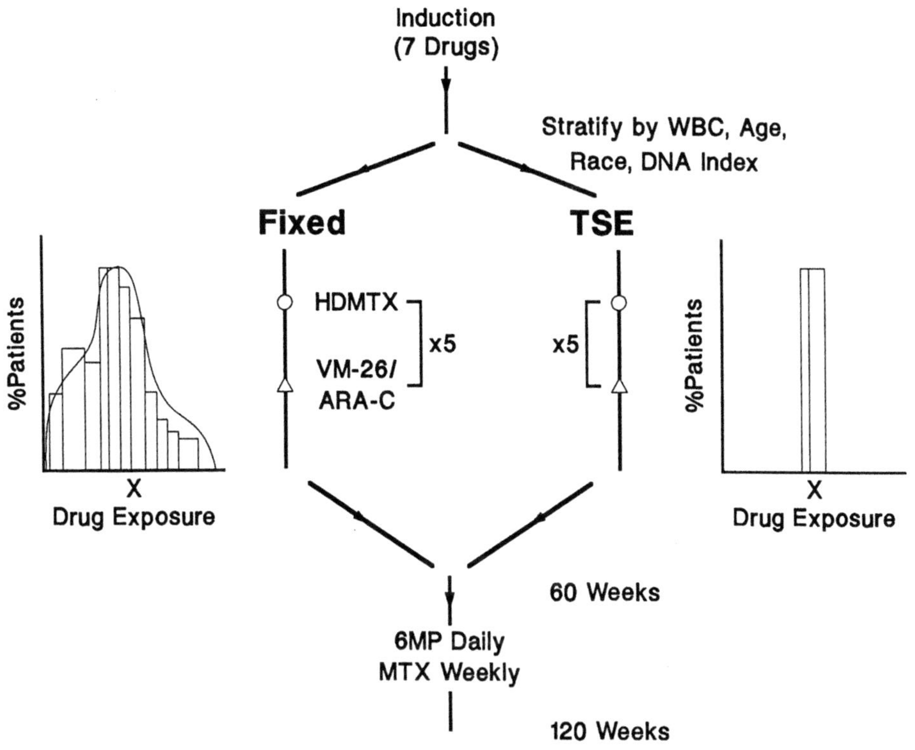

Fig. 3. Schema for planned Total XII study for children with non-B ALL showing general treatment plans and expected differences in systemic drug exposure between patient groups receiving fixed dosing versus those getting "modeled" drug delivery. *Fixed*, constant drug dosage; *TSE*, targeted systemic exposure

among successive courses administered to individual patients, and lower steady-state plasma drug concentrations may be associated with inferior treatment outcome [9, 10]. Our collective experience with the pharmacokinetic evaluation of each of these agents has permitted us to plan a study that will further test the potential of pharmacokinetically modeled therapy for improving the clinical outcome of children with non-B-cell ALL.

Therefore, in our Total XII study (Fig. 3), we will evaluate the impact of "modeled" administration of VM-26, ara-C, and high-dose MTX to maintain a narrow range of predetermined systemic exposure within one group of patients as compared with a fixed-dose approach in another group. Patients will be stratified by WBC count, age, race, and DNA index to assure prognostic uniformity of the groups being compared. Induction therapy will be given as in the Total XI study, and all biologic studies performed in Total XI will be repeated in Total XII. CNS prophylaxis will consist of periodic triple IT chemotherapy with CNS irradiation only in patients with high WBC levels. Periodic pulsed VM-26 and ara-C and high-dose MTX will be given to both groups using the aforementioned methods of drug administration, and 6-MP and IM MTX will be given between pulses during the first 60 weeks (Fig. 3). Thereafter, only the latter two drugs will be given (for 120 weeks). This research should provide data regarding the value of optimized systemic exposure to these effective agents as compared with the highly variable exposure seen with standard dosing. If significantly improved responses can be demonstrated in the former group, we will have obtained support for the concept that systemic exposure of leukemic cells to chemotherapy impacts favorably on cure rates in childhood ALL.

References

1. Dahl GV, Rivera GK, Look AT, Hustu HO, Kalwinsky DK, Abromowitch M, Mirro J, Ochs J, Murphy SB, Dodge RK, Pui C-H (1987) Teniposide plus cytarabine improves outcome in childhood acute lymphoblastic leukemia presenting with a leukocyte count $\geq 100 \times 10^9$/L. J Clin Oncol 5:1015–1021
2. Riehm H, Feickert H-J, Lampert F (1986) Acute lymphoblastic leukemia. In: Voute PA, Barrett A, Bloom J et al. (eds) UICC cancer in children: clinical management, 2nd edn. Springer, Berlin Heidelberg New York, pp 101–118
3. Steinherz PG, Gaynon P, Miller DR, Reaman G, Bleyer A, Finklestein J, Evans RG, Meyers P, Steinherz LJ, Sather H, Hammond D (1986) Improved disease-free survival of children with acute lymphoblastic leukemia at high risk for early relapse with the New York regimen – a new intensive therapy protocol: a report from the Children Cancer Study Group. J Clin Oncol 4:744–752
4. Clavell LA, Gelber RD, Cohen HJ, Hitchcock-Bryan S, Cassady JR, Tarbell NJ, Blattner SR, Tantravahi R, Leavitt P, Sallan SE (1986) Four-agent induction and intensive asparaginase therapy for treatment of childhood acute lymphoblastic leukemia. N Engl J Med 315:657–663
5. Rivera GK, Mauer AM (1987) Controversies in the management of childhood acute lymphoblastic leukemia: treatment intensification, CNS leukemia, and prognostic factors. Semin Hematol 24:12–26
6. Sullivan MP, Chen T, Dyment PG, Hvizdala E, Steuber CP (1982) Equivalence of intrathecal chemotherapy and radiotherapy as central nervous system prophylaxis in children with acute lymphatic leukemia: a pediatric oncology group study. Blood 60:948–958
7. Abromowitch M, Ochs J, Pui C-H, Fairclough D, Murphy SB, Rivera GK (1988) Efficacy of high-dose methotrexate in childhood acute lymphocytic leukemia: analysis by contemporary risk classifications. Blood 71:866–869
8. Pullen J, Boyett J, Frankel L, Iyer R, Van Eys J, Crist W, Harris M, Ravindranath Y, Sullivan M (1988) Extended triple intrathecal (T.I.T.) chemotherapy provides effective central nervous system (CNS) prophylaxis for both good and poor prognosis patients with non-T, non-B, acute lymphocytic leukemia (ALL); substitution of intermediate dose methotrexate (IDM for T.I.T. after consolidation provides less effective protection for the CNS). Proc ASCO 7:176 (abstr no 681)
9. Evans WE, Crom WR, Abromowitch M, Dodge R, Look AT, Bowman WP, George SL, Pui C-H (1986) Clinical pharmacodynamics of high-dose methotrexate in acute lymphocytic leukemia: identification of a relation between concentration and effect. N Engl J Med 314(8):471–477
10. Rodman JH, Abromowitch M, Sinkule JA, Hayes FA, Rivera GK, Evans WE (1987) Clinical pharmacodynamics of continuous infusion teniposide: systemic exposure as a determinant of response in a Phase I trial. J Clin Oncol 5:1007–1014

Results of Three Polychemotherapy Programs in Non-Hodgkin's Lymphomas

L. Konopka and S. Pawelski

A. Introduction

In the past 10 years, in spite of remarkable therapeutic progress with most hematological malignancies, non-Hodgkin's lymphoma (NHL) patients, particularly those with unfavorable histological types, remain a crux medicorum. Complete remissions are still comparatively rare and of short duration, and the disease is recurrent and progressive, with little hope of the cure so often observed in Hodgkin's disease.

Numerous salvage options based on "classical" cytostatics give variable results, and discrepancies between several published studies [1, 7, 8] may be explained at least partially by differences in patient characteristics, such as pathological type and clinical extent of non-Hodgkin's lymphoma.

Encouraged by the promising results of some investigators [3, 6, 10, 11, 16], we introduced in 156 cases three intensive multi-drug chemotherapy programs (CHOP or CHOP-Bleo, CHOMLA, CBVPM/AVBP; see Table 1) and now present the results.

B. Patients and Methods

Between 1979 and 1987, 156 patients with non-Hodgkin's lymphoma were treated at six hematological centers in Poland. All patients had a histological diagnosis of either intermediate (76 cases) or high-grade (80 cases) malignant lymphomas (Table 2).

Among 76 patients with intermediate-grade malignancy, 38 had large-cell lymphomas (31 diffuse and seven follicular), 24 had diffuse, small cleaved cell lymphomas, and 14 had diffuse mixed-cell lymphomas. Eighty patients classified as having high-grade malignancies were identified as follows: 33 cases, immunoblastic; 39, lymphoblastic, eight small, noncleaved cell lymphoma.[1] There were 95 men and 61 women with a median age of 46 years (range 18–65); 83 of these patients (31 in stage III and 52 in stage IV) had not received chemotherapy previously, and the remaining 73 patients (21 stage III and 52 stage IV) had been unsuccessfully treated with various chemotherapeutic regimens including COP, HCOP, LOP, MEV (A).

The extent of the disease was expressed in terms of criteria suggested at the Ann Arbor Conference [2] and included complete blood count with differential urinalysis, biochemical screening profile, serum protein electrophoresis, chest roentgenogram, and bone marrow aspiration or biopsy. In all patients a liver and spleen scan and abdominal ultrasonography were performed. In some patients, a CAT of the abdomen or a lymphanogiogram and spinal puncture for CSF cytology were done when the

Institute of Hematology, Chocimska 5, 00-957 Warsaw, Poland

Coordinators of Polish Multicenter Lymphoma Group: Institute of Hematology, Warsaw; Hemat. Dept., Poznań; Hemat. Dept., Szczecin, Hemat. Dept., Wrocław, Hemat. Dept., Lublin; Internal Med. Dept., Lódź

[1] In the present analysis the histology of lymph nodes was evaluated according to criteria of the Working Formulation of non-Hodgkin's lymphomas [15]

Table 1. Cytostatic programs

CHOP or CHOP-Bleo	
C – Cyclophosphamide	750 mg/m^2 i.v. on 1st day
H – Hydroxyrubicine	50 mg/m^2 i.v. on 1st day
O – Oncovine	1 mg/m^2 i.v. on 1st day
P – Prednisone	100 mg/m^2 p.o. on days 1–5
or	
B – Bleomycine	15 mg i.v. on 1st and 5th day
CHOMLA	
C – Cyclophosphamide	1000 mg i.v. on 4th day
H – Hydroxyurea p.o.	6 g/day/in 2 doses/on 1st and 7th day
O – Oncovine	1 mg i.v. on 2nd day
M – Methotrexate	120 mg i.v. on days 5, 10, and 15
L – Leucovorine	100 mg/in 4 doses/i.v. on days 6, 10, and 16
A – Adriblastine	40 mg/m^2 i.v. on 10th day
CBVPM/AVBP	
C – Cyclophosphamide	1200 mg/m^2 i.v. on 1st day
B – Bleomycine	13 mg i.m. on days 4–8
V – Vincristine	2 mg/m^2 (Max. dose 3 mg) i.v. on days 3 and 10
P – Prednisone	40 mg p.o. on days 4–17
M – Methotrexate	3 mg/kg i.v. on 20th day every 2 months
A – Adriblastine	30 mg/m^2 i.v. on days 2–4
V – Vincristine	1.4 mg/m^2 i.v. on days 1 and 8
B – Bleomycine	15 mg i.m. on days 2 and 8
P – Prednisone	40 mg p.o. on days 2–15 every 2 months

Table 2. Clinical and histological data

Program (no. of patients)		Histological grade of malignancy			
		Intermediate		High-grade	
		Previously		Previously	
		Untreated	treated	Untreated	treated
CHOP or CHOP-Bleo	(58)	20	10	18	10
CHOMLA	(46)	3	16	6	21
CBVPM/AVBP	(52)	20	7	16	9
Total	(156)	43	33	40	40
		76		80	

clinical presentation suggested possible involvement of the respective extranodal sites. Laparotomy was not routinely performed for staging purposes.

Patients were treated with three different chemotherapeutic regimens; CHOP or CHOP-Bleo (58 patients), CHOMLA (46 patients), and CBVPM/AVBP (52 patients) at the generally accepted doses of cytostatics included into these regimens [12–14]. Initial induction therapy with CHOP or CHOP-Bleo consisted of six cytostatic cycles, with CHOMLA of six to eight cycles, and with CBVPM/AVBP

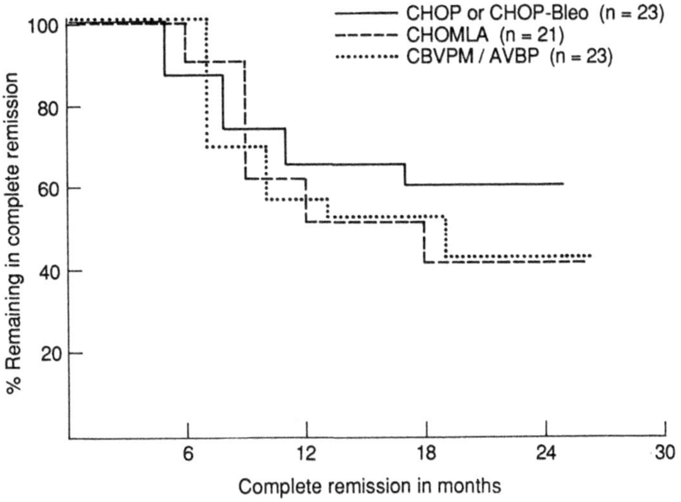

Fig. 1. Duration of complete remission for patients with intermediate and high-grade malignancy NHL. *Solid line* represents patients treated with CHOP or CHOP-Bleo, *broken line,* patients treated with CHOMLA, and *dotted line* patients treated with CBVPM/AVBP

Table 3. Results of treatment in all NHL patients

Program and number of patients		Evaluable patients (%)	Results of treatment (no. of patients) (%)		
			CR	PR	Failure or death
CHOP or CHOP-Bleo	(58)	51 (88)	23 (45.0)	7 (13.7)	12 (23.5) 9 (17.6)
CHOMLA	(46)	40 (87)	21 (52.5)	6 (15.0)	3 (7.5) 10 (25.0)
CBVPM/AVBP	(52)	47 (90)	23 (48.9)	11 (23.0)	9 (19.0) 4 (8.0)
Total		138 (88.5)	67 (48.5)	24 (17.4)	24 (17.4) 23 (16.6)
			91 (66.0)		

CR, Complete remission; PR, partial response

of four cycles each of the combined cytostatic regimen. After complete remission or partial response was achieved, two or three additional courses of the same cytostatics were given as consolidation. In some cases only adriamycin was omitted from the schedule when a total dose of 550 mg/m^2 had been reached.

After therapy (usually within 1–2 months after the last cycle of cytostatics) the patients were restaged again, particular attention being given to reassessing the initial involved site of the disease. Complete remission (CR) was defined as an absence of objective evidence of residual or subjective symptoms of persistent disease. Partial response (PR) required 50% reduction in tumor mass lasting at least 1 month.

Remission duration was ascertained from the date of the first objective complete remission to documented relapse. The data were analyzed by the life-table method [9], and differences among the curves were compared using the generalized Wilcoxon test modified by Gehan.

C. Results

One hundred thirty-eight (88.5%) of the 156 patients completed therapy and could be evaluated for response. Eighteen patients were not evaluable for response to therapy because of they did not complete follow-up. They were excluded from the survival curves.

Of the 138 evaluated patients, 67 (48.5%) achieved a complete remission and 24 (17.4%) a partial response (Table 3). Complete remissions were usually seen during induction cycles and at very similar rates (45%, 49% and 52.5%) after these three kinds of cytostatic regimens. Continuous CR in these three groups of treated patients has been documented in 42%–60% of patients.

The median disease-free survival for CHOP- or CHOP-Bleo-treated patients lasted 11.4 months and for those on the CHOMLA and CBVPM/AVBP programs 15 and 12.7 months respectively (Fig. 1).

The difference in the median disease-free survival between these three groups of patients was not statistically significant ($P=0.32$). There was no significant difference in the CR rate between the nontreated (37.2%) and the pretreated (38.2%) patients (Table 4).

Of all treated patients, 47 (34%) persons did not respond and 23 (16.6%) died within 4–11 months after starting therapy. The most common reasons for failure to achieve a complete response in these patients were persisting lymphadenopathy, bone marrow infiltration, and thrombocytopenic complications.

On evaluation of 138 patients under treatment in relation to the degree of ma-

Table 4. Results of chemotherapy in NHL patients who had received prior chemotherapy and in patients previously untreated

Program and past history of treatment (number of patients)			No. of evaluated patients	Response (no. of patients) (%)	
				CR	PR
CHOP or CHOP-Bleo	Untreated	(38)	51	19 (37.2)	5 (9.8)
	Refractory	(20)		4 (7.8)	2 (4.0)
CHOMLA	Untreated	(9)	40	7 (17.5)	1 (2.5)
	Refractory	(37)		14 (35.0)	5 (12.5)
CBVPM/AVBP	Untreated	(36)	47	18 (38.2)	1 (2.1)
	Refractory	(16)		5 (10.6)	10 (21.2)

CR, Complete remission; PR, partial response

Table 5. Results of chemotherapy according to grade of malignancy

Program and grade of malignancy (no. of patients)			No. of evaluable patients	Response (no. of patients) (%)	
				CR	PR
CHOP or CHOP-Bleo	Intermediate	(30)	28	12 (23.5)	5 (9.8)
	High	(28)	23	11 (21.5)	2 (4.0)
CHOMLA	Intermediate	(19)	18	12 (30.0)	2 (5.0)
	High	(27)	22	9 (22.5)	4 (22.5)
CBVPM/AVBP	Intermediate	(27)	23	5 (10.6)	9 (19.1)
	High	(25)	24	16 (38.2)	2 (4.2)

CR, Complete remission; PR, partial response

Table 6. Number of patients with NHL who relapsed after obtaining complete remission

Program (no. of patients)		Number of relapsing patients with NHL			
		to 3 months	to 6 months	to 12 months	above 12 months
CHOP or CHOP-Bleo	(28)	3	3	2	1
CHOMLA	(21)	2	6	2	2
CBVPM/AVBP	(23)	7	3	1	2

lignancy, it was found that the CBVPM/AVBP regimen was most effective in high-grade NHL (38.2% CR). In intermediate-grade NHL the best was the CHOMLA program (30% of CR, Table 5).

Approximately half of the patients who achieved an initial CR have relapsed within 3–15 months (Table 6). Relapse was most common at the previous site of involvement, i.e., lymph nodes (21 patients), spleen and liver (16 patients), and bone marrow (three patients).

D. Discussion

Although many agents are active in the treatment of NHL, only a few chemotherapy combinations have been shown to improve the survival of patients. The introduction of chemotherapy combinations containing adriamycin and bleomycin such as the CHOP and CHOP-Bleo regimens in particular represents an important advance in therapy. Several groups have reported plateaus in disease-free survival estimates, suggesting cure of 30%–35% of patients with intermediate or high-grade non-Hodgkin's lymphoma [4, 12, 14]. Unfortunately the majority of these patients do not achieve a plateau in *relapse*-free survival with CHOP or CHOP-Bleo or similar combination chemotherapy.

More recently, programs such as Pro-MACE-MOPP [5], COP-BLAM [11] M-BACOD [16], and MACOP-B [10] have achieved complete response rates of over 70% and appear to have extended the disease-free survival to more than 60%.

The results of treatment of our patients with three intensive and similar cytostatic programs were not satisfactory. Complete remissions were achieved in only 48.5% of all patients, and nearly half of these patients relapsed within 3–15 months after completing therapy.

Based on our experience with these three combined cytostatic programs, especially for resistant and relapsing NHL patients, a further modification of the present programs is necessary. But the real solution can be expected only with the introduction of potent new cytostatic drugs and the improvement of supportive care.

E. Conclusions

1. Of 138 advanced (clinical stage III or IV) NHL patients treated by three different cytostatic programs (CHOP or CHOP-Bleo, CHOMLA, CBVPM/AVBP), complete remission was obtained in 67 patients (48.5%) and partial response in 24 patients (17.4%).
2. Among the three cytostatic programs the highest rate of complete remission was observed after CHOMLA – 52.5%.
3. In untreated and high-grade malignant NHL patients the most effective cytostatic program was CBVPM/AVBP (38.2% complete remissions).
4. The results presented show that these three alternative chemotherapy programs are still only partly effective in the treatment of advanced intermediate and high-grade malignant NHL.

References

1. Bonadonna G (1985) Chemotherapy of malignant lymphomas. Semin Oncol 12:1–14
2. Carboni PP, Kaplan HS, Musshoff K, Smithers DW, Tubiana M (1971) Report of the committee on Hodgkin's disease staging classification. Cancer Res 31:1860–1861
3. Danieu L, Wong G, Koziner B, Clarkson B (1986) Predictive model for prognosis in advanced diffuse histiocytic lymphoma. Cancer Res 46:5372
4. Elias L, Portlock CS, Rosenberg SA (1978) Combination chemotherapy of diffuse histiocytic lymphoma with cyclophosphamide, adriamycin, vincristine and prednisone (CHOP). Cancer 42:1705–1710
5. Fisher RI, DeVita VT, Hubbard SM, Longo D, Wesley R, Chabner B, Young RC (1983) Diffuse aggressive lymphomas: Increased survival after alternating flexible sequences of ProMACE and MOPP chemotherapy. Ann Intern Med 98:304–309
6. Gaynor ER, Ultman JE, Golomb HM, Sweet DL (1985) Treatment of diffuse histiocytic lymphoma (DHL) with COMLA (cyclophosphamide, oncovin, methotrexate, leucovorin, cytosine arabinoside): a 10-year experience in a single institution. J Clin Oncol 3:1596
7. Horwich A, Peckham M (1983) "Bad-risk" non-Hodgkin lymphomas. Semin Haematol 20:35–65
8. Jelliffe AM, Vaughan Hudson G (1987) Multicentre studies and the treatment of lymphomas. Baillière's Clin Haematol 1:235–264
9. Kaplan EL, Meier P (1958) Nonparametric estimation from incomplete observation. J Amer Statis Assoc 53:457
10. Klimo P, Connors J (1985) MACOP-B chemotherapy for the treatment of diffuse large-cell lymphoma. Ann Intern Med 102:596
11. Laurence J, Coleman M, Allen S, Silver RT, Pasmantier M (1982) Combination chemotherapy of advanced diffuse histiocytic lymphoma with the six-drug COP-BLAM regimen. Ann Intern Med 97:190
12. McKelvey EM, Gottlieb JA, Wilson HE, Haut A, Talley RW, Stephens R, Lane M, Gamble JF, Jones SE, Grozea PN, Gutterman J, Cottman CA, Moon TE (1976) Hydroxyldaunomycin (adriamycin) combination chemotherapy in malignant lymphoma. Cancer 38:1484
13. Pawelski S, Konopka L, Pakulska J, Gutkowska J (1987) CBVPM/AVBP versus HCOMLA chemotherapy in non-Hodgkin lymphoma patients. In: Proc V Medit Congr Chemother Vol 6. 716–719 Eds. MS Sabboun, G Cocuzza (Florence, June 1987)
14. Rodriguez V, Cabanillas F, Burgess MA, McElvey EM, Valdivieso M, Bodey GP, Freireich EJ (1977) Combination chemotherapy (CHOP-Bleo) in advanced diffuse histiocytic lymphoma. Blood 49:325
15. Rosenberg SA, Berard CW, Brown BW, Burke J, Dorfman RF, Glatstein E, Hoppe RT, Simon R (1982) National Cancer Institute sponsored study of classifications of non-Hodgkin's lymphomas: summary and description of a working formulation for clinical usage. Cancer 49:2112
16. Skarin AT, Canellos GP, Rosenthal DS, Case DC, Mac Intyre JM, Pinkus GS, Moloney WC, Frei E (1983) Improved prognosis of diffuse histiocytic and undifferentiated lymphoma by use of high-dose methotrexate alternating with standard agents (M-BACOD). J Clin Oncol 1:91–98

Results of Treatment in Children with B-Cell Lymphoma: Report on the Polish Leukemia/Lymphoma Study Group *

J. Bogusławska-Jaworska [1], B. Kazanowska [1], B. Rodziewicz-Magott [1], W. Pietras [1], J. Armata [2], P. Daszkiewicz [2], A. Dłużniewska [3], M. Koehler [5], M. Matysiak [4], M. Ochocka [4], U. Radwańska [3], R. Rokicka-Milewska [4], D. Sońta-Jakimczyk [5], and I. Zmudzka [4]

A. Introduction

It is well established that B-cell lymphoma in children initially responds well to chemotherapy, and that prognosis depends upon the extent of the disease at diagnosis. However, the outcome of advanced B-cell lymphoma, particularly with initial bone marrow or central nervous system involvement remains far less favorable than that of localized disease [2–4]. Between 1983 and 1988 the Polish Children's Leukemia/Lymphoma Study Group used two protocols for disseminated B-cell lymphoma therapy, that of Murphy and Bowmann [2] and that of COAMP [1]. In this report the results of therapy with these regimens are described and are compared to those achieved with the LSA_2L_2 protocol [5].

B. Patients and Methods

A total of 152 children aged 1–15 years entered the study. Histologic classification was based on the Kiel system [3]. Clinical staging was done according to the criteria of Murphy et al. [2]. Initial characteristics of patients are presented in Table 1. Therapy protocols are outlined in Figs. 1 and 2.

C. Results

The overall results are summarized in Table 2. In patients treated according to the Murphy-Bowmann protocol and who achieved complete remission (CR), seven relapses were observed among the 18 cases of stage III (duration of CR, 2–10 months) and seven among the 10 cases of stage IV (duration of CR, 2–18 months). The life-table estimations for event-free survival after 62 months are 51% for stage III and 17% for stage IV (Fig. 3). In the patients with stage III and treated according to the COAMP regimen, all postremission relapses occurred between the 3th and 7th months after CR and in those with stage IV between the 2nd and 19th months after CR. The event-free survival after 61 months is 59% for those in stage III and 40% for those in stage IV (Fig. 4).

D. Concluding Remarks

Comparison of the efficacy of the three different therapy modalities for stage III of B-cell lymphoma indicates the superiority of the COAMP regimen. The results obtained in stage III are comparable with the best reported up to date [3, 4]. No particular toxicity of the COAMP protocol was observed; one patient died in CR due to septicemia. With the protocols used, no comparative advantage was achieved in treatment of stage IV B-cell lymphoma.

---

* Supported by grant CPBR 11.5 of the Institute of Oncology.

[1] Department of Pediatric Hematology, Medical School of Wroclaw, Wroclaw, Poland
[2] Department of Pediatric Hematology, Medical School of Krakow, Krakow, Poland
[3] Department of Pediatric Hematology, Medical School of Poznan, Poznan, Poland
[4] Department of Pediatric Hematology, Medical School of Warsaw, Warsaw, Poland
[5] Department of Pediatric Hematology, Medical School of Zabrze, Zabrze, Poland

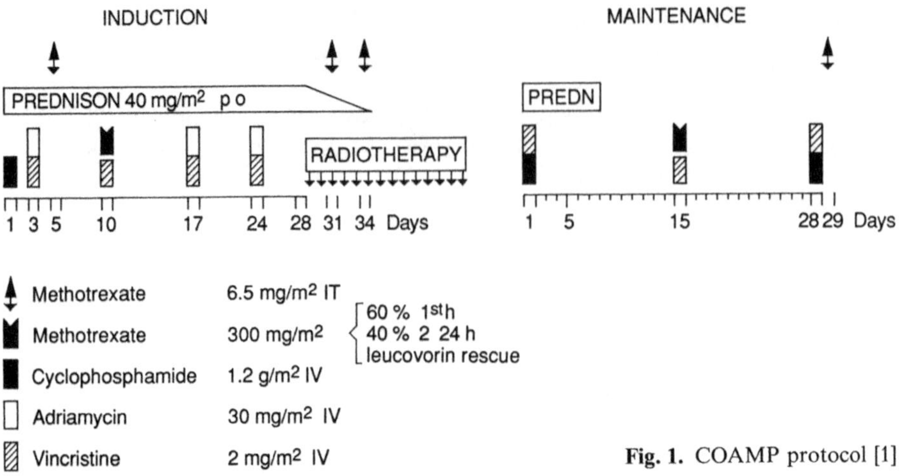

Fig. 1. COAMP protocol [1]

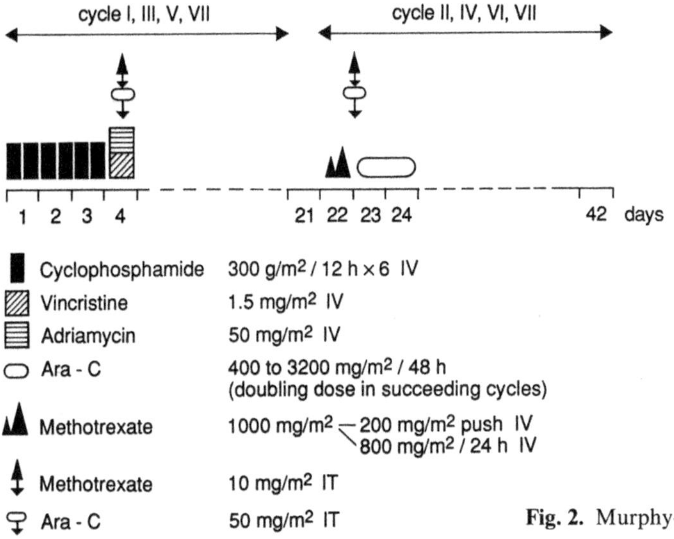

Fig. 2. Murphy-Bowmann protocol [2]

References

1. Anderson J, Wilson J, Jenkin D et al. (1983) Childhood non-Hodgkin's lymphoma. Results of a randomized therapeutic trial comparing a four-drug regimen (COAMP) with a ten-drug regimen (LSA_2L_2). N Engl J Med 308:559–565
2. Murphy S, Bowmann W, Hustu H, Berard C (1984) Advanced stage III–IV Burkitt's lymphoma and pharmacologic rationale for treatment and recent results (1979–1983). In: Lenoir G, O'Conor G, Olweny C (eds) Burkitt's lymphomas. IARC-WHO Publications, pp 405–418
3. Müller-Weihrich St, Henze G, Schwarze E, Budde M, Riehm H (1986) Childhood non-Hodgkin lymphoma: strategies for diagnosis and therapy. Monogr Paediat 18:167–186
4. Philip T, Pinkerton R, Biron P et al. (1987) Effective multiagent chemotherapy in children with advanced B-cell lymphoma: who remains the high risk patient? Br J Haematol 65:159–164
5. Wollner N, Exelby P, Lieberman P (1979) Non-Hodgkin's lymphoma in children. A progress report on the original patients treated with the LSA_2L_2 protocol. Cancer 44:1990

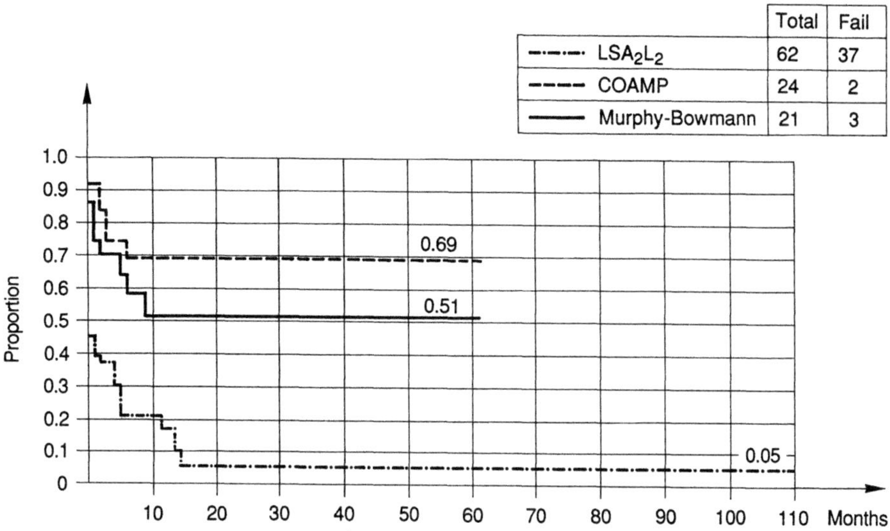

Fig. 3. Probability of event-free survival of children with B-cell lymphoma stage III with three different treatment protocols

Table 1. Initial characteristics of patients with B-cell lymphoma stages III and IV, treated according to three different protocols

	Protocol		
	LSA_2L_2	Murphy-Bowmann	COAMP
Total number admitted to the study	87	31	34
Sex			
Male	74	24	25
Female	13	7	9
Median age in years (range)	6 (1–15)	5 (1–12)	6 (1–14)
Clinical stage			
III	62	21	24
IV	25	10	10
BM	22	10	8
CNS	2	0	2
CNS + BM	1	–	–
Primary location:			
Intrathoracic	7	1	2
Abdomen	62	28	22
Head, neck	10	–	4
Peripheral nodes	8	2	6

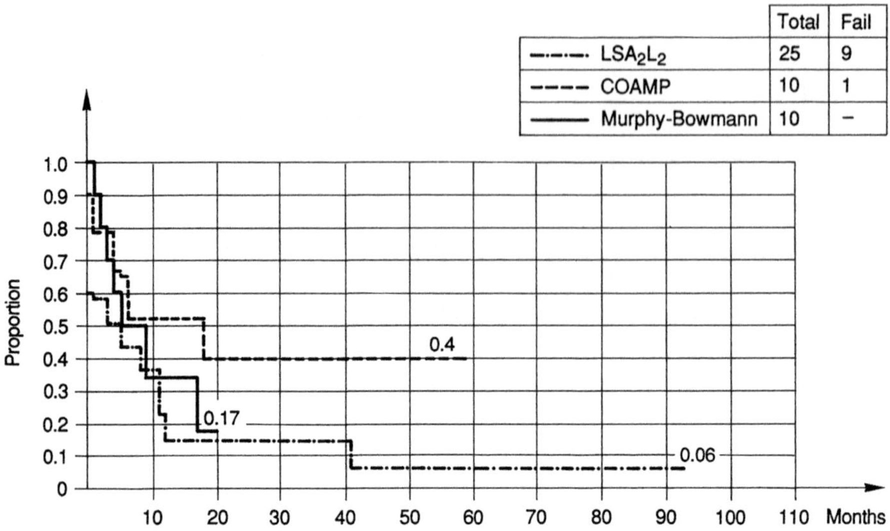

Fig. 4. Probability of event-free survival of children with B-cell lymphoma stage IV with three different treatment protocols

Table 2. Overall results achieved in children with B-cell lymphoma stages III and IV, treated according to three different protocols

	Protocol					
	LSA_2L_2		Murphy-Bowmann		COAMP	
	n	%	*n*	%	*n*	%
Total number admitted to the study	87	100	31	100	34	100
Achieved CR	51	58.6	28	90.3	31	91.2
No CR	36	31.4	3	9.7	3	8.8
Died in CR due to infection	6	11.8	1	3.6	1	3.2
Relapsed	25	49.0	13	46.4	11	35.5
Still in first CR	20	39.2	14	50.0	19	61.3
Median time of follow-up (months)	18		12		24	
Off therapy	20	39.2	12	42.9	15	48.4

Correlations of Clinical and Laboratory Data for Prognosis in Childhood Acute Lymphoblastic Leukemia

A. Petrakova [1], J. Stary [2], O. Hrodek [2], P. Goetz [3], K. Venclikova [1] and J. Srajer

A. Introduction

The prognostic importance of blast-cell DNA content and chromosomal findings in childhood acute lymphoblastic leukemia (ALL) is known [1–4, 6, 7, 9], but further correlations of hematologic, immunologic, and genetic data are needed for a better understanding of the biology of leukemic cells [5, 10]. Our pilot study is focused on the multivariate analysis of hematologic and genetic data from both a clinical and a laboratory point of view to be used for clinical prognosis in children with newly diagnosed ALL.

B. Material and Methods

Twenty-four clinical and laboratory parameters were analyzed in 65 children with newly diagnosed ALL, including age, sex, mediastinal mass, WBC and blast count in peripheral blood, blast count in bone marrow, FAB classification, cytochemical, histochemical, and immunological characteristics, chromosomal findings, karyotype index, DNA index, and familial incidence of cancers. All parameters were computerized and statistical analysis including multivariate analysis was done. Data from ALL patients were compared with control data from 60 healthy children without history of any malignancy. The prognosis of the disease was calculated in relation to the different risk factors.

C. Results

The absolute blast count in peripheral blood at the time of diagnosis proved to be the main prognostic risk factor in childhood ALL. Both karyotype and DNA index as well as the amount of nuclear DNA in bone marrow blasts correlated with absolute blast count in peripheral blood (Fig. 1) and are described as other important prognostic risk factors. A higher incidence of cancer in the ALL patients' families was found in comparison with the group of control children's families and seems to be another important genetic risk factor in the prognosis of childhood ALL. The significance of our findings was confirmed using multivariate analysis, which described an initial absolute blast count in peripheral blood as the most important determinant for the prognosis of ALL.

D. Discussion

The prognostic importance of laboratory and clinical data in relation to leukemia has recently been discussed in the literature [2, 4, 5, 7]. The published reports indicate that genetic factors contribute to the etiology of human leukemia. Current

[1] Research Institute of Child Development, Department of Genetics, Faculty of Pediatrics, Charles University, V uvalu 84, 15018 Praha 5, Czechoslovakia
[2] 2nd Pediatric Department, University Hospital Motol, Faculty of Pediatrics, Charles University, V uvalu 84, 15018 Praha 5, Czechoslovakia
[3] Department of Clinical Genetics, University Hospital Motol, V uvalu 84, 15018 Praha 5, Czechoslovakia

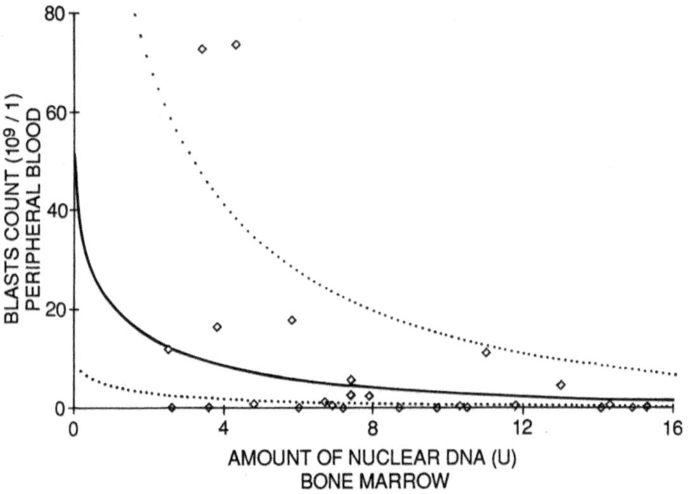

Fig. 1. Correlation of initial blast count in peripheral blood and DNA content in bone marrow blasts

epidemiologic knowledge supports the hypothesis that human leukemia is a disease arising from multiple causes. These complex causes include intrinsic factors (genetics and epidemiology) as well as environmental factors (irradiation, chemicals, and viruses). It would appear that no one cause is dominant and that the etiology of human leukemia is probably the result of various interactions producing a disturbance of cell division [8].

The first results of our pilot cooperative prognostic study in childhood ALL; presented here, indicate that there are different risk factors playing a role in the prognosis, but also in the cause of childhood ALL. The best direction for future prognostic investigations in childhood ALL may be the cooperation of clinical hematologists and oncologists with immunologists, geneticists, molecular biologists, and epidemiologists to clarify many different still unknown interactions of etiologic and prognostic importance.

E. Summary

Correlation of 24 different clinical and laboratory measures was performed for a calculation of the prognosis in 65 children with newly diagnosed acute lymphoblastic leukemia (ALL). The hematologic as well as genetic parameters were studied at the time of diagnosis. It was shown that the initial blast count in peripheral blood is the most important risk factor for the prognosis of ALL. Cytogenetic, DNA cytometric, and pedigree data correlate with blast count and also seem to be very important determinants for ALL prognosis. The statistically significant increase in the incidence of neoplasms in ALL families is reported.

References

1. Berneman ZN, De Bock R, Van Alsenoy L, Vingerhoets W, Van den Bergh M, Dumon J, Peetermans M (1985) Cytogenetic and DNA-flow cytometric studies of separated blasts. Leuk Res 9 (12): 1463–1466
2. Heath CW (1973) The epidemiology of leukemia. In: Schottenfield D (ed) Cancer epidemiology and prevention: current concepts
3. Heim S, Mitelman F (1987) Acute lymphoblastic leukemia. In: Cancer cytogenetics. Liss, New York, pp 141–174
4. Hiddeman W, Wormann B, Ritter J, Thiel E, Gohde W, Lahme B, Henze G, Schellong G, Riehm H, Buchner T (1986) Frequency and clinical significance of DNA aneuploidy in acute leukemia. Ann NY Acad Sci 458: 227–240

5. Knudson AG (1988) The genetics of childhood cancer. Bull Cancer (Paris) 75(1): 135–138
6. Look AT, Roberson PK, Wiliams OL, Rivera G, Bowman WP, Pul CHI, Ochs J, Abromowitch M, Kalwinsky D, Dahl GV et al. (1985) Prognostic importance of blast cell DNA content in childhood acute lymphoblastic leukemia. Blood 65(5): 1079–1086
7. Michael PM, Garson OM, Ekert H, Tauro G, Rennie GC, Pilkington GR (1988) Prospective study of childhood acute lymphoblastic leukemia: hematologic, immunologic and cytogenetic correlations. Med Pediatr Oncol 16: 153–161
8. Pui CH, Williams DL, Raimondi SC, Rivera GK, Look AT, Dodge RK, George SL, Behm FG, Crist WM, Murphy SB (1987) Hypodiploidy is associated with a poor prognosis in childhood acute lymphoblastic leukemia. Blood 70(1): 247–253
9. Suarez C, Miller DR, Steinherz PG, Melamed MM, Anreeff M (1985) DNA and RNA determination in 111 cases of childhood acute lymphoblastic leukemia (ALL) by flow cytometry: correlation of FAB classification with DNA stemline and proliferation. Br J Haematol 60(4): 677–686
10. Ueda T, Aozasa K, Tsujimoto M, Hamada H, Hayashi H, Ono K, Matsumoto K (1988) Multivariate analysis for clinical prognostic factors in 163 patients with soft tissue sarcoma. Cancer 62: 1444–1450

Central Nervous System Involvement in Non-Hodgkin's Lymphomas

K. Swencz-Szczepanik, M. Krämer, L. Konopka, M. Brodzki, S. Pawelski, and A. Deptala

A. Introduction

In the past 10 years the involvement of the central nervous system (CNS) in non-Hodgkin's lymphomas (NHL) has become apparent as the results of systemic chemotherapy have improved and complete remissions (CR) have increased in both frequency and duration [3, 7]. However, most studies report on cases in which CNS lymphomas were identified through the recognition of neurological symptoms and cytological examination of cerebrospinal fluid (CSF). In some cases, there are only a few malignant lymphoma cells in the CSF of asymptomatic patients at diagnosis, and they are of uncertain clinical importance.

To date there is no biochemical marker whose presence or elevation above normal concentration correlates with the presence of malignant lymphoma. Therefore, in this study we present the results of pulse-cytophotometry analyses, measurements of vitamin B_{12} level in CSF, and unsaturated vitamin B_{12} binding capacity ($UB_{12}BC$) by CSF protein to assess their roles as indices of CNS involvement in NHL.

B. Material and Methods

Between 1981 and 1987 at the Institute of Hematology in Warsaw, 199 NHL patients were treated with combined polychemotherapy. In 13 cases CNS involvement was suspected on the basis of neurological symptoms (Table 1). There were nine men and four women, aged between 19 and 59 years, in clinical stages III or IV.

In all of these patients and in another series of 17 cases with intermediate and high-grade NHL, the CSF was investigated as a routine procedure. Specimens were subjected to cytological examination after centrifugation with Cytospin 2-Shandon. Besides cytological examinations, DNA histograms (using PHYWE ICP 11 pulse-cytophotometer and computing according to the method described by Andreef [1], measurements of vitamin B_{12} levels in CSF (CSF was concentrated on Minicon CS-15), and assessment of $UB_{12}BC$ by CSF protein (using radioimmunological method described by Retief et al. [8]) were performed.

All patients with CNS involvement received methotrexate intrathecally (usually 15 mg every few days, for a minimum of six doses, or three beyond the first negative results on cytological examination), and in six cases of high-grade NHL additional cranial irradiation was performed.

C. Results

Malignant lymphoma cells were found in 21 specimens of CSF from 30 patients examined (including eight clinically asymptomatic cases). The histological classification of these patients is summarized in Table 2. The cytological changes of CSF in 21 cases were characterized by the presence of lymphoma cells which looked like lymphoblasts or parablasts and ranged in particular cases between

Institute of Hematology, 00-957 Warsaw, Poland

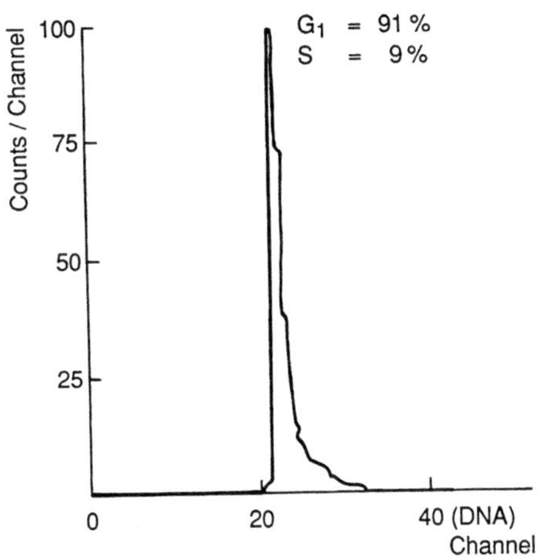

Fig. 1. DNA histogram of cerebrospinal fluid cells in patient F. M. without CNS involvement

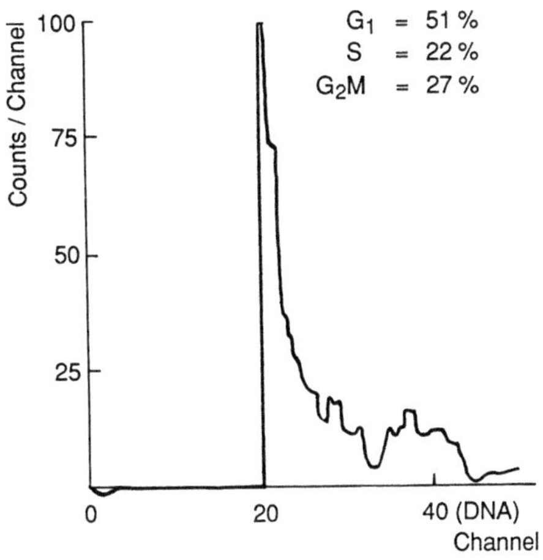

Fig. 2. DNA histogram of cerebrospinal fluid cells in patient S. R. with CNS involvement

Table 1. Neurological symptoms in 13 NHL patients with CNS involvement

Neurological symptoms	Number of patients
Headaches	5
Meningeal symptoms	2
Cranial nerve palsy	1
Peripheral nerve palsy	1
Polyneuropathy	7

Table 2. Histological classification of NHL patients with CNS involvement

Histological type of NHL (Kiel Classification)	Number of cases
1. High-grade lymphoblastic:	
– Burkitt type	1
– convoluted type	2
immunoblastic	3
2. Intermediate-grade centroblastic/centrocytic	5
3. Low-grade lymphoplasmocytoid	10
Total	21

6% and 93% of all presented in CSF cells.

In DNA histograms of patients without CNS involvement most cells were in the G_1 phase (Fig. 1). In DNA histograms of CSF in all the patients with CNS involvement, including the eight asymptomatic cases, increased numbers of cells in the S and G_2M phases were found (Fig. 2). During the successful treatment the values decreased proportionally to the number of lymphoma cells

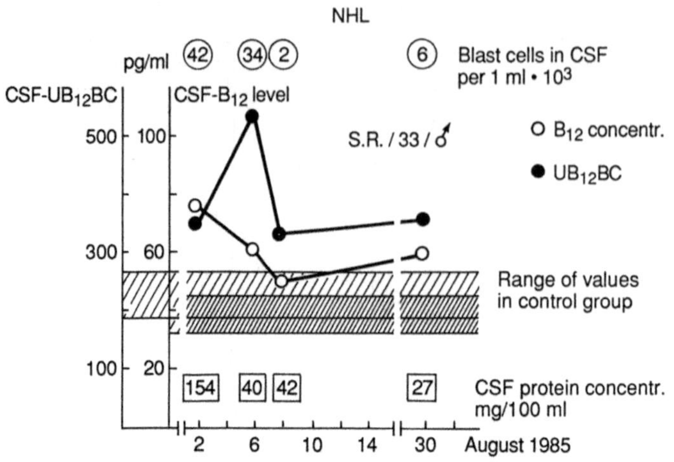

Fig. 3. The UB_{12} BC by CSF proteins and CSF vitamin B_{12} level in NHL patient (S. R.) (the same case as in Fig. 2) during intrathecal treatment with Mtx and in relapse

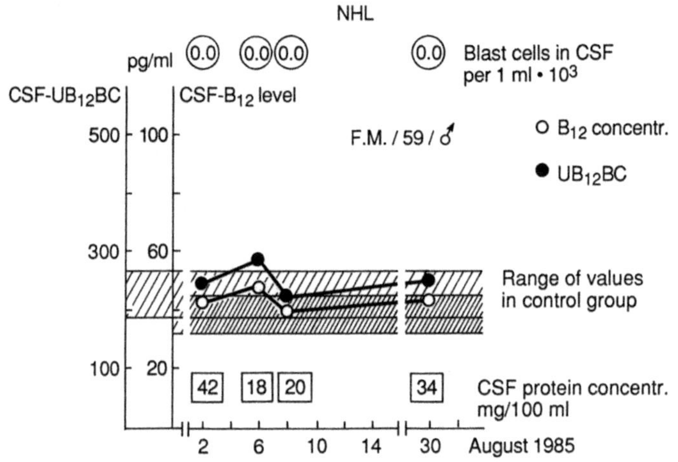

Fig. 4. The UB_{12} BC by CSF proteins and CSF vitamin B_{12} level in NHL patient F. M. (see case in Fig. 1) (studies repeated four times during one month observation).

and increased again in relapse. The $UB_{12}BC$ by CSF protein and the vitamin B_{12} level in CSF were increased only in patients with CNS involvement (Fig. 3), including the eight clinically asymptomatic cases. Both became normal after successful treatment and increased again in relapse. In patients without CNS involvement the $UB_{12}BC$ and the vitamin B_{12} level remained normal (Fig. 4).

D. Conclusions

CNS involvement in NHL was found in 21 of our cases (10.5%). Of these cases, eight were clinically asymptomatic. Cytophotometric analyses of DNA histograms and measurements of $UB_{12}BC$ and the vitamin B_{12} level proved the most accurate for confirmation of CNS involvement in clinically asymptomatic cases.

References

1. Andreef M (1980) Computer analysis of DNA histograms. University of Utrecht, p 36 (doctorate thesis)
2. Barligie B, Spitzer G, Hart JS, Johnson DA, Büchner T, Schumann J, Drewinko B

(1976) DNA histogram analysis of human hemopoietic cells. Blood 48:245–258
3. Ersbol I, Schultz HB, Thomsen BLR, Keiding N, Nissen N (1985) Meningeal involvement in non-Hodgkin's lymphomas, incidence, risk factors and treatment. Scand J Haematol 36:487–497
4. Glass JP, Melamed M, Chernik NL, Posner JB (1979) Malignant cells in cerebrospinal fluid (CSF). The meaning of a positive CSF cytology. Neurology 29: 1369–1375
5. Hansen M, Brynskov J, Christensen PA, Krinbel JJ, Gimsing P (1985) Cobalamin binding proteins (haptocornin and transcobalamin) in human cerebrospinal fluid. Scand J Haematol 34:209–212
6. Konopka L, Pawelski S, Brodzki M, Wegier-Filipiuk B (1987) Cerebrospinal fluid vitamin B_{12} concentration and its binding capacity in acute leukemias. Fourth international symposium on Therapy of Acute Leukemias, Rome, Feb 7–12 (program and abstr)
7. Recht L, Straus DJ, Cirrincione C, Tzvi Thaler H, Posner JB (1987) Central nervous system metastases from non-Hodgkin's lymphoma: treatment and prophylaxis. Am J Med 84:425–435
8. Retief FP, Gottlieb CW, Herbert V (1967) Delivery of $Co^{57}B_{12}$ to erythrocytes from alpha and beta globulin of normal B_{12}-deficient and chronic myeloid leukemia serum. Blood 29:837–851
9. Swencz-Szczepanik K (1982) Examination of DNA content in leukemia cells of cerebrospinal fluid. Doctoral dissertation. Institute of Haematology, Warsaw

Targeted Plasma Drug Concentration: A New Therapeutic Approach to Relapsed Nonlymphoblastic Leukemia in Children*

J. Mirro Jr.[1], W.R. Crom[2], D.K. Kalwinsky[1], V.M. Santana[1], D.K. Baker[2], and J. Belt[3]

A. Introduction

Chemotherapy produces approximately a 30% long-term disease-free survival in newly diagnosed children with acute nonlymphocytic leukemia (ANLL). Although 75%–80% of children will enter a complete remission (CR), most will relapse with resistant leukemia while receiving postremission chemotherapy.

Since 1976, investigators at St. Jude Children's Research Hospital have conducted a series of clinical trials in an attempt to improve the outcome of therapy for patients with previously untreated ANLL. In AML-76, we tested a cytokinetically based induction and nonmyelosuppressive maintenance therapy. Patients achieving CR were randomized to determine if splenectomy improved outcome. Although 72% of patients achieved a complete remission, the long-term survival was not satisfactory [1]. In 1980 we undertook an intensive chemotherapy trial (AML-80) modeled after the Dana-Farber VAPA study [2, 3]. AML-80 was quite toxic but did increase the median disease-free interval to approximately 18 months. Unfortunately, this approach did not improve long-term disease-free survival [4].

In 1983 we evaluated a new treatment strategy for childhood ANLL which included the introduction of additional drugs, such as etoposide (VP-16), early in therapy to maximize reduction in tumor burden during remission induction. The induction therapy was an intensive five-drug regimen utilizing conventional agents (ara-C, daunorubicin, 6-thioguanine) and newer drugs (VP-16, 5-azacytidine) [5, 6]. Postremission therapy included seven drugs given in sequentially rotated pairs. The objective of the postremission therapy was to expose leukemic cells to as many effective drugs as possible in an attempt to decrease the development of resistance. The induction schema for this protocol (AML-83) is outlined in Fig. 1. Despite this new treatment strategy the probability of event-free survival at 2 years in the AML-83 trial was 33% ± 7%, which was not significantly different from our two previous trials (AML-76 and AML-80) or many other reported trials [1–4, 7–11].

A	VP-16 200 mg/m^2 CI* (IV) d1-3 Ara-C 200 mg/m^2 CI (SQ) d4-8
B	Dauno 50 mg/m^2 (IV) d1,2 Ara-C 200 mg/m^2 CI (SQ) d1-5 6-TG 100 mg/m^2 (PO) d1-5
C	VP-16 250 mg/m^2 (IV) d1,2,3; 6,7 5-AZ 300 mg/m^2 (IV) d4,5

* Continuous infusion

Fig. 1. The induction schema of AML-83 included three combinations (*A*, *B*, and *C*) of antileukemic agents administered at standard dosages. Combination *A* and *B* of AML-83 contained similar drugs to cycle 1 and 2 of AML-R2 but at lower doses

Departments of Hematology/Oncology[1], Pharmacokinetics[2], and Biochemical Pharmacology[3], St. Jude Children's Research Hospital, Memphis, TN, USA

* Supported by NIH grant CA-20180 and by the American Lebanese Syrian Associated Charities (ALSAC). Dr. Mirro is a recipient of a Clinical Oncology Career Development Award from the American Cancer Society

B. Relapse Trial: AML-R2

Because these therapeutic trials for previously untreated patients failed to improve survival, we designed a completely novel approach to therapy. Our rationale for this new approach was that therapeutic failures may result because leukemic cells are inadequately exposed to effective agents in some patients. Likewise, some patients may develop unnecessary toxicity because of high drug plasma concentration and damage to normal tissues. Previous studies demonstrated marked interpatient variability for the plasma concentration of VP-16 and ara-C when administered in standard doses (per square meter). Therefore, we were interested in determining whether it was possible pharmacokinetically to control this variability by "targeting" the plasma concentration of these drugs to a predetermined level. In this novel approach to therapy, VP-16 and ara-C would not be administered at a standard dose per square meter, but at a dose that would produce a predetermined plasma concentration. This approach is the first step toward "individualizing" chemotherapy and permits evaluation of toxicity and efficacy with a standard systemic exposure to antileukemic agents. Such an approach also guaranteed that we would deliver more intensive therapy to all patients.

We piloted our targeted plasma drug concentration therapy on relapsed patients in the AML-R2 protocol. This protocol included antileukemic agents which patients received previously, but were known to be highly effective. The schema for the relapsed trial is shown in Fig. 2. The background data used for standardizing plasma concentrations was based on the previous plasma concentrations of VP-16 and ara-C measured in our AML-83 trial (Fig. 1). When VP-16 and ara-C are administered as a standard dose/m^2 per day there is a wide variability in plasma concentrations among patients. For example, there was a four- to five-fold variability in the plasma concentration of VP-16 when administered based on standard per square meter dosage (Fig. 3). A similar four-fold variability in the plasma concentrations of ara-C was also seen on AML-83 (data not shown). We selected

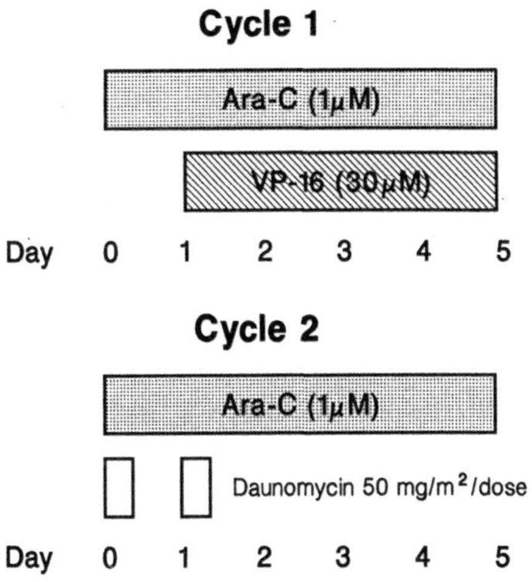

Fig. 2. Schema of AML-R2 which included targeted plasma concentrations of VP-16 and ara-C in cycle 1 and a targeted plasma concentration of ara-C with daunorubicin in cycle 2

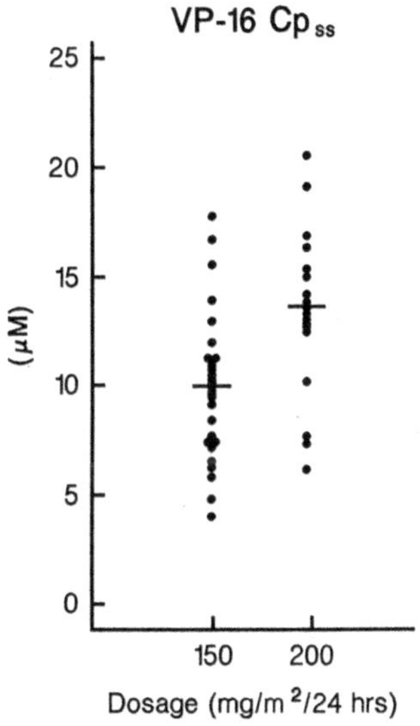

Fig. 3. The variability in the plasma steady-state concentrations of VP-16. Individual patients are marked and the median value is shown. All patients received VP-16 administered in a standard fashion at 150 or 200 mg/m^2 per dose

the maximum plasma concentrations of VP-16 and of ara-C which patients tolerated on AML-83 and used these as our planned target concentrations. If we could increase (maximize) the plasma concentration for all patients we would increase total exposure and standardize therapy.

In the first cycle of AML-R2 therapy (Fig. 2) the ara-C is administered as a continuous subcutaneous infusion for 5 days. The dose is adjusted within 12 h to achieve a plasma concentration of 1 μM. This dose adjustment is based on the plasma concentration determined by high-pressure liquid chromatography (HPLC) from samples obtained 1 and 6 h after the start of the ara-C infusion. A 4-day continuous intravenous infusion of VP-16 is started 24 h after starting ara-C and is administered to achieve a plasma concentration of 30 μM within 12 h based on the plasma concentrations assayed 1 and 6 h after the start of the VP-16 infusion. Cycle 2 of AML-R2 consists of ara-C, again administered to achieve a 1 μM plasma concentration in combination with daunorubicin 50 mg/m^2 per dose $\times$ 2 doses, given with the start of ara-C and 24 h later (Fig. 2).

All patients enrolled on this trial had previously received VP-16, ara-C, and daunorubicin at standard doses during initial induction therapy. Most relapsed while receiving the AML-83 regimen (Fig. 1), which contained almost identical induction therapy as AML-R2, but with much lower drug dosages of VP-16 and ara-C. Furthermore, with the targeted plasma concentrations on the AML-R2 relapse study all patients received the same total systemic exposure to VP-16 and ara-C despite being given different dosages.

Responses were evaluated by standard criteria. A CR was defined as a cellular marrow aspirate with $<5\%$ blasts cells, normal hemograms, and performance status for >1 month. Partial remission (PR) was an absence of peripheral blasts, $<25\%$ marrow blasts, and recovery from all toxicity. Toxicity was evaluated by the National Cancer Institute (NCI) common toxicity criteria.

C. Results

Nineteen children with relapsed ANLL were enrolled in the AML-R2 protocol (nine females and ten males). The median age at enrollment was 10.5 years (range 8 months to 17.1 years). Six patients had myeloblastic leukemic subtypes (FAB-M1 or M2), 1 patient had progranulocytic leukemia, and 12 patients had myelomonocytic or monocytic leukemic subtypes (FAB M4 or M5) (Table 1).

In cycle 1 of AML-R2 the median ara-C dose was 550 mg/m^2 per day (range 412–750 mg/m^2 per day). The median VP-16 dose was 500 mg/m^2 per day (range 350–750 mg/m^2 per day) (Table 2). The dose of ara-C administered in AML-R2 was approximately 2.5 times higher than on AML-83. Similarly the dose of VP-16 in AML-R2 was 2.5-fold higher than on AML-83. The "targeted" ara-C concentration of 1 μM on AML-R2 was readily achieved (Table 3). AML-R2 subjects also achieved the "targeted" VP-16 concentration of 30 μM with measured concentration of 32.44 ± 5.0 μM (median, $\pm$SD) (Table 3).

The therapeutic results of the AML-R2 protocol are encouraging. The overall complete response rate was 10 of 19 patients (53%) for patients who had previously received VP-16, ara-C, and daunorubicin therapy. There were 17 patients enrolled on this relapse trial (AML-R2) that relapsed after (or on) AML-83. Of the three patients enrolled on AML-R2 that achieved a complete remission after combination A of AML-83 with VP-16 and ara-C, all three also achieved a second complete remission after cycle 1 of AML-R2 containing the same two drugs. More interestingly, however, are the results of the 14 patients who were previously treated on AML-83 but had residual disease after combination A. Five of these 14 patients achieved a second complete remission using higher

Table 1. AML-R2 patient characteristics

No. patients	19
Sex	9 female/10 male
Age (median)	10.5 years
(range)	8 months to 17.1 years
FAB classification	
M1	4
M2	2
M3	1
M4	7
M5	5

Table 2. AML-R2 drug doses administered

AML-R2 cycle	Drug	Dose administered ($mg/m^2/day$)	
		Median	Range
1	Ara-C	550	412–750
1	VP-16	500	350–700
2	Ara-C	600	350–750

Table 3. Comparison of doses administered and plasma concentrations achieved on AML-83 and AML-R2

Protocol Phase of therapy	AML-83 Combination A	AML-R2 Cycle #1
Ara-C dose	200 $mg/m^2/day \times 5$	550 $mg/m^2/day \times 5$ (median)
Ara-C plasma concentration median (range)	0.1 μm (<0.1–0.28 μ*M*)	1.08 μ*M* (0.5–1.7μ*M*)
VP-16 dose	200 $mg/m^2/day \times 3$	500 $mg/m^2/day \times 4$ (median)
VP-16 plasma concentration median (range)	13.97 μ*M* (4–19.5 μ*M*)	32.44 μ*M* (25.7–43.9 μ*M*)

Table 4. AML-R2 toxicity

Toxicity	Cycle	Incidence
Sepsis	Cycle #1 and 2	6 Bacterial (32%) 3 Fungal (16%)
Fever	Cycle 1	100%
	Cycle 2	80%
Mucositis[a]	Cycle 1	90%
	Cycle 2	40%
Hepatotoxicity[a]	Cycle 1	26%
	Cycle 2	33%
Days hospitalized	Cycle 1	17.6 (5–35)
	Cycle 2	15.5 (7–39)

[a] Grade 3 or 4 by NCI common toxicity criteria

doses of VP-16 and ara-C on the AML-R2 study.

The AML-R2 protocol is extremely toxic producing prolonged pancytopenia and mucositis. One patient died from bacterial sepsis during drug-induced aplasia. Six patients had documented bacterial sepsis and three had documented fungal sepsis. In cycle 1, 100% of the patients had fever with neutropenia and were hospitalized for antibiotics, while in cycle 2, 80% required hospitalization for fever with neutropenia. Mucositis occurred predictably after cycle 1 in 90% of patients; in cycle 2 mucositis was much less of a problem and grade 3 or 4 mucositis occurred in approximately 40% of the patients (Table 4).

The incidence of abnormal liver enzymes was 26% during cycle 1 and 33% during cycle 2; alternations were usually mild and reversible. Cycle 1 resulted in skin toxicity with diffuse erythemia in 90% of the patients. No patient developed CNS toxicity on the AML-R2 protocol. The average hospitalization for cycle 1 was 17.6 days (range of 5–35 days) and for cycle 2 was 15.5 days (range of 7–39 days).

D. Discussion

The AML-R2 protocol was designed to determine if it is possible to "target" the plasma concentration and standardize the total systemic exposure for VP-16 and ara-C among all patients. In addition to achieving this objective, the protocol demonstrated that it was possible to increase the dose of VP-16 and ara-C by over two-fold. However, the dose was increased far less than the total systemic exposure. For ara-C the total systemic exposure on AML-R2 was two to ten times greater than on AML-83. Likewise, the VP-16 total systemic exposure was three to eight times greater on AML-R2 than on AML-83 (the VP-16 infusion was 1 day longer on AML-R2).

The AML-R2 protocol demonstrated the feasibility of adjusting drug dosage to achieve a target plasma concentration within 8–12 h of starting each agent. The approach of standardizing plasma concentration yielded a more uniform total drug exposure for all patients. This higher total exposure resulted in severe but very predictable toxicity with encouraging results.

The results of AML-R2 suggest a plasma concentration: response relationship for VP-16 and ara-C. Patients who achieved their initial CR with lower doses of VP-16 and ara-C on AML-83 achieved a second CR on AML-R2. More surprisingly, of the 14 patients who had residual disease on AML-83 after receiving VP-16 and ara-C in combination A, 5 achieved a second complete response after receiving higher dosages of the same drugs on cycle 1 of AML-R2. Unfortunately, the small number of patients treated on AML-R2 does not permit meaningful statistical analysis but our results support the hypothesis that the antileukemic effects of VP-16 and ara-C might be improved by increasing the total exposure to these agents. Our preliminary conclusion from the AML-R2 protocol is that more therapy may prove to be better for children with ANLL. This novel approach to antileukemic therapy will be used in our new front-line trial.

Acknowledgments. We thank the medical and nursing staff of St. Jude Children's Research Hospital for skilled patient care. We also thank Ms. M. Rafferty for data management and Ms. P. Vandiveer for typing the manuscript.

References

1. Dahl GV, Kalwinsky DK, Murphy SB, Look AT, Amadori S, Kumar M, Novack R, George SL, Mason C, Mauer AM, Simone JV (1982) Cytokinetically based induction chemotherapy and splenectomy for childhood acute nonlymphocytic leukemia. Blood 60: 856–863
2. Weinstein HJ, Mayer RJ, Rosenthal DS, Camitta BM, Coral FS, Nathan DG, Frei

E (1980) Treatment of acute myelogenous leukemia in children and adults. N Engl J Med 303:473–478
3. Weinstein HJ, Mayer RJ, Rosenthal DS, Coral FS, Camitta BM, Gelber RD (1983) Chemotherapy for acute myelogenous leukemia in children and adults: VAPA update. Blood 62:315–319
4. Dahl GV, Kalwinsky DK, Mirro J, Look AT (1987) A comparison of cytokinetically based versus intensive chemotherapy for childhood acute myelogenous leukemia. Hamatol Bluttransfus 30:83–87
5. Kalwinsky DK, Dahl GV, Mirro J, Jackson CW, Look AT (1986) Induction failures in childhood acute nonlymphocytic leukemia: etoposide/5-azacytidine for cases refractory to daunorubicin/cytarabine. Med Pediatr Oncol 14:245–250
6. Look AT, Dahl GV, Kalwinsky D, Senzer N, Mason C, Rivera G (1981) Effective remission induction of refractory childhood acute nonlymphocytic leukemia by VP-16-213 plus azacitidine. Cancer Treat Rep 65:995–999
7. Kalwinsky DK, Mirro J, Schell M, Behm F, Mason C, Dahl GV (1988) Early intensification of chemotherapy for childhood acute nonlymphocytic leukemia: improved remission induction with a five-drug regimen including etoposide. J Clin Oncol 6:1134–1143
8. Büchner TH, Urbanitz D, Hiddemann W, Rühl H, Ludwig WD, Fischer J, Aul HC, Vaupel HA, Kuse R, Zeile G (1985) Intensive induction and consolidation with or without maintenance chemotherapy for acute myeloid leukemia (AML): two multicenter studies of the German AML Cooperative Group. J Clin Oncol 3:1583–1589
9. Amadori S, Ceci A, Comelli A, Madon E, Masera G, Nespoli L, Paolucci G, Zaneoco L, Corelli A, Mandelli F (1987) Treatment of acute myelogenous leukemia in children: results of the Italian Cooperative Study AIEOP/LAM 8204. J Clin Oncol 5:1356–1363
10. Creutzig U, Ritter J, Riehm H, Langermann HJ, Henze G, Kabisch H, Niethammer D, Jürgen H, Stollman B, Lasson U (1985) Improved treatment results in childhood acute myelogenous leukemia: a report of the German cooperative study AML-BFM-78. Blood 65:298–304
11. Champlin R, Gale RP (1987) Acute myelogenous leukemia: recent advances in therapy. Blood 69:1551–1562

Use of All-trans Retinoic Acid in the Treatment of Acute Promyelocytic Leukemia

Huang Meng-er [1], Ye Yu-chen [1], Chen Shu-rong [1], Chai Jin-ren [1], Lu Jia-xiang [2], Zhao Lin [3], Gu Long-jun [1], and Wang Zhen-yi [1]

A. Introduction

Acute promyelocytic leukemia (APL) is considered to be a distinct entity among the acute myeloid leukemias (AMLs). Hemorrhagic diathesis often occurs and results in a rapid fatal outcome. The bleeding episodes are usually attributed to thrombocytopenia and/or disseminated intravascular coagulation (DIC), which is thought to result from the release of a procoagulant factor from the promyelocyte granules [1].

The use of daunorubicin in the induction therapy and improvement in the supportive therapy has greatly raised the rate of complete remission (CR) in APL [2, 3]. However, the increased mortality during induction therapy is higher in APL than in other forms of AML [3, 4] and there are still cases which continue to be refractory to induction chemotherapy. DIC remains a common lethal complication.

Induction of differentiation may be an alternative approach to the treatment of APL. Retinoic acid (RA), an analog of vitamin A, is one of the many agents which can induce differentiation and terminal cell division of leukemic cells in vitro [5]. At the present time, several cases of APL treated with 13-*cis* RA have been reported with encouraging results [6–9]. In this paper, we report the in vitro studies and therapeutic trials of 24 APL patients using all-*trans* RA.

B. Materials and Methods

I. Patients

The diagnosis of APL was made according to the criteria of the French-American-British (FAB) cooperative study group [10]. Every patient presenting to our hospitals since early 1986 with a diagnosis of APL was included in this study. The clinical characteristics of the 24 patients with APL are shown in Table 1. There were 11 females and 13 males, with a mean age of 35.5 years (range = 5–69 years). The total white blood cell counts ranged from 0.5×10^9/liter to 15.8×10^9/liter, including 20 cases (83.3%) with less than 3×10^9/liter, 3 cases (12.5%) with between 3×10^9 and 10×10^9/liter, and 1 case (4.1%) with more than 10×10^9/liter. The hemoglobin concentrations ranged from 41 g/liter to 121 g/liter including 8 cases (33.3%) with less than 60 g/liter, 12 cases (50%) between 60 and 90 g/liter and 4 cases (16.7%) with more than 90 g/liter. Platelet counts ranged from 10×10^9/liter to 337×10^9/liter including 15 cases (62.5%) with less than 50×10^9/liter, 7 cases (29.2%) between 50×10^9 and 100×10^9/liter and 2 cases (8.3%) with more than 100×10^9/liter. The percentage of promyelocytes in the marrow ranged from 15.6% to 94%, with 22 patients having more than 30% and the remaining 2 patients having between 15.6% and 30%. Of these 24 studied patients, 16 had never been treated. The

[1] Shanghai Institute of Hematology, Shanghai Second Medical University, Shanghai, People's Republic of China
[2] Shanghai Zhong-Shan Hospital
[3] Shanghai Children's Hospital, Shanghai, People's Republic of China

Table 1. Data of 24 patients with APL

Case	Sex	Age (years)	Previous therapy/ duration	PB WBC ($\times 10^9$/liter)	BM Prom (%)	Present therapy (mg/m^2/day)	PR (day)	CR (day)	BM Prom	Further therapy	Duration of CR (months)
1.	F	5	HOAP/7d HOP/9d H/20d	1.16	78	RA 80	31	68	1.5	(1)	8+
2.	M	6	HOP/26d	1.7	73.5	RA 100	18	28	2.5	(3)	–
3.	F	28	HOA/62d	1.8	89	RA 60	21	34	2.0	(3)	4[a]
4.	M	8	HOAP/10d OH/21d CR/3 months Relapse	2.2	33	RA 80	18	43	3.6	(2)	5[a]
5.	F	38	HOP, HOAP, H CR/1 month Relapse	7.7	15.5	RA 45	17	20	4.5	(1)	5+
6.	F	54	HOP, HOAP CR/30 months Relapse	4.0	28.5	RA 45	22	22	2.0	(1)	4[a]
7.	M	54	H/10d	1.4	94	RA 45	29	38	1.0	(1)	2[a]
8.	F	69	H/5d	0.5	48	RA 45	22	44	4.0	(3)	1+
9.	M	61	Untreated	1.0	74	RA 45–50	29	36	2.5	(2)	11+
10.	M	31	Untreated	1.6	76	RA 50	35	35	1.5	(2)	5[a]
11.	M	37	Untreated	1.4	65	RA 45	26	43	3.5	(1)	10+
12.	F	18	Untreated	0.9	81.5	RA 45–50	21	40	2.5	(4)	8+
13.	F	35	Untreated	2.2	70	RA 50	20	39	2.0	(2)	4[a]
14.	M	45	Untreated	2.1	84.5	RA 45	36	119	1.0	(4)	5+
15.	F	57	Untreated	1.9	88.5	RA 45	23	51	2.5	(4)	5+
16.	F	20	Untreated	0.9	85.5	RA 50	22	46	2.0	(1)	8+
17.	M	32	Untreated	1.1	78	RA 45	29	39	3.0	(4)	4[a]
18.	M	36	Untreated	1.1	89.7	RA 50	35	52	2.0	(4)	4+
19.	F	53	Untreated	15.8	75.5	RA 45	23	39	0	(3)	5+
20.	M	30	Untreated	6.5	90	RA 45	46	46	2.5	(4)	3+
21.	F	36	Untreated	1.7	91	RA 50	36	50	1.0	(3)	1+
22.	M	21	Untreated	1.7	90	RA 45	28	56	1.0	(4)	2+
23.	M	45	Untreated	1.4	78	RA 45	25	45	3.0	(4)	1+
24.	M	34	Untreated	1.1	30	RA 45 + ara-C20	60	98	1.5	(4)	3[a]

PB, peripheral blood count; BM, bone marrow; Prom, promyelocyte; PR (day), time to partial remission; CR (day), time to complete remission; –, lost to follow-up; +, "greater than", still under follow-up; d, days; (1), RA (20–30 mg/m^2/day); (2), RA (20–30 mg/m^2/day) + ara-C (10 mg every 12 h) or H (0.5 mg/m^2/day); (3), ara-C (10 mg every 12 h); (4), consolidated with HOAP, maintained by 6-mercaptopurine, methotrexate, or cyclophosphamide
[a] Relapse following CR

other eight (case #1–#8) had previously been treated with chemotherapy (HOAP[1], HOP, OH, COH, H). Of the eight treated patients, three were in relapse after 1–30 months of CR and five were resistant to or could not tolerate the chemotherapy (5–62 days of treatment). Twenty-two of the patients showed mild to moderate hemorrhagic manifestations (purpura, gingivorrhagia, gastrointestinal bleeding) but no laboratory evidence of DIC prior to treatment with RA except for a positive plasma protamin sulfate paracoagulation ($3P^+$) in three of the cases.

II. Marrow Preparation and Culture

A modification of the method of Flynn et al. [6] for short-term suspension culture was used. Marrow cells were aspirated from the iliac crest, layered onto Ficoll-Hypaque (specific gravity 1.077), and centrifuged at 800 *g* for 15 min. Interface cells were collected, washed with McCoy's 5A medium, and resuspended at a concentration of 5×10^5 cells/ml in McCoy's 5A medium containing 15% fetal calf serum. All-*trans* RA (Shanghai No. 6 Pharmaceutical Factory, Shanghai) was dissolved in absolute ethanol to a concentration of 1 m*M* and further diluted with the medium so that the final ethanol concentration in the cultures was 0.1% and the final RA concentration 1 μ*M*. Controls were cultured in medium alone. (It had been previously demonstrated that 0.1% ethanol had no effect on cell growth and on differentiation of HL-60 cells [11].) All cultures were incubated at 37° C in a 5% CO_2 atmosphere for up to 7 days. Cell density was determined by hemacytometer and cell viability by the trypan blue dye exclusion method. Aliquots of cells were removed for morphological examination on the 2nd, 4th, and 6th day of culture.

[1] H, Harringtonin (0.02–0.07 mg/kg/day); O, oncovin (0.02–0.03 mg/kg/day); A, cytosine arabinoside ara-C); P, prednisone; C, cyclophosphamide

III. Morphological Studies

Differential counts were performed on cell smears stained with Wright's solution. Chloroacetate esterase and alpha-naphthyl acetate esterase stains were performed using standard techniques [12]. Samples from four cases were prepared for transmission electron microscopic study. The nitroblue tetrazolium (NBT) reduction assay was performed as described by Francis et al. [13]. The percentage of cells containing intracellular blue-black deposits was determined in 200 cells on Wright's stained slide preparations.

IV. Colony Formation Assay

Blast cell colonies were grown as described by Minden et al. [14]. Conditioned medium was prepared from leukocytes (10^6 cells/ml) incubated at 37° C for 7 days in McCoy's 5A medium with 10% fetal calf serum and 1% (v/v) phytohemagglutinin-P (PHA-P) (DIFCO) and stored at 4° C until used. The preparation was termed PHA-LCM. Bone marrow cells were plated at 1×10^6 cells/ml using McCoy's 5A medium supplemented with 0.3% agar, 20% fetal calf serum, and 25% (v/v) PHA-LCM. GM-CFU was determined according to the technique of Pike and Robinson [15] for colony growth in agar. Briefly, 2×10^5 marrow cells were plated in 35-mm tissue culture dishes over a feeder layer of 1×10^6 leukocytes from healthy donors. The plates were incubated at 37° C in a humidified 5% CO_2 atmosphere. L-CFU colonies (more than 20 cells) were scored on day 8 and GM-CFU colonies (more than 40 cells) on day 10.

V. Treatment of Patients

The 24 patients in this series received all-*trans* RA (45–100 mg/m²/day) as the remission induction therapy. Informed consent was obtained from all patients (or their parents). Peripheral blood counts, bone marrow aspiration, and coagulation parameters (in 21 cases) including thrombin time, prothrombin time, plasma protamin sulfate paracoagulation test, euglobulin lysis test, and fibrinogen levels were determined before the start of therapy and at regular intervals thereafter. CR is defined as less than 5% blasts plus promyelocytes in a normal cellular marrow with a normal peripheral blood count and an absence of the signs and symptoms of leukemia on physical examination [16]. Partial remission (PR) is defined as less than 5% blasts plus promyelocytes in a normal cellular marrow but with a clinically moderate anemia. Blood transfusion and antibiotics were given as supportive treatment when necessary.

VI. Continuation Therapy Following Complete Remission

Twenty-three patients were followed after attaining CR. Further therapy was as follows: (1) Maintained by RA, 20–30 mg/m^2/day (six cases), (2) maintained by RA, 20–30 mg/m^2/day plus low-dose ara-C (10 mg i.m. every 12 h) or low-dose Harringtonin (0.5 mg/m^2 i.v. daily) in rotation (four cases), (3) maintained by low-dose ara-C, 10 mg i.m. every 12 h (five cases), (4) consolidated by chemotherapy (HOAP) and maintained by 6-mercaptopurine (2 mg/kg daily p.o.) and methotrexate (10 mg/m^2, i.v. weekly), or cyclophosphamide (200 mg/m^2, i.v. weekly) (nine cases).

C. Results

I. In Vitro Studies

Leukemic bone marrow cells derived from 15 patients and incubated for 7 days in suspension culture, with or without all-*trans* RA (1 μ*M*), showed little change in cell density. Viability of both control and RA-treated cells was consistently greater than 75%.

Leukemic promyelocytes from 14 patients showed morphologic and functional maturation when cultured in the presence of RA (Table 2), (Fig. 1 A, B). The

Table 2. Response of promyelocytes to RA in suspension culture

Case No.	Blasts (%)		Promyelocytes (%)		Myelocytes (%)		Mature[a] (%)		NBT (%)	
	Control	RA-treated	Control	RA-treated	Control	RA-treated	Control	RA-treated	Control	RA-treated
1.	3	2	68	7	15	34	4	48	ND	ND
3.	0	0	86	1	6	26	7	73	3.5	52
4.	0	1	47	4	14	48	18	36	ND	ND
9.	0	0	95	5	2	25	1	63	3	54
10.	0	0	98	2	0	24	2	74	5	38
11.	2	1	78	2	5	26	11	64	ND	ND
12.	2	1	81	4	9	27	6	62	2	39
13.	0	0	86	9	8	38	6	53	ND	ND
14.	0	0	86	3	11	29	3	68	ND	ND
16.	0	0	93	12	6	36	1	52	0	35.5
18.	1.5	0	77	1	20.5	42.5	1	55	12	43
19.	0	0	92	4	0	53	2	39	ND	ND
20.	0	0	90	8	8	38	2	52	ND	ND
21.	0	0	91.5	20	3.5	62	2	15	4	33
24.	0	0	84	80	2.5	4	6.5	7	ND	ND

Control, RA not added to the culture; NBT, nitroblue tetrazolium; ND, not done
[a] Metamyelocytes + bands + polymorphonuclear leukocytes

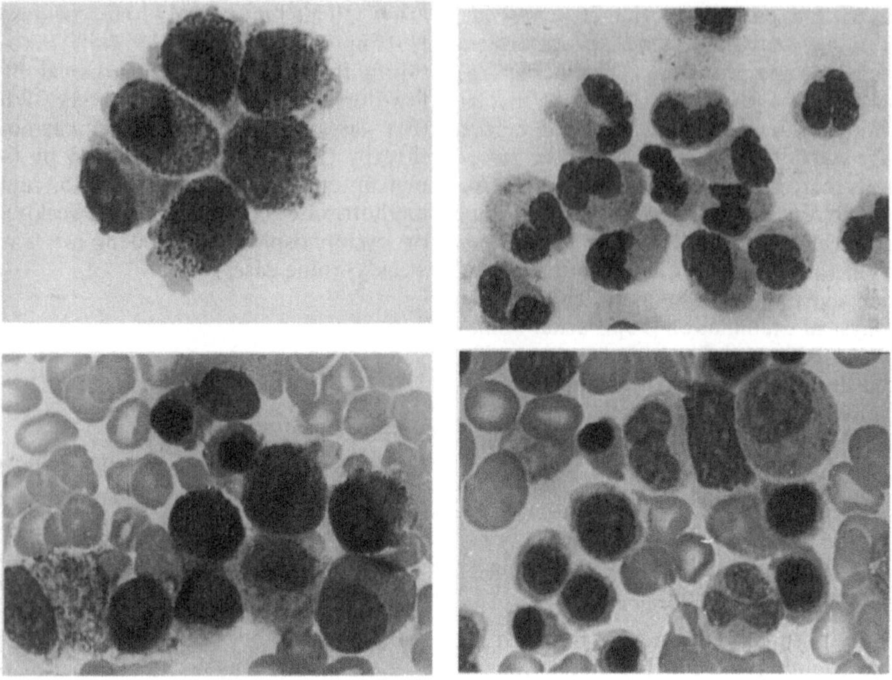

A B C D

Fig. 1 A–D. Morphological maturation of leukemic cells of case 10 in vitro and in vivo. **A** Cells cultured without RA, consisting of promyelocytes with characteristic cytoplasmic granules, ×1000. **B** Cells cultured with RA, showing maturation of granulocytes, ×1000. **C** Bone marrow before RA treatment. The predominance of promyelocytes (76%) indicates typical APL. **D** Bone marrow after 5 weeks of RA treatment. Promyelocyte level under 2%, and restoration of normal hemopoiesis without an aplasia phase are consistent with differentiation induction

percentage of promyelocytes in the control group versus the RA-treated group was 83.5% ± 12.8% [2] and 5.9% ± 5.0% [2] respectively. The percentage of mature cells (metamyelocytes + bands + PMNs) was 4.7% ± 4.5% [2] and 53.9% ± 15.4% [2] respectively. The rate of NBT reduction in RA-treated cells was 42.0% ± 7.5% [2], significantly higher than that of the control group (4.2% ± 3.5%) [2].

To examine the progression of cellular differentiation, we incubated cells from four patients with 1 μ*M* RA for various time intervals. After 48 h, morphologically recognizable changes in the promyelocytes could be observed. The nucleus became larger and fewer primary granules were observed in the cytoplasm. On the 4th day of culture, these cells gave rise to myelocytes which contained specific, or secondary, granules. The nuclear chromatin was more condensed and the nucleoli were either vague or no longer visible. There was an elevated population of metamyelocytes which had indented or horseshoe-shaped nuclei and cytoplasm filled with both primary and secondary granules appearing by day 6, as well as some band and fully mature granulocytes. When the cultures were continued for 7–8 days, the relative number of mature granulocytes increased.

Cytochemical analysis showed that in the control cells chloroacetate esterase activity varied from mildly to moderately

[2] Results represent the data from the patients studied and are expressed as mean % ± standard deviation.

positive, while in the RA-treated cultures intensely positive granules were seen, either diffusely scattered or accumulated in some portion of the cytoplasm. The majority of control cells showed weak nonspecific esterase activity while RA-treated cells had a stronger reaction.

Transmission electron microscopic examination of four cultures confirmed that in the presence of RA the cells had been differentiated to mature granulocytes. Condensation of the heterochromatin became evident and the nucleus had often been changed to a bean-shaped or even a segmented form. Neutrophilic granules were smaller and diffusely scattered throughout the cytoplasm. Azurophilic granules were markedly decreased.

II. Clinical Studies

Twenty-four patients were treated with all-*trans* RA as a single agent. All achieved both PR and CR except the one patient (# 24) whose cells were not inducible when cultured with RA in vitro. Subsequent bone marrow examination of this patient revealed a continuing proliferation of leukemic promyelocytes. When ara-C (10 mg) was added intramuscularly every 12 h, the patient achieved CR in 98 days (Table 1).

In the 12 patients studied who responded to the induction differentiation effect of RA, L-CFU growth was predominant (163.3 ± 129.0 colonies) and GM-CFU suppressed (0.63 ± 1.3 colonies) prior to treatment. GM-CFU reached normal levels (100.2 ± 55.1 colonies) with little or no growth of L-CFU after CR was achieved.

III. Pattern of Clinical Response to *trans*-Retinoic Acid

Systematic observation of the peripheral blood counts during RA treatment of the previously untreated patients revealed some specific patterns of change. There was a progressive rise in the total white blood cell count which started with initiation of treatment and which reached a peak between 7 and 14 days. After this, the white blood cell count fell with the progressive maturation of granulocytes. Increase in platelets was most prominent after 3 weeks. Elevation of the hemoglobin concentration appeared reluctant and slow. Bone marrow aspirate revealed that hypercellularity existed throughout the RA treatment. Partial remission could be expected within 1 month (Fig. 1C, D). Therapy with oral all-*trans* RA was accompanied by mild toxicity that consisted of dryness of the lips and skin (100%), headache (25%), nausea or vomiting (20.8%), moderate bone or joint pain (12.5%), and mild exfoliation (8.3%). Two patients had elevated SGPT. All of these side effects were well tolerated or alleviated when the dosage of oral RA was reduced.

IV. Disseminated Intravascular Coagulation

Coagulation parameters, including thrombin time, prothrombin time, plasma protamin sulfate paracoagulation test (3P), euglobulin lysis test, and fibrinogen levels, were measured simultaneously, in 21 patients, at the beginning of RA therapy and throughout the course of treatment. Of these patients, 18 who were normal in coagulation parameters prior to the start of RA therapy showed no changes during treatment. The other three patients who had been previously treated and who were 3P(+) became negative 7–10 days after RA. Therefore, DIC or other hemorrhagic complications did not occur when patients with APL were induced to remission with RA.

V. Duration of Clinical Remission

Twenty-three patients were followed after induction of CR (Table 1). Of the six patients maintained on RA alone, four were still in remission for a period of 5–10 months. Two patients relapsed in 2 and 4 months. Among the four patients maintained on RA with either low-dose ara-C or low-dose Harringtonin in rotation, three relapsed within a period of 4–5 months. Of the five patients maintained on low-dose ara-C alone, one case (#2) was lost to follow-up, one relapsed in 4 months, and the other three remained in CR for 1^+ to 5^+ months. Of the remaining nine patients who were consolidated by chemotherapeutic regimens and maintained on 6-mercaptopurine, methotrexate, or cyclophosphamide, two relapsed and seven have been in CR from 1^+ to 8^+ months. The new population of APL promyelocytes at relapse differed morphologically from those present at the start of treatment and were resistant to all-*trans* RA induction of differentiation in vitro.

D. Discussion

Recent approaches in treatment of leukemia include the use of "differentiation-inducing agents" such as RA, vitamin D_3, or low-dose ara-C [17–19]. Numerous studies both in vitro and in vivo have revealed that RA is a potent inducer of myeloid differentiation, both in the promyelocytic cell line HL-60 as well as in fresh promyelocytes from patients with APL, and at a concentration that was pharmacologically obtainable in man [11, 20]. 13-*cis* RA and all-*trans* RA were equally effective in induction of differentiation in vitro [5]. Our studies confirm that in vitro leukemic promyelocytes could be induced by all-*trans* RA to differentiate toward mature granulocytes. One exception was that the cells from patient #24 were resistant to RA induction. The morphological characteristics of these RA-resistant cells revealed a scanty cytoplasm with less-prominent coarse granulation. The differences in sensitivity to RA may be due to the heterogeneous entities of APL [21, 22].

In 1983, Flynn et al. [6] described the first case of APL treated with 13-*cis*-RA. Unfortunately, this patient died from disseminated candidiasis although there was a marked elevation in his peripheral granulocyte count after 2 weeks of treatment. Nilsson [7] reported a 30-year-old woman with APL in relapse for 10 months; she was treated by 13-*cis* RA (1 mg/kg) and began to respond after 1 month, and normal blood and bone marrow pictures continued for 11 months. Daenen et al. [8] reported a 33-year-old patient with refractory APL complicated by fibrinolysis and aspergillus pneumonia. He was treated with 13-*cis* RA (80 mg/day) alone, and attained a CR after 7 weeks. Recently, Fontana et al. [9] reported one case of refractory APL treated with 13-*cis* RA (100 mg/m^2) which resulted in CR after 13 days. In vitro studies of this patient's leukemic blasts showed differentiation in the presence of RA. Sampi et al. [23] reported a 58-year-old Japanese man who also had relapsed APL and failed to respond to etretinate (a form of retinoid) and dactinomycin, although the leukemic cells were sensitive to all-*trans* RA (10^{-6}–10^{-7} *M*) in vitro. We have treated our patients with all-*trans* RA and found that all-*trans* RA was not only effective in patients who had been refractory to chemotherapy, but also effective in those with "de novo" APL. Moreover, we were able to find predictive value in the in vitro differentiation studies. The single patient who was resistant to RA induction failed to show marrow improvement when treated with RA as the sole agent.

According to most authors, the main disquieting problem of APL is death during induction treatment [3, 4, 24], especially because of intracerebral hemorrhage. DIC is the most common complication of APL. Its severity and frequency are often aggravated by chemotherapy despite the use of heparin. In this study we report no aggravation of hemorrhagic manifestation or appearance of coagulation parameter abnormalities suggesting DIC during the course of RA treatment. This would be one of the striking advantages over aggressive chemotherapy which could destroy the leukemic cells and cause the release of procoagulant factors from the azurophilic granules into the circulation. It is possible that the leukemic cells are not destroyed during treatment of APL with RA, but that they have differentiated, undergone terminal cell division, and lost the capacity to release these coagulant factors during this process. The fact that there was no decrease, but rather an increase, of marrow cellularity during induction therapy supports this possibility.

The role of all-*trans* RA in the maintenance of remission is undetermined. Two cases of APL, reported by Daenen et al. [8] and Fontana et al. [9], relapsed in 6 and 12 months respectively. In our series, the patients were further treated with four different regimens after CR was induced, but it is too early to conclude which of these is the most effective. From the data obtained from both our clinical survey and the cytogenetic studies showing the persistence of abnormal clones (unpublished data), we suggest that intensive chemotherapy after CR may be beneficial.

The knowledge about the side effects of oral RA is mainly from the dermatological literature. Our data are compatible with other reports on the toxicity of oral all-*trans* RA [25]. The toxicity of 13-*cis* RA has been shown to be relatively lower than all-*trans* RA [25], but in our experience the side effects were well tolerated by the patients, some of whom have been taking RA for more than 10 months with no severe untoward effects.

Based on these observations, we conclude that all-*trans* RA is an effective agent for obtaining CR in APL. How to maintain and prolong the duration of the CR, however, requires further study.

Acknowledgments. We are grateful to Professor Samuel Waxman of Mount Sinai School of Medicine for his kind comments and suggestions in editing this manuscript, and to other physicians and hematologists at Shanghai Rui-Jin Hospital, Shanghai Zhong-Shan Hospital, Shanghai Chang-Zhen Hospital, and Shanghai Institute of Pediatrics for providing samples and care of patients studied. We are also grateful to Zhao Jin-chai for expert technical assistance.

References

1. Jones ME, Saleem A (1978) Acute promyelocytic leukemia: a review of literature. Am J Med 65:673–677
2. Bernard J, Weil M, Boiron M, Jacquillat C, Gemon MF (1973) Acute promyelocytic leukemia: results of treatment by Daunorubicin. Blood 41:489–496
3. Drapkin RL, Timothy SG, Dowling MD, Arlin Z, Mckenzie S, Kempin S, Clarkson B (1978) Prohylactic heparin therapy in acute promyelocytic leukemia. Cancer 41:2484–2490
4. Cordonnier C, Vernant JP, Brun B, Heilmann MG, Kuentz M, Bierling P, Farcet JP, Rodet M, Duedari N, Imbert M, Jouault H, Mannoni P, Reyes F, Dreyfus B, Rochant H (1985) Acute promyelocytic leukemia in 57 previously untreated patients. Cancer 55:18–25
5. Koeffler HP (1983) Induction of differentiation of human acute myelogenous leukemia cells: therapeutic implications. Blood 62:709–721
6. Flynn P, Miller W, Weisdorf D, Arthur D, Banning R, Branda R (1983) Retinoic acid treatment of acute promyelocytic leukemia: in vitro and in vivo observations. Blood 62:1211–1217
7. Nilsson B (1984) Probable in vivo induction of differentiation by retinoic acid of promyelocytes in acute promyelocytic leukemia. Br J Haematol 57:365–371
8. Daenen S, Vellenga E, Van Dobbenbugh OA, Halie MR (1986) Retinoic acid as antileukemic therapy in a patient with acute promyelocytic leukemia and aspergillus pneumonia. Blood 67:559–561
9. Fontana JA, Roger JS, Durham JP (1986) The role of 13-*cis* retinoic acid in the remission induction of a patient with acute promyelocytic leukemia. Cancer 57:209–217
10. Bennett JM, Catovsky D, Daniel MT, Flandrin G, Galton DAG, Gralnick HR, Sultan C (1976) Proposals for the classification of the acute leukemia. Br J Haematol 33:451–458
11. Breitman TR, Selonick SE, Collins SJ (1980) Induction of differentiation of the human promyelocytic leukemia cell line (HL-60) by retinoic acid. Proc Natl Acad Sci USA 77:2936–2940
12. Yam LT, Li CY, Crosby WH (1971) Cytochemical identification of monocytes and granulocytes. Am J Clin Pathol 55:283–290

13. Francis GE, Guimaraes JETE, Berney JJ, Wing MA (1985) Synergistic interaction between differentiation inducers and DNA synthesis inhibitors: a new approach to differentiation induction in myelodysplasia and acute myeloid leukemia. Leuk Res 9:573–581
14. Minden MD, Buick RN, McCulloch EA (1979) Separation of blast cell and T-lymphocyte progenitors in the blood of patients with acute myeloblastic leukemia. Blood 54:186–195
15. Pike BL, Robinson WR (1970) Human bone marrow culture in agar gel. J Cell Physiol 76:77–84
16. Vogler WR (1985) Post-remission therapy for acute myelogenous leukemia. In: Bloomfield CD (ed) Chronic and acute leukemias in adults. Martinus Nijhoff, Boston, p 209–228
17. Gold EJ, Mettelsmann RH, Itri LM, Gee T, Arlin Z, Kempin S, Clarkson B, Moore MAS (1983) Phase I clinical trial of 13-*cis* retinoic acid in myelodyplastic syndromes. Cancer Treat Rep 67:981–986
18. Koeffler HP, Hirji K, Itri L (1985) 1,25-Dihydroxyvitamin D_3: in vitro and in vivo effects on human preleukemic and leukemic cells. Cancer Treat Rep 69:1399–1407
19. Degos L, Castaigne S, Tilly H, Sigaux F, Daniel MT (1985) Treatment of leukemia with low-dose Ara-C: a study of 160 cases. Semin Oncol 12:196–199 [Suppl 3]
20. Breitman TR, Collins SJ, Keene BR (1981) Terminal differentiation of human promyelocytic leukemic cells in primary culture in response to retinoic acid. Blood 57:1000–1004
21. Golomb HM, Rowley JD, Vardiman JW, Testa JR, Butler A (1980) "Microgranular" acute promyelocytic leukemia: a distinct clinical, ultrastructural, and cytogenetic entity. Blood 55:253–259
22. Tomonaga M, Yoshida Y, Tagawa M, Jinnai I, Kuriyama K, Amenomori T, Yoshioka A, Matsuo T, Nonaka H, Ichimaru M (1985) Cytochemistry of acute promyelocytic leukemia (M_3): leukemic promyelocytes exhibit heterogeneous patterns in cellular differentiation. Blood 66:350–357
23. Sampi K, Honam Y, Hozumi M, Sakurai M (1985) Discrepancy between in vitro and in vivo induction of differentiation by retinoids of human acute promyelocytic leukemia cells in relapse. Leuk Res 9:1475–1478
24. Ruggero D, Baccarani M, Guarini A (1977) Acute promyelocytic leukemia: results of therapy and analysis of 13 cases. Acta Haematol (Basel) 58:108–119
25. Windhorst DB, Nigra T (1982) General clinical toxicology of oral retinoids. J Am Acad Dermatol 6:675–682

Simultaneous Presentation of B- and T-Cell Malignant Lymphoma

F.E. Cotter[1], P.A. Hall[2], B.D. Young[1], and T.A. Lister[1]

A. Introduction

Multiclonal NHL of the B-cell type has been described [1, 2], most often when two histological types of lymphoma have been noted in simultaneous or successive biopsies. This occurs predominantly when follicular centre cell (FCC) histology converts into a large cell diffuse histology [2]. Multiclonal disease of the B-cell type has been noted in heavily immunosuppressed patients [3]. Reports of composite lymphoma with diffuse T-cell histology arising from co-existant FCC have been described on morphological and immunological evidence [4–6]; however, no DNA analysis was reported. Recently, a multiclonal lymphoma, confirmed by DNA analysis, was reported with sequential B- and T-cell clones [7] in a patient with autoimmune disease who received chemotherapy for the initial B-cell lymphoma. Simultaneous presentation of B- and T-cell lymphomas, confirmed by DNA analysis, with no predisposing factors has not previously been reported.

B. Materials and Methods

I. Case Report

A fit 64-year-old man presented with lymphadenopathy in his left neck and right groin. He had no previous history of illness or exposure to carcinogens. Biopsy tissue from the neck lymph node was histologically a centroblastic/centrocytic follicular lymphoma. Staging investigations revealed no further sites of disease, and the patient was classified as having a stage-IIIA follicular lymphoma and treated with 6 weeks of oral chlorambucil at 10 mg per day. During the course of this treatment the swelling of neck glands disappeared, but the glands in the right groin continued to enlarge, leading to a biopsy of this second site of disease. The groin lymph node was histologically a diffuse T-cell lymphoma of the angioimmunoblastic lymphadenopathy type (AIL) with no evidence of a centroblastic/centrocytic lymphoma. Restaging revealed no new sites of disease. Treatment was commenced with weekly treatment of a regimen containing adriamycin, cyclophosphamide, vincristine, bleomycin and prednisolone, followed by a complete resolution of disease which has been sustained for 18 months.

II. Immunophenotype and Immunogenotype Studies

Fresh specimens were obtained from both biopsies and small portions were placed in air-tight plastic tubes, snap-frozen, and stored in liquid nitrogen. Immunostaining on paraffin sections (4 μm) and cryostat sections (6 μm), the latter air-dried and fixed in acetone, followed by periodate-lysine-paraformaldehyde as previously described [8], was performed using the indirect immunoperoxidase method with both B- and T-cell primary antibodies. Peroxidase-labelled second antisera were obtained from Dakopatts UK Ltd.

[1] Imperial Cancer Research Fund, Department of Medical Oncology and [2] Department of Histopathology, St. Bartholomew's Hospital, West Smithfield, London EC1 7BE, UK

III. Molecular Studies

DNA was extracted from the freshly frozen biopsy material by standard techniques [9]. Ten micrograms of DNA was digested with an appropriate restriction enzyme (Boehringer Mannheim, Federal Republic of Germany) and the fragments were studied by the Southern blot hybridization method [10] using Hybond-N filters (Amersham International PLC, Amersham, England). DNA probes were radiolabelled with the (32p)-dCTP random primer extension method [11]. Immunoglobulin heavy chain (JH) [12], pFL1 (kindly provided by Dr. M.L. Cleary) [13], and β-T-cell receptor gene (TCR) [14] probes were hybridized [15] and autoradiographed at −70° C.

C. Results

I. Histology

The histological features of the first biopsy from the left neck were normal nodal architecture effaced by an infiltrate of malignant lymphoma with a nodular architecture composed of an admixture of centroblasts and centrocytes. The appearance was that of centroblastic/centrocytic follicular lymphoma. The second biopsy, taken from the right groin, had a completely different appearance. The nodal architecture was effaced by a diffuse infiltrate of small and medium-sized lymphoid cells. Plasma cells and occasional eosinophils were present, and arborizing high endothelial venules were

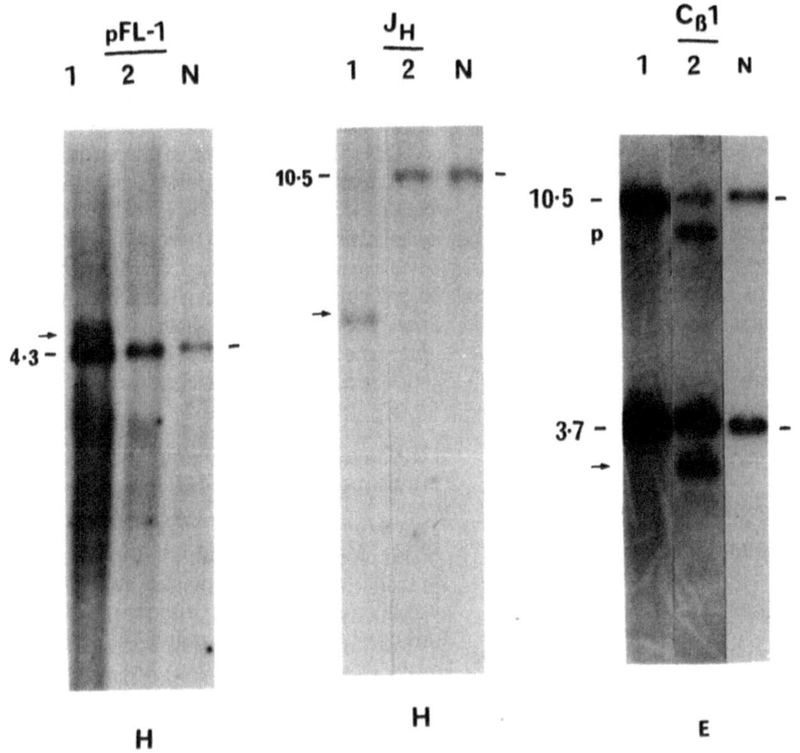

Fig. 1. Hybridization analysis of t(14;18) (*pFL · 1*), immunoglobulin (J_H), and T-cell receptor gene (C_β *1*) DNA. The probes used are indicated above each panel and the appropriate restriction enzyme used is shown below each panel. (*H*, *Hind*III, *E*, *Eco*RI). *Dashes* indicate germ line and *arrow* indicates clonal rearranged bands. Size makers are indicated in kilobases. (*P* represents in *Eco*RI site partly resistant to normal digestion; however, this is not a generalised partial digestion)

prominent. There was no evidence of centroblastic/centrocytic lymphoma, and a diagnosis of probable T-cell lymphoma of the AIL type was made.

II. Immunohistochemistry

The neoplastic cells seen in biopsy one (left neck) showed strong immunoreactivity with CD45, L26, MB1, MB2 and LN2 in paraffin section. Scattered cells with MT1 and UCHL1 immunoreactivity were present in the interfollicular areas, with occasional cells in the B-cell nodules. Cryostat section immunostaining showed an identical pattern using a large panel of antibodies. Kappa light-chain restriction was demonstrated. Staining with the antibody R4/23 (which recognises an antigen expressed on dendritic reticulum cells) accentuated the nodular pattern. In contrast, in biopsy two (right groin) occasional small clusters of lymphocytes with immunoreactivity with B-cell markers (L26, MB1, MB2) were seen, but the predominant population was T-lymphocytes (MT1, UCHL1-positive). Again, this pattern was repeated in the cryostat section immunostaining. The plasma cells and B-lymphocytes present were polyclonal. The pattern of staining with R4/23 seen in the first biopsy was not present, with only a few scattered cells staining.

III. Molecular Studies

Biopsy one showed immunoglobulin gene rearrangement with a clonal rearranged band, with the JH and pFL1 probes (Fig. 1) confirming the presence of a clonal B-cell neoplasm and the t(14:18) rearrangement found in the majority of follicular lymphomas (16,6). No gene rearrangement was seen with the TCR probe. Biopsy two confirmed the T-cell clonality by revealing a rearranged band with the TCR probe (Fig. 1). However, there was no rearrangement seen with the JH and pFL1 probes.

D. Discussion

This case revealed histological and immunophenotypical non-Hodgkin's lymphoma (NHL) with simultaneously presenting lymph nodes (LN) with both B- and T-cell clones. DNA analyses confirmed a clonal B-cell population in one LN together with a t(14:18) translocation (involving the BCL-2 gene) and a clonal T-cell population in the other LN. This has not previously been reported in a patient with no predisposing factors. This could be ascribed to independent development of two types of lymphoma. However, in view of previous evidence for multiclonal disease arising from FCC [1, 2, 4, 5, 7] this may not be the only possibility. The t(14:18) translocation was present in the FCC lymphoma but not in the T-cell lymphoma, suggesting that this translocation, present in nearly all FCC [16], may be a causative factor in the development of this histological type of disease. Subsequent or simultaneous development of diffuse histological types of lymphoma is reported to be associated with increasing absence of this translocation [17] and has been thought to have been lost with transformation of the lymphoma, and in addition is often associated with development of a second B-cell clone [1, 2]. A possible explanation for multiclonal lymphoma is the emergence from a malignant clone of pre-B-lymphocytes prior to B- or T-cell lineage commitment with the ability to develop into clonal B- or T-cell lymphoma. The presence of the BCL-2 gene rearrangement may be an indicator of a defective pre-B-lymphocyte and could have a bearing on the resistance of FCC to curative treatment [18, 19].

References

1. Sklar J, Cleary ML, Thielamanns K, Gralow J, Warnke R, Levy R (1984) Biclonal B-cell lymphomas. N Engl J Med 311:20–27
2. Siegelman MH, Cleary ML, Warnke R, Sklar J (1985) Frequent biclonality and Ig

gene alterations among B-cell lymphomas that show multiple histological forms. J Exp Med 161:850–863
3. Cleary ML, Sklar J (1984) Lymphoproliferative disorders in cardiac transplant recipients are multiclonal lymphomas. Lancet ii:489–493
4. York JC II, Cousar JB, Glick AD, Flexner JM, Stein R, Collins RD (1985) Morphological and immunologic evidence of composite B- and T-cell lymphomas. Am J Clin Pathol 84:35–43
5. Aisenberg AC, Block KJ, Wilkes BM (1981) Malignant lymphoma with dual B- and T-cell markers. J Exp Med 154:1709–1714
6. Jeannette JC, Reddick RL, Saunders AW, Wilkman AS (1982) Diffuse T-cell lymphomas preceded by nodular lymphoma. Am J Clin Pathol 78:242–248
7. Hu E, Weiss LM, Warnke R, Sklar J (1987) Non-Hodgkin's lymphomas containing both B- and T-cell clones. Blood 70:287–292
8. Hall PA, Stearn PM, Butler MG, d'Ardenne AJ (1987) Acetone periodate lysine paraformaldehyde (PLP) fixation and improved morphology for frozen-section immunohistochemistry. Histopathology 11:93–101
9. Cotter FE, Hall PA, Young BD (1988) Extraction of DNA from frozen section with simultaneous histological examination. J Clin Pathol 41:1125–1126
10. Southern EM (1975) Detection of specific sequences among DN fragments separated by gel electrophoresis. J Mol Biol 98:503–517
11. Feinberg AP, Vogelstein B (1983) A technique for radiolabelling DNA restriction endonuclease fragments to a high specific affinity. Anal Biochem 132:6–13
12. Nishida Y, Miki T, Hidsajima H, Honjo T (1982) Cloning of human immunoglobulin ε chain genes: evidence for multiple C_ε genes. Proc Natl Acad Sci USA 79:3833–3837
13. Cleary ML, Sklar J (1985) Nucleotide sequence of a t (14; 18) chromosomal breakpoint in follicular lymphomas and a demonstration of a breakpoint-cluster region near a transcriptionally active locus on chromosome 18. Proc Natl Acad Sci USA 82:7439–7443
14. Furley AJ, Mizutani S, Weilbaecher K, Dhaliwal HS, Ford AM, Chan LC, Molgaard HV, Toyonago B, Mak T, van den Elsen P, Gold D, Terhoist C, Greaves MF (1986) Developmentally regulated rearrangement and expression of genes encoding the T-cell receptor-T3 complex. Cell 46:75–87
15. Maniatis T, Fritsch EF, Sambrook J (1982) Molecular cloning: a laboratory manual. Cold Spring Harbor Laboratory, Cold Spring Harbor NY, p 387–389
16. Weiss LM, Warnke RA, Sklar J, Cleary ML (1987) Molecular analysis of the t(14; 18) chromosomal translocation in malignant lymphomas. N Engl J Med 317:1185–1189
17. Yunis JJ, Frizzera G, Oken MM, McKenna J, Theologides A, Arnesen M (1987) Multiple recurrent genomic defects in follicular lymphomas: a possible model for cancer. N Engl J Med 316:79–83
18. Gallagher CJ, Gregory WM, Jones AE, Stansfeld AG, Richards MA, Dhaliwal HS, Malpas JS, Lister TA (1986) Follicular lymphomas: prognostic factors for response and survival. J Clin Oncol 4:1470–1480
19. Qazi R, Aisenberg AG, Long JC (1976) The natural history of nodular lymphoma. Cancer 37:1923–1927

Phenotype Switch in Acute Leukemia Patients After Intensive Chemotherapy

R. Ihle, H. Matthes, and H. Ihle

A. Introduction

With the aid of monoclonal antibodies it has become possible to improve the correct diagnosis and classification of acute leukemias. From the literature and from our own experience we know that there is a distinct percentage of patients in whom the leukemic blasts express markers of different blood cell lineages. These so-called mixed-lineage leukemias are mostly associated with a poor prognosis in relation to the other cases. Besides the mixed lineage, we know that a lineage switch in the course of relapsing or resistant acute leukemia patients is also possible. This is perhaps important for the correct treatment of these patients.

B. Material and Methods

In the past few years we have analyzed our acute leukemia patients with a panel of 20 monoclonal antibodies which were kindly provided by Prof. Walter Knapp from the University of Vienna, Austria, with support of the International Society of Chemo- and Immunotherapy, Vienna. The antibodies used are listed in Table 1. The methods are described elsewhere [1, 2]. In a group of 64 patients with acute leukemia we performed simultaneously morphological, cytochemical, and immunocytological investigations in order to classify the leukemias. We diagnosed 49 as AML and 15 as ALL. The age ranged from 15–79 years. These investigations were also performed in the cases of relapsing or resistant disease.

The treatment of patients with acute myeloid leukemia was performed according to the TAD schedule after Gale and Cline, the treatment of ALL with a modified Hoelzer scheme. In cases of resistant or relapsing leukemia in patients under the age of 50 years we tried to give an intensive high-dose cytarabine AraC treatment, administering 3 g/m^2 twice daily over 6 days in combination with 45 mg/m^2 daunorubicin over 3 days.

C. Results

Of 64 patients with acute leukemia we found 49 to have AML and 15 to have ALL. In five patients (7.8%) there was a biphenotypical expression of myeloid and lymphoid markers. Three patients (4.7%) showed a lineage switch during the course of the disease (Table 2).

Following are the case reports in brief:

Case 1: Patient S. K., female, 17 years old. In April 1985 an M2 type of AML was diagnosed and treated according to the TAD schedule. A complete remission (CR) was achieved, but the patient refused an autologous bone marrow transplantation. Over a period of 6 months we performed a consolidation treatment,

Department of Internal Medicine, Division of Hematology, University Hospital Charité, Humboldt-Universität Berlin, German Democratic Republic

Abbreviation TAD = Thioguanine: 200 mg/sqm p.o. for 7 days Ara-C (Cytosinearabinoside): 200 mg/sqm i.v. for 7 days Daunorubicin: 60 mg/sqm for 3 days

Table 1. Our panel of monoclonal antibodies and their specificity (provided by Prof. W. Knapp, Vienna, with the support of the International Society of Chemo- and Immunotherapy, I.G.C.I., Vienna)

Cluster		Antibody	Specifity
B-cell marker	CD10	VIL-A1	Common-ALL antigen
	CD24	VIBC5	Pre-B, B cells, granulocytes
		VIBE3	Pre-B, B cells, granulocytes
T-cell marker	CD3	VIT3b	T cells
	CD1	VIT6	Cortical thymocytes
	CD4	VIT4	T-helper
	CD8	VIT8	T-suppressor
		VIT12	T cells
Myelomonocytic marker	CD15	VIMD5	Granulocytes
		VIM2	Granulocytes, monocytes
	CD11	VIM12	Granulocytes, monocytes
	CD14	VIM13	Monocytes
Erythroid marker		VIEG4	Glycophorin A
Platelet marker		VI-Pl1–3	Platelets, megakaryocytes
Proliferation marker		VIP1	
		VIP2b	
HLA-DR marker		VID1	HLA-DR

Table 2. Lineage switch in three patients with acute leukemia in the course of disease

Patient	Antibody (% positivity of blast cells)															Diagnosis
	CD10	VID1	CD24	CD1	CD3	CD4	CD8	VIP1	VIP2b	CD15	VIM2	CD11	CD14	VIEg4	VIPl1–3	
K. S.	–	10	30	–	–	–	–	–	–	70	40	25	20	–	–	M2-AML
	60	50	90	–	3	1	1	–	10	5	2	–	–	–	–	Common ALL
M. K.	–	20	60	–	10	10	10	5	3	90	90	10	8	–	–	M2-AML
	80	95	95	–	–	–	–	–	–	5	2	2	–	–	–	Common ALL
G. H.	–	90	70	8	15	10	10	40	35	20	24	3	20	–	–	B-ALL
	–	5	5	–	–	–	–	5	20	10	60	30	40	–	–	M4-AML

and then the treatment was stopped. In May 1986 a myeloid relapse was again treated with TAD. Again a complete remission was achieved. After the second relapse in October 1986 we administered a 6-day course of high-dose AraC. Only a partial remission was achieved. The blast cells were now morphologically undifferentiated, and an immunological diagnosis of common ALL was made. Despite continuous treatment the patient did not achieve complete remission. She died in April 1988.

Case 2: Patient M. K., female, 33 years old. In July 1985 an M2 type of AML was diagnosed. After three TAD courses a complete remission was achieved. She relapsed in November 1985, again with an M2 type. After repeated TAD a CR was achieved, continuing until September 1986. The morphologically undifferentiated blast cells now exhibited characteristics of a common ALL. No therapeutic benefit was possible, and the patient died 1 month later.

Case 3: Patient G. H., male, 65 years old. Between 1979 and 1983 a cyclophosphamide regimen for a glomerulonephritis was administered. Thereafter, the patient became pancytopenic. In November 1985 an acute lymphoblastic leukemia of the B-cell type was diagnosed. A modified Hoelzer therapy was performed, but there was only a partial response to the therapy. At the end of January, 1986, the cytomorphological pattern changed and we diagnosed AML of the M4 type morphologically as well as immunologically. A TAD therapy was not effective and the patient died in March 1986.

D. Summary and Conclusions

According to Stass et al. [4] the percentage of a lineage switch occurs in 6.7%–8.6% of patients with acute leukemia. Mostly, a conversion from the lymphoid to the myeloid phenotype is seen. In our three cases we found two switches from the myeloid to the lymphoid phenotype and only one from lymphoid to myeloid. This lineage switch is seen in relapsing and resistant leukemia cases [3].

Different hypotheses have been discussed concerning the phenotype switch. Cytostatic chemotherapy may eradicate one leukemic cell clone, allowing another one to proliferate. Otherwise, the leukemic transformed stem cell could be influenced by the chemotherapy, resulting in a change of the differentiation program of the cell and following with a switch of marker expression. Perhaps there is some clinical importance to monitoring the phenotype switch in order to administer the best treatment.

References

1. Holowiecki J, Lutz D, Krzemien S, Stella-Holowiecka B, Graf F, Kelenyi G, Schranz V, Callea V, Brugiatelli M, Neri A, Magyarlaki T, Ihle R, Jagoda K, Rudzka E (1986) CD-15 antigen detected by the VIM-D5 monoclonal antibody for prediction of ability to achieve complete remission in ANLL. Acta Haematol (Basel) 76:16–19
2. Knapp W (1985) Monoklonale Antikörper in der Leukämiediagnostik. Diagn Lab 35:12–22
3. Raghavachar A, Bartram CR, Gädicke G, Binder T, Heil G, Carbonell F, Kubanek B, Kleihauer E (1986) Conversion of acute undifferentiated leukemia phenotypes: analysis of clonal development. Leuk Res 10:1293–1299
4. Stass SA, Mirro J jr (1986) Lineage heterogeneity in acute leukaemia: acute mixed-lineage leukaemia and lineage switch. Clin Haematol 15:811–827

Expression of Cell Differentiation Antigens as a Prognostic Factor in Acute Leukemia

J. Hołowiecki, D. Lutz, V. Callea, M. Brugiatelli, S. Krzemień, B. Stella-Hołowiecka, A. Neri, R. Ihle, V. Schranz, F. Graf, G. Kelenyi, K. Jagoda, F. Zintl, T. Reves, G. Kardos, and S. Barceanu

A. Introduction

Much attention has been devoted to the study of prognostic factors in acute leukemia. Most of the published studies have been concerned with easily measured clinical and hematological parameters [6, 8] while studies requiring specialized techniques such as chromosome or molecular analysis [9] and immunophenotyping of leukemia [1–3, 7] have generally been rare.

In this report, based on the analysis of a large group of leukemia patients, we document that some cell differentiation antigens can be regarded as potential markers, predicting the ability to achieve complete remission and the duration of survival.

B. Materials and Methods

A total of 1054 untreated patients with acute leukemia were subclassified on the basis of the morphocytochemical FAB criteria and typed by means of a panel of 6–21 monoclonal antibodies (MoAb) of the VI series, donated by W. Knapp [5]. The mononuclear cells were isolated from the peripheral blood and/or bone marrow. Only samples with more than 70% blasts in the differential count were tested in indirect immunofluorescence [4, 5] using FITC-labeled goat Fab_2 antibodies against mouse IgG+IgM (Grubb) as the secondary reagent. The criterion for positivity was expression of the antigen by at least 15% of the blast cell population.

In this report special attention will be paid to the subgroup of 437 adult patients with acute nonlymphocytic leukemia (ANLL) – 236 men and 201 women, mean age 53 years (range 16–89 years) – and to the expression of antigens typical for the myelomonocytic line, such as VIM* 2, CD 15 (detected by VIM D5), CD 11_b (VIM 12), and CD 14 (VIM 13).

Patients were treated using one of the following remission-induction protocols: daunorubicin+Ara-C: 3+7 days or TAD, and those who reached complete remission (CR) received cyclic mainte-

* CDw 65

For the IGCI Cooperative Group: Vienna, Katowice, Reggio Calabria, Berlin, Budapest, Pecs, Jena, Bucharest

[1] Union des Societés, Des Sciences Médicales, 8 Rue Progresul, Bucarest, Romania (Prof. Dr. S. Barceanu)

[2] Department of Haematology, Silesian Medical Academy, 40-029 Katowice, Reymonta 8, Poland (Prof. Dr. J. Hołowiecki, S. Krzemień, B. Stella-Hołowiecka, K. Jagoda)

[3] Ludwig-Boltzmann-Institut für Leukämieforschung und Hämatologie, Hanusch-Krankenhaus, 11-40 Vienna, Austria (Doz. Dr. D. Lutz)

[4] Ospedale Riuniti, Div. di Ematologia, 89100 Reggio Calabria, Italy (Prof. Dr. A. Neri, V. Callea, M. Brugiatelli)

[5] Klinik für Innere Medizin d'Charité, Schumannstrasse 20/24, 104 Berlin, DDR (Prof. Dr. R. Ihle)

[6] Semmelweis Orvostudomanyi Egyetem, III Belklinika, Eötvös út. 12, 1121 Budapest, Hungary (V. Schranz, G. Kardos, F. Graf)

[7] Dept. of Pediatrics II, Semmelweis Univ. Medical School, Tözoltóúút. 7–9, 1094 Budapest, Hungary (T. Réves)

[8] Dept. of Pathology, Med. Univ. of Pecs, Szigeti út. 12, 7643 Pecs, Hungary (Prof. Dr. G. Kelényi)

[9] Univ. Kinderklinik, Kochstrasse 2, 6900 Jena, DDR (Prof. Dr. P. Zintl)

nance chemotherapy administered for periods of 12–36 months. The computer analysis was carried out using the BMDP1L and the Leukos 3 programs. The study was designed to correlate expression of differentiation antigens with CR rate and survival. Survival was measured from the date of entry into the study and included all cases with sufficient data regardless of the remission-induction result. Comparisons involving survival were based on the following tests; generalized Wilcoxon (Breslow), generalized savage (Mantel-Cox), and/or log-rank test.

C. Results

Using the FAB criteria and a panel of reagents detecting the determinants VIM 2, CD 15, 11, 14, glycophorin A, PL* 1–3, CD 24/19, CD 10, E-receptor, CD 7, CD 3, 4, 8, peroxidase (POX), unspecific esterase (ANAE) and PAS, 97.5% of the cases studied were classified as ALL or ANLL, while 2.5% remained unclassifiable.

At the time of this analysis data on the response to therapy of 437 adult patients with ANLL were available and form the basis for evaluation. The overall response rate in this group of patients (age 16–89 years) was 51% (42% CR, 9% PR). The correlation of CD 15 expression with the response to therapy was tested in all these cases. The remaining analyses were carried out in fewer cases, dependent upon the current availability of clinical data.

* CD 41–42

The relationship of antigen expression to CR rate is presented in Table 1. Significant differences were found only between the subgroups with different CD-15 expression.

The CR-rate was significantly higher in the CD 15-positive group than in the CD 15-negative one ($P<0.05$). A significant difference in the CR rate was also found between the subgroups with the proportions of CD 15-positive blasts above 50% and those with lower rates of expression ($P<0.02$). Expression of the other five antigens studied and of POX and ANAE did not correlate with the response to induction therapy.

An attempt was made to exclude the influence of other risk factors on the response. We found no statistically significant differences in age, WBC count, percentages of blasts and granulocytes in bone marrow, blast counts in peripheral blood, proportions of M_2-FAB subtype, platelet counts, and Hb level between the subgroups compared.

Table 1. Relationships of cell-differentiation antigen expression to CR rate in adults with ANLL

Antigen	Expression (% of blasts)	No.	CR (%)	Significance (chi^2)
CD15	<15%	170	34.7	4.4, $P<0.05$
	≥15%	267	45	
	<50%	323	38	5.4, $P<0.02$
	≥50%	114	51	
VIM2	<15%	64	42	n.s.
	≥15%	302	39.7	
CD14	<15%	126	44.7	n.s.
	≥15%	44	50	
CD11	<15%	131	48	n.s.
	≥15%	122	37.4	
Glycophorin A	<15%	221	43.3	n.s.
	≥15%	8	25	

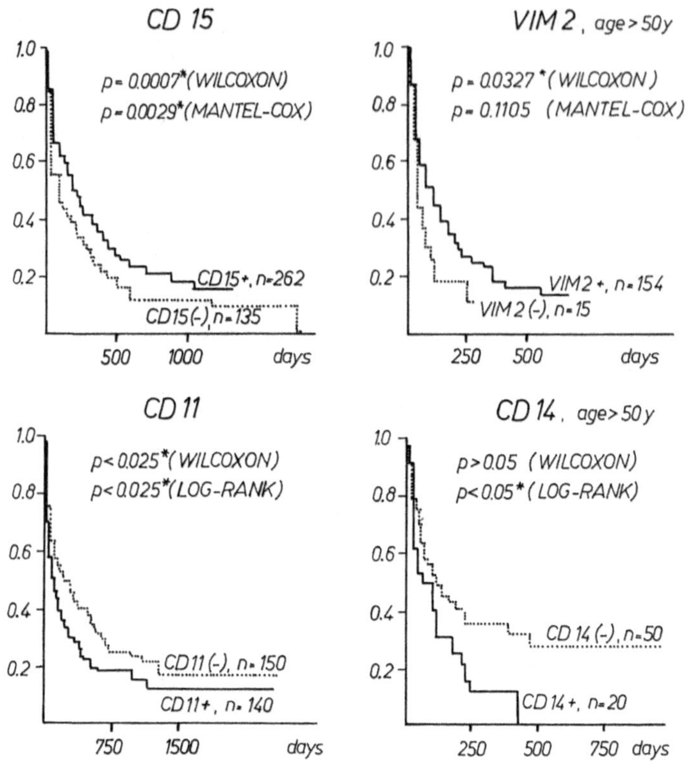

Fig. 1. Kaplan-Meier survival curves in patient groups defined by expression of selected antigens (the analyses of VIM 2 and CD 14 expression deal with subgroups of patients older than 50 years). *Asterisk*, Significant difference

Analysis of Kaplan-Meier survival curves (Fig. 1) shows that the CD 15-positive group had a significantly better outcome (mean 648 ± 353 days, median 218 days, 1-year survival = 41%) compared with the CD 15-negative one (mean 423 ± 279 days, median 76 days, 1-year survival = 29%), $P = 0.0007$ Wilcoxon, $P = 0.0029$ Mantel-Cox. Analogous analysis demonstrated that VIM 2 expression does not correlate significantly with survival, in spite of a higher median survival of the VIM 2-positive group. Significant differences were found when the effect of VIM 2 expression was studied in patients stratified into subgroups by age.

In the subgroup of those over 50 years of age, the VIM 2-positive patients had a better survival (median 101 days) compared with the VIM 2-negative one (median 26 days); $P = 0.03$ by Wilcoxon test.

In contrast to those "positive" interrelationships, the CD 11-positive group of ANLL patients had a shorter survival (median: 95 days, 1-year survival: 28%) than the CD 11-negative one (median: 219 days) 1-year survival: 43%); $P < 0.025$ by Wilcoxon test.

Similarly, a higher CD 14 expression appears to correlate with a shorter survival, but significant differences were found only in subgroups of patients older than 50 years (1-year survival equaled 10% and 35% respectively; $P < 0.05$ by log-rank test).

No association could be found between the survival and the expression of the following determinants; Ia (VID1), p45 (VIP 2b), transferrin (VIP 1), and glycophorin A (VIE G4).

D. Discussion

Today, when a number of significant therapeutic options are available, an attempt should be made to individualize therapy and to use new, more aggressive therapies mainly in prognostically poor groups of patients who are identified with the use of prognostic indices. The potential utility of cell differentiation antigens as prognostic factors has been investigated to a limited extent mainly in ALL [2, 7], but similar studies in ANLL are rare [1, 3, 4].

In our previous studies we showed that the expression of CD 15 provides prognostic information on the probability of achieving CR [4]. The present study confirms this observation in a larger group of 437 ANLL patients (Table 1) and shows that the predictive significance of CD 15 expression is not associated with other risk factors. As expected from our previous observations, the higher CD 15 expression was found to correlate with a better survival (Fig. 1). Thus, the expression of higher levels of CD 15 antigen is predictive of both a higher CR rate and a longer survival.

Of seven other antigens tested, the expression of CD 11_b, VIM 2, and CD 14 was found to correlate to some degree with survival. However, the significance of these interrelationships was lower. Moreover, VIM 2 and CD 14 antigen expression was predictive only in patients over the age of 50 years. It must be stressed that a high level of CD 15 and VIM 2 expression identified a group of patients with better outcome, whereas CD 11_b and CD 14 positivity were predictive of a shorter survival.

There are few publications referring to the prognostic relevance of immunophenotyping in ANLL. Griffin et al. [3] noted a correlation of My7 (CD 13) and My4 (CD 14) expression with a poor CR rate but found no significant relation to survival. In the same study no correlation could be shown between MCS1 (CDw15) expression and the response to therapy.

It is rather difficult to compare these results with our findings because of different reactivities of the monoclonal antibodies used in both studies. The frequencies of VIM D5 positivity in particular FAB subtypes in our study were as follows: M_1 – 60%, M_2 – 70%, M_3 – 65%, M_4 – 74%, M_5 – 71%. Corresponding frequencies obtained with MCS1 were lower, particularly in FAB subtypes M_1, M_2, and M_3 [3]. Thus, the reactivity of MCS1 is rather of the myelomonocytic type, whereas VIM D5 appears to be more specific for the granulocytic line [5].

Explanations for the higher CR rate and a better survival of the CD 15-positive group of ANLL patients are purely speculative. The CD 15 antigen can be found on normal granulocytic cells, from blasts to granulocytes [5]. It appears to be more specific for the granulocytic line than the VIM 2 antigen, which, in our experience, is expressed more strongly on both granulocytic and monocytic cells and may serve as a key marker of ANLL, being present in over 90% of cases. Assuming that patterns of antigen expression by normal hematopoietic cells are conserved by their malignant counterparts, it appears that CD 15 positivity reflects to some degree the maturity of leukemic cells. From this point of view, our findings are in agreement with the observations indicating a better prognosis in more mature subtypes of ANLL [6, 8].

Finally, the results presented here suggest that examination of the expression of CD 15, CD 11, VIM 2, and CD 14, and probably other myeloid antigens could be a useful addition to the existing systems of risk assignment in ANLL and could contribute to the improvement of therapy.

References

1. Dinndorf PA, Andrews RG, Benjamin D, Ridgway D, Wolff L, Bernstein ID (1986) Expression of normal myeloid-associated antigens by acute leukemia cells. Blood 67:1048–1052
2. Greaves M, Janossy G, Peto J, Kay H (1981) Immunologically defined subclasses of ALL in children: their relationship to presentation features and prognosis. Br J Haematol 48:179

3. Griffin JD, Davis R, Nelson DA, Davey FR, Mayer RJ, Schiffer C, McIntyre OR, Bloomfield CD (1986) Use of surface marker analysis to predict outcome of adult acute myeloblastic leukemia. Blood 68:1232–1241
4. Hołowiecki J, Lutz D, Krzemień S, Stella-Hołowiecka B, Graf F, Kelenyi G, Schranz V, Callea V, Brugiatelli M, Neri A, Magyarlaki T, Ihle R, Jagoda K, Rudzka E (1986) CD-15 antigen detected by the VIM-D5 monoclonal antibody for prediction of ability to achieve complete remission in acute nonlymphocytic leukemia. Acta Haematol (Basel) 76:16–19
5. Knapp W, Majdic O, Stockinger H, Bettelheim P, Liszka L, Köller U, Peschel C (1984) Monoclonal antibodies to human myelomonocyte differentiation antigen in the diagnosis of acute myeloid leukemia. Med Oncol Tumor Pharmacother 1:257–262
6. Mertelsman K, Thaler H, To L, Gee Ts, McKenzie S, Schauer P, Friedman A, Arlin Z, Cirrincione C, Clarkson B (1980) Morphological classification, response to therapy and survival in 263 adult patients with acute non-lymphoblastic leukemia. Blood 56:773–781
7. Ryan DH, Chapple CW, Kossover SA, Sandberg AA, Cohen HJ (1987) Phenotypic similarities and differences between CALLA-positive B-cell precursors. Blood 70:814–821
8. Swirsky DM, DeBastos M, Parish SE, Rees JKH, Hayhoe FGJ (1986) Features affecting outcome during remission induction of acute myeloid leukaemia in 619 adult patients. Br J Haematol 64:435–453
9. Yunis J, Brunning R, Howe R, Lobell M (1984) High-resolution chromosomes as an independent prognostic indicator in adult acute non-lymphocytic leukemia. N Engl J Med 311:812–818

Immunologic Subclassification of Acute Lymphoblastic Leukemia in Childhood and Prognosis (Modified BFM Protocol)*

F. Zintl and R. Häfer

A. Introduction

Acute lymphoblastic leukemias (ALLs) in childhood are clonal proliferations of lymphoid cells. It is now possible precisely to define stages of human lymphocyte differentiation using more traditional cell markers such as surface membrane immunoglobulin (SIg), sheep erythrocyte receptor (E), and cytochemical stains or highly specific monoclonal antibodies. The application of these immunologic methods to the study of ALL has further confirmed the heterogeneous nature of this disease. There appear to be clinical differences among the immunologic subtypes. It is as yet unclear whether T-cell disease is an independent prognostic variable. In the past B-cell ALL was characterized by an extremely poor prognosis. Relatively little information is available regarding the prognosis of the various stages of pre-B-ALL and T-ALL. In a modified Berlin-Frankfurt-Münster (BFM) study, children with non-T-ALL (except B-ALL) and T-ALL were treated according to the ALL VII/81 protocol. Our aim was to determine whether immunologic markers could define subgroups with distinctive clinical features and differing responses to standard chemotherapy.

Universitäts-Kinderklinik Jena, Abt. Hämatologie, Onkologie und Immunologie, Kochstraße 2, 6900 Jena, GDR

* This work was supported by Internationale Gesellschaft für Chemo- und Immunotherapie (IGCI) Vienna.

B. Material and Methods

One hundred and forty-three out of 525 untreated patients with childhood ALL were referred for immunophenotype determinations as part of a prospective multicenter study (1981–1987). ALL had been diagnosed in all patients by local and central review of cytologic and cytochemical features according to the French-American-British (FAB) criteria.

All patients were treated with first-line therapy for ALL (modified BFM protocol ALL/VII-81). Before starting cytostatic treatment, the individual risk factor for the patient was determined with the aid of a diagram. Three clinical values were important for the calculation of the risk factor (RF): the initial leukemic cell count and liver and spleen size. Patients with $RF < 1.2$ were considered to have a standard risk; those with $RF > 1.2$ and < 1.7 to have a medium risk; and those with $RF > 1.7$ to have a high risk [1–3]. Patients with confirmed B-cell features were treated completely differently [4]. A panel of monoclonal antibodies (W. Knapp, Vienna) to B-cell, T-cell, and myeloid antigens as well as E-rosetting, surface immunoglobulin, and acid phosphatase were used for phenotype determination.

C. Results

I. Immunologic Subgroups

In a modified BFM study (ALL VII/81) with a total number of 525 children with acute lymphoblastic leukemia, evaluation of 143 consecutively studied patients identified four major immunopheno-

Table 1. Major phenotypes of childhood ALL

		No. patients	%
Early Pre-B-ALL	($HLA\text{-}DR^+$, SIg^-, E^-, pT^-, $CD24^+$, $CD10^-$)	22	15.4
C-ALL	($HLA\text{-}DR^+$, SIg^-, E^-, pT^-, $CD24^+$, $CD10^+$)	82	57.3
B-ALL	($HLA\text{-}DR^+$, SIg^+, E^-, pT^-, $CD24^+$, $CD10^-$)	5	3.5
T-ALL	($HLA\text{-}DR^-$, SIg^-, $E^{+/-}$, $acP^{+/-}$, $CD1^{+/-}$, $CD3^{+/-}$, $CD10^{+/-}$)	34	23.8
Total		143	100.0

types of childhood ALL (Table 1). ALL of B-cell lineage (Ia^+, $CD24^+$, $CD10^+$, $SIg^{+/-}$) was observed in 109 patients (76.2%); T-cell lineage marker profiles ($E^{+/-}$, $acP^{+/-}$, $CD1^{+/-}$, $CD3^{+/-}$) were identified in 34 children (23.8%). The high proportion of T-ALL is caused by the fact that some centers included only patients with a high WBC in the immunophenotyping study. Four subgroups of B-cell lineage were defined (Fig. 1):

1. The first subgroup was Ia antigen positive, representing 6.4% of non-T-ALL.
2. Another subgroup expressed the Ia and CD24 antigen, representing 13.8% of cases.
3. The third subgroup expressed the Ia, CD24, and CALLA antigens, comprising 75.2% of the cases.
4. The final and most differentiated group represents SIg-positive B-ALL (3.5%).

The leukemic cells from 34 children with ALL expressed T-cell markers, including receptors for sheep erythrocytes, acid phosphatase and/or cell surface differentiation antigens specific for T cells.

Three subgroups of T-ALL could be identified (Fig. 1):

1. The early-T subgroup was CD1 and CD3 negative, representing 26.5% of T-ALL.
2. The second subgroup expressed $CD1^+$ and $CD3^{+/-}$ antigen, representing 53.0% of the cases (intermediate-T).
3. T-ALL cells of the mature-T subgroup (20.5%) lost the CD1 antigen and segregated into cells that had the phenotype of mature thymocytes and T-lymphocytes ($CD3^+$).

II. Immunologic Subgroups and Disease-Free-Survival

Results are reported as conventional product limit estimates by the Kaplan-Meier method [5]. The initial response to chemotherapy of patients with T-ALL

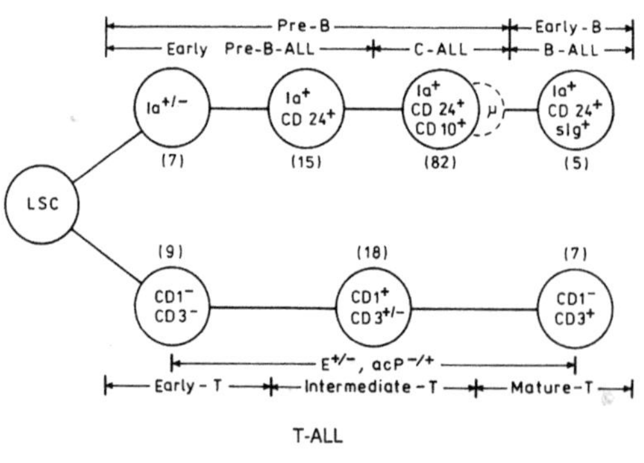

Fig. 1. Immunologic characteristics and subclassification of non-T- and T-ALL

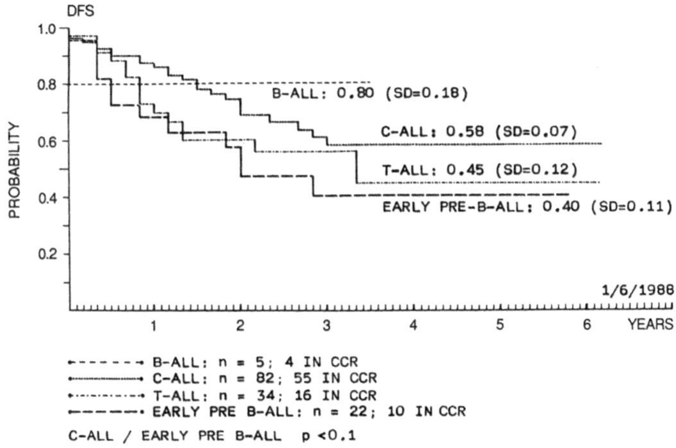

Fig. 2. Kaplan-Meier estimates for probability of disease-free survival for patients with C-ALL, T-ALL, early pre-B-ALL, and B-ALL

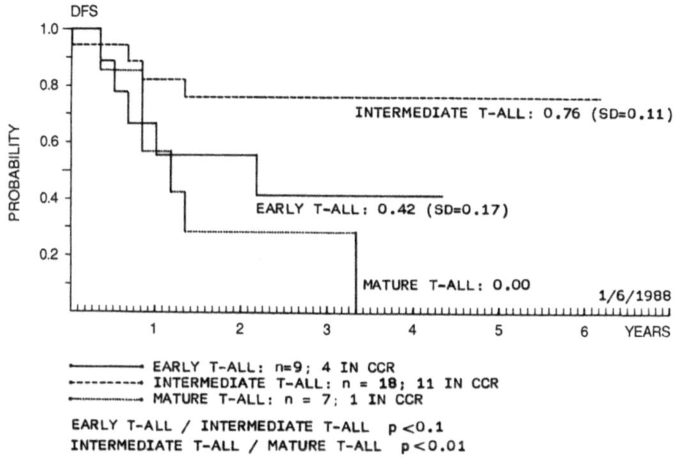

Fig. 3. Kaplan-Meier estimates for probability of disease-free survival for patients with T-ALL

did not differ from those with non-T-ALL.

The probability of disease-free survival (DFS) for 143 children was (Fig. 2):

1. C-ALL group, 0.58 ± 0.07
2. Early-pre-B-ALL group, 0.40 ± 0.11
3. T-ALL group, 0.45 ± 0.12
4. B-ALL, 0.80 ± 0.18

Although the immunologic subgroups of T-ALL were small, intermediate T-ALL patients fared significantly better than those with early T-ALL and mature T-ALL (Fig. 3).

B. Conclusions

Clinical studies have been performed using several series of monoclonal antibodies [6, 7]. In assessing the prognostic usefulness of antiboides in ALL, it is important to consider the influence of traditional unfavorable clinical factors (WBC, age, risk group, thymus tumor) on the patient's outcome:

1. Children with T-ALL had a poorer prognosis with a modified BFM therapy than those with C-ALL.
2. Patients with T-ALL were older and had higher white blood cell counts and organomegaly (Table 2).

Table 2. Comparative features of ALL subclasses

Characteristic	Early pre-B-ALL ($n=22$)	C-ALL ($n=82$)	T-ALL ($n=34$)	B-ALL ($n=5$)
Age (median in years)	$7^4/_{12}$	$4^4/_{12}$	$7^2/_{12}$	$6^5/_{12}$
WBC (median $\times 10^3$)	33.0	20.9	126.5	20.0
Risk group				
Standard	10 (45%)	45 (55%)	10 (29%)	2/4 (50%)
Medium	10 (45%)	34 (41%)	15 (44%)	1/4 (25%)
High	2 (10%)	3 (4%)	9 (27%)	1/4 (25%)
Mediastinal mass	3 (14%)	1 (1%)	17 (50%)	0 (0%)
HLA-DR positive	18 (82%)	79/80 (99%)	0 (0%)	5 (100%)
CALLA positive	0 (0%)	82 (100%)	4 (12%)	1 (20%)

3. Twenty-nine percent of children with T-ALL had standard risk characteristics versus 55% of those with C-ALL.
4. Early pre-B-ALL patients showed clinical factors like T-ALL patients (age, WBC, organomegaly) and had a poorer prognosis than C-ALL patients.
5. Children with intermediate T-ALL had a significantly better prognosis than those with early T-ALL and mature T-ALL.
6. The worse outcome of T-ALL was correlated with being older, higher white blood cell counts, and organomegaly but not with the "T" nature of leukemic cells.
7. B-ALL patients no longer have the worst prognosis of all ALL children when a B-ALL-tailored therapy is used.

References

1. Langermann HJ, Henze G, Wulf M, Riehm H (1982) Abschätzung der Tumorzellmasse bei der akuten lymphoblastischen Leukämie im Kindealter: prognostische Bedeutung und praktische Anwendung. Klin Pädiatr 194:209–213
2. Zintl F, Malke H, Plenert W (1985) Clinical experiences with a modified BFM protocol in childhood acute lymphoblastic leukemia. In: Neth R, Gallo RC, Greaves MF, Janka G (eds) Modern trends in human leukemia, vol VI. Springer, Berlin Heidelberg New York, pp 84–89
3. Riehm H, Henze G, Langermann HJ (1981) Multizentrische Therapiestudie BFM 81 zur Behandlung der akuten lymphoblastischen Leukämie im Kindes- und Jugendalter. Studienprotokoll
4. Müller-Weihrich S, Henze G, Schwarze EW, Budde M, Riehm H (1986) Childhood Non-Hodgkin's lymphoma strategies for diagnosis and therapy. In: Riehm H (ed) Malignant neoplasms in childhood and adolescence. Karger, Basel, pp 167–186 (Monogr. Paediatr, vol 18.)
5. Kaplan EL, Meier P (1970) Nonparametric estimation from incomplete observations. J Am Stat Assoc 53:457–481
6. Crist WM, Grossi CE, Pullen J, Cooper MD (1985) Immunologic markers in childhood acute lymphocytic leukemia. Semin Oncol 12:105–121
7. Bernstein I, Kersey J, Seeger R, Andrews R (1985) Immunodiagnostic and immunotherapy in childhood malignancies. Pediatr Clin North Am 32:575–599

Constellations of Genetic Abnormalities Predict Clinical Outcome in Childhood Malignancies*

A.T. Look

A. Introduction

A major thrust of tumor classification systems has been to recognize clinical and histological differences among tumors arising from the same or similar tissues, in the belief that such distinctions will provide a framework for the development of improved treatment. Clearly, this approach has been instrumental in advances toward uniformly curative treatment. However, the variable therapeutic responsiveness of many histopathologically classified tumors suggests the presence of biologically unique subgroups with prognostic importance. The challenge confronting pediatric oncologists is to devise classification schemes that will accommodate the biological diversity of childhood tumors. We need to be able to recognize, at diagnosis, those patients who will respond well to therapy despite having high-risk features by conventional criteria. Within so-called good-risk groups, we need to identify patients whose tumors have exceptional sensitivity to standard therapy, so that the severe acute toxicity and adverse late effects associated with intensive treatments can be avoided. Conversely, we need to recognize with greater reliability all exceptionally high risk patients for whom standard treatment approaches are likely to be futile.

The thesis of this article is that recognition of constellations of closely related genetic features may provide the predictive edge in treatment planning. Genetic abnormalities of tumor cells are in general acquired as somatic events and are probably tightly linked to the processes of malignant transformation and clonal evolution of aggressive growth properties within the malignant clone. Our studies indicate that certain profiles of genetic markers identify prognostically relevant subgroups within histopathologically defined childhood neoplasms. These genetic abnormalities include alterations of the ploidy (chromosome number) of malignant clones and the presence of structural cytogenetic abnormalities, such as translocations or deletions, that appear specific for subgroups of tumors. At the molecular level, gene amplification, rearrangement, mutation, and deletion indicate somatically acquired alterations that bear directly on the biological behavior of tumors. Because genetic abnormalities occur in defined patterns in human tumors, simple laboratory tests, such as the detection of abnormalities of ploidy by DNA flow cytometry, have proved especially useful for the prediction of prognosis in the clinical setting. My intention in this review is to present and discuss our findings in four different childhood neoplasms – unresectable neuroblastoma of infancy, acute lymphoblastic leukemia, Wilms' tumor, and osteogenic sarcoma of an extremity – to illustrate how discrete patterns of genetic abnormalities can be used as guides to optimal therapy.

Department of Hematology-Oncology, St. Jude Children's Research Hospital, and the Division of Hematology and Oncology, Department of Pediatrics, The University of Tennessee, Memphis, College of Medicine, Memphis, TN 38101, USA

* This work was supported in part by Cancer Center Support (CORE) grant CA-21765, by Leukemia Program Project grant CA-20180, by Solid Tumor Program Project grant CA-23099, and by the American Lebanese Syrian Associated Charities (ALSAC)

B. Disseminated Neuroblastoma of Infancy

We have shown that the ploidy of neuroblasts from infants with unresectable neuroblastoma, as detected at diagnosis by flow cytometric analysis of cellular DNA content, is closely linked to prognosis [1]. Although approximately 70% of infants with metastatic neuroblastoma are cured with five courses of cyclophosphamide and doxorubicin, the remaining 30% do not achieve a lasting response [2]. Infants whose tumors have diploid cellular DNA content are the ones who fail to respond to this therapy, while those with clonal hyperdiploid tumors have the best responses and are almost always cured of their disease [1, 3]. Our initial observations have now been extended to a total of 61 infants with unresectable disease who were treated with five courses of cyclophosphamide plus doxorubicin. In that study, conducted under the aegis of the Pediatric Oncology Group, approximately 90% of infants with hyperdiploid tumors responded to therapy, and approximately 80% have achieved durable remissions and are apparently cured. By contrast, of infants with diploid tumors, 60% failed to show an appreciable response to therapy, and only about 20% of this group are long-term disease-free survivors.

Comparison of ploidy findings with the presence or absence of N-*myc* gene amplification (determined by Dr. Garrett Brodeur) and structural karyotypic alterations (determined by Dr. Edwin C. Douglass) indicated that in neuroblastoma of infancy, diploid tumors are the ones with high levels of N-*myc* amplification [4–7] and abnormalities of chromosome structure, such as deletions of the short arm of chromosome 1 [8–11]. By contrast, hyperdiploid tumors tend to have simple additions of chromosomes with single copy numbers of the N-*myc* gene and no abnormalities of chromosome structure. The value of flow cytometry for classification of neuroblastoma in infants lies in its ability to serve as a marker for tumors with clinically relevant constellations of genetic abnormalities (Table 1). Whether any of the changes identified so far contribute directly to the observed patterns of drug sensitivity is unclear. It may be that the constellations we have recognized include other, as yet unidentified, genetic alterations that in fact mediate cellular sensitivity to chemotherapy. Whatever the explanation, it is now possible to use genetic measurements of neuroblastoma of infants to reliably predict response to therapy.

The above findings provided the impetus for a prospective clinical trial (conducted by Dr. Ann Hayes within the Pediatric Oncology Group) in which therapy for infants with neuroblastoma is being modified according to the patient's pretreatment DNA index. At the time of diagnosis, tumor cells are submitted for evaluation of DNA content, and patients are treated with one course of cyclophosphamide plus doxorubicin. The DNA index, defined as the ratio of the modal DNA content of G_0/G_1-phase tumor versus normal cells, is used to determine whether the patient will be given four additional courses of the drug combination (hyperdiploid tumors) or whether the therapy will be switched to platinum plus VM-26 (diploid tumors). The alternative treatment with platinum and VM-26 is both more effective and more toxic than therapy with cyclophosphamide and doxorubicin. In previous studies, the combination was shown to induce temporary remissions in approximately half of all infants who failed therapy with first-line agents [12]. If successful, this strategy should yield longer remissions in infants with high-risk genetic features while sparing others the added toxicity produced by cisplatin and VM-26.

We think it interesting that the well-defined genetic constellations in tumors of infants are not found in neuroblastomas in children over 1 year of age. N-*myc* gene amplification and structural abnormalities of chromosome 1 occur with equal frequency in tumors with diploid

Table 1. Constellations of clinically relevant genetic abnormalities in three childhood malignancies

Tumor	Genetic markers	
	Favorable prognosis	Unfavorable prognosis
Unresectable neuroblastoma in infants	Hyperdiploidy (DNA index >1) Absence of chromosomal structural changes or DMs/HSRs Single copy of N-*myc* gene	Near diploidy (DNA index = 1) Chromosome 1p deletions, DMS, or HSRs N-*myc* gene amplification
Acute lymphoblastic leukemia	Hyperdiploidy ≥53 chromosomes (DNA index ≥1.16) Simple additions of chromosomes with few translocations Trisomes preferentially involving chromosomes 4, 6, 10, 14, 17, 18, 20, 21 and X	Near diploidy (DNA index = 1) Specific chromosomal translocations [t(8;14), t(9;22), t(4;11)] *bcr-abl* fusion genes resulting from t(9;22)
Wilms' tumor	Near diploidy (DNA index = 1.0) Lack of complex chromosomal translocations	Near tetraploidy (DNA index = 2.0) Multiple complex chromosomal translocations

or hyperdiploid DNA content. Moreover, a group of older children with disseminated neuroblastoma and a potentially favorable outcome cannot be identified with any of these genetic markers, because the disease is almost uniformly fatal despite aggressive combination chemotherapy. New approaches to therapy are urgently needed for essentially all children more than a year of age who have disseminated disease at diagnosis.

C. Childhood Acute Lymphoblastic Leukemia

Improvements in the treatment of children with acute lymphoblastic leukemia (ALL) have galvanized efforts to identify features of leukemic cells that will reliably predict the clinical course of this disease in individual patients. It is important to identify patients who have a high risk of treatment failure and who would benefit from very intensified programs of therapy [13]. Concern over the immediate and long-term adverse sequelae of intensive chemotherapy, however, has stimulated attempts to identify children who respond well to less intensive treatment, so that they can be spared unnecessary toxicity.

It has been recognized for years that children with greater than 50 chromosomes in their leukemic cell karyotype have a more favorable prognosis than those with lower leukemia cell ploidies [14–19]. Careful analysis of the karyotypes of leukemic blasts with greater than 50 chromosomes has defined a constellation of cytogenetic features, such as trisomies of specific chromosomes and a relatively low frequency of translocations [19] (Table 1). About 20% of children have ploidy values in this range, and they comprise a subset of a larger common ALL antigen-positive group characterized by pre-B or B-cell precursor phenotypes [20]. We have shown that flow cytometric measurement of the DNA content of leukemic blast cells, expressed as a DNA index, is the method of choice for identifying patients in the hyperdiploid group who have a favorable prognosis. In a recent analysis of flow cytometric results for patients treated in St. Jude Total Therapy Study X, we found that the most favorable prognostic feature was a leukemic cell DNA index greater than or equal to 1.16, which corresponds to greater than or equal to 53 chromosomes per leukemic cell [21, 22]. Interestingly, the best results were obtained in the chemotherapy arm in which high-dose intravenous methotrexate plus intrathecal methotrexate was used for both intensification and central nervous system prophylaxis. The remainder of the treatment program included induction therapy with prednisone, L-asparaginase, and vincristine, and continuation therapy with mercaptopurine and low-dose methotrexate. With this treatment, an estimated 89% of children whose blasts had a DNA index greater than or equal to 1.16 will be in continuous remission for 3 years and 82% for 5 years. The overall treatment program is remarkable because of its low short-term toxicity and its reduced potential for producing adverse late effects, compared with other protocols that employ cranial irradiation and mutagenic agents such as cyclophosphamide and anthracyclines.

A group of patients with a much higher risk of treatment failure can also be defined by cytogenetic analysis (Table 1). These patients tend to have near-diploid chromosome complements in their leukemic blasts and an increased frequency of chromosomal translocations associated with a high risk of treatment failure, such as the t(8;14), t(9;22), and t(4;11) [19, 23]. Molecular genetic correlates of these specific translocations will undoubtedly be of increasing importance in the understanding of the pathogenesis of ALL and in the design of optimal therapy for this disease. For example, the t(9;22) (q34;q11) found in chronic myelogenous leukemia and some cases of ALL results in translocations of the c-*abl* gene from chromosome 9 to chromosome 22 [24–26]. The breakpoint on chromosome 22 in CML disrupts a gene called *bcr* (breakpoint cluster region) within a well-defined 5.8-kb region of genomic DNA [27–29]. Translocation results in fusion of the *bcr* and c-*abl* genes with production of a characteristic 8.5-kb mRNA [30–35] and a 210-kb hybrid protein that is activated as a tyrosine-specific kinase [36–39]. Although the 9;22 translocation that gives rise to the Philadelphia chromosome is cytogenetically identical in ALL to that found in CML, molecular studies have now revealed a potentially important difference. In ALL cells that harbor the Philadelphia chromosome, a 6.5- to 7.0-kb fusion transcript and a 185- to 190-kd hybrid protein are produced that are distinct from those of both the normal c-*abl* gene and the rearranged *bcr*/c-*abl* fusion gene found in CML [40–42]. The breakpoints of chromosome 22 in ALL cases with the 9;22 are not within the 5.8-kb region of *bcr* that contains the breakpoints in CML, but lie further upstream within the *bcr* gene. The ALL fusion protein includes aminoterminal determinants of the *bcr* gene but lacks internal *bcr* determinants that are found in the CML fusion protein near the *bcr-abl* junction [43–45]. It is now possible with molecular probes to detect rearrangements within the *bcr* fragment that is characteristically affected in CML, and it appears likely that probes will soon be defined that can identify the ALL abnormality. The molecular characterization of chromosomal translocations found in ALL promises to improve the understanding of the molecular basis of this disease, and to provide new means of identifying genetic rearrangements that have very specific implications for therapy.

D. Wilms' Tumor

In Wilms' tumor, the majority of patients have tumors with near diploid DNA content and karyotypes that lack major chromosomal rearrangements or translocations [46]. These patients also have favorable histological features, and a very high probability of cure with current treatment modalities. By contrast, a constellation of genetic features including cellular DNA content in the near-tetraploid range and multiple complex chromosomal translocations defines a group with anaplastic histological features and a very unfavorable treatment response (Table 1). As is the case in neuroblastoma and acute lymphoblastic leukemia, cellular DNA content measurements by flow cytometry appear to be the most practical means of identifying high-risk tumors in the clinical setting. It is important to note, however, that the correlation between very hyperdiploid tumors and a poor response to treatment is exactly opposite that observed in neuroblastoma in infancy and acute lymphoblastic leukemia. This emphasizes the fact that tumor-cell ploidy is probably not a primary determinant of drug responsiveness. Rather, it appears to be a marker of subgroups of tumors with unique cytogenetic and molecular genetic features. Thus, the correlation observed in Wilms' tumor between very high ploidy levels and complex chromosomal translocations may indicate that in this disease, high ploidy levels are associated with "genetic instability," which results in an increased rate of mutations leading to drug resistance. By contrast, in neuroblastoma of infants and acute lymphoblastic leukemia, hyperdiploid tumor cells tend to have the addition of whole chromosomes without structural rearrangements and thus may be less likely to have key genetic lesions, despite their higher ploidy levels.

E. Osteosarcoma of an Extremity

We have recently characterized the cellular DNA content of osteosarcoma cells obtained at diagnosis from patients with extremity lesions who lacked clinically evident metastases [47]. When treated with intensive combination chemotherapy as part of a large multi-institutional trial, patients with near-diploid tumor stem lines had a markedly improved disease-free survival by comparison with those whose tumors had only hyperdiploid stemlines. Thus, in this disease as in Wilms' tumor, hyperdiploidy correlated with an adverse prognosis. One important difference was observed, however. For osteosarcoma, the favorable influence of near diploidy was apparent even if additional hyperdiploid lines were present. Thus, in osteosarcoma there may be an interaction between stem lines of different ploidies that we have not observed in other tumors, possibly indicating that in osteosarcoma near-diploid stemlines may provide growth factors that are required by the hyperdiploid stem lines that accompany them.

F. Concluding Remarks

Our results indicate that for several types of childhood malignancies, it is possible to identify constellations of genetic abnormalities that appear important in pathogenesis and provide remarkable correlates with drug sensitivity. A goal of our studies has been to use such abnormalities to improve the design of clinical trials, which means that detection must be practical at the time of diagnosis for every patient's tumor in a clinical setting. One of the advantages of using flow cytometry to measure cellular DNA content is that reliable results can be obtained with samples shipped from participating institutions to a reference laboratory. Fortunately, it appears that simple measurements of tumor-cell ploidy can be used to identify subsets of patients with defined patterns of genetic abnormalities, including those that influence the responsiveness of tumor cells to chemotherapy. Careful clinical studies are needed to define important relationships between genetic abnormalities and therapeutic responsiveness for each tumor type. A prediction of these studies is that as molecular genetic abnormalities are more completely characterized in human tumors, the prognostic importance of these alterations will be increasingly useful for the design of optimal treatment programs.

Acknowledgements. I thank Dr. Edwin Douglass, Dr. Dorothy Williams, and Dr. Susana Raimondi for cytogenetic analysis, Dr. Garrett Brodeur for providing unpublished data on N-*myc* amplification, and John Gilbert for editing the manuscript.

References

1. Look AT, Hayes FA, Nitschke R, McWilliams NB, Green AA (1984) Cellular DNA content as a predictor of response to chemotherapy in infants with unresectable neuroblastoma. N Engl J Med 311:231–235
2. Green AA, Hayes FA, Hustu HO (1981) Sequential cyclophosphamide and doxorubicin for induction of complete remission in children with disseminated neuroblastoma. Cancer 48:2310–2317
3. Gansler T, Chatten J, Varello M, Bunin GR, Atkinson B (1986) Flow cytometric DNA analysis of neuroblastoma. Correlation with histology and clinical outcome. Cancer 58:2453–2458
4. Kohl NE, Kanda N, Schreck RR, Bruns G, Latt SA, Gilbert F, Alt FW (1983) Transposition and amplification of oncogene-related sequences in human neuroblastomas. Cell 35:359–367
5. Schwab M, Alitalo K, Klempnauer KH, Varmus HE, Bishop JM, Gilbert F, Brodeur G, Goldstein M, Trent J (1983) Amplified DNA with limited homology to *myc* cellular oncogene is shared by human neuroblastoma cell lines and a neuroblastoma tumor. Nature 305:245–248
6. Brodeur GM, Seeger RC, Schwab M, Varmus HE, Bishop JM (1984) Amplification of N-*myc* in untreated human neuroblastomas correlates with advanced disease stage. Science 224:1121–1124
7. Seeger RC, Brodeur GM, Sather H, Dalton A, Siegel SE, Wong KY, Hammond D (1985) Association of multiple copies of the N-*myc* oncogene with rapid progression of neuroblastoma. N Engl J Med 313:1111–1116
8. Brodeur GM, Green AA, Hayes FA, Williams KJ, Williams DL, Tsiatis AA (1981) Cytogenetic features of human neuroblastomas and cell lines. Cancer Res 41:4678–4686
9. Gilbert F, Feder M, Balaban G et al. (1984) Human neuroblastomas and abnormalities of chromosomes 1 and 17. Cancer Res 44:5444–5449
10. Kaneko Y, Kanda N, Maseki N et al. (1987) Different karyotypic patterns in early and advanced-stage neuroblastoma. Cancer Res 47:311–318
11. Hayashi Y, Inaba T, Hanada R, Yamamoto K (1988) Chromosome findings and prognosis in 15 patients with neuroblastoma found by VMA mass screening. J Pediatr 112:567–571
12. Hayes FA, Green AA, Casper J, Cornet J, Evans WE (1981) Clinical evaluation of sequentially scheduled cisplatin and VM26 in neuroblastoma: response and toxicity. Cancer 48:1715–1718
13. Rivera GK, Mauer AM (1987) Controversies in the management of childhood acute lymphoblastic leukemia: treatment intensification, CNS leukemia, and prognostic factors. Semin Hematol 24:12
14. Secker-Walker LM, Lawler SD, Hardisty RM (1978) Prognostic implications of chromosomal findings in acute lymphoblastic leukemia at diagnosis. Br Med J 2:1529
15. (1983) Chromosomal abnormalities and their clinical significance in acute lymphoblastic leukemia: Third International Workshop on Chromosomes in Leukemia. Cancer Res 43:868–873
16. Kaneko Y, Rowley JD, Variakojis D et al. (1982) Correlation of karyotype with clinical features in acute lymphoblastic leukemia. Cancer Res 42:2918
17. Secker-Walker LM, Swansbury GJ, Hardisty RM et al. (1982) Cytogenetics of acute lymphoblastic leukemia in children as a factor in the prediction of long-term survival. Br J Haematol 52:389
18. Williams DL, Tsiatis A, Brodeur GM et al. (1982) Prognostic importance of chromosome number in 136 untreated children with acute lymphoblastic leukemia. Blood 60:864
19. Williams DL, Harber J, Murphy SB et al. (1986) Chromosomal translocations play a unique role in influencing prognosis in childhood acute lymphoblastic leukemia. Blood 68:205
20. Look AT, Melvin SL, Williams DL et al. (1982) Aneuploidy and percentage of S-phase cells determined by flow cytometry correlate with cell phenotype in childhood acute leukemia. Blood 60:959
21. Look AT, Roberson PK, Williams DL et al. (1985) Prognostic importance of blast cell DNA content in childhood acute lymphoblastic leukemia. Blood 65:1079
22. Look AT, Roberson PK, Murphy SB (1987) The prognostic value of cellular DNA content in acute lymphoblastic leukemia of childhood (letter). N Engl J Med 317:1666
23. Bloomfield CD, Goldman AI, Berger AR et al. (1986) Chromosomal abnormalities

identify high-risk and low-risk patients with acute lymphoblastic leukemia. Blood 67:415
24. de Klein A, van Kessel AG, Grosveld G, Bartram CR, Hagemeijer A, Bootsma D, Spurr NK, Heisterkamp N, Groffen J, Stephenson JR (1982) A cellular oncogene is translocated to the Philadelphia chromosome in chronic myelocytic leukemia. Nature 30:765–767
25. Heisterkamp N, Stephenson JR, Groffen J, Hansen PF, de Klein A, Bartram CR, Grosveld G (1983) Localization of the c-*abl* oncogene adjacent to a translocation breakpoint in chronic myelocytic leukemia. Nature 306:239–242
26. Kozbor D, Giallongo A, Sierzega ME, Konopka JB, Witte ON, Showe LC, Croce CM (1986) Expression of a translocated c-*abl* gene in hybrids of mouse fibroblasts and chronic myelogenous leukemia. Nature 319:331–333
27. Groffen J, Stephenson JR, Heisterkamp N, de Klein A, Bartram Cr, Grosveld G (1984) Philadelphia chromosome breakpoints are clustered within a limited region, *bcr,* on chromosome 22. Cell 36: 93–99
28. Heisterkamp N, Stam K, Groffen J, de Klein A, Grosveld G (1985) Structural organization of the *bcr* gene and its role in the Ph[1] translocation. Nature 315:758–761
29. Stam K, Keisterkamp N, Reynolds FH jr, Groffen J (1987) Evidence that the *Ph1* gene encodes a 160000-dalton phosphoprotein with associated kinase activity. Mol Cell Biol 7:1955–1960
30. Gale RP, Canaani E (1984) An 8-kilobase *abl* RNA transcript in chronic myelogenous leukemia. Proc Natl Acad Sci USA 81:5648–5652
31. Collins SJ, Kubonishi I, Miyoshi I, Groudine MT (1984) Altered transcription of the c-*abl* oncogene in K562 and other chronic myelogenous leukemia cells. Science 225:72–74
32. Stam RS, Heisterkamp N, Grosveld G, de Klein A, Verma RS, Coleman M, Dosik H, Groffen J (1985) Evidence of a new chimeric *bcr*/c-*abl* mRNA in patients with chronic myelocytic leukemia and the Philadelphia chromosome. N Engl J Med 313:1429–1433
33. Romero P, Blick M, Talpaz M, Murphy E, Hester J, Gutterman J (1986) C-*sis* and c-*abl* expression in chronic myelogenous leukemia and other hematologic malignancies. Blood 67:839–841
34. Canaani E, Steiner-Saltz D, Aghai E, Gale RP, Berrebi A, Januszewicz E (1984) Altered transcription of an oncogene in chronic myeloid leukemia. Lancet 1:593–595
35. Shtivelman E, Lifshitz B, Gale RP, Canaani E (1985) Fused transcript of *abl* and *bcr* genes in chronic myelogenous leukemia. Nature 315:550–554
36. Kloetzer W, Kurzrock R, Smith L, Talpaz M, Spiller M, Gutterman J, Arlinghaus R (1985) The human cellular *abl* gene product in the chronic myelogenous leukemia cell line K562 has an associated tyrosine protein kinase activity. Virology 140: 230–238
37. Konopka JB, Watanabe SM, Witte ON (1984) An alteration of the human c-*abl* protein in K562 leukemia cells unmasks associated tyrosine kinase activity. Cell 37:1935–1942
38. Konopka JB, Watanabe SM, Singer JW, Collins SJ, Witte ON (1985) Cell lines and clinical isolates derived from Ph[1]-positive chronic myelogenous leukemia patients express c-*abl* protein with a common structural alteration. Proc Natl Acad Sci USA 82:1810–1814
39. Naldini L, Stacchini A, Cirillo DM, Aglietta M, Gavosto F, Comoglio PM (1986) Phosphotyrosine antibodies identify the p210 c-*abl* tyrosine kinase and proteins phosphorylated on tyrosine in human chronic myelogenous leukemia cells. Mol Cell Biol 6:1803–1811
40. Chan LC, Karhi KK; Rayter SI, Heisterkamp N, Eridani S, Powles R, Lawler SD, Groffen J, Foulkes JG, Greaves MF (1987) A novel *abl* protein expressed in Philadelphia chromosome-positive acute lymphoblastic leukemia. Nature 325:635–637
41. Clark SS, McLaughlin J, Crist WM, Champlin R, Witte ON (1987) Unique forms of the *abl* tyrosine kinase distinguish Ph[1]-positive CML from ALL. Science 235:85–88
42. Kurzrock R, Shtalrid M, Romero P, Kloetzer WS, Talpas M, Trujillo JM, Blick M, Beran M, Gutterman JU (1987) A novel c-*abl* protein product in Philadelphia-positive acute lymphoblastic leukemia. Nature 325:631–635
43. Hermans A, Heisterkamp N, von Linden M, van Baal S, Meijer D, van der Plas D,

Wiedemann LM, Groffen J, Bootsma D, Grosveld G (1987) Unique fusion of *bcr* and c-*abl* genes in Philadelphia chromosome-positive acute lymphoblastic leukemia. Cell 51:33–40
44. Walker LC, Ganesan TS, Dhut S, Gibbons B, Lister TA, Rothbard J, Young BD (1987) Novel chimaeric protein expressed in Philadelphia-positive acute lymphoblastic leukemia. Nature 329:851–853
45. Fainstein E, Marcelle G, Rosner A, Canaani E, Gale RP, Dreazen O, Smith SD, Croce CM (1987) A new fused transcript in Philadelphia chromosome-positive acute lymphocytic leukemia. Nature 330:386–388
46. Douglass EC, Look AT, Webber B, Parham D, Willimas JA, Green AA, Roberson PK (1986) Hyperdiploidy and chromosomal rearrangements define the anaplastic variant of Wilms' tumor. J Clin Oncol 4:975–981
47. Look AT, Douglass EC, Meyer WH (1988) Clinical importance of near-diploid tumor stem lines in patients with osteosarcoma of an extremity. N Engl J Med 318:1567–1572

Leukemia Cytogenetics in Children: Results of the German Therapy Studies

J. Harbott, J. Ritterbach, and F. Lampert

From January 1984 to June 1988, 1179 bone marrow samples from children with acute lymphoblastic (ALL) and non-lymphoblastic leukemia (ANLL) were received for chromosomal analysis. More than 90% were sent by mail from more than 60 hospitals all over the Federal Republic of Germany, all of them registered in one of the multicenter therapy studies CoALL or BFM.

After arrival, the bone marrow was washed twice in RPMI 1640 and afterwards cultured in RPMI + 20% FCS for 24 h, including a synchronization of the cell cycle by methotrexate for 17 h. After release of the MTX block, the cells were cultured again for 4.5 h and chromosome preparation was performed with KCl and methanol:acetic acid 3:1. The cell suspension was then dropped onto a cold, wet slide and stained by G-banding after drying for 2–6 days.

Chromosomal analysis was carried out in 219 samples of patients with ANLL, 127 at diagnosis and 20 at relapse, and was successful in 147 cases (67.1%). Bone marrow of children with ALL was derived from 960 patients and successfully analyzed in 554 cases (57.7%).

A normal karyotype was found in the bone marrow of only 30% (n=31) of children with ANLL, whereas the majority showed structural (n=38), numerical (n=16), or a combination of both types (n=19).

In contrast to ALL, numerical changes were very rare, and chromosome numbers of more than 47 appeared in only a small percentage. Most frequently, the loss or gain of a single chromosome was found, and chromosomes 8 and 7 were very often involved in a trisomy or monosomy, respectively. Both of them are typical aberrations of ANLL.

Further consistent chromosomal abnormalities were found in this group of leukemia patients, most of them translocations (Fig. 1). T(8;21) is a very common aberration of ANLL and appears mainly in patients with FAB-type M_2, but it was also found in M_1 and M_4. All of the patients with t(8;21) in this study showed a FAB-type M_2. Whereas each of the other consistent aberrations was found to be very specific for only one FAB-group – e.g., inv(16) in M_4 and t(15;17) in M_3 – abnormalities involving the long arm of chromosome 11 (11q23) were found in subtypes M_1, M_2, M_4, and M_5 (Fig. 2a). This band is a very interesting breakpoint in the cytogenetics of leukemia, for it is also detected in the bone marrow of patients with ALL, and was involved in translocations and deletions, e.g., t(9;11), del(11q). One patient (M_5) had a t(4;11) (q21;q23), which is very rare in ANLL and is found mainly in ppB-ALL or hybrid leukemia.

In 41% (n=167) of patients with ALL, whose bone marrow was analyzed at diagnosis, a normal karyotype was found; 24% (n=93) of the patients showed only structural abnormalities (pseudodiploid), whereas in 31% (n=118) a gain (hyperdiploid) and in 3% (n=12) a loss (hypodiploid) of chromosomes was detected. Hyperdiploidy with chromosome numbers of 47–49 was found in only 9% (n=34) of the ALL patients, whereas

University Children's Hospital, Feulgenstr. 12, D-6300 Giessen

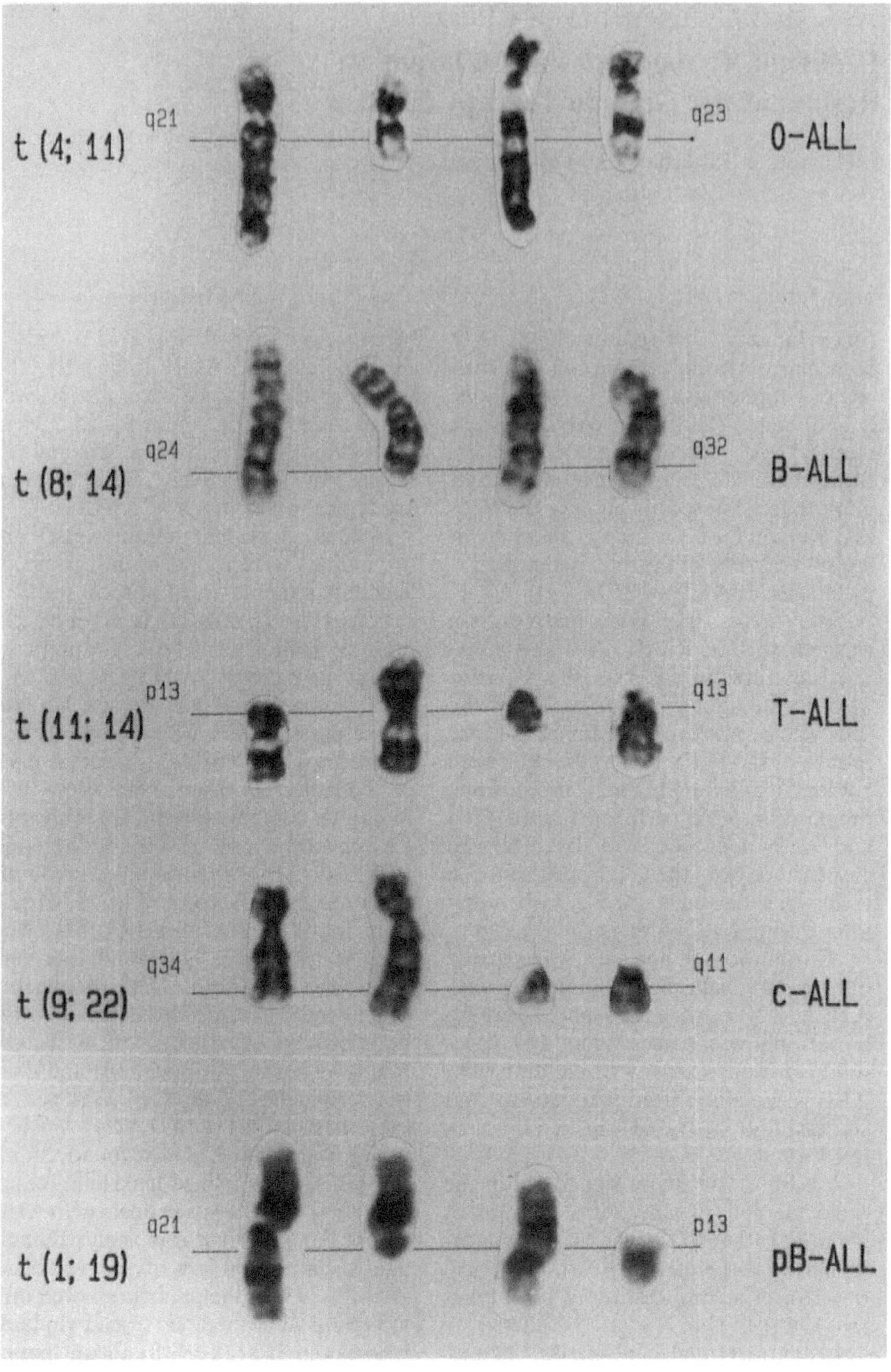

Fig. 1. Consistent aberrations of various immunophenotypes of ALL

Frequency of Consistent Aberrations in ALL

	B–cell ALL	T–cell ALL	common ALL	ppB–ALL	mixed
t(1;19)			8		
t(8;14)	14				
t(4;11) *				8	5
t(11;19)					1
t(9;22)			10		
t(11;14)		5			
t(1;14)		2			
t(10;14)		2			
t(12;14)		1			
inv(14q)		1		*) 1 patient with M5	

Frequency of Consistent Aberrations in ANLL

	M1	M2	M3	M4	M5
t(8;21)		15			
t(15;17)			4		
inv(16)				1	
11q23	1	2		2	9
+8	2	2		1	5

Fig. 2a, b. Frequency of consistent chromosomal abnormalities in ALL (**a**) and ANLL (**b**)

22% ($n=84$) had 50 and more chromosomes. The latter seems to be very typical, for it was found only in the bone marrow of patients with common ALL, a subtype in which pseudodiploidy is very rare.

In this immunophenotypical subgroup of ALL, however, two structural aberrations also appeared which were confined to this group: the Philadelphia chromosome, t(9;22), and the translocation (1;19) (Fig. 3). The t(8;14) were detected only in B-cell ALL, whereas the two variant forms of this aberration, namely t(2;8) and t(8;22), were never encountered. The t(4;11) is typical for very early stages of differentiation in hematopoiesis and was found in eight patients with ppB-ALL, five patients with a mixed leukemia, and, as mentioned above, in one child with ANLL FAB-type M_5. In T-cell ALL different aberrations were

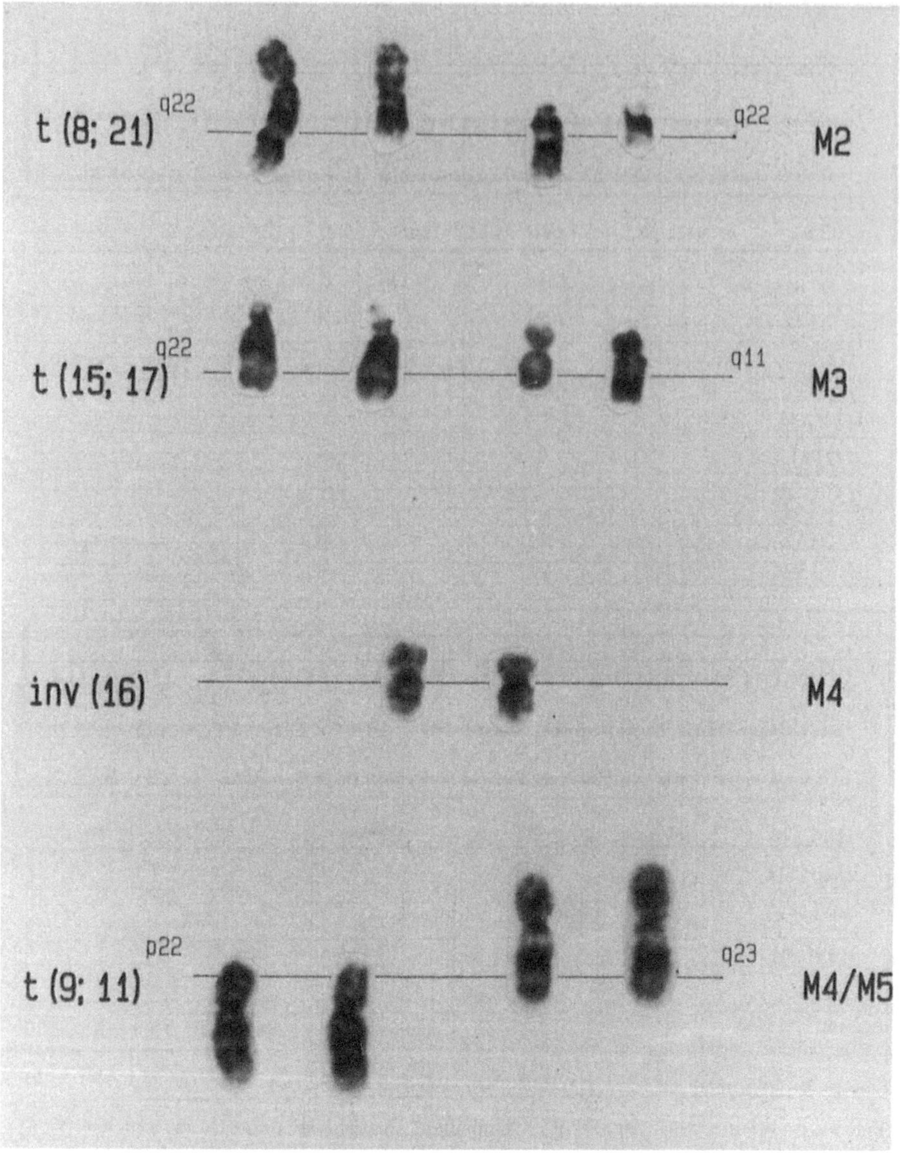

Fig. 3. Consistent aberrations of various FAB-groups of ANLL

found, all of them involving a single band of the long arm of chromosome 14 (14q11); t(11;14), t(1;14), t(10;14), t(12;14), and inv(14) (Fig. 2b). This band is the region where the TCR α and σ are located.

In order to propose prognostic meaning, we compared only patients who were treated by the same uniform therapy, as treatment is the most important prognostic factor. When the clinical outcome of patients with common ALL and different karyotypes was compared, a relapse became visible in five of 12 children with pseudodiploidy. In contrast, relapse was less often seen when a normal or hyperdiploid karyotype was diagnosed (Fig. 4). With regard to the meaning of

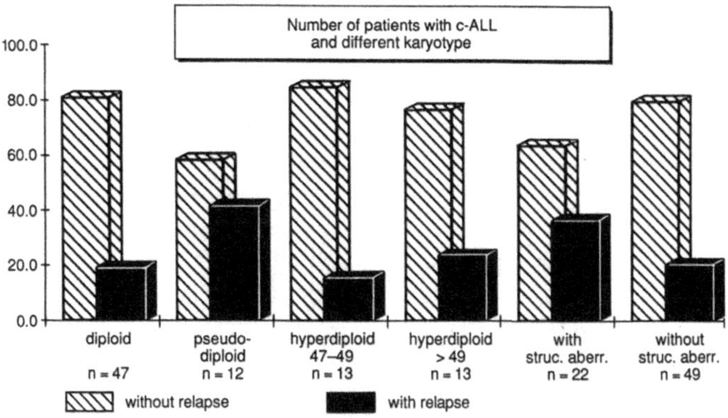

Fig. 4. Numbers of patients with and without relapse of common ALL. All of them were treated with the therapy protocol BFM-83, but different karyotypes were found in their leukemic cells at diagnosis of ALL

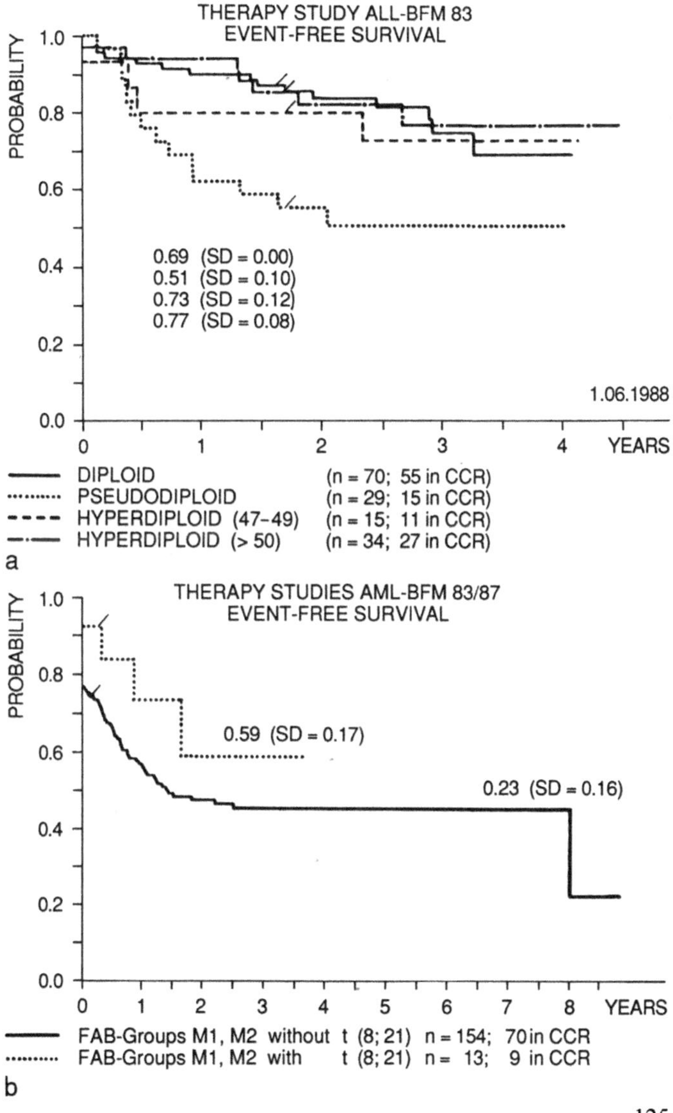

Fig. 5a, b. Life-table analysis according to Kaplan-Meier. (CCR = Continuous Complete Remission) **a** Patients with pseudodiploidy in their leukemic cells have a significantly poorer prognosis than those with other karyotypes. **b** Prognosis of patients with ANLL M_1 and M_2. Children with t(8;21) seem to have a better prognosis than those without this aberration structural abnormalities, the patient group with only numerical aberrations or normal karyotypes had only half the percentage of relapses compared with the patients with translocations, deletions, or other abnormalities (Fig. 4).

Life-table analysis by Kaplan-Meier also showed the poor prognosis of children with pseudodiploidy (Fig. 5a). By this analysis, however, it can also be demonstrated that patients with structural aberrations do not always have a poorer prognosis. The survival probability for children with translocation (8;21) in the leukemia karyotype seems to be significantly better than for the other patients with ANLL FAB-type M_2 (Fig. 5b).

By cytogenetic analysis of 554 children with ALL and 147 with ANLL it could be shown that chromosomal aberrations of acute leukemias are closely connected with single subgroups of leukemia and can be used for diagnostic classification. The comparison of relapse data showed a poorer prognosis for ALL patients with a pseudodiploid leukemia karyotype at diagnosis.

Cytostatic Influence of Thioproline on Peripheral Lymphocytes of Healthy Persons and Non-Hodgkin's Lymphoma Patients

T. Rozmysłowicz

A. Introduction

Thiazolidine-4-carboxylic acid (Thioproline, Norgamem) is an agent of possible antitumorigenic effect [1, 2, 7]. This drug seems to induce a "reverse transformation" of pathological cells by restoring contact inhibition in cell cultures [2–6]. No evidence of a cytotoxic effect of Thioproline has so far been observed [2, 4, 9]. In our previous work we demonstrated that Thioproline exerts strong inhibition of DNA synthesis in normal lymphocytes [8]. In this study we present experiments performed in normal as well as in non-Hodgkin's lymphoma (NHL) lymphocytes treated by Thioproline in vitro.

B. Materials and Methods

In vitro experiments on normal and NHL peripheral blood lymphocytes (PBLs) were carried out and the influence of different Thioproline concentrations was investigated by:

1. [^{3}H]Thymidine uptake into normal PBLs stimulated in culture with phytohemagglutinin (PHA)
2. [^{3}H]Thymidine uptake into pathological lymphocytes without adding PHA according to spontaneous incorporation of indicator
3. Suppressor activity of both groups of lymphocytes measured in the presence of Thioproline using ConA and [^{3}H]thymidine
4. Intracellular concentration of cAMP and cGMP
5. Trypan blue test evaluation of cell viability

We used the lymphocytes of 20 healthy persons (first-time blood donors) ranging from 18 to 40 years in age and lymphocytes obtained from 20 NHL patients (11 with lymphocytic lymphoma, 6 with centroblastic-centrocytic and immunoplasmocytic lymphoma, and 3 with acute lymphoblastic leukemia). Peripheral blood lymphocytes from healthy donors and PBLs from NHL patients were isolated on Lymphoprep (Nyegaard, Norway). A quantity of 2×10^5 PBLs were cultured for 72 h (37° C, 5% CO_2) with PHA (only normal PBLs) (Wellcome) together with Thioproline in four different concentrations calculated on the basis of the therapeutic doses according to blood volume (5, 20, 40, 80 μg) [2, 9]. The drug was diluted to 0.02 ml and incubated with examined lymphocytes (density of lymphocytes in suspension 0.2×10^6/ml) in a minimum essential medium (MEM) (0.2 ml-containing glutamine and antibiotics). Investigated lymphocytes were also incubated simultaneously with PHA 24 h before or after Thioproline intake. [^{3}H]Thymidine uptake was evaluated as counts per minute of isotope extracted from the lymphocytes and expressed as a percentage of the control value (without drug). Lymphocyte suppressor activity was defined as [^{3}H]thymidine uptake ratio of ConA lymphocyte culture stimulation and ConA stimulation in culture after 24 h. Intracellular level of cAMP and cGMP was measured using cAMP and cGMP RIA Kits (Amersham, UK).

Department of Internal Medicine, Institute of Hematology, Warsaw, Poland

C. Results

Of the normal and pathological PBLs used in the experiment, Thioproline caused dose-dependent inhibition of [^{3}H]thymidine uptake (except at the lower concentration of 5 μg), where increased uptake of [^{3}H]thymidine was observed (Fig. 1). Similar results were obtained by normal PBL incubation pretreated in culture with Thioproline 24 h before the addition of PHA and in a second group pretreated with PHA 24 h before the addition of Thioproline (Fig. 2). A viability test of PBLs after 24 and 48 h incubation with Thioproline showed a great decrease in viability of pathological lymphocytes. No significant decrease in normal lymphocyte viability was observed (Fig. 3). At a higher concentration, 80 μg, Thioproline caused a reduction in the intracellular level of cAMP and cGMP within normal PBLs as well as in NHL PBLs (Fig. 4). Thioproline has no effect on the suppression activity of normal and NHL PBLs (Fig. 5).

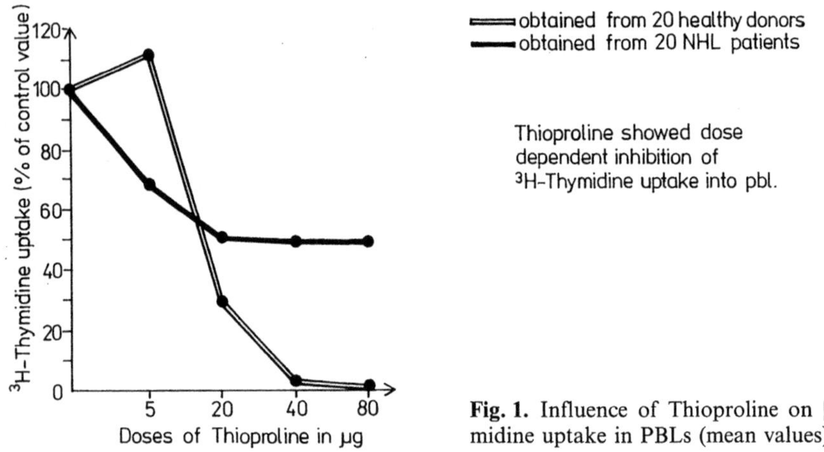

Fig. 1. Influence of Thioproline on [^{3}H]thymidine uptake in PBLs (mean values)

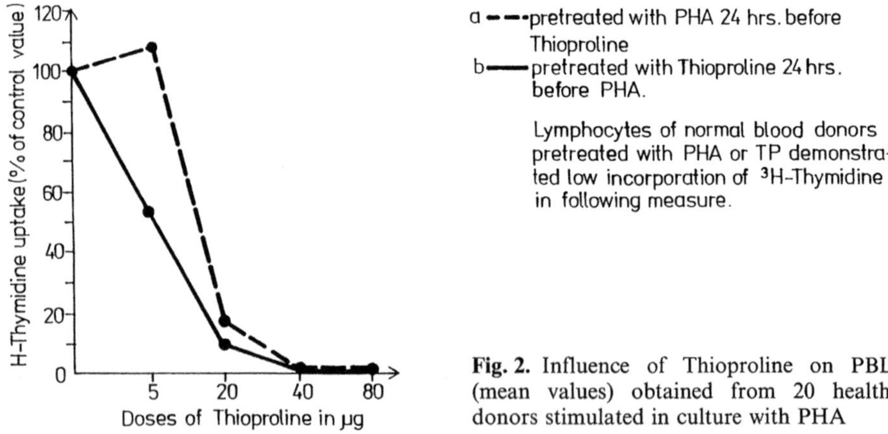

Fig. 2. Influence of Thioproline on PBLs (mean values) obtained from 20 healthy donors stimulated in culture with PHA

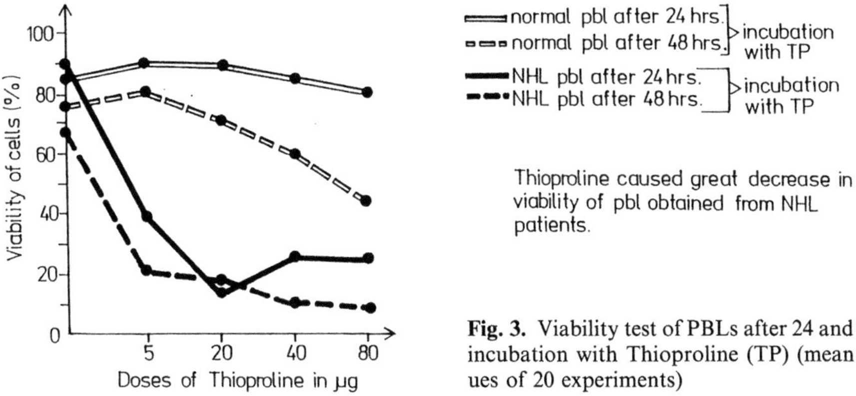

Fig. 3. Viability test of PBLs after 24 and 48 h incubation with Thioproline (TP) (mean values of 20 experiments)

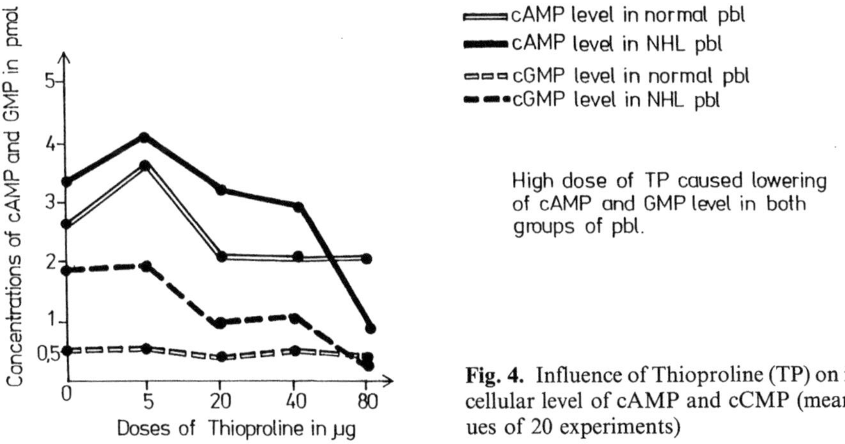

Fig. 4. Influence of Thioproline (TP) on intracellular level of cAMP and cCMP (mean values of 20 experiments)

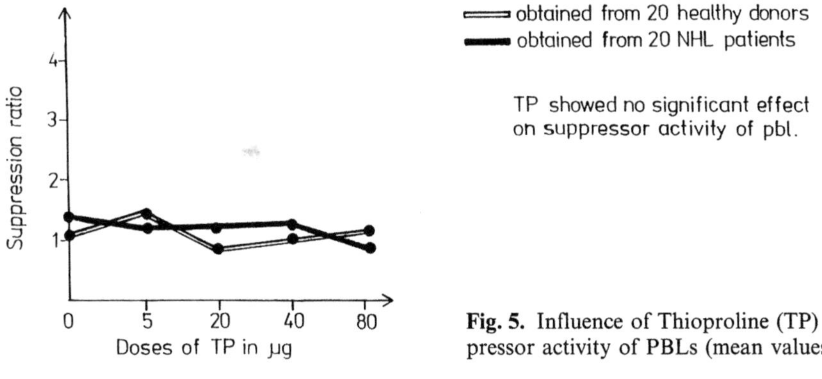

Fig. 5. Influence of Thioproline (TP) on suppressor activity of PBLs (mean values)

D. Conclusions

Thioproline caused in vitro dose-dependent inhibition of [^{3}H]thymidine uptake into normal and NHL PBLs. At the highest concentration Thioproline decreased the intracellular level of cAMP and cGMP in examined PBLs. Thioproline has no effect on the suppression activity of normal and NHL PBLs. Inhibition of [^{3}H]thymidine uptake into NHL PBLs seems to be caused by a cytotoxic mechanism.

References

1. Alberto P (1981) Thioproline (Norgamem) useless drug in the treatment of squamous cell carcinoma. Eur J Cancer Clin Oncol 17:1061–1062
2. Brugarolas A, Gosalvez M (1980) Treatment of cancer by an inducer of reverse transformation. Lancet 12:68–70
3. Diaz Gil J, Trilla C (1982) Diminution of the rate of growth of Hela cells caused by thioproline (Tp) and 2-aminothiazoline HC (2-AT). Effect of L-proline. Rev Esp Oncol 29:615–621
4. Gosalvez M (1983) Thioproline and reversal of cancer. Lancet 14:1108
5. Grier R, Merkley D, Roth Y (1984) Pilot study of the treatment with thioproline of 24 small animals with tumors. Am J Vet Res 45:2162–2166
6. McCarty M (1982) Cytostatic and reverse transformation therapies of cancer – a brief review and future prospects. Med Hypotheses 8:589–612
7. Parks R, Jones T, Banks A, Hessel E (1982) Thioproline: an inhibitor of chemical carcinogenesis. Neoplasma 29:535–537
8. Rozmysłowicz T, Konopka L, Takiel M, Pawelski S (1985) Influence of thioproline on ^{3}H-thymidine incorporation into the normal human lymphocytes. VIII Meet Int Soc Hem Eur Afr Div, Warsaw, Poland, p 396
9. Sahai Y, Imai K, Ibuka T, Sasaki T, Hayahawa M (1980) Toxicological study and phase I study of thioproline on inducer reverse transformation. Proc Congr Jap Soc Canc Therapy Tokyo, Japan, pp 595–596

Diversity of Molecular Phenotypes in Acute Leukemias

S. Mizutani [1], K. Nakamura [1], M. Ozaki [1], C. Tamura [1], M. Sasaki [2], Y. Tsunematsu [2], J. Fujimoto [1], J. Hata [1], N. Kobayashi [1]

A. Introduction

Recent advances in immunological studies of human leukemias have enabled us to address the lineage of leukemic cells. More recent molecular studies of leukemic cells have made it possible to study the regulation of genes encoding antigen receptors, indicating there is differentiation-linked rearrangement of the receptor genes with lymphoid leukemias. We have sought to evaluate the usefulness of molecular analyses of acute leukemias on the basis of independent morphological, cytochemical, and immunological studies. These studies have led us to conclude immunophenotypically classified acute leukemias carry divergent molecular phenotypes.

B. Materials and Methods

Thirty-nine acute leukemias including 16 B- and pre-B-cell leukemias (B group), 16 T-cell leukemias (T group), and 7 mixed lineage leukemias (MLLs) were studied. Leukemic cells were obtained from heparinized peripheral blood or bone marrow. Immunophenotyping was carried out as described previously [1] by flow cytometry using monoclonal antibodies CD10, CD19, CD15, CDw13, CD33, CD14, CD11 b, CD7, CD3, and CD2. Cytochemical and ultrastructural studies were also carried out according to standard methods. DNA was extracted from leukemic cells as has been described before [2]. Southern blot analysis of DNA was carried out after the cleavage of the DNAs with various restriction enzymes. DNA probes for this study include immunoglobulin heavy chain JH(IgJH, *Bam*HI/*Hind*III fragment), Ig μ-region genes (HM2), T-cell receptor (TCR) β (*Bgl*II/*Stu*I fragment of TCR β-constant region), and γ (pH60) region genes.

C. Results

Thirteen out of 16 B4 and/or CALLA-positive acute lymphocytic leukemias (stage II, III pre-B ALL, [3]) showed rearrangement of IgJH with or without those of TCR β and γ genes, which accord with the results obtained by others and ourselves previously. As 3 out of 16 B-group acute lymphoblastic leukemias (ALLs) showed no IgJH gene rearrangement, they were reprobed with Ig μ-chain gene. When reprobed with Cμ-region gene, one of these three with germ line IgJH showed two Ig μ-rearranged bands in addition to one germ line band, indicating this leukemia has a deletion of the IgJH region on both alleles. The other two cases of these three retained germ line configuration of the Ig μ region as well as the IgJH region (Fig. 1, Table 1). Densitometric analysis of these germ line bands has ruled out the possibility of the deletion of JH and Cμ region of Ig heavy

This work was supported by Grant-in-Aid for General Science Research from The Ministry of Education, Science and Culture to S.M. and N.K., and for Cancer Research from The Ministry of Health and Welfare to S.M. and J.H. in Japan

[1] The National Children's Medical Research Center and The National Children's Hospital, 3-35-31 Taishido, Setagaya-ku, Tokyo, Japan
[2] Urawa City Hospital, 2460 Mimuro, Urawa, Saitama, Japan

Table 1. Genomic diversity of precursor B ALLs

Case	Phenotype	Molecular type	Receptor genes			
			Ig JH	Ig μ	TCR β	γ
1, 2	Stage II/III	B-stem type	G/G	G/G	G/G	G/G
3	Stage III	Abortive pre B	D/D	R/R	G/G	G/G

Differentiation stage of precursor B ALLs was determined according to ref. [3]

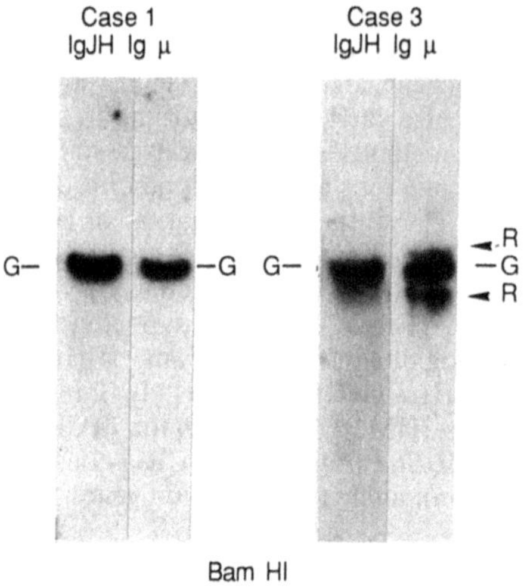

Fig. 1. Ig J and μ gene rearrangement in case 1 and 3. DNAs from case 1 and 3 were digested with Bam HI. After electrophoresis and Southern transfer to nitrocellulose paper the blot were hybridized successively with J and μ probes. The probes used are indicated above each panels. The unrearranged band is indicated by G and is 17 kb in this digest. Rearranged band is indicated by R.

chain locus in these leukemias (data not shown). Karyotypic analysis of one of them showed t(4;11) (q21;q23) acute leukemia, with which abnormality has been reported to carry a characteristic of progenitors common to B and monomacrophage. Careful examination of the cytochemical and ultrastructural features of this case showed granular positivity of β-glucuronidase, positive staining for TdT, and no ultrastructural peroxidase positivity, indicating lymphoid nature of this leukemia. Five cases among seven MLLs were identified as having B and myeloid phenotype. Molecular analyses with antigen receptor gene probes revealed two with germ line IgH and TCR β, γ genes, and three with rearrangement of the IgH gene. When referred to the morphological and cytochemical findings, these two with germ line antigen receptor genes showed Auer rods as well as strong peroxidase positivity by light microscopy (Tables 2, 3).

D. Discussion

We have investigated immunological and molecular features of 39 acute leukemias; 16 B-lineage leukemias, 16 T-lineage leukemias, and 7 MLLs were identified. As the results of this study we found one pre-B type ALL with deletion of the IgJH region on both alleles, suggesting a clonal expansion of abortive pre-B-lymphocytes. The immunophenotype of this leukemia, which is consistent with that of stage III pre-B cells, suggests deletion of the JH region might have taken place during the rearrangement of the IgH region gene. The other two cases with germ line Ig JH as well as the Ig μ gene have made it possible to identify the earliest B precursor cells. The immunophenotype of these leukemias was consistent with that of stage II/III pre-B-cell ALL. In the light of this study we suggest stage II/III pre-B lymphocytes may include those with germ line IgH genes and hence probably B stem cells. Thus the careful analyses of the IgH gene of pre-B ALLs may shed light on the diversity of molecular phenotypes in B precursor ALLs.

Table 2. Immunophenotypes of the cases presented in this study

	Pre B			MLLs with B & M				
Case	1	2	3	1	2	3	4	5
FAB	L2	L1	L2	M2	M2	L2	L2	L2
HLA DR	90	+	47	62	83	+	94	98
CD10 (CALLA)	–	62	46	–	–	87	–	17
CD19 (B4)	90	+	53	76	51	100	57	60
CD15 (LeuM1)	–	–	–	NT	15	NT	14	37
CDw13 (My7)	–	–	–	59	45	–	15	35
CD33 (My9)	–	–	–	–	16	NT	23	–
CD14 (My4)	NT	–	NT	NT	NT	34	NT	NT
CD11b (OKM1)	–	–	–	NT	–	37	–	NT
CD7 (Leu9)	–	–	19	–	20	NT	–	–
CD3 (Leu4)	–	–	NT	–	–	NT	–	–
CD2 (Leu2)	–	NT	15	23	–	–	–	–
SmIg	–	–	–	–	–	–	–	–
Tdt	+	+	+	NT	NT	70	NT	NT

NT, not tested; –, less than 10% positivity
Only the phenotype data of precursor B ALLs with germ line IgH and of MLLs with B and myeloid phenotype are presented. The data of the other pre-B ALLs, T ALLs, and MLLs with T and myeloid phenotype are omitted. Case No. 4 with B and myeloid phenotype was included in this study because of the transient appearance of light microscopic positivity or peroxidase.

Table 3. Molecular phenotypes in mixed lineage leukemias with B and myeloid phenotype

Case	1	2	3	4	5
IgJH	G/G	G/G	D/R	R/D	D/R
TCR γ	G/G	G/G	G/G	G/G	R/R
TCR β	G/G	G/G	G/G	G/G	G/G

Among five MLLs with B and myeloid phenotype, two cases showed no rearrangement of IgH or TCR genes. These two, with germ line antigen receptor genes, were positive for peroxidase by light microscopy, which is consistent with myeloid leukemias. Observation of Auer rods on these two leukemias made us confirm that they are bona fide myeloid leukemias. With the help of these findings, we speculate B4 antigen is not exclusively expressed on B precursor cells, but can be expressed on some myeloid leukemias. As to the three cases with rearrangement of the IgH gene with or without those of TCR genes, it is not yet clear whether they are lymphoid or myeloid. But the rearrangement of antigen receptor genes may reflect elevated recombinase activity, suggesting they are likely to be committed to lymphoid lineage rather than to myeloid lineage. Further study is necessary to determine the nature of these leukemias.

Acknowledgments. We thank Dr. Y. Kaneko for the karyotype study of the samples, Drs. J. Shimabukuro and Y. Hayashi for referring the patient's cells to us, and Dr. Enomoto for ultramicroscopic study of the samples. We are grateful to Drs. T. Mak, P. Leder, H. Molgaard, and T. Rabbitts for molecular probes TCR β, IgJH, Ig μ, and TCR γ respectively.

References

1. Chan LC, Pegram SM, Greaves MF (1985) Contribution of immunophenotype to the classification and differential diagnosis of acute leukemia. Lancet i: 475–479
2. Ford AM, Molgaard HV, Greaves MF, Could HJ (1983) Immunoglobulin gene organisation and expression in hematopoietic stem cell leukemia. EMBO J 2: 997–1001
3. Nadler LM, Korsmeyer SJ, Anderson KC et al. (1984) B cell origin of non-T cell acute lymphoblastic leukemia. J Clin Invest 74: 332–340

Treatment of Chronic Myelogenous Leukemia with Interferons: Hematologic, Cellular, and Genetic Investigations*

B. Opalka [1], O. Kloke [2], U. Wandl [2], R. Becher [2], and N. Niederle [2]

A. Introduction

Chronic myelogenous leukemia (CML) is a stem cell disorder characterized by the accumulation of immature precursors of the granulocytic and monocytic lineage. Most commonly, the disease is treated with chemotherapeutic agents, preferably busulfan and hydroxyurea. High-dose chemotherapy followed by allogeneic bone marrow transplantation may eradicate the malignant clone and offers the possibility of cure [1, 2].

In recent years, interferons (IFNs) have been successfully used as a novel therapeutic approach [3]. We have therefore started a study using IFN-alpha-2b alone or in combination with IFN-gamma. In addition to the clinical follow-up, the effects of IFNs were scored by molecular analysis and studies made on the growth of myeloid progenitor cells.

B. Materials and Methods

I. Patients' Characteristics

Two groups of patients with chronic-phase CML were treated. In the first group (48 patients), most patients were pretreated, while in the second group (24 patients) all patients were untreated. Of the first group all patients and in the second group 19 patients were evaluable for response.

II. Treatment Schedule

Recombinant human IFN-alpha-2b (Intron A°) with a specific activity of more than 1×10^8 units/mg protein was provided by Schering Inc., Kenilworth, New Jersey, United States. The drug was administered subcutaneously. During the introduction period, the daily dose was 4×10^6 units/m^2 body surface area. The dose was gradually reduced 2–4 weeks after initiation of IFN treatment. After normalization of white blood cell counts, the patients received individual doses ranging from 1×10^6 units every other day to 10×10^6 units daily. Some patients received a combination of IFN-alpha and -gamma at an initial dose of 4×10^6 units/m^2 IFN-alpha-2b and 50 μg IFN-gamma. IFN-gamma was provided by Biogen SA, Geneva, Switzerland, and had a specific activity of $2-4 \times 10^8$ units/mg protein.

III. Response Criteria

Response to IFN treatment was evaluated according to standard criteria [3]:

- Complete remission (CR): normalization of all clinical and hematologic parameters plus complete suppression of the Philadelphia (Ph') chromosome in all analyzable metaphases
- Hematologic remission (HR): a normalization of total and differential leukocyte counts, platelets, serum lactate dehydrogenase levels, and spleen size

[1] Institute for Molecular Biology, University of Essen (GHS), Hufelandstr. 55, D-43 Essen 1, FRG

[2] Department of Internal Medicine, University of Essen (GHS), Hufelandstr. 55, D-43 Essen 1, FRG

* This work was supported by the Meyer-Struckmann Foundation

- Partial hematologic remission (PHR): a more than 50% reduction of total leukocyte counts without normalization of differential counts
- Failure (PD): no reduction in leukocyte counts

IV. Preparation of DNA and Southern Blotting

Presence of the (Ph') chromosome [4] was analyzed by the Southern blot technique. Isolation of genomic DNA, digestion with restriction enzymes, and Southern type analyses were performed as described elsewhere [5, 6]. The probe used was a 5' *bcr*-specific 2.0-kb *Bgl*II/*Hind*III fragment [5].

V. Stem Cell Analysis

Isolation and culture conditions for bone marrow cells were performed as previously described [7, 8].

C. Results

I. Clinical Response to IFN-alpha-2b

In the first study, preferably pretreated patients received IFN-alpha alone. The results shown in the left of Table 1 demonstrate that most of the patients responded to the administration of IFN-alpha-2b. In 46% of the patients the leukocyte and differential counts normalized. Partial remissions were found in another 33%. However, no complete remission as defined by loss of the Ph' chromosome was obtained.

The therapeutic schedule was modified in the second group of patients, who had all received no previous therapy. As induction therapy, IFN-alpha-2b was given daily either alone or in combination with 50 μg IFN gamma. As maintenance, the highest tolerated dose was administered. The observation time ranged from 9 to 15 months. Preliminary data obtained with this regimen are presented in the right part of Table 1. Most patients evaluable for response ($n=19$) showed an objective

Table 1. Clinical response to IFN-alpha-2b or IFN-alpha-2b plus IFN-gamma

With prior therapy $n=48$		Without prior therapy $n=24$, evaluable 19 patients	
CR	0	CR	2
HR	22	HR	9
PHR	17	PHR	5
PHR (short) and PD	9	PHR (short) and PD	3

CR, complete remission; HR, hematologic remission; PHR, partial hematologic remission; PD, failure

remission to the IFN treatment. In two patients, moreover, cytogenetic analysis revealed the complete disappearance of Ph'-positive bone marrow cells.

II. Molecular Investigations

The rearrangements of the *BCR* gene caused by the Ph' translocation were routinely scored by the Southern blot technique [6]. In Fig. 1 data are presented from two patients. In the bone marrow DNA of patient 51 a very faint band was still present after 14 months of therapy, whereas no rearranged fragment could be found in the blood cell DNA of patient 62 after 9 months of treatment. The sensitivity of this assay was checked by mixing leukemic and normal DNA and was determined to be 1%–5% (Fig. 1).

With the same probe and *Pvu*II-digested DNA, patient 51 showed a polymorphic *BCR* allele. This allele was also found in 18 patients with different myeloproliferative disorders and their relatives, but not in a group of randomly selected control persons [9].

III. Stem Cell Analyses

In 13 patients analyzed, a reduction of the colony-forming progenitors CFU-GEMM, BFU-E, and CFU-C was observed during IFN treatment (Fig. 2). The most pronounced decrease was

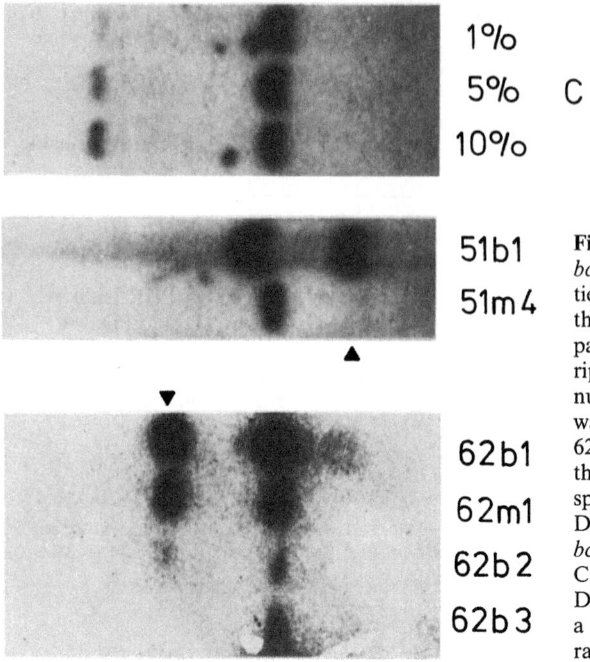

Fig. 1. Southern blot analysis of *bcr* rearrangements of two patients before and during IFN therapy. *First number*, number of patient; *m*, bone marrow; *b*, peripheral blood; *second number*, number of sample. Sample 51m4 was taken 14 months, sample 62b3 9 months, after initiation of therapy. *Arrows* indicate Ph′ specific bands. *Bgl*II-digested DNA was hybridized to the 5′ *bcr*-specific probe. For control, C, *Pvu*II-digested normal donor DNA was mixed with DNA from a leukemia patient at different ratios.

found in the erythroid compartment BFU-E. One patient showed initially an increase of BFU-E and CFU-GEMM. In this patient, however, all colony types did decrease later on during IFN treatment. Two patients with accelerated phase disease at the beginning of IFN-alpha-2b treatment did not respond to IFN administration. This failure correlated with an increase of granulocytic progenitor cell proliferation in vitro.

D. Discussion

The effects of IFN were evaluated in two groups of patients with chronic-phase CML. In the first group of patients, the minimal necessary dosage to keep patients in hematologic remission was administered as maintenance therapy after induction of remission. Most patients responded to IFN therapy. However, no disappearance or significant reduction of Ph′-positive metaphases was obtained with this therapeutic regimen (for details see [10, 11]). In a second study, therefore, higher maintenance doses were administered. So far, most patients have responded to IFN therapy. In two patients, moreover, a total loss of Ph′-positive bone marrow cells could be demonstrated by cytogenetic and molecular *bcr* analysis with a limit of detection of 1%–5% [3, 12]. Thus, the more aggressive therapy might perhaps be able to alter the course of the disease. So far, no superiority of combination therapy with IFN-alpha plus IFN-gamma over monotherapy with IFN-alpha has been proven. One has to keep in mind, however, that the number of patients tested is very low and the time of observation rather short.

The stem cell analyses showed a good correlation between the clinical response and the reduction of myeloid progenitor cell proliferation in vitro. Therefore, these in vitro determinations might serve as predictive tests for determining the in vivo response. The availability of the molecular analysis of the Ph′ chromosome allows a highly sensitive screening

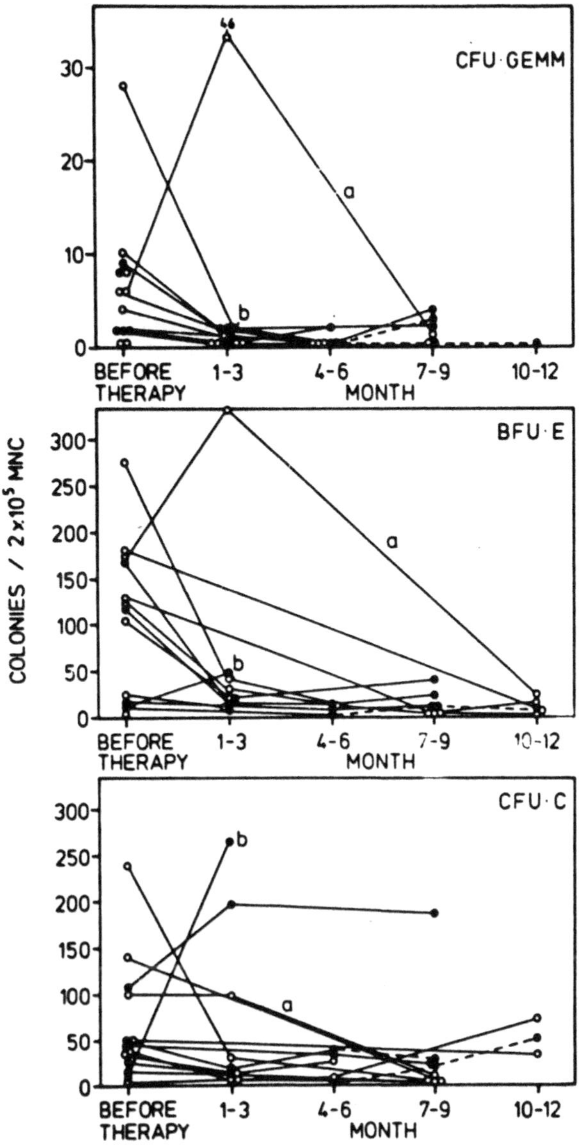

Fig. 2. Colony-forming potential in bone marrow of CML patients undergoing IFN therapy. (Redrawn from [8])

method which is independent of the presence of dividing cells and the preparation of metaphases. Moreover, the recently developed technique of PCR [13] will increase the sensitivity by about three to four orders of magnitude.

Acknowledgments. We thank Ms. C. Oberle, Ms. B. Muss, and Ms. B. Flöter for technical assistance, and Ms. C. Middendorf for preparing the manuscript.

References

1. Speck B, Gratwohl A, Osterwalder B, Nissen C (1984) Bone marrow transplantation for chronic myeloid leukemia. Semin Hematol 21:48–52
2. Schaefer UW, Beelen D, Graeven U, Kloke O, Niederle N, Quabeck K, Sayer H, Schmidt CG (1988) Allogenetic bone marrow transplantation in chronic myelogenous leukemia. In: Huhn D, Hellriegel

KP, Niederle N (eds) Chronic myelocytic leukemia and interferon. Springer, Berlin Heidelberg New York
3. Talpaz M, Kantarjian HM, McCredie K, Keating MJ, Trujillo J, Guttermann J (1987) Clinical investigation of human alpha interferon in chronic myelogeneous leukemia. Blood 69: 1280–1288
4. Nowell PC, Hungerford DA (1960) A minute chromosome in human chronic granulocytic leukemia. Science 132: 1497
5. De Klein A, Hagemeijer A, Bartram CR, Houwen C, Hoefsloot L, Carbonell F, Chan L, Barnett M, Greaves M, Kleihauer E, Heisterkamp N, Groffen J, Grosveld G (1986) *Bcr* rearrangement and translocation of the c-*abl* oncogene in Philadelphia positive acute lymphoblastic leukemia. Blood 68: 1369–1375
6. Opalka B, Wandl U, Kloke O, Koppe J, Niederle N (1988) Molecular biological investigation in chronic myelogenous leukemia patients undergoing interferon therapy. In: Huhn D, Hellriegel KP, Niederle N (eds) Chronic myelocytic leukemia and interferon. Springer, Berlin Heidelberg New York
7. Fauser AA, Messner HA (1979) Identification of megacaryocytes, macrophages, and eosinophiles in colonies of human bone marrow containing neutrophilic granulocytes and erythroblasts. Blood 53: 1023–1027
8. Wandl UB, Kloke O, Opalka B, Niederle N (1988) Suppressive effect of interferon alfa-2b on hematopoietic progenitor cells in patients with chronic myelogenous leukemia. In: Huhn D, Hellriegel KP, Niederle N (eds) Chronic myelocytic leukemia and interferon. Springer, Berlin Heidelberg New York
9. Opalka B, Wandl U, Kloke O, Oberle C, Koppe J, Niederle N, Schmidt CG (1989) A *Pvu* II polymorphism of the *bcr* region in patients with hemopoietic disorders and their families. Blood 73: 814–817
10. Niederle N, Kloke O, Osieka R, Wandl U, Opalka B, Schmidt CG (1987) Interferon alfa-2b in the treatment of chronic myelogeneous leukemia. Semin Oncol 14: 29–35
11. Niederle N, Kloke O, Osieka R, May D, Wandl UB, Becher R, Opalka B, Schmidt CG (1988) Treatment of chronic and acute phase chronic myelogenous leukemia with interferon-alpha-2b and interferon-gamma. In: Huhn D, Hellriegel KP, Niederle N (eds) Chronic myelocytic leukemia and interferon. Springer, Berlin Heidelberg New York
12. Yoffe G, Blick M, Kantarjian H, Spitzer G, Guttermann J, Talpaz M (1987) Molecular analyses of interferon-induced suppression of Philadelphia chromosome in patients with chronic myeloid leukemia. Blood 69: 961–963
13. Saiki RK, Scharf S, Faloona F, Mullis KB, Horn GT, Erlich HA, Arnheim N (1985) Enzymatic amplification of β-globin genomic sequences for diagnosis of sickle cell anemia. Science 230: 1350–1354

α Interferon in Myelodysplasia

D.W. Galvani

A. Patients and Methods

I. Myelodysplastic Patients

The FAB classification was employed in categorising MDS patients. Seventeen patients were studied initially – six with refractory anaemia (RA), six with RA with excess blasts (RAEB), one with transforming RAEB (trRAEB) and four with chronic myelomonocytic leukaemia. Of these, eight were suitable for α IFN therapy (3 RA, 3 RAEB, 1 trRAEB, 1 CMML). Treatment was commenced as 3 MU of Wellferon daily for up to 6 months. This dosage was reduced in the presence of thrombocytopenia.

II. Cell Phenotype

Peripheral blood mononuclear cells and neutrophils were separated using Ficoll/Hypaque centrifugation, and the neutrophils were further purified by dextran sedimentation.

Mononuclear antibody analysis was performed by an indirect immunofluorescent technique employing FITC-labelled goat anti-mouse immunoglobulin. Results were read on an FACS analyser. A spectrum of lymphocyte markers were used to characterise subsets, and several myeloid antibodies were used to stain neutrophils.

University Department of Haematology, Third Floor Duncan Building, Royal Liverpool Hospital, Prescot Street, Liverpool L69 3BX, United Kingdom

III. NK Assay

A standard ^{51}Cr-release assay was employed using K562 cells as targets. Various ratios of effectors-to-target cells were used, and cells were incubated at 37° C for 4 h. ^{51}Cr released from lysed cells was measured in the supernatant as a measure of cytotoxicity. The standard formula for % cytotoxicity was employed.

B. Results

I. Clinical Response to αIFN

One patient with RAEB showed a drop in marrow blast count from 18% to 0% over 6 months. The patient with CMML showed a fall in peripheral monocyte count from $>15 \times 10^9/l$ to $3 \times 10^9/l$ but became neutropenic. The other six patients showed no improvement in transfusion requirements or marrow. Interestingly, no infective episodes were observed, in contrast to a previous report on αIFN used in MDS.

II. Cell Phenotype

Percentage positivity of NK cells was not substantially different from normal before therapy and did not change during therapy (Table 1). Lymphocyte subsets were not found to be markedly abnormal before or after therapy. Mature neutrophils expressed the immaturity marker (CD33) in $18\% \pm 4\%$ (compared with $4\% \pm 1\%$ in normals).

III. NK Function

NK function was generally low in all 15 patients studied. The addition of αIFN in vitro enhanced this, however.

Table 1. Phenotypical NK cells in different forms of myelodysplasia

Diagnosis	No. of patients	IFN therapy[a]	% positive (±1 SEM)		Absolute no. of positive cells × 10^9/l (±1 SEM)	
			Leu7	CD16	Leu7	CD16
RA	4	–	13±3	17±3	0.1±0.1	0.3±0.1
	3	+	15±3	19±5	0.2±0.1	0.2±0.1
RAEB	6	–	25±4	19±4	0.3±0.1	0.2±0.1
	3	+	26±6	20±6	0.3±0.1	0.2±0.1
trRAEB	1	+	11	27	0.4	0.8
CMML	4	–	4±1	11±5	0.7±0.3	1.6±0.9
	1	+	9	8	0.5	0.4
Normal	10	–	14±2	21±3	0.4±0.1	0.6±0.1

[a] –Patient not receiving IFN; +patient receiving IFN

During αIFN therapy there was no consistent increase in NK cytotoxicity. When patients were studied sequentially, no consistent pattern emerged. One responder demonstrated increased activity, whilst the other demonstrated reduced activity. Nonresponders showed either enhancement or inhibition.

C. Discussion

A previous study of αIFN in MDS using 3 MU three times weekly revealed no clinical response. The present study suggests that a higher dosage may be effective occasionally in RAEB.

Although defective NK cytotoxicity was enhanced in vitro by αIFN, when αIFN was administered clinically it had no consistent effect on NK function. This phenomenon was independent of the percentage of NK cells present, which was in fact within normal limits. As NK cells may be implicated in haematopoiesis, we feel the effects of therapeutic αIFN are not mediated via these cells in MDS. Further, cold-target inhibition experiments using purified myeloid marrow cells showed no reduction in NK cytotoxicity.

We conclude that, although NK function is defective in MDS, the occasional beneficial effects of αIFN therapy are not mediated by this immunological mechanism.

Use of Investigational Drugs as Initial Therapy for Childhood Solid Tumors*

W.H. Meyer[1,3], P.J. Houghton[2], M.E. Horowitz[1,3], E. Etcubanas[1,3], C.B. Pratt[1,3], F.A. Hayes[1,3], E.I. Thompson[1,3], A.A. Green[1,3], J.A. Houghton[2], J.T. Sandlund[1,3], and W.M. Crist[1,3]

Solid tumors are relatively rare in children, but comprise about two-thirds of all malignancies that affect this age group. Most of these tumors respond well to initial treatment, and some (Wilms' tumor, low-stage Hodgkin's disease, low-stage rhabdomyosarcoma, and low-stage neuroblastoma) are readily cured with modern therapy. Still, many tumors that respond initially acquire clinical drug resistance, respond poorly to rechallenge with known active agents, and demonstrate a low level of responsiveness to experimental agents. This creates a major therapeutic dilemma for the pediatric oncologist. Although a critical need exists to identify new active agents for many solid tumors in childhood, current primary therapy is frequently quite active even in tumors which have a very high rate of relapse. Unfortunately conventional testing of anticancer drugs in previously treated patients can lead to ostensibly poor results when, in fact, the agent may have clinically significant activity. Several arguments can be marshaled against current phase II clinical trials. At relapse, patients may have tumors with multiple drug resistance and may tolerate therapy poorly due to previous treatment and the advanced state of their tumors. Finally, the attending physician, the patient, and the family may be reluctant to consider experimental chemotherapy, resulting in too few subjects in phase II trials to ensure adequate evaluation of all promising compounds.

One way to circumvent these difficulties is to test new phase II agents in previously untreated patients at high risk for failure on standard chemotherapy. This strategy is being implemented at St. Jude Children's Research Hospital by a team of investigators that includes pharmacologists who have developed models of human solid tumors in immune-deprived mice, clinical pharmacokineticists, and clinical oncologists. In this paper, we outline the conceptual framework of this investigative effort and the results of our initial efforts with three common pediatric solid tumors – rhabdomyosarcoma, osteosarcoma, and Ewing's sarcoma – emphasizing experience with melphalan in the treatment of patients with newly diagnosed rhabdomyosarcoma [1].

A. Conceptual Basis

Testing new agents in previously untreated patients yields the most reliable estimate of actual drug activity. Tumors are most sensitive to effective therapy at diagnosis, before the development of clinical drug resistance. When relapse occurs, the tumor is likely to have acquired resistance to some, if not all, of the agents used in primary therapy. In addition, the tumor may be clinically cross resistant to

[1] Departments of Hematology-Oncology, St. Jude Children's Research Hospital, Memphis, TN, USA
[2] Department of Biochemical and Clinical Pharmacology, St. Jude Children's Research Hospital, Memphis, TN, USA
[3] Department of Pediatrics, University of Tennessee, Memphis, College of Medicine, Memphis, TN, USA
* Supported in part by grants CA 23099 and CA 21765 from the National Cancer Institute and the American Lebanese Syrian Associated Charities (ALSAC)

new agents being tested in classic phase II trials. Such testing of new experimental agents in previously untreated children with typically advanced cancer is acceptable, both scientifically and ethically, *only* if care is taken in establishing appropriate criteria for the selection of patients and experimental agents.

B. Criteria for Patient Eligibility

The patient must be at high risk of ultimate treatment failure and, consequently, death from tumor. This does not mean that the tumor should be potentially resistant to all available therapy. In fact, in all the solid tumors selected for this approach at our center, combination chemotherapy exists which produces clinical responses in a significant proportion of patients. The key point is that all of the patients were judged to have a greatly increased risk of eventual treatment failure. Exactly how high this risk must be is difficult to ascertain; a minimum estimate would be probable failure and death in 40%–50% of patients. The greater the likelihood for cure with effective primary therapy, the greater the care that must be exercised in deciding who will be eligible for treatment with experimental agents as initial therapy.

C. Selection of Drugs

There must be a strong rationale for the selection of experimental agents. A drug could be selected for "up-front" testing if it demonstrates marked activity in a relevant model; e.g., the human tumor xenograft in immune-deprived mice. Melphalan and, to some extent, ifosfamide were selected for testing in untreated rhabdomyosarcoma following demonstration of very significant activity in xenografts of human rhabdomyosarcoma [2, 3]. An agent (or combination of agents) could also have shown activity in conventional phase II trials. Ifosfamide was selected on this basis for use in rhabdomyosarcoma [4–6] and in osteosarcoma [6, 7], as was the combination of ifosfamide/VP-16 in Ewing's sarcoma [8]. A compelling pharmacologic rationale with supporting laboratory data might be a third criterion for selection of a new agent. Although, to date, we have not based any choice of an agent solely on this reason, the demonstration of similar pharmacokinetics for melphalan in the xenograft model system and in children was a deciding factor in whether this agent would be used in children with previously untreated rhabdomyosarcoma [1].

The following sections present, in greater detail, implementation of this strategy in three common childhood solid tumors. The basic framework for drug testing is a phase II trial in which the experimental agent or drug combination is administered before any other therapy. This "window of opportunity" lasts for 6–9 weeks before initiation of "standard" treatment.

D. Rhabdomyosarcoma

Rhabdomyosarcoma is the most common soft tissue sarcoma in children [9]. Although patients with low-stage disease are frequently cured with surgery, combination chemotherapy, and radiation therapy [9], those with advanced tumors fare poorly [10, 11]. In addition, few new agents with significant antitumor activity have been identified by standard phase II testing.

The following categories of patients are eligible for testing with phase II agents before they receive standard chemotherapy: (a) patients with unresectable primary tumors excluding those in favorable sites (orbit, face, and cheek primaries) – IRS group III – and (b) patients with metastatic disease – IRS group IV. The outcome of aggressive multiagent chemotherapy, radiation therapy and surgery in these two categories has changed little over the past 10 years [10, 11], with the possible exception of parameningeal sites. At St. Jude Children's Research Hospital, the predicted disease-free survival for the last 96 consecutive patients in these two categories is about 30%.

The selection of new agents to be tested in rhabdomyosarcoma has been based primarily on the xenograft model, as discussed in the following paper by Houghton, and in a recent publication by Horowitz et al. [1]. Briefly, human rhabdomyosarcomas from previously untreated patients were grown as xenografts in immune-deprived mice and then were used to screen a spectrum of anticancer drugs for activity. The ranking of relative activity in the xenografts is essentially the same as in the human phase II trials. Of the agents tested to date, melphalan (L-phenylalanine mustard, L-PAM) emerged as the single most active agent in human rhabdomyosarcoma and ifosfamide the second most active. However, when moved into a conventional phase II trial, melphalan produced responses in only 1 of 15 previously treated patients. The drug would likely have been abandoned had not its plasma disposition, including total systemic clearance and AUC (area under the concentration vs. time curve), been similar in patients and in the xenograft model. This suggested that therapeutically adequate levels of melphalan were being attained in patients. Therefore, a phase II trial of melphalan in previously untreated patients was initiated. As recently published [1], melphalan proved highly active, producing responses in 10 of the first 13 patients with advanced rhabdomyosarcoma. Upon completion of this trial, ifosfamide has now been introduced for testing in previously untreated patients with rhabdomyosarcoma (RMS V Study). The outline of this protocol is shown in Fig. 1. Therapeutic results will be available when sufficient numbers of patients have been evaluated for response.

Undoubtedly, the significant activity of melphalan in rhabdomyosarcoma would have been undetected had we not tested this agent in previously untreated patients. In the initial phase II trial, all of the patients had received vincristine, doxorubicin, and cyclophosphamide and all but one had received dactinomycin and radiotherapy; hence, the early failure of melphalan was probably the result of tumors with resistance to multiple agents. Indeed, recent work by Houghton et al. [12] indicates that tumors resistant to vincristine are cross resistant to melphalan.

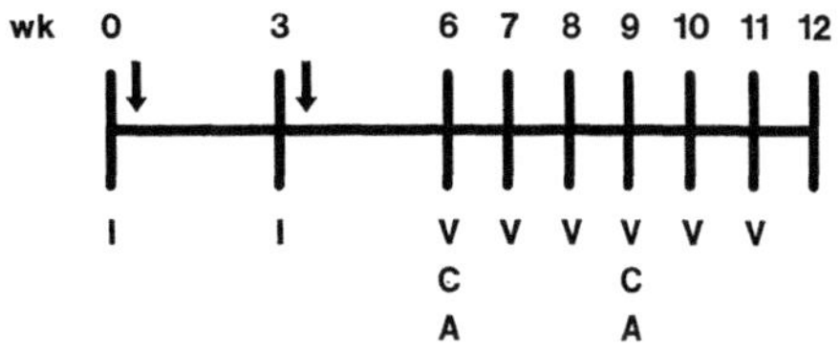

Fig. 1. Schema for ifosfamide trial (induction phase) on RMS V. *I*, ifosfamide; *V*, vincristine; *C* cyclophosphamide; *A*, Doxorubicin. *Arrows* indicate Mesna uroprotection

E. Osteosarcoma

First-line adjuvant and neoadjuvant treatment for osteosarcoma, the most common primary bone malignancy in the pediatric population [13], comprises relatively few agents: high-dose methotrexate, doxorubicin, cisplatin, and in some centers the combination of bleomycin, dactinomycin, and cyclophosphamide. In spite of aggressive multiagent protocols, over one-third of patients with nonmetastatic resectable primary tumors will relapse [14]. There has been little progress in the treatment of patients who present with metastatic disease at diagnosis or have unresectable primary tumors, groups that account for about one-third of all patients with osteosarcoma seen at our institution.

We have elected to enroll all patients with high-grade osteosarcoma in a modified phase II trial, because of the difficulty in predicting long-term, disease-free survival in this disease in the individual patient. With the possible exception of cellular DNA content [15] and serum lactate dehydrogenase levels [16, 17], there are no reliable prognostic indicators for patients with resectable osteosarcoma. Moreover, the clinical outcome of adjuvant therapy in patients treated at this center has not changed appreciably over the past 15 years [16]. This would appear to justify a high-risk classification in every case of high-grade osteosarcoma, and particularly in cases with metastatic lesions or unresectable primaries at diagnosis.

Ifosfamide was selected for up-front testing in the current osteosarcoma study (OS-86), because of results obtained in classic phase II trials conducted at this institution [6] and elsewhere [7]. Although xenograft models for osteosarcoma have recently been established [18], we have not used them to identify new agents. In its initial phase II trial, ifosfamide produced responses (complete plus partial responses) in 4 of 15 patients with relapsed osteosarcoma, including an unmaintained remission of over 4 years in one patient [6]. Although ifosfamide clearly has significant activity in relapsed osteosarcoma, it may increase toxicity when used in multiagent trials; thus, a better assessment of the level of activity in patients needs to be made before this agent is included in phase III protocols. Since ifosfamide caused consistent subclinical renal toxicity [19] as well as some instances of significant increases in serum creatinine [6], it is likely that it would add to the renal toxicity of combination chemotherapy, particularly that including cisplatin and high-dose methotrexate. Increased toxicity would be acceptable if ifosfamide demonstrated significant antitumor activity in more than 25% of previously untreated patients.

Figure 2 outlines the schema of therapy for the OS-86 trial. For the first 6 weeks on study, patients receive only ifosfamide. After complete radiologic and clinical evaluation to determine therapeutic efficacy, patients receive a third cycle of ifosfamide followed by high-dose methotrexate and doxorubicin. Following surgery at week 13, cisplatin is added to the schedule. This trial will clearly define the activity of this new agent in previously untreated patients and will begin to determine the spectrum of toxicities that are likely to occur when this agent is included in a multiagent bona fide phase III study.

F. Ewing's Sarcoma

Ewing's sarcoma is highly responsive to initial therapy. In the St. Jude EW-79 trial, over 90% of patients responded to low-dose oral cyclophosphamide and doxorubicin [20]. However, patients with metastatic disease at diagnosis and those with large primary lesions are known to have a high rate of relapse [21, 22]. Few patients with this tumor who relapse will be salvaged, particularly if relapse occurs

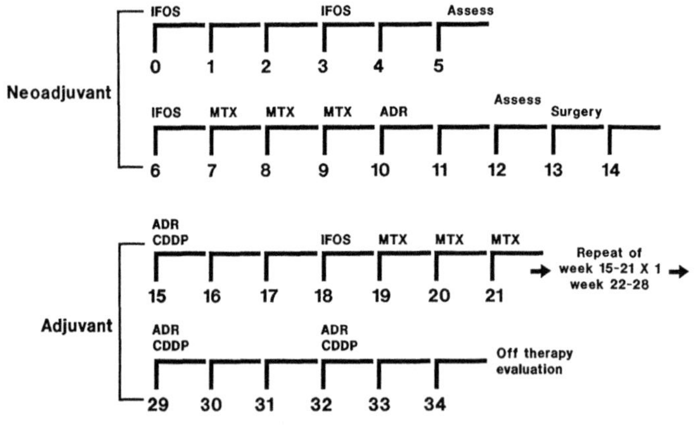

Fig. 2. Schema of chemotherapy for osteosarcoma (OS-86). *Ifos,* Ifosfamide; *MTX*, high-dose methotrexate with leucovorin rescue; *ADR,* doxorubicin; *CDDP*, cisplatin. All *numbers* indicate weeks of therapy

before or soon after cessation of chemotherapy. Unfortunately, only four chemotherapeutic agents are generally considered to have significant activity in Ewing's sarcoma: cyclophosphamide, doxorubicin, dactinomycin, and vincristine. Clearly, newer agents with significant activity against Ewing's sarcoma need to be identified. Recent phase II trials failed to identify new active agents; however, Miser et al. [8] showed that the combination of ifosfamide/VP-16 was very active in recurrent Ewing's sarcoma. Accordingly, an up-front trial of ifosfamide/VP-16 in patients at high risk for treatment failure has been initiated at St. Jude Hospital.

Patients with Ewing's sarcoma and clinical evidence of metastatic disease at diagnosis are at increased risk of treatment failure [21]. Recently, the size of the primary tumor has been shown to be an important prognostic indicator of treatment outcome [22]. At this institution, patients treated with metastatic disease at diagnosis or primary tumors >8 cm in largest dimension have about a 50% probability of disease-free survival, despite a very high initial response rate.

The combination of low-dose oral cyclophosphamide and doxorubicin shows marked activity in almost all patients with Ewing's sarcoma [20]. This has made selection of a new experimental drug for primary therapy quite difficult. The lack of data from human xenograft models and phase II trials that would suggest high levels of activity for any single phase II agent has made the selection of a single drug for this approach not feasible. However, initial reports from Miser et al. [8] showing marked activity of ifosfamide/VP-16 in relapsed Ewing's sarcoma (15 partial responses in the first 16 patients treated) indicated that this combination would be acceptable for trials in previously untreated patients. One could argue that these data are justification for immediate inclusion of ifosfamide/VP-16 in phase III trials in Ewing's sarcoma. Yet, the findings need to be confirmed, particularly in light of a recent follow-up report from Miser's group [23] that several patients have shown no response to this drug combination. In addition, ifosfamide/VP-16 will certainly add toxicity to any four-drug regimen presently used in Ewing's sarco-

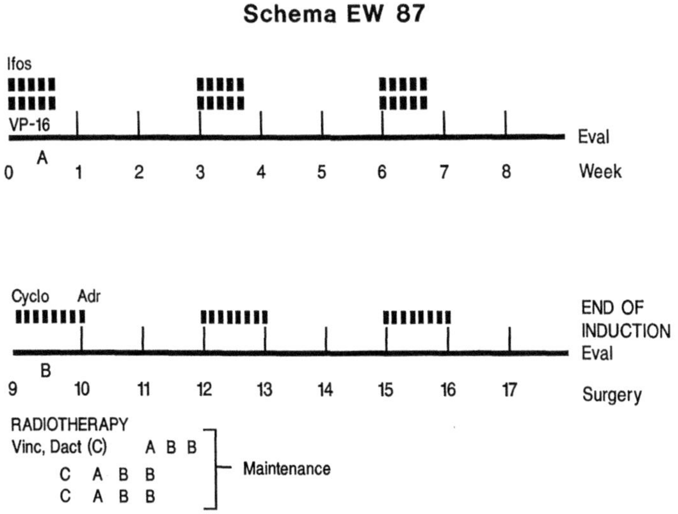

Fig. 3. Schema of chemotherapy for Ewing's sarcoma (EW-87). *Ifos*, ifosfamide; *VP-16*, etoposide; *Cyclo*, cyclophosphamide; *Adr*, doxorubicin; *Vinc*, vincristine; *Dact*, dactinomycin. Both the Ifos/VP-16 pair and the Cyclo/Adr pair are repeated times three at 21-day intervals during induction. *A*, Ifos/VP-16 drug pair; *B*, Cyclo/Adr drug pair; *C*, Vinc/Dact drug pair

ma. In this regard, it will be important to demonstrate the degree of activity of this combination in previously untreated patients, so that meaningful comparisons can be made with other known effective agents in Ewing's sarcoma.

The treatment schema for the present Ewing's sarcoma study for patients at high risk of relapse (EWI 87) is shown in Fig. 3. Three cycles of ifosfamide/VP-16 are given at 21-day intervals. After a complete clinical and radiologic assessment, the patient receives three cycles of low-dose oral cyclophosphamide and doxorubicin. The patient is then evaluated for response, including biopsy or resection of the primary site. Maintenance therapy consists of repetitive cycles of vincristine/dactinomycin, ifosfamide/VP-16, and cyclophosphamide/doxorubicin. High-dose (60 Gy) hyperfractionated radiotherapy is delivered to the primary tumor beginning at week 18.

G. Discussion

Development of new effective drugs is vital to the improvement of cure rates in childhood solid tumors. Conventional phase II studies will continue to be a useful tool for identifying potentially active new agents, but may underestimate the clinical value of many agents that would have significant activity against untreated tumors. For this reason, we have developed a program for testing selected new agents in previously untreated patients with solid tumors who are considered to have high risk of treatment failure. This approach is not unique to our institution. Indeed, earlier experience in similarly designed clinical trials suggests both the validity and inherent problems of up-front drug testing. Teniposide (VM-26), for example, was recently reported to be quite active in adults with small cell carcinoma of the lung [24] when used as primary therapy, despite its lack of activity in classic phase II trials. The Pediatric Oncology Group is also using this approach to identify new agents with activity in advanced-stage neuroblastoma.

Cullen and coworkers [25] tested idarubicin in previously untreated adults with small cell carcinoma of the lung; only 3 (14%) of 21 patients responded. At the completion of the trial, the authors concluded that responses to standard treatment with cyclophosphamide, vincristine, and etoposide following initial exposure to idarubicin seemed inferior to that previously seen in patients who received the same standard therapy without idarubicin. Although patients entered in that trial would clearly meet our criteria for a high-risk group (predicted median survival of approximately 8 months), and the experimental "window of opportunity" was short (16 of the 21 patients received only one or two courses of idarubicin), Cullen and coworkers used a rationale for drug selection that differs from ours. Their decision to begin up-front testing was based on the fact that idarubicin was an anthracycline analogue (doxorubicin is one the more active agents in small cell carcinoma of the lung), it can be given orally, and it is potentially less cardiotoxic than doxorubicin. They presented no data suggesting activity in previously treated patients nor data from relevant models of human small cell carcinoma of the lung. As one requirement for up-front testing of new compounds, Cullen et al. [26] recently suggested that "there should be evidence that the study drug is active in the disease in question," although they would accept activity of an analogue as sufficient evidence.

Kellie et al. [27] tested ifosfamide in previously untreated children with neuroblastoma, noting responses in 8 of 18 patients. Following exposure to ifosfamide, only four patients achieved a good partial or complete response to combination chemotherapy with vincristine, cyclophosphamide, cisplatin, and etoposide (OPEC) or a variant combination. Both a lower response rate and a shorter median survival were noted in these patients compared retrospectively with similar patients treated with OPEC from the time of diagnosis. The study population was appropriately selected and the phase II trial was sufficiently short. The primary rationale to test ifosfamide was that its analogue, cyclophosphamide, is active in neuroblastoma. In an earlier standard phase II trial of ifosfamide conducted by this group [28], only 2 of 25 patients responded. Thus, in both trials, the phase II agent was selected without clear evidence of activity in a relevant model or in classic phase II trials. Whether poorer responses to the upfront phase II agent will predict (or cause) poorer responses to standard treatment remains to be shown.

The identification of melphalan as a highly active agent in rhabdomyosarcoma (despite failure to define its activity in a standard phase II trial) provides strong support for the investigative approach we have described. However, safeguards must be in place to ensure an ethical study. As Cullen et al. [26] point out, careful clinical evaluation by experienced investigators, strict withdrawal criteria, continual protocol monitoring of response data, and reporting responses to both the phase II agent and standard therapy are essential. In our view, the two most critical factors required for implementation of this approach are: (a) careful selection of patients so that only those at high risk of treatment failure are included and (b) a strong rationale for selection of drug. Currently, we require that a prospective agent show either high levels of activity in the relevant human tumor xenograft or activity in classic phase II trials. The availability of the xenograft models and the collaboration of basic scientists, clinical pharmacokineticists, and clinical investigators makes this investigative effort unique. With appropriate care and diligence, primary testing of new agents in previously untreated patients should provide needed information regarding the actual level of antitumor activity of these agents and combinations. This in turn will speed the identification of clinically useful compounds and guide the future development of chemotherapy for childhood solid tumors.

Acknowledgment. We gratefully acknowledge John Gilbert for editing this manuscript.

References

1. Horowitz ME, Etcubanas E, Christensen ML et al. (1988) Phase II testing of melphalan in children with newly diagnosed rhabdomyosarcoma: a model for anticancer drug development. J Clin Oncol 6:308–314
2. Houghton JA, Houghton PJ, Green AA (1982) Chemotherapy of childhood rhabdomyosarcoma growing as xenografts in immune-deprived mice. Cancer Res 42: 535–539
3. Houghton J, Cook R, Lutz P, Houghton P (1985) L-Phenylalanine mustard (NSC 8806): a potential new agent in the treatment of childhood rhabdomyosarcoma. Cancer Treat Rep 69:91–96
4. Magrath I, Sandlund J, Rayner A et al. (1985) Treatment of recurrent sarcomas with ifosfamide. Proc ASCO 4:130
5. De Kraker J, Voute P (1984) Ifosfamide and vincristine in paediatric tumors: a phase II study. Eur Paediatr Haematol Oncol 1:47–50
6. Pratt CB, Horowitz ME, Meyer WH et al. (1987) Phase II trial of ifosfamide in children with malignant solid tumors. Cancer Treat Rep 71:131–135
7. Marti C, Kroner T, Remagen W et al. (1985) High-dose ifosfamide in advanced osteosarcoma. Cancer Treat Rep 69:115–117
8. Miser J, Kinsella T, Tsokos M et al. (1986) High response rate of recurrent childhood tumors to etoposide (VP-16), ifosfamide (IFOS) and mesna (MES) uroprotection. Proc ASCO 5:209
9. Green D, Jaffe N (1978) Progress and controversy in the treatment of childhood rhabdomyosarcoma. Cancer Treat Rep 5:7–27
10. Maurer HM, Beltangady M, Gehan EA et al. (1988) The intergroup rhabdomyosarcoma study-I. A final report. Cancer 61:209–220
11. Etcubanas E, Kun L, Pratt C et al. (1985) Patterns of failure in the treatment of

childhood rhabdomyosarcoma: a review of prospective studies at St Jude Hospital. Proceedings of the XVIIth Meeting of the International Society for Pediatric Oncology, Sept 30–Oct 4 Venice, p 95 (abstr)
12. Horton JK, Houghton PJ, Houghton JA (1987) Reciprocal cross-resistance in human rhabdomyosarcomas selected in vivo for primary resistance to vincristine and L-phenylalanine mustard. Cancer Res 47: 6288–6293
13. Bode U, Levine AS (1982) The biology and management of osteosarcoma. In: Levine AS (ed) Cancer in the young. Masson, New York
14. Link MP, Goorin AM, Miser AW et al. (1986) The effect of adjuvant chemotherapy on relapse-free survival in patients with osteosarcoma of the extremity. N Engl J Med 314:1600–1606
15. Look AT, Douglass EC, Meyer WH (1988) Clinical importance of near-diploid tumor stem lines in patients with osteosarcoma of an extremity. N Engl J Med 318: 1567–1572
16. Liddell RHA, Meyer WH, Dodge RK, Green AA, Pratt CB (1988) Prognostic indicators for patients with osteosarcoma (OS). Proc Am Assoc Cancer Res 29:226
17. Link MP, Shuster JJ, Goorin AM et al. (1988) Adjuvant chemotherapy in the treatment of osteosarcoma: results of the multi-institutional osteosarcoma study. In: Ryan JR, Baker LO (eds) Recent concepts in sarcoma treatment. Kluwer, Dordrecht, pp 283–290
18. Meyer WH, Houghton JA, Houghton PJ et al. (1987) Development and characterization of pediatric osteosarcoma (OS) xenografts. Proc Am Assoc Cancer Res 28:430
19. Goren MP, Wright RK, Horowitz ME, Pratt CB (1987) Ifosfamide-induced subclinical tubular nephrotoxicity despite mesna. Cancer Treat Rep 71:127–130
20. Hayes FA, Thompson EI, Hustu HO et al. (1983) The response of Ewing's sarcoma to sequential cyclophosphamide and Adriamycin induction therapy. J Clin Oncol 1:45–51
21. Pilepich MV, Vietti TJ, Nesbit ME et al. (1981) Radiotherapy and combination chemotherapy in advanced Ewing's sarcoma – intergroup study. Cancer 47:1930–1936
22. Göbel V, Jürgens H, Etspüler G et al. (1987) Prognostic significance of tumor volume in localized Ewing's sarcoma of bone in children and adolescents. J Cancer Res Clin Oncol 113:187–191
23. Miser J, Kinsella T, Triche T et al. (1988) Treatment of recurrent sarcomas in children and young adults: the use of a multimodality approach including ifosfamide (IFF) and etoposide (VP-16). Proc ASCO 7:258
24. Bork E, Hansen M, Dombernowsky, P et al. (1986) Teniposide (VM-26), an overlooked highly active agent in small-cell lung cancer. Results of a phase II trial in untreated patients. J Clin Oncol 4:524–527
25. Cullen MH, Smith SR, Benfield GFA, Woodroffe CM (1987) Testing new drugs in untreated small cell lung cancer may prejudice the results of standard treatment: a phase II study of oral idarubicin in extensive disease. Cancer Treat Rep 71:1227–1230
26. Cullen MH, Hilton C, Stuart NSA (1988) Evaluating new drugs as first treatment in patients with small-cell carcinoma: guidelines for an ethical approach with implication for other chemotherapy-sensitive tumors. J Clin Oncol 6:1356–1357 (letter)
27. Kellie SJ, De Kraker J, Lilleyman J et al. (1987) Ifosfamide in previously untreated disseminated neuroblastoma: results of study 3A of the European Neuroblastoma Study Group. Eur J Clin Oncol 24:903–908
28. De Kraker J, Pritchard J, Hartmann O, Ninane J (1987) Single-agent ifosfamide in patients with recurrent neuroblastoma (ENSG Study 2). Paediatr Hematol Oncol 4:101–104

Xenografts of Pediatric Solid Tumors: Predictive Intermediate Models? *

P.J. Houghton and J.A. Houghton

A. Introduction

Developing new therapy for treatment of solid tumors of childhood presents certain problems that are not encountered frequently with adult malignancies. In most instances, new agents are discovered through a process of serendipity, and certainly, design based upon metabolic characteristics of pediatric tumors has not been a major focus. Compounds showing potential against predominantly murine tumors and eliciting acceptable toxicity in preclinical toxicologic evaluations progress to traditional phase-I and phase-II clinical evaluation in adults. In the United States it is rare that a new anticancer drug will be tested in children simultaneously, or before adult phase-I trials have been completed. At this point in development, the new agent may fail to stimulate sufficient enthusiasm to further its evaluation in childhood malignancies and the compound may be discarded. However, there is no reason to assume that a failure to show activity against certain human tumors, predominantly of epithelial origin, will translate into a similar finding with childhood tumors of different origin. In fact, less than one in three new drugs evaluated in adults have received adequate testing against childhood malignancies. In part, this is a consequence of there being relatively few patients, precluding large-scale evaluation in randomized trials.

A second problem is that phase-II evaluation is conducted in children who have been exposed to multiple chemotherapeutic agents and have clearly resistant tumors. In addition, these patients can usually tolerate only reduced dose levels of a drug. Thus, a failure to demonstrate activity in these patients does not necessarily mean that the drug may not be active against the same tumor at diagnosis. However, evaluating new agents in previously untreated patients is in most cases not possible for childhood malignancies, because for many tumor types there is available, effective, and sometimes curative therapy. Thus, we find ourselves in a dilemma: clearly, we require new effective agents, but traditional means for identifying these are inadequate due to the reasons already given.

With this perspective we have approached the problem in the following manner. Our laboratories have been involved in developing preclinical models with the specific aim of ultimately developing histiotype-specific therapy. That is, instead of studying "generic cancer," we believe that new agents can be identified based upon the metabolic characteristics of particular histiotypes, or subtypes. The first step in testing such a hypothesis has been to develop specific models and to determine the relevance of these with respect to biologic, metabolic, and chemosensitivity characteristics. The approach we have taken is to establish human tumors in immune-deprived

---

Laboratories for Developmental Therapeutics, Dept. Biochemical and Clinical Pharmacology St. Jude Children's Research Hospital, Memphis, TN 38101, USA

* This work was supported in part by PHS awards CA 38933, CA 23099, CA 32613, CH-423 from the American Cancer Society and by the American Lebanese Syrian Associated Charities

mice. The models that will be dealt with here are rhabdomyosarcomas (RMS) of children and adult colon adenocarcinomas. These two types were chosen because they show quite different chemosensitivities and hence are of value for determining the relevance of this approach. In addition, we have established xenografts of rhabdomyosarcomas at relapse. Clearly, for the model to be of use these tumors should be significantly less sensitive than their counterparts heterotransplanted at diagnosis.

Results obtained with these models indicate that when used intelligently, human tumor xenografts may have a significant role in identifying agents that should receive priority for evaluation in childhood malignancies, and such data may be used further to justify evaluation of an agent in previously untreated children (with poor prognosis), even when it fails to demonstrate significant activity in classical phase-II evaluation against relapse tumors.

B. Material and Methods

I. Immune-deprived Mice

Four-week-old female CBA/CaJ mice were immune-deprived by thymectomy, followed at 3 weeks by total body irradiation (TBI; 925 cGy, ^{137}Cs source). Mice received either cytosine arabinoside (200 mg/kg) 48 h prior to TBI, or 3×10^6 nucleated marrow cells within 6 h of TBI [1, 2]. Mice were housed under conventional conditions in a humidity- and temperature-controlled environment, as described previously [1, 2].

II. Tumor Lines and Characteristics

Three models were used. Human colon adenocarcinomas were established from untreated primary lesions as described previously [2, 3]. These tumors retain morphologic, histologic, and karyotypic characteristics when grown in mice. For the first model of childhood RMS, lines were established from untreated tumor material, either primary or metastatic lesions. Characteristics of these tumors, histology, and karyotype have been presented in detail [1, 4]. In addition, five lines of RMS were established from patients with clearly progressive disease [5].

III. Experimental Design

Tumor material was transplanted into mice 2 weeks after TBI. Mice received a single administration of agent when the mean tumor diameter was >1 cm. Diameters were measured at 7-day intervals, and volumes were calculated as described previously [4]. Response criteria are presented in Table 1.

C. Results

The initial question that we posed was whether xenografts accurately represented the chemosensitivity of the tumor type from which they were derived. For this experiment we constructed three models, each comprising five or six tumor lines from untreated RMS, RMS at relapse, or adult colon adenocarcinoma. Colon adenocarcinoma xenografts were chosen, as this tumor type in man is essentially refractory to all conventional chemotherapeutic agents.

The responses of colon adenocarcinoma xenografts to single maximal tolerated dose (MTD) levels of seven agents are presented in Table 1. Clearly, these tumors in mice were poorly sensitive, with only methylCCNU, 5-fluorouracil, and cyclophosphamide showing marginal activity. Thus, the overall responsiveness of these tumors parallels the lack of response observed clinically.

In contrast, RMS at diagnosis is a relatively chemosensitive tumor, and this sensitivity is paralleled when these tumors are heterografted in immune-deprived mice. Responses to the MTD for known, clinically efficacious agents are presented in Table 2. Of particular note is the marked activity of vincristine, which at this dose level caused complete regres-

Table 1. Responses of colorectal xenografts

Agent/tumor	BR	AC_4	HC_1	GC_3	VRC_5	ELC_2
Cyclophosphamide	–	–	–	–	++	++
Actinomycin D		–	–	–		–
Doxorubicin		–	–	–	–	–
Vincristine			–	+	±	±
Fluorouracil	–	–	+	–	–	++
Methyl CCNU		+++	–	–	++	–
cis-DDP			–	–	–	–

Response criteria

Tumor response	Representation
No growth inhibition	–
Transient response, inhibition $<Td_2$[a]	±
Growth inhibition $\geq Td_2$	+
Growth inhibition $\geq 2 \times Td_2$	++
Growth inhibition $\geq 3 \times Td_2$	+++
Growth inhibition $\geq 3 \times Td_2$ + volume regression $\geq 50\%$	++++
Complete regression with subsequent regrowth	+++++
Complete regression with no regrowth of any tumors during the period of observation (≥ 84 days)	++++++

[a] Td_2, Mean time for tumor volume to double

Table 2. Responses of childhood RMS: diagnosis

Agent/tumor	Rh12	Rh18	Rh28	Rh30	Rh35	Rh39
Vincristine	++++++	+++	++++++	++++++	+++++	++
Cyclophosphamide	++	+++	++++	++	+	++++
Actinomycin D	–	++	++	–	±	–
Doxorubicin	++	±	+++	–	–	++

For response criteria see Table 1

Table 3. Ranking of agents in preclinical models and clinical trials

Drug	Xenograft[a] (%)	Clinic (%)
Vincristine	78	59
Cyclophosphamide	44	54
Doxorubicin	19	31
Actinomycin D	11	24

[a] Evaluated in six lines

sions in four of six tumor lines. A typical dose-response curve for RMS Rh12 treated with vincristine is shown in Fig. 1. It will be noted that the response curve is steep, and a fourfold reduction in dose of drug is sufficient to reduce its effect from "curative" to not significantly retarding tumor growth. This relationship has considerable importance when one is considering clinically relevant levels of drug resistance. The order of drug activity, both in the model and from clinical studies, appears to be in good agreement (Table 3).

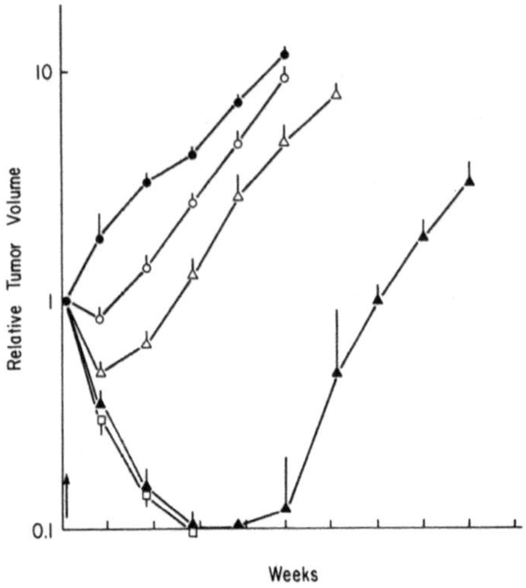

Fig. 1. Response of Rh12, childhood RMS, to increasing dose levels of vincristine (VCR). Mice received a single administration of VCR when tumors were ≥ 1 cm^3. Each curve represents the mean for 14 tumors. *Ordinate,* tumor volume relative to that at treatment; *abscissa,* time after treatment (●) No treatment control; (○) 0.375; (△) 0.75; (▲) 1.5, (□) 3.0 mg/kg vincristine

Table 4. Responses of childhood RMS: relapse

Agent/tumor	RD	LL	CD	Rh10
Vincristine	++	+	±	+
Cyclophosphamide	+	+	+	–
Actinomycin D	–	–	–	–
Doxorubicin	–	±	±	–

Response criteria as for Table 1

However, it was possible that the response of RMS xenografts was a consequence of host environment. Consequently, we established RMS lines from patients at relapse [5]. The responsiveness of these tumors to the standard clinically active agents is presented in Table 4. These tumors were significantly less sensitive to treatment than were the diagnosis specimens. These data therefore support the model as retaining chemosensitivity typical of the tumors from which it was derived.

Data derived from these models suggested that the RMS model for previously untreated tumors may be of value in identifying effective agents that had not been evaluated against this histiotype. Results with mitomycin C, *cis*-dichlorodiamminoplatinum, DTIC, and L-phenylalanine mustard (L-PAM) are presented in Table 5 and Fig. 2 [4]. Of these agents L-PAM showed very significant activity against five of six lines, causing complete regression of advanced tumor in four lines. Further, L-PAM had a fairly wide dose-response relationship [6].

Based on these data, a phase-I/II trial was initiated [7]. Initial studies showed that the MTD was 35 mg/m^2 given every 21 days. Pharmacokinetic profiles for L-PAM in children and in mice (Fig. 3) demonstrated that the area under concentration curve (AUC) and clearance were similar (Table 6). However, of 13 patients treated only one showed an objective response (Table 7). The difference between the clinical outcome and that predicted by the model (for diagnosis

Table 5. Responsiveness of xenografts of childhood rhabdomyosarcoma to DNA-reacting agents[a]

Agent/tumor	H × Rh12	H × Rh18	H × Rh28	H × Rh30	H × Rh35	H × Rh39
L-PAM	+++++	+++	++++++	+++++	++++++	++++++
Cyclophosphamide	++	+++	++++	++	+	++++
cis-DDP	+	++	++	+	+	++
Mitomycin C	–	+	++++	+	++++	+++++
DTIC	+	+++	+++++	++++	+++++	+++

[a] Agents administered as a single i.p. injection at equitoxic doses

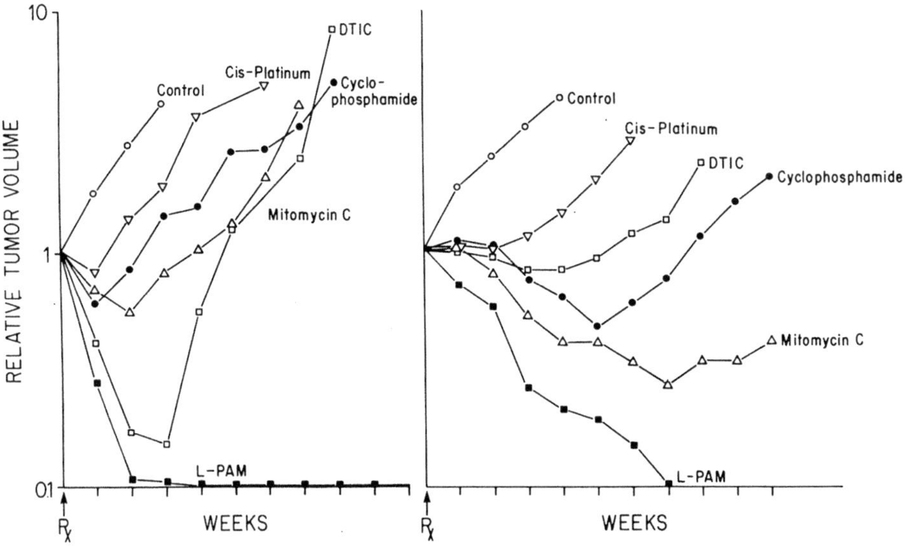

Fig. 2. Responses of Rh28 (*left*) and Rh39 (*right*) to DNA-interacting agents administered at the MTD. (From Houghton et al. [4])

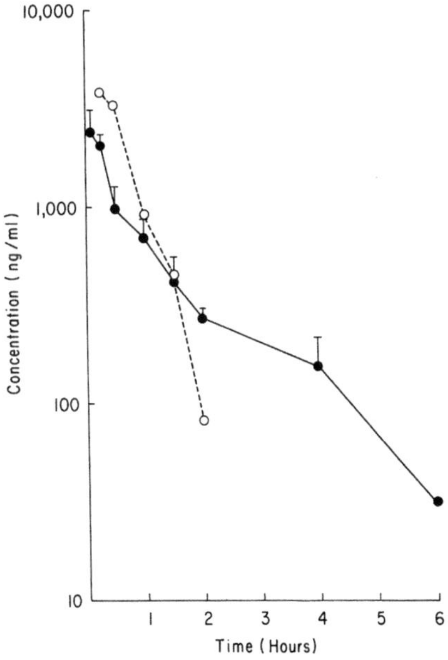

Fig. 3. Pharmacokinetic analysis of L-PAM in children and mice. Mice received 43 mg/m^2 as a single i.p. administration (o) and children received 35 mg/m^2 i.v. (●). *Bars* indicate standard deviation from the mean. (From Horowitz et al. [7])

Table 6. Pharmacokinetic parameters for L-phenylalanine mustard

	Child	Mouse
AUC	175000 μg/l/min	170000 μg/l/min
Clearance	232 mg/min/m^2	235 ml/min/m^2

RMS) was probably not due to different pharmacokinetics, as shown in Table 6, but was probably related to evaluation of L-PAM in relapse patients with drug-resistant tumors.

To test this hypothesis, a second clinical trial was undertaken in which L-PAM (45 mg/m^2) was administered to children diagnosed as having stage-4 (disseminated unresectable) RMS, for whom the prognosis was very poor. These patients have <15% long-term survival with conventional treatment. For these patients L-PAM demonstrated very significant activity, as shown in Table 7.

Table 7. Clinical features of 26 children with rhabdomyosarcoma who were treated with melphalan

Patient no.	Age (years) and sex	Histological classification	Prior therapy	Doses of melphalan	Treatment response	
					By site	Overall
Study I: Phase-II trial in previously treated patients						
1	11/F	Alveolar	VDCATR	1	Lung-PD	NR
2	21/M	Embryonal	VDCARER	4	Lung-CR, pelvis-PR	PR
3	12/M	Alveolar	VDCATR	2	Pelvis-SD, bone-PD	NR
4	19/M	Alveolar	VDCATR	2	Bone-PD, lung-PD	NR
5	11/F	Embryonal	VDCAPEMR	1	Brain-PD, lung-PD	NR
6	13/F	Embryonal	VDCATR	1	Chest wall-PD	NR
7	13/M	Embryonal	VDCATR	1	Pelvis-PD, lung-PD	NR
8	2/M	Embryonal	VACPEBR	1	Chest wall-PD	NR
9	12/M	Embryonal	VDCATR	2	Lung-SD	NR
10	15/F	Undifferentiated	VDCATR	1	Chest wall-PD	NR
11	1/F	Pleomorphic	VDCATR	1	Lung-PD	NR
12	15/M	Alveolar	VDCATR	3	Abdomen-OR, nodes-OR, pleural effusion-PD	NR
13	16/F	Alveolar	VDCA	1	Lung-PD	NR
Study II: Phase-II trial in newly diagnosed patients						
14	14/M	Alveolar	None	2	Hand-PR, extremity-CR, abdomen-CR	PR
15	<1/F	Undifferentiated	None	2	Extremity-PR, nodules-CR	PR
16	7/F	Embryonal	None	2	Head-PR	PR
17	8/F	Embryonal	None	2	Head-OR, lung-PD	NR
18	4/F	Embryonal	None	2	Pelvis-PR, pleural effusion-CR	PR
19	5/M	Embryonal	None	2	Head-PR	PR
20	2/M	Embryonal	None	2	Prostata-PR	PR
21	19/F	Alveolar	None	2	Pelvis-PR, CSF-CR bone marrow-CR	PR
22	9/F	Embryonal	None	2	Head-OR	NR
23	7/F	Embryonal	None	2	Head-PR	PR
24	3/M	Alveolar	None	1	Lung-SD	NR
25	14/F	Alveolar	None	2	Extremity-PR, bone-PR	PR
26	3/M	Embryonal	None	2	Head-PR	PR

Abbreviations: A, Doxorubicin; V, vincristine; D, dactinomycin; C, cyclophosphamide; P, cisplatin; E, etoposide; T, dacarbazine; M, methotrexate; B, dibromodulcitol; R, radiotherapy, OR, objective response; SD, stable disease; PD, progressive disease; PR, partial response; NR, no response; CR, complete response; CSF, cerebrospinal fluid

D. Discussion

The use of human tumor xenografts is now being accepted as a means by which phase-II preclinical evaluation may be used to prioritize and direct clinical testing of new agents. This may have considerable impact on the identification of new treatments for relatively rare tumors in adults and for most solid tumors of childhood. Extensive data from a series of models suggest that the following criteria may be useful in these studies: (a) models should be comprehensive and should attempt to encompass some of the heterogeneity observed in the clinical disease. We have routinely used six independently derived lines for each model, and this appears adequate; (b) response criteria should be similar to those used clinically, i.e., 50% tumor regression would be required at the MTD for a given agent; (c) to justify further evaluation of an agent in previously untreated, poor-prognosis patients, the agent should have demonstrated activity similar or superior to the most effective agent used in that disease, and in phase-I trials should have shown similar AUC and clearance to that in mice.

For most agents being evaluated in children, pharmacokinetic data from adults are often available. Thus, it is relatively easy to determine whether the mouse pharmacokinetics are similar to those that may be anticipated in children. Consequently, an agent that is tolerated in mice at dose levels that would be clearly toxic in man can be eliminated prior to trial in children. A more difficult decision is to move an "unknown" compound from the xenograft testing phase into a clinical trial. Under these conditions human tolerance has to be defined, and clearly, the mouse model may either overpredict or underpredict for efficacy in human beings. However, the criteria discussed above appear valid in this situation as well.

Acknowledgments. We thank Marc Horowitz, M.D., and Michael Christensen, Pharm. D., for fruitful collaboration.

References

1. Houghton JA, Houghton PJ, Webber BL (1982) Growth and characterization of childhood rhabdomyosarcomas as xenografts. JNCI 68:437–443
2. Houghton JA, Houghton PJ (1980) On the mechanism of cytotoxicity of fluorinated pyrimidines in four human colon adenocarcinoma xenografts maintained in immune-deprived mice. Cancer 45:1159–1167
3. Houghton JA, Taylor DM (1978) Maintenance of biological and biochemical characteristics of human colorectal tumours during serial passage in immune-deprived mice. Br J Cancer 37:199–212
4. Houghton JA, Cook RL, Lutz PJ, Houghton PJ (1984) Childhood rhabdomyosarcoma xenografts: response to DNA-interacting agents and agents used in current clinical therapy. Eur J Cancer Clin Oncol 20:955–960
5. Houghton JA, Houghton PJ, Green AA (1982) Chemotherapy of childhood rhabdomyosarcomas growing as xenografts in immune-deprived mice. Cancer Res 42: 535–539
6. Houghton JA, Cook RL, Lutz PJ, Houghton PJ (1985) L-Phenyl-alanine mustard (NSC 8806): a potential new agent in the treatment of childhood rhabdomyosarcoma. Cancer Treat Rep 69:91–96
7. Horowitz ME, Etcubanas E, Christensen M, Houghton JA, George SL, Green AA, Houghton PJ (1988) Predictability of pediatric rhabdomyosarcoma xenografts for melphalan activity in previously untreated patients: a model for development of cancer therapy. J Clin Oncol 6:308–314

Cell Biology

Stromal-Hematopoietic Interrelationships: Maximov's Ideas and Modern Models

A. Friedenstein

The idea of stromal-hematopoietic cell interactions was the essential part of Alexander Maximov's theory of hematopoiesis, which he proposed more than 60 years ago. According to Maximov (see Figs. 1–4), committed hematopoietic precursors descend from the hematopoietic stem cells due to local impacts generated by marrow stroma; this creates the conditions for hematopoietic cell differentiation [1]. Maximov's theory was far ahead of his time, and, though Maximov was highly respected in the scientific community, his concept of local "differentiation conditions" operative in hematopoiesis was met with particular skepticism. Today, Maximov's idea raises no doubt; in fact, it constitutes the essence of the problem of hematopoietic microenvironment (HME). What provokes discussions in modern hematology is the exact types of stromal cells responsible for HME and the mechanisms of stromal-hematopoietic cell interactions. Maximov assumed that the stromal cells in question were stromal fibroblasts (reticular cells), but for a long time many experimental hematologists denied this. Only recently has it been possible to apply two experimental models for checking the microenvironmental functions of marrow fibroblasts. The first model is the transfer of HME by heterotopic transplantation of marrow cells; the second is the establishment of HME in vitro by stromal cell underlayers in Dexter cultures.

The Gameleya Institute for Epidemiology and Morphology, Academy of Vet. Med. Sciences of USSR, Immunomorphol. Lab., Gameleya Street 18, Moscow D 98, USSR

Heterotopic transplantation of marrow cells results in the formation of marrow organs covered by a bone capsule [2–5]. Their hematopoietic cells are of the recipient origin [6], indicating that engraftment of some category of marrow cells results in the formation of bone and an HME suitable for population by hematopoietic cells and for their proliferation and differentiation. Heterotopic marrow can be retransplanted repeatedly with similar results, provided the recipients are compatible with H-2 antigens of the initial donor, not of the intermediate recipients [7–8]. This means that HME is transferred by engraftment of the marrow cells which remain unreplaced by the recipient cells. Chromosome typing of clonogenic stromal fibroblasts (CFUf) of the heterotopic marrow confirmed their donor origin [9, 10], and the problem was to check whether stromal fibroblasts were able to transfer HME when grafted heterotopically.

The in vitro descendents of CFUf after several passages compose diploid fibroblast cultures [11–13]. Tested by heterotopic transplantation, they were found to form bone marrow organs, while engraftment of cultured spleen fibroblasts (the descendents of spleen CFUf) produced lymphoid organs [14, 15]. Thus, cultured marrow fibroblasts appear to be able to transfer bone marrow HME. Depending on the origin of marrow fibroblast cultures (the source CFUf being from red or yellow marrow), their engraftment transferred not only the general pattern of HME, but also such details as the density of hematopoietic cells in a would-be marrow [16].

Cultured marrow fibroblasts produce hematopoietic growth factor (M-CSF, G-

CFS, GM-CFS, BFUf- and mixed-colony-CSF) which can be detected in the culture medium [17–20]. They regulate proliferation and differentiation of GM-CFU: their stimulatory effects were noted when the target marrow contained few spontaneous colonies, the inhibitory effects when large numbers of spontaneous GM-CFU were present [21]. Hematopoietic growth factors are also produced by cloned lines of marrow fibroblasts [22]. However, the direct proof of in vitro microenvironmental competence of marrow fibroblasts was their ability to establish HME in Dexter-type cultures. It has been shown [23] that when used as underlayers, the passaged murine marrow fibroblasts, free from macrophages and endothelial cells, supported hematopoiesis if seeded with stromal cell-depleted marrow suspensions.

Thus, cultured marrow fibroblasts transfer HME, release hematopoietic growth factors in vitro, and are capable of presenting them in a proper way to support hematopoiesis in cultures. This confirms Maximov's hypothesis of the role of marrow fibroblasts in hematopoiesis.

The population of marrow fibroblasts is probably a heterogeneous one, and there is no evidence that marrow fibroblasts which produce or present hematopoietic growth factors are the same cells which transfer HME, and vice versa. It may well be that there are several subpopulations of marrow fibroblasts with different microenvironmental functions. At present, fibroblasts including those from nonhematopoietic and hematopoietic organs look much alike, reminiscent of the situation with lymphocytes in Maximov's time. The main and most conclusive sine of fibroblasts (mechanocytes) is interstitial collagen types I and III synthesis, and few markers of their phenotype and genetic diversity have been so far ascertained. The diversity does exist, for instance, between marrow as compared with spleen fibroblasts, which is proved by the results of their heterotopic transplantation. The next question regarding HME seems to be the diversity of marrow fibroblasts including their clonogenic precursor cells.

In primary cultures of marrow cell suspensions the CFUf (CFCf) form adherent-cell colonies which are cell clones [24, 25]. The colonies are composed of fibroblasts which synthesize type-I and -III collagen and fibronectin and lack macrophage markers and VIII-factor-associated antigen [26–30]. Morphologically, the colonies are distinctly heterogeneous within each culture. Some are composed of elongated or blanket-like fibroblasts or of a mixture of both; the colonies may include fat cells or have a mineralized intercellular matrix [39]. These differences can hardly be regarded as markers of CFCf, the diversity not beeing stable at passaging and recloning.

In situ CFCf are outside the cycle arrested in Go [31]. Marrow fibroblasts possess PDGF receptors [32] and in medium with platelet-poor plasma their proliferation and the CFUf colony formation requires PDGF [33, 34]. It is believed that serum growth factors, which include PDGF, are sufficient for recruitment of CFCf into the cycle and that CFUf colony formation in serum-supplemented medium does not require additional growth stimulation. Yet this is probably not the case.

The efficiency of CFUf colony formation (CFEf) drops close to zero in low-density marrow cultures if they are depleted of nonadherent cells: 85% of CFCf do not proliferate at all or pass through one to three cell doublings (Fig. 1). On the other hand, the CFEf increases dramatically when such adherent marrow cell cultures are supplemented with irradiated marrow feeder cells or with platelets. This colony-stimulating activity is not replaced by additional PDGF and is expressed only in the serum-rich medium. Being stimulated by platelets each fibroblast precursor present in marrow cell suspensions turns out to be a clonogenic stromal cell (Fig. 1). Thus, nonstromal marrow cells which accompany CFCf in marrow cultures

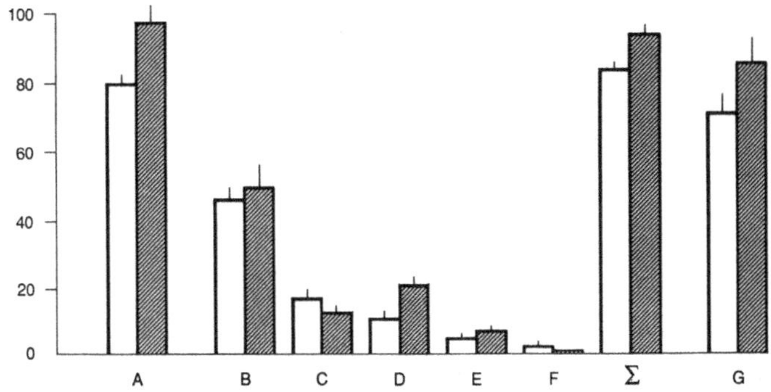

Fig. 1. CFUf colony formation in mice adherent marrow cell cultures. Cultures were initiated by injecting 5×10^5 mechanically (white colomns) or 5×10^4 trypzinised (black columns) marrow cells per culture flask (25 cm^2). Two hours after explantation the nonadherent cells were decanted from all cultures and further cultivation accomplished in aMEM medium plus 20% embryonal calf serum, part of the cultures (G) being additionally suplemented with 10^7 irradiated (60 Gy) marrow cells. Abscissa: *A–E* – fibroblast foci, fibroblast colonies and single fibroblasts in feeder non-supplemented cultures. *A* – single fibroblasts in one day cultures; *B–F* – 10 day cultures: *B* – single fibroblasts, *C* – two fibroblasts foci, *D* – three-eight fibroblasts foci, *E* – nine-forty nine fibroblasts foci, *F* – fibroblast colonies composed of 50 and more fibroblasts, $\sum$ – sum of *B, C, D, E* and *F* per culture. *G* –fibroblasts colonies in 10 days feeder-supplemented cultures. Ordinate: mean numbers ($M \pm m$) of single fibroblasts, fibroblast foci and fibroblast colonies for 3–5 cultures.

(probably megakaryocytes) provide growth-stimulating factors for CFUf colony formation. There are indications that CFCf are sensitive also to other growth-stimulating factors which induce the formation of fibroblast colonies with a different composition of matrix proteins. It has been reported [35] that marrow cells cultured in methylcellulose-clotted plasma with cortisone and PHA-stimulated leukocyte-conditioned medium produced fibroblast colonies with collagen type IV and laminin, in addition to collagen types I and III and fibronectin present in CFUf colonies, in liquid cultures with the serum-supplemented medium. The differences suggest either that there is a diversity of CFCf, which also require different colony-stimulating factors, or that the same CFCf can generate different descendents, depending on the stimulating factors used to induce colony formation.

Marrow CFCf diversity was demonstrated with regard to their proliferative and differentiative potencies. Only a small portion (10%) of single CFUf colonies transferred HME when grafted heterotopically, i.e., formed bone marrow organs [36]. At least 30% of CFCf appeared to be highly proliferative cells which provide single-colony-derived fibroblast cultures with 20–30 population doublings. When tested by transplantation of cells in diffusion chambers, 20% of these cultures formed simultaneously bone, cartilage, and reticular-like tissue, 30% formed only bone, and 27% only reticular-like tissue. The number of osteogenic units in late passages of cultured fibroblasts exeeded by far the total numbers of the initially explanted marrow cells, indicating that osteogenic precursors intensively multiplied within cultures [37]. There are reasons to consider CFCf with osteochondrogenic potencies as being osteogenic stem cells [38, 39]. One can assume that some of them are the progenitors of a marrow stromal lineage which includes committed osteogenic precursors, mature bone cells, and microenvironmentally competent fibroblasts (reticular cells). The assumption is backed up by the obligatory association

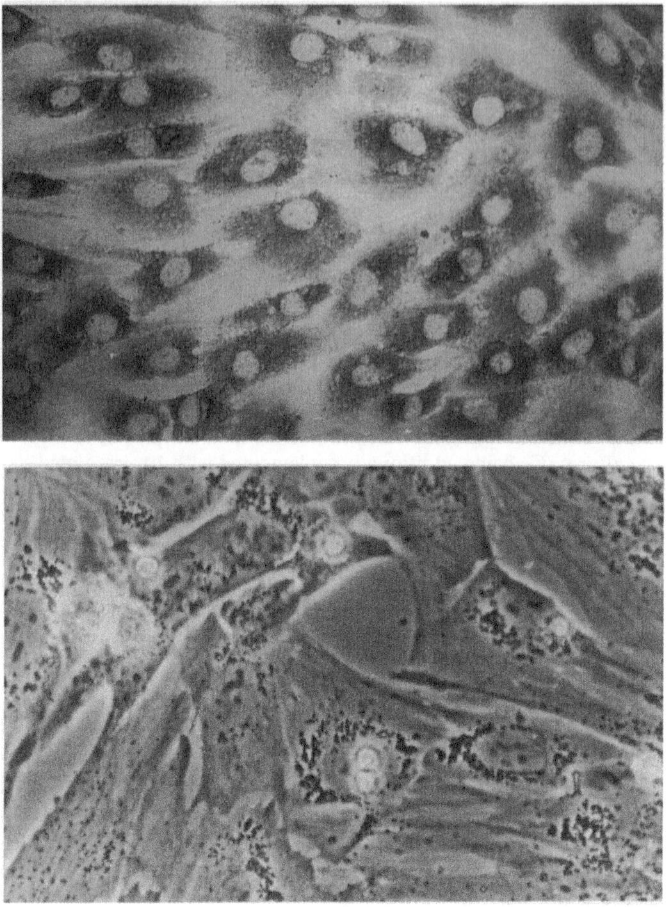

a

b

Fig. 2. Type I collagen in 12 day CFUf colony of guinea pig periferal blood leukocytes. Anticollagen antiserum, immunoperoxidase reaction (**a**). Live culture (**b**).

of HME transfer with bone formation, which applies to heterotopic transplantation of both freshly isolated marrow and single-CFUf-derived fibroblast colonies. In the heterotopic marrow the CFUf are of donor origin [9, 10], and it is reasonable to assume that the same applies to the microenvironmentally competent reticular cells. However, the ability of fibroblasts from single CFUf-colony-derived heterotopic bone marrow organs to support hematopoiesis in vitro, and their donor origin (which would be the proof of the above speculation) was not tested up to now. Anyway, the hierarchy of marrow precursors awaits further studies.

As far as Maximow's contribution to the problems of HME is concerned, it is impossible to omit his last work, entitled "Cultures of blood leukocytes. From lymphocyte and monocyte to connective tissue." [40]. It describes the formation of fibroblasts in plasma-clot cultures of guinea-pig blood cells. Subsequently, his results were put in question on the grounds of two possible objections, namely that the source of fibroblasts might be fragments of vessel walls which contaminate the blood during sampling, and that the cells in question were not fibroblasts (for references, see [41]). The first objection proved to be invalid when a CFUf colony assay was carried out

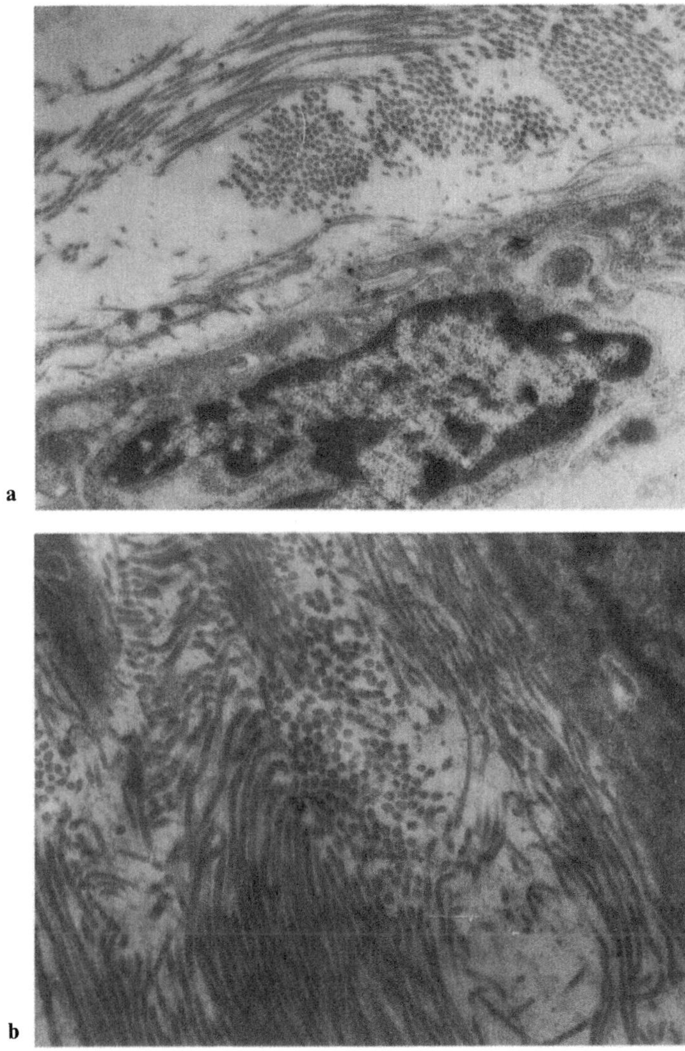

Fig. 3. Fibroblasts and collagen fibrils in 16 day CTUf colonies of rabbit periferal blood leukocytes. E. M.

with blood cells. It turned out that the incidence of CFUf colonies in guinea-pig and rabbit leukocyte cultures did not change with the number of punctures performed for blood sampling [42]. It has also been shown that fibroblasts in blood-derived CFUf colonies synthesize collagen type I [43] and lack VIII-factor-associated antigen and macrophage determinant *Mac*I [44], which confirms their fibroblast nature (Fig. 2, 3). It remains unknown from where CFUf migrate into blood, where they settle (if they do), and why blood-derived CFUf are not detectable in some mammals, including human beings. The presence of fibroblast precursors in blood discovered by Maximov is related to many unsolved problems of HME, in particular, to the possibility of CFUf repopulation; CFUf circulation in blood does not prove it at all.

Fig. 4. Professor Alexander Maximov

The story of the circulating fibroblast precursor cells demonstrates once again that not only Maximov's ideas, but also his experimental results are so topical that Professor Alexander Maximov almost remains a participant of present-day research (Fig. 5).

Fig. 5. Maximov in his tissue culture laboratory in the Military Medical Academy in Petersburg (1915)

Fig. 5a. Preparation of plasma for plasma-clot cultures

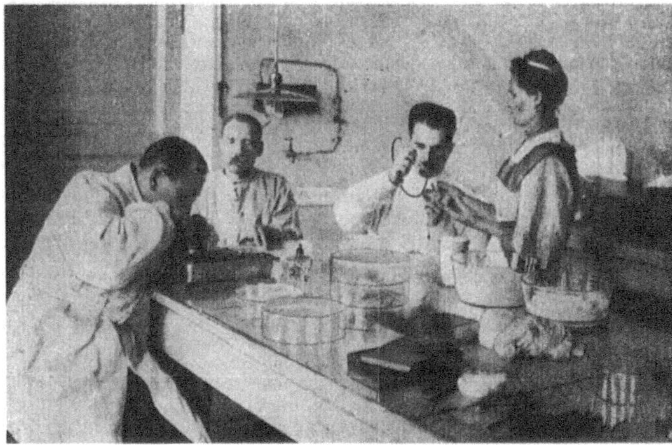

Fig. 5b. Placing tissue fragments in culture medium

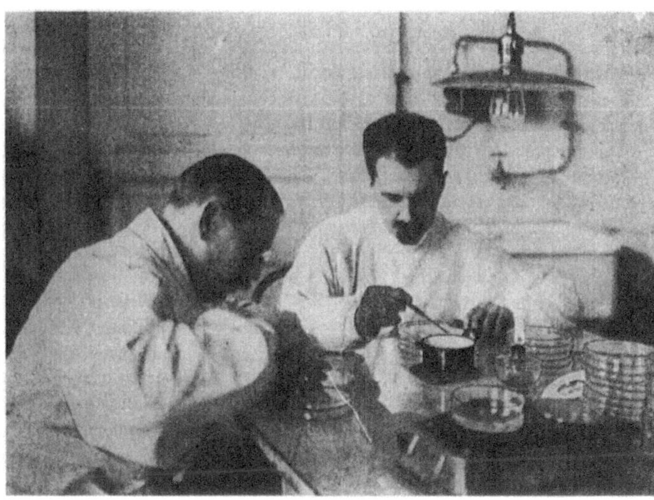

Fig. 5c. Kaissug hangrug-drop cultures in hallow-ground microscope slides.

References

1. Maximov AA (1906) Über experimentelle Erzeugung von Knochenmarks-Gewebe. Anat Anz 28:24–38
2. Tavassoli M, Crosby WH (1968) Transplantation of marrow to extramedullary sites. Science 161:54–56
3. Tavassoli M, Maniatis A, Binder RA, Crosby WH (1971) Studies on marrow histogenesis. Proc Soc exp Biol Med 138:868–870
4. Tavassoli M, Friedenstein A (1983) Hemopoietic stromal microenvironment. Am J Hemat 15:195–203
5. Friedenstein AJ, Latzinik NV, Grosheva AG, Gorskaya UF (1982) Marrow microenvironment transfer by heterotopic transplantation of freshly isolated and cultured marrow cells in porous sponges. Exp Hematol 10:217–227
6. Friedenstein AJ, Petrakova KV, Kuralesova AI, Frolova GF (1968) Heterotopic transplants of bone marrow. Analysis of precursor cells for osteogenic and hemopoietic tissues. Transplantation 6:230–247
7. Friedenstein AJ, Kuralesova A (1971) Osteogenic precursors of bone marrow in radiation chimeras. Transplantation 12: 99–108

8. Friedenstein A, Luria E (1980) Cellular bases of hemopoietic microenvironment. Moscow, Medicine (in Russian)
9. Friedenstein AJ et al. (1978) Origin of bone marrow stromal mechanocytes in radiochimeras and in heterotopic transplants. Exper Hematol 6:440–444
10. Friedenstein A (1976) Precursor cells of mechanocytes. Inter Rev Cytol 47:327–359
11. Friedenstein AJ, Chailakhjan RV, Lalykina KS (1970) The development of fibroblast colonies in monolayer cultures of guinea-pig bone marrow and spleen cells. Cell Tissue Kinet 3:393–403
12. Miskarova ED, Lalykina KS, Kokorin IN, Friedenstein AJ (1970) Osteogenic potencies of prolonged diploid cultures of myeloid cells. Bull Exp Biol Med 56:78–81
13. Friedenstein A (1973) Determined and inducible osteogenic precursor cells. Ciba Found Sympos (new series) 11:170–185
14. Friedenstein AJ et al. (1974) Stromal cells responsible for transferring the microenvironment of hemopoietic tissues. Cloning in vitro and retransplantation in vivo. Transplantation 17:331–340
15. Friedenstein A (1980) Stromal mechanocytes of bone marrow: cloning in vitro and retransplantation in vivo. In: Thierfelder S, Rodt H, Kolb H (eds) Immunology of bone marrow transplantation. Springer-Verlag 19–29
16. Patt HM, Maloney MA, Flannery ML (1982) Hemopoietic microenvironment transfer by stromal fibroblasts derived from bone marrow varying in cellularity. Exp Hematol 10:738–742
17. Akasaka M et al. (1987) Production of monoclonal antibody to adult bone marrow preadipocite line (H-I/A). Exp Hematol 15:610–618
18. Brockbank K, Van Peer CM (1983) Colony-stimulating activity produced by hemopoietic organ fibroblastoid cells in vitro. Acta Hematol 69:369–476
19. Zucali J et al. (1986) Interleukin I stimulates fibroblasts to produce granulocyte macrophage colony-stimulating activity and prostoglandin E_2. J Clin Invest 78:1306–1323
20. Lee M, Segal GM, Bagby GC (1987) Interleukin I induces human bone marrow-derived fibroblasts to produce multilineage hemopoietic growth factors. Exper Hematol 15:983–988
21. Brondy V et al. (1986) Monocytes stimulate fibroblastoid bone marrow stromal cells to produce multilineage hemopoietic growth factors. Blood 68:530–537
22. Greenberger BR, Wilson FD, Woo L (1981) Granulopoetic effects of human bone marrow fibroblastic cells and abnormalities in "granulopoietic microenvironment". Blood 58:557–563
23. Brockbank KCM, De Jong JP, Piersma AH, Voerman JSA (1986) Hemopoiesis on purified bone-marrow-derived reticular fibroblasts in vitro. Exper Hematol 14:386–394
24. Latzinik N et al. (1986) The content of stromal colony-forming cells (FCFC) in the mousebone marrow and the clonal nature of FCFC-derived fibroblast colonies. Ontogenesis 1:27–35
25. Friedenstein A et al. (1989) On clonality of CFUf-derived marrow stromal colonies (in press)
26. Latzinik NV, Sidorovich SY, Tarchanova IA (1980) Studies of surface receptors of the stromal mechanocytes of hemopoietic organs. Immunology 1:26–28
27. Castro-Malaspina H et al. (1980) Characterization of human bone marrow fibroblast colony-forming cells (CFUf) and their progeny. Blood 56:286–301
28. Wilson FD, O'Grady L, Mc Meil GJ, Munn SL (1974) The formation bone marrow-derived fibroblastic plaques in vitro. Exper Hematol 2:343–349
29. Friedenstein A et al. (1974) Precursors for fibroblasts in different populations of hemopoietic cells as detected by the in vitro colony assay method. Exper Hematol 2:83–92
30. Friedenstein AJ, Gorskaya UF, Kulagina NN (1976) Fibroblast precursors in normal and irradiated mouse hemopoietic organs. Exper Hematol 4:267–274
31. Keilis-Borok IV, Latzinik NV, Epichina SY, Friedenstein AJ (1971) Dynamics of the formation of fibroblast colonies in monolayer cultures of bone marrow, according to ^{3}H-thymidine incorporation experiments. Cytologia 13:1402–1409
32. Bowen-Pope DF, Seifert RA, Ross R (1985) The platelet-derived growth factor receptor. In: Boyton L, Leffert H (eds) Control of animal cell proliferation, Acad Press 1:281–312
33. McIntyre AP, Bjornson BH (1986) Human bone marrow stromal cell colonies: response to hydrocortisone and depen-

dence of platelet-derived growth factor. Exper Hematol 14:833–839
34. Wang OR, Wolf NS (1987) The effect of several growth factors on the in vitro growth of bone marrow stromal cells. Exper Hematol 15:610
35. Lim B et al. (1986) Characterization of reticulofibroblastoid colonies (CFU-RF) derived from bone marrow and long-term marrow culture monolayers. J Cellul Physiol 127:45–54
36. Chailakyan RK, Gerasimov YF, Friedenstein AJ (1984) Content of osteogenic precursor cells in the bone marrow and their proliferation in cultures. Bull Exp Biol Med 11:605–608
37. Gerasimov Y, Friedenstein AJ, Chajlakjan RK, Shiskova VV (1986) Differential potentiality of clonal strains of bone marrow fibroblasts. Bull Exp Biol Med 6:717–719
38. Friedenstein AJ, Chajlachyan RK, Gerasimov YF (1987) Bone marrow osteogenic stem cells in vitro cultivation and transplantation in diffusion chambers. Cell Tissue Kinet 20:263–272
39. Owen ME, Friedenstein AJ (1988) Stromal stem cells: marrow derived osteogenic precursors. In: Evered D, Harnett S. (eds) Cellular and molecular biology of vertebrate hard tissues. Ciba Found Symp 136:42–52
40. Maximov A (1928) Cultures of blood leucocytes. From leucocyte and monocyte to connective tissue. Arch exp Zellforsch 5:169–178
41. Jacoby F (1965) Macrophages. In: Willmer E (ed) Cells and tissues in culture, Acad Press 1:1–93
42. Luria EA, Panasyuk AF, Friedenstein AJ (1971) Fibroblast colony formation from monolayer cultures of blood cells. Transfusion 11:345–349
43. Luria E et al. (1989) Colony forming fibroblast precursors in circulating blood. Bull Exp Biol Med (in press)
44. Piersma AH, Ploemacher RE, Brockbank KG (1985) Migration of fibroblastoid stromal cells in murine blood. Cell Tissue Kinet 18:589–595

Bone Marrow Stromal Cells in Myelodysplastic Syndromes and Acute Nonlymphocytic Leukemia

E. Elstner, M. Wächter, and R. Ihle

A. Introduction

Despite some recent insights into the structural composition and function of hemopoietic stroma, there are still many open questions. One of these is whether there is any connection of stromal cells with the process of leukemic transformation in patients with acute leukemia.

We investigated whether bone marrow cells (BMCs) from patients with myelodysplastic syndromes (MDS) or acute nonlymphocytic leukemias (ANLL) are altered in their ability to form adherent stromal layers with active hematopoiesis in vitro and whether this depends on the stage of disease.

B. Material and Methods

BMCs were obtained from 24 normal volunteers, 28 patients with ANLL in different stages of the disease, and 9 patients with various forms of MDS: 4 with refractory anemia (RA), 4 with refractory anemia with excess of blast cells (RAEB), and 1 with RAEB in transformation. For studying the hematopoietic stroma we used modified Dexter liquid culture (Fig. 1).

C. Results

There were no differences between the stromal layers of patients with ANLL in complete remission (CR) and those of normal volunteers after 2 weeks of cultivation (Fig. 2). In most cases, however, both BMCs from patients with ANLL before treatment and BMCs from patients in relapse formed poorly adherent stromal layers. In 6 cases (4 with RA and 2 with RAEB) of 9 we observed the normal stromal grade in liquid culture of BMCs from patients with MDS (Fig. 3). It is of interest that the patient with RAEB in transformation showed no adherent layer in our system.

Growth characteristics of BMC from patients with ANLL and MDS in liquid culture are grouped in Fig. 4.

Bone marrow cultures from patients with ANLL in CR and those from patients with MDS, who established stromal layers, had quantitatively normal growth characteristics in liquid culture. However, there were qualitative differences in the nonadherent cell population between normal volunteers and ANLL patients in CR. In most cases of patients with ANLL in CR we found morphologically recognizable erythroid cells after 2 weeks culture in the non-adherent cell

Table 1. Growth characteristics of bone marrow cells in Dexter liquid culture

SG:	Stromal grade. Each Petri dish was assigned a score from 1 to 4, corresponding to a stromal layer covering from 25% to 100% of the area of the culture dish.
NAC:	Nonadherent cells (trypan-blue – negative)
F:	Fibroblast-like cells
CS:	"Cobble stone" (active hematopoietic areas)

Hämatologische Abteilung, Universitätsklinik für Innere Medizin, Bereich Medizin (Charité), Humboldt-Universität zu Berlin, GDR

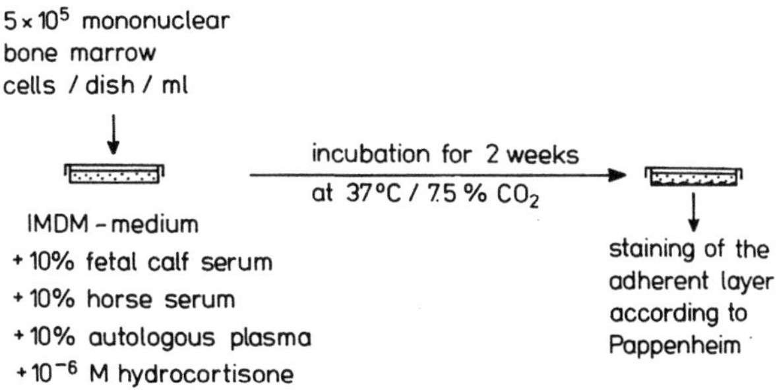

Fig. 1. Schematic representation of the method

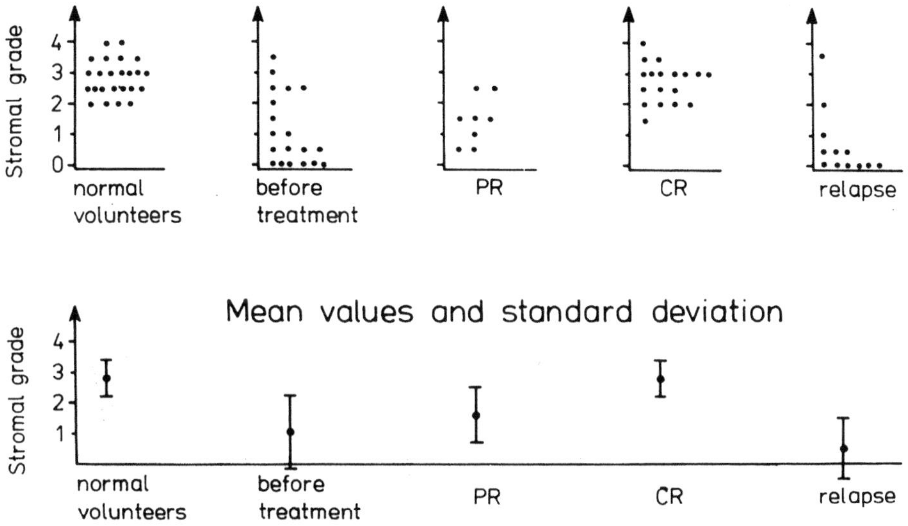

Fig. 2. Adherent layer in Dexter liquid culture (day 14) from bone marrow cells of patients at different stages of acute nonlymphoblastic leukemia

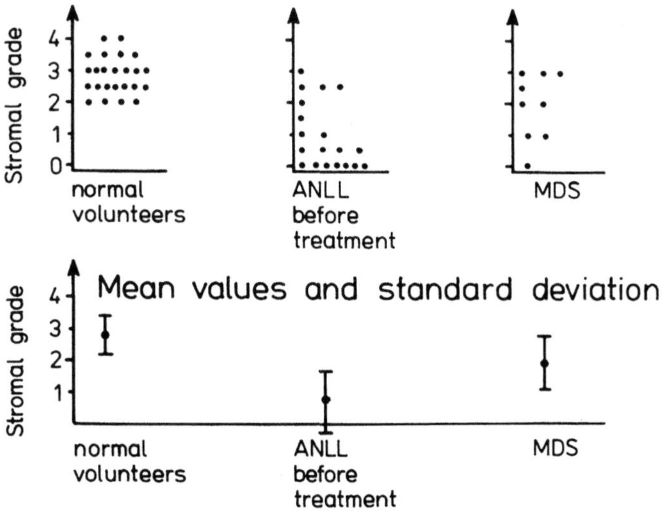

Fig. 3. Adherent layer in Dexter liquid culture (day 14) from bone marrow cells of patients with myelodysplastic syndromes (MDS) and acute nonlymphocytic leukemia (ANLL) before treatment

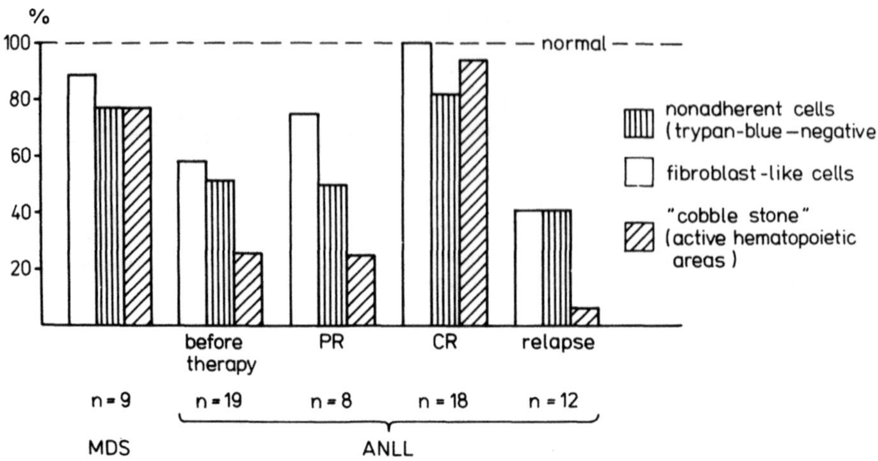

Fig. 4. Growth characteristics of bone marrow cells from patients with myelodysplastic syndromes (MDS) and acute nonlymphoblastic leukemia at different stages of disease in Dexter liquid culture (day 14)

population, which were not seen with normal volunteers. The growth characteristics of patients with ANLL in relapse and patients before therapy showed not only a poor formation of stromal layers, but there were also predominantly blast aggregations in the nonadherent cell population.

D. Discussion

Our results show that the ability of BMCs from patients with ANLL to form adherent stromal layers with active hematopoiesis depends on the stage of disease. Before therapy and in relapse we observed poor formation of stroma in most cases. In the nonadherent cell population there were predominantly blast aggregations in the 2-week culture. Nagao et al. reporting on CFU-F in patients with acute and chronic myelocytic leukemias, observed significant suppression of fibroblast colony formation at the time of diagnosis. The suppression was relieved during chemotherapy-induced remission. However, during relapse the level of CFU-F was again low. There is a close relationship between the leukemic disease and the functional state of bone marrow stroma [1]. At present, the cause of marrow stromal deficiency in leukemia remains unclear. Cocultivation of normal BMCs and leukemic cells reveals an inhibitory activity of leukemic blasts on CFU-F from normal marrow. Inhibition of CFU-F from normal marrow was also induced by leukemic-cell conditioned medium [2].

Bone marrow cells from ANLL patients before therapy are able to form a normal stromal layer with active hematopoiesis in more than four-week-old Dexter culture [3]. Normal stromal cells are present even in a marrow with 80%–90% leukemic blast cells. It is of interest that in our system BMCs showed a higher capacity to form stroma with active hematopoiesis before therapy (25%) than they did in relapse (6%). This finding suggests that in relapse of ANLL the normal stromal progenitors are either strongly reduced, absent, or changed, probably by the chemotherapy. It is known that chemotherapeutic agents impair stroma function [4–6]. We observed a qualitative difference between normal volunteers and some ANLL patients in CR in the nonadherent cell population over the stromal layer, in that morphologically recognizable erythroid cells were found. This is of interest, because normally in Dexter culture the development of erythroid progenitors is usually arrested at the stage of primitive burst-forming units (BFU-E) [7, 8].

The CR induction of acute leukemia is associated with the appearance of normal colony formation of bone marrow myeloid progenitor cells. However, some abnormalities of CFU-GM and CFU-E proliferation have been observed to persist in remission by several groups [9–13]. The erythroid differentiation to morphologically recognizable erythroid cells from BMCs of patients in CR in our system could be an indication of harmful effects on the balance of hematopoiesis caused by previous infiltration with leukemic cells and/or high-dose chemotherapy.

References

1. Nagao T, Komatsuda M, Yamanchi K, Arimori S (1981) Fibroblast colonies in monolayer cultures of human leukemia. J Cell Physiol 108:155–161
2. Ben-Ishay Z, Prindull G, Borenstein A, Shaton S (1984) Bone marrow stromal deficiency in acute myeloid leukemia in mice. Leuk Res 8:1057–1064
3. Coulombel L, Eaves CJ, Dubé ID, Kalousek DK, Eaves AE (1984) Variable persistence of leukemic progenitor cells in long-term CML and AML marrow cultures. In: Wright DG, Greenberger JS (eds) Long-term bone marrow culture. Liss, New York, pp 243–254
4. Adler SS, Kuznetsky RD (1984) Effects of chemotherapeutic agents on hemopoietic stromal tissue. 1. Effects of cyclophosphamide and bleomycin on hemopoiesis in subcutaneous femur implants. Exp Hematol 12:153–161
5. Elstner E, Goldschmidt H, Wächter M, Ihle R (1987) Sensitivity of stromal elements from human bone marrow cells to cytosine arabinoside in vitro. In: Neth R, Gallo RC, Greaves M, Kabisch H (eds) Modern trends in human leukemia. VII. Springer, Berlin Heidelberg New York, pp 177–180
6. Hays EF, Hale L, Villareal B, Fitchen JH (1982) "Stromal" and hemopoietic stem cell abnormalities in long-term cultures of bone marrow from busulfan-treated mice. Exp Hematol 10:383–392
7. Gregory CJ, Eaves AC (1978) In vitro studies of erythropoietic progenitor cell differentiation. Cold Spring Harbor Conf Cell Proliferation 5:179–192
8. Testa NG, Dexter TM (1977) Long-term production of erythroid precursor cells (BFU-E) in bone marrow cultures. Differentiation 9:193–195
9. Gustavsson A, Olofsson T, Olsson I (1979) Oscillations of marrow culture growth in acute myeloid leukemia during remission induction and remission. Acta Haematol 62:241–246
10. Mertelsmann R, Moore MAS, Broxmeyer ME, Cirrincione C, Clarkson B (1981) Diagnostic and prognostic significance of the CFU-C assay in acute nonlymphoblastic leukemia. Cancer Res. 41:4844–4848
11. Peschel C, Konwalinka G, Geissler D (1983) Studies of myelopoiesis in vitro on blood and bone marrow cell of patients with acute leukemia in long-term-remission. Leuk Res 7:397–406
12. Peschel G, Konwalinka G, Konwalinka DG, Geissler D, Braunsteiner H (1986) In vitro growth of erythropoietic progenitor cells in long-term remission of acute leukemia. Int J Cell Cloning 4:186–191
13. Tebbi K, Rubin S, Cowan DM, McCulloch EA (1976) A comparison of granulopoiesis in culture from blood and bone marrow cells of nonleukemic individuals and patients with acute leukemia. Blood 48:235–243

Growth and Development of Haemopoietic Cells: A Deterministic Process? *

C.P. Daniel, I.L.O. Ponting and T.M. Dexter

A. Introduction

I. Growth Factors

Haemopoietic colony stimulating factors (CSFs) are defined by their ability to stimulate the clonal expansion of bone marrow cells in semi-solid medium. Four main CSFs, interleukin-3 (IL-3), granulocyte macrophage-CSF (GM-CSF), macrophage-CSF (M-CSF) and granulocyte-CSF (G-CSF), have been described. Although all four CSFs are glycoproteins, they are otherwise unrelated. M-CSF is the only dimer, the others being monomers with disulphide bridges. Each factor binds to its own distinct receptor which, in the case of M-CSF, is similar to the product of the c-fms oncogene [1]. The responses mediated by these receptors in haemopoietic cells are survival, proliferation, lineage commitment and activation of end cell function.

Each of the four CSFs acts on a different, if overlapping, subset of haemopoietic cells called colony forming cells (CFCs). IL-3 stimulates the differentiation and expansion of a number of different progenitor cell types including CFC-GM, CFC-megakaryocyte, CFC-basophil, and CFC-eosinophil [2]. In addition, when combined with erythropoietin, IL-3 promotes the differentiation and growth of early erythroid progenitors (BFU-E). Colonies formed in the presence of IL-3 may also contain a mixture of cell types derived from multipotential cells (CFU-mix) [3]. Thus IL-3 acts both on early, multipotent, cells and on later, lineage committed cell types.

GM-CSF is more lineage restricted than IL-3 although their activities overlap. GM-CSF stimulates the bipotent progenitor cell GM-CFC but can also initiate the growth of BFU-E, CFC-Mix, CFC-Eos and CFC-Meg, although proliferation is not sustained [4]. G-CSF and M-CSF, as their names imply, are more lineage restricted still [5, 6].

In addition to their effects on progenitor cells CSFs can also activate specialised functions in their mature progeny. These include monokine production and phagocytosis in macrophages [7, 8], cell mobility in neutrophils [9] and antibody-dependent cytotoxicity in neutrophils and eosinophils [10].

A separate class of haemopoietic factors which are unable to support colony growth have been called synergistic factors [11]. Prominent among these are IL-1 (formerly haemopoietin-1) and IL-4. Both these factors can synergise with CSFs to promote colony growth in vitro. In theory a combination of CSFs and synergistic factors together with lymphokines (not discussed here) could regulate the entire spectrum of cell development from the multipotent progenitor to the activated end cell. In short-term colony-forming assays, however, growth is limited because the progenitor cells on which the growth factors act lack the capacity of self-renewal which is characteristic of normal haemopoiesis. In the bone marrow this capability is supplied by stem cells from which all other lineages are derived. The stem cell can, therefore, be

---

Department of Experimental Haematology, Paterson Institute for Cancer Research, Manchester, UK

* This work was supported by the Cancer Research Campaign

defined by its ability to regenerate the entire haemopoietic system including lymphoid cells. In practice, however, it is more usual to assay multipotential cells (CFU-S) by their ability to form colonies in the spleens of potentially lethally irradiated mice [12]. As well as containing all the haemopoietic lineages spleen colonies contain more CFU-S. Thus the CFU-S fulfill many of the criteria applied to stem cells. In order to reproduce these qualities in vitro a different approach to that of the colony forming assay has been adopted. The long-term culture of haemopoietic cells requires that they maintain intimate contact with their stromal microenvironment.

II. Interactions with Stromal Cells

Whereas the lifetime of haemopoietic cells in colony-forming assays may be measured in days, those maintained in the presence of a viable adherent layer of bone marrow stromal cells may persist for weeks [13] or even months [14]. Moreover, long-term bone marrow cultures (LTBMC) remain viable in the absence of added growth factors other than serum. Contact between the haemopoietic cells and those of the adherent layer is obviously important. Electron microscopy has shown, for instance, that erythroid progenitors from contacts with macrophages and sinusoidal epithelia. Similarly, granulocytic progenitors are found in contact with reticular cells and preadipocytes. Moreover, separation of the haemopoietic cells from the stroma prevents both self-renewal and differentiation [15].

The importance of stromal contact is further illustrated by experiments with multipotent haemopoietic cell lines (FDCP-Mix). These cells normally depend on the presence of IL-3 for both growth and survival [16]. When seeded onto a layer of marrow-derived stromal cells or mouse embryonic mesenchymal (3T3) cells, however, FDCP-Mix not only survive in the absence of IL-3 but are induced to differentiate. They do not, however, undergo significant self-renewal under these conditions [17, 18].

Long-term bone marrow culture and short-term colony forming assays indicate the potential importance of both stroma and growth factors to the regulation of haemopoiesis. Much remains to be discovered, however, about the way this process is actually controlled in vivo. This is especially important in view of the current use of recombinant CSFs as adjuvants in the treatment of various cancers [19]. In this article we attempt to clarify the possible role of growth factors and stromal elements in vivo in the light of recent results from our laboratory.

B. Mechanisms of Stromal Cell Dependent Haemopoiesis

Both contact with stromal cells and exposure to CSFs can induce progenitor cells to proliferate and differentiate. One mechanism, therefore, by which adherent cells might invoke these responses would be by the production of CSFs. A number of attempts have been made to assay CSF activity in bone marrow adherent cells. M-CSF has been detected by radioimmunoassay but, interestingly, no biological activity could be detected in conditioned medium from adherent cell cultures [20]. This could be due to the secretion of inhibitory substances which mask the activity of the CSF. In contrast, neither GM-CSF nor G-CSF are constitutively produced although both are secreted in response to treatment with IL-1 [21]. There is no evidence that IL-3 is produced by adherent cells.

Conditioned medium from LTBMC does not support the growth of haemopoietic cells which have been enriched for CFU-S [22]. When seeded onto an adherent layer, however, these cells both proliferated and differentiated [23]. This suggests that, if CSFs are present in stroma they must be retained in some way rather than secreted into the medium. Support for this concept is given by the experiments of Gordon et al. [24] who found that glycosaminoglycans (GAGs), extracted from bone marrow stromal extracellular matrix, could both bind GM-CSF and present it to haemopoietic cells so that it retained its biological activity. It seems possible, therefore, that GAGs could play a role in the retention and presentation of CSFs in stromal cell layers.

Earlier work in our laboratory had shown that modification of proteoglycan synthesis by β-D-xylosides in LTBMC increased the numbers of haemopoietic cells at all stages of development [25]. Thus it seemed increasingly probable that a component of the extra-cellular matrix (ECM) was important in haemopoiesis. In order to determine which component could be involved in this process, Matrigel, a basement membrane extract containing the various ECM components, was used. Matrigel was found to be capable of binding both GM-CSF and IL-3 and subsequently supporting haemopoietic cells. By selectively degrading the ECM using specific enzymes, it was shown that heparan sulphate was the component responsible for CSF binding [26]. Indeed, heparan sulphate alone, extracted from 3T3 cells, is sufficient to bind CSFs and stimulate haemopoietic cells [26]. Direct contact with adherent cells also seems to be important, however, since cell-free ECM does not support haemopoiesis.

These data support the hypothesis that haemopoiesis is regulated in vivo by growth factors produced and presented by the adherent bone marrow cells. We have attempted to confirm this using an in vitro model of haemopoietic differentiation. Factor-dependent FDCP-Mix cells were seeded onto a layer of 3T3 cells under conditions which have been shown to promote differentiation of the haemopoietic cells [17, 18]. At various times subsequently the RNA from both adherent and haemopoietic cells was probed for the expression of IL-3, GM-CSF and G-CSF by Northern analysis. No mRNA transcripts for any of these CSFs were detected in either cell type. Thus, if stromal-induced haemopoiesis is modulated by these CSFs, they must not only be cryptically retained by the cells, but also produced at very low levels, below the limits of detection for Northern analysis. The possibility that this is perhaps the case is indicated by recent results demonstrating a considerable degree of synergy between growth factors.

C. Synergistic Interactions

The direct action of growth factors can only be accurately interpreted by using pure proteins and homogeneous cell populations. We have met the first of these requirements by using recombinant IL-3, GM-CSF and G-CSF and highly purified M-CSF. Previous attempts to produce an enriched stem-cell population have depended on killing cycling cells using 5-fluorouracil (5-FU). This also spares mature cells, however, which are known to produce a number of haemopoietic factors. We have instead used fluorescence-activated cell sorting (FACS) to select a population highly enriched for CFU-Mix cells. FACS-purified cells are about 500 times more enriched for CFU-Mix than 5-FU cells and can be plated at a low concentration thus eliminating accessory cell interference. Thus, these cells can be used to study the direct effects of factors on multipotent cells.

Using FACS-separated cells we investigated the effects of the synergistic factor IL-1 on colony formation. IL-1 alone had no colony stimulating activity but, in combination with either GM-CSF or M-CSF, it induced both more and larger colonies than the CSFs alone. We interpret this to mean that the combination of IL-1 with these factors recruits a population of cells not responsive to the CSFs alone. These cells have a higher proliferation potential, hence the larger colonies, and are therefore probably more primitive than CFU-M or CFU-GM.

In 5-FU-treated bone marrow cells synergy has previously been reported between IL-1 and IL-3 [27]. As discussed above, however, 5-FU populations contain accessory cells which may mediate this effect. Alternatively FACS-separated cells may lack a sub-population of primitive progenitors present in the 5-FU population.

The absence of synergy between IL-1 and G-CSF seemed anomalous in view of the effects of IL-1 in combination with other CSFs. In further experiments, however, it was discovered that the spectrum of activity of G-CSF differed from that of the other CSFs. Like IL-1, G-CSF can synergise with M-CSF and GM-CSF to promote increased colony formation. The morphology of the colonies is determined not by G-CSF but by the other factor. Thus, a combination of G-CSF and M-CSF produces mostly macrophage colonies. These data strongly imply that G-CSF and IL-1 act on the same sub-population of cells, a conclusion which was validated by experiments with combinations of three factors. G-CSF and IL-1 plus GM-CSF, M-CSF or IL-3 did not stimulate more proliferation than combinations of two factors. Moreover, the population of cells recruited by IL-1 and G-CSF are probably primitive, IL-3-responsive cells. This can be inferred from the observation that the combination of IL-1 plus IL-3 plus M-CSF does not recruit any more cells than either IL-3 alone or IL-3 and IL-1.

While the details of these responses are complex, certain broad principles have emerged from the work. The response of the same population of cells to IL-3, IL-1 and G-CSF suggests that stem cells express receptors for all these factors. They may also have receptors for GM-CSF and M-CSF although these may only be functionally capable of transducing a proliferation response in the presence of IL-1 or G-CSF. Certainly these two factors are capable of recruiting cells to become responsive to GM-CSF and M-CSF. In the light of these findings we propose a model of stem cell development which is deterministic. The primitive multipotential cell has the capacity to respond to a range of factors. The lineage it eventually adopts will be determined by the exact combination of factors it encounters. Thus, a stem cell stimulated by IL-1 and M-CSF would give rise to monocytic end cells while one exposed to G-CSF and GM-CSF could produce both monocytes and granulocytes. A combination of factors rather than a single stimulus may be necessary to overcome an internal biochemical threshold which prevents unnecessary proliferation.

The attraction of our proposed model lies in the role it postulates for growth factors in the regulation of multipotent cells. We believe that the early involvement of factors in the determination of lineage commitment allows for both greater flexibility and speed of response than stochastic models. It remains to be seen, however, how far these results obtained in vitro reflect the regulation of haemopoiesis in vivo.

D. Summary and Conclusions

The in vitro methods used to study haemopoiesis fall into two distinct categories. Short-term colony forming assays have identified a number of potent soluble factors capable of maintaining survival, proliferation and differentiation of haemopoietic cells but not their self-renewal. In contrast, in long-term bone marrow culture, extensive self-renewal occurs in the absence of exogenous factors and direct physical contact between haemopoietic cells and cells of the adherent stromal layer seems to be important. Obviously, LTBMC more closely resembles the situation in haemopoietic tissues but the potency of growth factors imply that they too play a role. Our data suggest that this may be at an earlier stage of haemopoietic development than previously appreciated. Primitive multipotent cells have the potential to respond to CSFs which were previously thought to stimulate only committed progenitor cells. This response is only seen, however, when the cells are exposed to a combination of factors which include either IL-1 or G-CSF. Thus, a combination of factors is able to recruit cells which are not already committed and determine the lineage along which they will differentiate.

While it remains to be conclusively demonstrated that growth factors regulate normal "steady state" haemopoiesis in vivo it is clear that contact with stromal cells is important. The mechanisms by which the adherent layer influences haemopoietic development, however, are less obvious. We have shown that a component of stroma, heparan sulphate, is able to bind growth factors and present them to haemopoietic cells in a way that stimulates haemopoiesis. The adherent layer, therefore, has the potential to elaborate and present the factors which may control haemopoiesis. It is probably premature to speculate that different classes of adherent cells may present different combinations of factors. It is obvious, however, that the capacity of adherent cells to bind potent haemopoietic factors holds the potential to unify two hitherto separately regarded mechanisms for the regulation of haemopoiesis.

References

1. Sherr CJ, Rettenmier CW, Sacca R, Roussel MF, Look AT, Ek S (1985) The *c-fms* proto-oncogene product is related to the receptor for mononuclear phagocyte growth factor, CSF-1. Cell 41:665–675
2. Moore MAS (1988) Interleukin 3: an overview. Lymphokines 15:219–228
3. Prystowsky MB, Otten G, Naujokas MF, Vardiman J, Ihle JH, Goldwasser E, Fitch FW (1984) Multiple hemopoietic lineages are found after stimulation of mouse bone marrow precursor cells with interleukin 3. Am J Pathol 117:171–175
4. Metcalf D, Johnson GR, Burgess AW (1980) Direct stimulation by purified GM-CSF of the proliferation of multipotential and erythroid precursor cells. Blood 55: 138–143
5. Stanley ER, Heard PM (1977) Factors regulating macrophage production and growth. Purification and some properties of the colony stimulating factor from medium conditioned by mouse L cells. J Biol Chem 252:4305–4311
6. Nicola NA, Metcalf D, Johnson GR, Burgess AW (1979) Separation of functionally distinct human granulocyte-macrophage colony-stimulating factors. Blood 54:614–618
7. Moore RN, Oppenheim JJ, Farrar JJ, Carter CS, Waheed A, Shadduck RK (1980) Production of lymphocyte-activating factor (interleukin 1) by macrophages activated with colony-stimulating factors. J Immunol 125:1302–1308
8. Handman E, Burgess AW (1979) Stimulation by granulocyte-macrophage colony stimulating factor of Leishmania tropica killing by macrophages. J Immunol 122: 1134–1140
9. Gasson JC, Weisbart RM, Kaufman SE, Clark CS, Hewick RM, Wong GG, Golde GW (1984) Purified human granulocyte-macrophage colony-stimulating factor: direct action on neutrophils. Science 226: 1339–1343
10. Vadas MA, Nicola NA, Metcalf D (1983) Activation of antibody-dependent cell-mediated cytotoxicity of human neutrophils and eosinophils by separate colony stimulating factors. J Immunol 130: 795 pp.
11. Quesenberry PJ (1986) Synergistic hematopoietic growth factors. Int J Cell Cloning 4:3–15
12. Till JE, McCulloch EA (1961) A direct measurement of the radiation sensitivity of normal mouse bone marrow cells. Radiat Res 14:213–222
13. Dexter TM, Allen TD, Lajtha LG (1977) Conditions controlling the proliferation of haemopoietic stem cells in vitro. J Cell Physiol 91:335–341
14. Dexter TM, Spooncer E, Simmons P, Allen TD (1984) Long-term marrow cultures. An overview of techniques and experience. In: Wright DG, Greenberger JS (eds) Long term bone marrow cultures. Liss, New York
15. Bentley SA (1981) Close range cell:cell interaction required for stem cell maintenance in long-term bone marrow cultures. Exp Hematol 9:308–311
16. Spooncer E, Boettiger D, Dexter TM (1984) Continuous in vitro generation of multipotential stem cell clones from *src* infected cultures. Nature 310:228–230
17. Spooncer E, Heyworth CM, Dunn A, Dexter TM (1986) Self-renewal and differentiation of interleukin 3-dependent mul-

tipotent stem cells are modified by stromal cells and serum factors. Differentiation 31:111–118
18. Roberts RA, Spooncer E, Parkinson EK, Lord BI, Allen TD, Dexter TM (1987) Metabolically inactive 3T3 cells can substitute for marrow stromal cells to promote the proliferation and development of multipotent haemopoietic stem cells. J Cell Physiol 132:203–214
19. Bronchud MH, Scarffe JH, Thatcher N, Crowther D, Souza LM, Alton NK, Testa NG, Dexter TM (1987) Phase I/II study of recombinant human granulocyte colony-stimulating factor in patients receiving intensive chemotherapy for small cell lung cancer. Br J Cancer 56:809–813
20. Shadduck RK, Waheed A, Greenberger J, Dexter TM (1983) Production of colony-stimulating factor in long-term bone marrow cultures. J Cell Physiol 114:88–92
21. Fibbe WE, van Damme J, Billiau A, Goselink HM, Voogt PJ, van Eeden G, Ralph P, Altrock BW, Falkenburg JHF (1988) Interleukin 1 induces human marrow stromal cells in long term culture to produce granulocyte colony-stimulating factor and macrophage colony-stimulating factor. Blood 71:430–436
22. Lord BI, Spooncer E (1986) Isolation of haemopoietic spleen colony forming cells. Lymphokine Res 5:59–63
23. Spooncer E, Lord BI, Dexter TM (1985) Defective ability to self-renew in vitro of highly purified primitive haemopoietic cells. Nature 316:62–64
24. Gordon MY, Riley GP, Watt SM, Greaves MF (1987) Compartmentalization of haematopoietic growth factor (GM-CSF) by glycosaminoglycans in the bone marrow microenvironment. Nature 326:403–405
25. Spooncer E, Gallagher JT, Krisza F, Dexter TM (1983) Regulation of haemopoiesis in long term bone marrow cultures. IV. Glycosaminoglycan synthesis and the stimulation of haemopoiesis by B-D xylosides. J Cell Biol 96:510–516
26. Roberts R, Gallagher J, Spooncer E, Allen TD, Bloomfield F, Dexter TM (1988) Heparan sulphate bound growth factors: a mechanism for stromal cell mediated haemopoiesis. Nature 332:376–378
27. Stanley ER, Bartocci A, Patinkin D, Rosendaal M, Bradley TR (1986) Regulation of very primitive multipotent cells by hemopoietin-1. Cell 45:667–672

Limited Proliferative Potential of Primitive Hematopoietic Stem Cells: Hematopoiesis by Clonal Succession

J.L. Chertkov, E.I. Deryugina, N.J. Drize, and E.Yu. Sadovnikova

A. Introduction

The hierarchy of hematopoietic stem cells (HSC) is represented by several categories of maturating pluripotent progenitors. Most of them are the members of transitional cell populations and, obviously, have no capacity for self-maintenance, i.e., are not capable of giving rise to self-replicating offspring with the same proliferative potential as the parent had [1]. The foundator of this hierarchy has not yet been identified. The most probable candidate at present is the cells supporting long-term hematopoiesis in vivo after repopulation of lethally irradiated or genetically defective W-mutant mice or in vitro in long-term culture. However, the self-renewal is also not proven for these cells, and hematopoiesis, at least in culture, occurs by clonal succession [2]. The experimental data support the hypothesis that even primitive HSC (PHSC) exhibit high, though limited, proliferative potential. This cell category is usually identified by competitive repopulation assay using a mixture of tested and standard cells identifiable by biochemical, immunological, karyological, or other markers [3]. Limiting dilution analysis based on the ability of small numbers of +/+ hematopoietic cells to cure anemia of W-mutant mice has been also used for the determination of PHSC [4–6].

In the present study, the characteristics of PHSC were investigated in long-term bone marrow culture by both competitive repopulation and limiting dilution methods. Primitive HSC were defined as precursors responsible for the long-term maintenance of hematopoiesis, i.e., for the generation of mature as well as progenitor hematopoietic cells, particularly CFUs.

Hematological Scientific Center, Moscow, U.S.S.R., 125167

B. Materials and Methods

I. Mice

Eight- to 12-week-old male and female (CBA × C57BL/6)F_1 (CBF) and (C57BL/6 × DBA/2)F_1 (BDF) mice were used.

II. Irradiation

A ^{137}Cs-IPK irradiator was used. The recipient mice were irradiated with 13 Gy, the donor mice were sublethally irradiated with 2 or 4 Gy, at a dose rate of 20 cGy/min. The cultures were irradiated with 12 Gy at a dose rate of 4.5 Gy/min.

III. CFUs Determination

The spleen colony assay was used [7]. The recipient mice (8–10 per group) were injected with hematopoietic cells, and spleen colonies were counted 8 days later.

IV. Long-Term Bone Marrow Culture

The Dexter method of culture [8] was used as described elsewhere [9]. Briefly, the contents of a femur were flushed into a 25-cm^2 tissue culture flask (Lux) containing 10 ml of Fisher medium supplemented with 14% horse serum, 7% fetal calf serum (all Flow), and 10^{-6} M hydrocortisone sodium succinate (Sigma). The cultures were tightly sealed and placed in a dry 33° C incubator. One half of the medium with nonadherent cells was replaced weekly. In 3 weeks, after adherent cell layer formation, the cultures were irradiated and, after decanting of all medium with nonadherent cells, recharged with hematopoietic cells for competitive repopulation or for limiting dilution analysis.

1. Limiting Dilution Analysis

Irradiated 3-week cultures (at least 20 per point) were seeded with two doses of bone marrow cells: 0.01 or 0.0033 femur equivalent (f.e) per flask. At 5 (CBF mice) or 7 (BDF mice) weeks the individual cultures were scraped off with a rubber policeman and after repeated pipetting, both adherent and nonadherent cells from each culture were injected i.v. into two lethally irradiated mice for the verification of the presence of CFUs. The PHSC number (N), according to the equation $N = -\ln P_0$, was determined by evaluation of the fraction of cultures in which CFUs were not revealed (P_0).

2. Competitive Repopulation

The repopulating abilities of sublethally irradiated hematopoietic cells immediately after irradiation and those after regeneration were compared. The cultures were seeded with an equal mixture of bone marrow cells from mice irradiated just prior to the explantation and from mice of another genotype irradiated 10 or 19 weeks before explantation. Cell suspension was obtained by pooling equal amounts of bone marrow cells from CBF and BDF mice, and was used for the recharging of cultures (1–2 f.e./flask). The proportion of CFUs of each genotype was determined after 2–6 weeks in the nonadherent cell fraction. To discriminate the origin of the CFUs the mice of both genotypes were injected with the same mixture of pooled nonadherent cells. Previously, the recipients had been immunized i.p. more than twice, and not later than 2 months prior to the experiment with $2-3 \times 10^7$ spleen cells (CBF mice with DBA/2 spleen cells and BDF mice with CBA spleen cells). The syngeneic CFUs gave rise to the same spleen colony number in both immunized and nonimmunized recipients, whereas semiallogeneic CFUs did not generate spleen colonies in immunized mice (data not shown).

V. Chimeras

Irradiated recipients were injected with 0.35×10^6 or 35×10^6 syngeneic bone marrow cells. PHSC determination was performed by limiting dilution analysis 6 or 7.5 months after reconstitution.

VI. Cytostatics

Hydroxyurea (900 mg/kg) was injected i.p. six times at 6-h intervals, and donor mice were killed 2 h after the last injection. 5-Fluorouracil (150 mg/kg) was injected i.v., and donor mice were killed 2 or 4 days later.

C. Results

A significant correlation between the number of explanted bone marrow cells and the estimated amount of PHSC was observed, the linear regression line going through the origin (Fig. 1). Thus, the data suggest that the number of PHSC may be determined by the limiting dilution method. The content of PHSC in murine bone marrow as estimated by this method was 90 ± 20 per femur.

The treatment of donor mice with cytostatics revealed that PHSC are insensitive to the drugs used. Neither intensive treatment with hydroxyurea nor the injection of 5-fluorouracil essentially reduced the number of PHSC in bone marrow in spite of the fact that both drugs killed more than 99% of CFUs (Table 1).

Six or 7.5 months after the reconstitution, the number of PHSC in mice injected with a "small" dose (0.35×10^6) of bone marrow cells was lower than in

Table 1. The effect of phase-specific cytostatics on CFUs and PHSC content in murine bone marrow as measured by limiting dilution method

Drug	Time after treatment	CFUs/femur	PHSC/femur
–	–	7680	79
Hydroxyurea	2 hours	79	65
5-Fluorouracil	2 days	28	53
	4 days	191	45

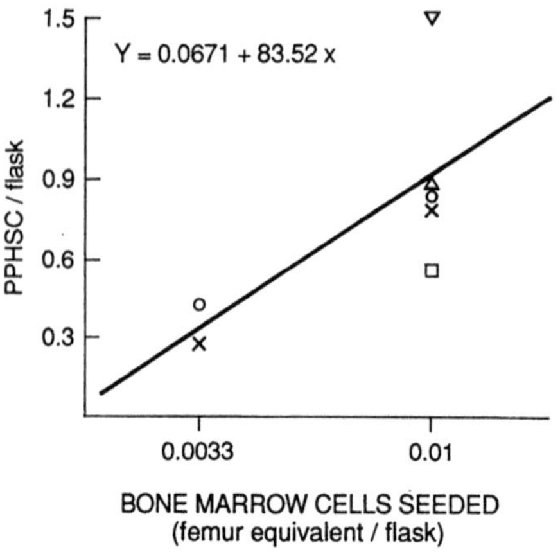

Fig. 1. Primitive hematopoietic stem cell content in normal mice as measured by limiting dilution analysis

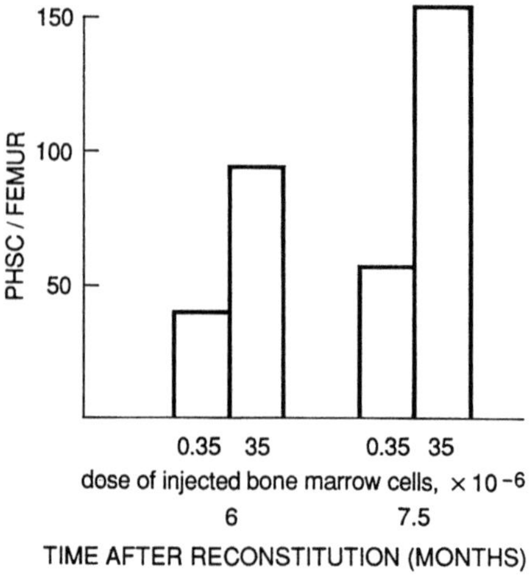

Fig. 2. Primitive hematopoietic stem cell content in mice reconstituted with different doses of bone marrow cells

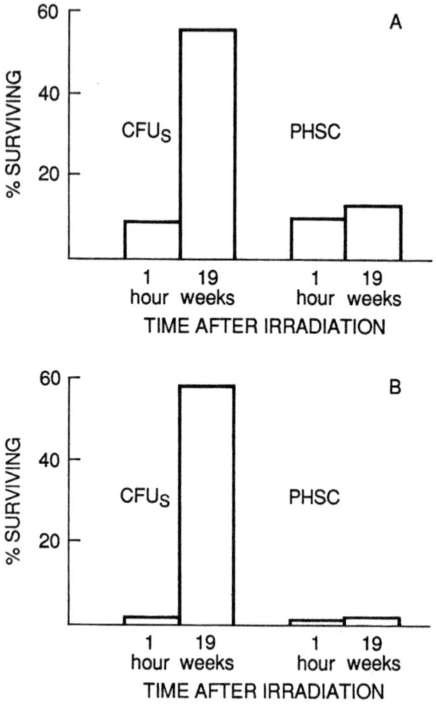

Fig. 3. Primitive hematopoietic stem cell content in bone marrow from mice immediately and 19 weeks after irradiation with a dose of 2 Gy (*A*) or 4 Gy (*B*)

recipients seeded with a "large" one (35×10^6) (Fig. 2), although the bone marrow cellularity and CFU content were approximately normal and were the same in both groups of mice (data not shown).

The content of PHSC in sublethally irradiated mice was estimated by competitive repopulation 19 weeks after 2 Gy irradiation in two experiments, and 10 or 19 weeks after 4 Gy irradiation in five experiments. The results obtained in a representative experiment are shown in Fig. 3. Immediately after irradiation with 2 Gy the number of CFUs was reduced ten times, and after 4 Gy, 100 times as compared with normal bone marrow. Two months later, the CFU population was restored to a subnormal level in both groups of mice. The competitive repopulation of cells irradiated with the same doses but harvested from mice immediately after irradiation or 10–19 weeks later revealed equal content of PHSC. Thus, the cells responsible for the maintenance of long-term hematopoiesis in culture do not possess the capacity for regeneration after irradiation.

D. Discussion

This paper presents the first estimation of PHSC in murine bone marrow as measured by limiting dilution analysis of long-term cultures. The frequency of PHSC is approximately 10 per 10^6 bone marrow cells. This number is in good agreement with data obtained in vivo by the limiting dilution method of reconstituted W-mutant mice [4–6]. The CFU population in bone marrow of studied genotypes reached 5000–8000 per femur. Taking the seeding efficiency factor as being 0.05–0.1 [10], the CFU population would be 50000–150000 per femur. Therefore, each pluripotent HSC is capable of producing a clone which includes 500–1500 CFUs and $1-4 \times 10^9$ differentiating cells (it should be borne in mind that an 8-day spleen colony may consist of 4×10^6 cells). Such enormous hematopoietic clones can obviously easily support the continuous production of hematopoietic cells during the whole life span of a mouse by the sequential expansion of a relatively small number of PHSC formed during embryogenesis. The hematopoietic tissue of an adult mouse is presented by approximately 4×10^8 cells. If the replacement of all hematopoietic cells in the bone marrow occurs in only 4 days, one PHSC capable of producing a clone of 10^9 cells would be enough for the maintenance of prolonged hematopoiesis for at least 10 days. Thus, only several hundred PHSC would be expended during the whole life span of a mouse. Therefore, although not yet conclusive, the results obtained support the hypothesis of hematopoiesis by clonal succession [11, 12].

The other important result obtained in this study suggests that PHSC are insensitive to cytostatics, in particular to phase-specific agents such as hydroxyurea which influence only cells synthesizing DNA. When the population of more mature precursors (CFUs) was nearly completely eliminated by cytostatics, the number of PHSC in bone marrow was unchanged. Therefore, PHSC apparently are in the G_0 phase of the cell cycle, and are members of a "hidden" reserve stem cell compartment. Their triggering into a state of proliferation could not be induced even by the strong hematopoietic stress. The data are in complete agreement with those obtained by competitive repopulation [3] and are not contradictory to the hypothesis that the population of "quiescent" primitive progenitors is formed during embryogenesis and is sequentially expended throughout postnatal development, producing one after another hematopoietic cell clones. If the latter were the case, then PHSC should not possess regeneration capacity. In the study performed, the increase of PHSC number until 19 weeks after 4 Gy irradiation was not really detected, although almost complete reconstitution of the CFU pool took place. The higher PHSC content in mice injected with a greater dose of bone marrow cells even several months after irradiation is also in accordance with this hypothesis because, if PHSC self-maintenance ability actually exists, the number of these cells would not depend on the injected cell dose.

On the whole, the data discussed above suggest that at present the hypothesis of hematopoiesis by clonal succession seems to be the most simple and requires the minimum number of additional assumptions. For the reliable discrimination between this hypothesis and the idea of the existence of immortal HSC, new data are necessary. Retrovirus-mediated gene transfer may be a very suitable system for the exploration of this intriguing problem.

References

1. Chertkov JL (1986) Early hemopoietic and stromal precursor cells. Int Rev Cytol 102:271–313
2. Chertkov JL, Deryugina EI, Drize NJ, Udalov GA (1987) Individual clones of hemopoietic cells in murine long-term bone marrow culture. Leukemia 1:491–496
3. Chertkov JL, Drize NJ, Gurevitch OA, Udalov GA (1986) Cells responsible for restoration of haemopoiesis in long-term murine bone marrow culture. Leuk Res 10:659–663
4. Boggs DR, Boggs SS, Saxe DF et al. (1982) Hematopoietic stem cells with high proliferative potential. Assay of the concentration in marrow by the frequency and duration of cure of W/W^v mice. J Clin Invest 70:242–253
5. Mintz B, Covarrabias L, Hawley RG (1987) Hematopoietic stem cells as potential vehicles for recombinant genes in prenatal mice. Haematologica 72 [Suppl]: 89–94
6. Nakano T, Naki N, Asai H, Kitamura Y (1987) Long-term monoclonal reconstitution of erythropoiesis in genetically anemic W/W^v mice by injection of 5-fluorouracil-treated bone marrow cells of $Pgk\text{-}1^b/Pgk\text{-}1^a$ mice. Blood 70:1758–1763
7. Till JE, McCulloch EA (1961) A direct measurement of the radiation sensitivity of normal mouse bone marrow cells. Rad Res 14:213–222
8. Dexter TM, Allen TD, Lajtha LT (1977) Conditions controlling the proliferation of hemopoietic stem cells in vitro. J Cell Physiol 91:335–344
9. Chertkov JL, Drize NJ, Gurevitch OA, Udalov GA (1983) Hemopoietic stromal precursors in long-term culture of bone marrow. I. Precursor characteristics, kinetics in culture, and dependence on quality of donor hemopoietic cells in chimeras. Exp Hematol 11:231–242
10. Siminovitch L, McCulloch EA, Till JE (1963) The distribution of colony-forming cells among spleen colonies. J Cell Comp Physiol 63:327–336
11. Kay HEM (1965) How many cell generations? Lancet II:418–419
12. Micklem HS, Lennon JE, Ansell JD, Gray RA (1987) Numbers and dispersion of repopulating hematopoietic cell clones in radiation chimeras as function of injected cell dose. Exp Hematol 15:251–257

The Effect of Human and Rat Fetal Bone on Hematopoiesis In Vitro and In Vivo

B.V. Afanasyev, V.A. Shatrov, and N. Balayan

Granulocyte-macrophage colony-stimulating factor (GM-CSF) is a regulatory glycoprotein necessary for the proliferation, maturation, and survival of myeloid cells [4]. Myelopoiesis is enhanced when recombinant human (rh) GM-CSF is given to mice, monkeys [6], and some patients with various forms of neutropenia [3, 5]. The clinical use of rhGM-CSF infusions is under intensive investigation now.

An alternative way of increasing the level of GM-CSF in the organism is by transplantation of CSF-producing cells, analogous to the transplantation of insulin-producing cells in patients with diabetes mellitus. It seems especially important to increase the intramedullary concentration of GM-CSF, as bone marrow is the central organ for regulating the proliferation and differentiation of hematopoietic cells. There is a positive correlation between the intramedullary concentration of GM-CSF and the prognosis for patients with leukemia [8]. One source of CSF-producing cells may be various embryonal cells which are available and genetically more tolerable.

The main objectives of this study were: (a) to evaluate the ability of various human fetal organs and tissues to develop colony-stimulating and -inhibiting activities (CSA and CIA); (b) to estimate their effect on colony-forming unit–granulocyte macrophage (FU-GM) and leukemic clonogenic cells (LCC); (c) to evaluate the effect of fetal bone (FB) on fibrous tissue of adult bone marrow in vitro; and (d) to evaluate the effect of rat FB transplantation into rats after lethal and sublethal doses of cyclophosphamide.

N.N. Petrov Institute of Oncology, Department of Bone Marrow Transplantation, Leningrad, USSR

A. Material and Methods

A total of 88 human fetuses (24 at 7–10 weeks; 38 at 11–15 weeks; 22 at 20–26 weeks) were used to evaluate CSA and CIA of various fetal organs and condition media (CM). Bone marrow cells of 87 hematologically healthy donors (control group), 92 patients with various hematological disorders, and 48 human fetuses (7 at 10–12 weeks; 10 at 13–15 weeks; 16 at 20–22 weeks; 15 at 23–26 weeks) were used as target cells. The cloning of hematopoietic cells was performed in agar drop-liquid medium [1]. CSA of fetal tissues (FT) or CM (CMFT) was estimated as: the number of aggregates in the plates with FB or CMFT divided by the number of aggregates in the plates with leukocyte feeder (F). IA was calculated as: CSA of F (100%) – CSA of (F + FT or CMFT).

Colonies and clusters were scored at day 7 or 8. The effect of FB on fibrous tissue was studied by morphometric assay of bone marrow trephine biopsy after its cultivation using the Marbrook system [2] with or without FB. The scheme of rat FB transplantation is presented in Fig. 1.

B. Results and Discussion

All human fetal tissues except from the liver were able to produce CSA in vitro

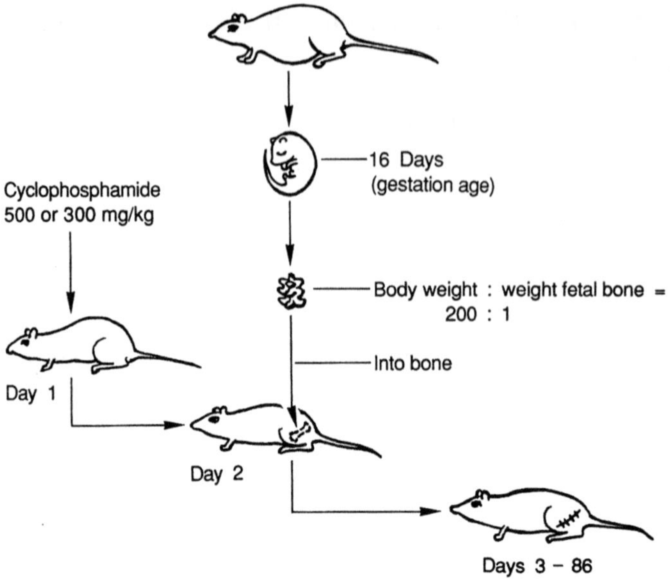

Fig. 1. Scheme of fetal bone transplantation in rat

Table 1. Colony-stimulating activity (CSA) and colony-inhibiting activity of human fetal bone (mean ± SD)

Target cells	Range of spontaneous colony formation	CSA of fetal bone (% leukocyte feeder)	CIA of fetal bone (Leukocyte feeder + fetal bone)
Control group	0 61.8 ± 27.4	62.8 ± 8.2 85.0 ± 10.4	100 – 80.6 ± 15.7 100 – 80.4 ± 11.3
Acute nonlymphocytic leukemia	0 52.7 ± 15.5	42.9 ± 11.1 67.8 ± 9.2	100 – 110.9 ± 19.1 100 – 115.1 ± 14.3
Fetal bone marrow	0 46.7 ± 15.5	50.2 ± 7.2 99.0 ± 11.4	100 – 52.0 ± 7.7 100 – 74.0 ± 7.9

Numerator, low level of spontaneous colony formation; denominator, high level of spontaneous colony formation

(Fig. 2), but FB was the only tissue which sustained this capacity for at least 4 weeks. The cocultivation of FB and AML (acute myeloid leukemia) or normal bone marrow in liquid culture for 2 weeks did not decrease the capacity of FB to produce CSA. The LCC were more sensitive to the endogenous stimulators and demonstrated less response to FB stimulation than normal CFU-GM (Table 1). With prolonged cultivation time of the AML bone marrow, morphometric analysis of trephine biopsy using the Marbrook system showed an increased proliferation of fibrous tissue and a disappearance of hematopoietic elements (Table 2). On the other hand, the cocultivation of trephine biopsy and FB inhibited the proliferation of fibrous tissue and prevented the destruction of hematopoietic cells. The cultivation of FB for 4 weeks in liquid medium or

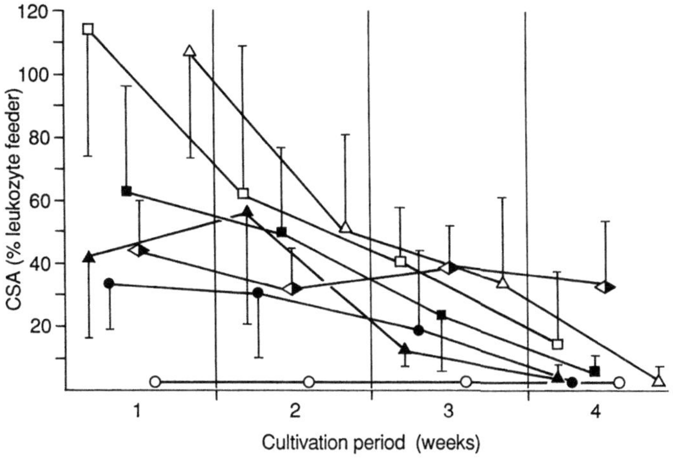

Fig. 2. Colony-stimulating activity (*CSA*) in various fetal tissues. ●, Kidney; ▲, spleen; ■, lung; ○, liver; △, muscle; □, thymus; ◁▶, bone

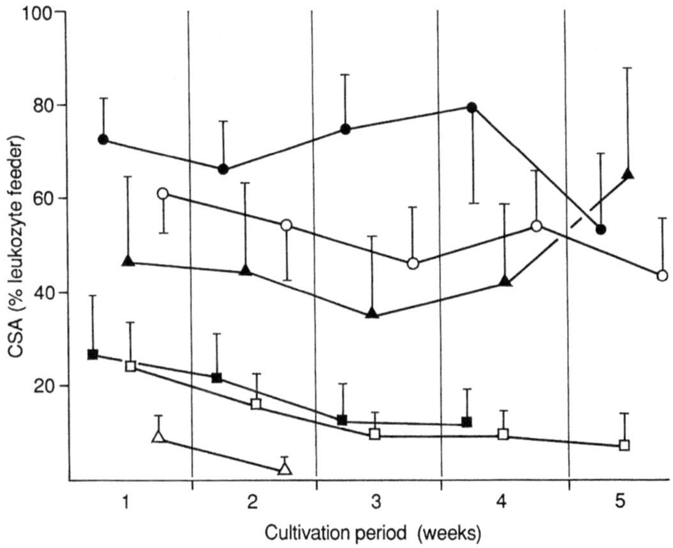

Fig. 3. Colony-stimulating activity (*CSA*) of human fetal bone (FB) under different perservation conditions. ●, Control; ▲, FB after cryopreservation; ■, crushed FB after cryopreservation; ○, FB preserved at +4° C for 1 week; □, FB preserved at +4° C for 2 weeks; △, FB preserved at +4° C for 4 weeks

cryopreservation at −196° C with 10% DMSO did not considerably affect its capacity for CSA production (Fig. 3). The transplantation of FB shortened the period of leukocytopenia after sublethal (300 mg/kg) or lethal (500 mg/kg) doses of cyclophosphamide in a rat model (Fig. 4) and increased the probability of survival (Fig. 5).

Thus, rat survival after FB transplantation was increased. This effect is probably related to the ability of FB to give long-term CSF production and the increase of the intramedullary CSF level. We have shown that FB induced the proliferation of LCC. FB has been suggested to exert a differentiated effect on LCC, which can return to their normal phenotype [7]. Moreover, FB inhibits fibrous tissue proliferation in human AML bone marrow. Further research is needed to establish the place of FB transplantation in clinical practice.

Table 2. Results of cultivation of bone marrow tissue in Marbrook system in the presence of human fetal bone (morphometric assay bone marrow trephine biopsies from the acute non-lymphocytic leukemia patients)

Cultivation period (weeks)	Type of tissue (%)			
	Hemopoietic (%)	Fat (%)	Bone (%)	Fibrous (%)
1	7–45	4–80	12–28	0–10
	17–38	26–54	12–19	0
2	0–20	11–70	15–40	8–42
	80	24–32	13–16	0
3	0	0–25	20–54	25–50
	8–28	18–24	23–28	0– 3
4–5	0	4– 9	16–32	30–60
	21–30	45–57	12–20	0– 3

Numerator, bone marrow trephine (BMT); denominator, BMT + fetal bone

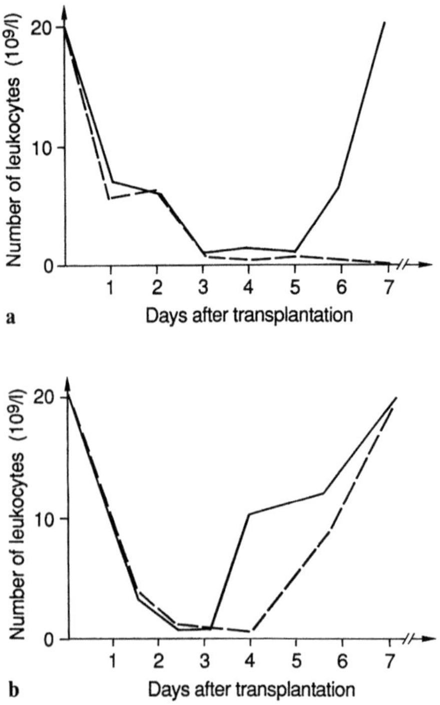

Fig. 4 a, b. Leukocyte count in peripheral blood after fetal bone transplantation. **a** Administration of 500 mg/kg cyclophosphamide. **b** Administration of 300 mg/kg cyclophosphamide. *Solid curve,* experimental group; *broken curve*, control group

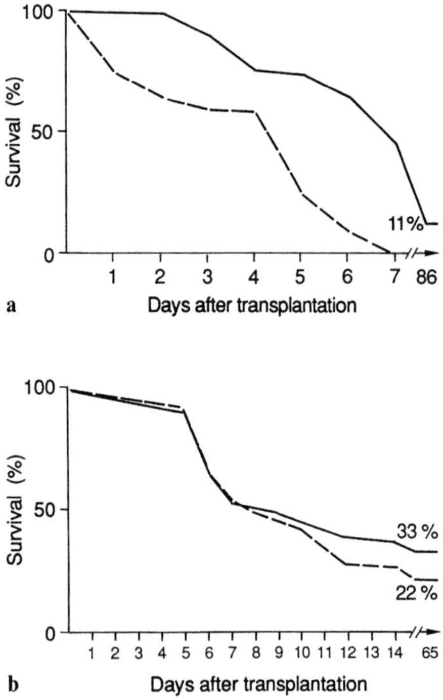

Fig. 5 a, b. Survival rate after fetal bone transplantation. **a** Administration of 500 mg/kg cyclophosphamide ($p=0.001$). **b** Administration of 300 mg/kg cyclophosphamide ($p=0.05$). *Solid curve,* experimental group ($n=63$); *broken curve,* control group ($n=42$)

References

1. Afanasyev BV, Tiranova SA, Kulibaba TG, Zubarovskaya LS, Bolshakova GD, Zabelina TS (1983) Cloning of human hemopoietic cells in the "agar drop-liquid medium" system. Ther Arch (USSR) 8:114
2. Golde DW, Cline MJ (1973) Growth of human bone marrow in liquid culture. Blood 41:45
3. Groopman JE, Mitsuyasu RT, DeLeo MJ, Oette DH, Golde DW (1987) Effect of recombinant human granulocyte macrophage colony stimulating factor on myelopoiesis in the acquired immunodeficiency syndrom. N Engl J Med 317:593
4. Metcalf D (1977) Haemopoietic colonies. Springer,Berlin Heidelberg New York
5. Nemunaitis J, Singer JW, Buckner CD, Hill R, Storb R, Thomas ED, Appelbaum FR (1988) Use of recombinant human granulocyte macrophage colony stimulating factor (rh GM-CSF) in autologous marrow transplantation for lymphoid malignancies. Blood 72:834
6. Nienhuis AW, Donahue RE, Karisson S, Clark SC, Agricola B, Antihoff N, Pierce JE, Turner P, Anderson WF, Nathan DG (1987) Recombinant human granulocyte macrophage colony stimulating factor (GM-CSF) shortens the period of neutropenia after autologous bone marrow transplantation in a primate model. J Clin Invest 80:573
7. Pierce GB (1983) The cancer cell and its control by the embryo. Am J Pathol 113: 117
8. Zubarovskaya LS, Afanasyev BV, Smirnova GA, Lukasheva TN, Patterson D, Shishkov AL, Balayan LN, Ostrovsky AM (1985) Colony-stimulating activity of bone marrow trephanobiopsies of patients with different pathological conditions of the blood system. Ther Arch (USSR) 7:22

Aberrant Expression of the Multi-CSF Gene in Hematopoietic Precursor and Stem Cell Lines Initiates Leukemogenic Progression

C. Laker[1], C. Stocking[1], N. Kluge[1], M.J. Franz[1], G. Beck-Engeser[1], E. Spooncer[2], M. Dexter[2], and W. Ostertag[1]

A. Introduction

The most common feature of many malignancies is the escape of transformed cells from normal growth requirements. In hemopoiesis, a series of well-characterized glycoproteins (colony-stimulating factors, CSFs) tightly control the survival, proliferation, differentiation, and functional activation of all blood cells [1]. In addition, the differentiation hierarchy of pluripotent stem cells, capable of self-renewal and differentiation into all lineages of myelopoiesis and lymphopoiesis, to terminally differentiated cells via multipotent precursor cells may be traced by surface markers and functional assays. Thus, the ability to determine the differentiation stage of a cell and an understanding of the molecules that trigger cell division make haemopoiesis a suitable system to study the initiation of malignant progression, i. e., leukemogenesis.

In the data presented here, we will focus on the following aspects:

1. Can aberrant expression of a CSF gene in a factor-dependent cell confer factor independence?
2. Is the acquisition of factor independence coincident with tumorigenesis?
3. If not, what are the other parameters that determine tumorigenic potential of such cells?
4. Can differences between stem and progenitor cells in tumorigenic progression following constitutive expression of CSF genes be detected?

To address these questions, we chose two different types of hemopoietic cell lines as model systems. The FDC-Pmix stem cell lines [2] retain a normal or nearly normal karyotype and their differentiation potential is close to that of normal myeloid stem cells. Indeed, some FDC-Pmix cells can repopulate the hematopoietic system of irradiated animals. The second cell types, FDC-P1 and FDC-P2, have a more restricted differentiation pattern, an abnormal karyotype, and most likely are representatives of myeloid progenitor cells [3].

B. Retroviral Transfer Into Hematopoietic Target Cells

The FDC-P progenitor cell lines were derived from normal long-term bone marrow cultures (Dexter-type cultures), whereas the FDC-Pmix stem cell lines were established from bone marrow cultures infected with Moloney murine leukemia virus (Mo-MuLV) and a recombinant retroviral vector carrying the avian v-src oncogene. The stromal cells of the long-term cultures express the src oncogene product pp60 v-src, whereas the FDC-Pmix cells do not. They do, however, express the ecotropic Mo-MuLV used as helper virus to provide viral spread of the replication defective src virus in the original cultures. In order to overcome retroviral interference by Mo-MuLV-infected FDC-Pmix cells, we used the amphotropic PA317 packaging cell line to assemble vector-carrying pseudotypes suitable for infection of the

[1] Heinrich-Pette-Institut für Experimentelle Virologie und Immunologie an der Universität Hamburg, Martinistrasse 52, 2000 Hamburg 20, FRG

[2] The Paterson Institute for Cancer Research, Manchester M20 9BX, UK

target cells. Since the FDC-P1 progenitor cell line does not express ecotropic retroviral sequences, the ecotropic ψ2 packaging cell line was used as a viral source.

The vectors used for transfer of either the granulocyte/macrophage (GM)-CSF or multi-CSF gene are based on the myeloproliferative sarcoma virus (MPSV), which has a broader host range as compared with Mo-MuLV due to mutations in the U3 region of the LTR [4]. All retroviral constructs contain the Tn5 drug resistance gene (Neo) and therefore allow selection with geneticin (G418). The CSF genes either replace the mos oncogene sequences of the native virus or are inserted into the viral genome coding for the gag leader just 3′ of the packaging site [5] (Fig. 1).

To test the transfer efficiency of MPSV-based vectors into hematopoietic cell lines, FDC-P1/2 progenitor cells or FDC-Pmix cells were infected with a limiting amount of virus particles, plated in the presence of exogenous growth factor and selected with G418 (1.5 mg/ml). The number of G418-resistant colonies was scored 9–16 days after infection and compared with the number of G418-resistant colonies of standard fibroblasts (NIH/3T3) obtained from a simultaneously performed infection by limiting dilution.

Table 1 shows the relative G418 transfer efficiency to hematopoietic target cells as compared with fibroblasts. The data demonstrate that the transfer efficiency into progenitor cells (FDC-P1/2)

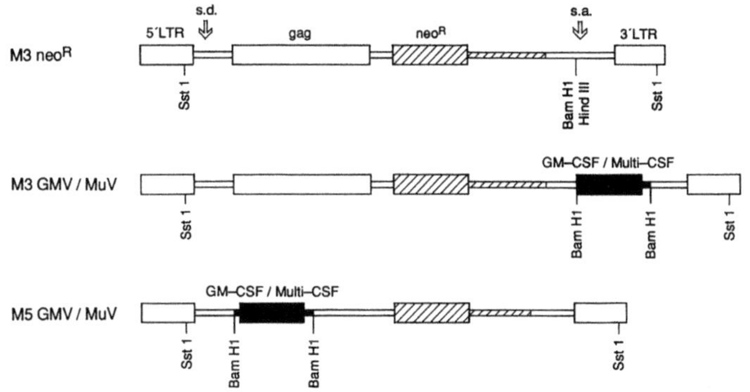

Fig. 1. Retroviral vectors. All vectors were based on the myeloproliferative sarcoma virus (MPSV) and have been described previously [4, 5]

Table 1. Transfer efficiency of Neo into hemopoietic target cells

Viral control regions	Helper function	Transfer efficiency: Fibroblasts[a] NIH/3T3	Hematopoietic precursor cell lines: FDC-P1	Hematopoietic precursor cell lines: FDC-P2	Hematopoietic stem cell lines[b]: A4	A7	15S
MPSV	ψ2	1	0.1–1.0	*	0	0	0
	PA317	1	10^{-3}–10^{-1}	10^{-3}–10^{-1}	10^{-5}–10^{-4}	10^{-7}–10^{-4}	10^{-5}–10^{-4}

[a] G418 selection at 0.4 mg/ml
[b] G418 selection at 1.5 mg/ml
* Expression of endogenous ecotropic sequences interferes with infection by ψ2 packaging cell lines

is significantly reduced, as compared with fibroblasts, and decreases dramatically in stem cell lines.

C. Introduction of a CSF Gene into Factor-Dependent Hemopoietic Target Cells Results in Two Types of Factor Independence

Infection of growth-factor dependent FDC-P1/2 progenitor cells with either multi-CSF or GM-CSF virus (M3MuV, M3GMV) conferred factor independence to the infected cells. However, the acquisition of growth autonomy after the autogenous production of CSFs appeared to be a consequence of two or more interdependent events. Uninfected cells or cell lines infected with control vectors containing the Neo gene but lacking the growth factor gene showed very low levels of spontaneously occurring growth factor independent mutants: $<10^{-7}$ or $<10^{-9}$ spontaneous mutants could be detected in FDC-P1 or FDC-Pmix cells, respectively. Immediately after introduction of the CSF gene, cells were independent of exogenously supplied CSF at high cell density. They were, however, not truly autonomous, as they still required exogenous CSF at low cell density for cell proliferation (nonautonomous factor independence). These cells often acquired true growth autonomy as a consequence of a second alteration. Growth of such cells was independent of cell density (autonomous). The first, but not the latter type of cells could be inhibited by specific antisera against the CSF.

Analysis of clones picked from primary infections and replated at several time points after infection with MuV demonstrates the shift from nonautonomous factor independence to true autonomy (Fig. 2a, b). Replated 4 weeks after infection in the absence of exogenous growth factor, cells display density-dependent clonability (Fig. 2a). However, after a culture period of several weeks under nonselective conditions (plus WEHI CM), replating of the clones gave rise to a more density-independent (linear) cloning pattern when plated in the absence of added growth factor (Fig. 2b). The shift from factor independence, where external stimulation is still required, to true autonomy was accelerated if cloned cultures were kept under selective conditions (minus WEHI CM).

Similar experiments performed by other groups have failed to distinguish the two types of factor independence [6–8], suggesting that infected cells could progress to growth autonomy in one step. Other groups obtained similar results as reported by us [9]. The most striking difference in our work was the use of retroviral vectors designed in our laboratory that express only low levels of the CSF gene product. We therefore designed an alternative construct (M5GMV) which permits a 100-fold increase in GM-CSF production, as compared with M3GMV. In the primary infection with the M5 construct, the infected cells selected for growth-factor independence (minus WEHI CM) showed a linear cloning pattern. The number of factor-independent clones obtained was, however, considerably lower than that obtained by selection for Neo colonies in the presence of WEHI-conditioned medium (Fig. 3). Analysis of clones picked from various cell densities plated under growth factor selection or G418 selection demonstrated that approximately 50% of the clones were already autonomous. These data thus suggest that the level of growth factor produced enhances the selective advantage of a second genetic event that leads to autonomous factor independence. If high levels of CSF production are necessary for the shift to growth autonomy, we would predict that autonomous clones should show high or increased release of growth factor. We were, however, unable to detect a significant increase of secreted CSF molecules accompanying the acquisition of growth autonomy.

Our results are, however, consistent with the hypothesis that specific alterations of the export of growth factors or

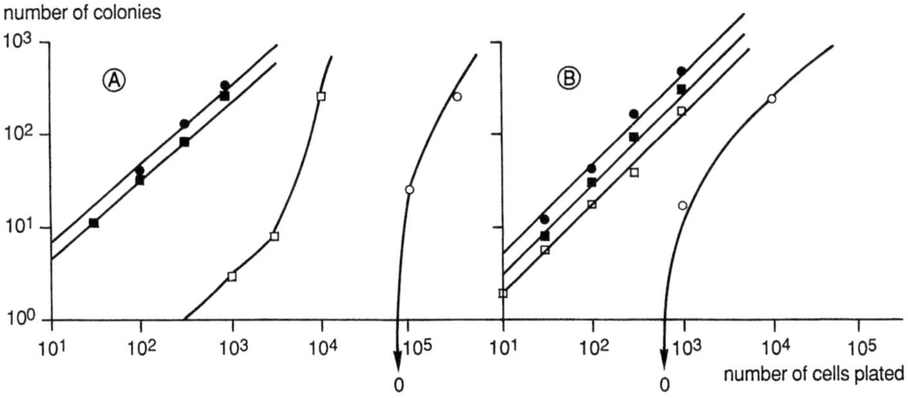

Fig. 2. Clonability of infected FDC-P1 progenitor cells as a function of cell density. The number of clones obtained at different cell densities is plotted on a log-log scale; thus, deviation from a slope of +1 reflects non-linear cloning efficiency. *Panel A:* M3GMV-infected FDC-P1 clone 6 (*squares*) and clone 4 (circles) recloned 4 weeks after primary infection under plus WEHI CM, plus G418 conditions (*closed symbols*) or minus WEHI CM selection (open symbols). *Panel B:* Recloning of the same clones after an additional tissue culture period of 12 weeks (=16 weeks after infection) depicts the shift to autonomous growth

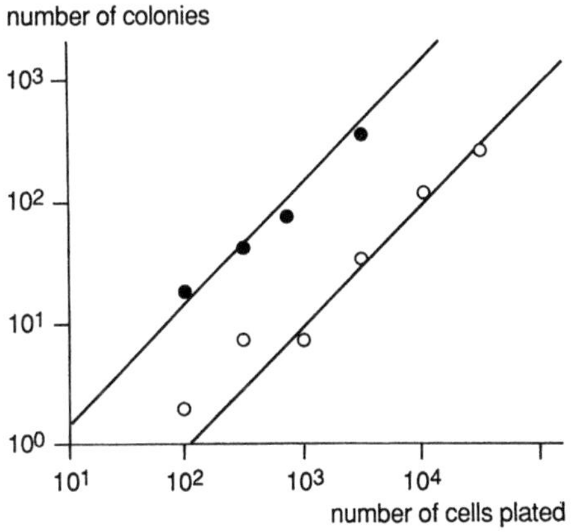

Fig. 3. Primary infection of FDC-P progenitor cells by M5GMV. Cells were infected for 6 h with supernatant of virus-producing $\psi 2$ fibroblasts, plated in the presence of WEHI CM and G418 (1.5 mg/ml; *closed symbols*) or absence of added growth factor (*open symbols*), and scored for colony formation 9 days after infection

the organization of the receptor molecules with respect to internal processing of the growth factor could lead to internal autocrine stimulation and growth autonomy.

To further investigate whether aberrant export of the constitutively expressed CSF gene product contributes to the acquisition of autonomous growth, we designed vectors containing two versions of the multi-CSF gene that differ in the constitution of the leader sequences of the mature protein. The sMuV contains the normal (short) leader sequence resulting in secretion of the mature CSF molecule, whereas the lMuV containing a cDNA clone with a longer leader sequence gives rise to a protein that is predominantly membrane-bound, the body of the protein remaining in the cytosol [10]. Only a small portion (~10%) of the molecules is cleaved and secreted.

A significant difference in cloning efficiency under growth factor selection (minus WEHI CM) was observed between FDC-P progenitor cells infected with either short or long leader MuV. FDC-P1 progenitor cells infected with the short leader MuV showed density-dependent colony formation in the absence of added growth factor when plated immediately after infection (Fig. 4a). In contrast, cells infected with the long leader MuV showed almost linear (autonomous)

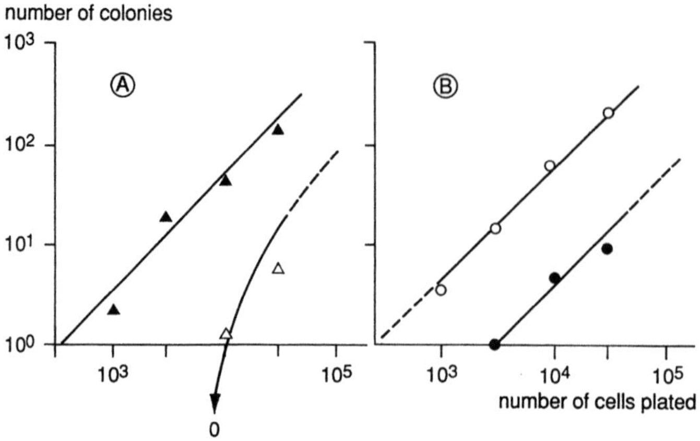

Fig. 4. Primary infection of FDC-P progenitor cells with either M3-sMuV or M3-lMuV. Cells were exposed to virus-containing supernatant for 6 h, washed three times, and plated in semisolid medium at various cell densities either in the presence of WEHI CM and G418 (*closed symbols*) or absence of WEHI CM (*open symbols*). *Panel A* depicts the cloning efficiency of FDC-P1 cells infected with M3-sMuV (short leader sequence) and *Panell B* the infection of FDC-P2 cells with M3-lMuV (long leader sequence). The proportionally higher clone numbers obtained after selection under minus WEHI conditions, as compared with G418 selection, is most likely due to the 2-day lag in cell death of FDC-P2 cells after removal of WEHI CM

clonability when plated under similar conditions (Fig. 4b).

Taken together, our data demonstrate that initial factor-independent growth requires receptor-ligand interaction at the cell surface. A second mutation leads to complete autonomous growth of the cell, and both the level and localization of the CSF molecule contribute to the frequency of obtaining such mutations.

D. The Shift to Growth Autonomy Occurs at a Lower Frequency in Stem Cell Lines

In contrast to the fast progression of infected FDC-P progenitor cell lines to autonomous factor independence (less than 16 weeks), the FDC-Pmix stem cell lines monitored over a 6-month period displayed a strikingly different behavior. First of all, in primary infections with either sMuV or lMuV, no clones could be isolated from plates under growth factor selection (cells plated without WEHI CM), even when cells were plated at the highest possible cell density (10^6 cells/ml). Follow-up of clones picked from G418 selection (in the presence of WEHI CM) revealed that only one out of 19 clones acquired growth factor autonomy during a 6-month observation period. Interestingly, this particular clone was of 15S origin, the parental cell line with the longest culture history and the most committed stem cell feature, having lost its responsiveness to differentiation induction. A4 and A7 stem cells infected with the M3MuV had to be maintained as bulk cultures under stringent conditions (limiting cell density and lack of added growth factor) to give rise to a few autonomous cell clones after more than 3 months of culture. A subline of the FDC-Pmix cell clone 15S that had been kept in culture continuously for 1 year, however, showed a similar pattern of growth-factor independence and autonomy after infection of lMuV as the progenitor lines FDC-P1 or FDC-P2. The lMuV-infected clones that grow density-dependent show an accelerated shift to growth factor autonomy (10^{-3} per cell generation).

Table 2. Tumorigenicity of M3MuV-infected FDC-P progenitor cell lines

MuV-infected FDC-P cell clones	Clones	Injected mice	No. of tumor-bearing mice (weeks after inoculation)		
(10^6 cells/mouse)	(*n*)	(*n*)	4	8	12
Non-autonomous clones	8	30	0/30 (0%)	3/30 (10%)	19/30 (63%)
Autonomous clones	2	9	4/9 (44%)	9/9 (100%)	9/9 (100%)
Controls					
Uninfected	–	3	0/3	0/3	0/3
Neo V-infected	8	30	0/30	0/30	0/30

However, infection of this subline with sMuV resulted in clones that shifted at a low rate (10^{-8} per cell generation) to growth autonomy. These experiments highlight that both the localization or processing of the CSF molecule and properties of the target cell are important in the acquisition of true autonomy after factor-independent growth by an autocrine mechanism. These properties may include the differentiation stage and mutations occurring during tissue culture adaptation, the latter involving mutations that cooperate with other genetic alterations leading to the final leukemogenic alterations.

E. Tumorigenicity of Factor-Independent Progenitor and Stem Cell Lines

Several groups have shown that tumorigenicity may be a property of cells that have acquired factor-independent growth by autonomous CSF production [6–9]. Having dissected two stages of factor independence in vitro (density dependent versus autonomous growth), we investigated whether acquisition of either type of factor-independent growth was sufficient to induce tumor formation in susceptible mice or whether growth autonomy is prerequisite for tumorigenicity.

Congenic mice were injected with 10^5 to 10^8 cells of density-dependent or autonomous M3MuV infected FDC-P1/-P2 progenitor cell clones. Injection of up to 10^8 control cells which were either uninfected or infected with a control vector into 33 individual mice did not induce tumors within a 4-month period of observation. In contrast, all mice injected with 10^8 M3MuV infected cells acquired tumors regardless of the nature of factor independence. However, the latency of tumor formation and progression in tumor weight was strikingly different from those tumors derived by nonautonomous clones as compared with autonomous clones. Four out of nine animals injected with as few as 10^6 cells of the autonomous clones showed tumor formation 4 weeks after inoculation and all animals (9/9) acquired tumors within 8 weeks time. The average tumor weight increased from 0.5 g at 2 weeks to more than 2.5 g just before death of the mice. The tumors induced by nonautonomous clones showed a significantly longer latency. Eight weeks after inoculation only 10% of injected mice (3/30) displayed tumors with an average tumor weight of 0.1 g and even after 12 weeks only 63% (19/30) of the animals showed tumors. To investigate whether the pronounced longer onset of tumor formation from nonautonomous clones was due to in vivo selection for autonomous variants, we either reestablished cell lines from tumors or plated cell suspensions from tumors directly in our in vitro assay in the absence of added growth factor. As expected, all cell lines established from tumors initiated by autonomous clones showed density-independent clonability when plated in the absence of WEHI

CM. The cells reestablished from tumors of density-dependent clones at the time point of inoculation gave rise to either autonomous or density-dependent subclones. Interestingly, reinjection of the in vivo passaged, density-dependent subclones resulted in accelerated tumor formation. Autonomous cell growth thus is only one parameter that may lead to rapid tumorigenicity; other and as yet unknown parameters unrelated to density-independent growth are also definitely involved in accelerating tumorigenicity in vivo.

In accordance with this hypothesis, neither density-independent nor autonomous clones of the M3MuV-infected FDC-Pmix stem cell clones (15S) induced tumor formation when injected into congenic animals. Multipotent stem cells thus require further alterations for acquisition of accelerated tumorigenesis.

F. Summary and Conclusions

Tumorigenesis of hemopoietic cells and acquisition of factor independence as a consequence of aberrant growth factor release are closely correlated [5–9]. In previous work we were able to dissect two stages leading to growth factor autonomy of cells: the first step requires the secretion of the constitutively expressed CSF gene product and extracellular interaction with its cognate receptor. This requirement for external stimulation is abrogated by a second step. We were interested in characterizing the parameters that influence the conversion from nonautonomous to autonomous growth properties of hematopoietic precursor cells. The frequency with which this alteration occurs varies and correlates with the level of growth factor production. However, a significant increase of CSF production accompanying the progression to autonomy could not be detected. We thus conclude that there is no direct link between level of CSF production and acquisition of true autonomy but an indirect influence enhancing the frequency of genetic alteration(s) that lead to growth autonomy.

Lang et al. [6] have suggested that the acquisition of autonomous growth occurs due to internal receptor-ligand interaction. Indeed, Keating and Williams [11] have claimed that PDGF may react with an intracellular PDGF receptor resulting in autocrine stimulation.

We therefore determined the importance of localization of the CSF molecules utilizing a pair of constructs that either result in expression of a secreted gene product (sMuV) or a mostly membrane-bound form (lMuv). In accordance with our expectation, we found a close to linear clonability of cells following infection with the construct conferring the membrane-bound form of the molecule, whereas infection with the secreted product vector showed density-dependent clonability under growth-factor selection. These data suggest that receptor-ligand interaction also may occur intracellularly, e.g., in the membranes of the endoplasmatic reticulum (ER) or at the inner face of the outer cell membrane.

Work by Browder et al. [12] is in accordance with this hypothesis. Infection of multi-CSF-dependent 32D cells with a construct encoding a modified multi-CSF protein that does not direct the CSF molecule into the ER for processing did not lead to factor independence, although the gene product localized in the cytosol proved to be biologically active.

In vivo studies performed to investigate the link between acquisition of factor independence and tumorigenicity of infected progenitor cells revealed that infected cell lines of autonomous as well as nonautonomous factor independence were tumorigenic. However, second mutations also accelerated the onset of tumor formation and progression.

The most intriguing result obtained from our experiments is the striking difference in initiation of leukemogenesis in haemopoietic stem cell lines as compared with the fast progression in precursor cells. Only one out of nine infected stem cell lines tested shifted from nonautonomous factor independence to true autonomy over a time period of 6 months. Inoculation of congenic mice did not induce tumor formation within the observation period when similarly infected FDC-P progenitor cells already elicited tumor formation. The stem cell lines thus appear to be more stable and less susceptible to genetic mutations than the more differentiated progenitor cell lines. Whether this is a characteristic of the undifferentiated stage or their close to normal karyotype remains to be clarified.

G. Outlook

Although it has long been clear that leukemogenesis is a multistep process, the genetic changes that are involved remain unclear. Analysis of leukemic cells from both AML and CML patients have provided evidence in support of the hypothesis that aberrant expression of growth factor genes (e.g., GM-CSF and G-CSF) may contribute to the uncontrolled growth of the leukemic cells [13, 14]. Indeed, two experimental studies have demonstrated the causal correlation of aberrant factor production with leukemogenic growth: unregulated expression of an IL-6 gene conjugated to the human Ig heavy chain gene enhancer (Eμ) in transgenic mice triggers generation of plasmacytoma/myeloma (Suematsu and Kishimoto, cited in [15]) and expression of IL-3 in a retroviral construct introduced in vivo via murine hematopoietic stem cells leads to a hematopoietic disorder similar to CML [12].

Although initial reports suggested that aberrant factor expression leads directly to tumorigenic growth [6, 8, 9], the studies presented here and elsewhere [5, 7, 17] have clearly demonstrated that secondary events are necessary for true autonomous growth and increased tumorigenic potential. These secondary events may be dependent on the autogenous factor production or may have obtained selective advantage in cells producing their own factor. The first step in the unmasking of these events has been the development of an in vitro model system, as has been described here. We hope that their ultimate genetic identification can be facilitated by insertional mutagenesis followed by selection for autonomous growth.

Acknowledgement. This work was supported by grants from the Deutsche Forschungsgemeinschaft (DFG Os 31-12) and the Deutsche Krebshilfe.

References

1. Metcalf D (1984) The haemopoietic colony stimulating factors. Elsevier, Amsterdam
2. Spooncer E, Heyworth CM, Dunn A, Dexter TM (1986) Self-renewal and differentiation of interleukin-3-dependent multipotent stem cells are modulated by stromal cells and serum factors. Differentiation 31:111-118
3. Dexter TM, Garland J, Scott D, Scolnick E, Metcalf D (1980) Growth of factor-dependent hemopoietic precursor cell lines. J Exp Med 152:1036-1047
4. Stocking C, Kollek R, Bergholz U, Ostertag W (1985) Long terminal repeat sequences impart hematopoietic transformation properties to the myeloproliferative sarcoma virus. Proc Natl Acad Sci USA 79:5746-5750
5. Laker C, Stocking C, Bergholz U, Hess N, DeLamarter JF, Ostertag W (1987) Autocrine stimulation after transfer of the granulocyte-macrophage colony stimulating factor gene and autonomous growth are distinct but interdependent steps in the oncogenic pathway. Proc Natl Acad Sci USA 84:8458-8462
6. Lang RA, Metcalf D, Gough NM, Dunn AR, Gonda TJ (1985) Expression of a hemopoietic growth factor cDNA in a factor-dependent cell line results in autonomous growth and tumorigenicity. Cell 43:531-542
7. Wong PMC, Chung S-W, Nienhuis A (1987) Retroviral transfer and expression of the interleukin-3 gene in hemopoietic cells. Genes Dev 1:358-365
8. Yamada G, Kitamura Y, Sonoda H, Harada H, Taki S, Mulligan RC, Osawa H, Diamantstein T, Yokoyama S, Taniguchi T (1987) Retroviral expression

of the human IL-2 gene in a murine T cell line results in cell growth autonomy and tumorigenicity. EMBO J 6:2705–2709
9. Hapel AJ, Vande Woude G, Campbell HD, Young IG, Robins T (1986) Generation of an autocrine leukaemia using a retroviral expression vector carrying the interleukin-3 gene. Lymphokine Res 5: 249–254
10. Haeuptle MT, Feint N, Gough NM, Dobberstein B (1989) A tripartite structure of the signals that determine protein insertion into the ER membrane. J Cell Biol 108:1227–1236
11. Keating MT, Williams LT (1988) Autocrine stimulation of intracellular PDGF receptors in v-*sis*-transformed cells. Science 239:914–916
12. Browder TM, Abrams JS, Wong PMC, Nienhuis AW (1989) The mechanism of autocrine stimulation in hematopoietic cells producing interleukin 3 after retroviral mediated transfer. Mol Cell Biol 9:204–213
13. Young DC, Griffin JD (1986) Autocrine secretion of GM-CSF in acute myeloblastic leukemia. Blood 68:1178–1181
14. Klein H, Becker R, Lindemann A, Oster W, Schleiermacher E, Souza L, Mertelsmann R, Herrmann F (1989) Synthesis of granulocyte colony-stimulating factor and its requirement for terminal divisions in chronic myelogenous leukaemia. J Exp Med (in press)
15. Hirano T, Kishimoto (1989) Interleukin-6. In: Sporn MB, Roberts AB (eds) Peptide growth factors and their receptors. Springer, Berlin Heidelberg New York, chap 14 (Handbook of Experimental Pharmacology vol 95, part 1)
16. Spooncer E, Katsuno M, Hampson I, Dexter TM, Just U, Stocking C, Kluge N, Ostertag W (1989) Biological effects of retroviral transfection of the murine interleukin-3 gene into FDC-P mix cells. In: Shen-ong GLC, Potter M, Copeland NG (eds) Mechanisms in myeloid tumor regenesis. Springer, Berlin Heidelberg New York (Current topics in microbiology and immunology, vol 149)
17. Johnson A, Betsholtz C, Heldin C-H, Westermark B (1986) The phenotypic characteristics of simian sarcoma virus-tranformed human fibroblasts suggest that the v-*sis* gene product acts solely as a PDGF receptor agonist in cell transformation. EMBO J 5:1535–1541

Infection of Multipotent IL-3-dependent Stem Cells With a Retroviral Vector Containing the IL-3 Gene Confers Density-dependent Growth Autonomy Without Blocking Differentiation*

U. Just[4], E. Spooncer[1], M. Katsuno[1], I. Hampson[1], M. Dexter[1], J.F. DeLamarter[3], C. Stocking[2], N. Kluge[2], and W. Ostertag[2]

A. Introduction

It is widely accepted that most naturally occurring leukemias are monoclonally derived from multipotent stem cells [5–7, 17], but the genetic changes leading to their transformation are poorly understood. A useful system in which to study the various processes occurring during leukemogenesis is offered by non-leukemic, multipotent stem cell lines (FDCPmix) established from murine long-term marrow cultures [20]. These cells grow continuously in vitro in the presence of Il-3, but they can also be induced to differentiate into mature granulocytes, macrophages, erythrocytes, and occasionally megakaryocytes, eosinophils, and mast cells by serum factors [20] or in association with marrow stromal cells [20] or certain embryonic mesenchymal cell lines [18]. Recent data have shown that hematopoietic growth-factor-dependent progenitor cell lines acquire growth-factor-independent growth and tumorigenicity when they are infected with retroviral vectors containing genes coding for Il-3 or GM-CSF [10, 11]. However, these studies have been restricted to cell lines which are blocked in differentiation and may therefore not reflect the alterations that occur in stem cells during leukemogenesis. To determine the effects of aberrant expression of Il-3 in differentiation-inducible stem cells we infected FDCPmix cells with a selectable retroviral vector carrying the cDNA of Il-3.

B. Materials and Methods

I. Vector Construction and Virus-producing Cell Lines

A cDNA clone of Il-3 (kindly provided by N. Gough, Melbourne) was subcloned into the MPSV-based M3neo vector [9, 10] and used for transfections into the amphotropic helper cell line, PA 317 [16], to produce infectious M3 MuV particles. Cell clone psi2 mos^{-1} no. 4 containing the neo MPSV *mos* deletion vector [22] was used to infect PA 317 in order to obtain amphotropic pseudotypes necessary for the infection of the ecotropic-virus producing FDCPmix cell lines [23]. Cell clones with titres of 10^3–10^5 for MuV and 10^5–10^8 GTU for mos^{-1} and with intact proviral genomes were used for co-cultivation experiments.

II. Cells

Virus-producing cell lines were kept in minimal essential medium supplemented

[1] Paterson Institute for Cancer Research, Christie Hospital and Holt Radium Institute, Manchester, M20 9BX, United Kingdom
[2] Heinrich-Pette-Institut für Experimentelle Virologie und Immunologie an der Universität Hamburg, Martinistraße 52, 2000 Hamburg 20, FRG
[3] Biogen, 46 Route des Acacias, 1227 Geneva, Switzerland
[4] Present address: Universitätskrankenhaus Eppendorf, II. Medizinische Klinik, Abteilung für Klinische Chemie, Martinistr. 52, D-2000 Hamburg 20, FRG

* This work was supported by the Cancer Research Campaign, United Kingdom, and the Deutsche Forschungsgemeinschaft. The Heinrich-Pette-Institut is financially supported by the Free Hanseatic City of Hamburg and the Bundesministerium für Jugend, Familie, Frauen und Gesundheit

with 10% fetal calf serum. Hematopoietic cell lines were maintained in Iscove's modified Dulbecco's medium, supplemented with 20% horse serum and Wehi 3BD-conditioned medium (WEHI CM) as a source of multi-CSF (Il-3) at a concentration that stimulated optimal cell growth.

III. Viral Infection and Selection Procedure

10^5 FDCPmix cells were inoculated onto subconfluent irradiated (20 Gy) virus-producer cell lines. Various FDCPmix cell lines were used. Two days later, the loosely adherent cells and cells in suspension were harvested, washed, and resuspended at about 10^5 cells/ml. After 2 days of culture, G418 was added to a final concentration of 1 mg/ml and the cells were subcultured as appropriate. Non-virus-infected cells died within 7 days but cells which had been co-cultured on the M3MuV and M3neo-producer cell lines continued to proliferate in the presence of the G418. About 2 weeks after selection with G418, the cells were cloned in soft agar in the presence of Il-3, and individual colonies were isolated and expanded from three different FDCPmix cell lines.

IV. Determination of Il-3 Activity

M3-MuV infected cells (10^6) were washed twice to remove residual Il-3 and incubated without WEHI-CM for 48 h. The supernatant was used as such or concentrated tenfold via Amicon filtration (exclusion mol. wt. <10000), dialyzed, and tested for stimulatory activity on indicator cell lines by determining [^{3}H] thymidine incorporation. Half-maximal stimulation of FDCP2 cells by either WEHI-CM or recombinant murine Il-3 was defined as 50 U/ml.

V. Growth Inhibition Assay

One of the clones of FDCPmix infected with M3MuV was grown at high density in the absence of Il-3. Cells were washed twice in medium without Il-3 and plated, 1×10^4 cells/well, into 96-well plates. Dilutions of the antiserum of preimmune rabbit serum ranged from 1:20 to 1:10240 final. 30 h after initiation, 0.5 μCi [^{3}H] thymidine was added for 14 h. Cells were harvested onto filters, using a cell harvester (Titertec), and counted.

VI. Colony Assay

10^3 control uninfected cells and cells infected with M3neo virus alone or M3MuV were plated in soft agar in culture conditions which allow the expression of multiple hematopoietic lineages [19]. Individual colonies were isolated after 10 days of growth, cytospin preparations made, and the cells stained with benzidine plus May-Grunwald Giemsa. At least 30 colonies were examined from each group.

VII. Marrow Stromal Cell Culture

Stromal cell cultures derived from bone marrow were irradiated [20, 21] and used as a supportive stroma for the growth of the FDCPmix cells, either uninfected, or M3MuV or M3neo infected. Between 2×10^6 and 10^7 FDCPmix cells were co-cultured with the marrow stromal cells. At various times after seeding of marrow stroma by the FDCPmix cells, cytospin preparations of the nonadherent cells were performed and the cells stained with May-Grunwald Giemsa.

VIII. Diffusion Chamber Culture

After two washes, aliquots of 5×10^5 FDCPmix cells, infected with either M3MuV or M3Neo, were inoculated into each diffusion chamber (DC). These were then inserted intraperitoneally into male CBA mice. After 7 days of culture the animals were killed; the chambers were removed and shaken for 40 min in a 0.5% Pronase solution (Merck). The resulting cell suspensions were counted for the total number of nucleated cells. Cytospin preparations were made and the cells were classified according to morphological criteria [12].

IX. Nucleic Acid Analysis

Cellular DNA was isolated and restricted by standard techniques and separated on agarose gels. Total RNA was isolated as previously described [2] and transferred to Gene Screen Plus (NEN) after denaturation with glyoxal and dimethylsulfoxide and electrophoresis through agarose gels [14]. Nucleic acids were transferred to Gene Screen Plus (NEN) and hybridized under the conditions recommended by the manufacturer, with probes labeled as previously described [4]. Probes used for analysis included an *Eco*RI-*Nco*I fragment of pMu21A containing the Il-3 cDNA clone (N. Gough, unpublished), the *Bgl*III-*Bam*HI fragment of pAG60 containing the coding region of the neo gene [3], and the PY80B probe specific for the murine Y-chromosome [1].

X. In Vivo Administration of FDCPmix Cells

Uninfected and A4/M3neo cells (cultured with Il-3) and A4/M3MuV cells (cultured without Il-3) were suspended in Fischer's medium at an appropriate cell concentration. The cells were injected i.v. in syngeneic B6D2Fl mice that had received 10 Gy, prior to inoculation of the cells.

C. Results

I. Virus Integration, Gene Expression, and Il-3 Secretion

Following infection and selection in liquid culture the cells were cloned in soft agar in the presence of G418. Individual clones were isolated and cultured for further analysis. Analysis of the virus insertion sites revealed that the resulting cell lines were monoclonal (data not shown). The Il-3 gene was expressed in the MuV-infected FDCPmix cells, as shown by Northern analysis (data not shown). Conditioned medium of the M3MuV-infected cells growing in the absence of Il-3 contained between 4 and 50 units of Il-3 activity per ml.

II. Density-dependent Growth Autonomy

All M3MuV infected cell lines could grow in high density without Il-3, whereas the uninfected FDCPmix cells and the M3neo-infected cells died in the absence of growth factor. Cloning of FDCPmix M3MuV-infected cells in soft agar resulted in nonlinear, density-dependent growth in the absence of Il-3 and in nearly linear growth in the presence of Il-3 (Fig. 1). Growth of the M3MuV-infected cells could be blocked by neutralizing antisera to Il-3 (Fig. 2).

III. Differentiation Induction

M3MuV-infected stem cell lines retained their capacity to undergo differentiation in response to serum factors or marrow

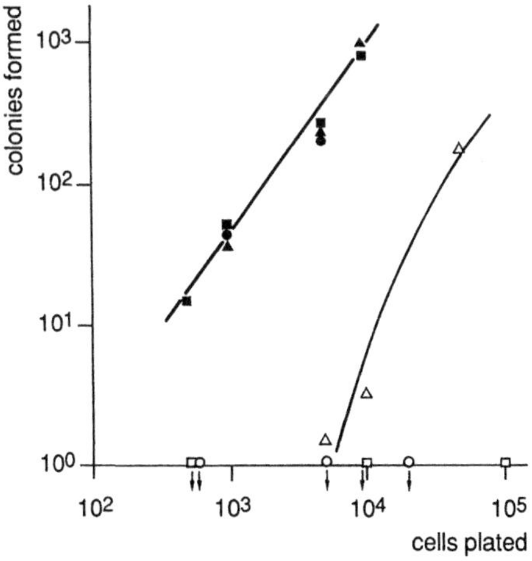

Fig. 1. Clonability of FDCPmix M3MuV-infected cells as a function of cell density. Uninfected FDCPmix cells with (■) and without (□) Il-3; M3neo-infected FDCPmix cells with (●) and without (○) Il-3; M3MuV-infected FDCPmix cells with (▲) and without (△) Il-3

Table 1. Colony formation by control and infected FDCPmix cells

	Colony morphology (%)					
	PE (%)	B	EG	LG	Mono	Mixed/Erythroid
Uninfected cells	6	9	4	63	8	16
$FDCP_{mix}$/M3neo	8	11	7	59	10	13
$FDCP_{mix}$/M3MuV	7	45	31	18	6	<1

PE (%), Plating efficiency, i.e., number of colonies formed per 100 cells plated; B, primitive blast cells; EG, promyelocytes and myelocytes; LG, metamyelocytes and mature granulocytes (including eosinophils); Mono, large mononuclear cells

Table 2. In vivo administration of FDCPmix cells

Group	Cells injected	No. of animals leukemic*	Spleen weight (mg, range)	Blood counts ($\times 10^{-3}$/mm^3, range)	Femur cellularity ($\times 10^{-7}$, range)
Uninfected	10^7	0/20	80– 120	3– 6	1.2–2.3
$FDCP_{mix}$/M3neo	10^7	0/10	70– 110	3– 7	1.3–2.2
$FDCP_{mix}$/M3MuV	5×10^6	15/15	530–1164	40–276	0.7–1.1

* Morbidity was first observed 6 weeks after injection of the cells, and the majority of the animals had developed clear evidence of hematopoietic disease (and were autopsied accordingly) within 12 weeks

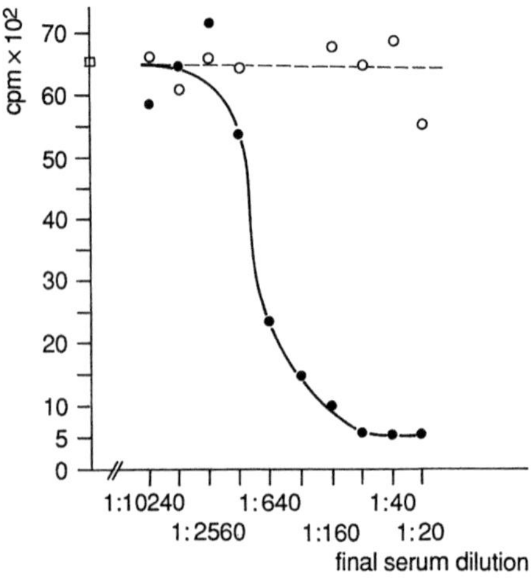

Fig. 2. Growth inhibition of FDCPmix M3-MuV-infected cells by anti-Il-3 antiserum. ● Growth medium with rabbit anti-Il-3 antiserum; ○ growth medium with rabbit preimmune serum; □ growth medium without antiserum or preimmune serum

stromal cells. In the mixed colony assay the plating efficiency was unaltered and the colonies produced contained maturing granulocytes and macrophages (Table 1). However, erythroid cells were rarely seen, and the balance between immature and mature granulocytes was changed in favor of immature cells (Table 1, Fig. 3). The same was true when the cells were co-cultured with marrow stromal cells (data not shown). Culture of M3MuV-infected FDCPmix cells in vivo in the DC led to an increase of immature and mature granulocytes and macrophages similar to the in vitro observations. In addition, erythroblasts were also found in the DC (data not shown).

IV. In Vivo Administration of FDCPmix MuV-infected Cells

When FDCPmix M3MuV-infected cells were injected into sublethally irradiated

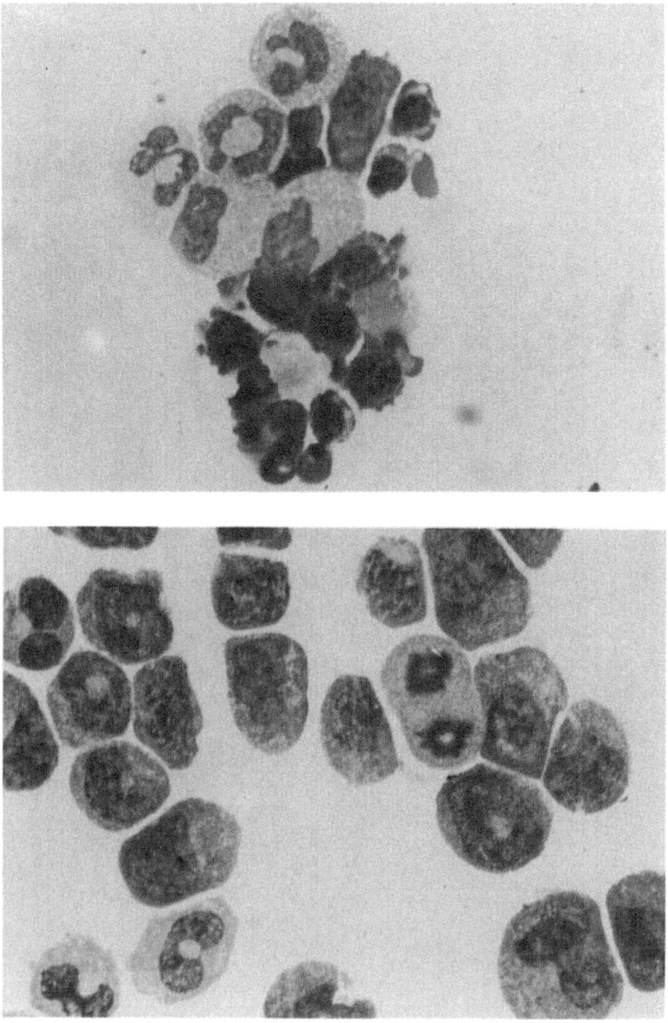

Fig. 3a, b. Mixed colony formation by control and infected FDCPmix cells. **a** Parental FDCP-mix cells; **b** FDCP-mix M3MuV-infected cells

syngeneic mice, the animals developed a five- to ten fold increase in the spleen weight, an increase in peripheral leukocytes, and a decreased hematocrit (Table 2). Morphological analysis of cells present in the spleen and peripheral blood of a representative mouse (Fig. 4) showed the following differential: spleen (blood) 16% (2)% blasts, 21% (10)% promyelocytes/myelocytes, 55% (75)% metamyelocytes and polymorphonuclear granulocytes, 5% (1)% nucleated erythroid cells, 3% (12)% other. The animals died within 2 months, whereas control mice, injected with stem cells containing the M3neo vector, showed no evidence of disease 6 months later. To determine the origin of the disease, cytogenetic analysis of cells in the spleen of the leukemic mice was performed. Initially, 80%–100% of the mitoses were of donor origin and possessed a normal (donor) male karyotype. Subsequently, however, spleen, bone marrow, and blood cells were of recipient origin, as revealed by Southern blotting (data not shown). Furthermore, the viral integration sites of cell lines recovered from leukemic animals were different

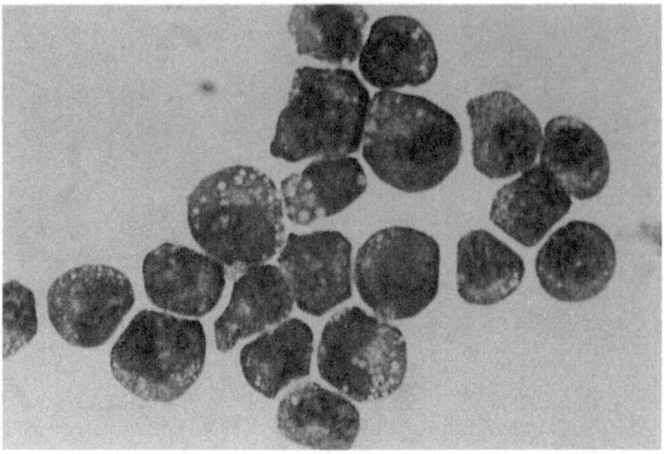

a

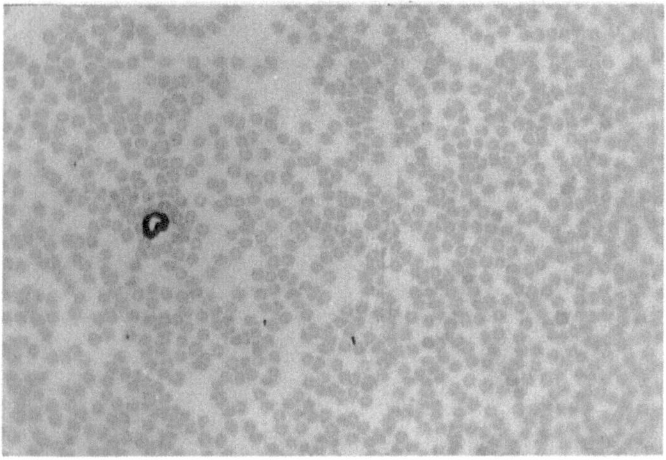

b

c

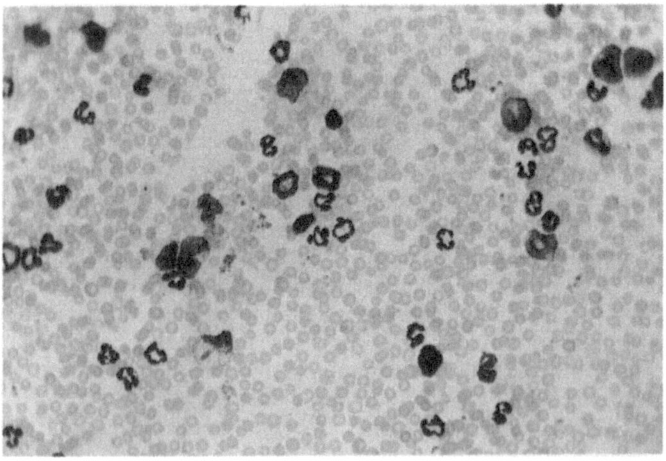

Fig. 4a–d. In vivo administration of FDCPmix M3MuV-infected cells. **a** FDCP-mix M3MuV-infected cells; **b** normal peripheral blood cells; **c** peripheral blood cells from leukemic mouse; **d** spleen cells from leukemic mouse

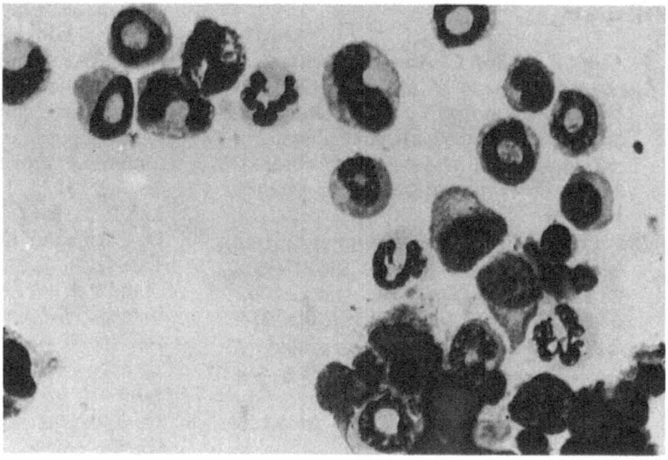

Fig. 4d

from the original viral insertion site of the donor FDCPmix cells (data not shown).

D. Discussion

The present results show that infection of multipotent Il-3-dependent stem cells with a retroviral vector containing the Il-3 gene confer density-dependent autocrine stimulation of growth without blocking differentiation, but with a change of the balance between differentiation and proliferation in favor of proliferation. From these data we conclude that inappropriate expression of Il-3 may play an important role in the multistep pathogenesis of leukemia. When the Il-3 infected cells were injected into sublethally irradiated syngeneic mice, the animals developed a myeloproliferative disease. However, the precise role of the injected cells remains to be determined. Analysis of spleen and blood cells of the leukemic mice revealed that the proliferating cells were initially derived from the transplanted stem cells but were subsequently of recipient origin. Furthermore, the viral integration sites in cell lines recovered from leukemic animals showed different bands as compared with the original injected cells, indicating infection of host cells. Since FDCPmix cells contain an ecotropic helper virus (MoMuLV) [23], it could have packaged the defective MuV vector to produce an infectious virus which may then have transformed host cells. However, when high levels of the original MuV virus are injected into mice no myeloproliferative disease is observed [9]. In the latter case, this may reflect a difficulty in the ability of the injected virus to "target" to the host cells in the sites of active hematopoiesis. This may not be the case for the MuV-infected FDCPmix cells, which can clearly lodge in the spleen and bone marrow and may be acting as "carriers" for infectious viral particles, thus facilitating infection of host hematopoietic cells. Also, it has been reported that injection of recombinant Il-3 into normal mice leads to an increase in spleen weight and content of CFU-S, as well as to an increase in progenitor cells of the myeloid lineage [8, 13, 15]. Therefore, Il-3 production by the infected cells (both donor and host) may have contributed to the disease by stimulating stem and progenitor cells from the recipient mice. Thus, the disease is probably multifactorial. Nonetheless, we have clearly shown that endogenous inappropriate expression of a growth factor gene can have profound biological effects and may well be a part of the process leading to leukemic transformation.

References

1. Avner P, Bishop C, Amar L, Cambrou J, Hatat D, Arnaud D, Mattei M-G (1987) Mapping the mouse X chromosome possible symmetry in the location of a family of sequences on the mouse X and Y chromosomes. Development-Camb 101 [Suppl]: 107–116
2. Auffray C, Rougeon F (1980) Purification of mouse immunoglobulin heavy-chain messenger RNAs from total myeloma tumor RNA. Eur J Biochem 107: 303–314
3. Colbere-Garapin F, Horodniceanu J, Kourilsky P, Garapin A-C (1981) A new dominant hybrid selective marker for higher eukaryotic cells. J Mol Biol 150: 1–14
4. Feinberg AP, Vogelstein B (1984) A technique for radiolabeling DNA restriction endonuclease fragments to high specific activity. Anal Biochem 137: 266–267
5. Fialkow PJ, Singer JW (1985) Tracing development and cell lineages in human hemopoietic neoplasia. In: Weissman IL (ed) Leukemia. Dahlem Workshop. Springer, Berlin Heidelberg New York, pp 202–222
6. Fialkow PJ, Singer JW, Adamson JW, Vaidya K, Dow LW, Ochs J, Moohr JW (1981) Acute nonlymphocytic leukemia: heterogeneity of stem cell origin. Blood 57: 1068–1073
7. Greaves MF, Delia D, Robinson J, Sutherland R, Newman R (1981) Exploitation of monoclonal antibodies: a "who's who" of haemopoietic malignancy. Blood Cells 7: 257–280
8. Kindler V, Thorens B, de Kossodo S, Allet B, Eliason JF (1986) Stimulation of hematopoiesis in vivo by recombinant bacterial murine interleukin 3. Proc Natl Acad Sci USA 83: 1001–1005
9. Laker C, Kluge N, Stocking C, Just U, Franz M-J, Ostertag W, Dexter M, Katsuno M, Spooncer E (1989) Leukemogenesis initiated by aberrant multi-CSF expression is a multi-step process and differs in hematopoietic stem and precursor cell lines. Mol Cell Biol (in press)
10. Laker C, Stocking C, Bergholz U, Hess N, DeLamarter JF, Ostertag W (1987) Autocrine stimulation after transfer of the granulocyte/macrophage colony-stimulating factor gene and autonomous growth are distinct but inter-dependent steps in the oncogenic pathway. Proc Natl Acad Sci USA 84: 8458–8462
11. Lang RA, Metcalf D, Gough N, Dunn AR, Gonda TJ (1985) Expression of a haemopoietic growth factor cDNA in a factor-dependent cell line results in autonomous growth and tumorigenicity. Cell 43: 531–542
12. Lau B, Jaeger G, Thiel E, Rodt H, Huhn D, Pachmann K, Netzel B, Boening L, Thierfelder S, Doermer P (1979) Growth of the Reh cell line in diffusion chambers. Evidence for differentiation along the T- and B-cell pathway. Scand J Haematol 23: 385–392
13. Lord BJ, Molineux G, Testa NG, Kelly M, Spooncer E, Dexter TM (1986) The kinetic response of haemopoietic precursor cells, in vivo, to highly purified, recombinant interleukin-3. Lymphokine Res 5: 97–104
14. McMaster GK, Carmichael GG (1977) Analysis of single- and double-stranded nucleic acids on polyacrylamide and agarose gels by using glyoxal and acridine orange. Proc Natl Acad Sci USA 74: 4835–4839
15. Metcalf D, Begley CG, Johnson GR, Nicola NA, Lopez AF, Williamson DJ (1986) Effects of purified bacterially synthesized murine multi-CSF (IL-3) on hematopoiesis in normal adult mice. Blood 68: 46–57
16. Miller AD, Buttimore C (1986) Redesign of retrovirus packaging cell lines to avoid recombination leading to helper virus production. Mol Cell Biol 6: 2895–2902
17. Rowley JD (1985) Significance of chromosome rearrangements in leukaemia and lymphoma. In: Weissman IL (ed) Leukemia. Dahlem Workshop. Springer, Berlin Heidelberg New York, pp 179–202
18. Roberts RA, Spooncer E, Parkinson EK, Lord BI, Allen TD, Dexter TM (1987) Metabolically inactive 3T3 cells can substitute for marrow stromal cells to promote the proliferation and development of multipotent haemopoietic stem cells. J Cell Physiol 132: 203–214
19. Spooncer E, Boettiger D, Dexter TM (1985) Continuous in vitro generation of mulitpotential stem cell clones from src-infected cultures. Nature 310: 228–230
20. Spooncer E, Heyworth CM, Dunn A, Dexter TM (1986) Self-renewal and differentiation of interleukin-3-dependent multipotent stem cells are modulated by stro-

mal cells and serum factors. Differentiation 31:111–118
21. Spooncer E, Lord BI, Dexter TM (1985) Defective ability to self-renew in vitro of highly purified primitive haematopoietic cells. Nature 316:62–64
22. Stocking C, Kollek R, Bergholz U, Ostertag W (1985) Long terminal repeat sequences impart hematopoietic transformation properties to the myeloproliferative sarcoma virus. Proc Natl Acad Sci USA 82:5746–5750
23. Wyke JA, Stoker AW, Searle S, Spooncer E, Simmons P, Dexter TM (1986) Perturbed hemopoiesis and the generation of mulitpotential stem cell clones in src-infected bone marrow cultures is an indirect or transient effect of the oncogene. Mol Cell Biol 6:959–963

Isolation of Revertants from a Factor-Independent Myeloid Cell Line

M. Kawai [1], T. Mori [1], C. Stocking [2], and K. Notake [1]

Mouse myeloid stem cell line D35 was obtained from a primary cell culture of C3H bone marrow cells. This cell grows completely dependent on the multi-lineage colony-stimulating factor (multi-CSF; IL-3) or the granulocyte macrophage colony-stimulating factor (GM-CSF).

By serial dilution of the medium every 3 days, autonomously growing mutants were obtained spontaneously at the rate of 10^{-8}–10^{-7}. Among 18 mutants, 11 produce GM-CSF by themselves, three produce multi-CSF, and two do not secrete any growth factors that stimulate the parental cell line. All of those mutants that were checked were tumorigenic in nude mice. Four of the GM-CSF-producing mutants turned out to be the result of the insertion of an intracisternal A particle (Dind 1), Rauscher murine leukemia virus (Dind 4), or Friend spleen focus-forming virus (F-SFFV; Dind 5, Dind 9) at either 3′ or 5′ of GM-CSF locus.

To clarify whether aberrant expression of a CSF gene is sufficient for autonomous growth and tumorigenicity, and to investigate possible mechanisms of gene activation, as well as insertion and excision of these retroviruses, we have attempted to obtain revertants that again have a growth dependency on factor. After cultivation of these autonomous mutants with multi-CSF, 10^6 cells were washed and incubated for 24 h without any growth factor. A final concentration of 10^{-5} *M* BrdU was then added to the culture. After another 21 h, Hoechst dye 33 258 (final 1 μg/ml) was added for 3 h to enhance the killing by light. All dishes were exposed to fluorescent light at a distance of 10 cm for 30 min. Cells were

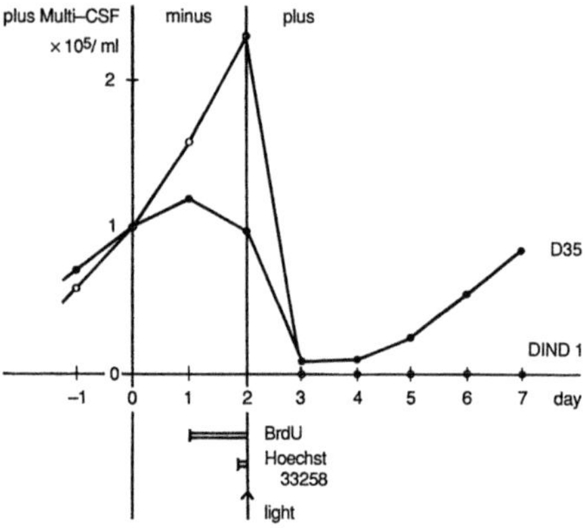

Fig. 1. Counter selection of Revertants from Dind 1 cells by BrdU-H33258-light killing

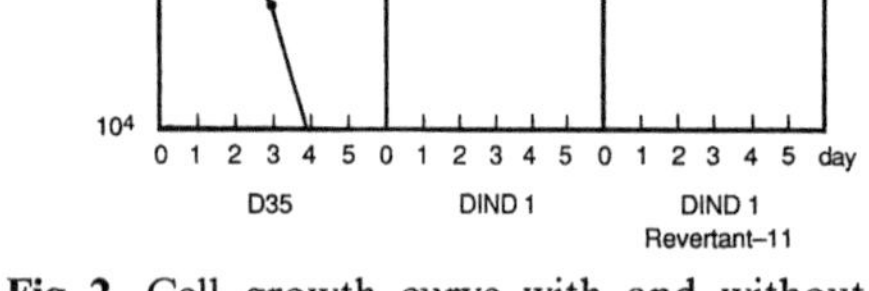

Fig. 2. Cell growth curve with and without multi-CSF. Open circles show the growth with multi-CSF (5% v/v of WEHI-3 cell conditioned medium), filled circles show the growth curve without multi-CSF

---

[1] Aichi Medical University, Aichi, Japan
[2] Heinrich-Pette-Institut, Hamburg, Federal Republic of Germany

subsequently washed three times with fresh media. The next day, as shown in Fig. 1, almost all the Dind 1 cells had been killed, but the control D35 cells had survived. After 7 days we saw 20–100 clonal clumps of the surviving Dind 1 cells in each dish. Clumps were directly transferred with Eppendorfer pipettes to the 96-well plates and cultivated with multi-CSF. After reaching confluency, each of three aliquots were shifted to 24-well plates to check their dependence of multi-CSF or GM-CSF. Among the 44 clumps, two were clearly dependent again on either factor. The other 42 were similar to normal or slow-growing Dind 1 cells. Figure 2 shows the growth curve of recloned revertant II in the presence or absence of multi-CSF, compared with D35 and Dind 1. As revertant II is shown to be a true revertant of Dind 1, it would be of interest to determine whether the IAP is still in the GM-CSF locus or not, and whether this revertant has lost the tumorigenicity in nude mice.

We have also shown here a more effective method for cell selection using a combination of H33258 with BrdU and light. This method decreases the cell survival rate to 10^{-5} from $\sim 10^{-2}$ obtained with standard systems using BrdU and light alone.

Comparison of Factors Which Induce Differentiation of the Murine Myeloid Leukaemic Cell Line M1 *

D.P. Gearing

A. Introduction

Naturally occurring regulatory molecules which can cause the terminal differentiation of leukaemic cells have recently been the subject of great interest, and to date, several protein factors have been described that can induce the differentiation of various myeloid leukaemic cells in vitro. The haematopoietic colony-stimulating factors, granulocyte-CSF (G-CSF) and granulocyte/macrophage-CSF (GM-CSF) [1, 2], which act as both proliferative and differentiative stimuli for normal myeloid progenitors, are able to induce the differentiation of certain murine and human myeloid leukaemic cells (including murine WEHI-3BD$^+$ and M1 and human HL-60) [3–5]. By contrast, a number of other activities have been described which are capable of inducing the differentiation of murine M1 leukaemic cells, yet which do not stimulate the proliferation of normal progenitor cells [6–8]. With the recent cloning of myeloid leukaemia inhibitory factor (LIF [22]; = D-factor [6]), they have all been cloned. In this report the genetic and polypeptide structures of LIF are compared with those of two growth factors also known to cause differentiation of M1 cells: G-CSF and interleukin-6 (IL-6; MGI-2).

The Walter and Eliza Hall Institute of Medical Research, Post Office, Royal Melbourne Hospital, Victoria, 3050, Australia

* This work was supported by the Anti-Cancer Council of Victoria, the National Health and Medical Research Council (Canberra), the AMRAD Corporation Ltd. (Melbourne) and the National Institutes of Health, Bethesda, MD, USA, Grant no. CA22556.

B. Genetic and Polypeptide Structures of M1 Differentiation-inducing Factors

I. G-CSF

Natural and recombinant G-CSF can induce the terminal differentiation of M1 leukaemic cells [9]. Genomic and cDNA sequences for both human and murine G-CSF have been reported [10, 11]. The genes are highly homologous and the encoded proteins display 76% identity with three insertions or deletions [11]. Murine G-CSF consists of 178 amino acids with a calculated Mr of 19061 and is O-glycosylated to an Mr of approximately 24000–25000. Comparison of the structures of human G-CSF cDNAs and the chromosomal gene revealed that there is a possibility of two different G-CSF molecules consisting of 177 or 174 amino acids [10], corresponding to a calculated molecular mass of 18671 for the smaller version. Native human G-CSF is O-glycosylated to approximately 21000 daltons [12]. Neither human nor murine G-CSF have any N-linked glycosylation sites, and each molecule is presumed to be intramolecularly linked via two disulphide bonds [11].

II. IL-6 [MGI-2]

Over the past few years, IL-6 has been variously known as B-cell stimulatory factor [13], interferon-β2 [14], 26-kD inducible protein [15], hybridoma/plasmacytoma growth factor [16] and interleukin-HP1 [17]. IL-6 also stimulates the differentiation of B cells [13] and has some activities as a GM-colony-stimulating factor [18]. To this variety of seemingly unrelated activities (and more; see [19]) can be added its ability to cause the terminal differentiation of M1 leukaemic cells (D. Metcalf, personal communication).

Human and murine IL-6 have been cloned [13–15, 20] and show low sequence conservation (42% between proteins, with five insertions or deletions [20]), especially in the amino terminal half. Human IL-6 consists of a 184-amino acid peptide of calculated mass 20781 which is presumed to be glycosylated to its observed native Mr of 26000 via two potential N-linked and four potential O-linked glycosylation sites [13]. Murine IL-6 consists of 187 residues with a calculated Mr of 21710 which is variably O-glycosylated to between 22000 and 29000 daltons [20, 21].

The primary sequence of the IL-6 protein revealed significant homologies with the sequence of G-CSF [13], including the conservation of four cysteine residues. Comparison of their gene sequences suggested that the two genes may be derived from a common ancestoral gene [13]. Recently, it has been demonstrated that the sequence of at least one form of purified MGI-2 [7] is the same as that of murine IL-6 (A. Zilberstein and L. Sachs, personal communication).

III. LIF [D-Factor]

LIF, a factor that we have recently purified [8] and cloned [25, 26] is able to induce the differentiation of M1 cells but, unlike G-CSF, not WEHI-3BD$^+$ cells [8]. LIF is of particular interest for leukaemia therapy, because, unlike the CSFs and IL-6, it does not appear to be a proliferative stimulus for either normal or leukaemic progenitor cells [8]. In addition, LIF is unique in that at high concentrations it inhibits M1 colony formation [8]. LIF is one of the most highly conserved of the regulators known to act within the myeloid system. The mature murine and human protein sequences share 78% identity (with no insertions or deletions) [23], and consist of 179 amino acids. Murine LIF is heavily glycosylated from its calculated protein mass of ~20000 daltons to approximately 58000 daltons, via mostly potential N-linked and some potential O-linked glycosylation sites [22]. The murine and human LIF molecules share six conserved cysteine residues. D-factor [6] has recently been purified and its determined amino acid sequence is the same as that of murine LIF (M. Hozumi, personal communication).

C. Is LIF Related to G-CSF and IL-6?

Given that G-CSF and IL-6 appear to be related in both polypeptide and gene arrangements [13, 24], it is interesting to consider whether LIF has any relationship to these two regulators.

I. Protein Sequence/Structure

LIF shares little detectable homology with G-CSF and IL-6 when their protein sequences are compared. Furthermore, the number and positions of the cysteine residues in the LIF proteins are not equivalent to those of G-CSF and IL-6 (Fig. 1), and whilst G-CSF and IL-6 are minimally glycosylated (mostly O-linked) [11–13, 20], LIF is heavily N-glycosylated [8, 22, 23]. However, murine and human LIF share two small patches of homology with murine and human G-CSF and minor similarities may be observed in the distribution of proline clusters in all three molecules, and each is of a similar length. Thus, at the polypeptide level LIF bears only little resemblance to G-CSF and even less to IL-6.

II. Gene Organization

Comparison of the gene organization of LIF with that shared by G-CSF and IL-6 reveals little similarity (Fig. 2). The LIF genes comprise three exons compared with the five exons of both G-CSF and IL-6 and, apart from the coding region of the first exon of murine LIF, which is the

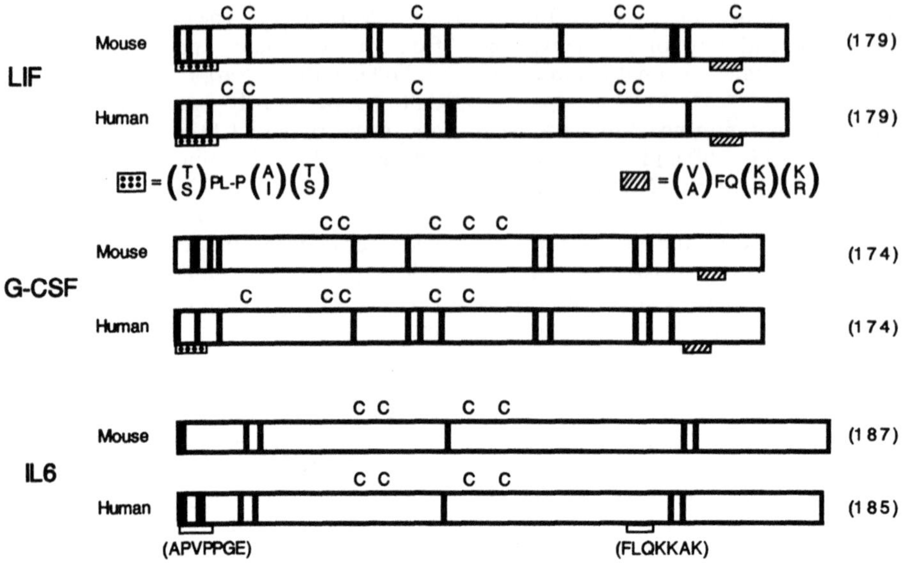

Fig. 1. Comparison of LIF, G-CSF and IL-6 polypeptides. Proline residues are marked by *vertical bars* and cysteine residues as *c*. Regions of homology are indicated by shaded boxes. (Based on reports in [10, 11, 13, 20, 22, 23]

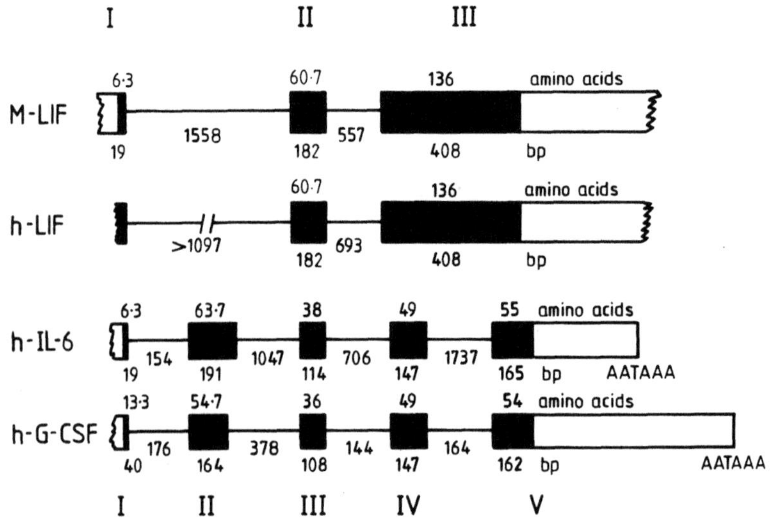

Fig. 2. Comparison of the gene organization between IL-6, G-CSF and LIF. *Boxes* represent exons. The coding and noncoding regions are shown by *closed* and *open boxes,* respectively. The *numbers above and below the boxes* indicate the numbers of amino acids of exons and those of nucleotides of exons for coding region, respectively. The *numbers below the lines* show the length of introns. (After the scheme presented in [24] and the data of [10] and [23]. Murine LIF gene organization: Gearing, King and Gough, Unpublished work)

same length as that of human IL-6 (6.3 codons), the lengths of the other exons are dissimilar. These data indicate that there is no reason to suggest that the three genes are related. However, it is interesting to note that the sum of the codons in the last three exons of the human G-CSF and IL-6 genes (139 and 142 codons, respectively) is approximately the same as the number of codons in the last exon of the LIF genes (136 codons). This suggests that if these genes are ancestrally related, then the LIF gene could have arisen by exon fusion of three exons into one, or that the G-CSF and IL-6 genes could have been derived by two introns inserting into a common larger last exon.

With the reasonable assumption that IL-6 is the molecule detected earlier as a plasmacytoma growth factor [25, 26], IL-6, G-CSF and LIF share two features in common: (a) when mice are injected with endotoxin, levels of each factor rise sharply in the serum within 1–3 h, and (b) all three molecules are produced in vitro by macrophages after priming with endotoxin [25–28] (D. Metcalf, personal communication), and G-CSF and LIF mRNAs are induced in macrophage lines following stimulation with endotoxin (N. Gough, personal communication). This suggests some intended biological coordination in their actions in vivo and raises the possibility of common transcriptional activation pathways.

D. Conclusions

From its primary protein sequences and its genetic organization, it can be concluded that LIF is a new structural entity which is probably not related to either G-CSF or IL-6, although the three molecules might in some ways be functionally related. At present, LIF appears unique in that it is not apparently a proliferative stimulus for either normal or leukaemic progenitor cells and it specifically inhibits colony formation and causes terminal differentiation of M1 leukaemic cells, this contrasts with G-CSF and IL-6, which can provide both a differentiative stimulus to M1 cells and a proliferative stimulus to normal cells.

Acknowledgements. Thanks to Nick Gough, Nic Nicola and Don Metcalf for their useful comments on this manuscript and to Iras Collins for typing.

Note added in proof:
Since this manuscript was submitted, LIF has been shown to be able to suppress the human leukemic cell lines HL60 and U937 when used in combination with GM-CSF or G-CSF [29]. Somewhat surprisingly, and by contrast to its differentiation-inducing activity on M1 leukaemic cells, LIF also has a powerful differentiation-inhibiting activity on embryonic stem [ES] cells [30, 31] and its use will aid in their propagation in vitro. In a further twist, COS cell conditioned-media containing LIF can also apparently act as a growth stimulus for the murine haemopoietic cell line DA-1a [32] so its postulated role as a differentiation-only factor may have to be reconsidered.

References

1. Metcalf D (1985) The granulocyte-macrophage colony-stimulating factors. Science 229:16–22
2. Metcalf D (1986) The molecular biology and functions of the GM-CSFS. Blood 67:257–267
3. Begley CG, Metcalf D, Nicola NA (1987) Purified colony stimulating factors (G-CSF and GM-CSF) induce differentiation in human HL60 leukemic cells with suppression of clonogenicity. Int J Cancer 39: 99–105
4. Metcalf D, Nicola NA (1982) Autoinduction of differentiation in WEHI-3B leukemia cells. Int J Cancer 30:773–780
5. Metcalf D (1979) Clonal analysis of the action of GM-CSF on the proliferation and differentiation of myelomonocytic leukemic cells. Int J Cancer 24:616–623

6. Tomida M, Yamamoto-Yamaguchi Y, Hozumi M (1984) Purification of a factor inducing differentiation of mouse myeloid leukemic M1 cells from conditioned medium of mouse fibroblast L929 cells. J Biol Chem 259:10978–10982
7. Lipton JH, Sachs L (1981) Characterization of macrophage- and granulocyte-inducing proteins for normal and leukemic myeloid cells produced by the Krebs ascites tumor. Biochim Biophys Acta 673: 552–569
8. Hilton DJ, Nicola NA, Metcalf D (1988) Purification of a murine leukemia inhibitory factor from Krebs ascites cells. Anal Biochem 173:359–367
9. Tomida M, Yamamoto-Yamaguchi Y, Hozumi J et al. (1986) Induction by recombinant human granulocyte colony-stimulating factor of differentiation of mouse myeloid leukemic M1 cells. FEBS Lett 207:271–275
10. Nagata S, Tsuchiya M, Asano S et al. (1986) The chromosomal gene structure and two mRNAs for human granulocyte colony-stimulating factor. EMBO J 5: 575–581
11. Tsuchiya M, Asano S, Kaziro Y, Nagata S (1986) Isolation and characterization of the cDNA for murine granulocyte colony-stimulating factor. Proc Natl Acad Sci USA 83:7633–7637
12. Nomura H, Imazeki I, Oheda M et al. (1986) Purification and characterization of human granulocyte colony-stimulating factor (G-CSF). EMBO J 5:871–876
13. Hirano T, Yasukawa K, Harada H et al. (1987) Complementary DNA for a novel human interleukin (BSF-2) that induces B lymphocytes to produce immunoglobulin. Nature 324:73–76
14. Zilberstein A, Ruggieri R, Kora JH, Revel M (1986) Structure and expression of cDNA and genes for human interferon -β-2, a distinct species inducible by growth-stimulatory cytokines. EMBO J 5:2529–2537
15. Content J, DeWit L, Pierard D et al. (1982) Secretory proteins induced in human fibroblasts under conditions used for the production of interferon β. Proc Natl Acad Sci USA 79:2768–2772
16. Van Damme J, Opdenakker G, Simpson RJ et al. (1987) Identification of the human 26-kD protein, interferon β2 (IFN-β_2), as a B cell hybridoma/plasmacytoma growth factor induced by interleukin 1 and tumor necrosis factor. J Exp Med 165:914–919
17. Van Snick J, Cayphas S, Vink A et al. (1986) Purification and NH_2-terminal amino acid sequence of a T-cell-derived lymphokine with growth factor activity for B-cell hybridomas. Proc Natl Acad Sci USA 83:9679
18. Wong GG, Witek-Giannotti JS, Temple PA et al. (1988) Stimulation of murine hemopoietic colony formation by human IL-6. J Immunol (1988) 140:3040–3044. J Immunol [in press]
19. Wong GG, Clark SC (1988) Multiple actions of interleukin 6 within a cytokine network. Immunol Today 9:137–139
20. Van Snick J, Cayphas S, Szikora J-P et al. (1988) cDNA cloning of murine interleukin-HP1: homology with human interleukin 6. Eur J Immunol 18:193–197
21. Van Snick J, Vink A, Cayphas S, Uyttenhove C (1987) Interleukin HP1, a T cell-derived hybridoma growth factor that supports the in vitro growth of murine plasmacytomas. J Exp Med 165:641
22. Gearing DP, Gough NM, King JA et al. (1987) Molecular cloning and expression of cDNA encoding a murine myeloid leukaemia inhibitory factor (LIF). EMBO J 6:3995–4002
23. Gough NM, Gearing DP, King JA et al. (1988) Molecular cloning and expression of the human homologue of the murine gene encoding myeloid leukemia-inhibitory factor. Proc Natl Acad Sci USA 85:2623–2627
24. Yasukawa K, Hirano T, Watanabe Y et al. (1987) Structure and expression of human B cell stimulatory factor-2 (BSF-2/IL-6) gene. EMBO J 6:2939–2945
25. Metcalf D (1973) Colony formation in agar by murine plasmacytoma cells: potentiation by hemopoietic cells and serum. J Cell Physiol 81:397–410
26. Metcalf D (1974) The serum factor stimulating colony formation in vitro by murine plasmacytoma cells: response to antigens and mineral oil. J Immunol 113:235–243
27. Burgess AW, Metcalf D (1980) Characterization of a serum factor stimulating the differentiation of myelomonocytic leukemic cells. Int J Cancer 26:647–654
28. Metcalf D, Nicola NA (1985) Synthesis by mouse peritoneal cells of G-CSF, the differentiation inducer for myeloid leukemia cells: stimulation by endotoxin, M-CSF and multi-CSF. Leukaemia Res 9:35–50

29. Maekawa T, Metcalf D (1989) Clonal suppression of HL60 and U937 cells by recombinant human leukemia inhibitory factor in combination with GM-CSF or G-CSF. Leukemia, in press
30. Williams RL, Hilton DJ, Pease S et al. (1988) Myeloid leukaemia inhibitory factor (LIF) maintains the developmental potential of embyronic stem cells. Nature 336:684–687
31. Smith AG, Heath JK, Donaldson DD et al. (1988) Inhibition of pluripotential embryonic stem cell differentiation by purified polypeptides. Nature 336:688–690
32. Moreau J-F, Donaldson DD, Bennett F et al. (1988) Leukaemia-inhibitory factor is identical to the myeloid growth factor, human interleukin for DA cells. Nature 336:690–692

Establishment of a Leukemic Cell Line MT-ALL With Multilineage Differentiation Potential *

F. Griesinger [1], J.M. Greenberg [2], D.C. Arthur [2], A.C. Ochoa [1], C. Hurvitz [3], R. Brunning [1], and J.H. Kersey [2]

A. Introduction

Multiple steps are involved in hematopoietic differentiation and maturation from multipotent progenitor cells to terminally differentiated cells of each lineage. The availability of recombinant growth factors and of various clonogenic precursor cell assays has provided useful information about hematopoietic precursors, notably of the myeloid lineage [1–3]. However, the direct evaluation of primitive human lymphohematopoietic progenitor cells has been hampered by their low frequency in normal donors and the difficulties involved in expanding them in in vitro culture [4, 5]. Therefore, little is known about the events and mechanisms that result in the irreversible commitment of multipotent uncommitted progenitors to a specific cell lineage. Leukemias have been traditionally used as a model for lymphohematopoietic development and have been studied extensively for the expression of differentiation antigens [6, 7] and recently for rearrangements of genes of the immunoglobulin supergene family [8, 9]. They are considered to represent clonal expansions of lymphoid and myeloid progenitor cells and therefore provide an opportunity to study homogeneous populations of lymphohematopoietic progenitor cells. However, leukemic cell lines with the potential for multilineage differentiation that would allow the study of early events of lymphohematopoietic development are very rare [10, 11]. We report a novel acute lymphoblastic leukemia (ALL), MT-ALL, with a predominantly mature $CD3^+$ TCR α/β^+ phenotype, which in response to various growth factors displays the potential for multilineage differentiation in vitro. This leukemia may be instrumental in elucidating mechanisms involved in early lymphohematopoietic development.

B. Materials and Methods

Leukemic cell cultures were set up as shown in Fig. 1 and as previously reported [12]. Southern analysis was performed according to standard methods [12, 13]. The genomic probe pδ7 detects the constant region of the T-cell receptor (TCR) δ gene (J.M. Greenberg, C.W. Wilkowski, J.H. Kersey, unpublished results). pTγ1 is a cDNA probe derived from the leukemic cell line HPB-mlt [14], kindly provided by R. Holcombe. The Tβ probe is a cDNA probe kindly provided by T.W. Mak. Methods for the lymphokine-activated (LAK) and natural killer (NK) cell assays as well as the oxygen production assay have been described elsewhere [15, 16].

---

[1] Departments of Laboratory Medicine and Pathology, and [2] Pediatrics, University of Minnesota, Minneapolis, MN 55455, USA
[3] Department of Pediatrics, Cedars-Sinai Medical Center, Los Angeles, CA 90048, USA

* Supported in part by the following grants for the National Institutes of Health: CA 25097 and CA 21737 and the University of Minnesota Children's Cancer Research Fund. Frank Griesinger is recipient of a fellowship from the Deutsche Forschungsgemeinschaft (Gr 301/1-1). James M. Greenberg is a Special Fellow of The Leukemia Society of America.

This work is published in the Journal of Experimental Medicine from which permission for the reproduction of figures and tables was obtained.

Clinical Course of the Patient. The 15-year-old caucasian male presented with hemorrhagic diathesis, hepatosplenomegaly, and mediastinal mass. White blood cell count was 759000/µl, containing 99% lymphoblasts, which were $CD2^+$ TdT^+ $CD4^-$ $CD8^-$. Complete remission was achieved lasting for 5 months. The patient relapsed and attained unsustained remissions during which peripheral blood was obtained at three different times after informed consent. The patient died 1 year after diagnosis.

C. Results and Discussion

The leukemia presented in the patient with a predominantly mature T-lymphocyte phenotype ($CD2^+$ $CD3^+$ $CD4^-$ $CD5^+$ $CD7^+$ $CD8^-$ $TCR\alpha/\beta^+$) and a morphology typical for lymphoblasts (Fig. 2A). In the presence of different recombinant growth factors, various leukemic cell lines belonging to different cell lineages were established (Fig. 1). In the presence of interleukin 3 (IL-3) and/or granulocyte-macrophage colony-stimulating factor (GM-CSF), myeloid and monocytoid cells grew out, including terminally differentiated neutrophilic granulocytes (Fig. 2B). The cells were myeloperoxidase positive and expressed the myelomonocytic differentiation antigens CD13, CD14, MY8, and CD33 [17], while they were devoid of T-lineage associated antigens. The oxygen radical production, which is a typical feature of mature functional granulocytes, was assessed by measuring the reduction of the reporter substrate nitroblue tetrazolium (NBT) in the presence of various stimuli [16]. After stimulation with phorbol myristate (5 ng/ml), but not with the chemotactic oligopeptide F-met-leu-phe [16], NBT was reduced, proving that the myeloid cells were at least partially functional (data not shown).

A growth factor independent (GFI) leukemic cell line was established from the IL-3 cultures. Morphologically, these were undifferentiated blasts (Fig. 2C), coexpressing T-lymphoid (CD2, CD3, CD5, CD7, TCRα/β) and myeloid lineage (CD13, CD14, CD33) associated differentiation antigens. In functional studies (oxygen radical production and NK and LAK activity; Fig. 3) these cells were inactive.

In the presence of IL-2, a cell line consisting of large lymphocytes with azurophilic granules (Fig. 2D) was established. In addition to the T-lineage associated antigens, the IL-2 dependent leukemic cells expressed CD8 and Leu-19, which can be found on cytotoxic T-lymphocytes [18, 19]. Functional studies confirmed that they represented the counterpart of MHC unrestricted cytotoxic T-lymphocytes [19]. The IL-2 de-

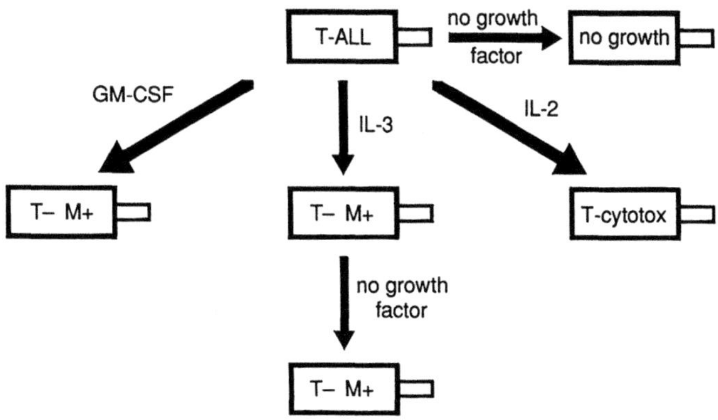

Fig. 1. Fresh peripheral blood leukemic blasts were cultured in the presence of rh GM-CSF (10–50 U/ml), IL-3 (10 U/ml), IL-2 (100 U/ml), or no growth factor. *T+/−*: positive/negative staining for T-lineage associated antigens; *M+/−*: positive/negative staining for myeloid lineage associated antigens and cytochemistry; *T-cytotox.*: expression of a phenotype consistent with cytotoxic T-lymphocytes (see text for details)

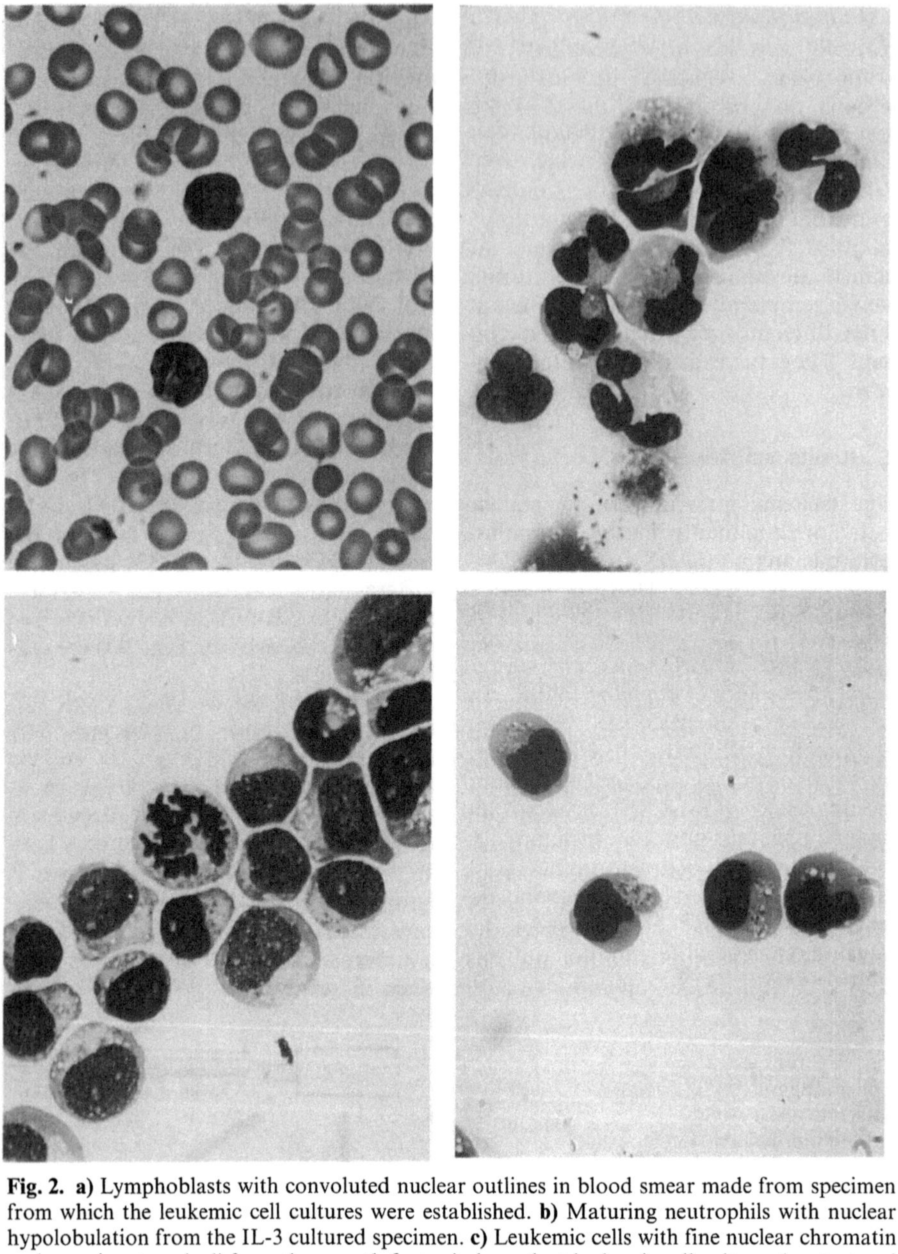

a b c d

Fig. 2. a) Lymphoblasts with convoluted nuclear outlines in blood smear made from specimen from which the leukemic cell cultures were established. **b)** Maturing neutrophils with nuclear hypolobulation from the IL-3 cultured specimen. **c)** Leukemic cells with fine nuclear chromatin and prominent nucleoli from the growth factor independent leukemic cell cultures (see text and Fig. 1). **d)** IL-2 dependent leukemic cell cultures. The majority of these cells have distinct nucleoli and azurophilic granulation. **(a–d**, Wright's Giemsa, ×1200) pendent leukemic cell cultures exerted high lytic activity against the NK sensitive target cell line K562 and the NK resistant cell line HL-60 [15] (Fig. 3). Surprisingly, the IL-2 dependent cell line coexpressed CD33 (data not shown).

Molecular genetic analysis demonstrated unambiguously that the various leukemic cell cultures were derived from the same malignant clone. Identical, unique rearrangements of the TCR δ (Fig. 4A, B), γ (Fig. 4C), and β (Fig. 4D) genes were demonstrated by Southern analysis. The TCR δ gene rearrangement, detected with the TCR δ constant region probe pδ7, most likely represents a Vδ3-Jδ1 rearrangement [12, 13, 19, 20]. Southern analysis using various restriction enzymes and a Jδ1 probe were consistent with this interpretation (data not shown). Cytogenetic analysis (Table 1) revealed a three way translocation (1;10;12) (q25;p13;p13) and a deletion of (6q15;q25) in all cell cultures analyzed, consistent with the molecular genetic findings.

In conclusion, we have characterized a novel leukemia which gives rise to cells belonging to different cell lineages, notably T-lymphoid and myeloid. We concluded that the cell lines were presumably derived from a putative multipotent leukemic progenitor. However, the exact pathway of multilineage differentiation remains to be elucidated. One possibility is that a minor population of multipotent leukemic progenitor cells which was present in the peripheral blood blasts of the patient, but undetectable by fluorescence-activated cell sorter (FACS) analysis, differentiated in response to various growth factors into the respective lineages. Alternatively, minor populations of lineage committed cells might have acquired a growth advantage under the different culture conditions which resulted in outgrowth of the different cell cultures. A third hypothesis suggests that due to posttransformational alterations of the genetic program, the mature $CD3^{+}TCR\alpha/\beta^{+}$ T-lymphoblasts were capable of undergoing lineage switch.

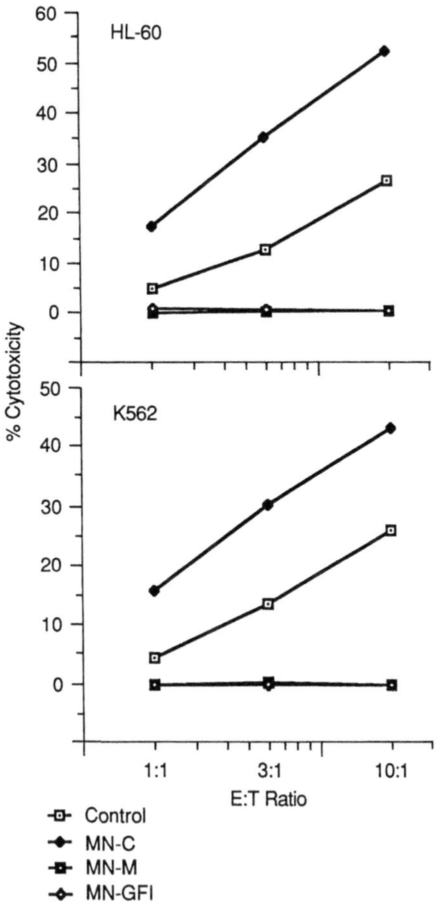

Fig. 3. Percent cytotoxicity of IL-2- (MN-C) and GM-CSF- (MN-M) dependent leukemic cell cultures, the growth factor independent leukemic cell cultures (MN-GFI) against the NK-sensitive target cell line K562 and the NK-resistant target cell line HL-60 in a chromium release assay. Control were peripheral blood mononuclear cells from a healthy donor which were cultured in the presence of IL-2 100 U/ml for 14 days. Mean values of triplicates are given, SEM was <5%. Representative data of one experiment (which were reproduced in two additional experiments) are shown

Such a mechanism has recently been described in B- and pre-B-cell lymphomas which were derived from Eμ-*myc* transgenic murine bone marrow cells. After coinfection with the viral oncogene v-*raf* these B-lineage tumors displayed "lin-

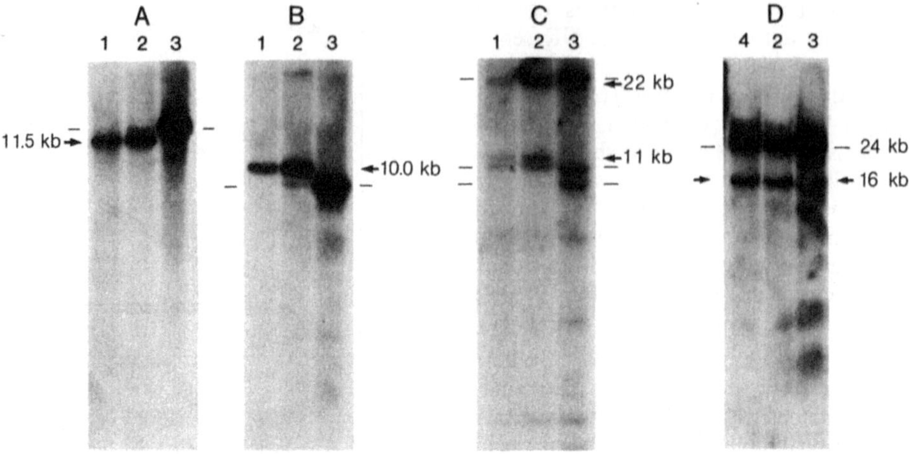

Fig. 4. High molecular weight DNA from IL-3-cultured (lane 1), IL-2-dependent (lane 4), uncultured leukemic cells (lane 2) and germline control (lane 3) were digested with Bam HI (Fig. 4 A, D) and Xba I (Fig. 4 B, C) and hybridized with the genomic TCR δ probe pδ7 (Fig. 4 A, B), the TCR γ cDNA probe pTγ1 (Fig. 4 C) and the TCR β cDNA probe Tβ. Germline bands are indicated with bars, rearranged bands with arrows. (Reproduced from the Journal of Experimental Medicine, 1989, vol. 169, pp. 1101–1122, by copyright permission of The Rockefeller University Press)

Table 1. Cytogenetic studies of cultured leukemic cells

Cultures	No. metaphases analyzed		G-banded karyotypes
	total	per clone	
Growth factor independent	10	7	47,XY,+19,del(6)(q15q25), t(1;10;12)(q25;p13;p13)
		2	47,XY,+19,del(6)(q15q25), dirdup(17)(q11.2→q23), t(1;10;12)(q25;p13;p13)
		1	48,XY,+19,del(6)(q15q25),+del(6)(q15q25), t(1;10;12)(q25;p13;p13)
GM-CSF	10	2	48,XY,+17,+del(6)(q15q25), t(1;10;12)(q25;p13;p13)
		8	49,XY,−7,+13,+19,+del(6)(q15q25), t(1;10;12)(q25;p13;p13),+der(7), t(7;17)(q36;21)
IL-3	10	5	47,XY,+19,del(6)(q15q25), t(1;10;12)(q25;p13;p13)
		2	48,XY,+19,del(6)(q15q25),+del(6)(q15q25), t(1;10;12)(q25;p13;p13)
		3	49,XY,+17,+19,+del(6)(q15q25), t(1;10;12)(q25;p13;p13)

(Reproduced from the Journal of Experimental Medicine, 1989, vol. 169, pp. 1101–1122, by copyright permission of The Rockefeller University Press).

eage instability" and underwent "lineage switch" to macrophages [22]. Whether similar mechanisms are operational in our novel leukemic cell line is currently under investigation. Further studies will specifically address possible mechanisms involved in the process of multilineage differentiation in this instructive leukemia. We hope that these studies will be useful for a better understanding of events of early lymphohematopoiesis, notably of the mechanisms that are involved in the irreversible commitment of multipotent progenitors to different cell lineages.

Acknowledgements. The authors thank Barbara Streifel for her expert technical assistance in performing the cytogenetic studies and Christopher W. Wilkowski for preparing pδ7. The authors are indebted to Celia Connoy and Lia Gore for editorial assistance.

References

1. Broxmeyer HE (1984) Colony assays of hematopoietic progenitor cells and correlations to clinical situations. CRC Crit Rev Oncol Hematol 1:227
2. Clark SC, Kamen R (1987) The human hematopoietic colony-stimulating factors. Science 236:1229
3. Metcalf D (1984) The hematopoietic colony stimulating factors. Elsevier, Amsterdam
4. Dexter TM, Allen TD, Lajtha LG (1977) Hematopoietic stem cells in vitro. J Cell Physiol 91:335
5. Whitlock CA, Witte ON (1982) Long-term culture of B lymphocytes and their precursors from murine bone marrow. Proc Natl Acad Sci USA 79:3608
6. Foon TA, Todd RF (1986) Review: Immunological classification of leukemia and lymphoma. Blood 68:1
7. Greaves MF, Chan LC, Furley AJW, Watt SM, Molgaard HV (1986) Lineage promiscuity in hematopoietic differentiation and leukemia. Blood 67:1
8. Minden MD, Mak TW (1986) The structure of the T cell antigen receptor. Blood 68:327
9. Toyonaga B, Mak TW (1987) Genes of the T-cell antigen receptor in normal and malignant T cells. Annu Rev Immunol 5:585
10. Hershfield MS, Kurtzberg J, Harden E, Moore JO, Whang-Pengh J, Haynes BF (1984) Conversion of a stem cell leukemia from a T-lymphoid to a myeloid phenotype induced by the adenosine deaminase inhibitor 2′-deoxycoformycin. Proc Natl Acad Sci USA 81:253
11. Kurtzberg J, Bigner SH, Hershfield MS (1985) Establishment of the DU.528 human lymphohematopoietic stem cell line. J Exp Med 162:1561
12. Griesinger F, Arthur DC, Brunning R, Parkin JL, Ochoa AC, Miller WJ, Wilkowski CW, Greenberg JM, Hurvitz C, Kersey JH (1989) Mature T-lineage leukemia with growth factor induced multilineage differentiation. J Exp Med 169:1101
13. Griesinger F, Greenberg JM, Kersey JH (1989) T cell receptor gamma and delta rearrangements in hematologic malignancies: Relationship to lymphoid differentiation. J Clin Invest 84:506
14. Dialynas DP, Murre C, Quertermous T, Boss JM, Leuden JM, Seidman JG, Strominger JL (1986) Cloning and sequence analysis of complementary DNA encoding and aberrantly rearranged human T-cell γ chain. Proc Natl Acad Sci USA 83:2619
15. Ochoa AC, Gromo G, Alter BJ, Sondel P, Bach FH (1987) Long term growth of lymphokine activated killer (LAK) cells: Role of anti CD3, β-IL-1, interferon-γ and -β. J Immunol 138:2728
16. Flynn PJ, Miller WJ, Weisdorf DJ, Arthur DC, Brunning R, Branda RF (1983) Retinoic acid treatment of acute promyelocytic leukemia: in vitro and in vivo observations. Blood 62:1211
17. Look AT, Peiper SC, Ashmun RA (1987) Binding of independently derived monoclonal antibodies to unique human myeloid differentiation antigens. In: McMichael AJ (ed) Leukocyte typing III. White cell differentiation antigens. Oxford University Press, Oxford, p 626
18. Lanier LL, Le AM, Civin CI, Loken MR, Phillips JH (1986) The relationship of CD16 (Leu 11) and Leu 19 (NKH-1) antigen expression on human peripheral blood NK cells and cytotoxic T-lymphocytes. J Immunol 136:4480
19. Lanier LL, Phillips JH (1986) Evidence for three types of human cytotoxic lymphocytes. Immunol Today 7:132
20. Hata S, Clabby M, Devlin P, Spits H, de Vries JE, Krangel MS (1989) Diversity and organization of human T cell receptor δ variable segments. J Exp Med 169:41
21. Griesinger F, Kersey JH (1988) Lineage restriction and diversity of T-cell receptor delta (TCR δ) rearrangements in hematopoietic malignancies. Blood 72 (Suppl 1): 196a
22. Klinken SP, Alexander WS, Adams JM (1988) Hematopoietic lineage switch: v-raf oncogene converts Eμ-myc transgenic B cells into macrophages. Cell 53:857

Establishment of a Hybrid Cell System Between Malignant Burkitt's Lymphoma Cells and Nonmalignant Lymphoblastoid Cells

J. Wolf[1], M. Pawlita[1], J. Bullerdiek[2], and H. zur Hausen[1]

A. Introduction

Burkitt's lymphoma (BL) is a high-grade malignant B-cell lymphoma found in a high-incidence endemic form in equatorial Africa and a rare sporadic form [1]. This tumor is strongly associated with a viral infection and specific chromosomal translocations. In the vast majority of endemic BL, Epstein-Barr virus (EBV) DNA has been demonstrated [2]. In all BLs one of three specific translocations is present, involving the cellular oncogene c-*myc* on chromosome 8 and loci for immunoglobulin genes on chromosomes 14, 22, and 2 [3]. The deregulation of c-*myc* caused by these translocations and the EBV infection are thought to be the critical steps in the development of EBV-positive BL, although the precise mechanisms are still unclear.

To investigate whether defects in cellular control genes may also play a causal role in BL pathogenesis, we established a fusion cell system between malignant BL cells and nonmalignant EBV-immortalized lymphocytes. The fusion of malignant and nonmalignant cells and the suppression of the tumorigenic phenotype in the resulting somatic cell hybrids, demonstrated for many tumors, but not yet for lymphoma-lymphocyte hybrids, has led to the concept of tumor suppressor genes [4, 5]. The activity of these genes is thought to prevent the malignant transformation of a normal cell and to suppress the malignant phenotype in somatic cell hybrids.

B. Results

The following cell lines (established and kindly provided by G. Lenoir, IARC, Lyon) were used: BL60, an EBV-positive BL cell line with a (8,22) translocation and highly tumorigenic in nude mice; and IARC 277, an EBV-immortalized nontumorigenic lymphoblastoid cell line originating from the same patient.

The neomycin resistance plasmid pSV2neo was introduced into BL60 cells by electroporation [6]. Several sublines resistant to 1200 μg/ml of the neomycin analogue G418 were obtained, which showed genomic integration of pSV2neo in Southern blot analysis. One of these sublines was cultured in medium containing 10^{-5} *M* 6-thioguanine (6TG) to select for spontaneous hypoxanthine-guanine-phosphoribosyltransferase (HGPRT)-negative mutants. A 6TG-resistant subline, named P7, that was stably HGPRT-negative (demonstrated by the consistent failure of growing in hypoxanthine, aminopterine, thymidine (HAT) medium) and in addition G418 resistant, was obtained. After ensuring that the high tumorigenicity of BL60 in nude mice was maintained in P7, this subline was used in fusion experiments with IARC 277.

The cells were attached to the bottom of plastic petri dishes by concanavalin A [7] and subsequently fused by PEG 1500. Thirteen hybrid clones growing in HAT medium containing G418 were obtained

[1] Institut für Virusforschung, Deutsches Krebsforschungszentrum, D-6900 Heidelberg, FRG
[2] Institut für Humangenetik, Universität Bremen, D-2800 Bremen 33, FRG

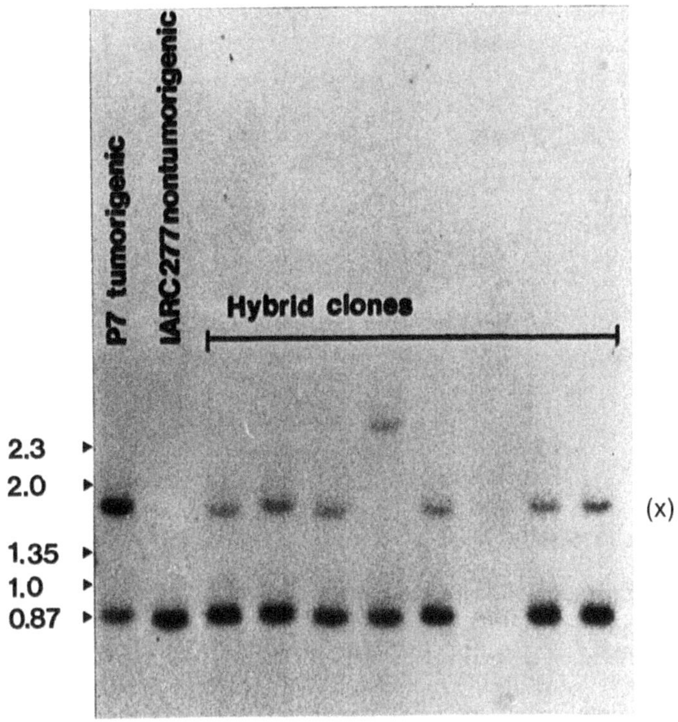

Fig. 1. Southern blot analysis for demonstration of the c-*myc* *Pvu*II-RFLP: presence of an additional DNA fragment (*X*) in the malignant BL60 subline P7 and in the hybrid clones in contrast to the nonmalignant cell line IARC 277, indicating the presence of the t(8;22) translocation. *Pvu*II-digested DNA was hybridized with c-*myc* exon 1 probe

in two independent experiments. The hybrid nature of these cells was verified by flow cytometry, restriction fragment length polymorphism (RFLP) analysis, and cytogenetics. Flow cytometric analysis of the DNA content revealed that the hybrid clones were near tetraploid by comparison with the diploid nontumorigenic and the near diploid tumorigenic parental line. For RFLP analysis cellular DNA was digested with the restriction enzyme *Pvu*II and probed in Southern blot analysis with radioactive-labeled c-*myc* exon 1. Because on chromosome $8q^+$, involved in the (8/22) translocation of BL60, one *Pvu*II site is abolished in c-*myc* exon 1, a new DNA fragment is created additional to the germline fragment [8]. This RFLP was used to demonstrate the presence of one chromosome $8q^+$ versus three normal chromosomes 8 in the hybrid cells (see Fig. 1).

These results were confirmed by cytogenetic analysis. Up to the present time seven hybrid clones have been analyzed. All show a near tetraploid karyotype with a modal range of 82–93 (peak 87) without any significant differences among each other (5 months after fusion). For all clones the presence of one copy of chromosome $8q^+$ could also be demonstrated in agreement with RFLP analysis (see Fig. 2).

We are currently investigating these hybrid clones with regard to the parameters of malignancy in vitro and in vivo. Preliminary results indicate a suppression of the BL phenotype in the hybrid cells demonstrating the same growth pattern as the nonmalignant IARC 277 cells, e.g., clumping in suspension culture, similar growth rate and maximal cell density. While the BL cells from large, progressively growing tumors without any sign of regression in nude mice, the grafts of the hybrid clones tested so far, as well as of the nonmalignant IARC 277 cells, stop growing after reaching a maximal size of 1 cm diameter, and then undergo complete regression.

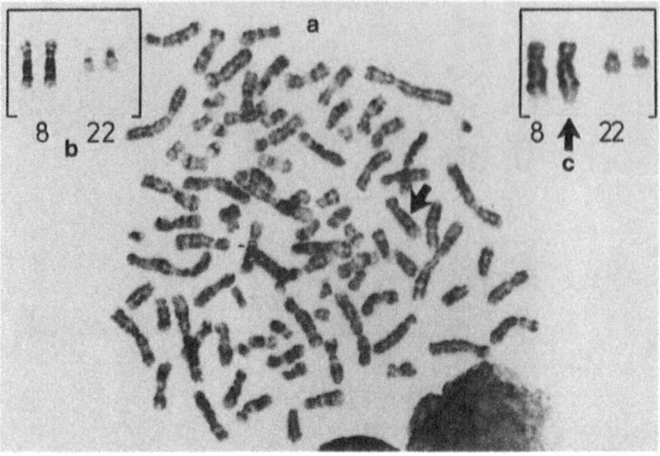

Fig. 2. a Cytogenetic demonstration of the t(8;22) translocation in the metaphase of one hybrid clone (*arrow*). **b** Normal chromosomes 8 and 22 of the nontumorigenic cell line IARC 277. **c** t(8;22) translocation of Burkitt's lymphoma subline P7 (*arrow*)

C. Conclusions

Malignant Burkitt's lymphoma cells and nonmalignant lymphoblastoid cells were fused. The resulting hybrid cells demonstrate a relatively complete chromosomal complement, including the presence of the (8,22) translocation, which is thought to play a causal role in the process of malignant transformation. First experiments indicate the suppression of the BL phenotype in the fusion cells. We hope that these hybrid cell system will be useful for further characterization of the mechanisms leading to the development of BL especially with regard to a possible causal role of defects in cellular suppressor genes.

References

1. De Thè G (1982) Epidemiology of Epstein-Barr virus and associated diseases in man. In: Roizman B (ed) The herpesviruses. Plenum, New York, pp 46–62
2. Zur Hausen H, Schulte-Holthausen H, Klein G, Henle W, Henle G, Clifford P, Santerson L (1970) EBV DNA in biopsies of Burkitt tumours and anaplastic carcinomas of the nasopharynx. Nature 228:1056–1058
3. Leder P, Battey J, Lenoir G, Moulding C, Murphy W, Potter H, Stewart T, Traub R (1983) Translocations among antibody genes in human cancer. Science 222:765–771
4. Klein G (1987) The approaching era of the tumor suppressor genes. Science 238:1539–1545
5. Harris H (1988) The analysis of malignancy by cell fusion: the position in 1988. Cancer Res 48:3302–3306
6. Potter H, Weir L, Leder P (1984) Enhancer-dependent expression of human k immunoglobulin genes introduced into mouse pre-B lymphocytes by electroporation. Proc Natl Acad Sci USA 81:7161–7165
7. Graessmann A, Wolf H, Bornkamm GW (1980) Expression of Epstein-Barr virus genes in different cell types after microinjection of viral DNA. Proc Natl Acad Sci USA 77:435–436
8. Szajnert MF, Saule S, Bornkamm GW, Wajcman H, Lenoir GM, Kaplan JC (1987) Clustered somatic mutations in and around first exon of non-rearranged c-*myc* in Burkitt lymphoma with t(8,22) translocation. Nucleic Acids Res 11:4553–4565

Effect of Natural and Synthetic Peptides on the Biological Function of Leukemic Cells *

B. Kazanowska [1], J. Bogusławska-Jaworska [1], and D. Konopińska [2]

A. Introduction

The goal of this work was to study the influence of some naturally occurring peptides and their analogs on the impaired biological properties of blood cells derived from leukemic patients. The effect of human tuftsin and of the insect neuropetide, proctolin, and their synthetic derivatives on the phagocytosis of granulocytes and the influence of proctolin and its analogs on the blastic transformation of lymphocytes were investigated.

B. Material and Methods

The phagocytic activity of granulocytes was determined in 30 patients with acute lymphoblastic leukemia (ALL). The lymphocyte transformation test was performed in 25 children with ALL and in a control group, consisting of 30 healthy children aged 2–12 years.

The phagocytic activity test was done according to the Steuden method, using ^{14}C-labeled *Staphylococcus aureus* 519 in the presence of tuftsin and its derivatives: Arg-tuftsin, Pro-Arg-tuftsin, Lys-Pro-Arg-tuftsin, and tuftsinyl-tuftsin [2–4]. The lymphocyte transformation test was performed using ^{14}C-thymidine in the presence of proctolin (Pr) and its derivatives: [Dopa^2]-Pr, (Homo-Arg)^1-Pr, [Phe(p-OMe)^2]-Pr, [Cha(-40Me)^2]-Pr, [Phe(p-NMe$_2$)^2]-Pr and [Phe(p-NO$_2$)^2]-Pr [1]. Both tests were performed in the initial phase of ALL and during complete remission on cytostatic therapy (Fig. 1).

A high stimulatory effect of tuftsin on the impaired PMN phagocytosis of patients in the initial phase of ALL was observed (67.4%). The influence of tuftsin on the phagocytic activity of granulocytes of ALL patients in remission was less striking. Tuftsin caused a decrease of phagocytosis of PMN in healthy children (Fig. 2).

The elongation of the tuftsin chain from the amino end resulted in a proportional decrease of the stimulatory activity of the subsequent peptides (Fig. 3).

It was found that proctolin restores the impaired phagocytosis of the ALL granulocytes to the normal level. A modification of the natural structure of proctolin by subsequent replacement of the -ON group in the aromatic ring of tyrosine by the -OMe, -NMe$_2$, and -ON$_2$ groups caused a lowering of this stimulatory effect. The proctolin analogs with 4-OMe-cyclohexyl and pNH$_2$-phenylalanyl did not influence the phagocytic activity of granulocytes (Fig. 4).

It was observed that the stimulatory effect of proctolin and its derivatives on the transformation rate of cultured lymphocytes was higher in those derived from patients who were in the initial stage of ALL than in those derived from patients in remission. Stimulation was higher in the presence of the analogs [Phe(p-NMe$_2$)^2]-Pr and (Homo-Arg)^1-Pr (63% and 59%). A less striking influence

[1] Department of Hematology, Medical Academy of Wrocław, Poland
[2] Institute of Chemistry, University of Wrocław, Poland
* Supported by grant CPBP 06.03 from the Institute of Pharmacology of the Polish Academy of Sciences

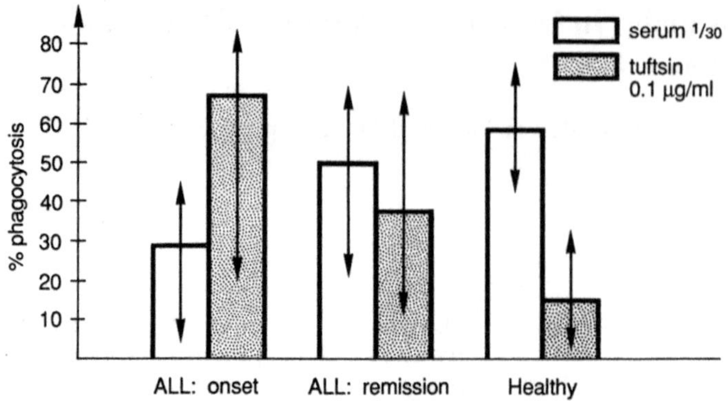

Fig. 1. Influence of tuftsin on the phagocytosis of ALL-affected and healthy PMN

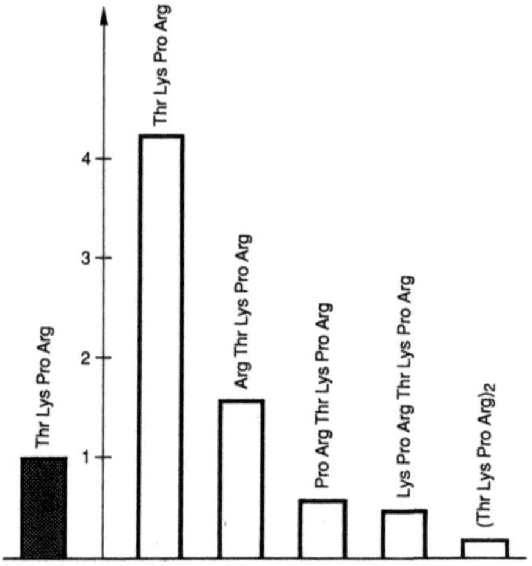

	ALL: onset	Remission	Healthy
Tuftsin	4.3	2.4	1.0
Arg-tuftsin	1.6	2.0	1.1
Pro-Arg-tuftsin	0.6	0.5	0.7
Lys-Pro-Arg-tuftsin	0.5	0.2	0.3
Tuftsinyl-tuftsin	0.2	0.1	0.0

Fig. 2. Stimulatory effects of tuftsin and its analogs on the phagocytosis of ALL-affected and healthy PMN

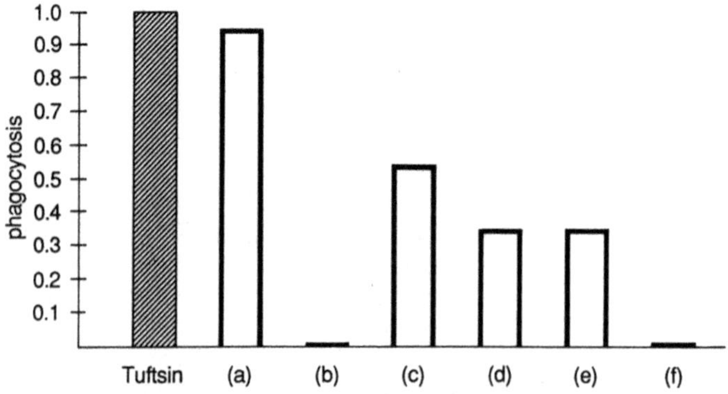

a – Proctolin (Pr)
b – [Phe – $(pNH_2)^2$] – Pr
c – [Phe – $pOMe^2$] – Pr
d – [Phe – $(pNMe_2)^2$] – Pr
e – [Phe – $(pNO_2)^2$] – Pr
f – [Cha – $(4\text{-}OMe)^2$] – Pr

Fig. 3. Influence of proctolin and its analogs, relative to that of tuftsin, on the phagocytic activity of ALL-affected PMN

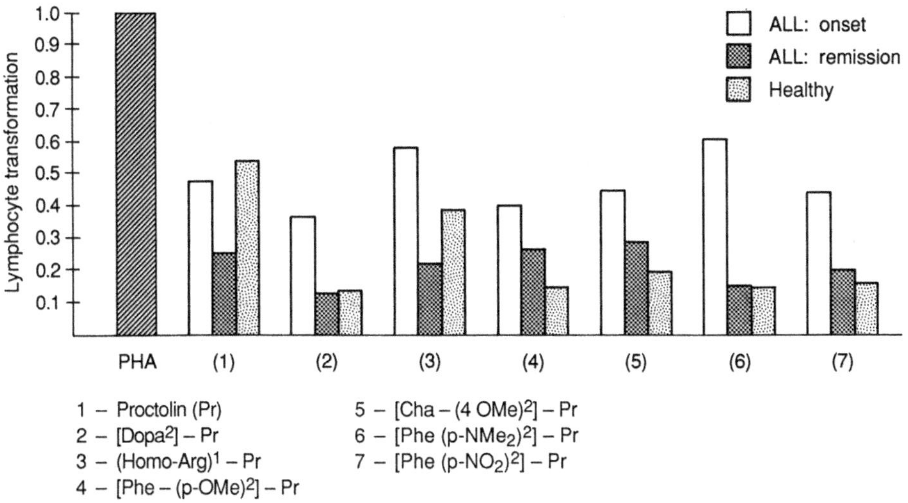

Fig. 4. Transformation of lymphocytes from ALL-affected and healthy children in the presence of proctolin and its analogs

on mitogenic activity was noticed when the other proctolin analogs were used (20%–30%).

C. Concluding Remarks

The high stimulatory effect of tuftsin and proctolin on the phagocytic activity of PMN derived from ALL patients was shown. A reduction in the phagocytosis of granulocytes of healthy subjects after preincubation with tuftsin, proctolin, and their analogs was observed, suggesting a regulatory effect of these peptides. The stimulating influence of tuftsin analogs on phagocytosis decreased with the elongation of the peptide chain at the N-terminal. A higher mitogenic effect of proctolin and its analogs was observed in cultured ALL lymphocytes compared with those of healthy children, which suggests a selective cytokinetic activity of the peptides studied.

References

1. Kazanowska B, Bogusławska-Jaworska J, Konopińska D, Rodziewicz-Magott B (1988) The influence of proctolin and its analogs on the blastic transformation of ALL lymphocytes. Arch Immunol Ther Exp (Warsz) (in press).
2. Kazanowska B, Steuden W, Bogusławska-Jaworska B, Konopińska D (1987) The influence of synthetic tuftsin and its analogs on the function of granulocytes of children with ALL. Arch Immunol Ther Exp (Warsz) 35:169
3. Konopińska D, Kazanowska B, Sobótka W, Bogusławska-Jaworska J (1986) An insect neuropeptide, proctolin. A new synthesis and its property of restoration of phagocytosis by defective human PMN leukocytes. Biull Pol Acad Sci 34:327
4. Konopińska D, Łuczak M, Wleklik M, Gumułka S, Kazanowska B (1983) Elongated tuftsin analogs – synthesis and biological investigation. Ann NY Acad Sci 419:35

Mechanisms of Glucocorticoid-Induced Growth Inhibition and Cell Lysis in Mouse Lymphoma Cells

K. Wielckens, S. Bittner and T. Delfs

A. Introduction

Glucocorticoid hormones are known to have a wide variety of molecular effects by induction or repression of proteins at the transcriptional level via a receptor system. Glucocorticoids, depending on the nature of the target tissue, can not only regulate carbonhydrate, protein and nucleic acid metabolism but can also accelerate or inhibit cellular growth or differentiation. Among the cell types that involute during prolonged exposure to steroids are certain lymphocytes and lymphoma cells. This effect is the basis of steroid therapy for malignant lymphomas. The steroid-induced death of lymphoma cells has morphological characteristics clearly distinct from necrosis, i.e., early alterations of the nuclear structure together with progressive reduction of the cellular volume but preservation of the integrity of cytosolic organelles [1] which are altered early during necrosis. This second type of cell death, called apoptosis, represents an active process involving an alteration of the pattern of active genes [2] in contrast to necrosis which is the consequence of environmental perturbations.

While the therapeutic potential of corticosteroids has been known for many years, the molecular basis of the steroid-triggered cytolysis of lymphoma cells has not been elucidated. In this contribution, an overview of the possible molecular events leading to the lysis of S49.1 murine lymphoma cells as well as the implications for the therapy of malignant lymphomas are presented.

Department of Clinical Chemistry, Medical Clinics, University of Hamburg, FRG

B. Two Phases of Glucocorticoid Action on Lymphoma Cells

S49.1 lymphoma cells exposed to the synthetic glucocorticoid dexamethasone respond to the steroid challenge by various metabolic alterations. Among the earliest effects is a reduction in the rate of glucose and amino acid transport [3], followed by a decrease in the rate of DNA, RNA and protein synthesis [4, 5]. These findings suggest that the first action of glucocorticoids is to switch cells from the proliferating to the nonproliferating state. This initial steroid effect is reflected by an accumulation in G_1 of the cell cycle [6]. This was confirmed by the experiment shown in Fig. 1. Glucocorticoid-induced cell death is initiated after a lag phase of about 24 h (Fig. 1A). During this time the steroid effect is fully reversible, i.e., when the glucocorticoid was removed and the cells cloned in soft agar no reduction in the colony forming ability was detectable before 24 h (Fig. 1B). In contrast, when lymphoma cells were treated with glucocorticoids, washed at various times to remove the steroid, and incubated further until day 3 after hormone addition, a pronounced decrease of the growth rate was detectable even when the cells were exposed to the steroid for only 6 h (Fig. 1C). Consequently, two phases of glucocorticoid action on lymphoma cells can be discriminated, a reversible cytostatic phase followed by an irreversible cytolytic phase. This finding sug-

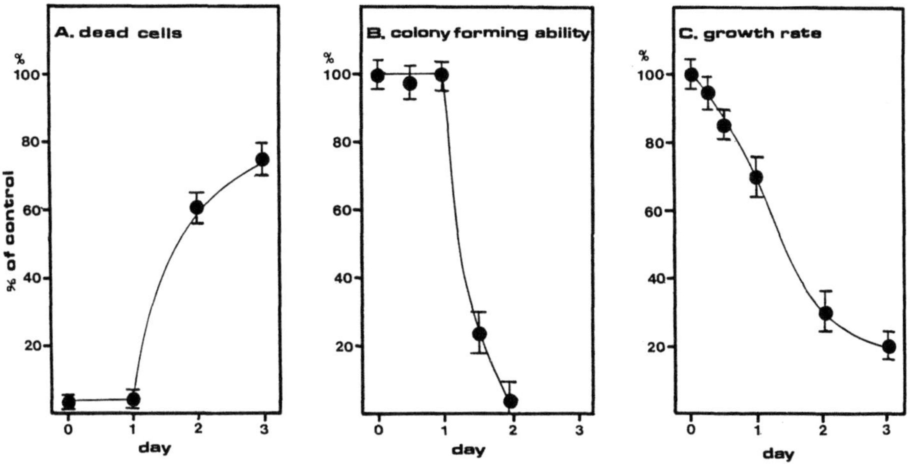

Fig. 1 A–C. Cell death (**A**), colony forming ability (**b**) and growth rate of dexamethasone-treated lymphoma cell cultures (**C**). S49.1 lymphoma cells were seeded and treated with 10^{-7} *M* dexamethasone. At the times indicated the cell viability (**A**) of dexamethasone-treated or control cultures was determined, or the cells were washed and reseeded at the same densities and the colony forming ability (**B**) was measured, or the number of living cells was monitored on the third day after steroid addition (**C**) as described in [7]

gests that the two processes are the result of distinct molecular mechanisms.

Since steroid "toxicity" depends on the growth rate, with a high number of lysed cells appearing under suboptimal growth conditions versus a lower number under optimal conditions [7], one can speculate that steroid-mediated growth inhibition may be brought about by interference with the growth-regulating system, either directly or indirectly by depression of the synthesis of autocrine growth factors.

C. Autocrine Growth factors of S49.1 Lymphoma Cells

Dilution of S49.1 cell cultures drastically reduces the growth rate, an effect which can be reversed by addition of conditioned medium and which points to the dependence of lymphoma cell proliferation on autocrine growth factors [7]. The nature of the factor(s), however, remains as yet unknown. Neither interleukin 2 nor other interleukins tested, including interleukins 1, 3 and 4, showed growth stimulating potential (Wielckens, unpublished observations).

Charcoal treatment of conditioned medium, however, revealed that one of the growth promoting factors is a small molecule with high affinity to charcoal. Since metabolites of arachidonic acid which bind strongly to activated charcoal have been shown to promote growth of cultured cells [8, 9] we first tested whether at least one of the autocrine growth factors is derived from arachidonic acid. Since the nonspecific lipoxygenase inhibitor nordihydroguaiaretic acid completely blocked S49.1 cell proliferation, in contrast to the cycloxygenase inhibitor indomethacin, the growth factor appeared to be a product of the lipoxygenase pathway. Of the metabolites tested (5-hydroxyeicosatetraenoic acid = 5-HETE, 8-HETE, 9-HETE, 12-HETE, 15-HETE, lipoxins A and B, leukotrienes B_4, C_4, D_4 and E_4), only leukotriene B_4 could restore the reduced growth rate in lymphoma cell cultures with low cell numbers, suggesting a critical role of this factor under the test conditions [7]. Even the precursor arachidonic acid did not significantly affect lymphoma cell proliferation. Furthermore, it was shown that S49.1 cells synthesize nanogram amounts of leukotriene B_4, arguing for it having a role as an autocrine growth factor [7].

As suggested by the lack of a pronounced effect of leukotriene B_4 on colony-forming ability in soft agar, however, S49.1 lymphoma cell growth appears not to depend only on this arachidonic acid metabolite. When cells are seeded at a still higher dilution than in the low density cell multiplication assays, the concentration of other essential autocrine growth factors may fall under a critical level. Therefore, it can be concluded that the diluted cell cultures still contain enough of other growth factors necessary to sustain lymphoma cell proliferation. Preliminary experiments suggest that at least two additional growth promoting factors are present in conditioned medium, one probably a protein, the other possibly a small molecule derived from arachidonic acid.

The antiinflammatory effect of corticosteroids has been attributed to inhibition of arachidonic acid release by induction of a phospholipase A_2 inhibitor, thereby blocking the formation of prostaglandins, leukotrienes and other hydroxyeicosatetraenoic acids [10]. The critical role of leukotriene B_4 suggests that the same mechanism is responsible or at least partially responsible for the antiproliferative effect on lymphoma cells. This notion is supported by the finding that dexamethasone abolished leukotriene B_4 synthesis almost completely [7]. In contrast to control cultures, where leukotriene B_4 was detectable at a constant level within 6 h, no leukotriene B_4 was detectable in dexamethasone-treated cultures. Since addition of leukotriene B_4 to dexamethasone-treated lymphoma cultures failed to reverse the growth inhibition, glucocorticoids appear to induce growth inhibition by more than one mechanism. It must be postulated that glucocorticoids also block the synthesis of other autocrine growth factors and/or the synthesis of growth factor receptors.

D. Glucocorticoid-Induced Lymphoma Cells Lysis

After accumulation of lymphoma cells in G_1 of the cell cycle by interference with the growth regulating system, glucocorticoids initiate cell lysis. The mechanism of corticosteroid-induced cell death has been subject to great controversy. It was proposed that it is the final response to an accumulation of various alterations of cellular metabolism [2]. This hypothesis, however, did not explain the early alterations of chromatin structure after corticosteroid challenge which are characteristic for apoptotic processes and which point to the nucleus having a particular role in the lytic event.

More recently it was postulated by Wyllie [11] that glucocorticoids activate an endonuclease in lymphatic cells, leading to the digestion of internucleosomal DNA and resulting in the discrete distribution of DNA fragments observed following extraction and separation on agarose gel [11]. The characteristic fragment pattern, corresponding to the DNA of oligo- and monosomes, has also been found in S49.1 lymphoma cells [7]. It cannot be ruled out, however, that DNA fragmentation occurs after cell death and merely represents a post-mortem phenomenon. This notion is supported by the finding that a comparable DNA fragment pattern is also detectable in cells treated by other toxic agents such as fluoride, azide, or cycloheximide ([12] and Wielckens, unpublished observations) and is therefore not related to death by necrosis or apoptosis. That a DNA fragmentation process could indeed be crucial for glucocorticoid-induced lymphoma cell lysis can be demonstrated by an indirect approach supporting Wyllie's postulate; however, the mechanism appears to be much more complicated than originally assumed.

The poly(ADP-ribosyl)ation reaction is a nuclear protein modification reaction utilizing NAD and catalyzed by a chromatin-bound, DNA-dependent enzyme, the poly(ADP-ribose)synthetase [13]. The enzyme polymerizes and transfers ADP-ribose moieties from NAD to chromatin proteins such as histone H2B or topoisomerase I [14, 15]. Moreover, the poly(ADP-ribose)synthetase is extensively automodified [16]. In intact DNA poly(ADP-ribose)synthetase is almost inactive but the enzyme is strongly activated by the introduction of strand breaks [17]. Therefore, it was proposed that the poly(ADP-ribosyl)ation reaction is involved in the DNA repair process [18]. The hypothesis of its involvement in DNA repair is now generally accepted, although the exact role of poly(ADP)-ribosyl)ation during DNA repair has yet to be elucidated.

Since DNA fragmentation either by alkylating agents or irradiation activates the poly(ADP-ribosyl)ation reaction [19, 20], an increase in poly(ADP-ribose)synthetase activity, a rise in the nuclear poly(ADP-ribose) content, and a decrease in the NAD level could reflect DNA fragmentation as well as activation of DNA repair. When S49.1 lymphoma cells were treated with dexamethasone and the poly(ADP-ribose)synthetase activity in permeabilized cells determined at various times after steroid addition, it became obvious that the poly(ADP-ribosyl)ation system was indeed activated during corticosteroid treatment. The amount of active enzyme increased up to fivefold in dexamethasone-treated cells (Fig. 2). By contrast, the amount of active enzyme in control cells was almost stable. The finding of activation of the poly(ADP-ribose)synthetase in dexamethasone-treated lymphoma cells was further supported by analysis of protein-bound poly(ADP-ribose) which revealed a significant increase in the amount of polymeric ADP-ribose residues [21] during the steroid challenge. Moreover, a decline in the NAD level, the substrate of the poly(ADP-ribosyl)ation reaction, was detectable many hours before cell lysis [22] and could be suppressed by addition of an inhibitor of poly(ADP-ribose) formation. The antagonistic effect of poly(ADP-ribosyl)ation in glucocorticoid cell death was also reflected by a strong potentiation of glucocorticoid toxicity by inhibitors of poly(ADP-ribosyl)ation such as benzamide [21, 22].

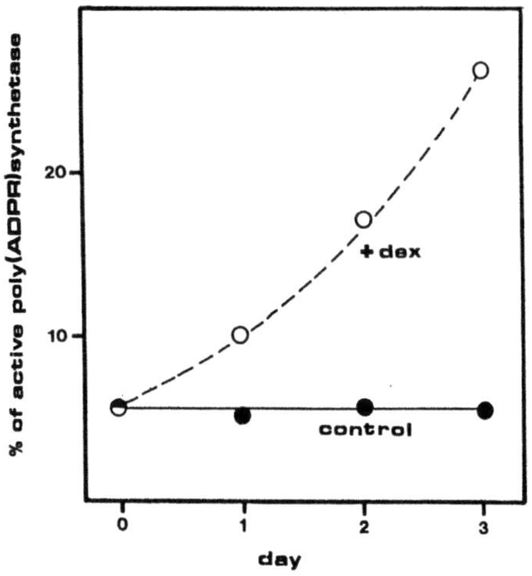

Fig. 2. Activation of poly(ADP-ribose)synthetase in dexamethasone-treated S49.1 lymphoma cells. Cells were incubated in the absence or presence of 10^{-7} *M* dexamethasone (+*dex*) for the times indicated, and the percentage of active poly(ADP-ribose)synthetase was determined as described in [21]

From these data it can be concluded that glucocorticoids really activate an endonuclease, which leads to DNA fragmentation. The activation of DNA repair mechanisms, however, antagonizes the DNA damage as long as sufficient NAD is present. Ultimately, the consumption of NAD limits the ability to repair DNA, and the consequent alterations of DNA repair capacity and breakdown of the energy metabolism together bring about cell lysis (Fig. 3).

Interestingly, benzamide not only enhanced the steroid toxicity but also shifted the initiation of cell death to earlier times. In S49.1 cell cultures treated only with glucocorticoids, cytolysis did not occur until 20–24 h following steroid addition [21]. When dexamethasone and

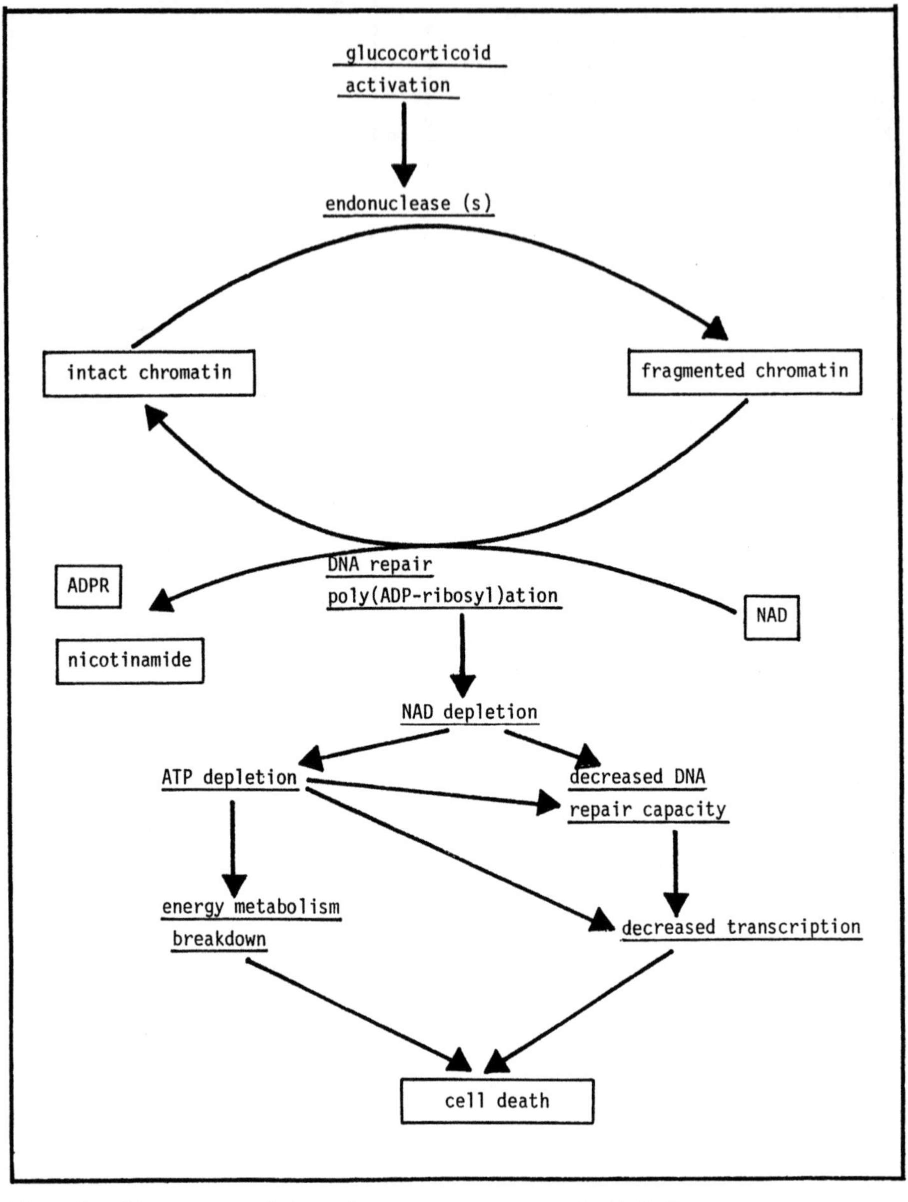

Fig. 3. Possible sequence of phases in response to glucocorticoid challenge

benzamide were combined, however, lysed cells appeared about 10 h earlier [22], a phenomenon easily explained by the idea that DNA repair, including the poly(ADP-ribosyl)ation reaction, antagonizes the glucocorticoid-induced DNA fragmentation. Taken together, the data suggest that three phases in the effect of glucocorticoids on lymphoma cells should be discriminated: a reversible cytostatic phase, a reversible precytolytic phase where endonuclease-mediated

1.	Normal tissue turnover
2.	Embryogenesis
3.	Atrophy of endocrine dependent tissue (e.g., prostate, adrenal cortex)
4.	Treatment with glucocorticoids (lymphocytes, lymphoma cells)
5.	Tumor regression
6.	X-irradiation or radiomimetic cytotoxic agents
7.	T-cell-mediated cell killing
8.	Cold shock
9.	Treatment with tumor necrosis factor
10.	Treatment with lymphotoxin

Table 1. Occurrence of apoptosis

DNA breakage is antagonized by DNA repair, and an irreversible cytolytic phase when the NAD pool is exhausted.

Cellular suicide processes comparable to those during glucocorticoid-induced lymphoma cell death have been demonstrated under a wide variety of conditions (Table 1). Therefore, a cellular suicide system could be essential for the development and homeostasis of a multicellular organism and it is possible that a system similar to that observed in lymphoma cells but triggered by other stimuli is present in nearly every eukaryotic cell. Moreover, the suicide system could be also involved in prevention of malignant transformation.

E. Conclusions

From our data it can be concluded that the therapeutic potential of glucocorticoids in the treatment of malignant lymphomas is not yet fully realized. The effectiveness of steroid therapy could be dramatically increased by combination with agents interfering with the cytostatic or cytolytic potential of the hormone, such as compounds which block the synthesis or binding of growth factors (e.g., inhibitors of the lipoxygenase pathway [7] or suramin [23]) or inhibitors of the poly(ADP-ribosyl)ation reaction, and this may cause no increase in undesirable side effects. Consequently, it is worth while investigating which growth factors are necessary for the proliferation of an individual type of lymphoma or leukemia and developing specific antigrowth factors such as protein growth factor analogs which have no intrinsic activity made by site-directed mutagenesis, antibodies against growth factor receptors, suramin analogs with decreased systemic toxicity, or inhibitors of certain lipoxygenases. The combination of corticosteroids with inhibitors of poly(ADP-ribosyl)ation could also provide a novel approach to the treatment of malignant lymphomas. To date, however, no clinically suitable inhibitor of poly(ADP-ribose)synthetase is available.

References

1. Wyllie AH (1980) Cell death: a new classification separating apoptosis from necrosis. In: Bowen ID, Lockshin RA (eds) Cell death in biology and pathology. Chapman and Hall, London, pp 9–34
2. Munck A, Crabtree GR (1980) Glucocorticoid-induced cell death. In: Bowen ID, Lockshin RA (eds) Cell death in biology and pathology. Chapman and Hall, London, pp 329–359
3. Rosen JM, Fina J, Milholland RJ, Rosen F (1972) Inhibitory effect of cortisol in vitro on 2-deoxyglucose uptake and RNA and protein metabolism in lymphosarcoma P1798. Cancer Res 32:350–355
4. Makman MH, Dvorkin B, White A (1966) Alterations in protein and nucleic acid metabolism of thymocytes produced by adrenal steroids. J Biol Chem 241: 1646–1648
5. Makman MH, Dvorkin B, White A (1968) Influence of cortisol on the utilization of precursors of nucleic acids and

protein by lymphoid cells in vitro. J Biol Chem 243:1485–1497
6. Harmon JM, Norman MR, Fowles BJ, Thompson EB (1979) Dexamethasone induces irreversible arrest and death of a human lymphoid cell line. J Cell Physiol 98:267–278
7. Bittner S, Wielckens K (1988) Glucocorticoid-induced lymphoma cell growth inhibition: the role of leukotriene B_4. Endocrinology 123:991–1000
8. Atluru D, Goodwin JS (1986) Leukotriene B_4 causes proliferation of interleukin 2-dependent T cells in the presence of suboptimal levels of interleukin 2. Cell Immunol 99:444–452
9. Palmberg L, Claesson HE, Thyberg J (1987) Leukotrienes stimulate initiation of DNA synthesis in cultured arterial smooth muscle cells. J Cell Science 88:151–159
10. Davies P, Bailey PJ, Godenberg MM, Ford-Hutchinson AW (1984) The role of arachidonic acid oxygenation products in pain and inflammation. Annu Rev Immunol 2:335–357
11. Wyllie AH (1980) Glucocorticoid-induced thymocyte apoptosis is associated with endogenous endonuclease activation. Nature 284:555–556
12. Vedeckis WV, Bradshaw HD (1983) DNA fragmentation in S49 lymphoma cells killed with glucocorticoids and other agents. Molec Cell Endocrinol 30:215–227
13. Ueda K, Hayaishi D (1985) ADP-ribosylation. Annu Rev Biochem 54:79–100
14. Adamietz P, Rudolph A (1984) ADP-ribosylation of nuclear proteins in vivo. J Biol Chem 259:6841–6846
15. Ferro A, Higgins P, Olivera B (1983) Poly(ADP-ribosyl)ation of a topoisomerase. J Biol Chem 258:6000–6003
16. Ogata N, Ueda K, Kawaichi M, Hayaishi O (1981) Poly-(ADP-ribose)synthetase, a main acceptor of poly(ADP-ribose) in isolated nuclei. J Biol Chem 256:4135–4137
17. Miller EG (1975) Stimulation of nuclear (polyadenosine diphosphate ribose) polymerase activity from Hela cells by endonucleases. Biochim Biophys Acta 395:191–200
18. Durkacs BW, Omidiji O, Gray DA, Shall S (1980) (ADP-ribose) participates in DNA excision repair. Nature 283:593–596
19. Wielckens K, Schmidt A, George E, Bredehorst R, Hilz H (1982) DNA fragmentation and NAD depletion. J Biol Chem 257:12872–12877
20. Jacobson EL, Antol KM, Juarez-Salinas H, Jacobson MK (1983) Poly(ADP-ribose) metabolism in ultraviolet irradiated human fibroblasts. J Biol Chem 258:103–107
21. Wielckens K, Delfs T (19869 Glucocorticoid-induced cell death and poly(adenosine diphosphate ribosyl)ation: increased toxicity of dexamethasone on mouse S49.1 lymphoma cells with the poly(ADP-ribosyl)ation inhibitor benzamide. Endocrinology 119:2382–2392
22. Delfs T, Wielckens K (submitted) Glucocorticoid-induced lymphoma cell death and poly(ADP-ribosyl)ation: further investigations on the mechanism of benzamide mediated dexamethasone toxicity increase.
23. Bittner S, Etavard N, Krenz G, Kraus U, Wielckens K (submitted) Suramin potentiation of glucocorticoid toxicity in murine S49.1 lymphoma cells.

Characteristics of 27 Human T-Cell Leukemia Cell Lines With/Without T-Cell Receptors of T3-Ti$\alpha\beta$ or T3-Ti$\gamma\delta$ Complex *

J. Minowada [1,2], K. Kohno [1], Y. Matsuo [1], H.G. Drexler [3], T. Ohnuma [4], W.J.M. Tax [5], and M.B. Brenner [6]

A. Introduction

The advances in the study of human leukocyte differentiation and its immunobiological function have been greatly facilitated by developments in four areas of methodology: first by the establishment of stable permanent leukemia cell lines of various origins, secondly by the development of numerous specific heterologous antibodies to various leukocyte differentiation antigens, thirdly by the introduction of many functional assays of both hematopoietic progenitor cells and mature leukocyte subsets, and fourthly by the development of recombinant DNA technology.

We have been interested in characterizing both permanent leukemia-lymphoma cell lines and fresh uncultured leukemia-lymphoma cells by means of multiple marker analysis which includes morphologic, immunobiologic, cytogenetic, enzymatic, virologic, functional, and molecular parameters [8, 9]. At present there are a total of 111 proven human leukemia-lymphoma cell lines which are being maintained and characterized in the laboratory. Of these cell lines, 35 were identified as those T-cell lines representing each of the 5 stages in the T-cell differentiation and maturation previously described, namely T-blast-I, -II, -III, -IV, and -V, respectively, in the order of maturation [9]. The advantages of utilizing leukemia-lymphoma cell lines are threefold: individual leukemia-lymphoma cell line presents an expanded monoclonal population, the marker profile reflects an arrested stage of various discrete points of hematopoietic cell differentiation, and stability and availability are high and unlimited. Furthermore, all characteristics except cytogenetic findings found in the leukemia-lymphoma cell lines are not tumor specific, but these characteristics appear to be the normal gene products often of vital significance in immunobiology [4, 8, 9].

The present report is a brief account in the expression of T-cell antigen receptor complex among 27 T-cell leukemia-lymphoma cell lines.

B. Materials, Methods, Results, and Discussion

As previously reported [8, 9, 12], the established leukemia cell lines and those cell lines transformed in vitro by HTLV-I infection were maintained in RPMI 1640 medium supplemented with heat-inactivated fetal calf serum in the standard procedure. Care was taken to maintain the cell cultures in an exponential growth phase in order to optimize the experiments. A standardized membrane immunofluorescence test using appropriate

---

[1] Fujisaki Cell Center, Hayashibara Biochemical Laboratories, Inc., Okayama, Japan
[2] Department of Microbiology, University of Illinois College of Medicine, Chicago, USA
[3] Department of Pediatrics, University of Ulm School of Medicine, Ulm, FRG
[4] Department of Neoplastic Diseases, Mount Sinai School of Medicine, New York, USA
[5] Department of Medicine, St. Radboudziekenhus, Nijmegen, The Netherlands
[6] Division of Tumor Virology, Dana-Farber Cancer Institute, Harvard Medical School, Boston, USA
* Supported in part by a grant from the Agency of Science and Technology, Japan

Table 1. T-cell leukemia-lymphoma cell lines with/without Ti-$\alpha\beta$ or Ti-$\delta\gamma$ (%)

Cell line	Cell stage	TdT	CD3 Leu-4	CD4 Leu-3a	CD8 Leu-2a	TcR WT-31	TcR Delta-1	CD10 BA3	CD10 NU-N1	CD34 MY10	CD5 Leu-1	CD7 3A1	CD25 TAC	CD28 Kolt-2	HLA-DR HLA-Dr	HLA-DP B7/21	HLA-DQ Leu-10
CD3$^+$ Ti-$\alpha\beta^+$ Ti-$\delta\gamma^-$																	
CCRF-CEM	T-II	100	100	100	0	100	0	100	100	90	100	100	0	100	0	0	0
HPB-ALL	T-II	100	100	100	100	100	0	100	100	0	100	100	0	90	0	0	0
HPB-MLT	T-II	100	100	100	100	100	0	100	100	0	100	100	0	100	0	0	0
HD-Mar-2	T-II	100	100	100	90	100	0	100	100	0	100	100	0	100	0	0	
TALL-1	T-III	100	100	100	100	50	0	0	0	0	100	100	0	90	0	0	
MOLT-16	T-III	100	100	10	0	100	0	0	0	0	100	100	0	90	0	0	0
JURKAT	T-III	60	100	50	0	100	0	0	0	0	90	90	0	50	0	0	
MAT	T-IV		70	0	0	60	0	0	0	0	90	100	0	100	0	0	0
H9	T-V	0	95	80	0	100	0	0	0	0	100	20	0	0	100	100	
ED-S	T-V	0	100	100	0	100	0		0		100	0	100	0	90	100	
ATL-35T°	T-V	0	100	100	0	50	0	0	0		100	100	100	0	100	100	100
CD3$^+$ Ti-$\alpha\beta^-$ Ti-$\delta\gamma^+$																	
DND-41	T-II	100	100	100	0	0	80	100	100	0	80	100	0	80	0	0	0
MOLT-13	T-III	100	90	0	0	0	100	0	0	100	90	90	20	100	0	0	
MOLT-14	T-III	100	90	0	0	0	100	0	0	100	100	90	30	100	0	0	
PEER	T-IV	0	90	90	0	0	100	0	20	0	100	100	0	100	0	0	
CD3$^+$ Ti-$\alpha\beta^-$ Ti-$\delta\gamma^-$																	
MKB-1	T-III	50	40	100	0	0	0	70	90	0		100	0	0	0	0	0
HUT-78	T-V	0	50	10	0	0	0	0	0	0	50	10	0	0	100	100	100
CD3$^-$ Ti-$\alpha\beta^-$ Ti-$\delta\gamma^-$																	
MOLT-3	T-III	100	0	100	75	0	0	0	0	70	100	100	0	100	0	0	0
MOLT-4	T-III	100	0	100	75	0	0	0	0	50	100	100	0	90	0	0	0
P12/ICHIKAWA	T-III	100	0	100	60	0	0	0	0	20	100	100	0	10	0	0	
SKW-3	T-IV	0	0	100	90	0	0	0	0	0	100	100	0	100	0	0	0
MOLT-15	T-IV	0	0	0	0	0	0	0	0	20	0	95	0	10		0	
ALL-SiL	T-IV	0	0	90	100	0	0	0	0	0	90	100	0	0		0	0
MT-1	T-V	0	0	0	0	0	0	0	0	0	100	0	100	100	100	100	100
HUT-102	T-V	0	0	100	0	0	0	0	0	0	100	0	100	0		100	100
C5/MJ	T-V	0	0	90	0	0	0	0	0	0	100	80	100	0	100	100	100
ATL-16T°	T-V	0	0	100	0	0	0	0	0		100	0	100	0	100	100	100

antibody reagents, morphologic test, and a few functional tests including IL-1, IL-2, and growth kinetic determinations were carried out. Numerous numbers of murine monoclonal antibody reagents for which the majority had been classified by the International Workshops [3, 7, 11] into the CD categories were used in the study.

Table 1 summarizes the percentage immunofluorescence test results with selected reagents relevant for this report. To determine expression of T-cell receptor (Ti$\alpha\beta$) and (Ti$\gamma\delta$), monoclonal antibodies, WT-31 [13] and TcR Delta-1 [2] were used respectively. In respect to the expression of CD3, four groups ($CD3^{+}$.Ti$\alpha\beta^{+}$; $CD3^{+}$.Ti$\gamma\delta^{+}$; $CD3^{+}$.TcR^{-}; $CD3^{-}$.TcR^{-}) were identified. Considering the stages of T-cell differentiation [8], T-cell receptor expression and CD3 expression occur in a relatively early stage. It was confirmed that Ti$\alpha\beta$ and Ti$\gamma\delta$ expressions are mutually exclusive and that CD3 expression is obligatory with T-cell receptor expression [1, 4, 14, 15]. Two cell lines with $CD3^{+}$ lack detectable expression of either forms of TcR. As expected all cell lines with $CD3^{-}$ were found to be negative for the TcR. Unlike reported "double-negative" T cells in respect to CD4 and CD8 [6, 10, 14], two T-cell lines with Ti$\gamma\delta$ (DND-41 and PEER) were positive for CD4. In view of the reported leukemia T-cell lines with IL-2 production [5], some T-cell lines with Ti$\alpha\beta$ (MOLT-16) or Ti$\gamma\delta$ (MOLT-14) were found to be capable of producing, specifically and nonspecifically, IL-2 with appropriate stimuli (data not shown).

The present study has therefore demonstrated again that these cell lines provide significant materials and possible models for basic research in human immunobiology.

References

1. Acuto O, Hussey RE, Fitsgerald KA, Protentis JP, Meuer SC, Schlossmann SF, Reinherz EL (1983) The human T cell receptor: appearance in ontogeny and biochemical relationship of α and β subunits on IL-2 dependent clones and T cell tumors. Cell 34:717–726
2. Band H, Hochstenbach F, McLean J, Hata S, Krangel MS, Brenner MB (1987) Immunochemical proof that a novel rearranging gene encodes the T cell receptor δ subunit. Science 238:682–684
3. Bernard A, Boumsell L, Dausset J, Milstein C, Schlossman SF (eds) (1984) Leucocyte typing. Springer, Berlin Heidelberg New York
4. Furley AJ, Mizutani S, Weilbaecher K, Dhaliwal HS, Ford AM, Chan LC, Molgaard HV, Toyonaga B, Mak T, van den Elsen P, Gold D, Terhorst C, Greaves MF (1986) Developmentally regulated rearrangement and expression of genes encoding the T cell receptor-T3 complex. Cell 46:75–87
5. Gillis S, Watson J (1980) Biochemical and biological characterization of lymphocyte regulatory molecules. V. Identification of an interleukin 2-producing human leukemia T cell line. J Exp Med 152:1709–1719
6. Lanier LL, Federspiel NA, Ruitenberg JJ, Phillips JH, Allison JP, Littman D, Weiss A (1987) The T cell antigen receptor complex expressed on normal peripheral blood $CD4^{-}$, $CD8^{-}$ T-lymphocytes. J Exp Med 165:1076–1094
7. McMichael A, Beverley P, Crimpton M, Milstein C (eds) (1987) Leukocyte typing III. Oxford University Press, New York
8. Minowada J (1978) Markers of human leukaemia-lymphoma cell lines reflect haematopoietic cell differentiation. In: Serrou B, Rosenfeld C (eds) Human lymphocyte differentiation: its application to cancer. INSERM symposium no 8. Elsevier/North Holland, Amsterdam
9. Minowada J (1985) Maker utility in the diagnosis and management of leukemias. Lab Med 16:305–309
10. Moingeon P, Jitsukawa S, Faure F, Troalen F, Triebel F, Graziani M, Forestier F, Bellet D, Bohuon C, Hercend T (1987) A γ-chain complex forms a functional receptor on cloned human lymphocytes with natural killer-like activity. Nature 325:723–726
11. Reinherz EL, Haynes BF, Nadler LM, Bernstein ID (eds) (1985) Leukocyte typing II, vols 1, 2, 3. Springer, Berlin Heidelberg New York

12. Sangster RN, Minowada J, Suciu-Foca N, Minden M, Mak TW (1986) Rearrangement and expression of the α, β, and γ chain T cell receptor genes in human thymic leukemia cells and functional T cells. J Exp Med 163:1491–1508
13. Spits H, Borst J, Tax W, Capel PJA, Terhorst C, deVries JE (1985) Characteristics of a monoclonal antibody (WT31) that recognizes a common epitope on the human T cell receptor for antigen. J Immunol 135:1922–1928
14. Van de Griend R, Borst J, Tax WJM, Bolhuis RLH (1988) Functional reactivity of WT31 monoclonal antibody with T cell receptor-γ expressing $CD3^+4^-8^-$ T cells. J Immunol 140:1107–1110
15. Weiss A, Newton M, Crommie D (1986) Expression of T3 in association with a molecule distinct from the T-cell antigen receptor heterodimer. Proc Natl Acad Sci USA 83:6998–7002

Intervention in Potential Leukemic Cell Migration Pathway Affects Leukemogenesis *

A. Peled and N. Haran-Ghera

A. Introduction

Several factors are involved in the high frequency of T-cell lymphomas of AKR mice, which appear mainly in the thymus at the age of 6–12 months [1]. The thymus is considered to play a major role in the disease since its removal prevents the development of T-cell lymphoma [2], while retransplantation of thymic epithelium to thymectomized AKR reconstitutes the high frequency of lymphoma [3]. Although the AKR/J strain has the predisposition to develop the disease since birth [4], the mean latent period is delayed until the age of 8 months. The long latent period has been attributed to the delayed formation of the leukemogenic dual tropic virus (DTV) with the MCF characteristic [5, 6], formed in the thymus as a consequence of recombination within the envelope gene of ecotropic and xenotropic murine leukemia virus (MuLV). DTVs are detected only in preleukemic thymus and leukemic tissues of strains of mice prone to develop high incidence of leukemia [7]. DTVs enhance leukemia development whereas endogenous ecotropic or xenotropic viruses are usually nontumorigenic. Exceptional is the ecotropic virus isolate SL_3 with the enhancing activity on T-cell lymphomagenesis [8]. These observations support the assumption that DTVs are proximal transforming agents of thymocytes and thereby responsible for high incidence of T lymphoma in AKR mice. Cloyd [9] proposed specific cellular tropism of two subclasses of MCF virus, and claimed that oncogenicity is closely linked to cellular differentiation. MCF isolated from lymphomatous thymus was replicating in the thymus and T peripheral cells, while nonlymphomagenic MCF isolated from leukemic spleen of NFS mice did not replicate in the thymus but rather in bone marrow cells, spleen, and lymph node B-lymphocytes.

Our previous studies showed that potential leukemic cells (PLCs) are initially detected among bone marrow cells rather than in the thymus of young AKR mice [10]. Infection of 14-day-old AKR mice with DTV did not change the spontaneous PLC distribution pattern in the host organs; however, it enhanced PLC transition to autonomous leukemic cells. A preferential cell tropism of DTV to cells among bone marrow and spleen cells rather than from thymocytes was also demonstrated [11]. We therefore considered DTV as a promoter of PLCs, triggering the natural progression and transition of PLCs into frank autonomous lymphoma. Very recently Buckheit et al. [12] proposed that a certain fraction of bone marrow cells in the AKR mouse enriched in prothymocytes is also high in ecotropic virus-producing population, seeding the spleen and thymus with infectious ecotropic virus. These cells may represent the PLCs in the bone marrow of AKR mice demonstrated by us [10, 11].

Removal of the thymus that prevents the emergence of T-cell lymphoma did

---

Department of Chemical Immunology, The Weizmann Institute of Science, Rehovot 76100, Israel
* This research was supported by grant from the "Ebner Family Leukemia Research Foundation at the Weizmann Institute of Science, in memory of Alfred Ebner."

not eliminate the presence of dormant PLCs among lymphoid organs of 8- to 12-month-old thymectomized AKR mice (thymectomy performed at the age of 50–60 days). Transplantation of lymphoid cells from these thymectomized mice into the appropriate recipients yielded 80%–100% incidence of B-cell lymphoma of AKR/J origin [13]. MCF-type viruses are probably not involved in the generation of these lymphomas. Most of the tested tumors were found negative to the monoclonal antibody 18-5, which recognizes MCF expression. Lack of recombinant virus formation was also observed in AKR strecker mice (athymic mutant) [14]. However, recent work by Fredrickson et al. [15] suggested a possible contribution of ecotropic MuLV in the development of B-cell lymphoma in NFS mice. Viral isolates from B-cell lymphomas of AKR origin were found to inhibit spontaneous T-lymphoma development [16]. One of the viral isolates CFC-666 was actually found to interfere with the spontaneous DTV formation in the thymus, thereby perhaps preventing the spontaneous T-cell lymphomagenesis in AKR mice. In the present study we further extended our analysis of the events occurring following infection with CFC-666 virus, and its effects on thymus differentiation antigens and on the occurrence of PLCs.

B. Results

The age-dependent susceptibility of AKR mice to the effect of CFC-666 shown previously to prevent spontaneous T-cell lymphoma development [16] was tested. Female and male newborn to 2-day-old AKR/J mice were injected i.v. (through the orbit plexus of the eye) with 0.1 ml CFC-666 or intrathymically into 14-, 60-, or 120-day-old AKR/J mice. The lymphoma incidence and the phenotype of the emerging tumors are shown in Fig. 1. A remarkable suppression in the development of T-cell lymphoma was observed in mice injected when newborn or 14 or 60 days old (27%, 16%, and 32%, respectively). A low incidence of B-cell lymphoma (about 20%) developed in the newborn and 14-day-old injected mice with CFC-666. The suppressive effect on T-cell lymphoma development was not observed when CFC-666 was injected into the thymus or spleen of 120-day-old mice. These mice yielded a 95% T-cell lymphoma incidence within 250 days, similar to the PBS intrathymic-injected mice or untreated controls.

Since CFC-666, when injected into young AKR mice, prevents leukemia development, it could also affect the occurrence of PLCs, their distribution, and their site. The presence of PLCs in thymus, bone marrow, or spleen of mice 12 months after CFC-666 injection into the thymus of 14-day-old AKR mice was demonstrated. Although mice injected with CFC-666 developed only 20% of lymphomas, 70%–90% of such infected mice were found to be carriers of PLCs. The majority of the lymphomas had pre-B-B characteristics, and sporadic occurrence of T- or null-cell lymphomas was also observed (Haran-Ghera et al., this issue).

These results resemble those of our previous observation of the existence of dormant PLCs in thymectomized AKR mice [13, 17], in spite of the fact of the presence of an intact thymus (although changed phenotypically). Thus, infection of young mice with CFC-666 might cause a "physiological thymectomy", thereby preventing PLC migration from the bone marrow to the intact thymus for further development into T-cell lymphoma.

The preleukemic thymus of AKR/J involves changes in thymocyte subpopulation and in viral expression observed at the age of 5–6 months [7]. The formation of the dual tropic virus (DTV) in the thymus within this age range has been related to those preleukemic changes. These changes could be accelerated by injection of DTV intrathymically to young AKR/J mice [11, 18]. The injection of CFC-666 into the thymus of 14-day-old mice pre-

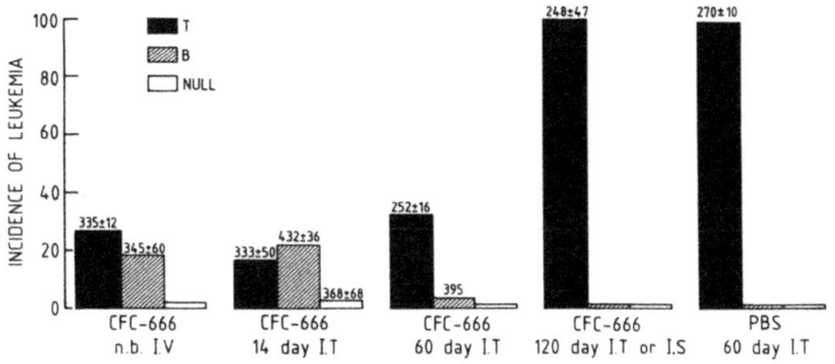

Fig. 1. Age-dependent susceptibility to CFC-666. Female and male new-born to 2-day-old AKR/J mice were injected i.v. (0.1 ml); 14-, 60-; or 120-day-old mice were injected intrathymically (0.02 ml). Lymphoma incidence and phenotypes

vented the changes occurring in the thymus of AKR/J spontaneously (or induced by DTV injection), namely the gradual elevation of class I histocompatibility antigens (H-2K and H-2D) and viral antigens, especially the expression of DTV [16]. In the present studies we further extended the investigation concerned with changes of thymus subpopulation in terms of quantitative expression of thymocyte differentiation antigens Thy-1, Lyt-2, and L3T4. The level of Thy 1.1 did not change during the first 9 months following CFC-666 inoculation and was similar to normal 2-month-old untreated mice, but dropped strikingly at 12 months post CFC-666 inoculation to about 36% ± 17% and a new population of Ig^+ cells appeared in these thymi (Fig. 2). The use of Lyt-2-FITC and L3T4-PE antibodies in FACS 440 analysis enabled us to calculate the level of the four populations of the thymus: $Lyt\text{-}2^+$ $L3T4^+$, $Lyt\text{-}2^+$ $L3T4^-$, Lyt^- $L3T4^+$, and $Lyt\text{-}2^-$ $L3T4^-$. In the normal adult mouse, Scolly et al. [19] found 81%, 5%, 9%, and 5%, respectively. In our studies (results presented in Fig. 3) we obtained the following: In (A) 2-month-old thymus and in (B) 9-month-old healthy thymus the four groups comprised 70%, 10%, 12%, and 8%, respectively. Although the pattern of distribution of cells within the group was different, the overall percent age of cells in each group was similar in both ages. These results are dramatically changed in the thymus of

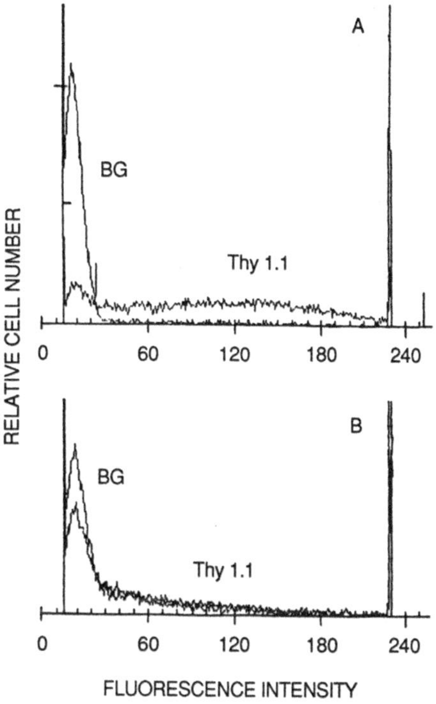

Fig. 2 A, B. FACS II analysis of AKR/J thymus for Thy 1.1 expression. **A** Normal thymus of 9-month-old AKR/J mice. **B** CFC-666-inoculated thymus (i.t. at 14 days) 12 months following virus inoculation

CFC-666-injected mice. (C) At 9 months following CFC-666 inoculation, the thymus comprised $L3T4^+Lyt\text{-}2^+ = 54\%$; $L3T4^+\text{-}Lyt\text{-}2 = 15\%$; $L3T4^-\text{-}Lyt\text{-}2^- = 14\%$; $L3T4^-\text{-}Lyt\text{-}2^+ = 17\%$. (D) Twelve months following CFC-666 inoculation the thymus changed its phenotype com-

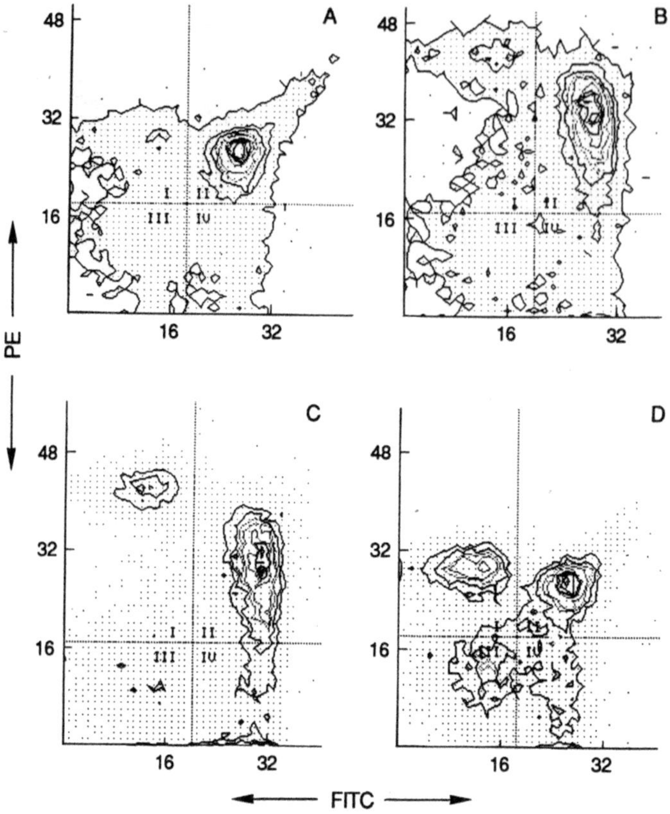

Fig. 3A–D. FACS 440 analysis of AKR/J thymus for Lyt-2 and L3T4 expression by double labeling using Lyt-2-FITC and L3T4-PE. **A** Two-month-old thymus. **B** Nine-month-old thymus. **C** CFC-666-inoculated thymus (i.t. at 14 days old) 9 months following virus injection. **D** Thymus 12 months following CFC-666 injection

pletely and the subgroups comprised 33%, 27%, 21%, and 19%, respectively. The thymi of CFC-666-injected mice remained normal in size but underwent changes in the expression of differentiation antigens.

C. Conclusion

In the studies described here we observed that the susceptibility to prevent spontaneous T-cell lymphomagenesis following injection of CFC-666 virus isolated from B-cell lymphoma of AKR origin is age dependent since the virus has to be injected early in life. The characteristic changes occurring spontaneously in the thymus of AKR mice during the preleukemic phase (at the age of 5–6 months) or following MCF injection were prevented by the inoculation of CFC-666. Namely, the amplified class I histocompatibility antigens, H-2K^k and H-2D^k, and especially the viral expression of MCF, were not observed. Nevertheless, dramatic changes in thymus differentiation antigens took place in the CFC-666-injected mice. Thus, CFC-666 interferes with MCF formation or replication in the thymus. We have suggested previously that DTV-induced changes in the thymus, including its "thymolytic" effect, trigger preexisting PLC migration from bone marrow into the thymus, thereby providing a suitable microenvironment for PLC progression to T-cell lymphoma. However, CFC-666 interference with this process might intervene also with the migration pathway of PLCs into the thymus, thereby allowing PLC dormancy. CFC-666 injection did not affect the presence of PLCs in the infected mice and only a low level of PLCs was observed in thymus of 12-month-old mice versus high PLC incidence observed in thymus of untreated 7- to 10-month-old mice. The PLCs demonstrated among bone marrow, spleen, and thymus cells eventually developed into B-cell lymphoma upon transplantation to new hosts (see Haran-Ghera et al., this issue). Reduced T-cell lymphoma development in AKR mice was demonstrated by the injection of DTV-SMX-1 into neonates [20]. Similarly, the prevention of spontaneous B-cell lymphoma development was observed by De Rossi et al. [21] following treatment of SJL/J (V^+) neonate mice with a DTV-SJL-151. In both cases the authors suggested viral interference as a plausible explanation for their observation.

References

1. Furth J, Seibold HR, Rathbone RR (1933) Experimental studies on lymphomatosis of mice. Am J Cancer 19:521–527
2. McEndy DP, Boom MB, Furth J (1944) On the role of thymus, spleen and gonads in the development of leukemia in high leukemic strain of mice. Cancer Res 4:377–383
3. Hays E (1968) The role of thymus epithelial reticular cells in viral leukemiogenesis. Cancer Res 28:21
4. Rowe WP, Pincus T (1972) Quantitative studies of naturally occurring murine leukemic virus injection of AKR mice. J Exp Med 135:429–436
5. Hartely JW, Walford MK, Old LJ, Rowe WP (1977) A new class of mouse leukemia virus associated with development of spontaneous lymphomas. Proc Natl Acad Sci USA 74:789–792
6. Elder JH, Gautsch JW, Jensen FC, Lerner RA, Hartely JW, Row WP (1977) Biochemical evidence that MCF murine leukemic viruses are envelope (ENV) gene recombinants. Proc Natl Acad Sci USA 74:4676–4680
7. Kawashima K, Ikeda H, Hartley JW, Stockert E, Rowe WP, Old LJ (1976) Changes in expression of murine leukemia virus antigens and production of xenotropic virus in the late preleukemic period in AKR mice. Proc Natl Acad Sci USA 73:4680–4684
8. Hays EF, Levy JA (1984) Differences in lymphogenic properties of AKR mouse retroviruses. Virology 138:49–57
9. Cloyd MW (1983) Characterization of target cells for MCF viruses in AKR mice. Cell 32:217–225
10. Haran-Ghera N (1980) Potential leukemic cell among bone marrow of young AKR/J mice. Proc Natl Acad Sci USA 77:2923–2926
11. Haran-Ghera N, Peled A, Hoffman AD, Leef F, Levy JA (1987) Enhanced AKR leukemogenesis by dual tropic viruses. I. The time and site of origin of potential leukemic cells. Leukemia 1:442–449
12. Buckheit RV Jr, Bolognesi DP, Weinhold KJ (1987) The effects of leukemosuppressive immunotherapy on bone marrow infectious cell centers in AKR mice. Virology 157:387–396
13. Peled A, Haran-Ghera N (1985) High incidence of B cell lymphomas derived from thymectomized AKR mice expressing TL.4 antigen. J Exp Med 162:1081–1086
14. Bedigian HG, Shultz LD, Meier H (1979) Expression of endogenous murine leukemic viruses in AKR/J Strecker mice. Nature 279:434–436
15. Fredrickson TM, Morse HC III, Rowe WP (1984) Spontaneous tumors of NFS mice congenic for ecotropic murine leukemic virus induction loci. J Nat Cancer Inst 73:521–524
16. Peled A, Haran-Ghera N (1988) Prevention of T cell lymphoma in AKR/J mice. Leukemia 2:125–131
17. Haran-Ghera N, Trakhtenbrot L, Resnitzky P, Peled A (1989) Preleukemic state in murine leukemogenesis. Leukemia (in press)
18. O'Donnell PV, Nowinski RC, Stockert E (1982) Amplified expression of murine leukemia virus (MuLV)-coded antigens on thymocytes and leukemic cells of AKR mice after infection by dual tropic (McF) MuLV. Virology 119:450–464
19. Scollay R, Bartlett P, Shortman K (1984) T cell development in the adult murine thymus: changes in the expression of the surface antigens Ly-2, L3T4 and B2A2 during development from early precursor cells to migrants. Immunol Rev 82:79–87

20. Stockert E, O'Donnell PV, Ohata Y, Old LJ (1988) Inhibition of AKR leukemogenesis by SMX-1, a dual tropic murine leukemia virus. Proc Natl Acad Sci USA 77:3720–3724

21. DeRossi A, D'Andrea E, Biasi G, Gollavo D, Chieco-Bianchi L (1983) Protection from spontaneous lymphoma development in SJL/J (v^+) mice neonatally injected with dual tropic SJL/151 virus. Proc Natl Acad Sci USA 80:2775–2779

Preleukemia in Experimental Leukemogenesis *

N. Haran-Ghera [1], L. Trakhtenbrot [1], P. Resnitzky [2], and A. Peled [1]

A. Introduction

The long latent period characteristic for both spontaneous and induced murine leukemias may reflect multiple steps in the leukemogenic process. The induction of preleukemia represents the initial phase in leukemogenesis. A sequence of further events (that could differ qualitatively from the initiating factor) is often required for the completion of the neoplastic transformation. The presence of potential leukemia-inducing cells (PLC) among bone marrow cells of mice shortly after treatment with different leukemogenic agents (which represents the preleukemic phase) was demonstrated by an in vivo transplantation bioassay. This method is based on the capacity of lymphoid cells transplanted into histocompatible hosts to give rise to lymphomas of donor origin [1]. A high incidence of PLC was demonstrated among bone marrow cells of intact or thymectomized BL/6 mice exposed to fractionated irradiation or treated with a potent chemical carcinogen. These treatments induced a high incidence of T-cell leukemia in intact mice, whereas thymus removal very markedly decreased the development of the disease [1-3]. The induction of PLC following intrathymic inoculation of two RadLV variants having H-2-associated high or low leukemogenic potential in adult mice has also been demonstrated [4, 5]. The most striking findings involved identification of PLC in mice resistant to overt leukemia development following virus inoculation. For example, a high incidence of PLC (in 80%-100% of the tested mice) was induced in BL/6 mice infected with D-RadLV or in B10S mice treated with A-RadLV, although overt T-cell leukemia incidence following these treatments was 0%-20% [6]. PLC have also been identified among fetal liver cells of 16-day-old AKR embryos as well as among bone marrow cells of intact or thymectomized AKR mice from the age of 14 days onwards [7]. Thus, resistance to leukemia development did not necessarily coincide with resistance to PLC induction.

In contrast to the leukemic cells in the experimental models described above that express the T-cell surface component Thy-1, PLC were shown to have the characteristics of prothymocyts [8, 9]. PLC were also shown to play the role of the immunogen in anti-RadLV-induced leukemia immune responses [6, 8, 10]. Thus, mice having H-2-linked resistance to RadLV leukemogenesis may be sensitized to their PLC, and further PLC differentiation into overt leukemia is suppressed by the PLC-induced antileukemia immune response. While leukemic cells grow progressively in any compatible host, further PLC proliferation and progression to leukemic cells is dependent on certain host conditions [9]. Thus, in T-cell leukemogenesis qualitative differences exist between PLC and leukemic cells.

---

[1] Department of Chemical Immunology, The Weizmann Institute of Science and

[2] Department of Internal Medicine "B" and Laboratory of Blood Morphology and Cytology, Kaplan Hospital, Rehovot, Israel

* This research was supported by a grant from the "Ebner Family Leukemia Research Foundation at the Weizmann Institute of Science, in memory of Alfred Ebner."

The present studies demonstrate the occurrence of a preleukemic phase during radiation-induced AML in SJL/J mice and the presence of dormant PLC in AKR mice following treatment that prevents their characteristic susceptibility to spontaneous T-cell leukemia development.

B. The preleukemic Phase in Radiation-induced Acute Myeloid Leukemia in SJL/J mice

Exposure of 3- to 4-month-old SJL/J mice to a single dose of 300 rads whole body irradiation results in AML development in about 20%–30% [11]. This incidence can be markedly increased by additional co-leukemogenic treatment with corticosteroids alone [12] and with additional cytophosphane treatment (Table 1). A single injection of different corticosteroids as co-leukemogenic treatment yielded about 50% AML; prolongation of the treatment to five weekly injections did not change this AML incidence. But five alternate weekly treatments of prednisone one week and cytophosphan the next week (the whole treatment lasting 10 weeks) increased AML incidence to 92% at a mean latency of 265 days. Neither agent without initial exposure to radiation had a leukemogenic effect. This combined co-leukemogenic treatment therefore seems to promote the proliferation and progression of the radiation preleukemic clone to overt AML.

Cytogenetic studies of AMLs induced in SJL/J mice by exposure to radiation with or without further co-leukemogenic treatment indicated that all tumors were characterized by a deletion of chromosome 2 [13]. This deletion therefore seems to be a necessary step in the radiation-induced tumorigenic process. Five types of chromosome-2 deletion were observed according to variations in size and formation, but each tumor had only one characteristic type of deletion in all cells involved. Control untreated SJL/J mice within the age range of 9–14 months had a normal diploid karyotype irrespective of the presence of spontaneous reticulum cell sarcoma [14], classified recently as B-cell neoplasms [15], observed in the older untreated tested mice.

Further studies revealed that chromosome-2 deletion represents the radiation-induced initiation of AML. Cells with this deletion were observed among bone marrow cells of SJL/J mice analyzed 4 months following exposure to 300 rads, although the cell donor mice looked normal, both on cytological and on histological examination. About 80% (14/17) of the tested irradiated mice had different levels of cells with deleted chromosome 2 among their bone marrow population (Table 2). It is interesting that among bone marrow cells from several mice tested, two or three clones with different chromosome-2 deletion types were observed [13], probably representing initial polyclonal transformation. In the final AML we always observed one type of chromosome-2 deletion in 100% of leukemic cells, suggesting the derivation of the tumor from a single altered clone. Among a small group of mice treated with both 300 rads and dexamethasone, analysis of bone marrow cells 4 months after treatment revealed cells with the chromosome-2 deletion in all treated mice (9/9). The percentage of cells with the deletion among the 50–70 karyotyped cells of each tested mouse was higher than in mice exposed to 300 rads alone (Table 2). A certain threshold level of cells with the chromosome-2 deletion, beyond 20%, seems to be essential for the actual overt AML development. In mice tested 4 months after receiving 300 rads, 28% (5/17) had more than 20% of deleted cells among those karyotyped versus 55% (5/9) with additional dexamethasone treatment. These levels match the final AML incidence observed in these two treated groups (Table 1).

A common missing segment D–G was observed in the different deletion types of chromosome 2 [13]. The loss of the genetic information located in this region may

Table 1. Induction of AML in SJL/J mice

Leukemogenic treatment [a]	AML incidence	Latency (days)
300 rads	3/16 – 18%	350±22
300 rads + Hydrocortisone acetate (2.5 mg i.p.)	14/30 – 47%	284±28
300 rads + Dexamethazone (0.5 mg s.c.)	16/26 – 60%	250±28
300 rads + Prednisone (1 mg s.c.)	16/32 – 50%	295±21
300 rads + 5 × (Prednisone $\xrightarrow{7\,d}$ cytophosphane) ↓ 1 mg s.c. ↓ 0.5 mg i.p.	12/13 – 92%	265±14
300 rads + 5 × Prednisone	7/14 – 50%	285±42
300 rads + 5 × Cytophosphane	2/10 – 20%	(340;365)
300 rads + 5 × (Prednisone $\xrightarrow{7\,d}$ cytophosphan)	0/18	–

[a] Three month-old female SJL/J mice were exposed to 300 rads whole body irradiation; the leukemogenic treatment was started 1–3 h after radiation.

Table 2. Incidence of chromosome-2 deletion among marrow cells of SJL/L mice, 4 months following exposure to 300 rads and dexamethasone treatment

Treatment of tested mice	Incidence of chromosome-2 deletion in tested mice	Percentage of cells per mouse with deletion 2 [a]				
		2%–5%	6%–10%	11%–20%	21%–30%	above 30%
300 rads	14/17 – 82%	6/17 – 35%	1/17 – 6%	2/17 – 11%	2/17 – 11%	3/17 – 17%
300 rads + dexamethasone	9/9 – 100%	0/9 –	2/9 – 22%	2/9 – 22%	1/9 – 11%	3/9 – 44%
SJL/J – 7 months old	0/10 –					

[a] 50–70 cells were karyotyped. The procedure used is described by Trakhtenbrot et al. [13].

be responsible for the generation of AML. The role of gene deletion as a predisposition in tumorigenesis has been extensively studied in connection with neuroblastoma and retinoblastoma development [16, 17]. Several investigators have suggested that loss or inactivation of negative regulatory genes through deletions may affect tumor development by derepressing other critical genes [18, 19]. The c-*abl* gene that has been implicated in the pathogenesis of human CML [20] was mapped by Goff et al. to the mouse chromosome 2 [21]. Perhaps some correlation between the localization of c-*abl* and the breakpoint in the deletion could also be implicated in radiation-induced AML in mice.

C. Preleukemia in AKR Mice

AKR mice are highly susceptible to spontaneous T-cell leukemia, arising predominantly in the thymus of 6- to 12-month-old mice. Leukemia development is associated with the expression of type-C retroviruses since birth, with increased levels of xenotropic virus titers, particularly in the thymus from 5 months onwards and in the formation of the dual tropic recombinant MCF type-C viruses that occur prior to leukemia development [22]. We have previously demonstrated that preleukemic cells (PLC), i.e., those having the potential to develop into T-cell leukemia, were present among bone marrow cells of intact and thymectomized young mice [7]. Thus, the occurrence of PLC in young AKR mice was shown to be thymus independent, but their progression to overt T-cell leukemia development was dependent on specific host conditions, including presence of the thymus. These findings suggested that the thymus provides a suitable environment for the promotion of PLC, contributing to the progression and differentiation of bone marrow-derived immortalized or preleukemic thymic precursors (prothymocytes) into T-cell leukemia. The promotion and progression of PLC into overt T-cell leukemia in AKR mice seems to be dependent on the delayed formation of the dual tropic virus (DTV) established in the thymus of 5- to 6-month-old AKR mice [22]. Indeed, recently we demonstrated that interference with DTV formation by injection of a viral isolate from a B-cell leukemia of AKR origin (designated 24-666) into the thymus of 14-day-old AKR mice very markedly reduced T-cell leukemia development. We observed a 10%–20% incidence at a mean latent period of 385 days. Some of these treated mice (10%–26%) developed B-cell leukemia beyond 400 days of age [24]. Since we suggested that the initial site of PLC occurrence in AKR mice is among bone marrow cells (and thymus independent), it seemed obvious that these virus 24-666-treated mice should be carriers of PLC. We therefore tested the leukemogenic potential of lymphoid cells from 12-month-old AKR mice infected when 14 days old with virus 24-666, using the transplantation bioassay method (transferring cells i.v. into [AKR × DBA/2]F1 mice). The results obtained (Table 3) clearly indicate that prevention of T-cell leukemia development by virus 24-666 does not affect the high incidence of PLC present among cells in the thymus, bone marrow, and spleen. The majority of the tumors originating from AKR donor cells had the characteristics of B-cell leukemias. Thus, dormant PLC were shown to be present in intact AKR mice that lacked DTV in their thymus.

Thymus removal in 1- to 3-month-old AKR mice markedly reduces the spontaneous tumor incidence (10%–20% extrathymic lymphoid tumors which appear late in life). The presence of PLC in old thymectomized AKR mice was previously demonstrated [25]. These PLC could be triggered to develop into B-cell leukemias after transplantation into syngeneic or hybrid (AKR × DBA/2)F1 intact or thymectomized recipients [25]. The majority of the leukemias expressed high levels of IgM, κ light chain, Ia molecules FcR, the Ly-1 marker, and the TL.4 antigen. Almost all thymectomized AKR mice were found to harbor PLC throughout their life span. These thymectomized AKR mice therefore represent an endogenous, spontaneously occurring preleukemic dormant state. Host inhibitory factors seem to contribute to the proliferation arrest of PLC in thymectomized mice, since only by transfer of lymphoid cells of old donor mice to young cell recipients could PLC proliferate and give rise to overt B-cell leukemia development. It is plausible to assume that prolonged thymectomy in AKR mice causes changes in regulatory mediators, including deficiency of T-cell factors necessary for B-cell growth turnover. It was therefore interesting to test whether grafting of a newborn AKR thymus subcutaneously into 10-month-old thymectomized AKR mice could affect the leukemia incidence in these thymus graft recipients. Since thymectomy prevents DTV formation, we also tested the possible effect of infection with DTV (from T-cell leukemia origin) or the virus 24-666 (isolated from a B-cell leukemia) on leukemia development and breakdown of PLC dormancy. Ten-month-old thymectomized AKR mice received a single i.v. injection of these virus isolates. The results obtained are summarized in Table 4. Indeed, the incidence of B-cell leukemia was markedly increased by these different experimental manipulations (66%–84% versus 18% in the untreated thymectomized control mice). Thus, PLC can be prompted to develop

Table 3. Leukemogenic potential of lymphoid cells from 12-month-old AKR mice infected when 14 days old with virus 24-666

Cells tested	Leukemia incidence	T-cell leukemia	B-cell leukemia	Null cell leukemia
Thymus	12/14 – 85% (86 ± 24)	3/14 – 21%	6/14 – 43%	3/14 – 21%
Bone marrow	12/14 – 85% (80 ± 16)	0/14 –	10/14 – 71%	2/14 – 14%
Spleen	10/13 – 77% (92 ± 27)	1/13 – 8%	7/13 – 54%	2/13 – 15%

Thymocytes, bone marrow cells, and spleen cells were injected i.v. from each individual mouse to (AKR × DBA/2)F1 recipients (1:1 transfer). The AKR donor origin of the developing leukemias and the tumor characteristics were done as previously described [25].

Table 4. Termination of PLC dormancy

Treatment[a]	B-Leukemia incidence	T-Leukemia incidence
–	5/28 – 18% (500 ± 20)	2/28 – 7% (385;430)
n.b. thymus s.c.	16/24 – 66% (480 ± 45)	2/24 – 8% (350;516)
DTV × 1 (i.v.)	22/26 – 84% (460 ± 74)	0/26
24-666 × 1 (i.v.)	21/26 – 80% (512 ± 58)	0/26

[a] AKR females thymectomized when 45–55 days old were further treated when 10 months old; both lobes of a syngeneic newborn (n.b.) thymus were grafted subcutaneously. The origin and preparation of the injected viruses are described in [7] and [24].

into overt lymphoma in the thymectomized mice by different factors, their mode of action being still an enigma.

The possible pathways in T- and B-cell leukemogenesis in AKR mice are summarized in Fig. 1. From birth, AKR mice carry "immortalized" PLC, identified among the bone marrow prothymocytes. In intact AKR mice the progression of PLC into overt T-cell lymphoma is dependent on the delayed formation of DTV in the thymus of 5- to 6-month old mice [23]. Thymus subpopulation changes due to DTV formation [26] may trigger stem cell migration, including PLC, into the "injured" thymus, and thereby provide the microenvironment for further PLC differentiation into T-cell leukemia. The capacity of DTV to impair certain T-cell functions [23] could also interfere with immune surveillance triggered by the immunogenic capacity of PLC and thereby contribute to the progression of PLC into overt T-cell leukemia. Interference with DTV formation in the thymus by a competitive virus (24-666) or by thymus removal abrogates the environment required for the transition of PLC to T-cell leukemia. These mice are carriers of PLC that can be triggered, by different experimental manipulations, to develop into B-cell leukemias. Different inhibitory factors may contribute to the arrest of PLC proliferation in these hosts. Most of the B lymphomas that develop express the Ly-1 antigen [25]. Since Ly-1$^+$ B cells were shown to secrete IgM autoantibodies [27], they may affect regulatory factors necessary for the proliferation of PLC into autonomous B-leukemic cells. Removal of PLC into young recipients might prevent the interaction of the autoantibodies with PLC and thereby facilitate their

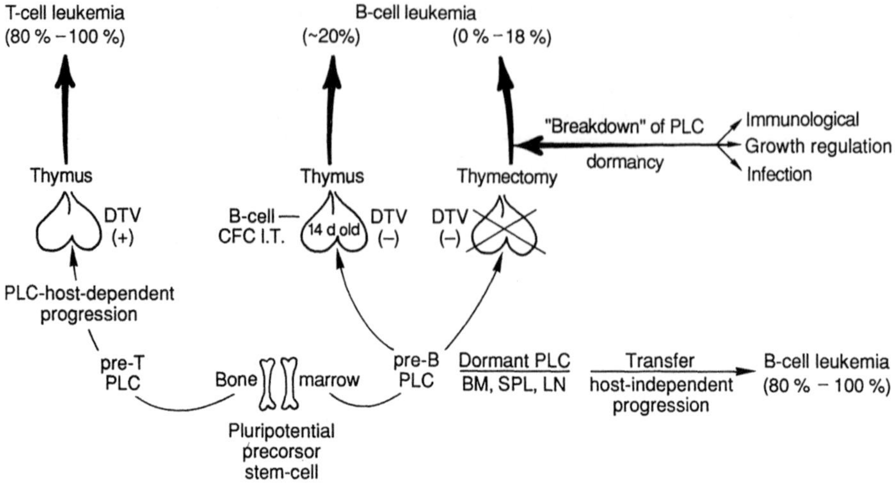

Fig. 1. T- and B-cell leukemogenesis in AKR/J mice

progression into B-cell leukemias. Breakdown of PLC dormancy in thymectomized mice by thymus grafts might act by providing the lacking hematopoietic growth factors. The viral infection might perhaps also act via lytic effects on certain lymphoid populations, thereby affecting immune integrity or regulatory factors mediated by these cells.

One obvious problem is whether PLC in intact and thymectomized AKR mice represent a pluripotential hematopoietic precursor cell sharing pre-T and pre-B properties or, alternatively, whether two different lymphoid progenitors restricted to T- or B-cell pathways are involved in T- and B-cell leukemia development in AKR mice. Our findings that B-cell leukemias of AKR origin express the T-cell antigen TL.4, found exclusively on T-cell leukemias, including the early-occurring ones in AKR mice, as well as on some lymphoid cells of 1- to 20-day-old AKR mice [25, 28], may suggest that a common progenitory $TL.4^+$ cell might serve as a precursor for both T- and B-cell leukemia in AKR mice.

References

1. Haran-Ghera N (1978) Spontaneous and induced preleukemia cells in C57BL/6 mice. JNCI 60:707–710
2. Haran-Ghera N (1973) Relationship between tumor cells and host in chemical leukemogenesis. Nature New Biol 240: 84–86
3. Haran-Ghera N (1976) Pathways in murine radiation leukemogenesis. In: Juhas JM, Tennant RW, Regan JD (eds) Biology of radiation carcinogenesis. Raven, New York, p 245–260
4. Haran-Ghera N, Ben-Yaakov M, Peled A (1977) Immunologic characteristics in relation to high and low leukemogenic activity of radiation leukemia virus variants, JM Yuhas, RW Tennant, JD Regand (eds). J Immunol 118:600–607
5. Lonai P, Haran-Ghera N (1980) Genetic resistance to murine leukemia virus variants; a comparative study on the role of the H-2 complex. Immunogenetics 11:21–30
6. Lonai P, Katz E, Peled A, Haran-Ghera N (1981) H-2-linked control of immunological resistance to viral leukemogenesis as a response to preleukemic cells. Immunogenetics 12:423–427
7. Haran-Ghera N, Peled A, Leef F, Hoffman AD, Levy JA (1987) Enhanced AKR leukemogenesis by the dual tropic viruses.

I. The time and site of origin of potential leukemic cells. Leukemia 1:442–449
8. Haran-Ghera N, Rubio N, Leef F, Goldstein G (1978) Characteristics of preleukemia cells induced in mice. Cell Immunol 37:308–314
9. Haran-Ghera N (1985) Prothymocytes as target cells of T-cell leukemogenesis in the mouse. Surv Immunol Res 4:19–26
10. Haran-Ghera N, Rubio N (1977) Immunologic characteristics in relation to high and low leukemogenic activity of radiation leukemia virus variants. II. Analysis of the immune response. J Immunol 118:607–614
11. Haran-Ghera N, Kotler M, Meshorer A (1967) Studies on leukemia development in the SJL/J strain of mice. JNCI 39:653–661
12. Resnitzky P, Estrov Z, Haran-Ghera N (1985) High incidence of acute myeloid leukemia in SJL/J mice after X-irradiation and corticosteroids. Leuk Res 9:1519–1528
13. Trakhtenbrot L, Krauthgamer R, Resnitzky P, Haran-Ghera N (1988) Deletion of chromosome 2 is an early event in the development of radiation-induced myeloid leukemia in SJL/J mice. Leukemia 2:545–550
14. Wanebo HJ, Gallmeier WM, Boyse EA, Old L (1966) Paraproteinemia and reticulum cell sarcoma in an inbred mouse strain. Science 154:901–903
15. Pattengale PK, Firth CH (1983) Immunomorphological classification of spontaneous lymphoid cell neoplasms occurring in female BALB/c mice. JNCI 70:169–179
16. Yunis JJ (1983) The chromosomal basis of human neoplasia. Science 221:227–230
17. Yunis J, Ransey M (1978) Retinoblastoma and sub-band deletion of chromosome 13. Am J Dis Child 132:161–170
18. Knudson AG Jr (1985) Hereditary cancer, oncogenes and antioncogenes. Cancer Res 45:1437–1443
19. Atkin NB (1985) Antioncogenes. Lancet ii:1189
20. Nowell PC, Croce CM (1986) Chromosomes, genes and cancer. Am J Pathol 125:8–12
21. Goff SP, D'Eustachio P, Ruddle FH, Baltimore D (1982) Chromosomal assignment of endogenous pro-oncogene c-*abl*. Science 218:1317–1319
22. Hartley JW, Walford MK, Old LJ, Rowe WP (1977) A new class of mouse leukemia virus associated with development of spontaneous lymphoma. Proc Natl Acad Sci USA 74:789–792
23. Peled A, Hoffman AD, Levy JA, Haran-Ghera N (1987) Enhanced AKR leukemogenesis by the dual tropic viruses. II. Effect on cell-mediated immune responses. Leukemia 1:450–456
24. Peled A, Haran-Ghera N (1988) Prevention of T-cell lymphoma in AKR mice. Leukemia 2:110, 12 Suppl. 1255–1315
25. Peled A, Haran-Ghera N (1985) A high incidence of B-cell lymphomas derived from thymectomized AKR mice expressing TL.4 antigen. J Exp Med 162:1081–1086
26. Kawashima K, Ikeda H, Hartley JH, Stockert E, Rowe WP, Old LJ (1976) Changes in expression of murine leukemia virus antigens and production of xenotropic virus in the late preleukemia in AKR mice. Proc Natl Acad Sci USA 73:4680–4684
27. Hayakawa K, Harday RR, Honda M, Herzenberg LA, Steinberg AD, Herzenberg LH (1984) Ly-1 B cells: functionally distinct lymphocytes that secrete IgM autoantibodies. Proc Natl Acad Sci USA 81:2494–2497
28. Peled A, Haran-Ghera N (1984) Age-related expression of TL antigen in AKR mice. Int J Cancer 34:121–126

RAS Mutations in Preleukaemias

R.A. Padua [1], G. Carter [1], D. Hughes [1], J. Gow [2], C. Farr [2], D. Oscier [3], F. McCormick [4], and A. Jacobs [1]

Activated *ras* genes have been implicated in a wide variety of neoplasms [2]. N-*ras* in particular has been shown to be involved in acute myelogenous leukaemia (AML) with activating mutations around codons 12/13 and 61 [3, 7]. The myelodysplastic syndromes (MDS) are a group of preleukaemias, a proportion of which will develop AML. Hirai et al. [9] described three MDS patients with N-*ras* mutations in codon 13 and Liu et al. [12] showed K-*ras* activations in two MDS patients. In this study we have screened DNA from peripheral blood or bone marrow of 50 MDS patients for *ras* mutations around codons 12/13 and 61 of H, K and N-*ras* and around codon 117 of N-*ras*. A mutation in position 117 of H-*ras* has been reported to be an activating mutation in vivo in chemically induced murine liver tumours [15]. Mutations in codons 116–119 of H-*ras* have been shown to reduce the ability of the H-*ras* p21 protein to bind and hydrolyse guanosine triphosphate (GTP) and some of these mutations are capable of activating the transforming potential of the normal gene [5, 20].

Using an amplification procedure called polymerase chain reaction (PCR) [16] and hybridisation with synthetic oligonucleotide probes [7, 21], *ras* mutations were detected in 20/50 (40%) MDS patients (Table 1). Fourteen N-*ras* mutations were observed (ten in codons 12/13 and four in codon 61), six K-*ras* mutations in codon 12 and two H-*ras* mutations (one in codon 12 and one in codon 61). Two patients had mutations in two different *ras* genes. This gives a total of 22 mutations from 50 individuals. Details of the patients and substitutions observed in these mutants have been reported elsewhere [14]. To date no mutations were observed with mutant-specific oligonucleotide probes to position 117.

Independent confirmation of these results were obtained with DNA from 11 patients with detectable *ras* mutations by using transformation assays (Table 2). A tumorigenicity assay identified another sideroblastic anaemia (RARS) patient with an N-*ras*-activated gene (*GD*). As this was not detected by the oligonucleotide screen, it is possible that the number of mutant *ras*-containing cells is very low. Another chronic myelomonocytic leukaemia (CMML) patient (RP86-CMML) with an N13 *Ala* substitution progressed approximately 1 year later to AML (RP87-AML). The mutation in the AML (RP87) stage was no longer detectable by PCR (Fig. 1 A). An N-*ras* clone pAT 8.8 with normal sequences around codon 12 [8] was used as a control. Other controls include HM-CMML, another N13 *Ala*-containing mutant and BM-PASA (sideroblastic anaemia or RARS). This observation was confirmed by directly hybridising the N13 *Ala* probe to unamplified DNA and showing differential hybridisation (Fig. 1 B). Using a minisatellite probe 33.6 [10], the DNA fingerprints confirmed that both samples were derived from the same patient (Fig. 1 C).

[1] Leukaemia Research Fund Preleukaemia Unit, University of Wales College of Medicine, Heath Park, Cardiff, UK

[2] Institute of Cancer Research, London, UK

[3] Royal Victoria Hospital, Bournemouth, UK

[4] Cetus Corporation, Emeryville, California, USA

	No. *ras* mutants/ No. samples	Substitution
Controls	0/10	
Myelodysplasia		
RAS	2/13	1 × H61 Arg (A→G)]
		1 × N12 Val (G→T)]
		1 × K12 Asp (G→A)
RA	4/13	1 × N12 Asp (G→A)
		1 × N12 Val (G→T)
		1 × N61 Leu (A→T)
		1 × K12 Asp (G→A)
RAEB	3/8	1 × N12 Ala (C→G)
		1 × N12 Asp (G→A)
		1 × K12 Arg (G→C)
CMML	11/16	1 × N12 Ala (G→C)
		1 × N12 Asp (G→A)
		3 × N13 Ala (G→C)
		3 × K12 Asp (G→A)]
		1 × H12 Val (G→T)]
		3 × N61 His (A→C)
Total	20/50 (40%)	

Table 1. *ras* mutations in MDS

Square brackets indicate mutations of the two patients with double mutations. RARS, refractory anaemia with ringed sideroblasts; RA, refractory anaemia; RAEB, refractory anaemia with excess blasts; CMML, chronic myelomonocytic leukaemia.

Both RP86-CMML and RP87-AML registered in a transformation assay (Table 2).

Several investigators have suggested that there is a heterogeneity of leukaemic cells with respect to the presence of an activated *ras* gene and that in some patients only a fraction of the malignant cells carry the mutant gene [18, 17, 7]. Similarly, heterogeneity in malignant melanoma has been described [1] where N-*ras* activation was detected in one out of five cultured tumour cell lines established from metastases of a melanoma patient. One explanation of these results is that mutations occur late, after a preleukaemic clone has already emerged, and give the premalignant clone an additional growth advantage. An alternative explanation is that the *ras* mutation occurs early in the preleukaemic process and there is later evolution with the emergence of a clone in which another gene has been activated. Our observations of activated *ras* genes in 21 preleukaemic patients argue for the latter explanation, though there is no evidence that this is an initiating event in leukaemogenesis. The possible reduction of a mutant *ras* gene with leukaemic transformation in one case also supports this argument. Similar observations have also been described in AMLs [7]. The evidence for such clonal evolution in leukaemogenesis is compelling [11, 6] and, in many cases, this may be clearly seen in serial karyotype studies [13, 19].

The incidence of *ras* mutations in the different FAB groups do not differ from those expected, with the exception of the CMMLs with a higher than expected frequency ($P=0.02$) (Table 3). To date, 8 of the 21 patients with mutant *ras* genes transformed to AML compared with 4 of the 29 patients with no detectable *ras* mutations. Out of the latter four, one of

Table 2. Transformation assays

		Tumorigenicity assay		
		No. tumours No. sites injected	Latency (days)	Transforming gene and substitution
Controls				
NIH3T3		2/30	33	–
EJ focus		16/16	21	Ha12 *val*
Myelodysplasia				
RARS	CN	8/8	38	N12 *val*+H-*ras*
	GD	6/8	19	N-*ras*
	PW	0/8	–	–
	ZG	0/8	–	–
RA	OB	8/8	21	N-*ras*
	GB	6/6	10	N61 *leu*
RAEB	KS	5/8	22	N12 *ala*
	TB	3/8	22	N12 *ala*
	JO	0/8	–	–
CMML	HM	3/8	36	N13 *ala*
	RP 1986	3/8	24	N13 *ala*
	DP	5/8	27	N-*ras*
	HE	4/8	26	K-*ras*
	HW	4/8	23	Non-*ras*
	JN	0/8	–	–
	AT	0/8	–	–
Acute myelogenous leukaemia				
	RP 1987	6/8	24	N-*ras*

		Focus formation	
		Transformation frequency Foci/μg DNA (No. foci/No. flasks)	
Controls			
NIH3T3		0 (0/5)	
EJ focus		0.2 (20.5)	Ha12 *val*
Myelodysplasia			
CMML	FB	0.06 (6/5)	N12 *asp*
CMML	MB	0.04 (8/10)	H12 *val*
RA	HW	0.017 (5/15)	Non-*ras*

Twenty micrograms of DNA was transfected onto NIH3T3 cells which were plated at a density of 3×10^5 24 h earlier. Precipitates were left overnight and media changed the next day. For tumorigenicity assays 1–2 μg PHSG272, a neomycin-resistance-containing cosmid [4], was cotransfected with genomic DNAs. Three days after transfection, cells were split into G418-containing media (800 μg/ml), resistant colonies were selected and 10^6 cells/site were injected in both of the hind flanks of athymic nude mice. Tumours with latency periods of less than 6 weeks were scored and analysed. For focus formation assays media were changed every 3 days in 5% serum. Foci were scored 3 weeks after transfection. One hundred micrograms of DNA of each sample was used per experiment and three to four mice were injected.

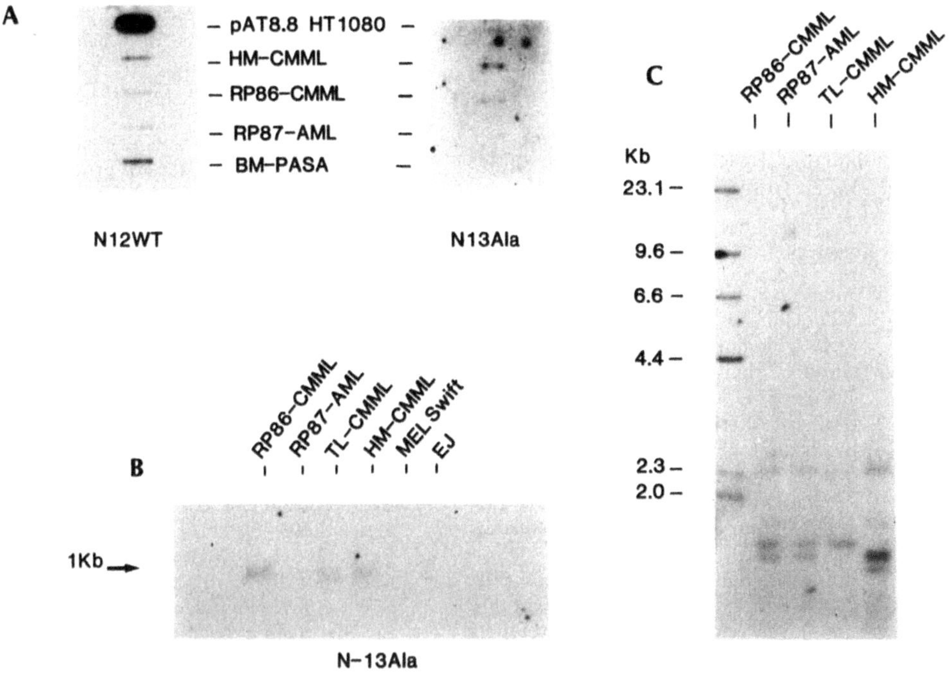

Fig. 1 A–C. Presence of N13 mutations in CMML and absence of mutation in AML. Amplified DNA was slot blotted and hybridised with N12/13 wild-type or N13 *ala* probes (**A**). Ten nanograms of cloned N-*ras* pAT 8.8 HT1080 [8] was used as a control. *Pvu*II-digested DNA was electrophoresed and dried gels were hybridised to the N13 *ala* probe (**B**). *Hinf*I-digested DNA was electrophoresed, transferred and filters hybridised to the minisatellite probe 33.6 [10] (**C**)

Table 3. Frequency of normal and mutant genes in RARS, RA, RAEB and CMML

	RARS	RA	RAEB	CMML	Total
Normal	10 (7.5)	9 (7.5)	5 (4.6)	5 (9.3)	29
Abnormal	3 (5.5)	4 (5.5)	3 (3.4)	11 (6.7)	21
	13	13	8	16	50

Figures in brackets are the frequencies that would be expected if the prevalence of mutant genes was the same in each group, and equal to the overall rate of 21/50 = 42%. A comparison of these expected frequencies with those actually observed by the exact test indicates that there are more CMML patients with mutant genes than would be expected ($\chi^2 = 5.39$, $P = 0.02$).

these patients (H.W.) had a detectable transforming gene which is not an activated *ras* gene (Table 2, RA-HW and CMML-HW). Transformants for both the refractory anaemia (RA) and CMML stages were anchorage independent and tumorigeneic in nude mice. DNA from the three non-mutated patients which progressed to AML have been found to be negative when tested in transformation assays (RAEB-JO; CMML-JN and CMML-AT). In agreement with Hirai et al. [9] and Liu et al. [12], our results show that MDS patients with mutations may evolve to AML more frequently than those without mutations. Thus, the presence of such mutations may be of prognostic value. However, clearly there are other factors which are important in progression.

This work was supported by the Leukaemia Research Fund. The tables are taken from Padua et al. [14].

References

1. Albino AP, Le Strange R, Oliff AI, Furth ME, Old LJ (1984) Transforming *ras* genes from human melanoma: a manifestation of tumour heterogeneity? Nature 308:69–72
2. Barbacid M (1987) *Ras* genes. Annu Rev Biochem 56:779–827
3. Bos JL, Verlaan-de Vries M, Van der Eb AJ et al. (1987) Mutations in N-*ras* predominate in acute myeloid leukaemia. Blood 69:1237–1241
4. Brady G, Jantzen HM, Bernard HU, Brown R, Schutz G, Hashimoto-Gotoh T (1984) New cosmid vectors for developing eukaryotic DNA cloning. Gene 27:223–232
5. Clanton DJ, Hattori S, Shih TY (1986) Mutations of the *ras* gene product p21 that abolish nucleotide binding. Proc Natl Acad Sci USA 83:5076–5080
6. Fearon ER, Philip BA, Burke PJ, Schiffer CA, Zohnbauer BA, Vogelstein B (1986) Differentiation of leukemia cells to polymorphonuclear leukocytes in patients with acute nonlymphocytic leukemia. N Engl J Med 315:15–24
7. Farr CJ, Saiki RK, Erlich HA, McCormick F, Marshall CJ (1988) Analysis of *ras* gene mutations in acute myeloid leukaemia by polymerase chain reaction and oligonucleotide probes. Proc Natl Acad Sci USA 85:1629–1633
8. Hall A, Marshall CJ, Spurr NK, Weiss RA (1983) Identification of transforming genes in two human sarcoma cell lines as a new member of the *ras* gene family located on chromosome 1. Nature 303:396–400
9. Hirai H, Kobayshi Y, Mano H et al. (1987) A point mutation at codon 13 of the N-*ras* oncogene in myelodysplastic syndrome. Nature 327:430–432
10. Jeffreys AJ, Wilson V, Thein SL (1985) Hypervariable 'minisatellite' regions in human DNA. Nature 314:67–73
11. Larson RA, Le Beau M, Vardiman JW, Testa JR, Golomb HM, Rowley JV (1983) The predictive value of initial cytogenetic studies in 148 adults with acute nonlymphoblastic leukemia: a 12 year study (1970–1982). Cancer Genet Cytogenet 10:219–236
12. Liu E, Hjelle B, Morgan R et al. (1987) Mutations of the Kirsten-*ras* proto-oncogene in human preleukaemia. Nature 300:186–188
13. Mecucci C, Rege-Cambrin G, Michaux J-L, Tricot G, Van den Berghe H (1986) Multiple chromosomally distinct cell populations in myelodysplastic syndromes and their possible significance in the evolution of the disease. Br J Haematol 64:699–706
14. Padua RA, Carter G, Hughes D et al. (1988) *Ras* mutations in myelodysplasia detected by amplification, oligonucleotide hybridisation and transformation. Leukemia 2:503–510
15. Reynolds SH, Stowers ST, Patterson RM, Maronpot R, Aaronson SA, Anderson MW (1987) Activated oncogenes in B6C3F1 mouse liver tumours: implications for risk assessment. Science 237:1309–1316
16. Saiki RK, Bugawan TC, Horn GT, Mullis KB, Erlich HA (1986) Analysis of enzymatically amplified beta-globin and HLA-DQalpha DNA with allele-specific oligonucleotide probes. Nature 324:163–166
17. Shen WP, Aldrich TH, Venta-Perez G, Granza BR Jr, Furth ME (1987) Expression of normal and mutant *ras* proteins in human acute leukemia. Oncogene 1:157–165
18. Toksoz D, Farr CJ, Marshall CJ (1987) *Ras* gene activation in a minor population of the blast population in acute myeloid leukaemia. Oncogene 1:409–413
19. Tomonaga M, Tomonaga Y, Kusano M, Ichimaru U (1984) Sequential karyotypic evolutions and bone marrow aplasia preceding acute myelomonocytic transformation from myelodysplastic syndrome. Br J Haematol 58:53–60
20. Walter M, Clark SG, Levinson AD (1986) The oncogenic activation of human *pn21 ras* by a novel mechanism. Science 233:649–652
21. Wood WI, Gitschier J, Lasky LA, Lawn RM (1985) Base composition-independent hybridization in tetramethylamonium chloride: a method for oligonucleotide screening of highly complex gene libraries. Proc Natl Acad Sci USA 82:1585–1588

Immunology

Logical Structure of the Immunoregulatory Compartment

N.A. Mitchison

The immune system as we know it in vertebrates is divided into a number of compartments, each of which contains a more or less complete set of lymphocytes with diverse receptors. Among them, by far the largest single compartment is devoted to immunoregulation and comprises the $CD4^+$, class II MHC-restricted set of T cells. Our invertebrate progenitors have molecules that are distantly homologous to antibody, but in this ancestral form they are believed to function in cell adhesion rather than defence, and are not diversified. At a stage in the evolution of vertebrates they developed a mechanism of diversification, and at that point must have been distributed on a single set of lymphocytes. This non-compartmentalized stage has not been found in any living vertebrates such as the primitive jawless fish. The rapid replacement of this missing link tells us that a compartmentalized immune system must provide significant evolutionary benefits.

The existence of a regulatory compartment permits:

1. Better antibodies to be made, by means of hypermutation
2. Coordination of the immunological attack on complex antigens, by means of epitope linkage
3. Control of hypersensitivity, by means of immunosuppression

Among these benefits, the third is the odd one out as it is needed only *after* the regulatory compartment has evolved; it corrects some of the problems that then arise as a result of overactivity within that compartment.

Each of these benefits needs further explanation. The first, that of hypermutation, is linked logically to the mechanism that the immune system has developed for avoiding reactivity with self. Tolerance of self, it is now generally agreed, results from deletion of self-reactive lymphocytes during development. This proposition needs minor qualification, in so far as mechanisms of suppression may supplement clonal deletion, but experience with unmanipulated systems (such as my group's work on F liver protein as an antigen in mice [1]) has underlined the primary role of deletion. If we accept this proposition, then it is also clear that hypermutation cannot be allowed to occur within the set of lymphocytes that is responsible for self-tolerance. Rajewsky's current estimate of the hypermutation rate in memory B-cells is $\sim 10^{-2}$/base pair per cell division; a rate as high as that would surely fill in any holes in the repertoire created by clonal deletion, and that would lead on to autoimmune disease. What actually happens is that the essential clonal deletion occurs only within the regulatory compartment (among helper T-cells), while hypermutation is confined to B cells, i.e., within one of the effector compartments. Clonal deletion may or may not also occur within the B-cell compartment, but that seems to be an optional extra that varies from one self macromolecule to another, depending mainly on concentration within body fluids. All this was known prior to the discovery of hyper-

Imperial Cancer Research Fund, Tumour Immunology Unit, University College London, Department of Biology, Medawar Building, Gower Street, London, WC1E 6BT, UK mutation; the real step forward is to understand that T-cell tolerance and hypermutation in B cells fit together logically, so that the two sets of observations mutually reinforce one another.

The second benefit, of epitope linkage, is of special interest because it has recently been discovered to work through two rather different mechanisms, and thus provides the first logical explanation of why T and B cells follow different traffic patterns within the immune system. Overall, linkage coordinates immunological attack in the following way: a regulatory lymphocyte (a helper T cell) recognizes an epitope (in effect a regulatory epitope) of a complex antigen, and then selectively activates effector lymphocytes (B cells or cytolytic T cells) that recognize other epitopes (effector epitopes) of the same antigen. In much the same way a suppressor regulatory cell (a suppressor T cell) can also selectively downregulate other cells (principally helper T cells) that recognize the same antigen. For both T and B cells the selective activation works via short-range lymphokines that are not antigen specific, and so the linkage depends exclusively on the regulatory and effector cells being brought into juxtaposition. There, however, the similarity stops. The T to B interaction depends on the formation of two-cell-type clusters, in which a B cell binds directly to T cells. The T-to-T interaction (helper to cytolytic T cell) depends rather on three-cell-type clusters, in which the two types of lymphocytes bind to a common antigen-presenting cell. Two crucial pieces of evidence establish that three-cell-type clusters do in fact form under physiological conditions in vivo [2]: (a) the two types of lymphocyte need not make a cognate interaction (in contrast to the T-B interaction, where such a requirement applies), and (b) with large numbers of antigenic particles (i.e. when each antigen-presenting cell can be calculated to pick up several particles) epitope linkage no longer operates (in contrast, again, to the T-B interaction, which saturates, if at all, only at much higher particle numbers).

Thus the immune system requires two quite different types of antigen-presenting cell. One, for B cells, must retain antigen in a conformationally intact form (most B cells recognize conformation epitopes); it must do so long term, in order to provide time for hypermutation to operate; and its dendrites need to be spaced together in a network dense enough to maximize the chances of contact with migrating B cells. The other, for T cells, must cleave antigen into a form able to bind to major histocompatibility complex (MHC) molecules; it need not retain antigen long term, for T cells do not hypermutate and their response plateaus early; and the presenting cells with their dendrites need to be spaced far enough apart to prevent the three-cell-type clusters, each with its own presenting cell, from interacting with one another and thus interfering with epitope linkage [3].

If one looks at the T-cell and B-cell area of a lymph node, two quite different (and possibly unrelated) types of antigen-presenting cell are evident in the two areas. Among T cells are to be found interdigitating dendritic cells, and among B cells follicular dendritic cells. Each has the appropriate combination of the three contrasting properties that have just been described. Furthermore the interdigitating dendritic cell, as well as being able to process antigen so that it associates with MHC molecule, is known also to bind T cells spontaneously in the absence of antigen and to stimulate them effectively in its presence. It is natural to conclude that the segregation of a lymph node into these two areas depends primarily on the need for the two types of antigen-presenting cell, and indeed that once the dendritic cells have sorted themselves out everything else in a lymph node will follow automatically as a consequence of selective aggregation of lymphocytes.

This account of lymph node structure is far from complete. It leaves unexplained the initial phase of the immune response, prior to localization of antigen on follicular dendritic cells in the form of antigen-antibody complexes; this early phase is still poorly understood. In addition, there is probably more to the structure of a follicle than simply aggregation of B cells around their antigen, for the germinal centre, the outer follicle, and the marginal zone around it are carefully arranged in a way that still require explanation.

The third benefit, of immunosuppression, raises the controversial issue of the suppressor T cell. Opinions among immunologists vary from those who regard this cell as playing a fundamental role in regulation of the immune response, to those who regard it as no more than an illusion. My own opinion, expressed in several recent and forthcoming reviews [4–7], lies somewhere between these two poles. I think it likely that a suppressor mechanism does operate, and that it has evolved primarily as a measure to counteract that threat of hypersensitivity. Over the course of evolution it is likely that the main threat of hypersensitivity has come from chronic infection: witness the extent to which chronic parasitic infection in the third world is usually well tolerated, except when it generates hypersensitivity. This line of thought finds support from studies in immunogenetics. On present evidence that MHC class II genes that mainly mediate suppression are HLA-DQ in man and H-2E in the mouse; as these are not homologous, this function must have flipped from one locus to another during mammalian evolution of the mammals. Furthermore mice, and possible rats too, fairly often lose expression of their suppression-mediating MHC class genes, as though the selective pressure of hypersentivity is diminished in these short-lived species.

As for the mechanism of suppression, the central questions remain unanswered pending the full deployment of molecular biology in this area. It is entirely possible that the phenomena of suppression can be accounted for by the known properties of suppression-mediating T cells: specialized restriction, surface markers such as CD45R, lymphokine-secretion profile, and high connectivity. Alternatively, these composite properties may eventually lead us to a set of T cells that have their own unique molecular mechanisms, such as a new set of receptors. The question is still open, and must surely occupy a high position on the agenda of contemporary immunology.

References

1. Griffiths JA, Mitchison NA, Nardi N, Oliveira DBG (1987) F protein. In: Sercarz E, Berzofsky J (eds) Immunogenicity of protein antigens: repertoire and regulation, Vol II. CRC, Boca Raton, pp 35–40
2. Mitchison NA, O'Malley C (1987) Three cell type clusters of T-cells with antigen presenting cells best explain the epitope linkage and non-cognate requirements of the in vivo cytolytic response. Eur J Immunol 17:579–583
3. Dexter M, Marvel J, Merkenschlager M, Mitchison NA, Oliveira D, O'Malley C, Smith L, Terry L, Timms E (1987) Progress in T cell biology. Immunol Lett 16:171–178
4. Mitchison NA, Oliveira DBG (1986) Chronic infection as a major force in the evolution of the suppressor T cell system. Parasitol Today 2:312–313
5. Mitchison NA, Oliveira DBG (1986) Epirestriction and a specialised subset of T helper cells are key factors in the regulation of T suppressor cells. In: Cinader B, Miller RG (eds) Progress in immunology, vol VI. Academic Press, London, pp 326–334
6. Marvel J, Mitchison NA, Oliveira DBG, O'Malley C (1987) The split within the CD4 (helper) T-cell subset, and its implications for immunopathology. Mem Inst Oswaldo Cruz [Suppl I] 82:260–273
7. Oliveira DBG, Mitchison NA (1989) Immune suppression genes. Clin Exp Immunol 75:167–177

Analysis of T Suppressor Cell-Mediated Tumor Escape Mechanisms Is Facilitated by the Selective In Vitro Activation of Tumor-specific Ts Cells

H.-D. Haubeck [1], I. Minkenberg [2], and E. Kölsch [2]

We have shown previously that tumor-specific T suppressor (Ts) cells were induced in vivo in BALB/c mice by the syngeneic plasmacytoma (PC) ADJ-PC-5 at very early stages of tumorigenesis [1, 2]. These Ts cells, which suppress a strong primary cytotoxic T cell response, have been characterized in detail [1–3].

There is evidence that Ts cell-inducing antigens (Ts-Ag) on ADJ-PC-5 plasmacytoma cells are expressed to some extent on normal BALB/c spleen cells and are therefore "self" antigens rather than tumor-specific neoantigens [4]. These data were subsequently confirmed by independent comparable studies using the EL4 thymoma of C57Bl/6 mice [5]. Thus, the induction of Ts cells by tumor-associated self antigens seems to be a more general rule and might be an important tumor escape mechanism.

To characterize Ts-Ag in more detail we have developed an in vitro system for the selective induction of tumor-specific Ts cells. Ts cell function would be masked in the in vitro Ts assay in the presence of activated cytotoxic T cells, which, like specific cytotoxic T cell clones, are not susceptible to suppression [2]. Activation of cytotoxic T cells is prevented by pretreatment of the ADJ-PC-5 stimulator cells with glutardialdehyd (GA) (Fig. 1). In contrast, specific Ts cells were activated by this approach which suppress the activation of specific cytotoxic T cells in the course of a primary mixed-lymphocyte tumor cell culture (MLTC) of BALB/c spleen cells against ADJ-PC-5

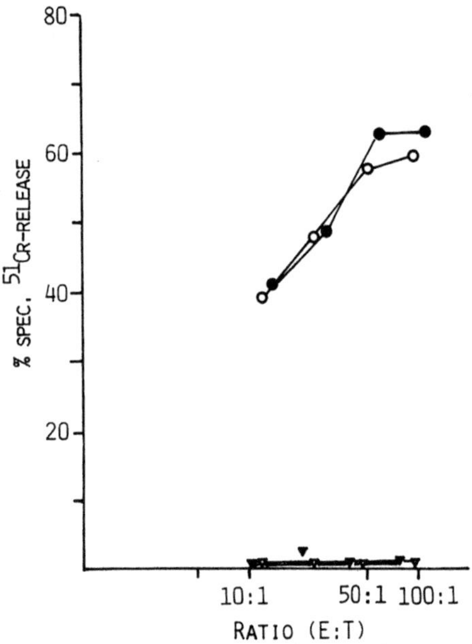

Fig. 1. Lack of induction of cytotoxic T (Tc) cells by glutaraldehyde (GA)-fixed ADJ-PC-5 plasmacytoma cells: 2×10^7 normal BALB/c spleen cells (SC) were incubated with 1×10^6 ADJ-PC-5 tumor cells in 10 ml MLTC medium containing 10% FCS in tissue culture flasks for 6 days. Thereafter, cells were harvested and tested for cytolytic activity against ADJ-PC-5 in a 6-h ^{51}Cr release assay. Several types of stimulator cells were used: (**a**) ADJ-PC-5 mitomycin-C-treated (○—○); (**b**) ADJ-PC-5 mitomycin-C-treated and subsequently fixed by GA (▼—▼); (**c**) a 1:1 mixture of ADJ-PC-5 mitomycin-C-treated and ADJ-PC-5 mitomycin-C-treated and GA-fixed (●—●). A control culture without stimulator cells is also shown (▽—▽)

[1] Department of Clinical Chemistry and Pathobiochemistry, Medical Faculty, University of Technology, Pauwelsstr., D-5100 Aachen, Federal Republic of Germany

[2] Department of Immunology, University of Münster, Domagkstr. 3, D-4400 Münster, Federal Republic of Germany

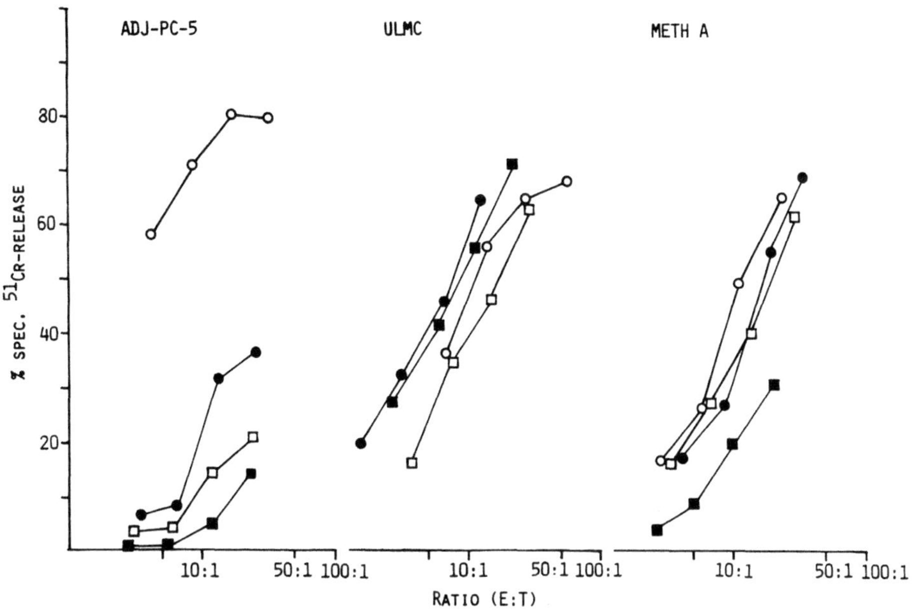

Fig. 2. Induction of specific Ts cells by stimulation of BALB/c SC with GA-fixed ADJ-PC-5 plasmacytoma cells: 4×10^6 normal BALB/c SC were incubated with 2×10^5 GA-fixed ADJ-PC-5 stimulator cells in MLTC medium containing 0.5% syngeneic normal mouse serum in Costar plates in a volume of 2 ml for 6 days. Subsequently, cells were harvested, 800 R x-irradiated, and preincubated with rIL2 (5 U/ml) for 1 h. They were then washed and used as a source for Ts cells. Graded numbers of Ts cells were added to primary mixed-lymphocyte tumor cultures (MLTC) of normal BALB/c SC against syngeneic tumor targets. After 6 days cells were harvested and tested for cytolytic activity. MLTC without Ts cells (○—○), with 2×10^5 (●—●), 6×10^5 (□—□), and 2×10^6 Ts cells (■—■). Tumor targets are ADJ-PC-5 (plasmacytoma), ULMC (lymphoma) and MethA (fibrosarcoma)

plasmacytoma cells, but not against the syngeneic control tumors ULMC (lymphoma) and MethA (fibrosarcoma) (Fig. 2). These Ts cells have been further characterized. Even in lectin-kill assays they have no cytolytic or NK-like activity, excluding a veto effect. In addition, suppression is not due to nonspecific effects like IL2 consumption, toxic effects by glutaraldehyde or PGE_2 release (data not shown).

The phenotype of these Ts cells was Thy 1.2^+, Lyt 2.2^+, $L3T4^+$, $I\text{-}A^{d-}$, $I\text{-}E^{d+}$ as evidenced by treatment with cytotoxic monoclonal antibodies and complement.

This in vitro system will be helpful for the isolation and characterization of Ts-Ag, but it also allows us to study in more detail the requirements for the induction of Ts cells and Ts-cell effector mechanisms.

References

1. Haubeck H-D, Kölsch E (1982) Regulation of immune responses against the syngeneic ADJ-PC-5 plasmacytoma in BALB/c mice. III. Induction of specific T suppressor cells to the BALB/c plasmacytoma ADJ-PC-5 during early stages of tumorigenesis. Immunology 47:503–509
2. Haubeck H-D, Kölsch E (1986) Regulation of immune responses against the syngeneic ADJ-PC-5 plasmacytoma in BALB/c mice. IV. Tumor-specific T suppressor cells, in-

duced at early stages of tumorigenesis, act on the induction phase of the tumor-specific cytotoxic T cell response. Immunobiology 171:357–363
3. Haubeck H-D, Kölsch E (1985) Isolation and characterization of in vitro and in vivo functions of a tumor-specific T suppressor cell clone from a BALB/c mouse bearing the syngeneic ADJ-PC-5 plasmacytoma. J Immunol 135:4297–4302
4. Kloke O, Haubeck H-D, Kölsch E (1986) Evidence for a T suppressor cell-inducing antigenic determinant shared by ADJ-PC-5 plasmacytoma and syngeneic BALB/c spleen cells. Eur J Immunol 16:659–664
5. Grooten J, Leroux-Roels G, Fiers W (1987) Specific suppression elicited by EL4 lymphoma cells in syngeneic mice. Specifity includes self-antigens. Eur J Immunol 17:605

Role of the Colony-Stimulating Factor-1 Receptor (c-*fms*) and Its Ligand in Oncogenesis

C.W. Rettenmier

A. Introduction

An increasing body of experimental evidence supports the concept that malignant cell transformation results from mutations causing structural alteration or inappropriate expression of products that serve critical control functions in normal growth and development. Among the genes for growth regulatory molecules that might be targets for such genetic damage are those encoding polypeptide growth factors or their cell surface receptors. The macrophage colony-stimulating factor CSF-1 is a glycosylated polypeptide homodimer that stimulates the proliferation, differentiation, and survival of mononuclear phagocytes [1]. CSF-1 also appears to enhance the specialized immune effector functions of terminally differentiated monocytes and macrophages. The receptor for CSF-1 is the c-*fms* proto-oncogene product [2], an integral transmembrane glycoprotein with intrinsic protein-tyrosine kinase activity. The receptor is oriented in the plasma membrane with its ligand-binding domain exposed at the cell surface and its tyrosine kinase domain in the cytoplasm. Binding of CSF-1 activates the receptor kinase, which in turn initiates intracellular signals leading to the transcription of genes that effect the mitogenic response.

The current availability of the cloned genes for both CSF-1 and its receptor has facilitated studies on their function in normal hematopoiesis and oncogenesis. The transforming potential of structurally altered CSF-1 receptors has been established, and complementary DNAs encoding the receptor and ligand have been introduced into a variety of cell types. The long-term goal of these investigations is to define the normal function of the growth factor and its receptor and to determine whether alterations in their expression contribute to human malignancy. These studies provide the basis for a possible role of the CSF-1 receptor and its ligand in myeloid leukemogenesis.

Department of Tumor Cell Biology, St. Jude Children's Research Hospital, 332 North Lauderdale, Memphis, Tennessee 38105, USA

Structural Alterations Constitutively Activate the CSF-1 Receptor

The product of the feline retroviral oncogene v-*fms* retains the general domain structure of its c-*fms*-encoded progenitor: an aminoterminal ligand-binding portion linked by a single membrane-spanning segment to a carboxylterminal tyrosine kinase. However, differences in the amino acid sequence of the two molecules enable the v-*fms* gene product to function as a constitutive kinase and generate growth-promoting signals in the absence of the ligand. Although the v-*fms* product includes the complete extracellular domain and is able to bind CSF-1, its higher basal level of tyrosine kinase activity is only slightly increased by the ligand. Moreover, when introduced into murine hematopoietic cell lines that require specific growth factors for proliferation and survival in culture, the v-*fms* gene induced factor-independent growth and tumorigenicity by a nonautocrine mechanism [3, 4]. The ability to transform a CSF-1-dependent macrophage cell line and interleukin-3-dependent myeloid cells did not result from autocrine production of the corresponding growth factors or transmodulation of their receptors by the v-*fms* gene product. Thus, the v-*fms* product is constitutively activated and can provide proliferative signals without binding the CSF-1 growth factor.

Introduction of v-*fms* into mouse bone marrow cells used to reconstitute lethally irradiated recipients resulted in proliferative disorders of multiple hematopoietic lineages [5]. The spleens of several primary recipients developed provirus-positive clones, which in some cases gave rise to clonal erythroleukemias or B-cell lymphomas when transplanted into secondary lethally irradiated hosts. Cells containing v-*fms* did not always have a proliferative advantage during serial transfers, suggesting that the development of these disorders was a multistep process in which expression of v-*fms* served as an initiating event. After a latency period, several other primary recipients developed myeloproliferative disorders which had clinical features reminiscent of chronic myelogenous leukemia but lacked evidence of clonality and were not efficiently transplanted to secondary hosts. Since diseases were observed in multiple hematopoietic lineages, the activity of this oncogene is not restricted to cells that ordinarily express CSF-1 receptors.

The v-*fms* gene efficiently transforms mouse NIH-3T3 fibroblasts to form colonies in semisolid medium and rendering them tumorigenic. Transduction of the human c-*fms* gene alone did not transform NIH-3T3 cells but did confer a CSF-1 responsive growth phenotype [6]. Cells expressing the normal CSF-1 receptor formed colonies in soft agar when plated in the presence of recombinant human CSF-1 and were morphologically transformed when cotransfected with a human CSF-1 cDNA. Thus, critical structural differences between the v-*fms* and c-*fms* gene products account for the transforming activity of the viral oncogene in the absence of the ligand.

The v-*fms* gene product differs from that of c-*fms* by several scattered amino acid changes and a carboxylterminal alteration that replaces 40 amino acids of the normal receptor with 11 unrelated residues in the oncogene product [7, 8]. The critical aspect of the C-terminal alteration appears to be the removal of a tyrosine residue at position 969, four amino acids from the receptor carboxylterminus. Chimeric v-*fms*/c-*fms* constructs demonstrated that this residue serves a negative regulatory function, possibly by limiting the activity of the receptor kinase in response to CSF-1. Substitution of the normal c-*fms*-encoded C-terminus for that of v-*fms* reduced the transforming efficiency of the oncogene more than tenfold [6, 9]. This inhibition was abrogated when the chimeric construct was prepared with a c-*fms* mutant encoding a phenylalanine residue in place of the wild-type tyrosine at position 969 [6]. Similarly, when NIH-3T3 fibroblasts were transformed by cotransfection of human c-*fms* and CSF-1 cDNAs, the efficiency of transformation was severalfold higher using a c-*fms* gene encoding the phenylalanine at position 969. However, the removal of tyrosine 969 by itself is insufficient to activate the transforming potential of the c-*fms* gene [6]. Rather, there appear to be one or more additional mutations in the v-*fms* gene that contribute to its oncogenic properties.

A chimeric construct, in which a major portion of the extracellular domain encoded by the v-*fms* gene was replaced by the corresponding region of the human c-*fms* gene, was markedly reduced in its transforming efficiency in the absence of CSF-1 [10]. However, like the normal c-*fms* gene product, the chimeric receptor efficiently induced transformation of NIH-3T3 cells when coexpressed with the ligand. Thus, despite the presence of the complete v-*fms*-coded kinase domain including its carboxylterminal truncation, the c-*fms*/v-*fms* chimera was to a large degree regulated by CSF-1. The simplest interpretation of these results is that the exchanged portion of the extracellular domain contains one or more alterations critical to activation of the v-*fms* gene. Since the mutation(s) maps to the ligand-binding domain of the receptor, it may induce a conformational change that simulates the effect of binding CSF-1 and thus constitutively activates the receptor kinase. Site-directed mutagenesis in the c-*fms* gene should allow the identification of the specific alteration(s) in the extracellular domain that unmasks the latent transforming potential of the CSF-1 receptor.

These studies demonstrate that genetic alterations are able to activate the normal CSF-1 receptor to an oncogenic protein that transforms a variety of cells including those that do not ordinarily express this receptor. Efficient activation probably involves two events, one of which removes a negative regulatory tyrosine residue near the receptor carboxylterminus and a second that appears to mimic a ligand-induced conformational change.

C. Aberrant Expression of CSF-1 and Its Receptor

Colony-stimulating factor-1 was first shown to be synthesized by mesenchymal cells, including stromal cells of the bone marrow, and to interact with its receptor on mononuclear phagocytes [1]. Based on subsequent studies, the growth factor may also function in placental development during embryogenesis. CSF-1 is produced by uterine glandular epithelial cells in response to estrogens and progesterone during pregnancy [11], and its receptor appears to be expressed on placental trophoblasts [12]. The concentration of uterine CSF-1 increases during pregnancy with the highest levels being detected at term. It thus seems likely that the growth factor plays a role in the formation and maintenance of the placenta.

Coexpression of human CSF-1 and its receptor transforms immortalized NIH-3T3 fibroblasts by an autocrine mechanism [6]. Similarly, rearrangement and expression of the murine CSF-1 gene has been implicated as a secondary transforming event in a CSF-1-dependent murine macrophage cell line immortalized by the c-*myc* gene [13]. The autocrine transforming activity of CSF-1 and its receptor may be dependent on the type of cell in which they are expressed. For example, introduction of a human CSF-1 gene into a factor-dependent murine macrophage cell line resulted in CSF-1 independence but not tumorigenicity [14]. Since the v-*fms* oncogene fully transforms this same cell line, there must be critical differences between the normal CSF-1-mediated response and that induced by a constitutively activated receptor in the absence of ligand. It has recently been demonstrated that peripheral blood monocytes are induced to express CSF-1 in response to phorbol esters, inflammatory mediators such as γ-interferon and tumor necrosis factor, and other cytokines such as granulocyte/macrophage CSF [15, 16]. These results raise the possibility that regulated production of CSF-1 may stimulate the function of terminally differentiated mononuclear phagocytes by autocrine or paracrine mechanisms.

Expression of the CSF-1 receptor within the hematopoietic compartment is normally restricted to mononuclear phagocytes. Synthesis of this receptor is one of the earliest markers of monocytic differentiation, and low numbers of CSF-1 receptors are present on committed bone marrow progenitors. The receptor number increases about tenfold during normal monocytic differentiation, and high numbers (ca. 50000/cell) are maintained on mature monocytes and macrophages. In the presence of interleukin-3 or hemopoietin-1 (the macrophage product, IL-1α), even more primitive bone marrow precursors are rendered responsive to the growth factor [17, 18]. Aberrant expression of the CSF-1 receptor in cells that do not ordinarily express c-*fms* might enable them to be stimulated by the ligand and confer a proliferative advantage.

The murine c-*fms* locus has been identified as a preferred integration site for a retrovirus that causes myeloblastic leukemia [19]. Friend murine leukemia virus (F-MuLV) induces this disease in mice with a latency of 6–12 months. High levels of expression of the complete CSF-1 receptor coding sequence in more than 20% of these cases is a consequence of proviral integration upstream of the first coding exon in the c-*fms* gene. These cells might be unusually sensitive to the ligand due to the high levels of CSF-1 receptor synthesis. A proliferative advantage could develop into frank leukemia by the accumulation of additional genetic alterations. Loss of the germ-line c-*fms* allele in some of the cases raises the possibility that these secondary transforming events might include mutations in the receptor gene itself that activate its transforming potential. Whatever the mechanism, expression of the c-*fms* gene as a consequence of proviral insertion in the system identifies this as an important initiating event in the leukemogenic process.

Recent surveys of human acute myelogenous leukemic (AML) cells have provided evidence for expression of CSF-1 or its receptor in some of these cases. Screening of AML blasts by in situ hybridization for mRNA revealed CSF-1 or c-*fms* transcripts in about one-half of the cases and expression of both mRNAs in cells from several patients [20]. A survey of human AML blasts by flow cytometry with monoclonal antibodies to the CSF-1 receptor demonstrated that approximately 30% of the pediatric cases expressed the c-*fms* gene product (R.A. Ashmun et al., manuscript submitted). CSF-1 receptors on the leukemic cells underwent downmodulation in response to the growth factor or phorbol esters, suggesting that they were functionally unaltered. As might be expected by the normally restricted distribution of CSF-1 receptors to mononuclear phagocytes, the highest percentage of positive cases was among leukemic cells that showed evidence of monocytic differentiation. However, in both studies c-*fms* expression was also observed in some AML cases that were undifferentiated or exhibited granulocytic differentiation. As in the F-MuLV-induced murine myeloblastic leukemia, aberrant CSF-1 receptor expression in early myeloid cells might be associated with the development of disease in these patients.

D. Discussion

Future investigations into a role of CSF-1 and its receptor in leukemogenesis must include studies on the regulation of their expression at the transcriptional and post-transcriptional levels. The soluble growth factor is derived by proteolytic cleavage of membrane-bound precursors, some of which are stably expressed at the cell surface [21]. Factors that influence the cleavage of CSF-1 precursors may represent an additional mechanism for regulation of CSF-1 function. In the human AML blasts that expressed CSF-1 transcripts, detectable quantities of the growth factor were secreted only when the cells were incubated with phorbol esters [20].

The long arm of human chromosome 2 contains the genes for several hematopoietic growth factors and cell surface receptors including CSF-1 at 5q33.1 and c-*fms* at 5q33.3–33.4. Acquired interstitial deletions of 5q are associated with a characteristic form of refractory anemia ("$5q^-$ syndrome") and AML that develops after exposure to toxic chemicals. A detailed molecular analysis of this portion of chromosome 5 will be required to determine whether genetic lesions contributing a human cancer involve either the coding or regulatory sequences of genes in this region.

References

1. Stanley ER, Guilbert LJ, Tushinski RJ, Bartelmez SH (1983) CSF-1 – a mononuclear phagocyte lineage-specific hemopoietic growth factor. J Cell Biochem 21:151–159

2. Sherr CJ, Rettenmier CW, Sacca R, Roussel MF, Look AT, Stanley ER (1985) The c-*fms* proto-oncogene product is related to the receptor for the mononuclear phagocyte growth factor, CSF-1. Cell 41:665–676
3. Wheeler EF, Rettenmier CW, Look AT, Sherr CJ (1986) The v-*fms* oncogene induces factor independence and tumorigenicity in CSF-1 dependent macrophage cell line. Nature 324:377–380
4. Wheeler EF, Askew D, May S, Ihle JN, Sherr CJ (1987) The v-*fms* oncogene induces factor-independent growth and transformation of the interleukin-3-dependent myeloid cell line FDC-P1. Mol Cell Biol 7:1673–1680
5. Heard JM, Roussel MF, Rettenmier CW, Sherr CJ (1987b) Multilineage hematopoietic disorders induced by transplantation of bone marrow cells expressing the v-*fms* oncogene. Cell 51:663–673
6. Roussel MF, Dull TJ, Rettenmier CW, Ralph P, Ullrich A, Sherr CJ (1987) Transforming potential of the c-*fms* proto-oncogene (CSF-1 receptor). Nature 325:549–552
7. Hampe A, Gobet M, Sherr CJ, Galibert F (1984) The nucleotide sequence of the feline retroviral oncogene v-*fms* shows unexpected homology with oncogenes encoding tyrosine-specific protein kinases. Proc Natl Acad Sci USA 81:85–89
8. Coussens L, Van Beveren C, Smith D, Chen E, Mitchell RL, Isacke CM, Verma IM, Ullrich A (1986) Structural alteration of viral homologue of receptor protooncogene *fms* at carboxyl terminus. Nature 320:277–280
9. Browning PJ, Bunn HF, Cline A, Shuman M, Nienhuis AW (1986) "Replacement" of COOH-terminal truncation of v-*fms* with c-*fms* sequences markedly reduces transformation potential. Proc Natl Acad Sci USA 83:7800–7804
10. Roussel MF, Downing JR, Ashmun RA, Rettenmier CW, Sherr CJ (1988) CSF-1 mediated regulation of a chimeric c-*fms*/v-*fms* receptor containing the v-*fms*-coded tyrosine kinase domain. Proc Natl Acad Sci USA 85:5903–5907
11. Pollard JW, Bartocci A, Arceci R, Orlofsky A, Ladner MB, Stanley ER (1987) Apparent role of the macrophage growth factor, CSF-1, in placental development. Nature 330:484–486
12. Rettenmier CW, Sacca R, Furman WL, Roussel MF, Holt JT, Nienhuis AW, Stanley ER, Sherr CJ (1986) Expression of the human c-*fms* protooncogene product (colony-stimulating factor-1 receptor) on peripheral blood mononuclear cells and choriocarcinoma cell lines. J Clin Invest 77:1740–1746
13. Baumbach WR, Stanley ER, Cole MD (1987) Induction of clonal monocyte-macrophage tumors in vivo by a mouse c-*myc* retrovirus: rearrangement of the CSF-1 gene as a secondary transforming event. Mol Cell Biol 7:664–671
14. Roussel MF, Rettenmier CW, Sherr CJ (1988) Introduction of a human colony stimulating factor-1 gene into a mouse macrophage cell line induces CSF-1 independence but not tumorigenicity. Blood 71:1218–1225
15. Horiguchi J, Warren MK, Kufe D (1987) Expression of the macrophage-specific colony-stimulating factor in human monocytes treated with granulocyte-macrophage colony-stimulating factor. Blood 69:1259–1261
16. Rambaldi A, Young DC, Griffin JD (1987) Expression of the M-CSF (CSF-1) gene by human monocytes. Blood 69:1409–1413
17. Stanley ER, Bartocci A, Patinkin D, Rosendaal M, Bradley TR (1986) Regulation of very primitive, multipotent, hematopoietic cells by hemopoietin-1. Cell 45:667–674
18. Mochizuki DY, Eisenman JR, Conlon PJ, Larsen AD, Tushinski RJ (1987) Interleukin 1 regulates hematopoietic activity, a role previously ascribed to hemopoietin 1. Proc Natl Acad Sci USA 84:5267–5271
19. Gisselbrecht S, Fichelson S, Sola B, Bordereaux D, Hampe A, André C, Galibert F, Tambourin P (1987) Frequent c-*fms* activation by proviral insertion in mouse myeloblastic leukaemias. Nature 329:259–261
20. Rambaldi A, Wakamiya N, Vallenga E, Horiguchi J, Warren MK, Kufe D, Griffin JD (1987) Expression of the macrophage colony stimulating factor and c-*fms* genes in human acute myeloblastic leukemia cells. J Clin Invest 81:1030–1035
21. Rettenmier CW, Roussel MF, Ashmun RA, Ralph P, Price K, Sherr CJ (1987) Synthesis of membrane-bound colony-stimulating factor-1 (CSF-1) and downmodulation of CSF-1 receptors in NIH 3T3 cells transformed by cotransfection of the human CSF-1 and c-*fms* (CSF-1 receptor) genes. Mol Cell Biol 7:2378–2387

Activation of CD4-Positive T Cells by Polysaccharide Fractions Isolated from the Cupressaceae Thuja occidentalis L. (Arborvitae) *

Sven H. Gohla [1], Hans-Dieter Haubeck [2], Silke Schrum [1], Helga Soltau [1], and Rolf D. Neth [1]

Various mitogens have been described which act on different subpopulations of mouse and/or human peripheral lymphocytes. In recent years it has been mainly Japanese and German research teams that have isolated polysaccharide fractions with immunomodulative properties [1–3, 17]. In addition to fungal polysaccharides such as lentinan, schizophyllan, and PSK polysaccharide, which are basically in the experimental step of clinical use [6, 10, 13], the focus of attention has mainly been on plants belonging to the asteracean family, such as *Echinacea purpurea* and *E. angustifolia,* where an increase in granulocyte phagocytosis is observable. *Thuja occidentalis* L., a Cupressaceae, has hardly been investigated up till now.

Several authors have demonstrated that allopathic extracts of this plant could be strong antiviral agents directed against plant and animal viruses [12, 3]. Polysaccharide fractions with molecular weights ranging between 5×10^5 and 1×10^6 and higher have been isolated from aqueous alkaline extracts of the herbal parts of *Thuja occidentalis* by ethanol precipitation and ultrafiltration using a Satorius Ultra Sart filtration cell [5].

A high molecular weight subfraction of *Thuja* polysaccharides (TPSg) proved to be highly mitogenic in peripheral blood leukocytes (PBL) (Figs. 1, 2). It was shown using alkalic phosphatase anti-alcalic phosphatase (APAAP) [7, 8, 9] and Pappenheim staining methods that the mitogenic and cluster-forming activity of TPSg causes T-cell induction rather than induction of B-cells [11]. In detail, more than 90% of all TPSg-induced blasts were shown to be Tpan-positive (T3, T11 and IOT1a marker); in contrast, Bpan-positive (CD22 marker) lymphoid cell induction occured in less than 4%. In particular, TPSg was identified as a potent T-cell mitogen which acts preferentially on the Okt4/Okt17-positive [14] T-cells (Fig. 3). As well as induction of the Okt4/Okt17 T-cell fraction, TPSg-induced generation of Okt16-positive immature T-cells/null cells was observed, whereas only a little stimulation of Okt8-positive T-cells was seen (Fig. 4). More than 75% of all TPSg-induced blast cells were shown to be Okt4/Okt17-positive, whereas less than 5% of all blasts generated were Okt8-positive. About 20%–25% of all TPSg-induced blasts were shown to be immature T-cells or null cells.

The Okt4/Okt17-positive T-helper/inducer cell generation is connected with increased production of interleukin-2. Furthermore, TPSg-dependent enhanced expression of the interleukin-2 receptor (75%) on the TPSg-triggered cells has been observed [15]. The cluster-forming ability and mitogenity of TPSg correlates well with [^{3}H]thymidine uptake and seems to be interleukin-1 and interferon-γ dependent, as was shown by blocking

[1] Department of Clinical Chemistry, Medical University Clinic, UKE, Martinistraße 52, D-2000 Hamburg, FRG
[2] Department of Clinical Chemistry, Medical University Clinic, Pauwelsstraße, D-5100 Aachen, FRG
* This study has been supported by the Karl- und Veronika Carstens-Stiftung im Stifterverband für die Deutsche Wissenschaft

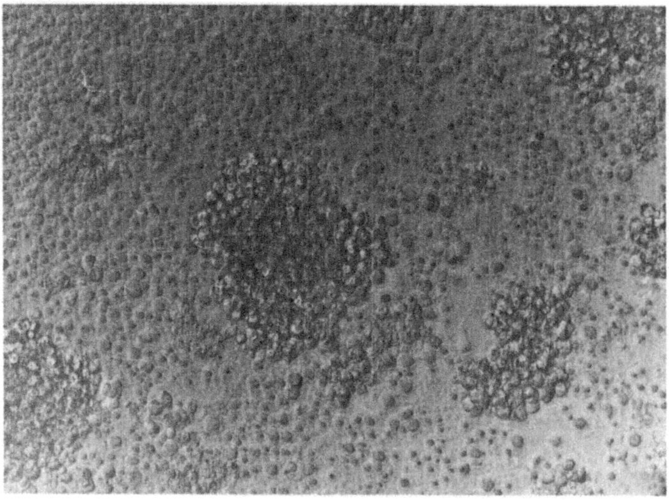

Fig. 1. Phase contrast micrograph of a 4-day-old culture of TPSg-triggered peripheral blood leukocytes. Magnification 550 ×. Peripheral blood leukocytes (PBL) were obtained by means of density gradient centrifugation (Ficoll-Hypaque, Pharmacia). The "buffy coat" was harvested after having been centrifugated for 20 min and washed 5 times with phosphate buffered saline pH 7.2. 5×10^5 cells were seeded out per well in a flat bottomed 96-well microtiter plate with a final volume of 100 µl Dulbeccois modified Eagle's medium (DMEM) +10% pooled human blood group AB serum. The cells were incubated with 100 µl/well of a TPSg solution in DMEM +10% pooled human AB serum with a final concentration of 1 mg/ml of TPSg

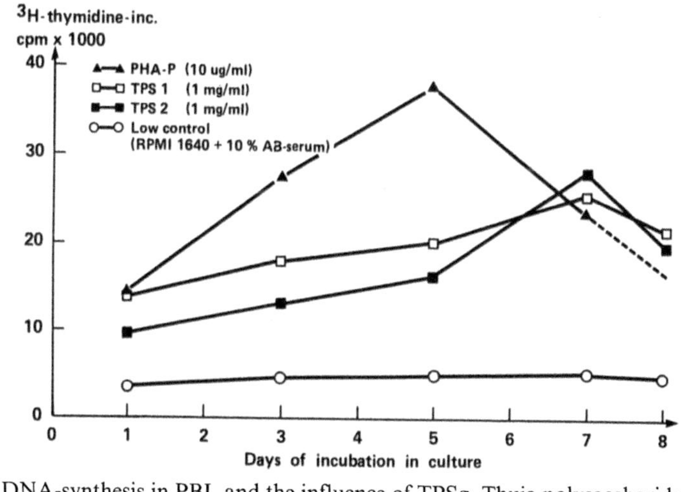

Fig. 2. Kinetics of the DNA-synthesis in PBL and the influence of TPSg. Thuja polysaccharide fraction. 5×10^5 PBL, resuspended in RPMI 1640, supplement with 6% fetal calf serum (FCS) and 4% human AB serum, were incubated at a final concentration of 1 mg/ml per well in a flat bottomed, microtiter plate. On days 0, 2, 4, 6, and 8, 0.5 µCi/50 µl RPMI 1640 was added for the last 12 h of incubation. Afterwards the cells were harvested (Scatron cell harvester) and the DNA synthesis rate was measured in a β-liquid scintillation counter. Supplemented medium was used as low control and Phytohemagglutinin. PHA-P, final concentration 10 µg/ml, was used as high control. Both controls were incubated under the same conditions as the TPSg cultures. All cultures were performed in triplicate. The results shown are the mean values of two experiments

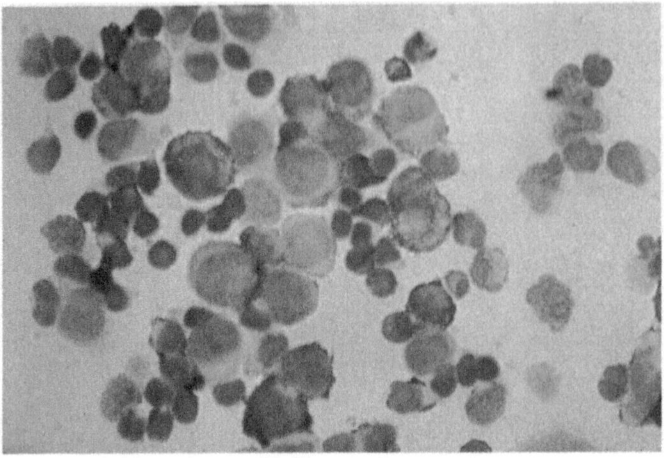

Fig. 3. Anti-Okt4 alkaline phosphatase anti-alkaline phosphatase (APAAP) staining of TPSg-triggered peripheral blood leukocytes. Magnification 630 ×. Air-dried cells were stained by the APAAP method (Erber et al.). The cells were fixed for 1 min with an acetone, methanol and formaldehyde (95:95:10) fixative. The following steps were carried out at room temperature (25 °C) in a humidified chamber. The first CD4+ surface antigen specific antibody (Dakko, cat. no. M716, Denmark) was incubated for 30 min. Subsequent washing was done using TBS (Tris-buffered saline, pH 7.6) unless otherwise indicated. After washing with TBS (1 min), the second incubation step with rabbit anti-mouse (RaM) bridge antibodies (Dakko, cat. no. Z259) was carried out for another 30 min. After washing, the third incubation with APAAP complex (Dakko, cat no. D651) was carried out for another 30 min. The quality of the staining can be improved if all incubation steps are done twice. The "fast red" method was used as a detection system. The sample was washed after APAAP incubation and was incubated with the substrate [dimethylformamide (200 µl); levamisole (30 µl); "fast red" (10 mg); 0.1 *M* Tris, pH 8.2; naphthol-AS-MX-phosphate (2 mg)] for a period of 30 min. After washing counterstaining with hematoxylin was done (1 min). All slides were placed in Apathy's mountian medium, Highman's modification

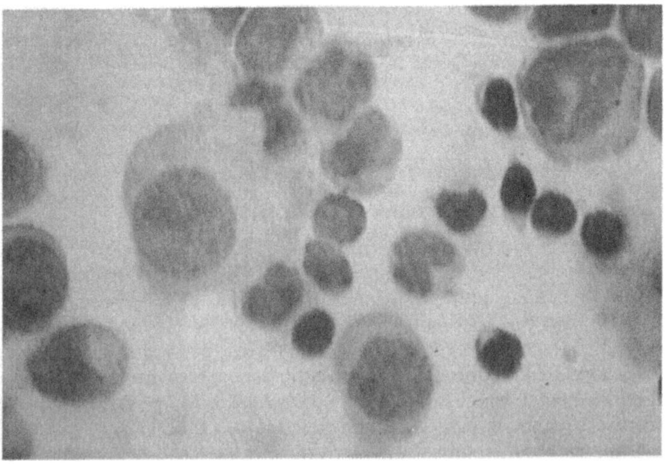

Fig. 4. Anti-T-suppressor/cytotoxic staining (OKT8 marker). Magnification 630 ×. For details of methods used see Fig. 3

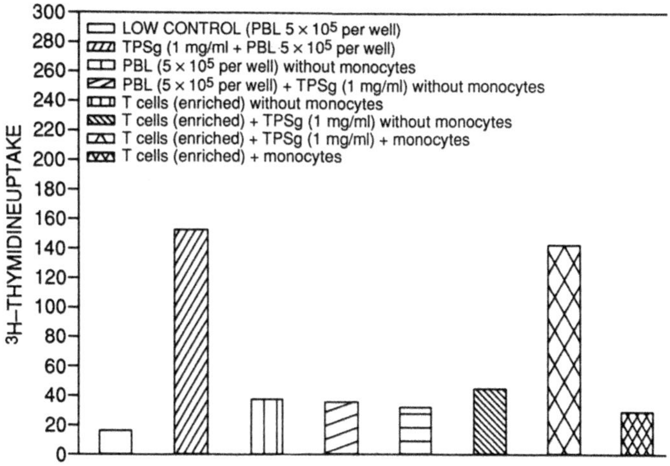

Fig. 5. The influence of TPSg and enriched T-lymphocytes in the presence or absence of peripheral blood monocytes. The peripheral blood monocytes were separated from the lymphocyte fraction by the monocyte adherence method described elsewhere and incubated separately under standard conditions. 5×10^5 PBL and 5×10^5 enriched T-cells were incubated with TPSg medium, final concentration 1 mg/ml, or with non-lectin-supplemented complete medium. The T-cell enrichement was performed according to the standard procedure using nylon wool as separation column. The T-cell fraction was obtained by rinsing the nylon wool 2 times with cold PBS. It could be shown that TPSg only works as a mitogen on PBL and enriched T-cell fractions in the presence of autologous monocytes. All results are mean values of two independent experiments. All concentrations per experiment were determined in triplicate

the mitogenic effect with interleukin-1- and interferon-γ-specific antibodies. It was also shown that Okt4-positive T-cell induction depends on the presence of autologous monocytes/macrophages (Fig. 5). Whether it is possible to use this polysaccharide fraction as an adjuvant in the therapy of immune deficiency syndromes and cancer must now be further investigated.

References

1. Beuscher N (1982) Über die medikamentöse Beeinflussung zellulärer und humoraler Resistenzmechanismen im Tierversuch. Arzneimittelforschung 32:134 ff
2. Beuscher N (1982) Über die medikamentöse Beeinflussung zellulärer und humoraler Resistenzmechanismen im Tierversuch. III. Steigerung der Leukozytenmobilisation bei der Maus durch pflanzliche Reizkörper. Arzneimittelforschung 30:821 ff
3. Beuscher N, Kopanski W (1986) Purification and biological characterization of antiviral substances from *Thuja occidentalis* L. In: 37th Annual congress on medical plant research, Hamburg, 22–27 Sept. 1986, Abstract of short lectures and poster presentation, Planta Medica, Thieme Verlag, p 75
4. Beuscher N, Beuscher H, Otto B, Schäfer B (1977) Über die medikamentöse Beeinflussung zellulärer und humoraler Resistenzmechanismen im Tierversuch. II. In vitro Untersuchungen an Peritoneal-Leukozyten und Seren der Ratte. Arzneimittelforschung 27:1655 ff
5. Caldes G, Prescott B, Thomas II CA, Bahr PJ (1981) Characterization of a polysaccharide from *Carthamus tinctorius* that cross reacts with type III pneumococcae polysaccharide. J Gen Appl Microbiol 27:157 ff
6. Chihara G, Maeda J, Hamuro J, Tahuma S, Fumiko F (1969) Inhibition of mouse sarcoma 180 by polysaccharides from *Lentinus etodes* (Berk.) Sing. Nature 222:687 ff
7. Cordell JL, Falini B, Erber WN (1984) Immunoenzymatic labeling of monoclon-

al antibodies using immune complexes of alkaline phospatase and monoclonal anti-alkaline phosphatase (APAAP complexes). J Histochem Cytochem 32:219 ff
8. Erber WN, Pinching AJ, Mason DY (1984) Immunocytochemical detection of T and B cell populations in routine blood smears. Lancet I:1042 ff
9. Erber WN, Mynheer LC, Mason DY (1986) APAAP labelling of blood and bone-marrow samples of phenotyping leukemia. Lancet I:761 ff
10. Fujimoto S, Takahashi M, Minami T, Ishigami H, Miyazaki M, Itoh K (1979) Clinical value of immunotherapy with OK432 or PS-K for stomach cancer patients. Jpn J Surg 9:190 ff
11. Gohla S, Haubeck H-D, Neth RD (1988) Mitogenic activity of high molecular polysaccharide fractions from the plant "Thuja occidentale L.". I. Monocyte-dependent induction of $CD4^+$ T-helper cells. Leukemia 2:528–533
12. Khurana PSM (1971) Effect of homoeopathic drugs on plant viruses. Planta Medica 20:142 ff
13. Maeda JY, Chihara G (1971) Lentinan, a new immuno-accelerator of cell-mediated responses. Nature 229:634 ff
14. Thomas Y, Rogozinski L, Rothman P, Rabbani LE, Andrews S, Irigoyen OH, Chess L (1982) Further dissection of the functional heterogeneity within the $OKT4^+$ and $OKT8^+$ human t-cell subsets. J Clin Immunol 2:85 ff
15. Uchiyama T, Nelson DL, Fleidher TA, Waldmann TA (1981) A monoclonal antibody (anti-TAC) reactive with activated and functionally mature human T-cells. J Immunol 126:1398–1404
16. Vollmar A, Schäfer W, Wagner H (1986) Immunologically active polysaccharides of *Eupatorium cannabinum* and *Eupatorium perfoliatum*. Phytochemistry 25:377 ff

Induction of Nonspecific Cell-Mediated Cytotoxicity: A Multisignal Event and its Cellular Regulation*

J. Atzpodien [1], C. Bührer [2], S.C. Gulati [3], D. Wisniewski [3], S. Öz [2], H. Kirchner [1], T. Benter [1], H. Poliwoda [1], K. Welte [2], and B. Clarkson [3]

A. Introduction

Nonspecific cytotoxic cells provide a major line of defense against tumor. While natural killer (NK) cells display spontaneous, non-MHC-restricted killing activity, both NK and non-NK lymphocytes can be induced by lymphokines to exhibit enhanced nonspecific cytotoxicity against tumor, including NK-resistant targets [1–3]. The latter activity is mediated by a heterogeneous cell population commonly termed lymphokine-activated killer (LAK) cells [2–4]. According to the initial concept, nonspecific LAK killing is generated via exposure to interleukin 2 (IL-2) of peripheral blood lymphocytes [5, 6]. Most of the LAK activity appears to be mediated by NK cells stimulated with IL-2; however, recent studies suggest that induction of MHC-unrestricted lymphokine-activated killing is a more complex phenomenon requiring a multitude of cellular and noncellular signals [2–4, 7–10].

In our study, we approached the LAK phenomenon by asking the following questions: (a) what are the cell types mediating LAK precursor and/or effector function? (b) What are the signals needed for the induction of nonspecific killing? Do different LAK precursor populations require different induction signals? (c) What is the regulatory role of various mononuclear cell subsets in the induction of nonspecific cytotoxicity? In order to answer these questions, we isolated and depleted NK cells, T lymphocytes, and monocytes, respectively, from fresh human peripheral blood. By culturing these cell populations under various conditions, we could define the role of both cellular and noncellular signals for the induction of MHC-unrestricted killing in human peripheral blood.

B. Methods

Production and characterization of monoclonal antibodies (mAb) used in this study have been described elsewhere [11–18]. NK cells, T-lymphocytes, and monocytes were selectively removed and purified from peripheral blood mononuclear cells (PBMC) using techniques that have been detailed previously [19–21]. In brief, for selective removal of NK cells and T cells, PBMC were preincubated with mAbs N901, and OKT3 plus H65, respectively. Reactive cells were then depleted by immunoadherence to plastic plates previously coated with affinity-purified goat anti-mouse IgG. NK-depleted fractions were additionally treated with the lysomotropic agent L-leucinemethylester (LeuOMe) [22]. Monocytes were removed from fresh PBMC by adherence to plastic petri dishes. After col-

[1] Department of Hematology and Oncology, Medizinische Hochschule Hannover, D-3000 Hannover 61, FRG
[2] Department of Pediatric Hematology and Oncology, Medizinische Hochschule Hannover, D-3000 Hannover 61, FRG
[3] Laboratory of Hematopietic Cell Kinetics, and Hematology/Lymphoma Service, Memorial Sloan-Kettering Cancer Center, New York, NY 10021, USA
* This work was supported by Deutsche Forschungsgemeinschaft grants At 12/1-1 and Bu 623/1-1 and by National Institutes of Health grants Ca-20194, Ca-08748, and Ca-19117

lecting the nonadherent cells, adherent fractions were harvested by incubation with medium plus EDTA. Fractions of monocyte-depleted PBMC were purified for T-lymphocytes and for NK cells. For isolation of T cells, the indirect immunoadherence technique was employed using mAbs B1, B4, MY4, MY8, MO-1, and N901; nonadherent cells were subsequently treated with LeuOMe to ensure complete removal of all NK cells and monocytes. For separation of NK cells from peripheral blood, PBMC were incubated with mAbs OKT3, H65, B1, B4, MY4, and MY8; positive cells were depleted by immune rosetting with bovine red blood cells that were coated with affinity-purified goat anti-mouse IgG by $CrCl_3$ linkage [23]. For immunofluorescence analyses [24], cells were incubated separately with mAbs and purified mouse IgG controls, then stained with fluorescein isothiocyanate conjugated $F(ab)_2$ fragments of goat anti-mouse IgG and analyzed on a FACScan flow cytometer. Phenotypic analyses of isolated cell fractions revealed that all purified and depleted populations contained under 2% (i.e., no detectable) inappropriate cells. To measure nonspecific NK and LAK cytotoxicity, a standard 4-h ^{51}Cr-release assay [25] was performed using two MHC class I negative tumor cell lines (K562, Daudi) [26, 27] as targets.

C. Results

I. NK Cells as Nonspecific Precursor and Effector Cells

As shown in Fig. 1, fresh thoroughly purified NK cells lysed approximately 50% of the NK-sensitive K562 cell line, as compared to 0% lysis of NK-resistant, LAK-sensitive Daudi targets (50:1 E:T ratio). Following 5-day stimulation with recombinant IL-2 (1000 U/ml), cytotoxicity mediated by isolated NK cells was measured at 100% against Daudi and 16% against K562 (50:1 E:T ratio). When compared to unfractionated mononuclear cells, fresh unstimulated NK cells exhibited an approximately tenfold higher non-specific killing activity, while day 5 IL-2 induced cytotoxicity was measured at equal levels in both purified NK and unseparated PBMC cultures. Thus, stringent isolation of fresh NK cells did not result in a detectable enrichment for nonspecific lymphokine-activated killing as defined by day 5 Daudi cell lysis. However, rigorous depletion of NK cells from PBMC prior to IL-2 induction completely abrogated spontaneous as well as IL-2 activated non-MHC-restricted cytotoxicity (Fig. 2).

II. T-Lymphocytes as LAK Precursor and Effector Cells

Figure 3 A gives the phenotypic analyses of highly purified T-lymphocytes that were initially depleted of NK cells and grown in short-term culture in the presence of IL-2 or IL-2 plus mitogen (PHA-M). While IL-2 by itself was unable to induce the generation of nonspecific cytotoxic cells, costimulation of IL-2 induced CD3 lymphocytes via the T3/Ti receptor antigen resulted in the generation of T-lymphocytes that expressed NK-related surface molecules (N901, H25) [17, 18] and concomitantly acquired both NK and LAK cytotoxicity (Fig. 3). As shown in Fig. 2B, when using unseparated peripheral blood mononuclear cells, a stringent removal of T cells on day 0 resulted in a partial, though significant, depletion of IL-2 induced nonspecific cytolysis of Daudi. This suggested that a proportion of N901-negative T-lymphocytes represents a precursor population to MHC-unrestricted lymphokine-activated killer cells. While mediating LAK precursor functions, $CD3^+$ T cells were also demonstrated to account for approximately 30%–50% of IL-2 induced MHC-unrestricted cytotoxicity generated in unfractionated cultures comprising NK cells, T-lymphocytes, and monocytes; evidence supporting this was provided in cell depletion experiments whereby NK cells

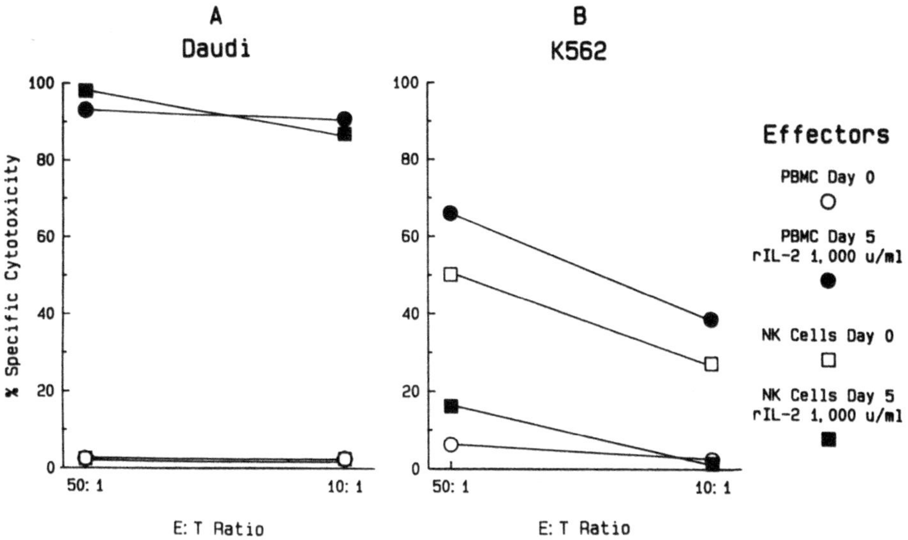

Fig. 1 A, B. Comparison of cytotoxic activities against Daudi (**A**) and K562 (**B**) mediated by unseparated PBMC before and after 5-day IL-2 activation, or by purified NK cells before and after 5-day stimulation with IL-2

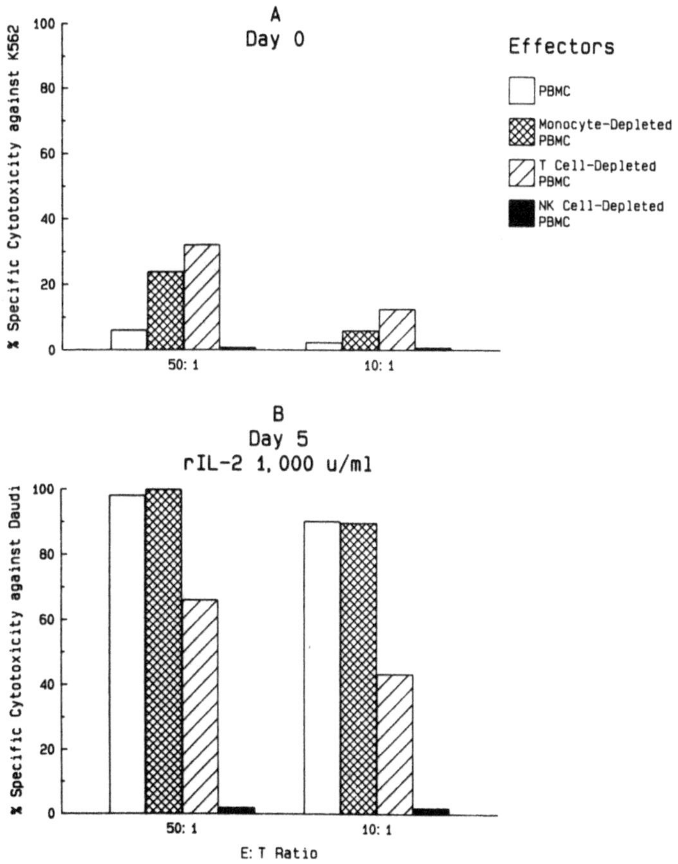

Fig. 2 A, B. Cytotoxicity before (**A**) and after (**B**) 5-day IL-2 incubation mediated by unseparated PBMC and PBMC that were depleted on day 0 of NK cells, T cells, or monocytes

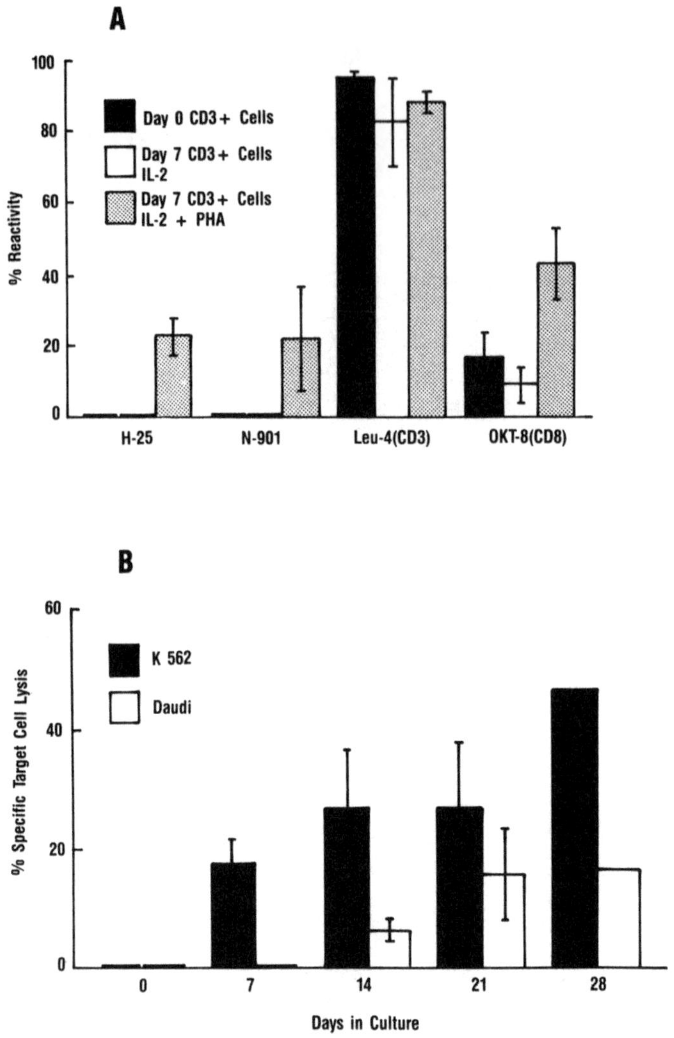

Fig. 3 A, B. Phenotypic (**A**) and functional (**B**) analyses of purified CD3$^+$ T-lymphocytes following induction with IL-2 in the absence (**A**) or presence (**A, B**) of PHA mitogen

and monocytes were stringently removed from day 5 IL-2 induced cultures (Fig. 4).

III. Monocytes as Regulators of Nonspecific Cytotoxicity

PBMC, initially depleted of all detectable monocytes, showed an increased relative cytotoxicity against K562 while lymphokine-activated killing of Daudi was essentially unchanged (Fig. 2). In short-term culture with IL-2, monocytes did not exhibit any nonspecific cytotoxicity against tumor cells. Nonetheless, proliferation of MHC-unrestricted killer cells was augmented significantly when purified NK cells were cultured with IL-2 in the presence of autologous monocytes (Table 1). Consistently, the cumulative nonspecific cytolytic activity induced in NK cultures in response to IL-2 was increased approximately twofold with autologous monocytes present (Fig. 5).

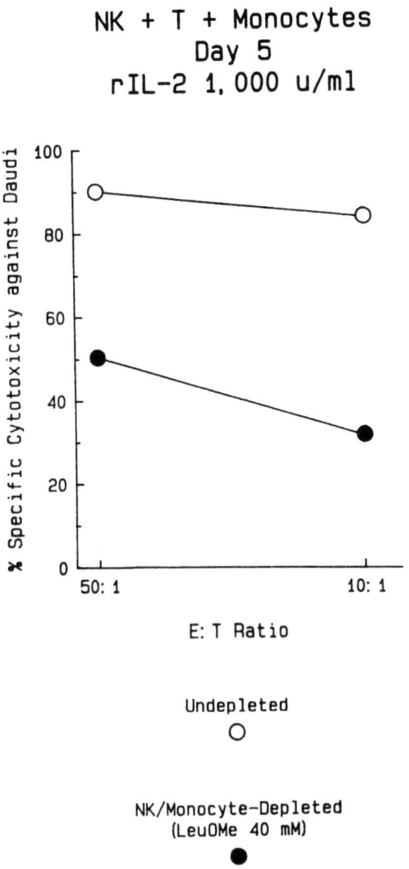

Fig. 4. IL-2 induced non-MHC-restricted cytotoxicity mediated by day 5 undepleted, and NK cell plus monocyte depleted effector populations that were generated in purified NK cultures in the presence of equal numbers of autologous T cells and monocytes

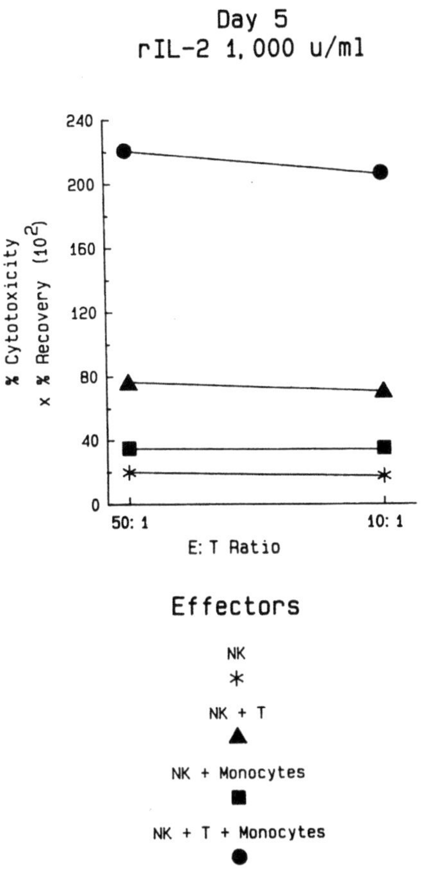

Fig. 5. Cumulative lymphokine-activated cytotoxicity (percent cytolysis × percent recovery) generated in IL-2 cultures of highly purified NK cells that were incubated alone, or in the presence of T cells, monocytes, or T cells plus monocytes

IV. T Cells and Monocytes as Coregulators of MHC-Unrestricted Lymphokine-Activated Killing

When added to NK cells on day 0, autologous T-lymphocytes could induce an approximately fourfold increase in the absolute number of nonspecific effector cells generated upon induction with IL-2 (Table 1). While percent cytotoxicity against Daudi was essentially unchanged (data not shown), addition of T-lymphocytes to freshly isolated NK cells resulted in a significant (i.e., more than threefold) enhancement of cumulative non-MHC-restricted killing (percent cytotoxicity × percent recovery) following 5-day IL-2 activation (Fig. 5). As shown in Table 1, costimulation of NK cells, T cells, and monocytes with IL-2 had an even more pronounced effect on the expansion of cytolytic effectors in response to cytokine. Compared to highly purified NK cells alone, approximately 15- and

Table 1. Proliferation and cell recovery after 5-day incubation with rIL-2 of highly purified subpopulations isolated from human peripheral blood

Peripheral blood subpopulation	[^{3}H]-Thymidine incorporation [a] (cpm; mean ± 1 SD)	Cell recovery [b] (%)
NK cells	914 ± 301	20
T cells	211 ± 107	68
Monocytes	565 ± 162	10
NK cells plus T cells [c]	1482 ± 316	80
NK cells plus monocytes	3162 ± 802	35
T cells plus monocytes	3466 ± 860	60
NK cells plus T cells plus monocytes	14041 ± 2718	245

[a] 10^4 cells each were pulsed with [^{3}H]-thymidine for 4 h.
[b] Cell recovery as compared to day 0 was assessed by trypan blue dye exclusion test.
[c] After purification, the various subpopulations were admixed at equal ratios prior to rIL-2 culture.

12-fold higher levels of proliferation and cell recovery, respectively, were measured in NK cultures grown in the presence of equal numbers of T cells plus monocytes (Table 1). Thus, autologous T-lymphocytes and monocytes, when added to fresh NK cells, were able to induce a dramatic (i.e., more than tenfold) increase in total nonspecific cytolytic activity generated upon 5-day stimulation with IL-2.

D. Discussion

Both NK and LAK cells are functionally defined by their non-MHC-restricted cytotoxicity against tumor [1–3]. In the present study, the main findings about the induction of nonspecific cell-mediated cytotoxicity in human peripheral blood are the following: (a) NK cells (N901$^+$ CD3$^-$) and T-lymphocytes (CD3$^+$ N901$^-$) can function separately as precursor cells to MHC-unrestricted LAK cells; (b) non-NK (N901$^-$) T cells can give rise to a subset of NK-like (N901$^+$) T-lymphocytes that coexpress NK (N901) and T-cell (CD3) specific antigens and concomitantly acquire NK and LAK cytotoxicity in response to IL-2; (c) while in NK cells, IL-2 provides a sufficient signal for the induction of nonspecific lymphokine-activated cytotoxicity, additional activation via the T3/Ti receptor antigen is needed in order to induce non-MHC-restricted LAK activity in a proportion of mature N901$^-$ T-lymphocytes; (d) T cells do not only mediate but also regulate LAK precursor and effector functions whereby cumulative nonspecific cytolysis (percent cytotoxicity × recovery) is increased more than threefold in IL-2 induced NK cultures grown in the presence of CD3$^+$ cells; (e) monocytes, when coincubated with NK cells, augment the cumulative LAK activity generated in NK cultures upon stimulation with IL-2; and (f) monocytes and T cells strongly synergize in enhancing the generation and activation of NK-derived MHC-unrestricted killer cells, thus producing a dramatic (i.e., more than tenfold) increase in the total nonspecific killing induced by IL-2.

While the ontogeny of IL-2 activated killer cells continues to be controversial [2, 3, 8–10], there is pertinent evidence that the phenomenon of LAK cells defines a cellular function rather than a distinct cell type [2–4, 6]. In this study, we show that a variety of different mononuclear cell subsets contribute to the generation and activation of non-MHC-restricted IL-2 induced killer cells. Based on the observation of various cellular and noncellular signals for the induction in human peripheral blood of nonspecific cell-mediated cytotoxicity, we propose that activation of unrestricted cytolytic cells constitutes a multisignal event tightly regulated at the cellular level. Our data indicate that application of IL-2 has both direct and indirect effects, whereby NK cells, T-lymphocytes, and monocytes may be induced to synergize in generating nonspecific killing activity in a subset of cells presenting with NK or NK/T cell phenotype. As to the ontogeny of NK and LAK cells, the present results also suggest that a proportion of NK-like peripheral blood mononuclear cells can be derived from mature T-lymphocytes.

Acknowledgements. We are indebted to Dr. Robert Knowles, New York, for providing mAb H65, and to Dr. James Griffin, Boston, for the supply of mAbs B1, B4, MY4, MY8, MO-1, and N901.

References

1. Ortaldo JR, Reynolds CW (1987) Natural killer activity: definition of a function rather than a cell type. J Immunol 138:4545
2. Phillips JH, Lanier LL (1986) Dissection of the lymphokine activated killer phenomenon. J Exp Med 164:814–825
3. Ortaldo JR, Mason A, Overton R (1986) Lymphokine-activated killer cells. Analysis of progenitors and effectors. J Exp Med 164:1193–1205
4. Lotzová E, Herberman RB (1987) Reassessment of LAK phenomenology: a review. Nat Immun Cell Growth Regul 6:109–115
5. Lotze MT, Grimm EA, Mazumder A, Strausser JL, Rosenberg SA (1981) Lysis of fresh and cultured autologous tumor by human lymphocytes cultured in T-cell growth factor. Cancer Res 41:4420–4425
6. Grimm EA, Mazumder A, Zhang HZ, Rosenberg SA (1982) Lymphokine-activated killer cell phenomenon: lysis of natural killer resistant fresh solid tumor cells by interleukin-2 activated autologous human peripheral blood lymphocytes. J Exp Med 155:1823–1841
7. Talmadge JE, Wiltrout RH, Counts DF, Herberman RB, McDonald T, Ortaldo JR (1986) Proliferation of human peripheral blood lymphocytes induced by recombinant human interleukin-2: contribution of large granular lymphocytes and T lymphocytes. Cell Immunol 102:261–272
8. Roberts K, Lotze MT, Rosenberg SA (1987) Separation and functional studies of the human lymphokine-activated killer cell. Cancer Res 57:4366–4371
9. McMannis JD, Fisher RI, Creekmore SP, Braun DP, Harris JE, Ellis TM (1988) In vivo effects of recombinant IL-2. I. Isolation of circulating Leu-19$^+$ lymphokine-activated killer effector cells from cancer patients receiving recombinant IL-2. J Immunol 140:1335–1340
10. Grossman Z, Herberman RB (1986) Natural killer cells and their relationship to T-cells: hypothesis on the role of T-cell receptor gene rearrangement on the course of adaptive differentiation. Cancer Res 46:2651–2658
11. Knowles RW (1986) Immunochemical analysis of the T cell-specific antigens. In: Reinherz ER, Haynes BF, Nadler LM, Bernstein ID (eds) Human T lymphocytes. Springer, Berlin Heidelberg New York, pp 259–287
12. Kung PC, Goldstein G, Reinherz EL, Schlossman SF (1979) Monoclonal antibodies defining distinctive human T cell surface antigens. Science 206:347–349
13. Stashenko P, Nadler LM, Hardy R, Schlossman SF (1980) Characterization of a human B lymphocyte-specific antigen. J Immunol 125:1678–1685
14. Nadler LM, Anderson KL, Marti G, Bates M, Park E, Daley JF, Schlossman SF (1983) B4, a human B lymphocyte-associated antigen expressed on normal, mitogen-activated, and malignant B lymphocytes. J Immunol 131:244–250
15. Griffin JD, Ritz J, Nadler LM, Schlossman SF (1981) Expression of myeloid differentiation antigens on normal and malignant myeloid cells. J Clin Invest 68:932–941
16. Todd RF, Nadler LM, Schlossman SF (1981) Antigens on human monocytes by monoclonal antibodies. J Immunol 126:1435–1442
17. Griffin JD, Hercend T, Beveridge R, Schlossman SF (1983) Characterization of

an antigen expressed by human natural killer cells. J Immunol 130:2947–2951
18. Bai Y, Beverly PCL, Knowles RW, Bodmer WF (1983) Two monoclonal antibodies identifying a subset of human peripheral blood mononuclear cells with NK and K cell activity. Eur J Immunol 13:521–527
19. Wysocki LJ, Sato VL (1978) Panning for lymphocytes: a method for cell selection. Proc Natl Acad Sci USA 75:2844–2848
20. Wisniewski D, Strife A, Wachter M, Clarkson B (1985) Regulation of human peripheral blood erythroid burst-forming unit growth by T lymphocytes and T lymphocyte subpopulations defined by OKT4 and OKT8 monoclonal antibodies. Blood 65:456–463
21. Atzpodien J, Gulati SC, Shimazaki C, Bührer C, Öz S, Kwon JH, Kolitz JE, Clarkson BD (1988) Ewing's sarcoma: ex-vivo sensitivity towards natural (NK) and lymphokine-activated (LAK) killing. Oncology 45:437–443
22. Thiele DL, Lipsky PE (1986) Leu-leu-ome sensitivity of human activated killer cells: delineation of a distinct class of cytotoxic T lymphocytes capable of lysing tumor targets. J Immunol 137:1399–1406
23. Goding JM (1976) The chromic chloride method of coupling antigens to erythrocyte: definition of some important parameters. J Immunol Methods 10:61–66
24. Herzenberg LA, Herzenberg LA (1977) Analysis and separation using the fluorescence activated cell sorten (FACS) In: Weir EM (ed) Handbook of experimental immunology. Blackwell, Oxford, pp 22.1–22.21
25. Pross HF, Maroun JA (1984) The standardization of NK cell assays for use in studies of biological response modifiers. J Immunol Methods 68:235–249
26. Lozzio CB, Lozzio BB (1975) Human chronic myelogenous leukemia cell line with positive Philadelphia chromosome. Blood 45:321–334
27. Klein E, Klein G, Nadkarni JS, Nadkarni JJ, Wigzell H, Clifford P (1968) Surface IgM-kappa specificity on a Burkitt lymphoma cell in vivo and in derived culture lines. Cancer Res 28:1300–1310

Signal Transduction Mechanisms in Human Natural Killer Cells Mediating Antitumor Immunity

P.J. Leibson[1,2], K.P. Windebank[1,2], T.J. Barna[1], and R.T. Abraham[3]

A. Introduction

The spontaneous cytotoxicity of normal human peripheral blood mononuclear cells (PBMCs) is mediated by a subset of lymphocytes called natural killer (NK) cells. Unlike other cytotoxic lymphocytes, which require prior sensitization by the specific antigen to which they are programmed to respond, single NK cells are directly cytotoxic to a wide variety of malignant cells without prior exposure. Neither the NK cell receptor responsible for activation during this interaction nor its target cell cognate have been identified biochemically. NK cells can also be activated for lysis by the Fc region of IgG antibodies which have coated a malignant target cell (antibody-dependent cellular cytotoxicity, ADCC). In contrast to the situation in direct cytotoxicity, the NK cell surface Fc receptor (CD16) has been well characterized and can be identified by monoclonal antibodies (mAbs) such as 3G8 and Leu-11.

In this study, we used cloned, human NK cell lines to characterize the intracellular signal transduction pathways that are used during NK cell activation. NK cells activated by direct binding of sensitive tumor cells or by Fc receptor ligation by anti-CD16 antibody demonstrated a rapid increase in phosphoinositide hydrolysis. Moreover, this response was modulated in a heterologous manner by the cAMP second messenger pathway, a system known to exert a significant regulatory action on NK cell cytolytic activity.

Mayo Clinic and Foundation, Departments of Immunology[1], Pediatrics[2] and Pharmacology[3], Rochester, MN 55905, USA

B. Materials and Methods

I. Isolation, Passage, and Characterization of NK Cell Lines

Adherent cell-depleted human PMBCs were stained with fluoresceinated anti-Leu-11a (CD16) and sorted on a FACS IV cell sorter. Using a modification of the procedure by van de Griend et al. [1], the separated populations (>98% $CD16^+$ upon reanalysis) were plated in limiting dilution (0.3–3.0 cells/well) with irradiated (4000 R) autologous PBMCs, irradiated (10000 R) allogeneic EBV-transformed cells, human recombinant IL-2 (20 units/ml), and 20% human sera. After 7 days, fresh IL-2 (10 units/ml) was added. On day 13, the replicate wells were scored for cell proliferation and clonal cell lines were selected based on the Poisson distribution. The cell lines were passaged weekly and their phenotype was monitored using fluorescent antibodies and flow cytometry.

II. Cytotoxicity Assay

The ^{51}Cr-release assay was performed using a procedure previously described [2]. Results are expressed as lytic units/10^6 cells, where 1 lytic unit is the number of cells required to give 20% specific chromium release [3]. Lytic units were calculated using computer software generously provided by Hugh F. Pross (Queens University, Kingston, Ontario).

III. Measurement of Inositol Phosphates

Inositol phosphate generation was evaluated using a modification [4] of the procedure previously described [5]. Briefly, NK cells were prelabeled with myo[^{3}H]inositol and then incubated with either target cells (E/T ratio = 1/1) or monoclonal antibodies (1 μg/10^6 cells) in media containing 10 m*M* lithium chloride, an inhibitor of inositol-1-phosphatase. After various time intervals (1–60 min), the reactions were terminated by the addition of methanol:chloroform:HCl (200:100:2). After the addition of chloroform and water, the aqueous phase was collected, added to a 60 m*M* sodium formate-5 m*M* sodium tetraborate buffer, and loaded onto AG1-X8 anion-exchange columns. IP1 and IP2 were eluted from the column with 0.5 *M* ammonium formate – 100 m*M* formic acid. The remaining IP3 or, alternatively, the total inositol phosphates were eluted with 1.2 m*M* ammonium formate – 100 m*M* formic acid. [^{3}H]-labeled inositol phosphates were quantitated by liquid scintillation counting.

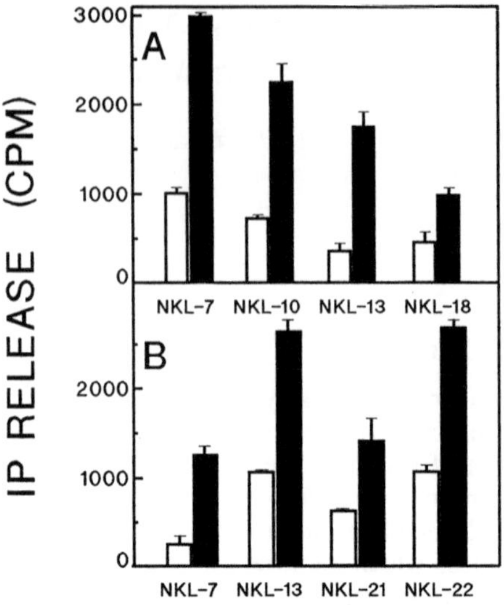

Fig. 1 A, B. Inositol phosphates are generated during binding of NK-sensitive targets (**A**) or during Fc receptor ligation (**B**). Different human CD16$^+$/CD3$^-$ NK cell lines were prelabeled with [^{3}H]myoinositol and incubated for 30 min with medium alone (*open bars* in **A**), K562 leukemic cells (*solid bars* in **A**), control antibody MKD6 (*open bars* in **B**), or with anti-CD16 antibody 3G8 (*solid bars* in **B**). Inositol phosphates were then extracted, partitioned by anion-exchange chromatography, and quantitated by liquid scintillation counting

C. Results and Discussion

In order specifically to assess signal transduction in homogeneous populations of human NK cells, cloned CD16$^+$/CD3$^-$ cell lines were selected and characterized. When these NK cell lines were exposed to NK-sensitive tumor targets for 30 min, the level of inositol phosphates rose two to five times above background (Fig. 1 A). To determine whether the target cell-induced inositol phosphate generation correlated with sensitivity to NK cell-mediated lysis, a panel of tumor target cell lines were comparatively evaluated in both inositol phosphate release and cytotoxicity assays (Fig. 2). There was complete concordance between target cell sensitivity to lysis and stimulation of phosphoinositide hydrolysis.

Since ADCC is generated by the crosslinking of Fc receptors on the NK cell surface, we also evaluated the direct effect of Fc receptor ligation using anti-Fc receptor antibodies (3G8). Incubation of the [^{3}H]inositol-labeled NK cell lines with 3G8 mAb consistently increased total inositol phosphate levels two to five times over those present in unstimulated cells (Fig. 1 B). Kinetics experiments demonstrated that stimulation by either susceptible target cells or by Fc receptor ligation led to rapid (1 min) generation of the Ca^{2+}-mobilizing second messenger inositol trisphosphate (IP3), with slower accumulation of inositol bisphosphate (IP2) and inositol monophosphate (IP1).

Previous studies have demonstrated that activation of the cyclic AMP-dependent second messenger pathway strongly inhibits NK cell-mediated cytotoxic functions. Treatment of NK effector cells with forskolin to elevate intracellular cAMP levels resulted in a concentration-dependent inhibition of phosphoinositide hydrolysis induced by both NK-sen-

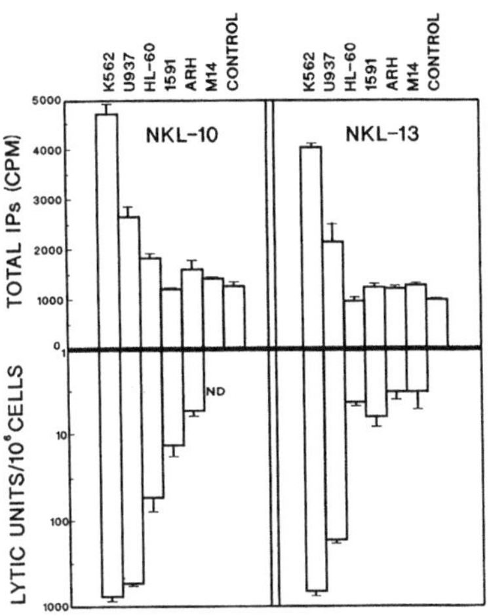

Fig. 2. Phosphoinositide metabolism is differentially stimulated by targets of varying sensitivity to NK lysis. Two human $CD16^+/CD3^-$ NK cell lines were incubated for 30 min with medium alone or with six different tumor targets. In parallel, the sensitivity of each tumor to NK cell-mediated lysis was measured in a 4-h ^{51}Cr-release assay

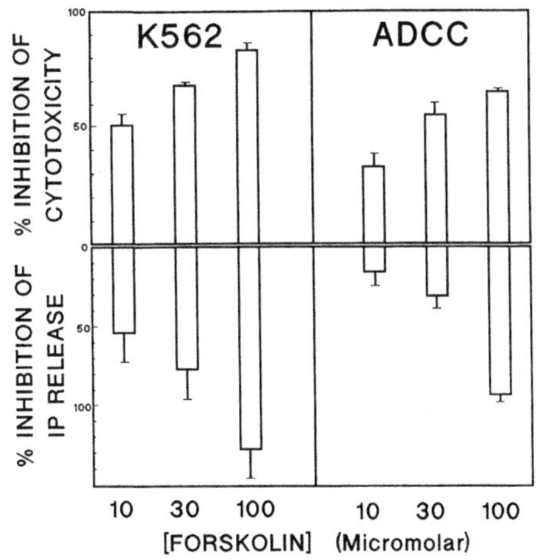

Fig. 3. Coordinate inhibition of NK cytotoxicity and inositol phosphate formation by intracellular cAMP elevations. The human NK cell line NKL-22 was preincubated for 10 min with various concentrations of forskolin and then assayed for direct and antibody-mediated cytotoxicity (*upper panels*). In parallel, forskolin-treated NK cells were tested for inositol phosphate release after stimulation by K562 cells or after Fc receptor ligation by antibody 3G8 (*lower panels*)

sitive targets and 3G8-mediated Fc receptor ligation (Fig. 3).

These results suggest that phosphoinositide turnover represents a critical early event in the human NK cell cytolytic process. Moreover, the potent inhibitory effect of cAMP on NK cell cytotoxicity may be explained by the uncoupling of NK receptors from phospholipase C-mediated phosphoinositide hydrolysis.

References

1. Van de Griend RJ, Van Krimpen BA, Bol SJ, Thompson A, Bolhuis RL (1984) Rapid expansion of human cytotoxic T cell clones: growth promotion by a heat-labile serum component and by various types of feeder cells. J Immunol Methods 56:285–298
2. Leibson PJ, Hunter-Laszlo M, Douvas GS, Hayward AR (1986) Impaired neonatal natural killer cell activity to herpes simplex virus: decreased inhibition of viral replication and altered response to lymphokines. J Clin Immunol 6:216–224
3. Pross HF, Bain MG, Rubin P, Sharagee P, Patterson MS (1981) Spontaneous human lymphocyte-mediated cytotoxicity against tumor target cells. IX. The quantitation of natural killer cell activity. J Clin Immunol 1:51
4. Abraham RT, Ho SN, Barna TJ, McKean DJ (1987) Transmembrane signaling during interleukin-1 dependent T cell activation. J Biol Chem 262:2719
5. Bijsterbasch MK, Meade CJ, Turner GA, Klaus GG (1985) B lymphocyte receptors and phosphoinositide degradation. Cell 41:999

Tumor-Specific Antigens and Tumor-Specific Mutant Proteins in Mouse and Man *

H. Schreiber, H. Koeppen, and P.L. Ward

A. Introduction

The modern era of cancer immunology began with the discovery that inbred mice could be immunized against cancers that had been induced by chemical carcinogens such as the polycyclic hydrocarbon methylcholanthrene (MCA) [1–4]. Particularly, studies of Prehn and Main in 1957 made it highly unlikely that the antigens on the cancers were also widely expressed on normal tissue. It was shown that normal tissue of the host from which the tumor had been isolated did not immunize the recipient to reject the tumor challenge; furthermore, mice immunized against the tumor still accepted normal skin grafts from the mouse of tumor origin. Thus, these antigens were seemingly tumor specific. Another important aim of the experiments using MCA-induced murine tumors was the search for antigens that were tumor specific as well as shared among different independently induced cancers. The identification of such antigens would allow the same antigen to be utilized for the therapy and diagnosis of different types of cancers occurring in different individuals. The existence of such antigens would have great significance in medical praxis. However, very extensive transplantation experiments showed that the tumor-specific rejection antigens on these cancers were unique, i.e., individually specific for a particular tumor even when compared to other tumors of the same histologic type induced in the same organ system with the same carcinogen in supposedly genetically identical mice. In fact, careful studies searching for cross-reaction among ten tumors expressing unique antigens showed no repeatable protective immunity except when immunization and challenge involved the same tumor [5]. Thus, it appears from these studies that the antigenic repertoire is, in fact, very large. Tumors induced with other chemical and physical carcinogens and even spontaneous cancers also display unique (individually specific) antigens that can elicit tumor rejection [6–11]. Finally, a single cancer cell may display multiple independent unique antigens, so that the diversity of unique antigens may be greater than previously anticipated [12].

B. Genetic Origin of Murine Unique Tumor Antigens

The seemingly endless diversity of unique tumor antigens on experimentally induced and spontaneous cancers has stimulated the interest of many immunologists. Burnet, for example, postulated that the unique antigens might be the result of clonal expansion of single cells expressing the particular (preexistent) antigen [13]. This situation would be similar to the idiotype of B- and T-cell malignancies that are individually distinct and are immunogenic in the host of origin [14, 15]. The nonmalignant clone carrying the idiotype is, under normal circumstances, present in too low a frequency to be detected by the immune system or the

Department of Pathology, University of Chichago, 5841 S. Maryland Ave., Chicago, IL 60637, USA

* This work was supported by grants CA-19266, CA-22677, and CA-37156 from the National Institutes of Health

scientist trying to prove the absolute restriction of the antigen to malignant cells. Burnet suggested that gene families known to allow enormous antigenic diversity, such as the receptors for antigens on T and B cells or MHC class I antigens, could represent the genes encoding tumor antigens [13]. In fact, certain experiments pointed at the possibility that immunoglobulin genes or MHC class I genes can encode unique tumor-specific rejection antigens [16]. The question of whether amplification of preexistent normal clonal antigens is the basis for the uniqueness of tumor-specific antigens has been addressed experimentally. In two such studies, a nonmalignant fibroblast line was cloned, then expanded, and subclones were malignantly transformed [17, 18]. Immunological studies indicated that all had individually distinct antigens even though all tumors had been derived from the same precursor cell. At face value, these experiments seem to indicate that the appearance of the antigens followed the carcinogen exposure and that these are, therefore, new antigens or neoantigens that were not previously expressed on the precursor cell. However, normal cells can generate considerable diversity of surface molecules during clonal expansion from a single precursor [19, 20], and the transformation event caused by the carcinogen may simply fix a particular antigenic phenotype [21]. Alternatively, it is possible that normal previously nonexpressed genes are randomly activated by the carcinogen [22]. Obviously both mechanisms could produce considerable antigenic diversity with apparent tumor specificity even though these antigens are expressed on normal cells. Sometimes only restricted populations of normal cells express these antigens, so the fact that they are not tumor specific may be difficult to recognize since the appropriate control cells expressing this antigen may not have been tested [22]. Together, the previous experiments cannot prove the possibility that the so-called tumor-specific antigens are tumor specific in the strictest sense since they might be encoded by normal genes and even be expressed on an unrecognized normal cell population.

C. Are Unique Tumor Antigens Encoded by Tumor-Specific Mutations?

Since most, if not all, carcinogens are mutagens, it appears quite logical to hypothesize that tumor-specific antigens may commonly arise from tumor-specific mutations of structural genes. The extreme uniqueness of transplantation antigens induced by chemical carcinogens would be consistent with the fact that mutagenic chemicals randomly affect genes. However, to date there is no genetic evidence that a cancer-specific mutation and not normal genes encoded in the germline encode unique tumor antigens. Recent work in animal tumors led to the development of cytotoxic T-lymphocyte (CTL) and antibody probes that can be used to unravel the genetic origins of unique tumor antigens. However, there are serious questions whether previously isolated tumors can be used for a meaningful genetic analysis of the origin of unique antigens, since none of the previously generated tumors were isolated along with nonmalignant control cells and DNA. Without such controls one cannot prove that a particular abnormal gene was not already present in normal DNA of the host in which the tumor originated. This is particularly relevant since subtle germline mutations, residual heterozygosity, contaminations of the strain of tumor origin during breeding [23] would easily be distinguished from tumor-specific mutations if autochthonous normal DNA was available for each tumor analyzed [16]. Our laboratory has previously used ultraviolet light (UV)-induced murine skin tumors as an experimental model to study the host's immune responses against a cancer [24, 25]. These tumors often exhibit such a strong immunogenicity that they are rejected by syngeneic animals. We recently generated a new series of UV-induced tumors [32]; these tumors were isolated with all necessary controls, such as cells and DNA from normal tissues of each tumor-bearing animal. This material should enable us to unravel the genetic origin of unique tumor antigens and finally answer the question of whether these antigens are tumor specific in the strictest sense, in that they are encoded by tumor-specific genes not present in normal somatic cells of the host of tumor origin.

D. Do Tumor-Specific Mutant Proteins Encode Tumor-Specific Antigens?

It must be expected that chemical and physical carcinogens mutate intracellular as well as surface proteins. Many, or most, of these mutations are probably a disadvantage to the cell and are, therefore, selected against during the clonal evolution of cancer [16]. In contrast, specific mutational changes that favor the malignant process would be retained. An example is a highly selected point mutation caused by the chemical carcinogen nitrosomethylurea in the cellular *ras* oncogenes [26]. This mutation favors malignant growth and is, therefore, found regularly in certain tumors, such as mammary tumors induced by this carcinogen. Other examples of mutations leading to fusion of exons between distinct genes that are brought together by tumor-specific translocations are found in certain types of human leukemias ([27–30], also see J.D. Rowley, this volume). Thus, fusion between the *BCR* and *ABL* genes leads to several types of fusion proteins that must clearly be expected to generate a new antigenicity. Since these fusion genes caused by the translocations are not observed in normal cells, one can assume that these genes may well encode truly tumor-specific antigens. The mutant *ras* genes, as well as the *BCR-ABL* fusion genes, encode intracellular proteins. Until recent years, it was postulated by immunologists that CTL could only recognize cell surface proteins. However, previous and recent evidence demonstrating CTL recognition of the nuclear SV40 virus T antigen and influenza virus nuclear protein made it clear that intracellular proteins are indeed recognized by CTL (for review see [31]). The explanation for this enigmatic finding is that CTL can recognize peptides of enzymatically cleaved antigens which are then "expressed" on the cell surface in association with MHC class I molecules.

E. Conclusions

Although we lack conclusive evidence, it is certainly possible that tumor-specific mutant proteins can be recognized by CTL or helper T cells as tumor-specific antigens. Interestingly, mutant genes such as *BCR-ABL* represent mutations that are shared by leukemias of the same type but independently induced in different patients. Thus, these changes represent common or shared tumor-specific mutations that may encode yet common tumor-specific antigens in man. This is important since the search for common yet tumor-specific antigens in experimental tumors has been without convincing success. At present, we do not know how regularly tumor-specific mutant proteins are found in human cancer cells, or whether they indeed encode tumor antigens that can be exploited therapeutically and diagnostically. However, it is likely that more tumor-specific mutant proteins will be discovered in human cancers in the future and that cancer development as a multistep process is probably dependent upon several rather than a single mutational event. Certainly, several of these mutations, such as the *BCR-ABL* fusion gene, may be essential for maintaining the malignant phenotype. Such mutant proteins, if they act as tumor-specific antigens would be ideal targets since the cancer cell could not escape therapy directed at this target by gene loss or down-regulation. Thus, discovery of these mutant proteins that are truly tumor specific and genetically defined needs the most serious evaluation by tumor immunologists.

References

1. Gross L (1943) Intradermal immunization of C3H mice against a sarcoma that originated in an animal of the same line. Cancer Res 3:326–333
2. Foley EJ (1953) Antigenic properties of methylcholanthrene-induced tumors in mice of the strain of origin. Cancer Res 13:835–837
3. Prehn RT, Main JM (1957) Immunity of methylcholanthrene-induced sarcomas. J NCI 18:769–778
4. Old LJ, Boyse EA, Clarke DA, Carswell EA (1962) Antigenic properties of chemically-induced tumors. Ann NY Acad Sci 101:80–106
5. Basombrio MA (1970) Search for common antigenicities among twenty-five sarcomas induced by methylcholanthrene. Cancer Res 30:2458–2462
6. Globerson A, Feldmann M (1964) Antigenic specificity of benzo(a)pyrene-induced sarcomas. J NCI 32:1229–1243
7. Pasternak G, Graffi A, Horn K-H (1964) Der Nachweis individual-specifischer Antigenität bei UV-induzierten Sarkomen der Maus. Acta Biol Med Ger 13:276–279
8. Kripke ML (1974) Antigenicity of murine skin tumors induced by ultraviolet light. J NCI 53:1333–1336
9. Vaage J (1968) Nonvirus-associated antigens in virus-induced mouse mammary tumors. Cancer Res 28:2477–2483
10. Carswell EA, Wanebo HJ, Old LJ, Boyse EA (1970) Immunogenic properties of reticulum cell sarcomas of SJL/J mice. J Natl Cancer Inst 44:1281–1288
11. Morton DL, Miller GF, Wood DA (1969) Demonstration of tumor-specific immunity against antigens unrelated to the mammary tumor virus in spontaneous mammary adenocarcinomas. J Natl Cancer Inst 42:289–301
12. Wortzel RD, Philipps C, Schreiber H (1983) Multiplicity of unique tumor-specific antigens expressed on a single malignant cell. Nature 304:165–167
13. Burnet FM (1970) A certain symmetry: histocompatibility antigens compared with immune receptors. Nature 226:123–126
14. Lynch RG, Graff RJ, Sirisinha S, Simms ES, Eisen HN (1972) Myeloma proteins as tumor-specific transplantation antigens. Proc Natl Acad Sci USA 69:1540–1544
15. Lampson LA, Levy R (1979) A role for clonal antigens in cancer diagnosis and therapy. J NCL 62:217–219
16. Schreiber H, Ward PL, Rowley DA, Stauss HJ (1988) Unique tumor-specific antigens. Annu Rev Immunol 6:465–483
17. Basombrio MA, Prehn RT (1972) Studies on the basis of diversity and time of appearance of chemically-induced tumors. NCI Monogr 35:117–124
18. Embleton MJ, Heidelberger C (1972) Antigenicity of mouse prostate transformed in vitro. Int J Cancer 9:8–18
19. Moscona AA (1974) Surface specifications of embryonic cells: lectin receptors, cell recognition, and specific cell ligands. In: Moscona AA (ed) The cell surface in development. Wiley, New York, pp 67–99
20. Hood L, Huang HV, Dreyer WJ (1977) The area-code hypothesis: the immune system provides clues to understanding the genetic and molecular basis of cell recognition during development. J Supra Str 7:531–559
21. Srivasta PK, Old LJ (1988) Individually distinct transplantation antigens of chemically induced mouse tumors. Immunol Today 9:78–83
22. Old LJ (1981) Cancer immunology: the search for specificity – G.H.A. Clowes Memorial Lecture. Cancer Res 41:361–375
23. Bailey DW (1982) How pure are inbred strains of mice? Immunol Today 3:210–214
24. Koeppen H, Rowley DA, Schreiber H (1986) Tumor-specific antigens and immunologic resistance to cancer. In: Steinman RM, North RJ (eds) Mechanisms of host resistance for infectious agents, tumors and allografts. Rockefeller University Press, New York, pp 359–386
25. Urban JL, Schreiber H (1988) Host-tumor interactions in immunosurveillance against cancer. Prog Exp Tumor Res 32:17–68
26. Sukumar S, Notario V, Martinzanca D, Barbacid M (1983) Induction of mammary carcinomas in rats by nitrosomethylurea involves malignant activation of H-*ras*-1 locus by single point mutations. Nature 306:658–661

27. Rowley JD (1973) A new consistent chromosomal abnormality in myelogenous leukemia identified by quinacrine fluorescence and Giemsa staining. Nature 243:290–293
28. deKlein A, Geurts van Kessel A, Grosveld G, Batram C, Hagemeijer A, Bootsma D, Spurr NK, Heisterkamp N, Groffen J, Stephenson JR (1982) A cellular oncogene is translocated to the Philadelphia chromosome in chronic myelocytic leukemia. Nature 300:765–767
29. Shtivelman E, Lifshitz B, Gale R, Canaani E (1985) Fused transcripts of *abl* and *bcr* genes in chronic myelogenous leukemia. Nature 315:550–554
30. Ben-Neriah Y, Daley G, Mes-hasson A, Witte O, Baltimore D (1986) The chronic myelogenous leukemia-specific p210 protein is the product of the *bcr/abl* hybrid gene. Science 233:212–214
31. Braciale TJ, Morrison LA, Sweetser MT, Sambrook J, Gething M, Braciale V (1987) Antigen presentation pathways to class I and class II MHC-restricted T lymphocytes. Immunol Rev 98:95–113
32. Ward PL, Koeppen H, Hurtean T, Schreiber H (1989) Tumor antigens defined by cloned immunological probes are highly polymorphic and are not detected on autologous normal cells. J Exp Med 170:217–232

Modification of HLA Expression as a Possible Factor in the Pathogenesis of Burkitt's Lymphoma

R. Voltz, W. Jilg, and H. Wolf

Introduction

According to the Kiel classification, Burkitt's lymphoma (BL) is a high-grade malignant non-Hodgkin's lymphoma of the B-cell lineage. In all the BL cells one finds a gene translocation which can activate the c-*myc* oncogen. In addition, 95% of the endemic-type BLs are infected with the Epstein-Barr virus (EBV). In vitro infection with EBV can transform peripheral blood B-cells into immortalized nonmalignant lymphoblastoid cell lines (LCLs). LCLs can also grow out spontaneously from peripheral blood lymphocytes of EBV-positive donors after the removal of T cells or their inhibition by cyclosporin A in vitro. In vivo EBV-infected cells are normally controlled by the immune system. Immunosuppression leads to a higher risk of EBV-associated oligoclonal lymphoma. In vitro cytotoxic chromium release assays performed with pairs of BL cells and LCLs of the same BL patient have shown that BLs are not recognized by cytotoxic T cells (CTLs) in contrast to LCLs [1]. Most of the EBV-specific CTLs are HLA class I restricted; there are, however, up to 30% of class II restricted CTLs in the EBV system. One possibility for the BL cells not being recognized is the altered surface expression of EBV-specific structures which have been functionally termed lymphocyte-detected membrane antigen (LYDMA). One of the possible candidates for LYDMA is the latent membrane protein LMP of the reading frame BNLF1, which has been shown to be differentially expressed on BL cells and LCLs [2]. The specific recognition may also be influenced by an altered HLA class I or class II expression as suggested by Torsteindottir et al. [3]. We therefore studied the HLA expression on eight pairs of BL cells and LCLs.

Max von Pettenkofer Institute, University of Munich, Pettenkoferstr. 9a, D-8000 München 2, FRG

Results

For measuring the HLA density we used a radioimmunoassay with monoclonal antibodies against a framework determinant of class I (clone W6/32) or class II (clone L243) antigen and ^{125}I-labeled protein A. To correct for differences in surface area, cells were also incubated with saturating amounts of a mixture of polyclonal antisera against BL cells and LCLs; the cpm values obtained with the monoclonal antibodies were divided by the cpm values of the polyclonal serum mixture, leading to values of relative density of class I and II antigen. Most of the pairs tested showed a significantly lower HLA class I (Fig. 1) and class II density (data not shown) on the BL cells. EBV-positive LCLs showed a much higher HLA density than peripheral blood B-cells of the same donor (data not shown). In contrast to the corresponding LCLs, both in EBV-positive and -negative BL cells the HLA expression could be stimulated by a factor 3–5 by IFN-γ (Fig. 2) and TNF (data not shown). TNF did not show a direct toxic effect against BL cells or LCLs (data not shown).

After immunoprecipitation we performed a biochemical separation of HLA class I and II antigens by one-dimension-

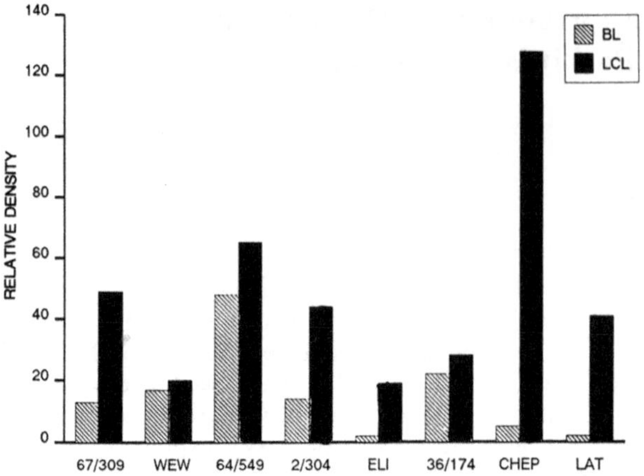

Fig. 1. Expression of class I antigens in BL cells and corresponding LCLs

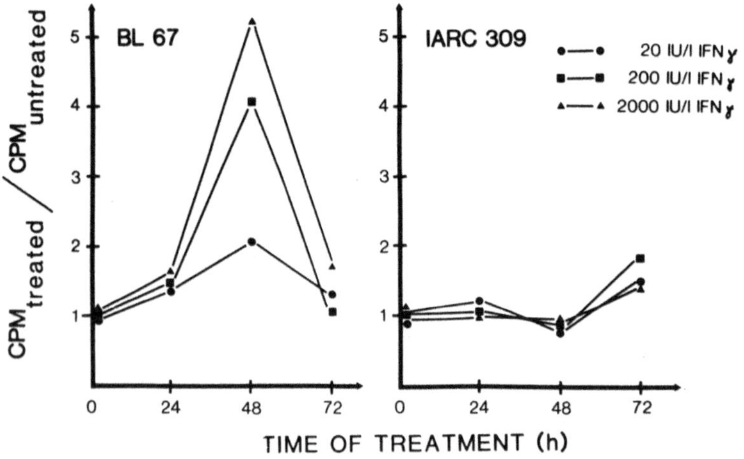

Fig. 2. HLA class I expression under stimulation with IFN-γ

al SDS-PAGE and two-dimensional electrophoresis which separates according to the isoelectric point and molecular weight. BL cells and LCLs of the same patient showed a different pattern: some of the HLA specificities expressed on LCLs were not found – or some only in traces – on the corresponding BL cells (Fig. 3). After incubation with tunicamycin preventing glycosylation of proteins, these differences were still detected, indicating that they are not due to differences in glycosylation in BL cells and LCLs (data not shown). These differences could not be affected by IFN-γ or TNF (data not shown). EBV infection itself did not alter HLA expression qualitatively. This was shown by a comparison of EBV-negative pokeweed mitogen-stimulated B-blasts and corresponding EBV-positive LCLs and a comparison of an EBV-negative BL-cell line before and after infection with EBV (data not shown).

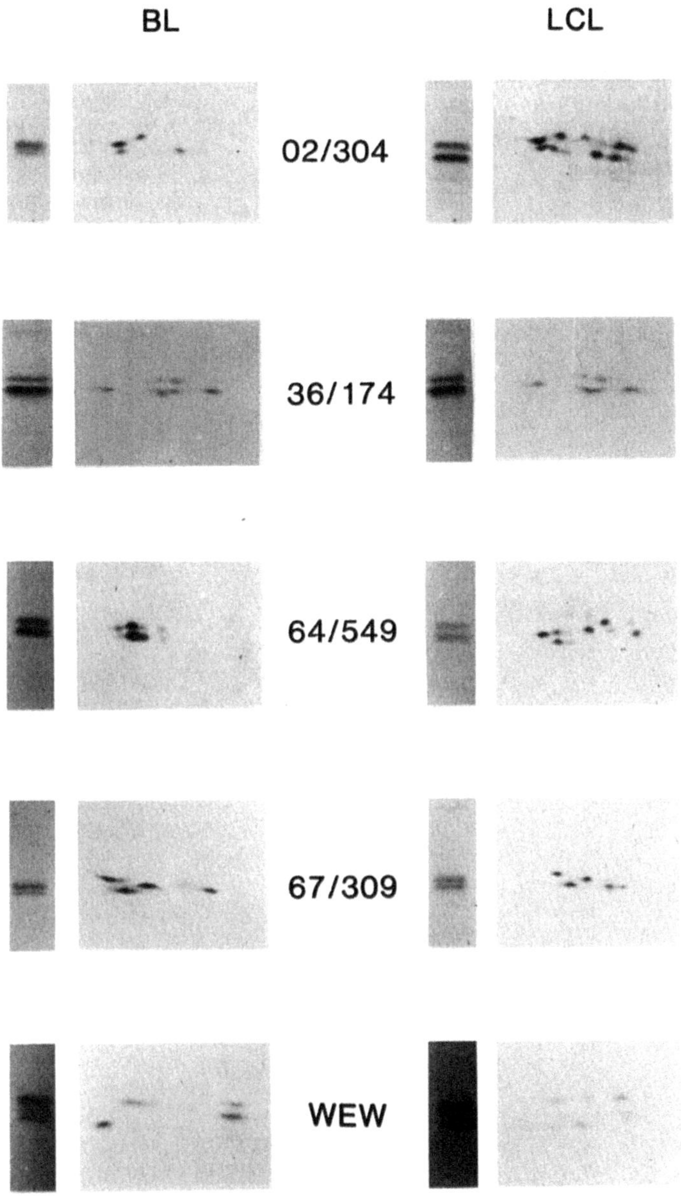

Fig. 3. SDS PAGE and two-dimensional gel electrophoresis of HLA class I immunoprecipitates of five BL/LCL pairs

Summary and Conclusions

Our data show that

- Most BL cells express significantly less HLA class I and II than the corresponding LCLs.
- Lymphoblastoid cell lines have a much higher HLA class I and II density than normal peripheral blood B cells.
- There are qualitative differences in class I and II expression in most BL/LCL pairs.
- These qualitative differences are not caused by EBV infection. They might be due to a selective downregulation of HLA specifities by c-*myc* activation.

We conclude that these modifications of HLA expression may well play a role in the reduced specific immune recognition of Burkitt's lymphoma cells.

References

1. Rooney CM, Rowe M, Wallace LE, Rickinson AB (1985) Epstein-Barr virus-positive Burkitt's lymphoma cells not recognized by virus-specific T-cell surveillance. Nature 317:629–631
2. Modrow S, Wolf H (1986) Characterization of two related Epstein-Barr virus-encoded membrane proteins that are differentially expressed in Burkitt lymphoma and in vitro-transformed cell lines. Proc Natl Acad Sci USA 85:5703–5707
3. Torsteinsdottir S, Masucci MG, Ehlin-Henriksson B, Brautbar C, Ben Bassat H, Klein G, Klein E (1986) Differentiation-dependent sensitivity of human B-cell-derived lines to major histocompatibility complex-restricted T-cell cytotoxicity. Proc Natl Acad Sci USA 83:5620–5624

Disorders of the Expression of the Multichain IL-2 Receptor in HTLV-I-Associated Adult T-Cell Leukemia

T.A. Waldmann

A. Introduction

The activation of T cells requires two sets of signals from cell surface receptors to the nucleus. The first signal is initiated when appropriately processed and presented foreign antigen interacts with the 90-kD polymorphic heterodimeric T-cell surface receptor for the specific antigen. Following the interaction of antigen presented in the context of products of the major histocompatibility locus and interleukin-1 or interleukin-6 with the antigen receptor, T cells synthesize interleukin-2 (IL-2) [1, 2]. To exert its biological effect, IL-2 must interact with specific high-affinity membrane receptors. Resting T cells do not express high-affinity IL-2 receptors, but receptors are rapidly expressed on T cells after activation with an antigen or mitogen [3, 4].

Progress in the analysis of the structure, function, and expression of the human IL-2 receptor was greatly facilitated by the production by Uchiyama et al. [5] of a monoclonal antibody (termed anti-Tac) that was shown to recognize the human IL-2 receptor [6].

We have utilized the anti-Tac monoclonal antibody and radiolabeled IL-2 in cross-linking studies to: (a) define multiple IL-2-binding peptides that participate in the human receptor for IL-2; (b) molecularly clone cDNAs for the 55-kD peptide of the human IL-2 receptor; (c) determine the immunological events that require the interaction of IL-2 with its receptor; (d) analyze disorders of IL-2 receptor expression in leukemia, especially those forms of leukemia associated with the retrovirus HTLV-I; and (e) develop protocols for the therapy of patients with IL-2 receptor-expressing adult T-cell leukemia and T-cell-mediated autoimmune disorders, and for individuals receiving organ allografts.

The Metabolism Branch, National Cancer Institute, National Institutes of Health, Bethesda, Maryland 20892, USA

B. Structure of the Multisubunit IL-2 Receptor

The high-affinity IL-2 receptor consists of multiple distinct IL-2-binding peptides. The IL-2-binding receptor peptide identified by the anti-Tac monoclonal on PHA-activated normal lymphocytes is a 55-kD glycoprotein [6]. We and others have defined a second non-Tac IL-2-binding peptide with an M_r of 68–76 kD (p75) [7, 8]. Using cross-linking methodology, we demonstrated the p75 peptide on MLA 144, a gibbon T-cell line that does not express the Tac antigen but manifests a few thousand relatively low-affinity ($K_d = 14$ nM) IL-2-binding sites per cell. The p75 peptide was also identified in addition to the Tac peptide (p55) in cell populations that express both high- and low-affinity receptors. We proposed a multichain model for the high-affinity IL-2 receptor in which an independently existing Tac or p75 peptide would represent low- and intermediate-affinity receptors, respectively, whereas high-affinity receptors would be expressed when both peptides are expressed and associated in a receptor complex [7]. To test this working hypothesis, we fused cell membranes from a low-affinity IL-2-binding cell line bearing the Tac peptide alone (MT-1) with membranes from a cell line bearing the p75 peptide alone (MLA 144) and generated hybrid membranes bearing high-affinity receptors [9]. These studies support the multichain model for the high-affinity IL-2 receptor [7].

There is evidence suggesting a more complex subunit structure that involves peptides in addition to the p55 and the p75 IL-2-binding peptides. Two monoclonal antibodies, OKT27 and OKT27b, were produced that react with distinct epitopes of a 95-kD peptide. The OKT27b antibody inconsistently coprecipitated the 55-kD Tac peptide as well as the 95-kD peptide [10]. A flow cytometric energy transfer technique was used to demonstrate a close nonrandom proximity between the p55 Tac and 95-kD T27 peptides [10]. In addition, fluorescence photobleaching recovery measurements suggest that the Tac and T27 peptides physically interact in situ in HUT 102 membranes [11]. In independent chemical cross-linking studies with radiolabeled IL-2, Herrmann and Diamantstein [12] and Saragovi and Malek [13] presented evidence for an independent 100–115-kD IL-2-binding peptide in mice associated with the p55- and p75-kD chains of the high-affinity form of the IL-2 receptor on mouse T-cell blasts, CTLL-16 cells, and sublines of EL-4 transfected with the gene encoding the p55 peptide. This 100- to 115-kD peptide was not precipitated by an anti-p55-specific antibody. Taken together, these studies suggest that three IL-2-binding peptides (p55, p75, and p95–115) are associated in the multisubunit high-affinity IL-2 receptor.

The three-dimensional structure of the 133 amino acid lymphokine IL-2 has been defined [14]. These studies, taken in conjunction with studies using site-specific mutagenesis of IL-2 and monoclonal antibodies directed toward defined regions of IL-2 in neutralization and binding assays [15, 16], have aided in the analysis of the structure-function relationships of human IL-2. Furthermore, they have led to the identification of the amino acid residues required for binding to the different IL-2 receptor peptides and for biological activity. IL-2 has an α-helical tertiary structure involving six α helices that suggests that certain portions of the molecule form a structural scaffold that underlies the receptor-binding facet of the molecule [14]. A short helical segment (helix A, amino acid residues 11–19) is required for biological activity and appears to be involved in binding to the p75 IL-2-binding peptide. The second helix on the structural scaffold helix is an extended loop involving residues 33–56 that form a helix interrupted in the middle by Pro^{47}. These two segments are referred to as B and B′. This segment appears to be required for binding to the p55 Tac peptide. An additional α helix E (amino acids 107–113) is also positioned on the binding plane and could theoretically bind the proposed 95–115-kD IL-2-binding peptide. However, no extensive studies of this region of IL-2 have been made. Finally, the carboxy terminal residues 121–133 and two of the three cysteine residues (58 and 105) are required for full biological activity and binding [15].

C. Lymphocyte Functions That Are Regulated by the Interaction of IL-2 with Its Receptor

The anti-Tac monoclonal antibody has been used to define those lymphocyte functions that require an interaction of IL-2 with the 55-kD inducible receptor on activated T- and B-lymphocytes. The addition of anti-Tac to cultures of human peripheral blood mononuclear cells inhibited the proliferation of T-lymphocytes stimulated by soluble antigens and by cell surface antigens (autologous and allogeneic mixed lymphocyte reactions) [17]. Anti-Tac was also shown to inhibit the generation of both cytotoxic and suppressor T-lymphocytes in allogeneic cell cultures, but did not inhibit their action once generated. In contrast to the action on T cells, anti-Tac did not inhibit the IL-2-induced activation of large granular lymphocytes into effective MK and LAK cells. As noted above, large granular lymphocytes express the p75 but not the 55-kD Tac peptide. Furthermore, upregulation of the expression of Tac mRNA and Tac peptide by IL-2 has been demonstrated for a number of cell types (e.g., large granular lymphocytes, B cells, and resting T cells), including some that initially express few if any Tac molecules [18, 19]. The addition of IL-2 to such Tac-negative cells, including large granular lymphocytic leukemia cells, augmented transcription of the Tac gene and induced the expression of the Tac peptide [20]. Neither the IL-2-induced activation of large granular lymphocytes nor the upregulation of Tac gene expression was inhibited by the addition of anti-Tac. These results strongly suggest that the p75 peptide is responsible for IL-2-induced activation of large granular lymphocytes and that the p75 peptide can mediate an IL-2 signal without coexpression of the Tac peptide. Thus, the p75 peptide may play an important role in the IL-2-mediated immune response not only by participating with the Tac peptide in the formation of the high-affinity receptor complex on T cells but also by contributing to the initial triggering of large granular lymphocyte activation so that these cells become efficient NK and LAK cells.

D. Disorders of IL-2 Expression in Adult T-Cell Leukemia

A distinct form of mature T-cell leukemia was defined by Takasuki and coworkers [21] and termed adult T-cell leukemia (ATL). T-cell leukemias, such as ATL, that are caused by HTLV-I, as well as all T-cell lines infected with HTLV-I, express large numbers of IL-2 Tac receptor peptides. An analysis of this virus and its protein products suggests a potential mechanism for this association between HTLV-I and IL-2 receptor expression. The complete sequence of HTLV-I has been determined by Seiki and colleagues [22]. In addition to the presence of typical long terminal repeats (LTRs), *gag, pol,* and *env* genes, retroviral gene sequences common to other groups of retroviruses, HTLV-I and -II were shown to contain an additional genomic region between *env* and the 3′ LTR referred to as pX that encodes at least three peptides of 21, 27, and 40–42 kD. Sodroski and colleagues [23] demonstrated that one of these, a 42-kD protein they termed the tat protein, is essential for viral replication. The mRNA for this protein is produced by a double splicing event. The tat protein acts on a 21-bp enhancer-like repeat within the LTR of HTLV-I, stimulating transcription [24, 25]. This tat protein also appears to play a central role in directly or indirectly increasing the transcription of host genes such as the IL-2 and especially the IL-2 Tac receptor genes involved in T-cell activation and HTLV-I-mediated T-cell leukemogenesis [26–28].

E. IL-2 Receptor as a Target for Therapy in Patients with ATL

The observation that ATL cells constitutively express large numbers of IL-2 receptors identified by the anti-Tac monoclonal antibody, whereas normal resting cells and their precursors do not, provided the scientific basis for therapeutic trials using agents to eliminate the IL-2 receptor-expressing cells. Such agents could theoretically eliminate Tac-expressing leukemic cells or activated T cells involved in other disease states while retaining the Tac-negative mature normal T cells and their precursors that express the full repertoire for T-cell immune responses. The agents that have been used or are being developed include: (a) unmodified anti-Tac monoclonal; (b) toxin (e.g., a chain of ricin toxin, *Pseudomonas* toxin, truncated *Pseudomonas* toxin) conjugates of anti-Tac; (c) alpha- and beta-emitting isotopes (e.g., bismuth-212 and yttrium-90) chelated to anti-Tac; (d) "humanized" recombinant antibodies that combine the variable or hypervariable domains of mouse anti-Tac associated with the constant domains of human immunoglobulin kappa light and IgG-1 or IgG-3 heavy chains; and (e) interleukin-2 toxin fusion proteins (e.g., IL-2-truncated *Pseudomonas* toxin).

We have performed a clinical trial to evaluate the efficacy of intravenously administered anti-Tac monoclonal antibody in the treatment of patients with ATL [29]. None of the ten patients treated suffered any untoward reactions, and only one, a patient with anti-Tac-induced clinical remission, produced antibodies to the anti-Tac monoclonal. Three of the patients had a mixed, partial, or complete remission following anti-Tac therapy. These patients may have represented an early autocrine stage of ATL, wherein the leukemic T cells still require IL-2 for their proliferation. Alternatively, the clinical responses may have been mediated by host cytotoxic cells reacting with the tumor cells bearing the anti-Tac mouse immunoglobulin on their surface.

These therapeutic studies have been extended in vitro by examining the efficacy of toxins coupled to anti-Tac selectively to inhibit protein synthesis and viability of Tac-positive ATL lines. The addition of anti-Tac antibody coupled to *Pseudomonas* exotoxin inhibited protein synthesis by Tac-expressing HUT 102-B2 cells, but not that by the acute T-cell line MOLT-4, which does not express the Tac antigen [30]. Anti-Tac conjugated with unmodified *Pseudomonas* toxin (PE) was hepatotoxic. Subsequent functional analysis of deletion mutants of the PE structural gene has shown that the 26-kD domain I of the whole 66-kD PE is responsible for cell recognition; domain II for translocation of the toxin across membranes; and domain III for ADP-ribosylation of elongation factor 2, the step actually responsible for cell death [31]. A PE molecule from which domain I has been deleted (PE40) has full ADP-ribosylating activity but extremely low cell-killing activity because of the loss of the cell recognition domain. Anti-Tac PE40 conjugates retained the capacity of unmodified PE to kill Tac-expressing T cells but were two logs less toxic to Tac-non-expressing cells.

PE40 was also used in IL-2 PE40 constructs to provide an alternative (lymphokine-mediated) method of delivering PE40 to the surface of IL-2 receptor, Tac-positive, cells [32]. These constructs were effective in inhibiting protein syntheses and in killing IL-2 receptor-expressing cells but not the cells that did not display the cell surface IL-2 receptor.

The action of toxin conjugates of monoclonal antibodies depends on their ability to be internalized by the cell and released into the cytoplasm. Anti-Tac bound to IL-2 receptors on leukemic cells is internalized slowly into coated pits and then endosomic vesicles. Furthermore, the toxin conjugate does not pass easily from the endosome to the cytosol, as required for its action on elongation factor 2. To circumvent these limitations, an alternative cytotoxic reagent was developed that could be conjugated to anti-Tac and that was effective when bound to the surface of leukemic cells. It was shown that bismuth-212 (^{212}Bi), an alpha-emitting radionuclide conjugated to anti-Tac by use of a bifunctional chelate, was well suited for this role [33]. Activity levels of 0.5 μCi or the equivalent of 12 rad/ml of alpha radiation targeted by ^{212}Bi-labeled anti-Tac eliminated greater than 98% of the proliferative capacity of the HUT 102-B2 cells, with only a modest effect on IL-2 receptor-negative lines. This specific cytotoxicity was blocked by excess unlabeled anti-Tac, but not by human IgG. Therefore, ^{212}Bi-labeled anti-Tac is a potentially effective and specific immunocytotoxic agent for the elimination of IL-2 receptor-positive cells.

In addition to its use in the therapy of patients with ATL, IL-2 receptor-directed therapy is being attempted in other clinical states. Specifically, therapeutic studies have been initiated using monoclonal antibodies directed toward the IL-2 receptors expressed on autoreactive T cells of certain patients with autoimmune disorders, on host T cells responding to foreign histocompatibility antigens on organ allografts, and on leukemic T and B cells.

References

1. Morgan DA, Ruscetti FW, Gallo RC (1976) Selective in vitro growth of T-lymphocytes from normal human bone marrows. Science 193:1007–1008
2. Smith KA (1980) T-cell growth factor. Immunol Rev 51:337–357
3. Robb RJ, Munck A, Smith KA (1981) T-cell growth factor receptors. J Exp Med 154:1455–1474
4. Waldmann TA (1986) The structure, function, and expression of interleukin-2 receptors on normal and malignant T cells. Science 232:727–732
5. Uchiyama T, Broder S, Waldmann TA (1981) A monoconal antibody (anti-Tac) reactive with activated and functionally mature human T cells. I. Production of anti-Tac monoclonal antibody and distribution of Tac^{+} cells. J Immunol 126:1393–1397
6. Leonard WJ, Depper JM, Uchiyama T, Smith KA, Waldmann TA, Greene WC (1982) A monoclonal antibody that appears to recognize the receptor for human T cell growth factor; partial characterization of the receptor. Nature 300:267–269
7. Tsudo M, Kozak RW, Goldman CK, Waldmann TA (1986) Demonstration of a new (non-Tac) peptide that binds interleukin-2: a potential participant in a multichain interleukin-2 receptor complex. Proc Natl Acad Sci USA 83:9694–9698
8. Sharon M, Klausner RD, Cullen BR, Chizzonite R, Leonard WJ (1986) Novel interleukin-2 receptor subunit detected by cross-linking under high-affinity conditions. Science 234:859–863
9. Tsudo M, Kozak RW, Goldman CK, Waldmann TA (1987) Contribution of a p75 interleukin-2 binding peptide to a high affinity interleukin-2 receptor complex. Proc Natl Acad Sci USA 84:4215–4218
10. Szollosi J, Damjanovich S, Goldman CK, Fulwyler M, Aszalos AA, Goldstein G, Rao P, Talle MA, Waldmann TA (1987) Flow cytometric resonance energy transfer measurements support the association of a 95-kDa peptide termed T27 with the 55-kDa Tac peptide. Proc Natl Acad Sci USA 84:7246–7251
11. Eddidin M, Aszalos A, Damjanovich S, Waldmann TA (1988) Lateral diffusion measurements give evidence for association of the Tac peptide of the IL-2 receptor with the T27 peptide in the plasma membrane of HUT-102-B2 T cells. J Immunol 141:1206–1210
12. Herrmann F, Diamantstein T (1987) The mouse high affinity IL-2 receptor complex. I. Evidence for a third molecule, the putative γ-chain associated with the α- and/or β-chain of the receptor. Immunobiology 175:145–158
13. Saragovi H, Malek TR (1987) The murine interleukin-2 receptor: irreversible cross-linking of radiolabeled interleukin-2 to high affinity interleukin-2 receptors reveals a non-covalently associated subunit. J Immunol 139:1918–1926
14. Brandhuber BJ, Boone T, Kenny WC, McKay DB (1987) Three-dimensional structure of interleukin-2. Science 238: 1707–1709
15. Ju G, Collins J, Kimberlee LK, Tsien W-H, Chizzonite R, Crowl B, Bhatt R, Kilian PL (1987) Structure-function analysis of human interleukin-2: identification of amino acid residues required for biological activity. J Biol Chem 262:5723–5731
16. Kuo L, Robb RJ (1986) Structure-function relationship for the IL-2 receptor system. I. Localization of a receptor binding site on IL-2. J Immunol 137:1538–1543
17. Depper JM, Leonard WJ, Waldmann TA, Greene WC (1983) Blockade of the interleukin-2 receptor by anti-Tac antibody: inhibition of human lymphocyte activation. J Immunol 131:690–696
18. Waldmann TA, Goldman CK, Robb RJ, Depper JM, Leonard WJ, Sharrow SO, Bongiovanni KF, Korsmeyer SJ, Greene WC (1984) Expression of interleukin-2 receptors on activated human B cells. J Exp Med 160:1450–1466
19. Reem G, Yeh N-H (1984) Interleukin-2 regulates expression of its receptor and synthesis of gamma interferon by human T lymhocytes. Science 255:429–430
20. Tsudo M, Goldman CK, Bongiovanni KF, Chan WC, Winton EF, Yagita M,

Grimm EA, Waldmann TA (1987) The p75 peptide is the receptor for interleukin-2 expressed on large granular lymphocytes and is responsible for the interleukin-2 activation of these cells. Proc Natl Acad Sci USA 84:5394–5398
21. Takasuki K, Uchiyama T, Sagawa K, Yodoi J (1977) Adult T cell leukemia in Japan. In: Seno S, Takaku F, Irino S (eds) Topics in hematology. Excerpta Medica, Amsterdam, pp 73–77
22. Seiki M, Hattori S, Hirayama Y, Yoshida M (1983) Human adult T-cell leukemia virus: complete nucleotide sequence of the provirus genome integrated in leukemia cell DNA. Proc Natl Acad Sci USA 80:3618–3622
23. Sodroski JG, Rosen CA, Haseltine WA (1984) *Trans*-acting transcriptional activation of the long terminal repeat of human T lymphotrophic viruses in infected cells. Science 225:381–385
24. Paskavis H, Felber BK, Parlakis GN (1986) *Cis*-acting sequences responsible for the transcriptonal activation of human T-cell leukemia virus type I constitute a conditional enhancer. Proc Natl Acad Sci USA 83:6558–6562
25. Shimotohno K, Miwa M, Slamon DJ, Chen ISY, Hoshino H, Takano M, Fujino M, Sugimara T (1985) Identification of new gene products coded with X-regions of human T-cell leukemia viruses. Proc Natl Acad Sci USA 82:302–306
26. Inoue J, Seiki M, Taniguchi T, Tsuru S, Yoshida M (1986) Induction of interleukin-2 receptor gene expression by $p40^x$ encoded by human T-cell leukemia virus type I. EMBO J 5:2883–2888
27. Cross SL, Feinberg MD, Wolf JB, Holbrook NJ, Wong-Staal F, Leonard WJ (1987) Regulation of the human interleukin-2α chain promoter: activation of a nonfunctional promoter by the transactivation gene of HTLV-I. Cell 49:47–56
28. Maruyama M, Shibuya H, Harada H, Haitakeyama M, Seiki M, Fujita T, Inoue J, Yoshida M, Taniguchi T (1987) Evidence for aberrant activation of the autocrine loop by HTLV-I encoded $p40^x$ and T3-Ti complex triggering. Cell 48:343–350
29. Waldmann TA, Goldman CK, Bongiovanni KF, Sharrow SO, Davey MP, Cease KB, Greenberg SJ, Longo D (1988) Therapy of patients with human T-cell lymphotrophic virus I-induced adult T-cell leukemia with anti-Tac, a monoclonal antibody to the receptor for interleukin-2. Blood 72:1805–1816
30. FitzGerald D, Waldmann TA, Willingham MC, Pastan I (1984) *Pseudomonas* exotoxin-anti-Tac: cell-specific immunotoxin active against cells expressing the human T-cell growth factor receptor. J Clin Invest 74:966–971
31. Huang J, FitzGerald DJT, Adga S, Pastan I (1988) Functional domains of *Pseudomonas* exotoxin identified by deletion analysis of the gene expressed in *E. coli.* Cell 48:129–136
32. Lorberboum-Galski H, Kozak R, Waldmann TA, Bailon P, FitzGerald D, Pastan I (1988) IL2-PE40 is cytotoxic to cells displaying either the p55 or p75 subunit of the IL-2 receptor. J Biol Chem 263: 18650–18656
33. Kozak RW, Atcher RW, Gansow OA, Friedman AM, Hines JJ, Waldmann TA (1986) Bismuth-212 labeled anti-Tac monoclonal antibody: alpha-particle emitting radionuclides as modalities for radioimmunotherapy. Proc Natl Acad Sci USA 83:474–478

Monocyte Interleukin-1 Secretion Is Regulated by the Sequential Action of γ-Interferon and Interleukin-2 Involving Monocyte Surface Expression of Interleukin-2 Receptors*

F. Herrmann [1], A. Lindemann [1], S. A. Cannistra [2], M. Brach [1], W. Oster [1], J. D. Griffin [2], and R. Mertelsmann [1]

A. Introduction

Interleukin-1 (IL-1) is a polypeptide synthesized as a high-mol.-wt. precursor and subsequently secreted after proteolytic cleavage to 17500-dalton active forms in murine [1] and human cells [2]. Recently, cDNAs for murine [3] and for two distinct human IL-1 species, IL-1α and IL-1β, have been isolated, sequenced, and cloned [2, 4].

IL-1 is produced by a variety of cell sources including macrophage-containing tissues, such as peripheral blood, bone marrow, dendritic cells, pulmonary alveolar cells, Kupffer cells, astrocytes, and glial cells (reviewed in [5, 6]) and non-macrophage cells such as B-lymphocytes [7], large granular lymphocytes [8], epidermal cells [9], and mesangial cells [10], and it exerts a multiplicity of nonspecific biological activities amplifying inflammatory reactions and modulating the immune response. IL-1 activity results in the release of acute phase reactants such as serum amyloid P component, fibrinogen and C-reactive protein [11, 12], increases the production of collagenase and prostaglandins by rheumatoid synovial cells and chondrocytes [13], induces synthesis of prostacyclin [14] and procoagulant activity [15] in vascular cells, acts on vascular endothelial cells to increase adhesion of granulocytes and monocytes [16], stimulates fibroblast growth [17], release of γ-interferon (IFN) [18], and osteoclast-mediated bone reabsorption [19], and mediates tumor cytostasis [20, 21].

IL-1 has been shown to provide a signal for T-cell proliferation, in response to both antigen-specific and polyclonal T-cell stimulation [22, 23], but it is not required for the recognition of an exogenous antigen provided by an Ia identical cell [24–26] and cannot substitute for accessory cells in MHC-restricted antigen presentation.

It has been proposed that IL-1 induces T cells to produce mitogenic lymphokines such as interleukin-2 (IL-2) and to express their respective receptors on T cells [27–31]. The synergistic action of IL-1 with IL-2 and IFN in boosting NK activity has also been reported [32]. Furthermore, an in vitro role of IL-1 in enhancing the proliferative response of activated B cells to B-cell growth factors [33–36] and in modulating antibody production [37, 38] has been suggested.

In an immune response, IL-1 is released by macrophages after activation by two pathways: one is the genetically unrestricted direct challenge of macrophages by various compounds including lipopolysaccharides, purified tuberculin protein derivatives, muramyl dipeptide, phorbol myristate acetate, silica particles, and the glucocerebroside GL-1 [6]. A second pathway for the secretion of IL-1 involves activated T cells that stimulate monocytes to produce IL-1 by a cell contact-dependent, genetically

[1] Department of Hematology, University of Mainz, Mainz, Federal Republic of Germany
[2] Division of Tumor Immunology, Dana-Farber Cancer Institute, Department of Medicine, Harvard Medical School, Boston, MA 02115, USA
* This work was supported by grant He 1380-2/1 from the Deutsche Forschungsgemeinschaft and grant 0319015 from the Bundesministerium für Forschung und Technologie

restricted pathway which requires Ia antigen identity [39, 40]. Recently, ligand binding to monocyte membrane structures as provided by monoclonal antibodies to Ia antigens has been reported to stimulate monocyte IL-1 secretion [41]. Also, lymphokines released by T cells, such as the granulocyte/monocyte colony-stimulating factor, have been implicated in monocyte IL-1 secretion [42].

However, little is known about mechanisms underliying T-cell/monocyte interactions that lead to monocyte IL-1 secretion.

We have previously shown that T-cell-derived γ-IFN induced binding sites for IL-2 on monoblast line U 937 and the promyelocyte line HL 60 that bound biologically active IL-2 [43]. In addition, we have presented evidence for IL-2-R expression on γ-IFN treated human peripheral blood monocytes [43]. We have now shown, by means of biochemical and molecular identification and bioassays, that cultured human peripheral blood monocytes display binding sites for IL-2. Stimulation of monocytes with γ-IFN, lipopolysaccharide (LPS), or phytohemagglutinin (PHA) enhanced surface IL-2-R expression up to threetimes. Binding of IL-2 to monocyte IL-2-R resulted in IL-1 secretion that could be enhanced five- to sixfold when monocytes were costimulated with LPS. Moreover, the progressive loss of monocyte IL-1 secretion upon monocyte aging in vitro was reversed by either IL-2 or γ-IFN alone, but was most effective in the presence of both factors. IL-2 effects on IL-1 secretion by highly purified monocyte preparations could be partially blocked by the addition of anti-IL-2-R antibodies to the cultures.

These results provide insights into the T-cell/monocyte interactions that lead to monocyte IL-1 secretion, suggesting that the T-cell lymphokines IL-2 and γ-IFN may act on monocytes to amplify the immune response by establishing a positive feedback loop.

B. Materials and Methods

I. Separation of Monocytes

Peripheral blood mononuclear cells (PBMC) were isolated from healthy volunteer donor buffy coats by Ficoll-Hypaque (density 1.077 g/cm) gradient centrifugation. T cells were recovered by rosetting with AET-treated SRBC (5% vol/vol solution); monocytes were separated by repeated adherence steps of the E-rosette negative fraction [44]. Purity was determined by Wright-Giemsa staining, ANAE staining, and immunofluorescence analysis with monoclonal antibodies to the Mo-2, T-11, B-1, and NKH1-A antigens [45–48].

II. Culture Conditions

In some experiments monocytes were incubated at 1×10^6 cells/ml in 60×15 mm Petri dishes (Falcon, Oxnard, CA) at 37° C in a 5% CO_2 atmosphere and cultured for a period of 12–72 h in the presence or absence of 50–1000 U/ml ($SA = 6 \times 10^7$ U/mg) of purified *Escherichia coli*-derived recombinant γ-IFN (Biogen Research Corporation, Cambridge, MA). The culture medium was RPMI 1640 supplemented with 5% low-endotoxin FCS, 2 m*M* L-glutamine, 100 U/ml penicillin, 100 µg/ml streptomycin, and 1% sodium pyruvate (Gibco, Grand Island, NY) (=standard culture medium).

In selected experiments other compounds potentially affecting monocyte functions were added, either alone or in combination, for 48 h to monocyte cultures, including prostaglandin E_2 (10^{-6} *M*), LPS; *E. coli* 026: B6 (15 µg/ml, Sigma, St. Louis, MO), PHA (2 µg/ml, Burroughs-Wellcome, Greenville, NC), 1.25 (OH) vitamin D_3 (10^{-8} *M*), recombinant αA-IFN, β-IFN (500 U/ml; $SA = 2 \times 10^8$ U/mg, Hoffman-La Roche, Nutley, NJ) and recombinant TNFα (150 U/ml; $SA = 7.2 \times 10^7$ U/mg, Genentech, San Francisco, CA). Cells derived from cultures as described above were subjected to immunofluorescence staining and analysis.

In other experiments designed to examine the effects of γ-IFN and IL-2 on modulation of IL-1 secretion by monocytes, monocytes were cultured in 24-well, 16-mm flat-bottom plates (Falcon) at 2.5×10^5 cells/ml and 1 ml/well for 24–168 h in the presence or absence of γ-IFN (250 U/ml), recombinant IL-2 (2.5–500 U/ml; SA = 10^6 U/mg, Biogen) or a combination of both lymphokines, IL-1-inducer LPS (15 µg/ml), anti-IL-2-R 1 mcAb [49] (diluted 1:100–1000), or control IgG2a mcAbs, that were either binding or nonbinding to monocytes (diluted 1:100). Supernatants were harvested after various culture times and subjected to IL-1 and IL-2 bioassays.

In some experiments (indicated in the table headings) supernatants were treated with 1500 nU/ml of a neutralizing mcAb to γ-IFN, 3C11C8 [44], and mcAb to IL-2 (DMSl, 0.5 mg/ml; kindly provided by Dr. K. A. Smith) for 2 h at 27° C. In other selected experiments supernatants were additionally absorbed with 1×10^7 cloned murine IL-2-dependent CTLL cells at 4° C for 24 h to remove IL-2 activity that was potentially not neutralized. Culture medium in this culture type was standard culture medium supplemented with indomethacin (10^{-6} *M*, Sigma) to prevent the endogenous secretion of prostaglandins, known to inhibit IL-1 release [50].

III. Immunofluorescence Staining

For one-color staining, 1×10^6 fresh or cultured monocytes were incubated for 30 min at 4° C with the appropriate mcAbs to monocyte-, T-cell-, B-cell-, NK-cell antigens (Mo-2, T-11, B-1, NKH1-A) or several antibodies to IL-2-R, including anti-IL-2-R 1 [49]), anti-TAC [51], and anti-7G7/B 6 [52] (kindly provided by Drs. T. Waldmann and D. Nelson) or irrelevant isotype-identical control antibodies. If antibodies were not directly fluoresceinated, a second incubation was performed for 30 min at 4° C with fluorescein-conjugated goat anti-mouse IgG + IgM (Tago, Burlingame, CA). The washing medium was minimal essential medium (MEM; Gibco), containing 10% pooled human AB serum. For dual fluorescence studies, monocytes that had been treated with γ-IFN, PHA, or LPS were incubated with biotin-conjugated anti-Mo-2 and a fluoresceinated anti-IL-2-R 1. After washing, cells were developed with avidin-conjugated Texas red (Molecular Probes, Junction City, OR). Cells were analyzed using a dual-laser flow cytometer (EPICS V; Coulter Electronics, Hialeah, FL).

IV. Assay for IL-1

Single thymocyte suspensions from C3H/HeJ mice (female, 6–8 weeks old) (Jackson Laboratory, Bar Harbor, ME) were cultured at 1.5×10^6 cells/150 µg/well in standard culture medium supplemented with 2.5×10^{-5} *M* 2 mercaptoethanol (2 ME; Sigma) and submitogenic concentrations of PHA (0.5 µg/ml) in the presence or absence of supernatant conditioned by monocytes (25% vol/vol) for 72 h at 37° C, 5% CO in 98-microwell flat-bottom plates (Falcon) as described [53]. Eighteen hours before harvesting, cultures were pulsed with 1 mCi/ml = 37 kBq/ml of tritiated-thymidine (^{3}H-Td; Schwartz-Mann, Spring Valley, NY). The incorporated radioactivity was collected onto fiberglass filters and assayed using a liquid scintillation counter (Packard Instruments, Downer's Grove, IL). The levels of IL-1 activity in supernatants tested are expressed as cpm values of ^{3}H-Td incorporated by thymocytes (mean of triplicate cultures). In selected samples IL-1 activity present in the supernatants was detected using the LBRM 33-IA5B6 conversion assay [54].

V. Assay for IL-2

Interleukin-2 activity was determined by assaying the growth of IL-2 dependent murine CTLL 2 cells as described [55]. CTLL 2 cells were cultured at 5×10^6

cells/100 µl/well at 37° C, 5% CO_2, in 98-microwell flat bottom plates (Falcon) in standard culture medium supplemented with 5×10^{-5} *M* 2 ME in the presence or absence of monocyte conditioned supernatants (25% vol/vol) for 48 h. The cells were pulsed with 3 H-Td for the final 18 h before harvesting. The incorporated radioactivity was measured in the same manner as described for the IL-1 assay. IL-2 levels are expressed as cpm values of 3 H-Td incorporated by CTLL 2 cells in response to IL-2 (mean of triplicate cultures).

VI. Iodination, Immunoprecipitation, and SDS-PAGE

Monocytes treated for 48 h with 250 U/ml γ-IFN were externally labeled with 125 I using the lactoperoxidase-catalyzed method. Immunoprecipitates with anti-IL-2-R 1 antibody were analyzed by sodium dodecyl–10% polyacrylamide gel electrophoresis (SDS-PAGE) [43].

VII. Northern Blot Analysis

Total cellular RNA was prepared employing the guanidinium isothiocyanate/cesium chloride method [23]. Monocytes were lysed in situ, adherent on the Petri dishes. After glyoxylation, 10-µg samples of RNA were size-fractioned by agarose gel electrophoresis and transferred onto synthetic membranes. Filters were hybridized with a full-length IL-2-R cDNA (kindly provided by Dr. W. C. Greene), radiolabeled, and exposed to Kodak XAR5 using Dupont Cronex intensifying screens. To exclude contamination of monocyte RNA with RNA derived from T cells possibly contaminating monocyte culture, filters were reprobed with a cDNA specific for the T-cell receptor β-chain gene (kindly provided by Dr. H. D. Royer).

C. Results

I. Purification of Monocyte Preparation

Isolation of monocytes by two sequential adherence steps of E-rosette-negative PBMC resulted in cell preparations consisting of >98% monocytes by morphology and cytochemistry (Wright Giemsa and ANAE staining). Cytofluorographic analysis of these cells employing mcABs to Mo-2 (monocytes), T-11 (T cells), B-1 (B cells), and NKH1-A (NK cells) revealed 96%–99% Mo-2-positive cells. Reactivity with anti-T-11, -B-1, and -NKH1-A was below the background fluorescence. The possibility that the few remaining T cells, present in the monocyte cultures and not detectable by means of morphology or immunofluorescence, produced endogenous IL-2 was examined by assaying samples of IL-1-containing conditioned medium for their effects on IL-2-dependent CTLL 2. A rat IL-2 sample used for maintenance of CTLL 2 cells served as a positive control. None of the conditioned media tested exhibited any activity on the CTLL 2 cells (data not shown).

II. Constitutive and Modulated IL-2-R Expression on Peripheral Blood Monocytes

IL-2-R expression is negligible in freshly isolated monocytes (<2%). However, using mcAb anti-IL-2-R 1 and immunofluorescence, surface IL-2-R were detectable in monocytes after 24 h of culture, with a maximum (25%) after 48–60 h. Exposure of monocytes to r γ-IFN (50–1000 U/ml) resulted in enhanced expression of binding sites for anti-IL-2-R 1; they were at a maximum when 100 U/ml r γ-IFN was present during 60 h of culture (Table 1). Similar results were obtained using other anti-IL-2-R antibodies, such as anti-TAC and anti-7G7/B6 (data not shown).

Identification of these binding sites as IL-2-R was confirmed by immunoprecipitation of the same 60- to 65-kD protein

Cultures of monocytes	IL-2-R1 Number of stained cells (%)
Nontreated	25[a]
Treated with	
γ-IFN	63
αA-IFN	25
β-IFN	24
TNFα	24
1.25 (OH) vitamin D_3	26
PGE2	23
LPS	52
PHA	55
γ-IFN + TNFα	74
γ-IFN + PGE2	39
γ-IFN + LPS	64
γ-IFN + αA-IFN	64
γ-IFN + β-IFN	60

See text for concentrations used. Recovery of viable cells in each fraction was >78%.
[a] Data of one representative experiment.

Table 1. Effect of various compounds (48-h exposure) on IL-2-receptor expression of cultured peripheral blood monocytes

from surface-labeled PHA-activated T cells (72 h, 2 μg/ml) as from γ-IFN-induced cultured peripheral blood monocytes (60 h, 100 U/ml; Fig. 1). Other species of IFN, such as r α-A-IFN or r β-IFN, or compounds that are known to be involved in the modulation of monocyte/macrophage functions, such as PGE_2, 1.25 (OH) 2 vitamin D_3, and TNF α, failed to enhance monocyte surface IL-2-R expression. However, when various combinations of these compounds were assayed, it was demonstrated that PGE_2 partially inhibited the γ-IFN-induced enhancement of IL-2-R on monocytes, whereas TNF α – although it demonstrated no enhancing effect on IL-2-R expression by itself – synergized with γ-IFN to increase the number and fluorescent intensity of IL-2-R 1+monocytes (data not shown), while cocultures of γ-IFN together with αA-IFN, β-IFN, and LPS did not alter the effects observed with γ-IFN alone (Table 1).

PHA and LPS also increased the number of IL-2-R 1+monocytes. This was confirmed by two-color immunofluorescence using biotin-conjugated anti-Mo-2 and fluoresceinated anti-IL-2-R 1 antibody (Fig. 2). Of the cells used for two-color immunofluorescence studies, 97.2%–99.4%) were stained by biotinylated Mo-2; 53.5%, 59.2%, and 62.5%, respectively, of cells from PHA-, LPS-, and γ-IFN-treated cultures were double-stained by biotinylated Mo-2 and fluoresceinated anti-IL-2-R 1. To determine whether monocyte surface expression of IL-2-R was associated with induced expression of the IL-2-R gene, cytoplasmic mRNA of monocytes treated with LPS, PHA, or γ-IFN was extracted and hybridized to an IL-2-R gene-specific cDNA. As shown in Fig. 3, unstimulated monocytes failed to accumulate transcripts for IL-2-R. In the presence of the stimulatory compounds, however, the IL-2-R message of 3.5 and 1.5 kb in size, became detectable 12 h after initiation of cultures. These mRNA were similar in size to the corresponding messengers seen in PMA/PHA-activated normal T cells. Lack of T-cell receptor-β-chain transcripts in RNA derived from monocyte cultures excluded the possibility that T-cell contamination was responsible for the IL-2-R messenger (data not shown).

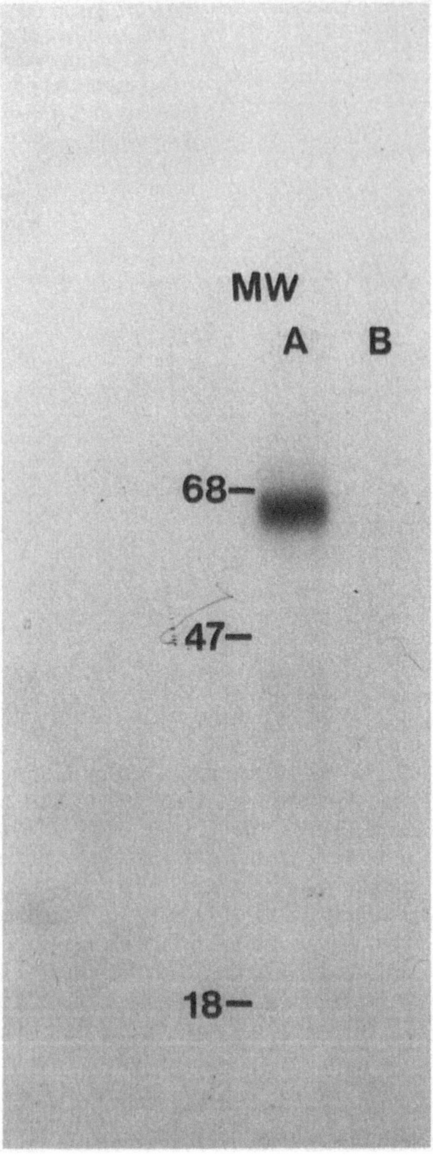

Fig. 1. SDS-PAGE analysis of immunoprecipitates obtained with anti IL-2-R 1 antibody from γ-IFN-treated (48 h, 250 U/ml) monocytes (*lane A*) under reducing conditions. *Lane B* represents control immunoprecipitates from the same cells using an unreactive control antibody (anti-glycophorin A)

III. γ-IFN and IL-2 Enhance LPS-induced and Maintain LPS-inducible IL-1 Release by Cultured Peripheral Blood Monocytes

When monocytes were cultured for 24 h, subsequent stimulation with LPS (15 μg/ml) resulted in the release of IL-2 activity in the supernatants. Costimulation with 250 U/ml r γ-IFN at initiation of the cultures yielded a twofold increase in the level of IL-1 activity when monocytes were induced with LPS. As seen in Table 2, when r IL-2 (500 U/ml) was present during the first 24 h of culture and was thoroughly washed off before inducing with LPS, an up to 30% increase in detectable IL-1 activity was observed, as compared with untreated LPS-induced cultures.

However, IL-2 synergized with γ-IFN to yield a sixfold enhancement of LPS-inducible secretion of IL-1 activity, whereas LPS-uninduced cells did not secrete significant levels of IL-1. None of the supernatants assayed for IL-1 activity showed any mitogenic stimulation of IL-2-dependent CTLL 2 cells (data not shown).

Monocytes cultured for more than 84 h were diminished in their ability to support LPS-inducible IL-1 production. This loss of IL-1 secretory potential is complete after 214 h, but it can be overcome by the addition of IL-2 or γ-IFN, and more dramatically by a combination of both lymphokines (Table 3) when introduced at the time of initiation of cultures. When cultures were performed in the presence of IL-2, γ-IFN, or both, the ability of LPS to induce the release of IL-1 was largely maintained and the mitogenic activity in supernatants remained comparable to that in fresh cultures. Again, when cultures were washed thoroughly before LPS was introduced no IL-2 activity could be detected in the supernatants to be assayed for IL-1 activity (not shown).

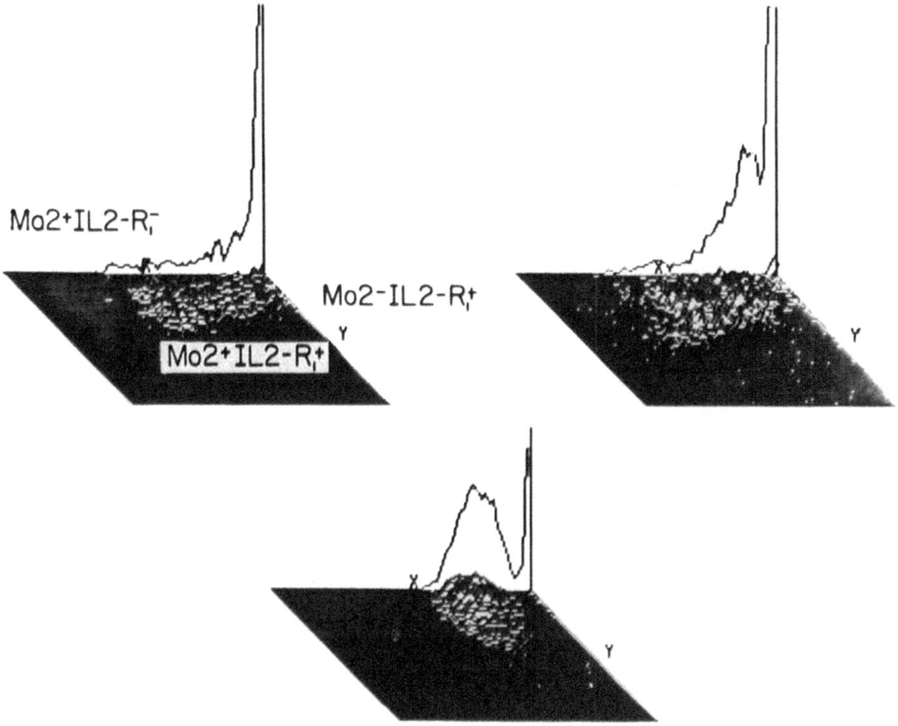

Fig. 2. Dual-color fluorescence analysis of IL-2-R expression on peripheral blood monocytes cultured for 48 h in the presence of LPS (15 μg/ml; *top left panel*), PHA (2 μg/ml; *top right panel*), or γ-IFN (250 U/ml; *bottom panel*). Mo-2 antigen was detected with biotin-conjugated anti-Mo-2 (*x axis*), and IL-2-R was detected with fluoresceinated IL-2-R 1 antibody (*y* axis)

Table 2. γ-Interferon and interleukin-2 enhancement of LPS-induced interleukin-1 release by cultured monocytes

Culture of monocytes in:	^{3}H-Td Incorporation[a] of murine thymocytes induced by monocyte-conditioned media[b] (MCM) stimulated with:					
	Culture medium			LPS		
	Expm. 1	Expm. 2	Expm. 3	Expm. 1	Expm. 2	Expm. 3
Culture medium	368	127	309	6687	9859	5392
γ-IFN	446	141	287	12036	18329	11007
IL-2	439	213	308	7914	9997	6985
γ-IFN + IL-2	471	242	357	36875	33341	29890

Monocytes (2.5×10^{5}/ml) were cultured in culture medium in the absence or presence of γ-IFN (250 U/ml), IL-2 (500 U/ml), or a combination of both for 24 h. After several washings, cultures were reincubated in fresh culture medium with or without LPS (15 μg/ml).
After further 24 h, cell-free supernatants were harvested and assayed for IL-1 biological activity.
[a] Values are expressed as cpm of triplicate cultures. SD never exceeded 7%. ^{3}H-Td Incorporation of thymocytes stimulated with PHA (0.5 μg/ml) or PHA + LPS (15 μg/ml) was 264 ± 67 and 339 ± 71.
[b] The result given correspond to a dilution of MCM of 25% vol./vol.

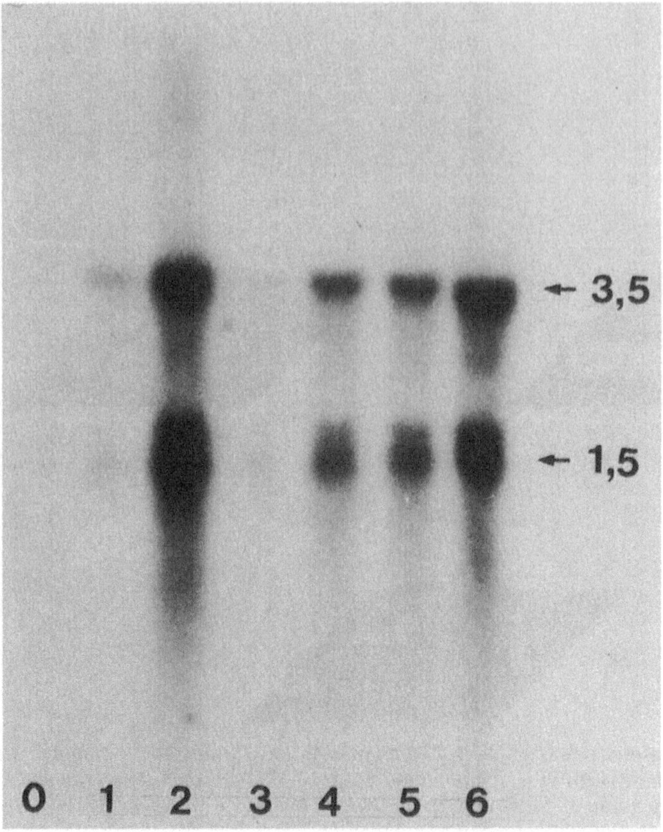

Fig. 3. Detection of IL-2-R mRNA in monocytes cultured for 12 h in the presence of LPS (15 µg/ml; *lane 4*), PHA (2 µg/ml; *lane 5*), or γ-IFN (250 U/ml; *lane 6*). Monocytes cultured in the presence of medium alone (*lane 3*) failed to display IL-2-R mRNA. Lanes 0, 1, and 2 represent control lanes of T cells cultured for 12 h in medium alone (*lane 0*), in the presence of PHA (2 µg/ml; *lane 1*), or PHA plus PMA (10^{-9} *M*; *lane 2*)

IV. Effect of γ-IFN and IL-2 on the Induction of IL-1 Release by Cultured Monocytes in the Absence of LPS

In an attempt to examine whether cultured monocytes were able to release IL-1 activity in response to a cascade of lymphokines, independently from induction by LPS, experiments were performed in which γ-IFN- or IL-2-treated or untreated control cultures (0–144 h) received a second lymphokine pulse, either with γ-IFN or with IL-2. As seen in Tables 4 and 5, when monocytes were cultured in the presence of γ-IFN (250 U/ml), they could be induced to release IL-1 activity by IL-2; this was maximally detectable when γ-IFN was present in the primary culture for 60 h, requiring a minimal IL-2 concentration of ≥250 U/ml (Table 5). When primary cultures were performed in the absence of γ-IFN and with or without IL-2, γ-IFN was unable to induce any IL-1 release (Table 4). Supernatants from secondary cultures, induced with either γ-IFN or IL-2, were treated with neutralizing concentrations of mcAb to γ-IFN (3C11C8) or to IL-2 (DMS1), to avoid a carry-over of antiproliferative effects (although not detectable in control experiments) of γ-IFN or mitogenic activities of IL-2 into the final IL-1 assay. In selected experiments supernatants were additionally absorbed with 1×10^7 IL-2-dependent cloned CTLL cells (24 h, 4° C), to remove any possible residual exogenous IL-2 activity (although not detectable in the CTLL 2 proliferation assay). A possible mechanism by which IL-2 could induce IL-1 secretion, particularly in γ-IFN-primed cultures, would be that γ-IFN enhances the monocyte ability to respond to the presence of IL-2. Therefore, further experiments were designed to examine whether IL-1-enhancing effects of IL-2 were mediated via γ-IFN-induced receptors for IL-2.

V. Effect of Monoclonal Antibody to IL-2-R (anti-IL-2-R 1) on IL-1 Release from γ-IFN-pretreated and IL-2-induced Cultures of Peripheral Blood Monocytes

Since γ-IFN was shown to increase threefold the number of monocytes expressing binding sites for anti-IL-2-R 1 antibody, possibly resulting in an enhancement of the ability of monocytes to respond to IL-2, we investigated the effect of anti-IL-2-R 1 antibody on the IL-2-induced IL-1 release from γ-IFN-primed monocyte cultures. Under optimal experimental conditions for release of IL-1 activity in IL-2 (500 U/ml)-stimulated cultures of γ-IFN (250 U/ml)-pretreated monocytes (60 h), the presence of anti-IL-2-R 1 antibody (diluted 1:100) during the last 24 h resulted in a reduction of detectable IL-1 activity in the supernatants by 60%. A 20% reduction of released IL-1 activity was observed when IL-2-induced supernatants were tested that had been generated in the absence of γ-IFN (Table 6), suggesting that IL-1 induction by IL-2 might be mediated at least in part via IL-2-R on cultured monocytes, either induced by γ-IFN or (to a much lower extent) constitutively expressed. However, cultures that were performed in the absence of IL-2, or those in which anti-IL-2-R 1 antibody was replaced by isotype-identical binding and nonbinding control mcAbs were free of detectable IL-1 activity or exhibited unchanged levels of IL-1 release, respectively.

D. Discussion

The interaction of amplifying soluble messenger molecules may be the prereq-

Table 3. γ-Interferon and interleukin-2 maintenance of LPS-induced interleukin-1 release by cultured monocytes

Monocytes in culture (h)	^{3}H-Td Incorporation of murine thymocytes induced by monocyte-conditioned media stimulated with											
	Culture medium			γ-IFN			IL-2			γ-IFN + IL-2		
	Expm. 1	Expm. 2	Expm. 3	Expm. 1	Expm. 2	Expm. 3	Expm. 1	Expm. 2	Expm. 3	Expm. 1	Expm. 2	Expm. 3
48	6384	8173	8064	12987	13112	16507	7801	9231	9201	27340	34107	32998
108	2911	3702	3322	13017	13004	15743	8016	9746	9384	39108	41693	43821
192	897	1965	1713	13223	13897	15789	8092	9787	9402	29974	33342	34502
240	312	ND[a]	ND	11314	ND	ND	7314	ND	ND	21117	ND	ND

Monocytes (2.5×10^5/ml) were cultured in culture medium in the absence or presence of γ-IFN (250 U/ml), IL-2 (500 U/ml) or a combination of both for 24, 84, 168, and 216 h. After several washings, cultures were reincubated with fresh culture medium supplemented with LPS (15 μg/ml) for an additional 24 h and cell-free supernatants were harvested and assayed for IL-1 biological activity. For further details see legend to Table 2.

[a] Not done.

Table 4. Effect of γ-interferon and interleukin-2 on the induction of interleukin-1-release by cultured monocytes in the absence of LPS

Induction culture with γ-IFN	^{3}H-Td Incorporation of murine thymocytes induced by monocyte conditioned media stimulated with								
	Culture medium			Culture medium + γ-IFN			Culture medium + IL-2		
Monocytes in culture (h)	Expm. 1	Expm. 2	Expm. 3	Expm. 1	Expm. 2	Expm. 3	Expm. 1	Expm. 2	Expm. 3
24	293	366	298	246	353	342	258	359	317
84	316	357	353	270	377	359	278	384	339
168	208	199	ND[a]	188	301	ND	243	292	ND

Induction culture with IL-2	^{3}H-Td Incorporation of murine thymocytes induced by monocyte conditioned media stimulated with								
	Culture medium			Culture medium + IL-2			Culture medium + γ-IFN		
Monocytes in culture (h)	Expm. 1	Expm. 2	Expm. 3	Expm. 1	Expm. 2	Expm. 3	Expm. 1	Expm. 2	Expm. 3
24	441	429	407	473	427	421	791	886	809
84	609	594	706	701	600	733	19384	18247	17003
168	489	383	ND	493	394	ND	5604	4041	ND

Monocytes (2.5×10^5/ml) were cultured for various time in culture medium in the absence or presence of γ-IFN (250 U/ml) or IL-2 (500 U/ml). At time point 0, after 60 h and 144 h, cultures that were washed several times received fresh culture medium supplemented with either γ-IFN (250 U/ml) or IL-2 (500 U/ml). After a further incubation period of 24 h, cell-free supernatants were harvested, treated with neutralizing concentrations of monoclonal antibody to γ-IFN (3C11C8; 1500 nU/ml) and monoclonal antibody to IL-2 (DMS 1; 0.5 mg/ml) (2 h at 27 °C) and assayed for IL-12 biological activity. For further details see legend to Table 2.

[a] Not done.

Table 5. Effect of graded concentrations of IL-2 on IL-1 secretion from γ-IFN-treated monocytes

IL-2 (U/ml)	^{3}H-Td Incorporation of murine thymocytes induced by monocyte conditioned medium		
	Expm. 1	Expm. 2	Expm. 3
0	256	293	301
2.5	249	289	317
25	436	301	314
250	5332	4736	6004
500	15937	13889	16019

Monocytes (2.5×10^5/ml) were cultured in the presence of γ-IFN (250 U/ml). After 60 h, graded concentrations of IL-2 (2.5–500 U/ml) were added to the cultures. After further 24 h, cell-free supernatants were harvested, treated with antibodies 3C11C8 and DMS1 (as described in Table 4), absorbed with CTLL cells 24 h, 4 °C), and assayed for IL-1 biological activity. For further details see legend to Table 2.

uisite for the events leading to effector functions of cells participating in an immune response [6]. Production of γ-IFN is closely associated with the production of IL-2 in IL-2-R-bearing T cells, which is regulated at the transcriptional level [56–60]. IL-2-R expression in T cells and IL-2 production is amplified by IL-1 and requires the presence of appropriate MHC products on accessory cells [31] that are in turn regulated by γ-IFN [43, 61–63]. More recently, the T-cell lymphokine TNF-β and monokine TNF-α have been implicated in this cascade: it was demonstrated that both molecules are involved in the cytotoxic effector function of the respective cell type when induced with γ-IFN or IL-2 and when acting synergistically with γ-IFN [60, 64, 65]. Therefore, the study of the regulation of the functional interrelationship of these lymphocyte/monocyte-derived activities may represent an interesting model for the investigation of intercellular communications in the immune response.

Whereas the association of the interactions of the humoral regulator molecules during the intial steps of the cascade of an immune response – i.e., the IL-1-mediated enhancement of IL-2 production, leading to IL-2-mediated enhancement of γ-IFN production, leading to secretion of TNF-α and -β, and the final downregulation of macrophage Ia-expression, IL-1 secretion, and T-cell IL-2 and γ-IFN release by prostaglandins or other cAMP agonists – is well documented [50, 66–68], relatively little is known about mechanisms that contribute to the preservation and perpetuation of these activities.

The studies in this article present evidence for the surface expression of IL-2-R in cultured monocytes and detail their possible implication in the secretion of IL-1. IL-2-R were detected by several anti-IL-2-R mcAbs and were indistinguishable from T-lymphocyte IL-2-R by immunoprecipitation. The prerequisite for monocyte IL-2-R expression was contact with plastic surfaces during culture, since monocyte culture in hydrophobic dishes did not result in IL-2-R expression (data not shown). The time course of monocyte IL-2-R expression that was first detectable after 24 h of culture suggested de novo synthesis rather than the unmasking of cryptic receptors. A several-fold increase in the number of IL-2-R-bearing monocytes was obtained when monocyte cultures were performed in the presence of γ-IFN (100–250 U/ml) or, to a lower extent, in the presence of LPS (15 μg/ml) and PHA (2 μg/ml). This effect was not seen when monocytes were cultured in the presence of purified recombinant preparations of αA-IFN, β-IFN, TNF-α, or other modulators of monocyte function such as PGE_2, and 1.25 (OH) 2 vitamin D_3. However, culture combinations of γ-IFN with TNF-α or PGE_2 lead to a further increase (TNF-α) or decrease (PGE_2) in the number and fluorescence intensity of IL-2-R-expressing monocytes as well as in the levels of surface HLA-DR expression (data not shown), confirming previous observations of synergistic or antagonistic actions of both

Table 6. Effect of monoclonal antibody to interleukin-2-receptor (IL-2-R1) on interleukin-1 by γ-IFN-pretreated monocytes cultured in media containing interleukin-2

Induction culture in	^{3}H-Td Incorporation of murine thymodytes induced by monocyte conditioned media stimulated with							
	Control medium γ-IFN		Control medium + γ-IFN		Control medium IL-2		Control medium + γ-IFN	
	Expm. 1	Expm. 2	Expm. 1	Expm. 2	Expm. 1	Expm. 2	Expm. 1	Expm. 2
Culture medium	282	279	280	286	1433	1291	17355	15137
IL-2-R1 (1:1000)	291	268	301	289	1352	1284	14997	13989
IL-2-R1 (1:100)	286	272	284	289	997	1017	8569	6146
Control antibody 1	274	273	270	278	1439	1276	17158	15206
Control antibody 2	259	268	267	271	1418	1272	17204	15217

Monocytes (2.5×10^{5}/ml) were cultured in culture medium in the absence or presence of γ-IFN (250 U/ml). After 26 h, cultures received a second γ-IFN (250 U/ml) or an IL-2 (500 U/ml) pulse. After an additional 24 h, monoclonal antibody IL-2-R1 (1:1000 and 1:100 final dilution) or isotype-.identical control binding (1) or nonbinding (2) antibodies were added. After further 12 h of culture, cell-free supernatants were harvested, treated with antibodies 3C11C8 and DMS1 (as described in Table 4), and assayed for IL-1 biological activity. In Expm. 2 supernatants were additionally absorbed with cloned murine CTLL cells (24 h, 4 °C) before assay for IL-1 activity. γ-IFN and antibody to γ-IFN (3C11C8), IL-2 (DMS1), IL-2-R (anti IL-2-R1), and control antibodies had no effect on the thymocyte assay for IL-1 by themselves, nor did they modify the activity of a standard IL-1 preparation (purified IL-1, Genzyme, Norwalk, CT) (not shown). For further details see legend to Table 2.

components with γ-IFN in terms of other macrophage functions [50, 64, 69, 70]. IL-2-R expression in nonlymphoid cells has now been repeatedly documented in transformed cells including monoblast line U 937 [43] and promyelocytic line HL 60, when differentiation along the monocytic axis was induced with γ-IFN [43, 71], whereas myeloblast line KG 1 [43] and a newly established line derived from a human eosinophilic leukemia expressed IL-2-R [72] in the absence of exogenous inducing agents. Although we were unable to detect inducible or constitutively expressed IL-2-R on mouse peritoneal macrophages using antimouse IL-2-R antibody AMT 13 (T. Diamantstein and F. Herrmann, unpublished observations), in line with previous reports that functions of macrophages derived from different sites or sources may be regulated differentially [73–75], the recent finding of low-affinity IL-2-R on murine IL-3-dependent myelomonocytic cell lines [76] demonstrates the lack of possible interspecies differences.

To examine the biological consequences of IL-2-R expression by human peripheral blood-derived monocytes, in particular when upregulated by γ-IFN, we assessed the effects of the interactive cascade of both lymphokines on the regulation of monocyte IL-1 secretion, since it has been shown that T-cell products may induce IL-1 secretion in the absence of cell-to-cell contact [39, 40, 53]. Every attempt was made, and a large array of control experiments were introduced to rule out any contaminants that would cause mitogenic activity other than that of IL-1, including the use of low-endotoxin serum, highly purified preparations of recombinant lymphokines, and highly enriched monocyte preparations. In addition, we excluded the presence of exogenous or endogenously generated biologically active lymphokines in the final IL-1 assay, using neutralizing mcAbs to γ-IFN and IL-2 as well as absorption techniques to purify the supernatants to be assayed for IL-1 from IL-2 activity by using an IL-2-dependent murine T-cell clone.

Under these experimental conditions our results can be summarized as follows: Monocytes do not secrete IL-1 in the absence of inducers such as LPS, IL-2, or γ-IFN, and, more effectively, the synergistic action of both lymphokines results in an up to sixfold enhancement of LPS-inducible IL-1 secretion. The loss of monocytes to release IL-1 due to aging in culture was overcome when monocytes were cultured in the presence of γ-IFN, IL-2, and, more effectively, in the presence of *both* T-cell products before induction with LPS. These results confirm previous observations by other groups in some aspects [26, 63, 77]. Whereas γ-IFN seems not to have any IL-1-inducing property, an IL-1-inducing capacity was demonstrated for IL-2, although concentrations of more than 250 U/ml were necessary.

The capacity of IL-2 to induce IL-1 secretion required at least in part the presence of IL-2-R on monocytes, as demonstrated by blocking experiments with anti-IL-2-R mcAbs. Prior enhancement of the number of IL-2-R-bearing monocytes by γ-IFN resulted in an about 30fold increase of IL-2-mediated secretion of IL-1 activity, suggesting a fundamental role for the sequential action of both lymphokines in establishing a positive feedback loop to generate IL-1 secretion, and thus to preserve and perpetuate biological activities required for immune responses.

Current studies are underway to address the role of IL-2 itself in modulating levels of IL-2-R in monocytes. In addition, it has been suggested that subpopulations of cultured monocytes may be unable to secrete IL-1 upon induction [73], in particular, those that are involved in counter-regulatory monocyte functions. Expression of IL-2-R on these cells, which could lead to adsorption of IL-2 activity and thus serve as a signal to limit cytotoxic events at inflammatory sites, needs further investigation.

References

1. Giri JG, Lomedico PT, Mizel SB (1985) Studies on the synthesis and secretion of interleukin-1. I. A 33000 molecular weight precursor for interleukin-1. J Immunol 134:343–349
2. March CJ, Mosley B, Larsen A, Ceretti DP, Braedt G, Price V, Gillis S, Henney, CS, Kronheim SR, Grabstein K, Conlon PJ, Hopp TP, Cosman D (1985) Cloning, sequence and expression of two distinct human interleukin-1 complementory DNAs. Nature 315:641–647
3. Lomedica, PT, Gubler U, Hellman CP, Dunkovich M, Giri JG, Pan YE, Collier K, Saminow R, Chua AO, Mizel SB (1985) Cloning and expression of murine interleukin-1 cDNA in *Escherchia coli*. Nature 312:458–462
4. Auron PE, Webb AC, Rosenwasser LJ, Mucci SF, Rich A, Wolfe SM, Dinarello CA (1984) Nucleotide sequence of human monocyte interleukin-1 precursor DNA. Proc Natl Acad Sci USA 81:7907–7912
5. Oppenheim JJ, Mizel SB, Meltzer MS (1979) Biological effects of lymphocyte- and macrophage-derived mitogenic "amplification" factors. In: Cohen S, Pick E, Oppenheim JJ (eds) Biology of the lymphokines. Academic Press, New York, p 291–302
6. Oppenheim JJ, Gery I (1982) Interleukin-1 is more than an interleukin. Immunol Today 3:113–119
7. Matsushima K, Procopio A, Abe H, Scala G, Ortaldo JR, Oppenheim JJ (1985) Production of interleukin-1 activity by normal human peripheral blood B-lymphocytes. J Immunol 135:1132
8. Scala G, Allavena P, Djeu J, Kasahara T, Ortaldo J, Herberman R, Oppenheim JJ (1984) Human large granular lymphocytes are potent producers of interleukin-1. Nature 309:56–59
9. Luger TA, Stadler BM, Katz SI, Oppenheim JJ (1981) Epidermal cell-derived thymocyte-activating factor. J Immunol 127:1493–1498
10. Lovett DH, Ryan JL, Sterzel RB (1983) Stimulation of rat mesangial cell proliferation by macrophage interleukin-1. J Immunol 131:2830–2836
11. Sztein MB, Vogel SN, Sipe JD, Murphy PA, Mizel SB, Oppenheim JJ (1981) The role of macrophages in the acute phase response: SAA inducer is closely related to lymphocyte-activating factor and endogenous pyrogen. Cell Immunol 63:164–176
12. Dinarello CA (1984) Interleukin-1 and the pathogenesis of the acute phase response. N Engl J Med 311:1413–1418
13. Mizel SB, Dayer JM, Krane SM, Mergenhagen HG (1981) Stimulation of rheumatoid synovial cell collagenase and prostaglandin production by partially purified lymphocyte-activating factor (interleukin-1). Proc Natl Acad Sci USA 78:2474–2477
14. Rossi V, Breviario F, Ghezzi P, Deyana E, Mantovani A (1985) Prostacyclin synthesis induced in vascular cells by interleukin-1. Science 229:174–176
15. Bevilacqua MP, Pober JS, Majeau GR, Cotran RS, Gimbrone MA (1984) Interleukin-1 induces biosynthesis and cell surface expression of procoagulant activity in vascular endothelial cells. J Exp Med 160:618–623
16. Bevilacqua MP, Pober JS, Wheeler ME, Cotran RS, Gimbrone MA (1985) Interleukin-1 acts on cultured human vascular endothelial cells to increase the adhesion of polymorphonuclear leukocytes, monocytes and related leukocyte cell lines. J Clin Invest 76:2003–2011
17. Schmidt JA, Mizel SB, Cohen D, Green I (1982) Interleukin-1 a potential regulator of fibroblast proliferation. J Immunol 128:2177–2182
18. Van Damme J, De Ley M, Opdenakker G, Billiau A, de Sommer P, Van Beekmen J (1985) Homogenous interferon-inducing 22 k factor is related to endogenous pyrogen and interleukin-1. Nature 314:266–268
19. Gowen M, Meikle MC, Reynolds JJ (1983) Stimulation of bone resorption in vitro by a nonprostanoid factor released by human monocytes in culture. Biochim Biophys Acta 762:471–474
20. Onozaki K, Matsushima K, Aggarwal BB, Oppenheim JJ (1985) Human interleukin-1 is a cytocidal factor for several tumor cell lines. J Immunol 135:3962–3968
21. Lovett D, Kozan B, Hadam M, Resch K, Gemsa D (1986) Macrophage cytotoxicity: interleukin-1 as a mediator of tumor cytostasis. J Immunol 136:340–347
22. Farrar WL, Mizel SB, Farrar JJ (1980) Participation of lymphocyte-activating factor "interleukin-1" in the induction of cytotoxic T-cell response. J Immunol 124:1371–1377

23. Herrmann F, Oster W, Meuer SC, Lindemann A, Mertelsmann RH (1988) Interleukin-1 stimulates T-lymphocytes to produce granulocyte-monocyte colony-stimulating factor. J Clin Invest 81:1415–1419
24. Chu E, Rosenwasser LJ, Dinarello CA, Larean M, Geha RS (1984) Role of interleukin-1 in antigen-specific T-cell proliferation. J Immunol 132:1311–1316
25. Chu E, Gesner M, Gorga J, Geha RS (1985) Role of Ia antigens and interleukin-1 in T-cell proliferation to phytohemagglutinin. Clin Immunol Immunopathol 36:70–80
26. Haq AU, Mayernik DG, Orosz C, Rinehart JJ (1984) Interleukin-1 secretion is not required for human macrophage support of T-cell proliferation. Cell Immunol 87:517–527
27. Shaw J, Caplin B, Paetkan V, Pilarski M, Delovitch Tl, McKenzie IFC (1980) Cellular origins of co-stimulator and its activity in cytotoxic lymphocyte responses. J Immunol 124:2231–2239
28. Ruscetti FW, Mier JW, Gallo RC (1980) Human T-cell growth factor: parameters of production. J Supramol Struct 13:229–241
29. Stadler BM, Sougherty SF, Farrar JJ, Oppenheim JJ (1981) Relationship of cell cycle to recovery of IL-2 activity from human mononuclear cells, human and mouse T-cell lines. J Immunol 127:1936–1941
30. Kaye J, Gillis S, Mizel SB, Shevach EM, Malek TR, Dinarello CAA, Lachman BL, Janeway CA (1984) Growth of a cloned helper T-cell line induced by a monoclonal antibody specific for the antigen receptor. Interleukin-1 is required for the expression of receptor for interleukin-2. J Immunol 133:1339–1345
31. Schwab R, Crow MK, Russo C, Weksler ME (1985) Requirements for T-cell activation by OKT 3 monoclonal antibody: role of modulation of T-3 molecules and interleukin-1. J Immunol 135:1714–1718
32. Dempsey RA, Dinarello CA, Mier JW, Rosenwasser LJ, Allegretta M, Brown TE, Parkinson DR (1982) Differential effects of human leukocytic pyrogen/lymphocyte-activating factor, T-cell growth factor and interferon on human natural killer activity. J Immunol 129:2504–2510
33. Booth RJ, Prestidge RL, Watson JD (1983) Constitutive production by the WEHI-3 cell line of B-cell growth and differential factor, that co-purifies with interleukin-1. J Immunol 131:1289–1293
34. Booth RJ, Watson JD (1984) Interleukin-1 induces proliferation of two distinct B-cell subpopulations responsive to two different murine B-cell growth factors. J Immunol 133:1346–1349
35. Howard M, Mizel SB, Lachman LB, Ansel J, Johnson B, Paul WE (1983) Role of interleukin-1 in anti-immunoglobulin-induced B-cell proliferation. J Exp Med 157:1529–1536
36. Hoffmann MK, Mizel SB, Hirst JA (1984) IL-1 requirement for B-cell activation revealed by use of adult serum. J Immunol 133:2566–2568
37. Falkoff RJM, Maruguschi A, Hong JX, Butler JL, Dinarello CA, Fauci AS (1983) The effects of interleukin-1 in human B-cell activation and proliferation. J Immunol 131:801–810
38. Lipsky PE, Thompson PA, Rosenwasser LJ, Dinarello CA (1983) The role of interleukin-1 in human B-cell activation: inhibition of B-cell proliferation and the generation of immunoglobulin-secreting cells by an antibody against human leukocyte pyrogen. J Immunol 130:2708–2714
39. Unanue ER, Kiely JM, Calderon J (1976) The modulation of lymphocyte functions by molecules secreted by macrophages. II. Conditions leading to increased secretion. J Exp Med 144:155–161
40. Unanue ER (1980) Cooperation between mononuclear phagocytes and lymphocytes in immunity. N Engl J Med 303: 977–981
41. Palacios R (1985) Monoclonal antibodies against human Ia antigens stimulate monocytes to secrete interleukin-1. Proc Natl Acad Sci USA 82:6652–6656
42. Moore RN, Oppenheim JJ, Farrar JJ, Carter CS, Waheed A, Shadduck RK (1980) Production of lymphocyte-activating factor (interleukin-1) by macrophages activated with colony-stimulating factors. J Immunol 125:1302–1305
43. Herrmann F, Cannistra SA, Levine H, Griffin JD (1985) Expression of interleukin-2 receptors and binding of interleukin-2 by gamma interferon induced human leukemic and normal monocytes. J Exp Med 162:1111–1116
44. Herrmann F, Cannistra SA, Griffin JD (1986) T-cell-monocyte interactions in the production of humoral factors regulating human granulopoieses in vitro. J Immunol 136:2856–2863

45. Todd RF, Nadler LM, Schlossman SF (1981) Antigens on human monocytes identified by monoclonal antibodies. J Immunol 126:1435–1441
46. Reinherz EL, Kung PC, Goldstein G, Levy RH, Schlossman SF (1980) Discrete stages of human intrathymic differentiation: analysis of normal thymocytes and leukemic lymphoblasts of T lineage. Proc Natl Acad Sci USA 77:1588–1592
47. Stashenko P, Nadler LM, Hardy R, Schlossman SF (1980) Characterization of a human B-lymphocyte-specific antigen. J Immunol 125:1678–1685
48. Hercend T, Griffin JD, Bensussan A, Schmidt RE, Edson MA, Brennan A, Murray C, Daley JF, Schlossman SF, Ritz J (1985) Generation of monoclonal antibodies to a human natural killer clone. Characterization of two natural killer-associated antigens, NKH IA and NKH 2, expressed by subsets of large granular lymphocytes. J Clin Invest 75:932–943
49. Fox DA, Hussey RE, Fitzgerald A, Bensussan A, Daley JF, Schlossman SF, Reinherz EL (1984) Activation of human thymocytes via the 50 kD T-11 sheep erythrocyte binding protein induces the expression of interleukin-2 receptors on both T-3 (+) and T-3 (–) populations. J Immunol 134:330–337
50. Kunkel SL, Chensue SW, Phan SH (1986) Prostaglandins as endogenous mediators of interleukin-1 production. J Immunol 136:186–192
51. Uchiyama T, Broder S, Waldmann TA (1981) A monoclonal antibody (anti-TAC) reactive with activated and functionally mature human T cells. J Immunol 126:1393–1399
52. Rubin LA, Kurman CC, Biddison WE, Goldman ND, Nelson DL (1985) A monoclonal antibody, 7G7/B6, binds to an epitope on the human interleukin-2 receptor that is dinstinct from that recognized by IL-2 or anti-TAC. Hybridoma 4:91–102
53. Mizel SB, Oppenheim JJ, Rosenstreich DL (1978) Characterization of lymphocyte-activating factor produced by macrophage cell line P388D1. I. Enhancement of LAF production by activated T-lymphocytes. J Immunol 90:1497–1502
54. Conlon PJ (1983) A rapid biological assay for the detection of interleukin-1. J Immunol 131:1280–1285
55. Gillis S, Ferm MM, Ou W, Smith KA (1978) T-cell growth factor, parameter of production and a quantitative microassay for activity. J Immunol 120:2027–2034
56. Kasahara T, Hooks JJ, Dougherty SF, Oppenheim JJ (1983) Interleukin-2-mediated immune interferon (IFN-gamma) production by human T cells and T-cell subsets. J Immunol 130:1784–1792
57. Reem GH, Yeh NH (1984) Interleukin-2 regulates the expression of its receptor and synthesis of gamma-interferon by human T-lymphocytes. Science 225:429–430
58. Vilcek J, Henriksen-Destefano D, Siegel D, Klion A, Robb RJ, Le J (1985) Regulation of IFN-gamma induction in human peripheral blood cells by exogenous and endogenously produced interleukin-2. J Immunol 135:1851–1856
59. Palacios R (1984) Production of lymphokines by circulating human lymphocytes that express or lack receptors for interleukin-2. J Immunol 132:1833–1840
60. Svedersky LP, Nedwin GE, Goeddel DV, Palladino MA (1985) Interferon-gamma enhances production of lymphotoxin in recombinant interleukin-2-stimulated peripheral blood mononuclear cells. J Immunol 134:1604–1608
61. Kelley VE, Fiers W, Strom TB (1984) Cloned human interferon-gamma, but not interferon-beta or interferon-alpha, induces expression of HLA-DR determinants by fetal monocytes and myeloid leukemic cell lines. J Immunol 132:240–245
62. Sztein MB, Steeg PS, Johnson HM, Oppenheim JJ (1984) Regulation of human peripheral blood monocyte DR antigen expression in vitro by lymphokines and recombinant interferons. J Clin Invest 73:556–565
63. Newton RC (1985) Effect of interferon on the induction of human monocyte secretion of interleukin-1 activity. Immunology 56:441–449
64. Williamson BD, Carswell EA, Rubin BY, Prendergast JS, Old LJ (1983) Human tumor necrosis factor produced by human B-cell lines: synergistic cytotoxic interaction with human interferon. Proc Natl Acad Sci USA 80:5397–5402
65. Nedwin GE, Svedersky LP, Bringman TS, Palladino MA, Goeddel DV (1985) Effects of interleukin-2, interferon-gamma, and mitogens on the production of tumor necrosis factor alpha and beta. J Immunol 135:2492–2497

66. Synder DS, Beller DI, Unanue ER (1982) Prostaglandins modulate macrophage Ia expression. Nature 299:163–164
67. Walker C, Kristensen F, Bettens F, Deweck AL (1983) Lymphokine regulation of activated (G-1) lymphocytes. I. Prostaglandin E_2-induced inhibition of interleukin-2 production. J Immunol 130:1770–1778
68. Chouaib S, Welte K, Mertelsmann R, Dupont B (1985) Prostaglandin E_2 acts at two distinct pathways of T-lymphocyte activation: inhibition of interleukin-2 production and downregulation of transferrin receptor expression. J Immunol 135:1172–1179
69. Stone-Wolff DS, Yip YK, Kelker HC, Le J, Henriksen-Destefano D, Rubin BY, Rinderknecht E, Aggarway BB, Vilcek J (1984) Interrelationship of human interferon-gamma with lymphotoxin and monocyte cytotoxin. J Exp Med 159:828–843
70. Russel SW, Pace JL (1984) Gamma-interferon interferes with the negative regulation of macrophage activation by prostaglandin E_2. Mol Immunol 21:249–255
71. Dubreuil P, Mannoni P, Olive D, Winkler-Lowen B, Mawas C (1985) Expression of T-cell-related antigens on cells from the myelomonocytic lineage. In: Reinherz EL, Haynes BF, Nadler LM, Bernstein ID (eds) Leukocyte typing II, vol 1. Springer, Berlin Heidelberg New York, p 335
72. Saito H, Bourinbaiar A, Ginsburg M, Minato K, Ceresi E, Yamada K, Machover D, Breard J, Mathe G (1985) Establishment and characterization of a new human eosinophilic leukemia cell line. Blood 66:1233–1241
73. Khansari N, Chou YK, Fudenberg HH (1985) Human monocyte heterogeneity: interleukin-1 and prostaglandin E_2 production by separate subsets. Eur J Immunol 15:48–51
74. Hayari Y, Kubulansi T, Globerson A (1985) Regulation of thymocyte proliferative response by macrophage-derived prostaglandin E_2 and interleukin-1. Eur J Immunol 15:43–47
75. Koretzky GA, Elias JA, Kay SI, Rossman MD, Nowell PC, Daniele RP (1983) Spontaneous production of interleukin-1 by human alveolar macrophages. Clin Immunol Immunopathol 29:443–450
76. Koyasu S, Yodoi J, Nikaido I, Tagaya Y, Taniguchi Y, Honjo T, Yahara I (1986) Expression of interleukin-2 receptors on interleukin-3-dependent cell lines. J Immunol 136:984–987
77. Arenzana-Seisdedos F, Virelizier JL, Fiers W (1985) Interferons as macrophage-activating factors. III. Preferential effects of interferon-gamma on the interleukin-1 secretory potential of fresh or aged human monocytes. J Immunol 134:2444–2451

The Expression of T-Cell Receptor-Associated Proteins in Normal and Leukaemic Immature T Cells

D. Campana [1], E. Coustan-Smith [1], L. Wong [2], and G. Janossy [1]

A. Introduction

T lineage cells express different T-cell receptor (TCR) heterodimers, $\alpha\beta$ and $\gamma\delta$, in association with the same set of monomorphic CD3 polypeptides [1–6]. In the post-natal thymus and the peripheral blood of humans [7, 8], rodents [9, 10] and avians [11, 12] TCR$\alpha\beta$ cells are predominant. The distribution of these subpopulations during ontogeny, however, might deviate from this rule. For example, in 12- to 14-day chicken embryos, cells expressing TCR$\gamma\delta$ are higher in number than cells with membrane TCR$\alpha\beta$ [11, 12].

The origin, proliferative activity and expression of CD3 and TCR molecules of the cells which migrate into the thymic epithelial rudiment are not yet fully documented. Several studies have suggested that the rearrangement of the genes encoding for the TCRβ, -γ and -δ chains occurs early in ontogeny, preceding that of the TCRα locus [9, 13–16]. In mice, approximately 75% of thymocytes have rearranged the β chain locus and transcribe the corresponding mRNA at the 17th day of fetal life, when the activation of the TCRα chain gene is initiated [15, 16].

The putative evidence for the rearrangement of the TCR$\gamma\delta$ genes in immature thymocytes derives mainly from the investigation of their configuration in selected CD4$^-$, CD8$^-$ ("double negative") cells, regarded as T-cell precursors [5, 9, 17]. Since such "double negative" thymocytes also include TCR$\gamma\delta^+$ cells (together with TCR-germline non-T cells), the finding of rearranged TCR$\gamma\delta$ loci in such populations is not entirely unexpected, but the suggestion that these clones are derivatives of immature cells is not fully convincing. On the basis of these experiments it has been hypothesized that the rearrangement of TCR$\gamma\delta$ is attempted first and is followed, when not successful, by that of β and α genes. However the possibility that the two lineages follow separate pathways is still open [18]. It is relevant here that the Cδ locus is situated between the Vα and Jα loci [19]; the rearrangement of α genes deletes the Cδ locus, rendering the simultaneous $\alpha\delta$ or successive $\alpha\rightarrow\delta$ expression unlikely.

It is known that once in the thymus, immature T cells actively divide and accumulate CD3 molecules in the cytoplasm [20–22]. This is an early sign of T-cell commitment which is reflected by the consistent expression of cytoplasmic CD3 (cCD3) in thymic acute lymphoblastic leukaemia (T-ALL), the malignant counterpart of early thymic progenitors [21, 22]. It is also known that these malignant cells monoclonally rearrange their TCRβ and TCRγ genes and transcribe TCRβ mRNA [23, 24]. Nevertheless, it has not yet been explored whether such T-ALL blasts are capable of expressing the various TCR-associated chains in their cytoplasm or in their membrane.

Monoclonal antibodies (MAbs) to human TCR proteins have been recently developed. One of these MAbs, βF1, has a framework reactivity against the TCR$\alpha\beta$

[1] Department of Immunology, Royal Free Hospital School of Medicine, London NW3 2QG, UK

[2] Fetal Tissue Bank, Royal Marsden Hospital, London SW3 6JJ

and recognizes the separated β subunit in Western blotting. This antibody does not bind to the surface of viable T cells and requires membrane permeabilization for its use [25]. The MAb βF1 can also be used in an informative combination with WT31, a MAb which recognizes an epitope jointly formed by the assembled α and β chains of the TCR complex [26]. Cells expressing βF1 but no WT31 contain free β chains without membrane TCR. The third reagent, TCRδ-1, identifies one determinant of the human TCRδ chain on the surface of a T-lymphocyte subpopulation [27].

In our study [27a, b], we first applied MAbs to CD3 and TCR chains in double- and triple-colour staining methods in order to investigate the expression of the TCR$\alpha\beta$ and δ proteins during fetal and post-natal T-cell differentiation. Secondly, we used the same reagents to analyse acute leukaemia blasts and compared these findings to the TCR expression in normal immature T cells.

B. Materials and Methods

I. Handling of Samples

Human fetal tissues were obtained following legal termination of pregnancy. The gestational age ranged from 7 to 20 weeks. This was calculated from the foot length compared to a standard curve constructed on the basis of a comparative analysis of this parameter, menstrual records and, when possible, of crown/rump length in 3000 samples studied at the same institution (L. Wong, manuscript in preparation). Infant and adult thymus samples were from patients undergoing cardiac surgery, aged 3–10 years and 17–25 years, respectively. Cell suspensions prepared from parts of the liver and the thymus samples were washed twice in RPMI-1640 with L-glutamine (Gibco, Paisley, UK) and 10% fetal calf serum.

Normal peripheral blood (PB) samples were obtained from healthy volunteers (aged 23–37 years). PB and bone marrow (BM) samples of patients with acute and chronic leukaemia were received by the laboratory for immunological diagnosis. Mononuclear cells from PB, BM and cell suspensions were separated after centrifugation on Ficoll-Hypaque density gradient and washed three times in phosphate-buffered saline (PBS).

II. Staining of Cells

Monoclonal antibodies used in this study are listed in Table 1. Double-colour immunofluorescence (IF) was performed by using second-layer antisera to different mouse Ig isotypes and to Ig of other species. These were conjugated to fluorochromes such as fluorescein isothyocyanate (FITC) and tetramethylrhodamine isothyocyanate (TRITC) or to colloidal gold. Biotin-conjugated βF1 was used with the corresponding second-layer streptavidin (Sigma Chemicals, Poole, UK; cat. no. S. 4762) labelled with TRITC in the laboratory.

Cells at a final concentration of 2×10^7/ml were resuspended in PBS containing 0.2% bovine serum albumin and 0.2% sodium azide (PBSA). The viability, assessed with trypan blue, was >95% in the adult and infant samples and >80% in the fetal samples studied. Fifty milliliters of these suspensions was distributed into LP3 tubes (Luckham, West Sussex, UK), and MAbs were added at saturating concentrations ranging from 0.1 µg to 1.0 µg as determined by titration. After 10 min incubation at room temperature cells were washed twice in PBSA and incubated with second-layer antisera for a further 10 min. After two washes in PBSA a small droplet from the pellet was transferred onto a microscope slide, covered with a coverslip and observed on a Zeiss fluorescence microscope equipped with selective filters for FITC and TRITC and a polarizing filter for the detection of colloidal gold [28]. Cells were also studied on an Epics V Cell Sorter (Coulter Electronics, Hialeah, Florida, USA) using the following settings: laser 200 mW at

Table 1. Monoclonal antibodies and heterologous antisera

	CD	Class	Source
Monoclonal antibodies			
βF1	–	IgG1	TCS cat no. TA1051
TCRδ-1	–	IgG1	Dr. M.B. Brenner, Boston, MA, USA
WT31	–	IgG1	BD cat no. 7770
UCHT1	3	IgG1	Dr. P. Beverley, London, UK
T10B9	3	IgM	Prof. J. Thompson, Lexington, KY, USA
MEM57	3	IgG2a	Dr. I. Hilgert, Prague, Czechoslovakia
NA1/34	1	IgG2a	Prof. A. McMichael, Oxford, UK
RFT11	2	IgG1	RFH
RFT1	5	IgG1	RFH
RFT2	7	IgG2a	RFH
RFT4	4	IgG1	RFH
RFT8	8	IgM	RFH
Leu11b	16	IgM	BD cat no. 7530
RFDR1	anti-HLA-DR	IgM	RFH
2D1	45	IgG1	Dr. P. Beverley, London, UK
RFAL3	10	IgM	RFH
RFB7	37	IgM	RFH
RFB4	22	IgG1	RFH
Ki67	–	IgG1	Dakopatts, Denmark, cat no. M722
Heterologous antisera			
Rabbit anti-TdT			SL cat no. CT-004
G-anti-M Ig FITC/TRITC			RFH
G-anti-R IgG FITC			SL cat no. CT-008
G-anti-R Ig TRITC			SL cat no. SBA-4010-03
G-anti-M IgM TRITC			SL cat no. SBA-1020-03
G-anti-M IgG FITC			SL cat no. SBA-1030-02
G-anti-M IgG1 TRITC			SL cat no. SBA-1070-03
G-anti-M IgG2a FITC			SL cat no. SBA-1080-02
G-anti-M IgG gold-conjugated			Janssen, Belgium, cat no. 23.688.20
Streptavidin TRITC			RFH

TCS, T Cell Sciences Inc., Cambridge, MA, USA; BD, Becton Dickinson, Mountain View, CA, USA; RFH, Royal Free Hospital, London, UK; SL, Sera-Lab Ltd., Sussex, UK.

480 nm; photomultiplier tube 700 V; 515 nm dichroic mirror to collect 90° light scatter and 515 nm blocking filter.

In order to visualize intracellular antigens, cytocentrifuge preparations of unlabelled and labelled mononuclear cells were made. For TdT and Ki67 staining cytospins were fixed in cold methanol for 30 min or in a mixture of cold acetone and methanol for 15 min. For visualizing cytoplasmic antigens slides were fixed in acetone for 5–10 min at 20° C. Cytospins and tissue sections were incubated with MAbs for 45 min and washed in PBS. Second-layer antisera were added for the same length of time. After washing in PBS, the slides were mounted in an equal mix of glycerol and PBS.

Double-colour IF with βF1 and another MAb of IgG_1 class (e.g. UCHT1) was performed as follows. First, cells were stained with UCHT1, followed by goat antiserum to mouse Ig-FITC (G-anti-M Ig FITC). Of these cells cytocentrifuge preparations were made. To these normal mouse serum was added in order to saturate free combining sites of the previous second layer. Finally, the smears were incubated with βF1 conjugated to biotin and streptavidin TRITC. It is known that βF1 labels β chains only in permeabilized cells. In the triple-staining experiments, cell suspensions were first incubated in suspension with WT31 + G-anti-M IgG conjugated to colloidal gold, and cytocentrifuge preparations were made. After fixation in acetone:methanol 1:1 and incubation for 30 min with diluted normal mouse serum, rabbit antiserum to TdT (R-anti-TdT) and βF1-biotin were both added. The second layers were goat antisera to rabbit Ig conjugated to FITC (G-anti-R Ig FITC) and streptavidin TRITC, respectively. Controls, including MAbs of identical subclass (e.g. CD22; IgG_1) were studied in parallel samples in place of WT31 or βF1.

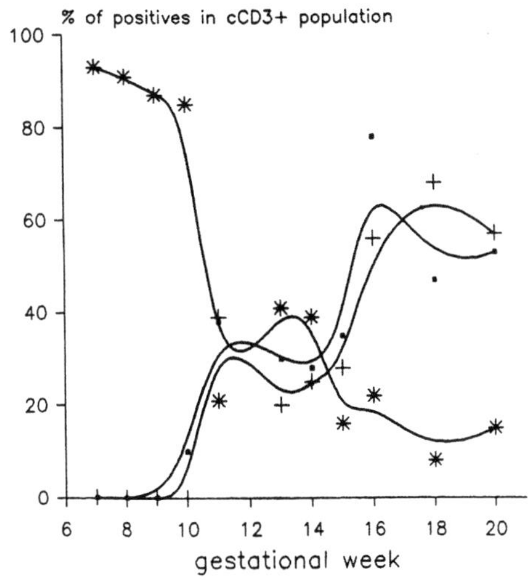

Fig. 1. Phenotypic and proliferative changes of $CD3^+$ cells in the fetal liver during ontogeny. In samples obtained from the 7th to the 9th week of gestation, $cCD3^+$ cells are $mCD3^-$ (+) and $\beta F1^-$ (●). These cells show a high proliferative activity, indicated by Ki67 expression (*). After the 10th week of gestation $mCD3^+$ and $\beta F1^+$ cells appear, and the proliferative activity decreases [27a]

C. Results

I. Phenotype of T Lineage Cells in the Fetal Liver

Fifteen samples of fetal liver of 7–20 weeks gestation were studied [27a]. Lymphoid cells labelled by CD3 (UCHT1) in cytocentrifuge preparations were found in all the samples analysed and represented 0.6%–12% of the mononuclear cells. When the staining was performed in suspensions of viable cells, no membrane CD3 ($mCD3^+$) cells were seen in the liver samples of 7- to 9-week-old fetuses, but a few $mCD3^+$ cells appeared at 10.5 weeks (39% of $cCD3^+$ cells). These cells then persisted in samples from 11- to 20-week-old fetuses. Nevertheless, the percentages of cells labelled by UCHT1 in cytospins (1.0%–12%) remained higher than those labelled in cell suspension (0.1%–7%), suggesting that a considerable proportion (0.8%–5.2%) of cells expressed cCD3 but no mCD3. These findings were confirmed by double-colour IF staining using a CD3 MAb of IgM class (T10B9) in suspension in combination with one of IgG_1 class (UCHT1) on cytospins. In the earliest samples at the 7th to 9th weeks no $cCD3^+$ cells had mCD3 ($cUCHT1^+$, $mT10B9^-$), while in those from the 11th to 20th weeks 9%–68% of $cCD3^+$ cells expressed mCD3 ($cUCHT1^+$, $mT10B9^+$). Thus T-lineage cells of $cCD3^+$, $mCD3^-$ phenotype persist in fetal liver samples taken after the 11th gestational week, i.e. even after the emergence of $mCD3^+$ lymphocytes (Fig. 1). From these findings we conclude that immature T-lineage cells appear in the fetal liver prior to the emergence of the thymus and occur together with more mature cells after the establishment of the thymic gland.

The phenotypic analysis of $cCD3^+$, $mCD3^-$ cells taken at 7–9 weeks of age showed that >98% were $CD7^+$ ($RFT2^+$), $CD45^+$ ($2D1^+$) and $D8^{+/-}$ ($RFT8^{+/-}$) without any detectable staining with CD1 (Na1/34), CD2 (RFT11), CD5 (RFT1) or CD4 (Leu3). Cytoplasmic CD^{3+} cells were also negative with CD16

(Leu11b), anti-HLA-DR (RFDR1), CD10 (RFAL3) and did not show nuclear TdT staining, although other $CD3^-$ cells including B-cell progenitors were labelled by the same reagents in all the samples studied. The majority (>90%) of $CD7^+$ ($RFT2^+$) cells were $cCD3^+$. These findings confirm the immaturity of these precursors.

In the following experiments the TCRβ chain expression was studied using βF1 in fixed cytocentrifuge preparations. In the liver samples obtained at weeks 7–9 $cCD3^+$ cells failed to express TCRβ chains ($\beta F1^-$). The $\beta F1^+$ cells first appeared after the 11th gestational week and then persisted (Fig. 1). Double-colour IF showed that $\beta F1^+$ cells acquired mCD3 only gradually. In samples taken at weeks 14–16 15%–32% of $\beta F1^+$ cells were still $mCD3^-$. In samples at weeks 17–20 virtually all (>98%) $\beta F1^+$ cells were also $mCD3^+$.

The proliferative activity of $CD3^+$ cells was investigated using a MAb Ki67 which identifies a nuclear antigen expressed during the G_1, S, G_2 and M phases of the cell cycle [29]. In the fetal liver at weeks 7–9 the proportion of $CD3^+$ cells in cycle was 87%–93%. In the 11- to 20-week samples the proportion of $CD3^+$ $Ki67^+$ progressively declined, and only 8%–39% of $CD3^+$ cells were $Ki67^+$ (Fig. 1). This proliferative activity was seen exclusively in the $cCD3^+$, $mCD3^-$ population, for <5% of $mCD3^+$ cells were $Ki67^+$. However, in the 14- to 16-week samples a significant proportion (10%–25%) of $\beta F1^+$ cells were $Ki67^+$, indicating that some TCRβ^+, $mCD3^-$ cells are still in cycle. These findings are compatible with the possibility that $cCD3^+$ cells proliferate locally in the liver and may even develop TCRβ positivity during this proliferative cycle in situ.

II. TCR$\alpha\beta$ Expression in the Fetal and Infant Thymus

In the 15 thymus samples studied from gestational weeks 10.5–20 a high proportion of thymocytes (91%–98%) were $cCD3^+$ ($UCHT1^+$), but only 49%–72% of cells were labelled by βF1, and 19%–42% were $mCD3^{++}$. Double-labelling experiments indicated that, similarly to the fetal liver population, 17%–46% of $cCD3^+$ cells lacked TCRβ chains identifiable by the βF1 MAb. Most of these $cCD3^+$, $\beta F1^-$ cells were blasts with a high nuclear/cytoplasmic ratio and nucleoli. On the other end of the spectrum, the reactivity of WT31, detecting an epitope formed by the assembled $\alpha+\beta$ chain, was seen on 67%–91% of $mCD3^+$ cells. These cells were regarded as maturing T-lymphoid cells [27a].

In the thymocytes obtained at 10.5–18 weeks of gestation nuclear TdT was absent, as reported previously [30]. Nevertheless, positive TdT staining appeared in the four samples studied at the latest gestational periods taken at the 19th to 20th weeks. In these samples TdT^+ cells represented 53%–61% of the total thymocytes, and double-colour IF showed that the large TdT^+ blasts comprised particularly high proportions (31%–80%) of $cCD3^+$, $\beta F1^-$ putative precursor cells. The TdT positivity was absolutely restricted to the $cCD3^+$ cells, of which 67%–79% were $\beta F1^+$, but no TdT^+ cells expressed strongly WT31 among the more than 10^4 analysed [27a].

In seven samples of infant thymus five main differentiation-related markers such as TdT, cCD3, intracellular TCRβ chains (βF1), membrane-associated TCR$\alpha\beta$ (WT31) and mCD3 were studied. Four major categories of cells could be distinguished on the basis of their phenotype defined by double and triple markers (Fig. 2). Firstly, 18% (range 8%–26%) of cells were TdT^+ and expressed cCD3 only, without any detectable TCRβ ($\beta F1^-$). The majority of TdT^+ blasts (60%–80%) were amongst these $\beta F1^-$ cells (category I). The rest of TdT^+ cells (42%; range 22%–62%) were $\beta F1^+$. These were $WT31^{+/-}$ or $WT31^-$ and showed the morphology of typical small cortical thymocytes (category II). With triple-marker analysis a population

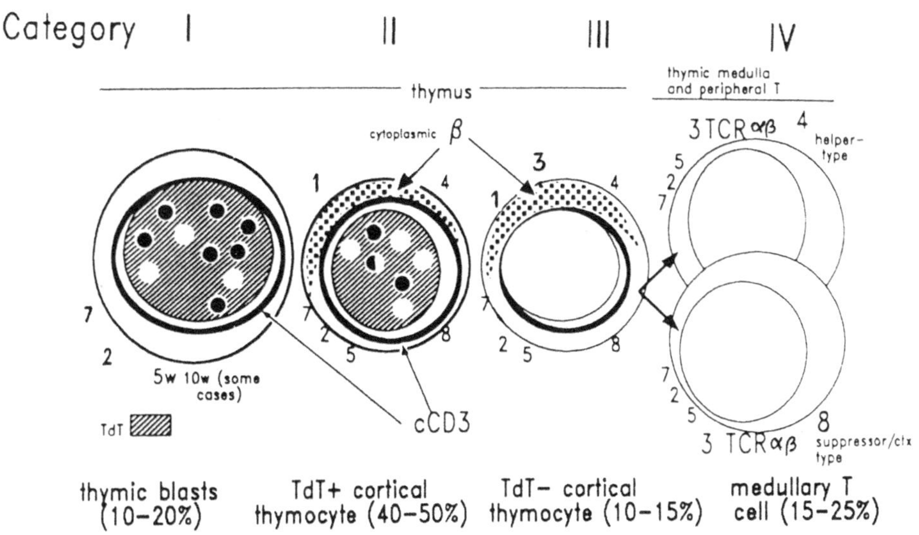

Fig. 2. Stages of T-cell differentiation defined by MAb to TCR associated molecules and TdT. CD3 expression in the cytoplasm (cCD3) and in the membrane (mCD3) is shown by *continuous circles*. The TdT labelling of the nucleus is depicted by a *mesh*. Cytoplasmic TCRβ chains are shown by *small dots*. Cells of categories I and II have a high proliferative activity, indicated by nuclear Ki67 positivity (*white dots; one dot,* 10% positive cells) and bromodeoxyuridine incorporation (*black dots; one dot,* 10% positive cells). The majority of T-ALL show features of proliferating category I (TdT$^+$, cCD3$^+$, βF1$^-$) and category II (TdT$^+$, cCD3$^+$, βF1$^+$) or asynchronous development of TCR molecules [27a]

of βF1$^+$ cells (12.5%; range 3%–30%) could be identified which were TdT$^-$ but still lacked strong WT31 (category III). Finally, 23% (range 17%–27%) of cells had features of mature T cells, signified by the membrane expression of assembled TCR$\alpha\beta$ as detected by WT31 (category IV). The intensity of staining with βF1 on cells of categories III and IV was stronger than that observed on TdT$^+$ category II thymocytes [27a].

In these infant thymic suspensions, 42% (range 28%–52%) mCD3^{++} cells were seen, including the WT31$^+$ population (category IV) together with most of the TdT$^-$, βF1$^+$ cells (category III). In addition, a few TdT$^+$ cells (2%–27%) were also weakly mCD3$^+$. These results, taken together, suggest that the insertion of CD3 into the membrane precedes the assembly of TCR$\alpha\beta$ as detected by WT31. Although WT31 binding was visualized with a very sensitive detection method, the immunogold second layer, these findings do not exclude the possibility that the cell type referred to as category III already expresses, in addition to mCD3, small amounts of TCRβ and/or TCRα on its surface. This possibility will need to be further investigated with additional MAbs to TCRβ and TCRα chains.

III. Reactivity of TCRδ-1 in Fetal and Post-natal Tissues

Fifteen samples of fetal liver (7–20 weeks of gestation) and seven samples of fetal thymus (10.5–20 weeks of gestation) were investigated for the expression of TCRδ chain in cell suspension, cytocentrifuge preparations and in cryostat section. No TCRδ-1$^+$ cells were seen in the liver samples. Very low proportions (0.02%–0.7%) of TCRδ-1$^+$ cells were observed in the cytospins of fetal thymus, irrespective of whether these were TdT$^-$ or TdT$^+$, i.e. before or after the 18th week of gestation. The rare TCRδ-1$^+$

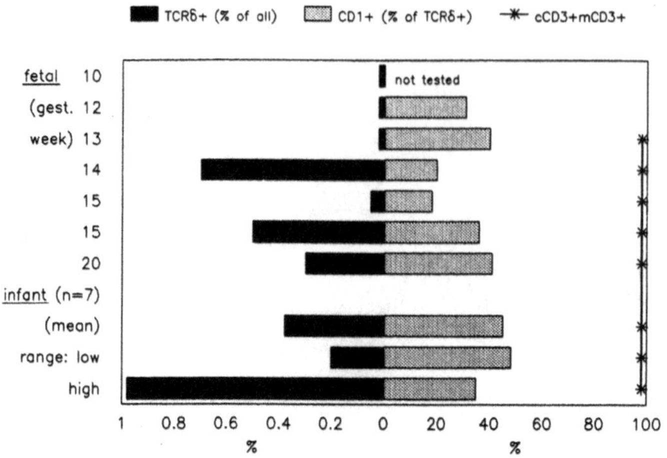

Fig. 3. Expression of TCRδ chains in fetal and infant thymus. TCRδ-1$^+$ cells are invariably CD3$^+$, class II$^-$ and TdT$^-$, but the thymic origin of a proportion of these cells is indicated by CD1 positivity [27a]

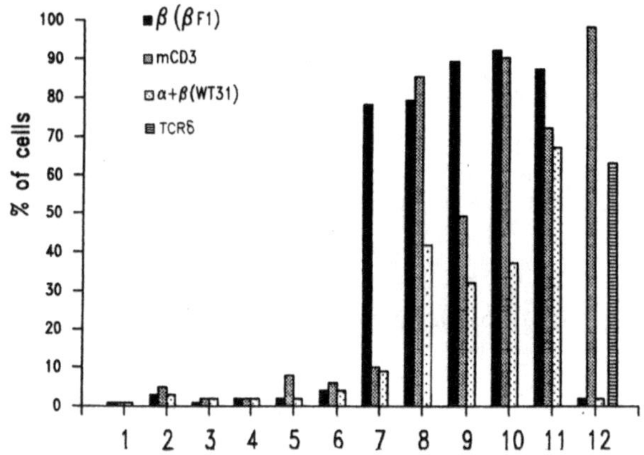

Fig. 4. Heterogeneous expression of TCR proteins in T-ALL. Half of the cases illustrated are βF1$^-$ with few residual normal T cells (TdT$^-$) that are βF1$^+$. Cases 7–11 show βF1 positivity, and in cases 8–11 some of these blasts are also WT31$^+$. Such TdT$^+$, WT31^{++} double-stained cells are not found amongst normal thymocytes. In case 12 mCD3 is associated with TCRδ chains [27b]

thymic cells together with the controls from selected donors with high proportions (9%–13%) of TCRδ-1$^+$ cells were, however, intensely labelled. In infant thymus 0.1%–0.8% cells were TCRδ-1$^+$. Similarly low values (0.1%–0.9%) were observed in 8/10 adult thymus samples, but in 2/10 samples the proportion of TCRδ-1$^+$ cells was somewhat higher – 3% and 4%. These thymic TCRδ-1$^+$ cells were invariably TdT$^-$, class II$^-$ and CD3$^+$. The thymic origin of a proportion (18%–59%) of these cells was indicated by their expression of CD1 (Fig. 3) [27a].

IV. TCR Protein Expression in Malignant T Cells

The 20 cases of T-ALL studied had rearrangements of the TCRβ and TCRγ genes, and contained >70% strongly CD7$^+$ and TdT$^+$ blasts. Cytoplasmic CD3 was detected by UCHT1 in 75%–99% of blasts in all cases studied. These T-ALL cases fell into two categories on the basis of their βF1 positivity, as illustrated by representative samples in Fig. 4. In 12 samples <5% blasts were βF1$^+$ (see e.g. cases 1–6). These βF1$^-$ cases were also WT31$^-$ (<5%) and predominantly $CD1^-$, $CD4^-$, $CD8^-$. In eight cases of T-ALL 78%–92% $\beta F1^+$ blasts were detected (see e.g. cases 7–11), and four of these had mCD3 on 49%–85% of blasts and WT31 positivity of 32%–67% of blasts (see e.g. cases 8–11). In the same subgroup of T-ALL some blasts were also $CD1^+$ (5%–63%), and large proportions of blasts (60%–92%) exhibited CD4 and/or CD8 antigens. The same samples were reinvestigated with the TCRδ1 MAb. In two $\beta F1^-$ cases TCRδ was detected in 48% and 63% of T-ALL blasts (see e.g. case 12). These $TCR\delta^+$ T-ALL were the only tow cases in our series expressing mCD3 without WT31 positivity [27b].

Finally, the lineage specificity of βF1, WT31 and TCRδ1 was investigated in non-T acute leukaemias. Ten cases of common ALL with IgH gene rearrangements were analysed, but none of these showed TCR chains on the membrane or in the cytoplasm in spite of the rearrangements of both TCRγ and TCRβ in two cases, TCRγ in another two cases and TCRβ alone in one more case. All cases of acute myeloblastic leukaemia and of B-cell chronic lymphocytic leukaemia tested showed no positivity with cCD3, βF1, WT31 and TCRδ1 [27b].

D. Discussion

It is known that haemopoietic progenitors reside in the fetal liver from the 5th gestational week, even before they appear in the BM around the 10th week [31]. It is also documented that the fetal liver contains TdT^+, $CD10^+$ and $c\mu^+$ B cell progenitors between weeks 11 and 26 [32–35]. In this study we have first investigated the possibility that T-cell progenitors might also be present in the liver in these early stages of development before the formation of the thymus, which occurs at around the 10th week.

The expression of CD3 molecules in the perinuclear area is an early event in T-cell differentiation, and the detection of cCD3 is also a reliable marker for T-cell commitment in T-ALL while other leukaemias of B and myeloid origin are invariable cCD3 negative [21, 22, 24]. Immature cells of the T lineage could indeed be identified in the fetal liver by virtue of their cCD3 positivity, and such precursors appeared to encompass two phases of successive development steps. First, at the 7th to 9th gestational weeks $cCD3^+$ cells were observed with no identifiable mCD3 or β chain ($\beta F1^-$). These cells expressed CD7 and CD45 without class II or nuclear TdT, and without any other known features of cortical thymocytes such as CD1 and CD4 with CD8. The next cell type, however, carried TCRβ and mCD3, and such cells could be seen from the 11th week onwards. In the samples from weeks 14–16 the double-labelling experiments have identified a few cells showing only TCRβ ($\beta F1^+$) without mCD3. There are two possible explanations for this finding: $cCD3^+$ cells may undergo development in the liver in situ, resembling the features of T-cell colony-forming cells in vitro [36]. Alternatively, $cCD3^+$ cells may seed to the thymus from where they rapidly disseminate, and some may return to the liver as both "transitional" ($TCR\beta^+$, $mCD3^-$) and more mature ($TCR\beta^+$, $mCD3^+$) cells [27a].

The investigation of the proliferative capacity of these liver-borne immature T cells may give some tentative clues to distinguish between these two possibilities. In the early liver samples (7th to 9th weeks) the majority (>90%) of $cCD3^+$ cells are $Ki67^+$ cells within their cell cycle while the $mCD3^+$ liver lymphocytes are virtually all $Ki67^-$ (>95%). In the 14- to 16-week samples the $\beta F1^+$ population includes 10%–25% of $Ki67^+$ cells, indicating that the $TCR\beta^+$, $mCD3^-$ cells are in cycle [27a]. Thus, at least some of the $mCD3^+$ cells might originate locally from these immature T cells. Lobach et al. have already described $CD7^+$, $CD45^+$ cells in the sinusoids of fetal liver samples and the perithymic parenchyma at the 7th week [37]. In their study the

CD7$^+$ cells appeared to be CD3$^-$ when tested with the Leu4 MAb. The discrepancy between these and our findings might be due to technical difficulties in detecting low antigen levels in fetal tissues.

Other cell types with the putative prothymocyte features have also been described, and future studies may indeed reveal that among these are the cells which most efficiently home to and proliferate in the thymus. Nevertheless, the question of these cells' commitment to the T-cell lineage prior to their arrival to the thymus remains unanswered. Hokland et al. reported that after the seeding of CD10$^+$ fetal BM cells onto irradiated thymic cells in vitro lymphoid cells of CD1$^+$, CD4$^+$, CD8$^+$ phenotype develop in the presence of interleukin-2 [38]. Van Dongen et al. have characterized a rare CD7$^+$, TdT$^+$, Class II$^+$ cell type in the adult BM and suggested that it might fit the description of prothymocytes [39]. These features are also seen in a group of acute non-lymphoid leukaemias which show no signs of TCRβ or IgH gene rearrangements [40]. In contrast, the detection of cCD3 can be taken as a reliable expression of T-cell commitment [21, 22, 24, 41]. This is demonstrated in malignant disease such as T-ALL as well as in the progenitor cell line KG-1. This line lacks CD3 antigens and exhibits a germline configuration of the TCRβ genes together with myeloid antigens; a subclone of this cell line, KG-1a, however, expresses cCD3, TCRβ gene rearrangement and CD7, paralleled by the disappearance of the myeloid associated antigens [42]. These observations support the view that CD3 antigens accumulate in the perinuclear area at the earliest stages of T-cell differentiation and imply that cCD3$^+$ cells in the fetal liver include a class of TdT$^-$ prothymocytes.

These data taken together suggest a model of human T-cell ontogeny in which cCD3$^+$, CD7$^+$, TdT$^-$ T-cell progenitors originate and proliferate in the liver before thymic generation. The capacity of these progenitors to colonize thymic epithelium needs to be investigated. The question of a second TdT$^+$ prethymic cell also remains open because we find that cCD3$^+$, TdT$^+$ cells are restricted to the thymus and are absent from the fetal liver as well as in the fetal and adult BM.

In the infant thymus four major subpopulations have become identifiable with the help of reagents to TCR-associated proteins [27a]. The large TdT$^+$ blasts show progenitor features, such as a high proliferative activity, and expression of cCD3 [20–22, 42]. Many of these thymic cells lack cytoplasmic TCRβ chains (category I). In these mCD3$^-$, CD1$^-$ blasts a germline configuration of the TCRβ genes was reported [21], but the DNA analysis in such a small polyclonal population presents considerable technical difficulties. The majority of TdT$^+$ cells are smaller and heterogeneously exhibit mCD3, TCRβ, CD1, CD4 and CD8 antigens. The expression of TCRβ chains appears to be initiated at this cortical thymocyte stage (category II). The gradual change into the third cell category is signified by the loss of TdT and the presence of cells which contain TCRβ chains but do not as yet exhibit a fully assembled TCR$\alpha\beta$ that would be detectable by WT31 (category III). This minority population is similar to the TdT$^-$ pre-B cells seen in the BM which show cytoplasmic μ heavy chains without membrane Ig. These findings also confirm observations in murine neonatal thymus where free cytoplasmic β chains are synthesized at high rate but are rapidly degraded in thymocytes with no detectable surface receptor [43]. Finally, the α chains are produced, and the assembly of membrane bound TCR is initiated. Our study suggests that the insertion of CD3 into the membrane might precede that of the fully assembled TCR$\alpha\beta$ complex.

On the basis of studies indicating the early rearrangement and expression of TCR$\gamma\delta$ genes in T-cell ontogeny [4–6, 9, 11], we anticipated that a particularly high proportion of TCRδ^+ precursors might be detectable in the early fetal liver and thymus, perhaps coinciding with the TdT^- stage of development. The results do not confirm this hypothesis because the output of $TCR\delta^+$ cells remains low at the time of the highest TdT^- proliferative activity among $cCD3^+$ cells in the fetal liver and thymus [27a].

Studies in rodents [10] and avians [11, 12] have demonstrated that the development of at least a proportion of the $TCR\gamma\delta$ lineage cells occurs in the thymus. The expression of CD1 in 18%–59% of $TCR\delta1^+$ fetal and post-natal thymic cells, shown in our investigation, provides evidence for such a process. Despite CD1 positivity, we could not find $TCR\delta1^+$ thymic cells with nuclear TdT. There are two possible explanations for this observation. The generation of diversity of the $TCR\delta$ chains might be totally TdT independent, or TdT might be lost as soon as the synthesis of $TCR\delta$ chains is initiated, similar to the events occurring during B-cell development: in B-cell progenitors TdT expression ceases with the accumulation of μ heavy chains [44]. The possibility that TdT does contribute to the generation of diversity of $TCR\delta$ is also suggested by the observation that N-region nucleotide addition in the $TCR\delta$ gene is seen in murine thymic clones when these are derived from TdT^+ post-natal samples, but not when derived from TdT^- fetal thymic samples [9].

The paucity of $TCR\delta1$ positivity on thymocytes is surprising in view of the fact that some cases of T-ALL successfully synthesize $TCR\delta$ [27b] and $TCR\gamma$ [45] chains. The comparative analysis of normal and leukaemic thymic cells suggests that most cases of T-ALL reflect the features of the two earliest stages of TdT^+ development such as $cCD3^+$, $TCR\beta^-$ (category I) and $cCD3^+$, $TCR\beta^+$ (category II). Nevertheless, in a substantial number of samples asynchronous, aberrant expressions of TCR receptor proteins are also seen. An example is the combined presence of both TdT and bright WT31, a phenomenon not seen in normal cells [27a, b].

A further point of interest is as follows. In 6 of the 20 T-ALL cases studied, blasts expressed mCD3. This was associated with membrane $TCR\alpha\beta$ chains in four cases and with $TCR\delta$ chains in the remaining two. The fact that both patients with $TCR\delta^+$ T-ALL had enlarged mediastinal mass suggests thymic involvement and indirectly confirms the occurrence of a thymic development of $TCR\gamma\delta$ cells in humans, as already indicated by the CD1 expression on $TCR\delta1^+$ thymic cells. In addition, the TdT positivity of the blasts in the two $TCR\delta^+$ T-ALL cases further demonstrates the asynchronous phenotypic expression of leukaemia [27b). It remains to be investigated whether this heterogeneity of T-ALL is reflected in their clinical course and response to treatment.

Acknowledgments. We thank Dr. M. B. Brenner (Dana-Farber Cancer Institute, Boston, Massachusetts, USA), Dr. P. C. L. Beverley (University College Hospital, London, UK), Dr. S. Ip (T Cell Sciences, Cambridge, Massachusetts, USA), Prof. J. Thompson (University of Kentucky, Lexington, Kentucky, USA), Dr. I. Hilgert (University of Prague, Czechoslovakia) and Prof. A. J. McMichael (Nuffield Department of Medicine, Oxford, UK) for the gift of antibodies. We also thank Dr. P. Kung (T Cell Sciences, Cambridge, Massachusetts, USA) and Dr. W. Tax (Nijmegen, The Netherlands) for discussions.

References

1. Reinherz EL, Meuer BC, Fitzgerald KA, Hussey RE, Hodgdon JC, Acuto O, Schlossman SF (1983) Comparison of T3 associated 49/43kd cell surface molecules on individual human T-cell clones: evidence for peptide variability in T-cell receptor structure. Proc Natl Acad Sci USA 80:4104–4108
2. Borst J, Alexander S, Elder J, Terhorst C (1983) The T3 complex on human T lymphocytes involves four structurally distinct glycoproteins. J Biol Chem 258:5135–5141
3. Brenner MB, Trowbridge IS, Strominger JL (1985) Cross-linking of human T cell receptor proteins: assocation between the

T cell idiotype subunit and the T3 glycoprotein heavy subunit. Cell 40:183–190
4. Brenner MB, Mclean J, Dialynas DP, Strominger JL, Smith JA, Owen FL, Seidman JG, Ip S, Rosen F, Krangel MS (1986) Identification of a putative T-cell receptor. Nature 322:145–149
5. Bank I, DePinho RA, Brenner MB, Cassimeris J, Alt FW, Chess L (1986) A functional T3 molecules associated with a novel heterodimer on the surface of immature human thymocytes. Nature 322:179–181
6. Loh EY, Lanier LL, Turck CW, Littman DR, Davies MM, Chien YH, Weiss A (1987) Identification and sequence of a fourth human T cell antigen receptor chain. Nature 330:569–572
7. Lanier LL, Weiss A (1986) Presence of Ti (WT31) negative T lymphocytes in normal blood and thymus. Nature 324:268
8. Borst J, Van Dongen JJM, Bolhuis RLH, Peters T, Haller DA, De Vries E, Van de Griend RJ (1988) Distinct molecular forms of human T cell receptor $\gamma\delta$ detected on viable T cells by a monoclonal antibody. J Exp Med 167:1625–1634
9. Chien YH, Iwashima M, Wettstein DA, Kaplan KB, Elliott JF, Born W, Davies MM (1987) T-cell receptor δ gene rearrangements in early thymocytes. Nature 330:722–727
10. Pardoll DM, Fowlkes BJ, Lew AM, Maloy WL, Weston MA, Bluestone JA, Schwartz RH, Coligan JE, Kruisbeeg AM (1988) Thymus-dependent and thymus-independent developmental pathways for peripheral T cell receptor $\gamma\delta$ bearing lymphocytes. J Immunol 140:4091–4096
11. Chen CH, Cihak J, Losch U, Cooper MD (1988) Differential expression of two T cell receptors, TcR1 and TcR2, on chicken lymphocytes. Eur J Immunol 18:539–543
12. Sowder JT, Chen CH, Ager LL, Chan MM, Cooper MD (1988) A large subpopulation of avian T cells express a homologue of the mammalian T gamma/delta receptor. J Exp Med 167:315–322
13. Hayday AC, Saito H, Gillies SD, Kranz DM, Tanigawa G, Elsen HN, Tonegawa S (1985) Structure organization and somatic rearrangement of the T cell gamma genes. Cell 40:259–269
14. Born W, Yague J, Palmer E, Kappler J, Marrack P (1985) Rearrangement of T-cell receptor β chain genes during T-cell development. Proc Natl Acad Sci USA 82:2925
15. Snodgrass HR, Kisielow P, Kiefer M, Steinmetz M, Von Boehmer H (1985) Ontogeny of the T cell antigen receptor within the thymus. Nature 313:592
16. Snodgrass HR, Dembic Z, Steinmetz M, Von Boehmer H (1985) Expression of T cell antigen receptor genes during fetal development in the thymus. Nature 315:232
17. Crispe LN, Moore MW, Husmann LA, Smith L, Bevan MJ, Shimonkevitz RP (1987) Differentiation potential of subsets $CD4^-8^-$ thymocytes. Nature 329:336–339
18. Davis MM, Bjorkman PJ (1988) T cell antigen receptor genes and T cell recognition. Nature 334:395–402
19. Chien YH, Iwashima M, Kaplan KB, Elliott JF, Davis MM (1987) A new T-cell receptor gene located within the alpha locus and expressed early in T-cell differentiation. Nature 327:677–682
20. Campana D, Janossy G (1988) Proliferation of normal and malignant human immature lymphoid cells. Blood 71:1201–1210
21. Furley AJ, Mizutani S, Weilbaecher K, Dhalival HS, Ford AM, Chan LC, Molgaard HV, Toyonaga B, Mak T, Van der Elsen P, Gold D, Terhorst C, Greaves MF (1986) Developmentally regulated rearrangement and expression of genes encoding the T cell receptor-T3 complex. Cell 46:75–87
22. Campana D, Thompson JS, Amlot P, Brown S, Janossy G (1987) The cytoplasmic expression of CD3 antigens in normal and malignant cells of the T lymphoid lineage. J Immunol 138:648–655
23. Sangster RN, Minowada J, Suciu-Foca N, Minden M, Mak TW (1986) Rearrangement and expression of the α, β and γ chain T cell receptor genes in human thymic leukemia cells and functional T cells. J Exp Med 163:1491
24. Furley AJW, Chan LC, Mizutani S, Ford AM, Weilbaecher K, Pegram SM, Greaves MF (1987) Lineage specificity of rearrangement and expression of genes encoding the T cell receptor-T3 complex and immunoglobulin heavy chain in leukemia. Leukemia 1:644
25. Brenner MB, Mclean J, Scheft H, Warnke RA, Jones N, Strominger JI (1987) Characterization and expression of the human $\alpha\beta$ T cell receptor by using a framework

monoclonal antibody. J Immunol 138:1502–1509
26. Spits H, Borst J, Tax W, Capel PJA, Terhorst C, De Vries JE (1985) Characteristics of a monoclonal antibody (WT-31) that recognizes a common epitope on the human T cell receptor for antigen. J Immunol 135:1922–1928
27. Band H, Hotchstenbach F, McLean J, Hata S, Krungel MS, Brenner MB (1987) Immunological proof that a novel rearanging gene encodes the T cell receptor $\gamma\delta$ subunit. Science 238:682
27a. Campana D, Janossy G, Coustan-Smith E, Tian W-T, Ip S, Wong L, Amlot PL (1989) The expression of T cell receptor-associated proteins during T cell ontogeny in man. Immunol 142:57–66
27b. Campana D, Janossy G et al. (1989) Heterogeneous T cell receptor expression in T cell acute lymphoblastic leukemia. (submitted)
28. Van Dongen JJM, Hooijkaas H, Comans-Bitter WM, Benne K, Van Os TM, De Jong JDJ (1985) Triple immunological staining with colloidal gold, fluorescein and rhodamine as labels. J Immunol Methods 80:1–6
29. Gerdes J, Lemke H, Baisch H, Wacker HH, Schwab U, Stein H (1984) Cell cycle analysis of a cell proliferation-associated human nuclear antigen defined by the monoclonal antibody Ki-67. J Immunol 133:1710–1715
30. Bodger MP, Janossy G, Bollum FJ, Burford GD, Hoffbrand AV (1983) The ontogeny of terminal deoxynucleotidyl transferase positive cells in the human fetus. Blood 61:1125–1131
31. Kelemen E, Calvo W, Fliedner TM (1979) Atlas of human hemopoetic development. Springer, Berlin Heidelberg New York
32. Hokland P, Rosenthal P, Griffin JD, Nadler LM, Daley J, Hokland M, Schlossman SF, Ritz J (1983) Purification and characterization of fetal hematopoietic cells that express the common acute lymphoblastic leukemia antigen. J Exp Med 157:114
33. Rosenthal P, Rimm IJ, Umiel T, Griffin JD, Osanthanondh R, Schlossman SF, Nadler LM (1983) Ontogeny of human hematopoietic cells: analysis utilizing monoclonal antibodies. J Immunol 31:232
34. Bonati A, Delia D, Starcich B, Buscaglia M (1984) Phenotype of the terminal transferase-positive cells in human foetal liver and bone-marrow: analysis with monoclonal antibodies. Scand J Haematol 33:418
35. Bofill M, Janossy G, Janossa M, Burford GD, Seymour GJ, Wernet P, Kelemen E (1985) Human B cell development. II. Subpopulations in the human fetus. J Immunol 134:1531–1538
36. Mossalayi MD, Lecron JC, Goube de Laforest P, Janossy G, Debre P, Tanzer J (1988) Characterization of prothymocytes with cloning capacity in human bone marrow. Blood 71:1281–1287
37. Lobach DF, Hensley LL, Ho W, Haynes B (1985) Human T cell antigen expression during early stages of fetal thymic maturation. J Immunol 135:1752–1759
38. Hokland P, Hokland M, Daley J, Ritz J (1987) Identification and cloning of a prethymic precursor T lymphocyte from a population of common acute lymphoblastic leukemia antigen (CALLA)-positive fetal bone marrow cells. J Exp Med 165:1749–1754
39. Van Dongen JJM, Hooijkaas H, Comans-Bitter M, Hahlen K, De Klein A, Van Zanen GE, Van't Veer MB, Abels J, Benner R (1985) Human bone marrow cells positive for terminal deoxynucleotidyl transferase, HLA-DR, and a T cell marker may represent prothymocytes. J Immunol 135:3144–3150
40. Norton JD, Campana D, Hoffbrand AV, Janossy G, Coustan-Smith E, Jani H, Yaxley JC, Prentice HG (1987) Rearrangement of immunoglobulin and T cell antigen receptor genes in acute myeloid leukemia with lymphoid associated markers. Leukemia 1:757–761
41. Janossy G, Coustan-Smith E, Campana D (1989) The reliability of cytoplasmic CD3 and CD22 antigen expression in the immunodiagnosis of acute leukemia – A study of 500 cases. Leukemia 3:170–181
42. Furley AJ, Reeves BR, Mizutani S, Altass LJ, Watt SM, Jacob MC, Van der Elsen P, Terhorst C, Greaves MF (1986) Divergent molecular phenotypes of KG1 and KG1a myeloid cell lines. Blood 68:1101
43. Hannum C, Marrack P, Kubo R, Kappler J (1987) Thymocytes with the predicted properties of pre-T cells. J Exp Med 166:874
44. Janossy G, Bollum F, Bradstock KF, Ashley J (1980) Cellular phenotypes of normal and leukaemic haemopoietic cells deter-

mined by analysis with selected antibody combinations. Blood 56:430–441
45. Elliott JF, Rock EP, Patten PA, Davis MM, Chien YH (1988) The adult T-cell receptor δ chain is diverse and distinct from that of fetal thymocytes. Nature 331:627–631
46. Van Dongen JJM, Wolvers-Tettero ILM, Seidman JG, Ang SL, Van de Griend RJ, De Vries EFR, Borst J (1987) Two types of gamma T cell receptors expressed by T cell acute lymphoblastic leukemias. Eur J Immunol 17:1719

Molecular Genetics

Molecular Analysis of the Translocation Breakpoint in a Philadelphia-positive, bcr-negative ALL Patient *

M. J. M. van der Feltz, M. K. K. Shivji, G. Grosveld, and L. M. Wiedemann

A. Introduction

Translocation t(9; 22) (q34; q11) occurs in 90% of patients with chronic myeloid leukaemia (CML) [13] and in 5% of children and 10%–20% of adults with acute lymphoblastic leukaemia (ALL) [16]. It is easily identified by the $22q^-$, or Philadelphia, chromosome. In CML all chromosome 22 breakpoints are located within the 5.8-kb breakpoint cluster region (bcr) in the 3′ part of the *phl* gene [6]. In ALL, however, only some breakpoints are in bcr. Others are more 5' in the *phl* gene, as suggested by observations made at the RNA and protein level [3, 4, 9, 17] and at the DNA level by pulsed-field gel electrophoresis [14]. Both in CML and in ALL the breakpoint on chromosome 9 is within the *abl* oncogene upstream of the common exon [7].

Depending on the breakpoint on chromosome 22, a hybrid 8.5-kb mRNA [15] and a 210-kDa fusion protein [10] are formed (in Ph^+ bcr^+ CML and ALL), or a hybrid 7-kb mRNA and a 190-kDa fusion protein (in Ph^+ bcr^- ALL) [1–4, 9, 11, 17]. Both protein products seem to have an enhanced tyrosine kinase activity compared with the 145-kDa *abl* protein when assayed in vitro.

In order to characterize a p190 ALL breakpoint at the genomic level we have cloned and sequenced the breakpoint of the $9q^+$ chromosome from a Philadelphia-positive p190 ALL patient.

* This work was supported by the Leukaemia Research Fund of Great Britain

B. Results and Discussion

We have shown previously that a c-*abl*-related 190-kDa protein is expressed by the leukaemic cells of patient F. Y. investigated here [1]. When DNA from the leukaemic cells was analyzed by Southern blotting and hybridization with a probe spanning the bcr region, no rearrangements were found. After hybridization with a c-*abl* probe, however, a rearranged band was detected in *Bgl*II and *Bam*HI digested DNA from the leukaemic cells which was not present in the DNA from a skin biopsy of the same patient [1, 5]. Using the same c-*abl* probe (0.52 E in Fig. 1 a) the rearranged 13.4-kb *Bgl*II fragment was isolated from an EMBL3 library made from DNA from the patient's leukaemic cells [5]. The 11.2-kb *Bgl*II fragment spanning the corresponding germline chromosome 9 region was isolated from the same library. Comparison of the restriction maps of the two *Bgl*II fragments showed we had cloned the $9q^+$ junction fragment (Fig. 1 b). As in CML, the breakpoint on chromosome 9 is upstream of the common exon a2, here between exon Ia and a2. Probes FY1 and FY2 from the non-chromosome 9 part of the clone (Fig. 1 b) were used to hybridize to Southern blots of DNA from several rodent-human somatic cell hybrids. In this way we found that the non-chromosome 9 sequences are derived from chromosome 22. Comparison with a genomic clone of the *phl* gene (Fig. 1 d, λbcr 5.7) which contains the second exon and 11.7 kb of the 3′ portion of the adjacent intron, revealed that the cloned $9q^+$ junction fragment contains part of this *phl* intron. There-

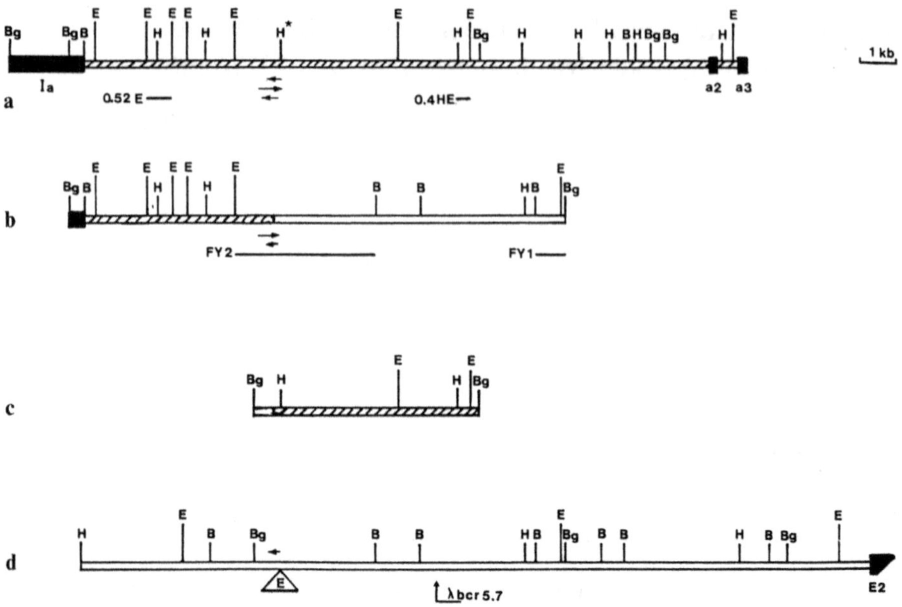

Fig. 1a–d. Restriction enzyme maps of the junction regions of the DNA of the Ph-positive p190 ALL compared with the normal regions of chromosomes 9 and 22. Abbreviations of restriction enzymes: *B, Bam*HI; *Bg, Bgl*II; *E, Eco*RI; *H, Hind*III. The *Hind*III site in (**a**) which is marked by an asterisk is described in the text. Horizontal arrows in (**a, b, d**) denote the sequencing strategy. *Hatched* and *white regions* represent chromosome 9 and chromosome 22 respectively. **a.** Germline c-*abl*, adapted from [7]. Probes 0.52 E and 0.4 HE are indicated below the map. Exons are indicated by *black boxes*. **b.** The cloned *Bgl*II fragment from the $9q^+$ chromosome. Probes FY1 and FY2 are shown below the map. **c.** The size of the chromosome $22q^-$ *Bgl*II fragment is derived from the $9q^+$ map in comparison with the germline c-*abl* and *phl* restriction enzyme maps. **d.** The *phl* map is compiled from a genomic clone [9], the $9q^+$ junction fragment and data from Southern blots of DNA from several patients after hybridization to probes FY1 and FY2. The exact size of *phl* exon 2 (E2) is not known; its location has been determined by hybridization with a cDNA probe. The *arrow* indicates the position of the 5′ end of the λbcr 5.7 genomic clone. The *triangle* below the map denotes the approximate location of a 1-kb deletion polymorphism containing an *Eco*RI site. Additional *Bam*HI and *Bgl*II sites may exist

```
                   10        20        30        40        50        60        70        80        90       100
chrom. 9    ACAAAAATCTCATTTCTTTTTTTTTTTGAGACAAGAGTCTCACTCTGTCGCCCAGACTGGAGTGCAGTTGCACGATCTCAGCTCACTGCAAGCTCCGCCT
chrom. 9q+  ACAAAAATCTCATTTCTTTTTTTTTTTGAGACAAGAGTCTCACTCTGTCGCCCAGACTGGAGTGCAGTTGCACGATCTCAGCTCACTGCAAGCTCCGCCT
chrom. 22   TCAGATTTTTTCAAGGAGGGGTGTCTTAGTCACTTCAGGCTGCTATAACAAAAATGCCATAAACTGGGTACCTTAAACAACAAACACTTACTCCTCACAG

              ↓        110       120       130       140       150       160       170       180       190       200
chrom. 9    CCTGGGTTCACGCCATTCTCCTGCCTCAGCCTCCCTAGTAGAGGGTACTACAGGCGCCCGGGGGTTTCACCATGTTAGCCAGGATGGTCTCCATCTCCTG
chrom. 9q+  CCCTGGAGGCAGGAAGTCAGTATCAAGGAGCTGGAACATTTGGAGTCTGCCAAGGGCCCACTTCCTGGTTCTTAGCCATCTTCTTGCTGTGTTCTCACCT
chrom. 22   TGCTGGAGGCAGGAAGTCAGTATCAAGGAGCTGGAACATTTGGAGTCTGCCAAGGGCCCACTTCCTGGTTCTTAGCCATCTTCTTGCTGTGTTCTCACCT
              ↑

                   210       220       230       240       250       260       270
chrom. 9    ACCTCGTGATCCGCCCACCTCGGCCTCCCAAAGCCCTGGGATTACAGGCGTGAGCCACTGCACCGGGCCA
chrom. 9q+  GGTGTGGAAAGGATGGGGGCTGTCTCTGGGGTTTCAAAAGGGCACTAATCCCCTTCATGAGGGTTCCGTC
chrom. 22   GGTGTGGAAAGGAT
```

Fig. 2. Nucleotide sequence of the chromosome 9, $9q^+$, and 22 regions around the breakpoint. Normal chromosome 9, and the $9q^+$ sequences are from patient F.Y.; chromosome 22 sequences are from non-ALL DNA. The breakpoint is indicated by arrows between 102 and 103. Sequences showing homology to Alu sequences are underlined. Between 160 and 161 there is a 40-bp stretch of Alu sequences missing. Numbers are arbitrary. The *Hind*III site marked with an asterisk in Fig. 1 a is not shown here; it would be at number 368 fore, we conclude that the breakpoint on chromosome 22 in this patient is located in the putative first intron of the *phl* gene, approximately 16.5 kb upstream from exon 2. This location is in agreement with results from the analysis of the RNA and the protein which result from this type of translocation [3, 4, 9, 17]. The chromosome 22 breakpoint in the ALL cell line SUP-B13 [14] lies within the same *Bam*HI fragment as the breakpoint in patient F.Y. described here. Using probes FY1 and FY2 (Fig. 1 b) we detected rearrangements in the DNA from two of 12 additional Ph-positive p190 ALL patients.

We determined the nucleotide sequence around the breakpoint from double-stranded templates using the Sequenase kit (United States Biochemical Corporation). Oligonucleotide primers were prepared on an Applied Biosystems 381A DNA synthesizer. The sequencing strategy is indicated in Fig. 1 a, b and d; part of the sequence is shown in Fig. 2.

The chromosome $22q^-$ junction fragment has not been cloned; its restriction map in Fig. 1 c has been predicted from the chromosome 9, $9q^+$, and 22 maps. When c-*abl* probe 0.4 HE (indicated in Fig. 1 a) was hybridized to a Southern blot of *Bgl*II-digested DNA from the patient's leukaemic cells, a rearranged band of 6.2 kb (the predicted size in Fig. 1 c) was observed in addition to the expected germline 11.2-kb fragment [5]. The nucleotide sequence of the chromosome $22q^-$ junction fragment will be obtained after amplification from the patient's DNA by the polymerase chain reaction. So far, there is no evidence for large insertions or deletions at this breakpoint. The breakpoint on chromosome 9 is 265 bp 5′ of the *Hind*III site, which has been marked with an asterisk in Fig. 1 a. There is 88% homology between the c-*abl* sequences around the breakpoint and a consensus Alu repeat sequence, when a 40-bp stretch of Alu which is missing between nucleotide 160 and 161 is omitted. Homology to Alu sequences is occasionally observed near translocation breakpoints, e.g., in the CML t (9; 22) translocation [8, 12]. In chromosome 22, however, there is no Alu homology in the region described here (Fig. 2), which makes recombination between Alu sequences as a mechanism for translocation unlikely.

C. Conclusions

We have shown by cloning of the genomic DNA from the leukaemic cells of a Ph^+ p190 ALL patient that the translocation breakpoint on chromosome 22 is located in the first putative intron of the *phl* gene. It is clear that in Ph^+ p190 ALL the chromosome 22 breakpoints are not located in such a small region that they can be detected with one probe, as is the case in Ph^+ bcr^+ CML. The nucleotide sequence around the breakpoint described here does not reveal any obvious indication for the mechanism of the translocation.

Acknowledgments. We thank Dr. A. Geurts van Kessel for sending us DNA from somatic cell hybrids, Drs. K. Merrifield and M. Osborne for providing us with oligonucleotides, and Dr. J. Groffen for the chromosome 22 5.3 kb *Bam*HI fragment.

References

1. Chan LC, Karhi KK, Rayter SI, Heisterkamp N, Eridani S, Powles R, Lawler SD, Groffen J, Foulkes JG, Greaves MF, Wiedemann LM (1987) A novel abl protein expressed in Philadelphia chromosome positive acute lymphoblastic leukaemia. Nature 325:635–637
2. Clark SS, McLaughlin J, Crist WM, Champlin R, Witte ON (1987) Unique forms of the abl tyrosine kinase distinguish Ph′-positive CML from Ph′-positive ALL. Science 235:85–88
3. Clark SS, McLaughlin J, Timmons M, Pendergast AM, Ben-Neriah Y, Dow LW, Crist W, Rovera G, Smith SD, Witte ON (1988) Expression of a distinctive BCR-ABL oncogene in Ph′-positive acute lymphocytic leukemia (ALL). Science 239:775–777

4. Fainstein E, Marcelle C, Rosner A, Canaani E, Gale RP, Dreazen O, Smith SD, Croce CM (1987) A new fused transcript in Philadelphia chromosome positive acute lymphocytic leukaemia. Nature 330:386–388
5. van der Feltz MJM, Shivji MKK, Grosveld G, Wiedemann LM (1988) Characterization of the translocation breakpoint in a patient with Philadelphia positive, bcr negative acute lymphoblastic leukaemia. Oncogene 3:215–219
6. Groffen J, Stephenson JR, Heisterkamp N, de Klein A, Bartram CR, Grosveld G (1984) Philadelphia chromosomal breakpoints are clustered within a limited region, bcr, on chromosome 22. Cell 36:93–99
7. Heisterkamp N, Stephenson JR, Groffen J, Hansen PF, de Klein A, Bartram CR, Grosveld G (1983) Localization of the c-abl oncogene adjacent to a translocation break point in chronic myelocytic leukaemia. Nature 306:239–242
8. Heisterkamp N, Stam K, Groffen J, de Klein A, Grosveld G (1985) Structural organization of the bcr gene and its role in the Ph′ translocation. Nature 315:758–761
9. Hermans A, Heisterkamp N, von Lindern M, van Baal S, Meijer D, van der Plas D, Wiedemann LM, Groffen J, Bootsma D, Grosveld G (1987) Unique fusion of bcr and c-abl genes in Philadelphia chromosome positive acute lymphoblastic leukemia. Cell 51:33–40
10. Konopka JB, Watanabe SM, Witte ON (1984) An alteration of the human c-abl protein in K562 leukemia cells unmasks associated tyrosine kinase activity. Cell 37:1035–1042
11. Kurzrock R, Shtalrid M, Romero P, Kloetzer WS, Talpaz M, Trujillo JM, Blick M, Beran M, Gutterman JU (1987) A novel c-abl protein product in Philadelphia-positive acute lymphoblastic leukaemia. Nature 325:631–635
12. Rogers J (1985) Oncogene chromosome breakpoints and Alu sequences. Nature 317:559
13. Rowley JD (1973) A new consistent chromosomal abnormality in chronic myelogenous leukaemia identified by quinacrine fluorescence and Giemsa staining. Nature 243:290–293
14. Rubin CM, Carrino JJ, Dickler MN, Leibowitz D, Smith SD, Westbrook CA (1988) Heterogeneity of genomic fusion of BCR and ABL in Philadelphia chromosome-positive acute lymphoblastic leukemia. Proc Natl Acad Sci USA 85:2795–2799
15. Shtivelman E, Lifshitz B, Gale RP, Canaani E (1985) Fused transcript of abl and bcr genes in chronic myelogenous leukaemia. Nature 315:550–554
16. Third international workshop on chromosomes in leukaemia. 1980. (1981) Cancer Genet Cytogenet 4:101–110
17. Walker LC, Ganesan TS, Dhut S, Gibbons B, Lister TA, Rothbard J, Young BD (1987) Novel chimaeric protein expressed in Philadelphia positive acute lymphoblastic leukaemia. Nature 329:851–853

Note added in proof:

More extended sequence data and the exact restriction map around the breakpoint have been described in: van der Feltz et al. (1989) Nucl Acids Res 17:1–10.

Cloning of the Breakpoint Junction of the Translocation 14;19 in Chronic Lymphocytic Leukemia

T. McKeithan, H. Ohno, J. Rowley, and M. Diaz

Human B-cell lymphocytic neoplasms are often associated with specific cytogenetic abnormalities that correlate with their histological and immunologic phenotypes. The genes located at the breakpoints of these recurring chromosomal translocations appear to be integrally involved in the pathogenesis of the corresponding B-cell neoplasms. Our laboratory has recently reported that t(14;19) (q32;q13.1) is a recurring translocation in chronic lymphocytic leukemia (CLL). We have analyzed the leukemic cells from two such patients in detail using various probes from the very complex immunoglobulin heavy chain locus (IGH). In both cases, the t(14;19) was part of a three-way translocation with loss of the derivative chromosome containing the q terminus of the affected chromosome 14. Using Southern blot analysis, numerous rearrangements and deletions were found within IGH in these two cases; however, in both, rearrangements were found involving one of the two α constant regions (Cα), which are components of IGH. As described below, these rearrangements result from the 14;19 translocation.

In addition, two internal deletions within IGH have occurred in case D.B., as confirmed by molecular cloning: the μ switch region is largely deleted, and another deletion extends from the switch region of $\alpha 1$ to the switch region of $\gamma 2$. Several additional rearrangements appear to involve the γ gene segments and may be related to a translocation 2;14, which is also present in this patient's malignant cells. Despite the involvement of the α regions in the translocations, the remaining J region, presumably responsible for any immunoglobulin heavy chain produced, is found associated with other constant region segments – Cμ in one patient and C$\gamma 4$ in the other.

In each case, clones containing the rearranged Cα sequences were isolated from a bacteriophage lambda library made using complete *Bgl*II digests. In patient J.L., the rearrangement involved C$\alpha 2$ while C$\alpha 1$ was affected in patient D.B. In both cases, the break in chromosome 14 appears to involve one of the α switch regions. Detailed restriction mapping was used to identify the portion of the clones which was not derived from chromosome 14. Subclones free of repetitive sequences were, in each case, localized to chromosome 19 using a panel of 31 mouse/human somatic cell hybrids. These results were confirmed by in situ hybridization to metaphase chromosomes. By hybridization to somatic cell hybrids containing fragments of chromosome 19, the probes were sublocalized to the q arm of chromosome 19, bands q12 to 13.2.

We have used a human library, prepared from the DNA of a patient without abnormalities of chromosome 19, to isolate clones from the normal sequences on chromosome 19. Restriction mapping of these clones, which encompass about 30 kb of DNA, demonstrates that the breakpoints in the two patients are about 19 kb apart. Such variability is common in other chromosome translocations associated with leukemia or lymphoma. Thus, it is likely that the same gene is affected by the t(14;19) in both patients.

Deparatments of Pathology and Medicine, University of Chicago, Chicago, IL, USA

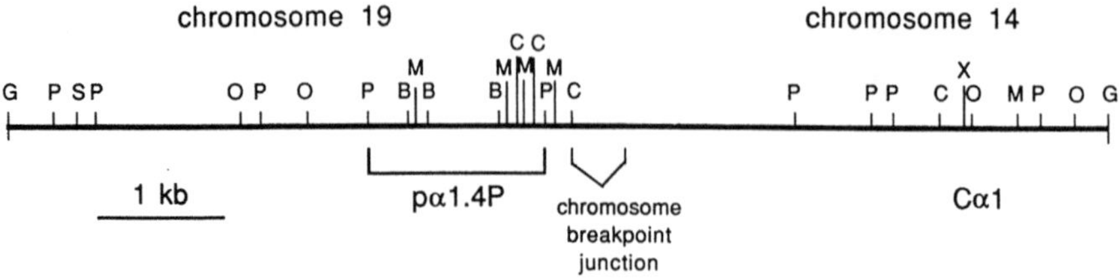

Fig. 1. Restriction map of the cloned *Bgl*II fragment from case 2. The 3′ portion contains Cα1 sequences. A subclone, pα1.4P, was mapped to chromosome 19 using in situ hybridization and hybridization to a panel of somatic cell hybrids. This probe detects a 2.3-kb message in several hematological cell lines. Restriction enzyme sites are illustrated as follows: *B, Bam*HI; *C, Sac*II; *G, Bgl*II; *M, Sma*I; *O, Oxa*NI; *P, Pst*I; *S, Sal*I; *X, Xho*I. Note the cluster of sites for *Sma*I and *Sac*II which partly overlaps pα1.4P. These sites are rare in human DNA except in "CpG islands," which are regions with high C+G content and unusually high frequency of the dinucleotide CpG. These islands are associated with the 5′ ends of many genes. The presence of a CpG island was confirmed by sequencing the pα1.4P segment

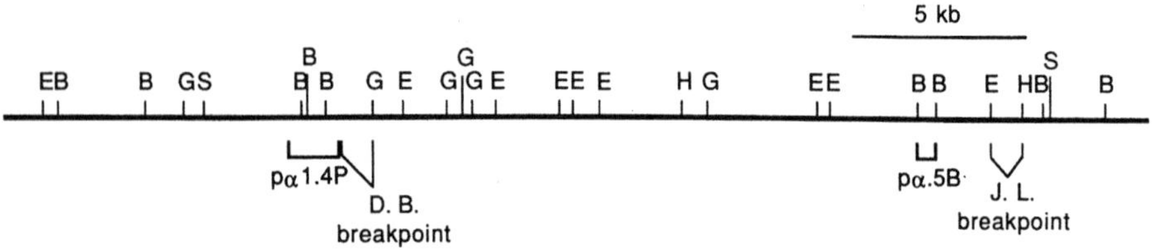

Fig. 2. Composite map of the region on chromosome 19 affected by the t(14:19). This map is based on overlapping bacteriophage clones which were isolated from a normal library, using the two probes pα.5B and pα1.4P adjacent to the translocation breakpoint junctions of the two patients. The locations of the two probes and of the breakpoint junctions are indicated. Restriction sites are as follows: *B, Bam*HI; *G, Bgl*II; *E, Eco*RI; *H, Hind*III; *S. Sal*I

Very recently, we have obtained evidence for a break in the same region in a third patient with the t(14;19).

The region on chromosome 19 near the breakpoint in patient D.B. has an unusually large number of sites for restriction enzymes containing the dinucleotide "CG" as part of their recognition sequence. For example, three *Sma*I sites (one CG) and three *Sac*II (two CGs) are found within a region of about 600 basepairs; *Sac*II sites are generally much less frequent than this, occurring on average about 200 kb apart. Most such restriction sites are associated with "CpG islands," areas of high C+G content and unusually high abundance of the dinucleotide CG, which is otherwise highly depleted in the human genome. CpG islands are found associated with the 5′ portions of all known "housekeeping" genes and many tissue-specific genes as well.

The suspicion that this region is a CpG island was confirmed by DNA sequencing. A 400-basepair region was found to have a high abundance of CG dinucleotides and a C+G content of 80%, which is twice the value for most human sequences. Hybridization of a probe from this region to a 2.3-kb RNA transcript in several lymphoid cell lines has provided further evidence that a gene on chromosome 19 is adjacent to the chromosome breakpoint. We are currently attempting to clone and analyze this gene.

Variant Breakpoint Positions on Chromosome 22 in Ph′-Positive Chronic Myelogenous Leukemias

A. Guerrasio [1], C. Rosso [1], A. Tassinari [2], A. Zaccaria [2], and G. Saglio [1]

As a consequence of the molecular events associated with the presence of a Philadelphia (Ph′) chromosome, two different types of abnormal c-*abl* proteins, termed P210 and P190 according to their molecular weight, have been found in Ph′ positive leukemic cells [1, 2]. P210, expressed in almost all chronic myelogenous leukemias (CML) and in approximately half of the Ph′ positive acute lymphoblastic leukemias (Ph′$^+$ ALL), is codified by a hybrid gene arising from a rearrangement between the 5′ of c-*abl* oncogene and a restricted region called ("m-*bcr*" – major breakpoint cluster region) of a gene on chromosome 22, also denominated *bcr* [3–5]. By contrast, P190, present in the remaining half of the Ph′$^+$ ALLs, derives from a rearrangement involving the same two genes, but with a different breakpoint on chromosome 22, mapping more 5′ and within the first large intron of the *bcr* gene [6, 7]. The relationship between the two types of *bcr*/c-*abl* rearrangement leading to P210 or P190 protein production and to the acute or chronic leukemic phenotype of the Ph′$^+$ cells, is still matter of debate.

In order to further elucidate this problem, we have investigated the breakpoint position on chromosome 22 in a large series of Ph′$^+$ CML patients, both in chronic and in the blast phase of the disease. We studied 102 Ph′$^+$ CML patients: 79 were in chronic, 6 in accelerated, and 17 (9 lymphoid, 8 myeloid) in blast phase. Using a set of restriction endonucleases (*Bgl*II, *Bam*HI, *Eco*RI, *Bcl*I, *Kpn*I and *Hind*III) and probes corresponding to different parts of the *bcr* gene (see Fig. 1), in 96 out of the 102 cases classical rearrangements were shown to occur within the m-*bcr* region of the *bcr* gene. However, in seven cases of this group, m-*bcr* rearrangements were detected only with a probe (probe B in Fig. 1) corresponding to *bcr* sequences remaining on chromosome 22 after the t(9;22) translocation, whereas they were not detected with a probe (probe A in Fig. 1) corresponding to *bcr* sequences moving to chromosome 9. This finding is mainly due to small deletions of sequences corresponding to probe A occurring during the translocation to chromosome 9, as we never observed in our cases the loss of the entire chromosome 9q$^+$.

The most surprising result, however, was that six cases in chronic phase showed breakpoints mapping outside the m-*bcr* area: two presented breakpoints located approximaely 12 and 10 kb upstream (positions 1 and 2 in Fig. 1) and one about 20 kb downstream to the m-*bcr* region (position 3 in Fig. 1), whereas three Ph′$^+$ CML patients were apparently lacking any rearrangement of the *bcr* gene detectable with the presently available probes, which allow exploration of the entire coding part of the gene with the exception of the first large intron. The latter finding is similar to that in the majority of the Ph′$^+$ ALLs expressing P190. Although protein expression cannot easily be assessed in CML chronic phase, a Northern analysis of one of these cases showed the presence of c-*abl* transcripts of 7 and 6 kb but no hybrid 8.5-kc *bcr*/c-*abl* messages. Taken together all these findings are compatible with the presence either of a rearrange-

[1] Dipartimento di Scienze Biomediche e Oncologia Umana, University of Turin, Turin, Italy

[2] Istituto di Ematologia, L. and A. Seragnoli, University of Bologna, Bologna, Italy

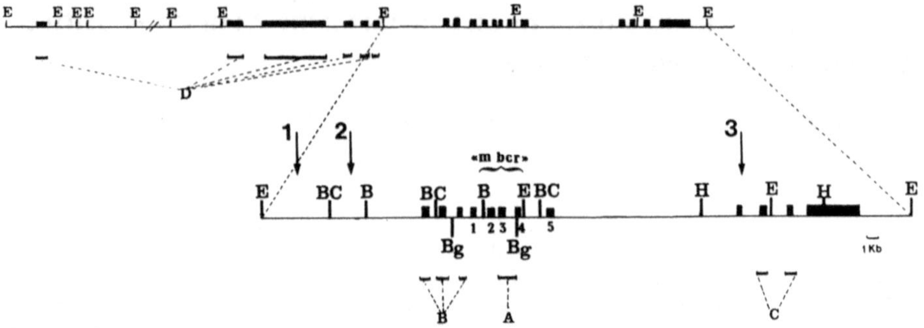

Fig. 1. Schematic map of the *bcr* gene on chromosome 22. *Black boxes* indicate exons. *A, B, C, D* represent the four probes used in this study. The major breakpoint cluster region where rearrangements usually take place in CML is marked as m-*bcr*. *Numbered arrows* point to three different variant breakpoint positions found in CML patients (see text). Abbreviations for restriction endonucleases: *E, Eco*RI; *B, Bam*HI; *BC, Bcl*I; *H, Hind*III; *Bg, Bgl*II

ment on chromosome 22 similar to that of Ph'$^+$ *bcr*- ALLs or of a rearrangement occurring outside the *bcr* gene.

Although further studies are needed to characterize completely these variant cases, they, as others recently described [8], show that a variability of the breakpoint position on chromosome 22 is present also in a minority (5%–6%) of the cases in chronic phase. At this regard it is interesting to note that the clinical presentation and disease course in patients showing variant breakpoint positions on chromosome 22 do not seem to differ from those in patients with the common type of *bcr* rearrangement. This point raises important questions concerning the relationship between the type of *bcr-abl* rearrangement, the type of abnormal *abl* protein expressed, and the clinical presentation of the disease. At present, no simple models can be proposed, but further correlations between clinical and molecular data will help to solve this problem as well as to clarify the role of the c-*abl* proto-oncogene in human leukemogenesis.

References

1. Konopka JB, Watanabe SM, Witte ON (1984) An alteration of the human c-*abl* protein in K562 leukemia cells unmasks associated tyrosine kinase activity. Cell 37:1035
2. Chan LC, Karhi KK, Rayter SI, Heisterkamp N, Eridani S, Powles R, Lawler SD, Groffen J, Foulkes JG, Greaves MF, Wiedemann LM (1987) A novel *abl* protein expressed in Philadelphia chromosome positive acute lymphoblastic leukemia. Nature 325:635
3. Groffen J, Stephenson JR, Heisterkamp N, de Klein A, Bartram CR, Grosveld G (1984) Philadelphia chromosome breakpoints are clustered within a limit region, *bcr*, on chromosome 22. Cell 36:93
4. Heisterkamp N, Stamm K, Groffen J, de Klein A, Grosveld G (1985) Structural organization of the *bcr* gene and its role in the Philadelphia translocation. Nature 315:758
5. Shtivelman E, Lifshitz B, Gale RP, Canaani E (1985) Fused transcript of *abl* and *bcr* genes in chronic myelogenous leukemia. Nature 315:550
6. Hermans A, Heisterkamp N, von Lindern M, van Baal S, Meijer D, van der Plas D, Wiedemann LM, Groffen J, Bootsma D, Grosveld G (1987) Unique fusion of *bcr* and c-*abl* genes in Philadelphia chromosome positive acute lymphoblastic leukemia. Cell 51:33
7. Fainstein E, Marcelle C, Rosner A, Canaani E, Gale RP, Dreazen O, Smith SD, Croce CM (1987) A new fused transcript in Philadelphia chromosome positive acute lymphocytic leukemia. Nature 330:386
8. Selleri L, Narni F, Emilia G, Colo A, Zacchini B, Venturelli D, Donelli A, Torelli U, Torelli G (1987) Ph' positive chronic myelogenous leukemia with a chromosome 22 breakpoint position outside the *bcr* region. Blood 70:1659

Expression of Cellular Oncogenes in Primary Cells from Human Leukemias

Guo-wei Rong and Shi-shu Chen

Alterations of cellular oncogene expression have been reported in numerous neoplasias, particularly of the hematopoietic system. However, a cause-effect relationship between cellular oncogene abnormalities and oncogenesis is still unclear. In order to explore the involvement of cellular oncogenes in human cancer cells, we studied the levels of the expression of three cellular protooncogenes in the primary leukemic cells from 53 leukemic patients with different types and stages.

A. Materials and Methods

Thirty-five patients with acute myeloblastic leukemia (AML), 14 with acute lymphoblastic leukemia (ALL), 2 with chronic myelocytic leukemia (CML), 2 with chronic lymphocytic leukemia (CLL), and 8 normal individuals as controls were studied. Diagnosis was based on the morphological evaluation of bone marrow smears according to the French-American-British (FAB) criteria. Leukemic cells, obtained from peripheral blood after diagnosis and before initiation of chemotherapy, were enriched up to 90% by centrifugation on Ficoll-Hypaque density gradient [1]. HL-60, a human promyelocytic leukemia cell line, was used as reference.

Department of Biochemistry and Laboratory of Molecular Biology, Shanghai Second Medical University, Shanghai, People's Republic of China

I. Probes

The following cellular or viral oncogene probes were used: a 1.6-kb DNA sequence (pMC41) of human c-*myc*; 1.0 kb *Pvu*I-*Pvu*I fragment (pA T 8.8) of N-*ras*, and 5.8 kb of *Hind*III-*Hind*III fragment (pFBJ-2) of v-*fos*. These probes were all obtained as gifts from Dr. Land H.

II. mRNA Analysis

The mRNA of the corresponding protooncogene in all fresh leukemic cells from patients or control cells was detected by Quick-blot assay [2]. Briefly: cells → deproteinized by proteinase K → add detergents (Brij-35, Doc) and NaI → filter through BA-85 membrane → wash membrane consecutively in H_2O, $EtOH/H_2O$, dilute acetic anhydride → hybridization with corresponding radioisotope-labeled probes → autoradiography → detect intensity of spots by densitometer (Shimadzu CS930).

B. Results

I. Storage of Samples

In order to repeat the experiment, we stored the cell samples in 1 × VRC (vanadyl ribonucleaside complex) solution in liquid nitrogen for 1–2 months. There was no difference in mRNA content between 15 cases of fresh and stored samples (Fig. 1).

II. Expression Levels of Protooncogenes in Normal Leukocytes

The expression levels of c-*myc*, N-*ras*, and c-*fos* in peripheral leukocytes from

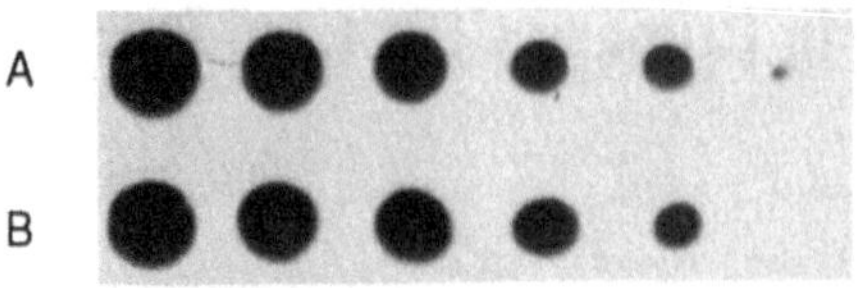

Fig. 1. Comparison of the mRNA content in fresh and stored cells: *A*, fresh leukemic cells from AML-M3; *B*, sample from the same patient but stored in liquid nitrogen for 45 days. Cell number from left to right spot: 2.0×10^6, 1.0×10^6, 5.0×10^5, 2.5×10^5, 1.25×10^5, 7.5×10^4

Table 1. Oncogene expression in HL-60 and WBCs from eight normal individuals

Cell source	c-*myc*	N-*ras*	c-*fos*
WBCs	< +1 [a]	+	–
HL-60	+5	+1 ~ +2	–

[a] Using "+5" to indicate the expression level of c-*myc* in HL-60, and one-fifth of the level as "+1," and so on.

eight different normal individuals, and the HL-60 cell line were detected. The level of the c-*myc* transcripts in HL-60 was denoted as "+5" [3], and the one-fifth of the level as "+1" and so on. Figure 2 and Table 1 show that the *myc* gene was slightly expressed and N-*ras* was marginally expressed, whereas the expression of c-*fos* was undetectable in the normal leukocytes.

III. Expression Levels of Protooncogenes in Leukemic Cells

In ten AML-M3, four AML-M2, four AML-M4, and ten ALL patients, before chemotherapy, the c-*myc* was obviously expressed in almost all leukemic cells irrespective of the cell types, while N-*ras* and c-*fos* were unconstantly expressed. However, the c-*fos* was expressed in all four cases of AML-M4 (Tables 2, 3). The c-*myc* transcripts were detected but the N-*ras* and c-*fos* were not in four chronic

Table 2. Protooncogene expression levels in 18 AML [a] patients

	No. studied	No. positive	Degree of expression				
			+1	+2	+3	+4	+5
AML-M3							
c-*myc*	10	7	4	2		1	
N-*ras*	9	4	1		2	1	
c-*fos*	8	2	1	1			
AML-M2&M4							
c-*myc*	8	6	4	1	1		
N-*ras*	8	3	2	1			
c-*fos*	8	4	3	1			

[a] Before chemotherapy.

Table 3. Protooncogene expression levels in eight ALL patients

c-*onc*	No. studied	No. positive	Degree of expression				
			+1	+2	+3	+4	+5
c-*myc*	8	6	2	3	1		
N-*ras*	8	3	3				
c-*fos*	6	1	1				

[a] Before chemotherapy.

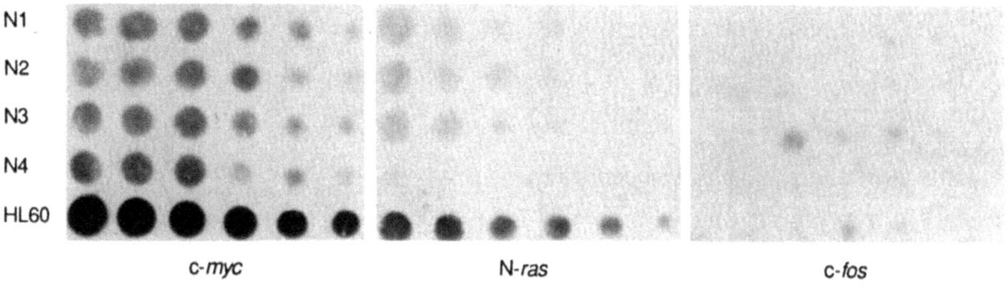

Fig. 2. Oncogene expression in normal leukocytes. N1–N4: four different normal individuals. HL-60 used as a reference. Cell number the same as in Fig. 1

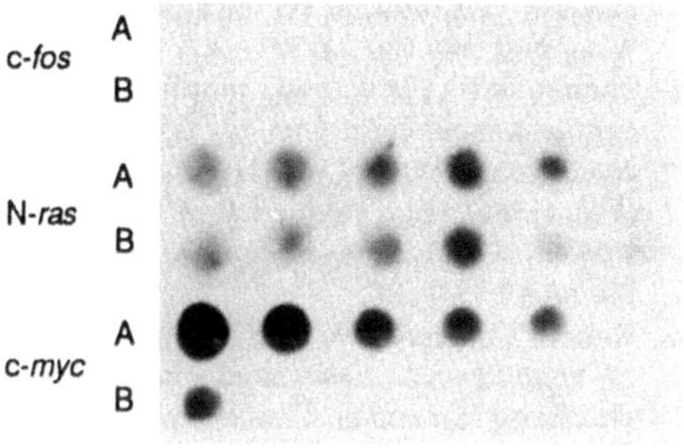

Fig. 3. Protooncogene expression in remitted AML-M2 patient: *A*, before chemotherapy; *B*, remission phase

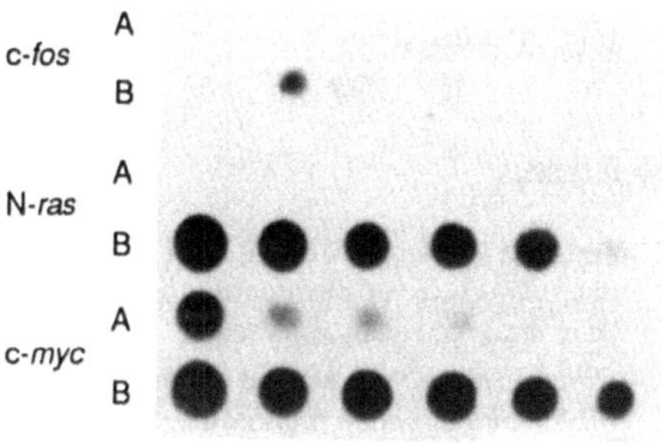

Fig. 4. Protooncogene expression in a CLL patient in blast crisis: *A*, before blast crisis; *B*, in blast crisis

leukemic cases. The samples were obtained from the three AML patients 2–3 months after initial therapy, while the patients were in remission and on maintenance chemotherapy. The results showed that the levels of c-*myc* mRNA decreased obviously in one of three remission samples (Fig. 3), while the levels of c-*myc* mRNA remained essentially unchanged in the others. Figure 4 demonstrates that c-*myc* and N-*ras* mRNA increased dramatically in CLL during blast crisis.

C. Discussion

In this experiment, we selected three cellular protooncogenes – c-*myc* and c-*fos* gene coding for nuclear proteins, the former relating to early cell differentiation and the latter linking to monocytic differentiation and N-*ras* for putative intermediate transducers of mitogenic signals – for detecting their expression levels in primary cells from human leukemia. Most previous studies on the expression of cellular oncogenes in leukemias have been carried out on neoplastic lines, which may not faithfully represent the primary cancer cells [4–6]. So in this study we observed the expression of protooncogenes in primary cells from human leukemias. Our results showed that only c-*myc* gene expressed slightly in normal leukocytes while expressed obviously in almost all leukemic cells irrespective of the cell types. The c-*myc* expression level is higher in acute leukemia than in chronic leukemia. But the levels of N-*ras* and c-*fos* transcripts were variable (Tables 2, 3). An interesting finding is that the c-*fos* gene was expressed in all four AML-M4. The recent data indicate that the N-*ras* gene may be activated by mutation in AML [7]. Our results showed the N-*ras* was strictly expressed in acute leukemia, but there is no correlation between the quantitative expression in AML and ALL. So the level of N-*ras* expression may not be significantly changed in leukemic cells.

Until now there have only been a few articles written about oncogene expression during remission and after chemotherapy [8]; we analyzed three cases of AML and one of CLL, but no definite suggestion can be put forward. We think that it is important to study the changes of oncogene expression in the course of remission.

References

1. Ferrante A, Thong YH (1978) A rapid one-step procedure for purification of mononuclear and polymorphonuclear leukocytes from human blood using a modification of the Hypaque-Ficoll technique. J Immunol Methods 24:389–393
2. Bresser J, Doering J, Gillespie D (1983) Quick-blot: selective m-RNA or DNA immobilization from whole cells. DNA 2:243–254
3. Blick M, Westin E, Gutterman J et al. (1984) Oncogene expression in human leukemia. Blood 64:1234–1239
4. Chen SS, Cheng F, Rong GW et al. (1989) The molecular biological properties of a transplantable human primary gastric cancer in nude mice (THPGC-1). Chin J Oncol 11(3):171–174
5. Tang JQ, Chen SS (1988) Changes of expression and structure of some protooncogenes in TPA induced HL-60 and Raji cells. Acta Biol Exp Sin 21:201–208
6. Shimke RT (1986) Gene amplification in culture animal cells. Cell 37:705–713
7. Needleman SW, Kraus MH, Srivastava SK et al. (1986) High frequency of N-*ras* activation in acute myelogenous leukemia. Blood 67:753–757
8. Rossen D, Tereba A (1983) Transcription of hematopoietic associated oncogenes in childhood leukemias. Cancer Res 43:3912–3916

c-myc and *c-myb* Oncoproteins During Induced Maturation of Human Myeloid and Erythroid Leukaemic Lines

M. A. Bains, P. Pedrazzoli, T. G. Hoy, and A. Jacobs

c-myc and c-*myb* mRNA have been found to be tightly regulated during haemopoietic differentiation [1]. While the expression of both oncogenes is induced in cells stimulated to proliferate, c-*myc* in the transition from G_0 to G_1 [2], and c-*myb* in the transition from G_1 to S phase [3], expression ceases in terminally differentiating haemopoietic cell lines. This takes the form of a progressive monotonic mRNA decline during macrophage (MΦ) and granulocyte (GN) differentiation and a biphasic mRNA decline during erythroid (E) differentiation [4–6]. Although this decline may reflect the cessation of proliferation which accompanies the differentiation process, recent reports that constitutive expression of either oncogene inhibits differentiation in murine [7] and human [8] leukaemic lines imply a causative role for each in haemopoietic maturation.

We have studied nuclear c-*myc* and c-*myb* proteins through the cell cycle during MΦ, GN, E and megakaryocytic (MK) differentiation of KG1, HL60 and HEL leukaemic cells. The p62$^{c\text{-}myc}$ and p75$^{c\text{-}myb}$ content of propidium-iodide stained nuclei was quantitated by flow cytometry using fluoresceinated antibodies CT14-G4 and MB4.3, respectively (gifts of G. Evan, Cambridge, UK), following our published method [9].

Figure 1 is a dot display of p62$^{c\text{-}myc}$ and p75$^{c\text{-}mcb}$ fluorescence in uninduced HL60, demonstrating a less than two-fold increment in both oncoproteins over the cell cycle.

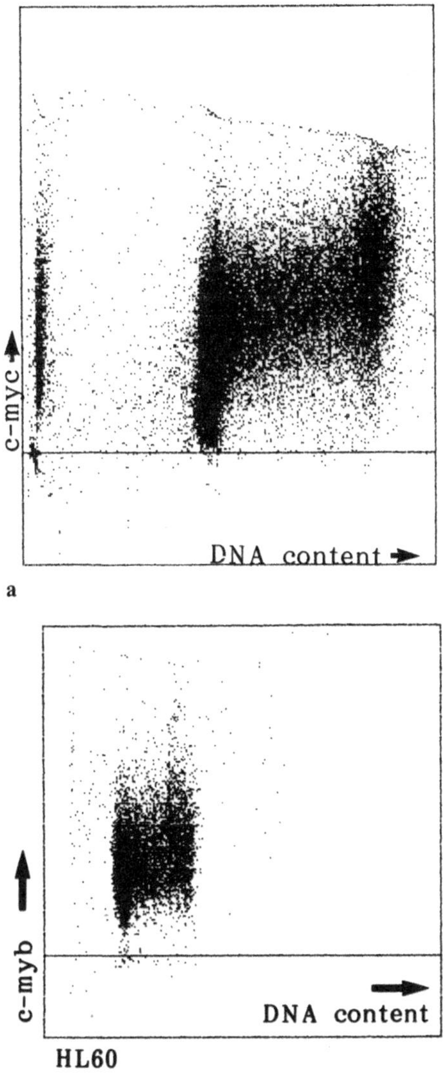

Fig. 1. Dot display of p62$^{c\text{-}myc}$ (**a**) and p75$^{c\text{-}myb}$ (**b**) fluorescence in uninduced HL60 in terms of DNA content

Department of Haematology, University of Wales College of Medicine, Cardiff, UK

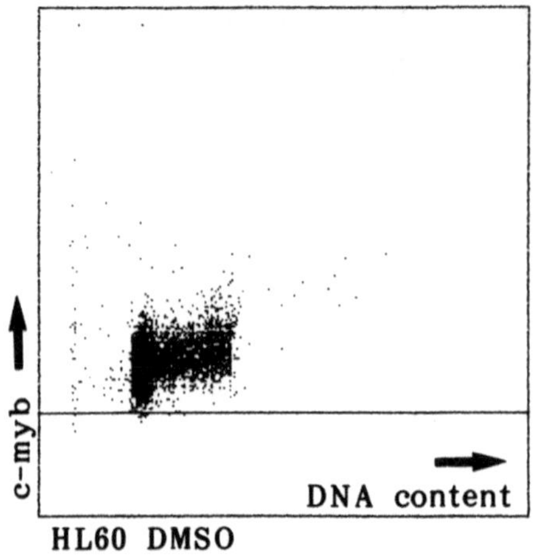

Fig. 2. Reduced level of p75$^{c\text{-}myb}$ in HL60 with DMSO-induced maturation but unchanged cell cycle distribution

Table 1. Oncoprotein levels of HL60, HEL and KG1

Cells	p62$^{c\text{-}myc}$	p75$^{c\text{-}myb}$
HL60	1.0	1.0
HEL	0.9	1.1
KG1	0.68	0.9

Table 1 shows the oncoprotein levels of different leukaemic lines relative to HL60. Figure 2 demonstrates that while p75$^{c\text{-}myb}$ levels declined in HL60 with DMSO-induced maturation, the cell cycle distribution did not change. MΦ induction of different leukaemic lines resulted in an early increase in both oncoproteins, followed by a decline simultaneous with the reduction of S-phase cells and appearance of α-naphthyl-acetate esterase positive cells, as shown in Table 2.

Different patterns of oncoprotein change were found when different inducing agents were used for GN differentiation of HL60, as shown in Table 3. Growth arrest and phenotypic maturation preceded a decline in p62$^{c\text{-}myc}$ in retinoic acid treated HL60; the opposite occurred in DMSO-treated cells.

Hemin-induced E differentiation of HEL resulted in biphasic p62$^{c\text{-}myc}$ and p75$^{c\text{-}myb}$ kinetics, without significant change in growth fraction, while DMSO-induced MK differentiation caused an early and steady decline of both oncoproteins, as shown in Table 4.

Thus although there are early transient increases, generally c-*myc* and c-*myb* proteins decline with differentiation, well before proliferation ceases in some lineages. The kinetics of oncoprotein decline resemble closely those published for the mRNA of these oncogenes. The patterns of decline differ between the two oncogenes and vary with the lineage induced and with the inducer used. There is no simple relationship of either oncogene to proliferation during induced maturation, and the cell cycle distribution of their proteins does not change during the differentiation process. These data support disparate roles for c-*myc* versus c-*myb* during human haemopoietic differentiation and suggest that multiple signal transduction pathways exist for down-regulation of these genes.

In summary, we have quantitated, for the first time, c-*myb* protein over the cell cycle in human haemopoietic cells, using our original method of flow-cytometric assay of nuclear bound c-*myc* protein product [9]. As for c-*myc* mRNA, c-*myb* RNA has been found to be tightly regulated during haemopoietic differentiation, and aberrant expression has been related to failure of maturation induction in both human and murine leukaemic cell lines [1, 7]. We have considered in detail the changes in c-*myc* and c-*myb* proteins during the induced maturation of human myeloid and erythroid leukaemic cell lines and correlated these changes with the differentiative and proliferative status of these cells. We observed that, as in normal cells [9], both oncoproteins decline in differentiating leukaemia cells, well before proliferation ceases in some lineages. Although oncoprotein levels and distribution over the cell cycle are similar among the uninduced cells of different leukaemic lines, oncogene-, cell-

Table 2. Relative $p62^{c\text{-}myc}$ and $p75^{c\text{-}myb}$ and cell cycle status during TPA induction of leukaemic cells to macrophages

Cells	Time 0	30 min	2 h	6 h	24 h	48 h
HL60						
C-*myc*	1.0	1.5	2.0	1.4	0.7	0.5
C-*myb*	1.0	1.3	0.95	1.3	1.0	0.85
% S	34.7	39.5	34.1	39.8	15.0	15.3
KG1						
C-*myc*	1.0	1.0	1.3	0.55	0.35	0.4
C-*myb*	1.0	1.1	1.3	0.7	1.4	0.4

Table 3. Relative $p62^{c\text{-}myc}$ and $p75^{c\text{-}myb}$ and cell cycle status during induction of HL60 to granulocytes

Agent	Time 0	30 min	2 h	6 h	24 h	48 h	120 h
DMSO							
C-*myc*	1.0	0.7	0.5	0.7	0.5	0.5	0.5[a]
C-*myb*	1.0	1.3	1.1	1.0	0.8	0.6	0.2
% S	32.0	34.1	31.2	31.6	30.1	19.1	7.7
Retinoic acid							
C-*myc*	1.0	1.2	1.2	1.4	1.4	1.2	0.7[b]
C-*myb*	1.0	0.7	0.6	0.8	0.7	0.6	0.3
% S	36.3	35.4	40.0	35.8	29.1	30.9	10.6

[a] 80% NBT-positive cells
[b] 92% NBT-positive cells

Table 4. Relative $p62^{c\text{-}myc}$ and $p75^{c\text{-}myb}$ and cell cycle status in HEL after induction

Agent	Time 0	30 min	2 h	6 h	24 h	48 h	72 h	120 h
Hemin								
C-*myc*	1.0	0.8	0.7	2.8	1.1	0.6	0.5	0.6
C-*myb*	1.0	1.1	0.7	1.4	1.0	0.7	0.5	–
% S	39.6	52.0	39.0	40.0	38.0	40.1	39.2	37.3
DMSO								
C-*myc*	1.0	0.7	0.4	0.6	0.3	0.3	0.3	0.2
C-*myb*	1.0	0.7	0.4	0.8	0.4	0.4	0.4	0.3
% S	45.1	37.3	37.3	32.9	30.1	26.5	28.5	32.4

line-, lineage-, and inducer-specific kinetics occur in this decline. The cell cycle distribution of the oncoproteins does not change during maturation. Our data (a) suggest that there is no simple relationship of either oncoprotein to proliferation, (b) confirm other reports [10] that multiple metabolic cascades exist in leukaemic cells to down-regulate and upregulate genes important in the differentiation process, and (c) support disparate roles for c-*myc* versus c-*myb* during haemopoietic differentiation.

Acknowledgment: This work was supported by the Leukaemia Research Fund, U.K. The Figures and Table 1 are reproduced from Pedrazzoli et al. [11].

References

1. Kirsh IR, Bertness V, Silver J, Hollis G (1986) Regulated expression of the c-*myb* and c-*myc* oncogenes during erythroid differentiation. J Cell Biochem 32:11–21
2. Heikkila R, Schwab G, Wickstrom E, Loke SL, Pluznik DH, Watt R, Neckers LM (1987) A c-*myc* antisense oligodeoxynucleotide inhibits entry into δ phase but not progress from G_0 to G_1. Nature 328:445–449
3. Thompson C, Challoner PB, Nieman PE, Groudine M (1986) Expression of the c-*myb* proto-oncogene during cellular proliferation. Nature 319:374–380
4. Dalla Favera RD, Westin EH, Gelmann EP, Martinotti S, Bregni M, Wong-Staal F, Gallo RC (1983) The human oncogene c-*myc*: Structure, expression, and amplification in the human promyelocytic leukemic cell line HL60. Hamatol Bluttransfus 28:247–253
5. Yen A, Guernsey D (1986) Increased r-*myc* RNA levels associated with the precommitment state during HL60 myeloid differentiation. Cancer Res 46:4156–4161
6. Prochownik E, Kukowska J (1986) Deregulated expression of c-*myc* by murine erythroleukaemia cells prevents differentiation. Nature 322:848–850
7. Clarke M, Kukowska-Latallo J, Westin E, Smith M, Prochownik E (1988) Constitutive expression of a c-*myc* cDNA blocks friend murine erythroleukaemia cell differentiation. Mol Cell Biol 8:884–892
8. Larsson L, Ivhed I, Gidlund M, Patterson U, Vennstrom B, Nilsson K (1988) Phorbol-ester induced terminal differentiation is inhibited in human U937 monoblastic cells expressing a V-*myc* oncogene. PNAS 85:2638–2642
9. Bains MA, Hoy TG, Baines P, Jacobs A (1987) Nuclear c-*myc* protein, maturation, and cell cycle status of human haemopoietic cells. Br J Haematol 67:293–300
10. Yen A, Brown D, Fishbaugh J (1987) Control of HL60 monocytic differentiation: Different pathways and uncoupled expression of differentiation markers. J Exp Cell Res 168:247–254
11. Pedrazzoli P, Bains MA, Watson R, Fisher J, Hoy TG, Jacobs A (1989) c-*myc* and c-*myb* oncoproteins during induced maturation of myeloid and erythroid human leukaemic cell lines. Cancer Res (in press)

Isolation and Characterisation of a Myeloid Leukaemia Inducing Strain of Feline Leukaemia Virus

T. Tzavaras [1], N. Testa [2], J. Neil [1], and D. Onions [3]

A. Introduction

Feline leukaemia virus (FeLV) is the aetiological agent of a wide range of neoplastic and degenerative conditions [1]. The predominant, naturally occurring, FeLV-induced tumours are T-cell lymphomas, and recently some of the viral events in their pathogenesis have been elucidated. Both transduction and insertional mutagenesis of the *myc* gene are frequent concomitants of T-cell transformation by FeLV. In addition, a possible role for the T-cell antigen receptor gene in leukaemogenesis has been revealed with the discovery of an FeLV-mediated transduction of the β-chain of the T-cell antigen receptor [2, 3].

In contrast, the viral aetiology of myeloid leukaemia has received less attention. In these studies we have isolated a virus complex, FeLV-GM1 from a naturally occurring case of myeloid leukaemia corresponding to stage M6 in the FAB classification. This isolate contained both subgroups A and B of FeLV, and on passage into kittens it produced a spectrum of myeloproliferative disease including myeloid leukaemia. In vitro colony assays of bone marrow early erythroid precursor (BFU-E) and granulocytic macrophage precursor cells (GM-CFC) indicated that two stages in the development of disease could be recognised. In the first stage, in which no histopathological abnormalities were observed, there was a gross expansion in the GM-CFC compartment. At a later stage in those cats that developed myeloid leukaemia, large numbers of small clusters were superimposed on a residual normal GM-CFC colony pattern.

FeLV-GM1 contains both subgroup A and B components which were not separable by endpoint titration. Molecular analysis and cloning of the FeLV-GM1 isolate has revealed that the subgroup B component is defective for replication, and from recent pathogenesis experiments with cloned viruses it is now evident that B component was not required to induce the early proliferative events. However, only in cats inoculated with both components has full leukaemia development been observed so far.

We hypothesise that the viral events leading to myeloid leukaemia can be divided into two discrete stages. In the first stage the virus induces a polyclonal expansion of myeloid precursor cells with altered response to, and/or production of, growth factors. This proliferating cell population may now become a target for further viral events necessary for complete transformation.

B. Cellular Events in FeLV-GM1 Leukaemia

Initial experiments with FeLV-GM1 involved the passage of virus from the original tumour into newborn kittens. Of 30 cats challenged with this virus 6 died of myeloid leukaemia within 8–40 weeks, 17 developed aplastic anaemia and 7 remained clinically normal for 1 year. In order to characterise the effects of the

[1] Beatson Institute for Cancer Research, Glasgow, UK
[2] Department of Experimental Haematology, Paterson Laboratories, Manchester, UK
[3] LRF Human Virus Research Centre, Glasgow, UK

Table 1. Colony-forming units of bone marrow precursor cells in control and FeLV-GM1 infected cats

Cat no.	Status	Weeks after infection	GM-CFC per 10^5 bone marrow cells	GM-CFC colony morphology	Pathology
1	Control	1.5	65	Normal	Normal
2	Infected	1.5	201	Normal	Normal
3	Infected	1.5	360	Normal	Normal
4	Control	4	48 ± 4	Normal	Normal
5	Infected	4	6 ± 2	Normal	Preleukaemic
6	Infected	4	42 ± 6	Normal	Preleukaemic
7	Control	5	57 ± 5	Normal	Normal
8	Infected	5	30	Plus 10^3 clusters	Myeloid Leukaemia
9	Infected	5	200	Plus 10^3 clusters	Myeloid Leukaemia
10	Control	8	79 ± 11	Normal	Normal
11	Infected	8	13 ± 2	Plus 10^3 clusters	Myeloid Leukaemia
12	Infected	8	ND	ND	Myeloid Leukaemia
13	Control	20	172 ± 7	Normal	Normal
14	Infected	20	792 ± 76	Normal	Normal
15	Infected	20	764 ± 11	Normal	Normal

Bone marrow from cats infected with FeLV-GM1 was prepared, plated at densities ranging from 10^4 to 10^5, and assayed as described elsewhere [7]. Results are expressed as the means ± SD of four plates; results without standard deviations are the means of two wells.

virus on bone marrow precursor cells, ten neonatal kittens were infected with ca. 10^4 ffu/ml FeLV-GM1. Starting at 10 days post-infection and at intervals thereafter infected cats with their age-matched uninfected controls were sampled for the assay of plasma viraemia and bone marrow colony-forming cells (Table 1).

Appropriate tissue was examined histologically so that correlations could be made between the extent of the disease and the pattern of myeloid colony formation (Table 1 and Fig. 1). As early as 10 days after infection a dramatic effect on the myeloid lineage was noted with a three- to six-fold increase in the number of GM-CFC colonies. At this stage both the morphology of the colonies and the histology of the marrow remained normal. In cats examined at 5 and 8 weeks after infection, myeloid leukaemia was present. Three of these cats had myeloid leukaemia with little differentiation, typical of stage M1 disease in the FAB classification, while a further cat showed clear evidence of stage 4, myelomonocytic leukaemia. A striking feature of the GM-CFC colony pattern in all of the leukaemia cats examined was the presence of several thousand small cell clusters which were superimposed on a pattern of few remaining colonies, a finding which has been recorded in some human patients with acute myeloblastic leukaemia and which is pathognomic for that disease [4]. In one of the leukaemic cats the morphologically normal GM-CFC population was expanded about four times above the control value while in two others (cats 14 and 17) the GM-CFC colonies were reduced below control values. The likeliest interpretation of these findings is that, in the first stage of the disease the virus induces an expansion of the GM-CFC population which is eventually replaced by a leukaemic population arising from these cells.

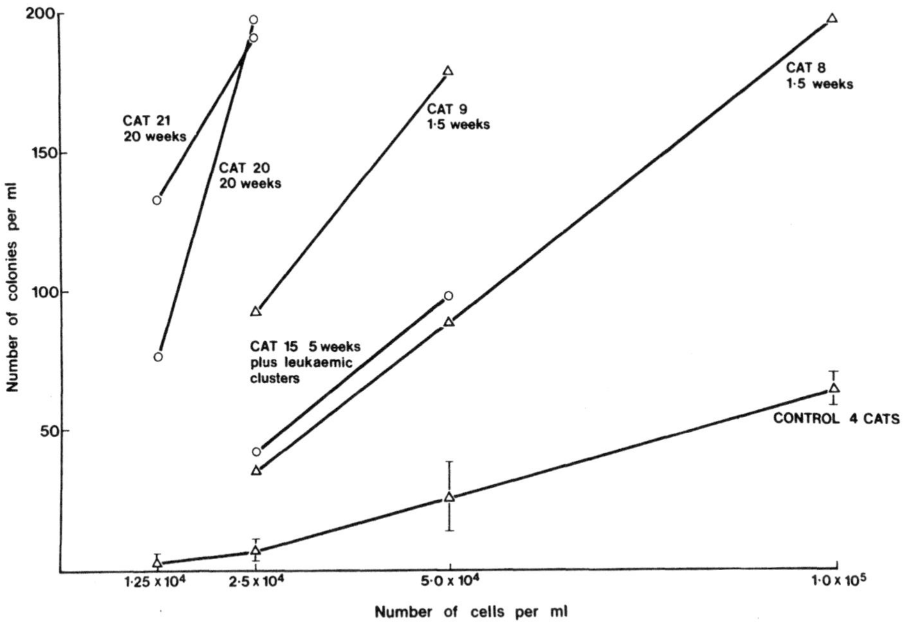

Fig. 1. Colony-forming units of granulocyte macrophage precursors (GM-CFC) determined at varying cell densities of nucleated bone marrow cells in the absence of exogenous colony-stimulating factor. The control values are the means of four cats (±SE). Only colonies with normal morphology were scored. Note in this figure cats 8, 9, 15 and 20 correspond to cats 2, 3, 9, 14 and 15 in table 1

C. Characterisation of FeLV-GM1 Virus

A distinguishing feature of cats that developed leukaemia following FeLV-GM1 infection was the presence of both subgroups A and B in the plasma. Cats that remained clinically normal were usually viraemic with subgroup A virus, while remaining latently infected with subgroup B. A temporal association between the onset of subgroup B viraemia and leukaemia was also observed. In one cat 9 months post-infection GM-CFC colony numbers remained within normal limits, and the cat was viraemic with subgroup A alone. A month later the cat had developed myeloid leukaemia and was viraemic with both subgroups. However this association is complex in that cats viraemic with subgroup A alone often had titres of less than 10^3 ffu/ml, whereas cats viraemic with both subgroups had titres of ca. 10^5 ffu/ml. Consequently the presence of the B subgroup in plasma could have been a consequence of efficient subgroup A replication rescuing the B virus from latency.

In order to resolve the roles of the individual components of FeLV-GM1 in disease induction, we undertook to molecularly clone and analyse the biological effects of the subgroup A virus alone and the reconstituted A plus B virus complex. The subgroup B virus was found to have a 1.5-kb deletion with *gag-pol* and was therefore defective for replication.

In Table 2 the result of GM-CFC colony assays plated at limiting dilution without exogenous colony-stimulating factor are presented for cats infected with either subgroup A or subgroup AB. All the AB infected cats displayed an expansion of GM-CFC at 4 weeks, as did three of the six subgroup A infected cats, indicating that the subgroup B virus was not an absolute requirement for this event.

At 38 weeks cat 22 developed myeloid leukaemia and, as in other leukaemic

Table 2. GM-CFC colony formation at limiting dilution 4 weeks after infection

Cat no.	Virus challenge	GM-CFC per 10^4 BM cells without exogenous CSF
16	Control	0.7 ± 1.1
17	AB	10 ± 4
18	AB	12 ± 1
19	AB	11 ± 3
20	AB	14 ± 1
21	AB	18 ± 4
22	AB	16 ± 3
23	AB	15 ± 4
24	Control	1 ± 1
25	A	2 ± 0
26	A	1 ± 1
27	A	3 ± 1
28	A	20 ± 3
29	A	8 ± 4
30	A	16 ± 6

GM-CFC colony numbers of uninfected control cats and cats infected with molecularly cloned FeLV-GM1. Cells were plated at limiting dilution without the addition of exogenous CSF.

cats, a clonal proviral integration pattern was observed.

D. Discussion

The study of retrovirus-induced leukaemias has been of value in revealing the multistage pattern of leukaemogenesis and in implicating specific cellular genes in haemopoietic transformation. In long-term bone marrow cultures Friend virus infection permits the establishment of autonomously proliferating cell lines which may grow independently of exogenous growth factors but respond to differentiation factors and are non-leukaemic in vivo. At a later stage cells with only a limited capacity to differentiate develop and these cells are leukaemic in vivo and are often aneuploid [5]. These observations are parallelled by the in vivo experiments with FeLV-GM1 in which the initial step appears to be an increase in granulocyte macrophage cells which have an altered response to CSF. This in turn is followed by the development of leukaemic cells which do not differentiate normally in response to CSF.

The mechanism responsible for the initial GM-CFC expansion is not clear. Both the rapidity of the expansion and the absence of a clonal retroviral integration pattern suggests that this process is polyclonal. Retrovirus infection has been associated with enhanced CSF production [6], and FeLV infection of feline embryo cells can induce the production of factors with burst promoting activity (J. Abkowitz, personal communication). However, preliminary experiments in which marrow from infected cats was used as a source of CSF indicate that a minimal increase in CSF production occurs in FeLV-GM1 infection. A hypothesis worth further investigation is that infection of bone marrow precursor cells can lead to autocrine stimulation of GM-CFC.

Studies with the molecularly cloned viruses indicted that the subgroup A component of FeLV-GM1 could reproduce the early proliferation of GM-CFC. Similarly we have previously shown that another subgroup A virus, FeLV-Glasgow/1, can produce a lesser but significant increase in GM-CFC numbers soon after infection [7]. FeLV-Glasgow/1 is less rapidly oncogenic than FeLV-GM1 but has produced myeloid leukaemia in some cases.

In those cats that developed myeloid leukaemia a clonal pattern of proviral integration was observed in the bone marrow. In Friend virus-induced myeloblastic leukaemias three distinct proviral integration sites, the *fim* loci, have been identified, one of which, *fim*-2, spans the 5′ end of the c-*fms* gene [8]. In the leukaemic cats we have not detected rearrangement of the c-*fms* gene, but we cannot preclude a proviral integration some distance from this locus.

The role of the subgroup B virus in the secondary leukaemogenic events remains unresolved. One possibility is that it is involved in overcoming viral interference in preleukaemic cells that are already infected with subgroup A [9]. FeLV subgroups are defined by viral interference so that a cell infected by subgroup A cannot be superinfected with the same subgroup but is susceptible to subgroup B infection. Consequently the initial events may be dependent on subgroup A virus while secondary events require the presence of the subgroup B virus.

References

1. Neil JC, Onions DE (1985) Feline leukaemia virus: Molecular biology and pathogenesis. Anticancer Res 5:49–64
2. Neil JC, Hughes D, McFarlane R, Wilkie NM, Onions DE, Lees G, Jarrett O (1984) Transduction and rearrangement of the *myc* gene by feline leukaemia virus in naturally occurring T-cell leukaemias. Nature 308:814–820
3. Onions D, Lees G, Forrest D, Neil JC (1987) Recombinant feline viruses containing the *myc* gene rapidly produce clonal tumours expressing T-cell antigen receptor gene transcripts. Int J Cancer 40:40–45
4. Moore MAS, Spitzen G, Williams N, Metcalf D, Buckley J (1974) Agar culture studies of 127 cases of untreated acute leukaemia: the prognostic value of reclassification of leukemia according to the *in vitro* growth characteristics. Blood 44:1–18
5. Testa NG, Dexter TM, Scott D, Teich NM (1980) Malignant myelomonocytic cells after *in vitro* infection of marrow cells with Friend Leukaemia Virus. Br J Cancer 41:33–39
6. Koury MJ, Pragnell IB (1982) Retroviruses induce granulocyte macrophage colony stimulating activity in fibroblasts. Nature 299:638–640
7. Testa NG, Onions D, Jarrett O, Frassoni F, Eliason JF (1983) Haemopoietic colony formation (BFU-E GM-CFC) during the development of pure red cell hypoplasia induced in the cat by feline leukaemia virus. Leuk Res 7:103–116
8. Bordereaux D, Fichelson S, Sola B, Tambourin PE, Gisselbrecht S (1987) Frequent involvement of the *fim*-3 region in friend murine leukemia virus-induced mouse myeloblastic leukemias. J Virol 61:4043–4045
9. Neil JC, Forrest D (1987) Mechanisms of retrovirus-induced leukaemia. Selected aspects. Biochim Biophys Acta 907:70–79

Cell Lineage Specificity of Chromatin Configuration and Gene Expression in Haemopoietic Progenitor Cells

A. M. Ford [1], L. E. Healy [1], S. M. Watt [1], E. Spooncer [2], T. M. Dexter [2], H. V. Molgaard [1], and M. F. Greaves [1]

Haemopoietic stem cells have the potential to activate up to eight distinct cell lineage specific genetic programmes [1]. The mechanisms of cell lineage choice or commitment are fundamental to developmental biology in general and are beginning to be unravelled at least in invertebrates (e.g., Drosophila, Nematodes, slime moulds). Transacting DNA binding proteins that directly or indirectly regulate gene transcription are central players in the game [2], as are inductive cellular interactions [3]. In haemopoietic differentiation, it is clear that selective progenitor-stromal cell interactions involving both adhesive and growth factor recognition [4–7] are playing a role in early decision-making but the basic mechanisms whereby uni-lineage adoption is made are still obscure. A sample scheme might incorporate the following possibilities:

1. That when mesodermal cells become committed to blood cell formation [1] as opposed to say, muscle, then this specification must be reflected in inheritable changes in DNA structure and should logically include changes in genes that are functionally coupled to the individual lymphoid and myeloid lineage programmes. These genes could encode transcription factors, growth factor receptors, adhesion receptors and other functionally important molecules.
2. Alterations in genes that register pan-haemopoietic commitment plus potentiality for specific haemopoietic lineages could involve active transcription and/or alterations in chromatin structure. It is known that active genes are preferentially sensitive to DNAse I and that transcriptional control regions are hypersensitive [8–10]. Accessibility of chromatin to DNAse I is necessary but not sufficient for transcription and may provide therefore a convenient marker for genes that have become primed for activation.
3. That lineage programmes are expressed as a coordinated cascade of transcriptional control and the commitment process itself involves only a small number of genes, i.e. those that are expressed earliest in the developmental programme. Recent studies on muscle cell differentiation have provided an instructive precedent, indicating that single genes coding for regulatory (DNA binding) proteins can initiate a full programme of striated muscle lineage specific gene expression [11, 12].

With this background, we have asked whether any of the known haemopoietic lineage-associated genes are either transcriptionally active or have DNAse I hypersensitive sites in multi-potential cells. Immunoglobulin heavy chain and T-cell receptor (γ, β) genes were obvious choices as they have (sterile) transcripts (prior to gene rearrangement) in the earliest identifiable B or T lymphocyte percursors and their genomic structure is known in considerable detail [13, 14].

[1] Leukaemia Research Fund Centre, Institute of Cancer Research, Chester Beatty Laboratories, 237 Fulham Road, London SW3 6JB, UK and

[2] Paterson Laboratories, Christie Hospital and Holt Radium Institute, Wilmslow Road, Withington, Manchester M20 9BX, UK

We first screened a series of human lymphoid and myeloid leukaemia cell lines and normal tissue for DNAse I hypersensitivity of the IgH enhancer region, methylation of Hha I sites flanking this region and sterile mRNA. These results have been published in detail elsewhere [15].

As tested by DNAse I hypersensitivity, the chromatin structure of the IgH enhancer region in human B-cell precursor cell lines was in an open or accessible conformation. All T-cell lines, with either germline or rearranged IgH genes, were also hypersensitive to DNAse I but in contrast to B-cell precursors showed no detectable $C\mu$ expression. Normal thymocytes similarly had a hypersensitive IgH enhancer site. In contrast to lymphoid cells, all myeloid cell lines tested, as well as normal granulocytes, were not DNAse I hypersensitive and did not express $C\mu$. Two Hha I restriction sites on either side of the IgH enhancer were not methylated in all $C\mu$-expressing lines but methylated in non-expressing cell lines. A putative lympho-myeloid progenitor cell line KG1 [16], although having a germline configuration of Ig genes, produced $C\mu$ transcripts (and TCR γ mRNA) and was hypersensitive to DNAse I in the IgH enhancer region. After induction of myeloid differentiation the Ig enhancer region of KG1 cells is no longer hypersensitive or transcriptionally active. These results show that an open chromatin structure around the heavy chain enhancer is necessary but insufficient for initiating transcription from unrearranged IgH genes and further suggests that this region may be in an open or accessible configuration prior to lineage commitment and close following adoption of the myeloid lineage.

To pursue this possibility further, we have performed similar analyses in murine IL3-dependent cell lines established from long-term bone marrow culture [17, 18]. These lines have the considerable advantage over human cell lines (such as KG1) that they retain multilineage differentiation potential in vivo and in vitro, are nonleukaemogenic and have a normal diploid karyotype.

Four independent cell lines have been tested with similar results; we show here results with one line, A4. The cells maintained in medium supplemented with IL3 have a DNAse I hypersensitive IgH enhancer site as revealed by the presence of a 1.8 kb DNAse I digest product (EHS in Fig. 1, lane 2). There is however no stable μ mRNA detectable. TCR γ but not β genes are transcribed and in vitro hybridization analysis with a ^{35}S-DNA TCR γ probe indicates that $>95\%$ of A4 cells contain TCR γ mRNA. When A4 cells are induced to differentiate by removing IL3 and placing the cells in contact with either normal bone marrow stroma or 3T3 fibroblasts, then they differentiate into various types of myeloid cells (but predominantly granulocytes). Analysis of the differentiated progeny of A4 cells reveals that the IgH enhancer region is now resistant to DNAse I (i.e. closed; Fig. 1) and no TCR γ mRNA is detectable (Fig. 2) compared with an actin mRNA control.

We interpret these results to indicate that normal primitive myeloid progenitors have active or "primed" lymphoid genes but that these are closed down when definitive myeloid differentiation occurs. Since we also find that primitive embryonic stem cells (ES cells; [19]) and the primitive pan-mesodermal cell line 10T1/2 [20] have a DNAse I resistant IgH enhancer region and no TCR γ transcripts, we consider it likely that these features are characteristic of haemopoietic stem cells.

Experiments are in progress with other haemopoietic genes that are expressed very early in haemopoietic differentiation, e.g. CD3, CD2, CD19, λ 5, β-spectrin, CD33, and MPO, to see if they are transcriptionally active or are primed with DNAse I hypersensitive enhancer regions.

One speculative interpretation of these data is that haemopoietic stem cells register their multi-lineage potential by activating a small set of regulatory genes

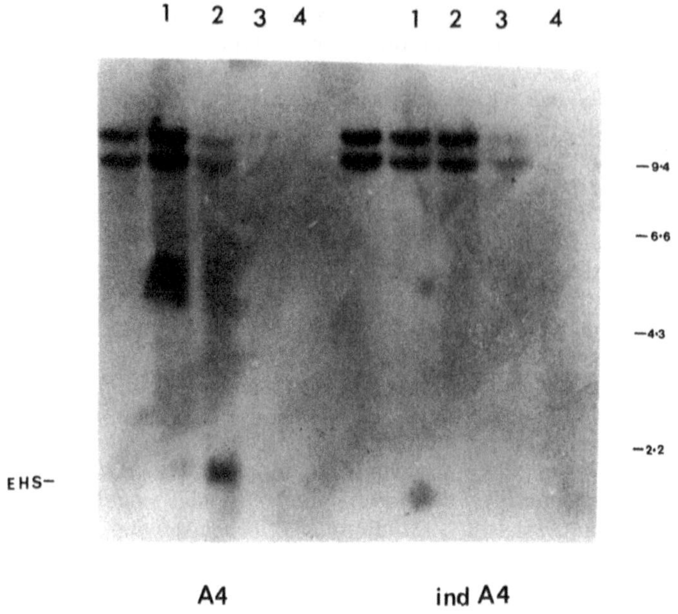

Fig. 1. DNAse I hypersensitive site analysis of the μ gene in myeloid-induced and non-induced A4 cells. *Lane 0:* nuclei incubated without added DNAse I. *Lanes 1–4:* nuclei incubated with 1.2, 1.8, 2.7 and 4 µg/ml DNAse I respectively (each lane contains 10 µg DNA restricted with BamHI and hybridized to a JH-enhancer intron probe). EHS, enhancer hypersensitive site; molecular weight markers are given in kb

whose protein products confer a DNAse I hypersensitive configuration on genes that play a pivotal and possibly initiating role in uni-lineage commitment. Interaction of such cells with appropriate stroma-associated environmental ligands [5, 6] might then selectively up-regulate particular transcription factor(s) and so initiate a cascade of selective gene expression for T cells or granulocytes, etc. Adoption of one lineage also then involves closing down the availability of other previously accessible lineage restricted genes.

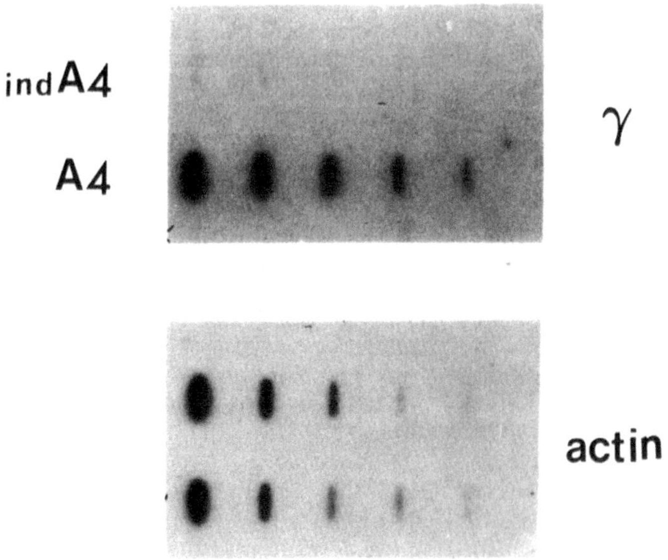

Fig. 2. RNA slot blot analysis of cytoplasmic RNA extracted from induced and non-induced A4 cells. Each lane contains samples of 2, 1.0, 0.5, 0.25 and 0.125 µg total cellular RNA, hybridized to murine TCR γ and actin probes respectively

Clearly many more experiments are required to verify these ideas. In particular, we need to establish that the results obtained to date are not unique to the rearranging lymphoid genes or to the cell lines used.

Finally, these data have potentially important implications for an understanding of lineage specific gene expression in human leukaemia cells. A proportion of acute leukaemias (5%-10%) display multilineage gene expression in individual blast cells [21]. It was suggested that this phenotypic pattern reflected either (a) infidelity of gene expression or genetic misprogramming arising as a direct consequence of gene rearrangements in leukaemia, or (b) an origin of such leukaemias in multi-lineage progenitor cells with effective maturation arrest in a proliferating mode such that their intrinsic capacity to activate early components of the lineage specific programme is revealed. This latter interpretation accords with the data we report in this paper and elsewhere [15]. The two interpretations are not exclusive however; certainly it is possible that activated or mutated protooncogenes could have profound effects on the regulation of lineage specific gene programmes [22].

Acknowledgements. The work described in this paper was supported by the Leukaemia Research Fund of Great Britain and the Cancer Research Campaign. We are grateful to Barbara Deverson for preparation of the manuscript.

References

1. Metcalf D, Moore MAS (1971) Haemopoietic cells. North-Holland, Amsterdam, p 45
2. Maniatis T, Goodbourn S, Fischer JA (1987) Regulation of inducible and tissue-specific gene expression. Science 236:1237
3. Harland R (1988) Growth factors and mesoderm induction. Trends Genet 4:62-63
4. Dexter TM (1982) Stromal cell associated haemopoiesis. J Cell Physiol [Suppl] 1:87-94
5. Gordon MY, Riley GP, Watt SM, Greaves MF (1987) Compartmentalization of a haemopoietic growth factor (GM-CSF) by glycosaminoglycans in the bone marrow microenvironment. Nature 326:403-405
6. Roberts R, Gallagher J, Spooncer E, Allen TD, Bloomfield F, Dexter TM (1988) Heparan sulphate bound growth factors: a mechanism for stromal cell mediated haemopoiesis. Nature 332:376-378
7. Sieff CA (1987) Hematopoietic growth factors. J Clin Invest 79:1549-1557
8. Mathis D, Oudet P, Chambon P (1980) Structure of transcribing chromatin. Prog Nucleic Acid Res Mol Biol 24:1-55
9. Burch JBE, Weintraub H (1983) Temporal order of chromatin structural changes associated with activation of the major chicken vitellogenin gene. Cell 33:65-76
10. Elgin SCR (1981) DNAase I-hypersensitive sites of chromatin. Cell 27:413-415
11. Lassar AB, Paterson BM, Weintraub H (1986) Transfection of a DNA locus that mediates the conversion of 10T1/2 fibroblasts to myoblasts. Cell 47:649-656
12. Pinney DF, Pearson-White SH, Konieczny SF, Latham KE, Emerson CP (1988) Myogenic lineage determination and differentiation: evidence for a regulatory gene pathway. Cell 53:781-793
13. Alt FW, Blackwell TK, DePinho RA, Reth MG, Yancopoulos GD (1986) Regulation of genome rearrangement events during lymphocyte differentiation. Immunol Rev 89:5-30
14. Marrack P, Kappler J (1987) The T cell receptor. Science 238:1073-1079
15. Ford AM, Watt SM, Furley AJW, Molgaard HV, Greaves MF (1988) Cell lineage specificity of chromatin configuration around the immunoglobulin heavy chain enhancer. EMBO J 7:2393-2399
16. Furley AJ, Reeves BR, Mizutani S, Altass LJ, Watt SM, Jacob MC, van den Elsen P, Terhorst C, Greaves MF (1986) Divergent molecular phenotypes of KG1 and KG1a myeloid cell lines. Blood 68:1101-1107
17. Spooncer E, Boettiger D, Dexter TM (1984) Continuous in vitro generation of multipotential stem cell clones from src-infected cultures. Nature 310:228-230
18. Spooncer E, Heyworth CM, Dunn A, Dexter TM (1986) Self-renewal and differentiation of interleukin-3-dependent multipotent stem cells are modulated by stromal cells and serum factors. Differentiation 31:111-118

19. Hooper M, Hardy K, Handyside A, Hunter S, Monk M (1987) HPRT-deficient (Lesch-Nyhan) mouse embryos derived from germline colonization by cultured cells. Nature 326:292–295
20. Taylor SM, Jones PA (1979) Multiple new phenotypes induced in 10T1/2 and 3T3 cells treated with 5-azacytidine. Cell 17:771–779
21. Greaves MF, Chan LC, Furley AJW, Watt SM, Molgaard HV (1986) Lineage promiscuity in hemopoietic differentiation and leukemia. Blood 67:1–11
22. Klinken SP, Alexander WS, Adams JM (1988) Hemopoietic lineage switch: v-raf oncogene converts Eμ-myc transgenic B cells into macrophages. Cell 53:857–867

Chromosomal Translocations Involving the T-Cell Receptor δ Chain Locus and Two Loci on the Short Arm of Chromosome 11 *

T. Boehm [1], R. Baer [1], L. Buluwela [1], A. Forster [1], I. Lavenir [1], E. Nacheva [2], J. Waters [3], L. White [4], D. Williams [5], and T. H. Rabbitts [1]

Chromosomal abnormalities in T-cell acute lymphoblastic leukaemia frequently involve chromosome 14q11, the site of the T-cell receptor (TCR) δ/α chain locus [1–3]. We have previously described the molecular analysis of two such translocations involving the short arm of chromosome 11, t(11; 14)(p15; q11) and t(11; 14)(p13; q11) [4, 5]. Here, we briefly summarise these findings with emphasis on the mechanism by which these translocations arise.

A. The Translocation t(11; 14)(p15; q11) in RPMI 8402

The DNA sequence analysis of chromosomal junction of $11p^+$ and $14q^-$ chromosomes, respectively, indicated that chromosome 11 sequences were joined to a DDJ element of the TCRδ chain gene (chromosome $11p^+$). A typical signal sequence (heptamer/nonamer with 12-bp spacer), presumably derived from the 5′ end of Dδ1, is joined, back to back, to a heptamer-like sequence derived from chromosome 11 at the chromosome $14q^-$ junction. It is noteworthy that there is no nonamer-like sequence on either side of this chromosome 11 derived heptamer. This suggests that the actual translocation involved a break [1] at the 5′ end of a Dδ element, which had previously joined to a DJ segment, and [2] at the heptamer sequence of chromosome 11. It is likely that the recombinase involved in the physiological process of antigen-receptor gene rearrangements attempted to join to a Vδ gene to the DDJ element, but mistakenly utilised the chromosome 11 derived heptamer to cause the translocation.

B. The Translocation t(11; 14)(p13; q11)

The molecular cloning of breakpoints of two tumours carrying the t(11; 14)(p13; q11) translocation [5] showed a DDJ join at one $11p^+$ breakpoint (tumour #8511) and a D-D join at another $11p^+$ junction (tumour LALW-2). The analysis of chromosome 11 germ-line sequences at these respective breakpoints revealed the presence of a heptamer-like sequence in one instance (tumour #8511) but not in the other tumour (LALW-2). In the latter case, the translocation process seemed to have ignored the heptamer used in the #8511 tumour, although both breakpoints in that region occur within only 800 bp [5]. Figure 1 A summarises the events leading to the translocation in tumour LALW-2 and indicates that the translocation most likely takes place after a D-D join during an attempt to join a Vδ element. Figure 1 B schematically shows the spatial relation-

[1] Medical Research Council Laboratory of Molecular Biology, [2] Addenbrooke's Hospital, Department of Haematological Medicine, and [3] Cytogenetics Department, University Clinical School, Hills Road, Cambridge, CB2 2QH, UK
[4] Division of Haematology/Oncology, The Prince of Wales Children's Hospital, High Street, Randwick, NSW 2031, Australia
[5] Division of Pathology and Laboratory Medicine, St Jude Children's Hospital, Memphis, TN 38101, USA
* T. B. was supported by a Mildred-Scheel fellowship

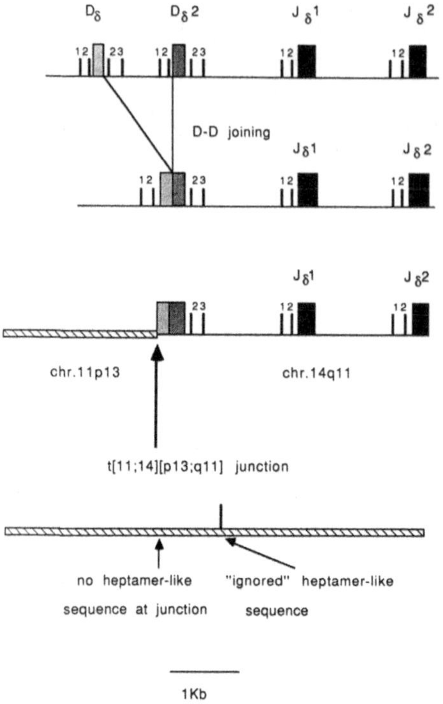

Fig. 1. **A** Schematic illustration of a D-D join at the TCRδ locus and preceding the translocation in tumour LALW-2 [5] and the structure of the $11p^+$ chromosome (*third line*) after translocation. **B** Schematic illustration of the chromosome 11 germ-line breakpoint cluster region, indicating the breakpoint in the LALW-2 tumour about 0.8 kb centromeric to a hepatamer-like sequence

ship of the heptamer to the translocation breakpoint of tumour LALW-2.

C. Conclusion

The contribution of the t(11; 14) (p15; q11) translocation to tumour formation in the RPMI 8402 cell line is as yet unknown. The 11p15 sequences near the breakpoint are transcriptionally active, so the possibility exists that the translocation aberrantly activated this gene in one way or another with pathogenic consequences for the afflicted cell. However, an alternative view would be that the translocation occurred only because of the presence of the heptamer-like sequence near the breakpoint mistakenly utilised by the recombinase while attempting to rearrange a Vδ gene to the DDJ segment. The translocation could thus have been innocuous with respect to tumour formation. As such, the RPMI 8402 provides an excellent example to test the central dogma of cancer cytogenetics about the pathogenic significance of chromosomal abnormalities.

A completely different situation is encountered in the t(11; 14)(p13; q11) translocation. In this case, a breakpoint cluster region can be defined, strongly suggesting that a disrupted 11p13 locus is important for tumour formation. Furthermore, the variable presence of heptamer sequences at the breakpoints within that region suggests that a sequence-specific cut is not necessary on chromosome 11 for the translocation process. This observation distinguishes a two- and a one-site recognition model for sequence-specific recombinase involvement in these translocations [6].

The 11p13 breakpoint may prove to be useful in another respect. Due to its apparent proximity to the WAGR complex, it might help to refine the map of this important region, thought to contain tumour suppressor genes.

References

1. Mengle-Gaw L, Willard HF, Smith CIE, Hammärstrom L, Fischer P, Sherrington P, Lucas G, Thompson POW, Baer R, Rabbitts TH (1987) Human T-cell tumours containing chromosome 14 inversion or translocation with breakpoints proximal to immunoglobulin joining regions at 14q32. EMBO J 6:2273–2280
2. Williams DL, Look AT, Melvin SL, Robertson PK, Dahl G, Flake T, Stass S (1984) New chromosomal translocations correlate with specific immunophenotypes of childhood acute lymphoblastic leukemia. Cell 36:101–109
3. Baer R, Boehm T, Yssel H, Spitz H, Rabbitts TH (1988) Complex rearrangements

within the human Iδ-Cδ/Iα-Cα locus and aberrant recombination between Iα segments. EMBO J 7:1661–1668
4. Boehm T, Baer R, Lavenir I, Forster A, Waters JJ, Nacheva E, Rabbitts TH (1988) The mechanism of chromosomal translocation +(11; 14) involving the T-cell receptor C locus on human chromosome 14q11 and a transcribed region of chromosome 11p15. EMBO J 7:385–394
5. Boehm T, Buluwela L, Williams D, White L, Rabbitts TH (1988) A cluster of chromosome 11p13 translocations found via distinct D-D rearrangements of the human T-cell receptor δ chain gene. EMBO J 7:2011–2017
6. Rabbitts TH, Boehm T, Mengle-Gaw L (1988) Chromosomal abnormalities in lymphoid tumours: mechanism and role in tumour pathogenesis. Trends Genet 4:300–304

Genetic Markers for Oncogenes, Growth Factors, and Cystic Fibrosis *

M. Dean [1], C. Stewart [1], A. Perry [1], B. Gerrard [1], T. Beck [2], U. Rapp [2], M. Drumm [3], M. Iannuzzi [3], F. Collins [3], S. O'Brien [2]

A. Introduction

The techniques of molecular biology have had a dramatic effect on the advancement of human genetics. In particular, the development of restriction fragment length polymorphisms (RFLPs) has allowed researchers to generate genetic markers for virtually any region of the human genome. Most RFLPs occur when a mutation creates or deletes a recognition site for a restriction enzyme, generating a DNA fragment of altered size. In the simplest case this will create two alleles. A DNA probe which hybridizes to this fragment will detect the presence of these alleles in the DNA from different individuals. Probes used to detected RFLPs have been derived from both cloned genes and randomly isolated DNA segments. Thus, each RFLP is a genetically inherited marker for a precise location on a chromosome.

By analyzing the inheritance of RFLPs in families, linkage analysis techniques can be used to determine the order and genetic distance between different polymorphic probes [1]. When this information is combined with data on the physical location of the probes, a genetic map of a chromosome can be constructed [2]. Genetic maps are powerful tools, useful in the detailed molecular analysis of biological problems.

By studying the inheritance of polymorphisms in families afflicted with a genetic disease, the approximate chromosomal location of the gene responsible for the disease can be determined. Linkage with RFLP markers has been detected for several human disease genes including Huntington's disease [3], muscular dystrophy [4], and cystic fibrosis [5–7]. Linkage is often a critical step in the eventual isolation of the gene itself [8]. RFLPs have also provided critical data in the analysis of specific chromosomal abnormalities in human tumors. DNA markers can be used to distinguish between the two homologous chromosomes in the normal cells of a patient and, when compared with DNA from a tumor, can detect the loss of a specific chromosome or chromosomal region. Such analysis helped to identify the region containing the gene responsible for retinoblastoma [9] and regions containing putative genes involved in Wilm's tumor and renal cancer [10, 11] among others. In this report we present data on several newly detected RFLPs in biologically important genes (Table 1), describe in detail the identification of new RFLPs in three loci, and discuss the potential usefulness of genetic polymorphisms in the analysis of human disease.

[1] BCDP, Program Resources Inc., NCI-Frederick Cancer Research Facility, Frederick, MD 21701, USA

[2] Laboratory of Viral Carcinogenesis, NCI-Frederick Cancer Research Facility, Frederick MD 21701, USA

[3] Howard Hughes Medical Institute, University of Michigan, Ann Arbor, MI, USA

* This project has been funded at least in part with Federal Funds from the Department of Health and Human Services under contract number N01-C0-74102 with Program Resources, Inc. The content of this publication does not necessarily reflect the views or policies of the Department of Health and Human Services, nor does mention of trade names, commercial products, or organizations imply endorsement by the U.S. Government

B. Results and Discussion

The c-*raf*-1 gene is a member of the family of serine protein kinases and is the cellular homologue of the v-*raf* gene. v-*raf* was identified as the transforming gene of the murine 3611 acute transforming retrovirus [12]. It is believed that part of the function of the c-*raf*-1 gene is to act as an intracellular signal transmission molecule for certain hormones or growth factors [13]. The human c-*raf*-1 gene has been localized to chromosome 3 at bands p24–25 [14]. Using cDNA probes from the human c-*raf*-1 gene, we have searched for the presence of RFLPs in this locus. The pTuc-8 probe detects a polymorphism with the enzyme *Eco*RI, with alleles of 15 and 12 kb (Fig. 1). In an analysis of 71 unrelated Caucasians, we found a frequency of 0.87 for the 15-kb allele and 0.13 for the 12-kb allele (Table 1). In addition, the *Eco*RI poly-

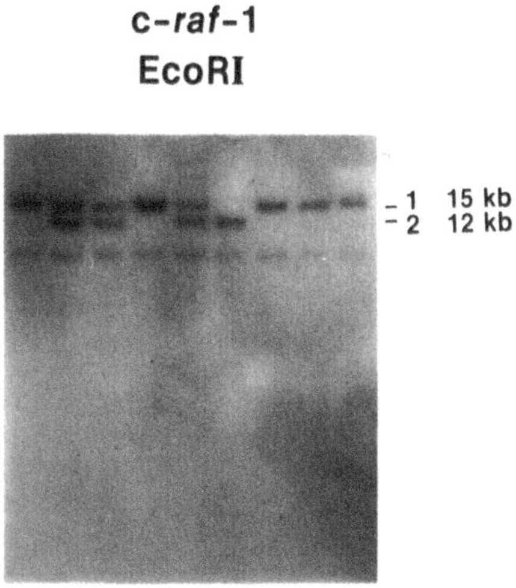

Fig. 1. *Eco*RI RFLP in the c-*raf*-1 gene. DNA from nine individuals was hybridized with the pTUC-8 cDNA probe. Size of the alleles in kilobases (*kb*)

Table 1. Newly described RFLPs

Gene	Location	Enzyme	Alleles	Size (kb)	Frequency
IL-1α	2q13–21	*Taq*I	>5 VNTR	5.5–6.0	
		*Taq*I	1	5.5–6.0	0.74
			2	4.2	0.26
RAR	17q21.1	*Pst*I	1	3.0	0.18
			2	2.6	0.82
c-*raf*-1	3p24–25	*Eco*RI	1	15	0.87
			2	12	0.13
pCF29	7q31	*Eco*RV	1	12	0.40
			2	9.5	0.60
		*Pvu*II	1	9.0	0.39
			2	6.0	0.61
CF7032	7q31	*Sac*I	1	15	0.39
			2	6	0.61
T4	12pter -p12	*Taq*I	1	7.5	0.80
			2	7.0	0.20
TCR zeta		*Bam*HI	>5 VNTR[a]	5.5–6.5	
ovc-2		*Pst*I	1	6.6	0.44
			2	5.0	0.56
		*Rsa*I	1	1.7	0.53
			2	1.2	0.57
c-*jun*		*Kpn*I	1	10.5	0.50
			2	10.0	0.50

[a] VNTR = variable number of tandem repeats

morphism was shown to segregate properly in three large three-generation pedigrees (data not shown).

Specific deletions in the short arm of chromosome 3 have been described in patients with small-cell lung cancer (SCLC) and renal cell carcinoma [15, 16]. Using RFLPs within the c-*raf*-1 locus, we have examined DNA from a total of 83 human lung carcinomas (Sidthansen et al., unpublished). In an analysis of ten paired (normal versus tumor) SCLC DNA samples, five informative cases showed loss of heterozygosity at this locus. A *Bgl*I polymorphism which shows 50% heterozygosity in a normal population was used to analyze 73 unpaired lung carcinoma DNAs. Fifteen of 31 non-SCLC samples showed heterozygosity; however, none of the 42 SCLC samples were heterozygous. This striking apparent loss of heterozygosity at the c-*raf*-1 locus in SCLC provides evidence that the c-*raf*-1 locus is deleted in small-cell lung carcinoma.

Recently, Siezinger et al. [17] used RFLPs in the c-*raf*-1 gene to detect linkage between this gene and the gene responsible for Von Hippel-Lindau syndrome, an inherited cancer syndrome. The polymorphism described here can be used to further study families with this disease in order to pinpoint the location of the gene.

We used probes from the interleukin-1-alpha (IL-1α) gene to screen for RFLPs at this locus. IL-1α is a secreted protein that is involved in the stimulation of the growth of lymphocytes. While secreted principally from macrophages, the protein is also produced from keratinocytes and lymphocytes and plays a role in the stimulation of lymphocytes and fibroblasts [18]. Using an IL-1α cDNA probe we detected a two-allele polymorphism with the enzyme *Taq*I, and found that these alleles have frequencies of 0.74 and 0.26 (Table 1). In addition, we found that a fragment of this gene, detected by several different enzymes, showed a high degree of variability between individuals. This type of variation is characteristic of

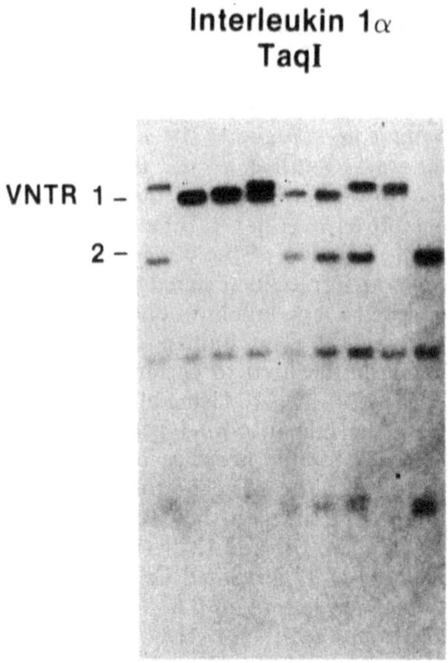

Fig. 2. Polymorphisms in the interleukin-1-alpha gene. *Taq*I detects a two-allele site polymorphism as well as a high polymorphic VNTR

polymorphisms known as VNTRs (variable number of tandem repeats). VNTRs are regions of DNA containing a group of tandem-repeated sequences [19]. The number of repeats at a given locus is often highly variable between different individuals, making these polymorphisms very informative for genetic analyses. Examination of the published genomic sequence of the genomic IL-1α gene [20] revealed the presence of a group of tandem repeats in the intervening sequence between the fifth and sixth exons (Fig. 2). The sequence of these repeats is typical of those found in other VNTRs which have been characterized. The IL-1α gene has been mapped to chromosome 2q13-21 and the polymorphisms we have described will be useful genetic markers for this region. In addition, these polymorphisms can be used to test whether the IL-1α gene is linked to any human genetic disease for which family pedigrees are available.

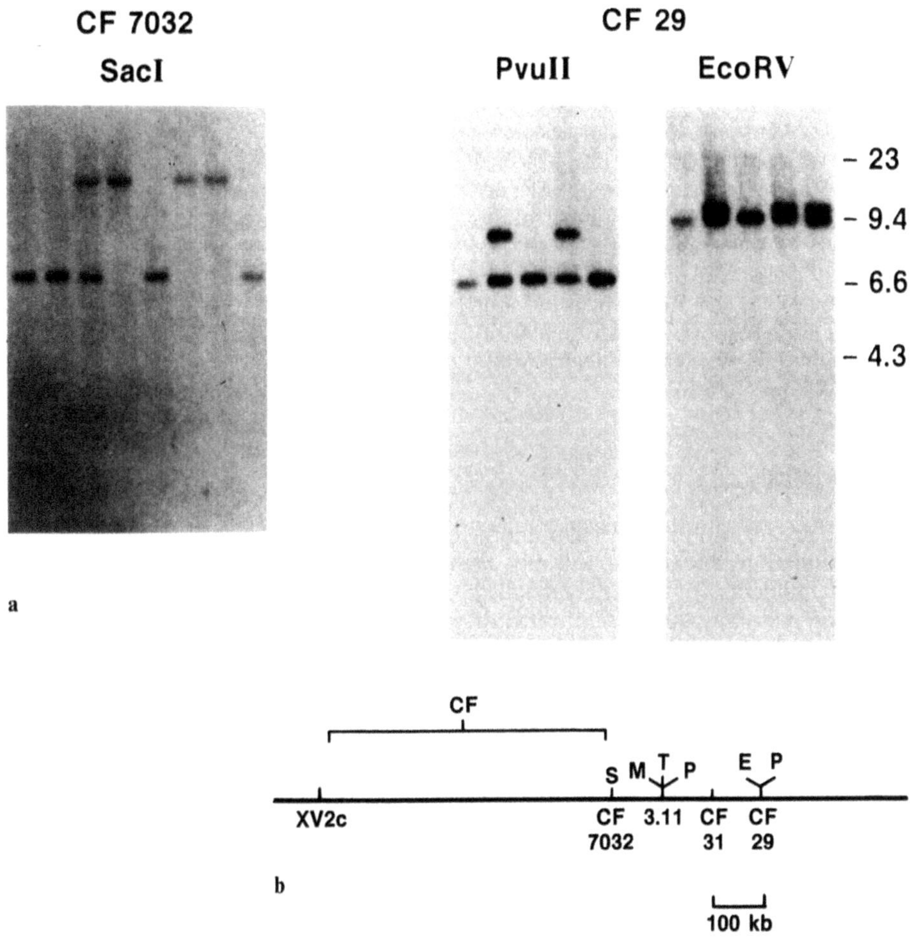

Fig. 3. Polymorphisms for probes tightly linked to the CF gene. The positions of the chromosome-jumping clones CF 70-32, CF 31, and CF 29 are shown in relation to the pJ3.11 probe. RFLPs detected by these clones are displayed with sizes in kilobases. *S, Sac*I; *M, Msp*I; *T, Taq*I; *P, Pvu*II; *E, Eco*RV

Cystic fibrosis (CF) is the most common fatal genetic disease among Caucasians; it is caused by a recessive mutation at a single locus. The gene responsible for cystic fibrosis has been linked to polymorphic markers that map to chromosome 7q31 [21]. Recently, the location of the gene has been narrowed to a 700-kb region between the probes XV2c and pJ3.11 [22]. In order to clone and characterize additional sequences from this region, we have used chromosome jumping [23, 24] to isolate sequences surrounding the pJ3.11 locus. Briefly, chromosome jumping involves the circularization of large DNA fragments and the cloning of the ends of such fragments. By generating a library of chromosome-jumping clones and screening with a probe from the pJ3.11 locus, we have been able to isolate sequences spanning 300–400 kb surrounding pJ3.11 (Fig. 3). Using probes from this region we have searched for RFLPs and have discovered three new polymorphisms to date (Table 1). Two of these

RFLPs are detected with the CF29 probe, which is located approximately 150 kb from pJ3.11, away from the CF gene. The third RFLP is detected with CF70-32, a jumping clone that lies closer to the CF gene (Fig. 3; unpublished data). While all three of these RFLPs are fairly informative, in preliminary analysis they all show a high degree of linkage disequilibrium with the *Msp*I RFLP detected by pJ3.11. In other words, all four of these RFLPS reveal very little new information and increase the heterozygosity only from 53% to 58%.

C. Conclusion

Further progress in the molecular analysis of human genetic disease will rely on the development of increasingly detailed chromosome maps. At present, most of the maps of human chromosomes were constructed principally with anonymous DNA markers, not known to be part of genes. We have begun to characterize RFLPs in biologically important genes, many of which have been mapped to specific chromosomal locations by in situ hybridization. The list of genes in which we have detected RFLPs includes the proto-oncogenes and growth factors c-*raf*-1, c-*jun*, c-*ovc*-2, and IL-1α, and cell surface molecules and receptors such as the retinoic acid receptor, the T-cell receptor zeta gene, and the T4 gene, receptor for HIV. In addition to providing markers for specific chromosomal locations, the RFLPs in these genes will allow them to be tested as candidate genes for human genetic diseases.

References

1. Botstein D, White R, Skolnick M, Davis RW (1980) Construction of a genetic linkage map in man using restriction fragment length polymorphisms. Am J Hum Genet 32:314–331
2. White R, Leppert M, Bishop DT, Barker D, Berkowitz J, Brown C, Callahan P, Holm T, Jerominski L (1985) Construction of linkage maps with DNA markers for human chromosomes. Nature 313:101–105
3. Gusella JV, Wexler, NS et al. (1983) A polymorphic DNA marker genetically linked to Huntington's disease. Nature 306:234
4. Davies KE, Pearson PL et al. (1983) Linkage analysis of two cloned DNA sequences flanking the Duchenne muscular dystrophy locus on the short arm of the human X chromosome. Nucleic Acids Res 11:2303–2312
5. Tsui L-C, Buchwald M, Barker D, Braman JC, Knowlton R et al. (1985) Cystic fibrosis locus defined by a genetically linked polymorphic DNA marker. Science 230:1054–1057
6. Wainwright BJ, Scambler PJ, Schmidtke J, Watson EA, Law H-Y, Farrall M, Cooke HJ, Eiberg H, Williamson R (1985) Localization of cystic fibrosis locus to human chromosome 7cen-q22. Nature 318:384–386
7. White R, Woodward S, Leppert M, O'Connell P, Hoff M, Herbst J, Lalouel J-M, Dean M, Vande Woude G (1985) A closely linked genetic marker for cystic fibrosis. Nature 318:382–384
8. Monaco AP, Neve RL, Colletti-Feener C, Bertelson CJ, Kurnit DM, Kunkel LM (1986) Isolation of candidate cDNA's for portions of the Duchenne muscular dystrophy gene. Nature 323:646–650
9. Cavenee WK, Dryja TP, Phillips RA, Benedict WF, Godbout R, Gallie BL, Murphree AL, Strong LC, White RL (1983) Expression of recessive alleles by chromosomal mechanisms in retinoblastoma. Nature 305:779–784
10. Koufos A, Hansen MF, Copeland NG, Jenkins NA, Lampkin BC, Cavenee WK (1986) Loss of heterozygosity in three embryonal tumors suggests a common pathogenetic mechanism. Nature 316: 330– 334
11. Zbar B, Brauch H, Talmadge C, Linehan M (1987) Loss of alleles of loci on the short arm of chromosome 3 in renal cell carcinoma. Nature 327:721
12. Rapp UR, Goldsborough MD et al. (1983) Structure and biological activity of v-*raf*, a unique oncogene transduced by a retrovirus. Proc Natl Acad Sci USA 80:4218
13. Rapp UR, Storm SM, Cleveland JL (1987) Oncogenes: Clinical Relevance.

Hematology and Blood Transfusion 31:450–459
14. Bonner T, O'Brien SJ, Nash WG, Rapp UR (1984) The human homologs of the raf (mil) oncogene are located on human chromosomes 3 and 4. Science 223:71–74
15. Whang-Peng J, Kao-Shan CS, Lee EC (1982) Specific chromosome defect associated with human small-cell lung cancer: deletion 3p(14-23). Science 215:181–182
16. Kok K, Osinga J, Carritt B, Davis MB, van der Hout AH, van der Veen AY, Landsvater RM et al. (1987) Deletion of a DNA sequence at the chromosomal region 3p21 in all major types of lung cancer. Nature 330:578–581
17. Seizinger BR, Roleau GA et al. (1988) Von Hippel-Lindau disease maps to the region of chromosome 3 associated with renal cell carcinoma. Nature 332:268–269
18. Oppenheim JJ, Kovacs E, Matsushima K, Durum SK (1986) There is more than one interleukin 1. Immunol Today 2:45–56
19. Nakamura Y, Leppert M, O'Connell P, Wolfe R, Holm T, Culver M, Martin C, Fujimoto E, Hoff M, Kumlin E, White R (1987) Variable number of tandem repeat (VNTR) markers for human gene mapping. Science 235:1616–1622
20. Furutani Y, Notake M, Fukui T, Ohue M, Normura H, Yamada M, Nakamura S (1986) Complete nucleotide sequence of the gene for human interleukin 1 alpha. Nucl Acids Res 14:3167–3179
21. Dean M (1988) Molecular and genetic analysis of cystic fibrosis. Genomics 3:93–99
22. Drumm ML, Smith CL, Dean M (1988) Physical mapping of the cystic fibrosis region by pulsed-field gel electrophoresis. Genomics 2:346–354
23. Collins FS, Weissman SM (1984) Directional cloning of DNA fragments at a large distance from an initial probe: A circularization method. Proc Natl Acad Sci USA 81:6812–6816
24. Collins FS, Drumm ML, Cole JL, Lockwood WK, Vande Woude FG, Iannuzzi MC (1987) Construction of a general human chromosome jumping library, with application to cystic fibrosis. Science 235:1046–1049

Search for Genes Critical for the Early and/or Late Events in Carcinogenesis: Studies in *Xiphophorus* (Pisces, Teleostei)*

C. Zechel, U. Schleenbecker, A. Anders, M. Pfütz, and F. Anders

A. Introduction

I. Historical Background

The concept of genes that code for neoplastic transformation, called "oncogenes," originates from two sources, virology and animal genetics. The virological source can be traced back to the year 1910 when Peyton Rous discovered the virus that causes sarcoma in chickens. It took, however, about 60 years until evidence was produced that the cancer determinants located in the genome of this and related viruses (retroviruses) are truly genes [1, 2]. The source in animal genetics dates from 1929, when Myron Gordon, Georg Häussler, and Curt Kosswig independently discovered that the F_1 hybrids between certain domesticated ornamental breeds of the Central American fish species *Xiphophorus maculatus* (platyfish) and *X. helleri* (swordtail) spontaneously develop melanoma that is inherited in the hybrid generations like the phenotype of any normal Mendelian gene located in the genome of the fish. The basic idea in both the retrovirus and the *Xiphophorus* model is that oncogenes present in the genome of animals are activated by changes in structure (point mutation, translocation, truncation) and/or changes in expression (ectopic expression, unscheduled expression), and that products of the activated genes mediate the neoplastic transformation of a target cell [3–6].

In addition, a recent extension of the oncogene hypothesis is that "tumor-suppressor genes" or "antioncogenes" control the expression of oncogenes and the manifestation of a tumor phenotype [7, 8]. Such "oncostatic genes" have been identified in several systems: firstly, the retinoblastoma gene in humans [9, 10], secondly, the lethal giant larvae gene in *Drosophila* [11, 12], and thirdly, the differentiation gene *Diff* in *Xiphophorus* [3, 13].

While many investigators have focussed their attention on the role of retroviral oncogenes (symbolized by v-*onc*s) in neoplastic transformation, we concentrated on cellular oncogenes (symbolized by x-*onc*s) and the oncostatic genes that might be involved in the early and late events in carcinogenesis in *Xiphophorus*.

Our research on oncogenes in *Xiphophorus* began in 1957 with systematic crossings between populations, races, and species and with mutagenesis studies in purebred and hybrid fish. *Xiphophorus* from wild populations of the natural habitat and *Xiphophorus* bred from wild populations in the laboratory are almost completely insusceptible to neoplasia, i.e., insensitive to mutagenic carcinogens and tumor promotors, whereas hybrids derived from crossings between different wild populations develop neoplasms spontaneously or after treatment with carcinogens [14–16]. Subsequently we found that melanoma and a large variety of other neoplasms developing either spontaneously or after treatment with carcinogens can be as-

Genetisches Institut der Universität Giessen, D-6300 Giessen, FRG

* This work was supported by the University of Giessen, Deutsche Forschungsgemeinschaft (grant 17/26-1–3 and SFB 103), Bundesminister für Forschung und Technologie, and the Umweltbundesamt, FRG

signed to a particular Mendelian gene located on particular sex chromosomes [3, 14, 16, 17]. This gene is an "oncogene" by definition, and was designated as "tumor gene," *Tu*. The phenotypic expression of *Tu* is regulated by systems of modifying genes that may stimulate (*S* genes) or repress (*R* genes) *Tu* activity [18]. The genetic make-up of the modifying gene system is highly different among the species of the genus *Xiphophorus*, less different among the races of a certain species, but still different among populations of a certain race. Evidence for this assumption comes from the observation that the degree of malignancy of melanomas in interspecific hybrids depends on the parental genomes [18, 19].

Studies on transformed pigment cells in purebreds and interspecific hybrids led to the characterization of the *Tu* and the *R* genes by means of classical pheno- and cytogenetic methods, however little is known about the *S* genes. The *Tu* gene, although detected in fish with transformed melanocytes (Tr melanocytes) is also present in all specimens of wild populations of *Xiphophorus*, irrespective of the phenotypic expression of Tr melanocytes. It is postulated that *Tu* fulfills an essential, so far unknown function (indispensable *Tu*), and that its function in fish developing tumors is an accessory one (accessory *Tu*). In the wild populations malignant expression of *Tu* is under stringent control exerted by *R* genes that are organized mainly as members of three interrelated *R* gene systems: (a) *Tu*-linked tissue-specific *R* genes (*R-mel, R-neu, R-epi, R-mes*) which, if impaired, lost, or translocated, permit the *Tu*-encoded tumor formation in the pigment cell system and neurogenic, epithelial, and mesenchymal tissues [3]; (b) *Tu*-linked compartment-specific *R* genes, which restrict spots and melanomas to distinct compartments of the body; 14 compartment-specific *R* genes (*R-co*) have been identified which, if impaired, correspond to sites of the body where the spots in the purebreds and the melanomas in the hybrids occur [3]; (c) the *Tu*-nonlinked modifying genes (*R* and *S* genes), which control proliferation and differentiation of the transformed pigment cells, e.g., the prominent *R* gene *Diff* which was disclosed by the clear-cut 1:1 segregation between the BC hybrids bearing malignant melanomas and those bearing benign melanomas [13, 20, 21]. *Tu, R-mel,* and *R-co* are closely linked to each other and form a Mendelian entity designated as "tumor gene complex" (*Tu* complex). *Tu* complexes that are accessory in the fish and determine Tr melanophore patterns are the subject of the first part of the present article and modifying genes that of the second part.

II. Approach

Table 1 shows the gross constitution of the *Tu* complex-containing region of the sex chromosomes of *Xiphophorus* in wild-type order and after X-ray induced structural changes. Most of the mutants used were genetically and phenogenetically analyzed in 1973 (for photographs see [14]). Since then, more and new mutants have been isolated and studied (for photographs see [22]). The respective Mendelian genes on the chromosomes are arranged in a uniform order with (a) the sex-determining region proximal to the centromer, followed by (b) the pterinophore loci (*Dr, Ar, Ye, Or, Br*) and (c) the melanophore loci (*Sd-Tu, Sr-Tu,* etc.). Purebred animals containing a *Tu* complex exhibit spots. If, however, the *Tu* complex is present in *X. maculatus*/*X. helleri* hybrids, melanoma develops either spontaneously or following treatment with carcinogens. Animals lacking the *Tu* complex are almost completely incapable of developing spots or melanoma [15, 16]. Although the *Tu* complexes are rather well understood in terms of Mendelian genetics, studies undertaken at the molecular level have failed so far to characterize them and to distinguish between those that are indispensable and those that are accessory. Not only have no gene products of the *Tu* complexes been identified, but also no

Table 1. Wild-type and structural mutants in *Xiphophorus*

					1973	1988
A	FM..........	*X. helleri,*	wild type,	Rio Lancetilla		
X	F *Dr, Sd-Tu*	*X. maculatus,*	wild type,	Rio Jamapa		
X	M *Ar, Sr-Tu*	*X. maculatus,*	wild type,	Rio Jamapa		
X	F *Ye, Li-Tu*	*X. variatus,*	wild type,	Rio Panuco		
Y	M *Or, Pu-Tu*	*X. variatus,*	wild type,	Rio Panuco		
Z	M *Br, Ni-Tu*	*X. maculatus,*	wild type,	Belize River		
Z	M	*X. maculatus,*		Rio Usomacinta		
X	F,	Deletions,	no *Tu* complex left		1	14
Y	M,	Deletions,	no *Tu* complex left			3
X	F *Dr,*	Deletions,	pterinophore locus left		2	48
Y	M *Ar,*	Deletions,	pterinophore locus left			4
Z	M *Br,*	Deletions,	pterinophore locus left			3
X/Y	M *Dr, Sr-Tu*	Crossovers			1	2
A/X	FM..., *Sd-Tu*	Interspecific translocations			1	11
A/Y	FM..., *Sr-Tu*	Interspecific translocations			1	3
X/X	F *Dr, Li-Tu*	Interspecific crossovers			1	2
X+Y	F *Ye, Li-Tu, Or, Pu-Tu*	Unequal crossovers, duplications			4	5
A/X	FM..., *Dr,*	Interspecific translocation and deletion			1	
Y+X	M *Ar, Sr, Sd-Tu*	Unequal crossover			1	4
X+Y	F *Dr, Ar, Sr-Tu*	Deletions and crossover			1	1
X/Y	F *Dr, Ar,*	Deletions/deletions/crossovers				3
Total					14	103

A, Autosome of *X. helleri* homologous to the sex chromosomes of *X. maculatus*; X, Y, Z, sex chromosomes; F, M, female- and male-determining regions. Pterinophore loci: *Dr*, dorsal red; *Ar*, anal red; *Ye*, yellow; *Or*, orange; *Br*, brown; *Ni*, nigra. Melanophore loci: *Sd*, spotted dorsal; *Sr*, striped; *Li*, lineatus; *Pu*, punctatus.

alterations have been observed which would be expected to appear when the *Tu* complexes switch over to tumorigenic potency. Moreover, the *Tu*-nonlinked *R* genes and the *S* genes are poorly defined in terms of Mendelian and molecular genetics. The approach was to identify and to map sequences strictly correlated with the inheritance of the tumor phenotype, that is to say, of the *Tu* complexes and the modifying genes, and to study their expression.

We had probes specific for 15 molecularly defined viral oncogenes at our disposal when we started our search for genes structurally and functionally related to the genetic factors determining neoplasia in *Xiphophorus*. Southern blot analyses and some sequence data revealed that almost all oncogenes corresponding to the probes are present in the genome of all individuals of *Xiphophorus* tested so far [16, 23–28]; only *ros* and *mos* could not be identified in the fish. Some of the xiphophorine cellular oncogenes (x-*onc* genes) show restriction fragment length polymorphisms (RFLP), the patterns of which have evolved differently in the various taxonomic groups of fish [22, 27, 28]. For instance, the pattern of the lengths of the restriction fragments of x-*sis* is specific to each of the different species but there is no RFLP within each of the species; actually these species show a monomorphism of the restriction fragment length of x-*sis* [27]. In contrast, the pattern of lengths of the restriction fragments of x-*erb*A and x-*erb*B is species nonspecific but is specific to the different races and populations of the species. The lengths of certain fragments of x-*erb*B are even different in females and males of the same population [22, 28]. We used the RFLP phenomenon as an indicator for the Mendelian inheritance of the x-*onc*s through the purebred and hybrid generations. If a certain oncogene fragment is inherited independently from the inheritance of spot or melanoma formation, one can conclude that the respective oncogene is not "critical" for the first step of melanoma formation. This is not to say that such an oncogene is not involved in melanoma formation at all; for instance, x-*src*, x-*sis*, x-*ras*, x-*myc* are expressed in the melanoma [27, 29, 30, 31] and are certainly implicated in tumor growth or tumor progression, but they are contributed by *X. helleri* to the hybrid while the appearance of spots or melanoma is contributed by *X. maculatus*. These genes, therefore, are not candidates for the primary event leading to melanoma.

B. Results and Discussion

I. Oncogenes that Might Be Considered "Critical" for the Early Events in Carcinogenesis

1. x-*erb*B Restriction Fragment Length Polymorphism

In the following paragraphs we concentrate on certain viral *erb*B (v-*erb*B) homologous DNA fragments because they are so far the only fragments that show the same inheritance as the susceptibility to melanoma. These fragments correspond to the x-*erb*B gene that represents a xiphophorine epidermal growth factor (EGF) receptor gene ([22]; for an overview on v-*erb*B and c-*erb*B see [32]).

Figures 1 and 2 show that Southern analyses revealed a different distribution of several *Eco*RI fragments of x-*erb*B in the different purebred and hybrid genotypes. Fragments of 3.5 kb and 4.3 kb, and some bands larger than 12 kb, are distributed in all populations and species of the fish genus without any as yet detectable pattern. Two fragments comprising 5.5 and 7.5 kb are constantly present in all individuals of all populations of *Xiphophorus* tested and, therefore, appear to be located on an autosome and to be structurally unrelated to the *Tu* complexes which determine the spot pattern and the melanoma formation. Three fragments, however, comprising 4.9, 6.7, and 11.5 kb are restricted to individuals exhibiting the sex-chromosome-linked spot patterns or melanomas. The latter fragments claimed our special interest.

The 4.9-kb *Eco*RI fragment is restricted to all individuals of *X. maculatus* from Rio Jamapa (female XX, male XY; Fig. 1, lanes a and b) exhibiting the X-chromosomal *Dr Sd-Tu* complex (dorsal red, spotted dorsal). The 6.7-kb *Eco*RI fragment is linked to the Y-chromosomal *Ar Sr-Tu* (anal red, stripe sided) from the same population and to the Z-chromosomal *Br Ni-Tu* (brown, nigra) from *X. maculatus* from Belize River (female WZ, male ZZ; lane d). The 11.5-kb fragment is specific to the X-chromosomal *Ye Li-Tu* (yellow, lineatus) of *X. variatus* from the Rio Panuco lane c. *X. maculatus* from Rio Usomacinta (lane e) and *X. helleri* from Rio Lancetilla (lane f) that lack both the sex-chromosomal pterinophore locus and the *Tu* complex (no spot patterns occur) lack also the sex-chromosomal restriction fragments. The latter results confirm that the sex-chromosomal spot patterns and the sex-chromosomal v-*erb*B related *Eco*RI fragments are linked to each other.

To assign the X-, Y- and Z-specific fragments more specifically to melanoma we introduced normal and structurally mutated sex chromosomes of *X. maculatus* and a normal X chromosome of *X. variatus* into the genome of *X. helleri* by introgression comprising mostly more than five backcrosses. The BC hybrids, which as expected segregated into equal portions of siblings exhibiting or lacking melanomas, were examined in parallel to the purebreds (Fig. 2; compare with Fig. 1). The 4.9-kb fragment was found in the normal and tumorous tissues of the melanoma developing BC segregants that carry the X-chromosomal *Dr Sd-Tu*

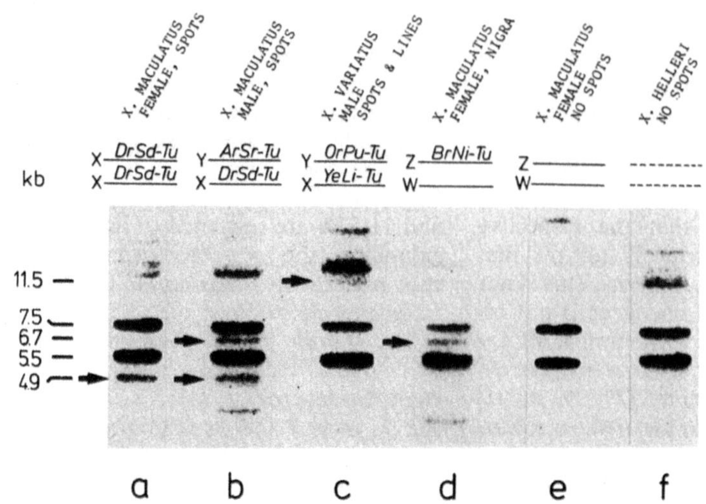

Fig. 1. Assignment of three v-*erb*B homologous fragments (*arrows*) to the spot determining sex chromosomes (Southern blot) of purebred xiphophorine fish. *X,Y,W,Z,* sex chromosomes. *Dashes* indicate autosomes of *X. helleri* that are homologous to the sex chromosomes of *X. maculatus* and *X. variatus*. For gene symbols and phenotypes of the animals see Table 1

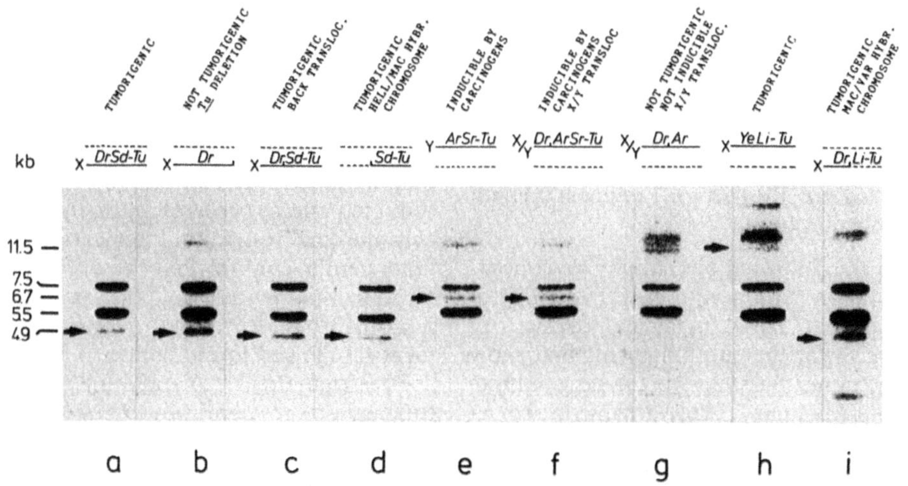

Fig. 2. Assignment of the three v-*erb*B homologous EcoRI fragments detected in the purebreds (see Fig. 1), to the melanoma-determining region of the tumor gene complexes in *X. maculatus*/*X. helleri* and *X. variatus*/*X. helleri* hybrids carring normal and structurally changed chromosomes. (For symbols and phentoypes see Table 1)

complex but was not found in the melanoma-free segregants (siblings) lacking this X chromosome (not shown in Fig. 2; the restriction fragment pattern was identical to that of the purebred *X. helleri*). The 6.7-kb fragment was identified in all BC hybrids exhibiting the Y-chromosomal *Ar Sr-Tu* complex but was not detected in the siblings not having inherited this Y chromosome. The Z-chromosomal 6.7-kb fragments were not investigated in the hybrids. The 11.5-

kb fragment was identified in the normal tissues of all melanomatous BC hybrids displaying the X-chromosomal *Ye Li-Tu*. These results confirm the linkage of the three sex-chromosomal fragments to the respective sex chromosomes of *X. maculatus* and *X. variatus* that transmit the capability of the spontaneous or induced development of melanoma through the hybrid generations.

2. Location of the Xiphophorine *erb*B Restriction Fragments

Cytogenetic observations on 88 X-ray-induced and spontaneous structural chromosome changes have shown that the *Tu* complexes and the adjacent pterinophore loci are terminally located on the sex chromosomes [3, 14]. Our present study on the three sex-chromosome-specific *Eco*RI Southern fragments disclosed new details of the genetic make-up of the region where the information for pigment cell transformation is located.

a) The 4.9-kb Fragment. A melanoma-free red individual of *Dr Sd-Tu* BC hybrid segregants was propagated to a red substrain which turned out to be completely incapable of developing melanoma spontaneously and almost completely incapable of developing melanoma following treatment with mutagenic carcinogens and tumor promoters. The breeding and treatment experiments suggest that the red substrain had lost the *Sd-Tu* complex but had retained its *Dr* locus on its X chromosome (Fig. 2, lane b). Southern analysis showed that the X-chromosomal 4.9-kb band is still present in the mutant. Provided that the terminal chromosome deletion was created by an unequal crossing over, the finding of the *Sd-Tu* deletion suggests the existence of the corresponding translocation of the platyfish *Sd-Tu* to a chromosome of *X. helleri* thus forming an interspecific hybrid chromosome. A total of 11 translocation events of this type have been observed (see Table 1). Our studies on such a *Sd-Tu* translocation substrain revealed a quasi-purebred greyish green strain of the swordtail that has acquired both the 4.9-kb fragment and the capacity to develop melanoma spontaneously (Fig. 2, lane d). Since, on the one hand, the reddish deletion animals described above have retained the 4.9-kb fragment, and, on the other hand, the greyish green translocation animals have gained this fragment, we conclude that the breaking points of both structural mutation events were different, one proximal to the 4.9-kb fragment and the other distal. The retranslocation of the *Sd-Tu* chromosome fragment (lane d) to the *Dr*-deletion chromosome (lane b) did not affect the 4.9-kb band but restored the capacity to develop melanoma spontaneously (lane c) on the *Dr*-mediated reddish skin.

b) The 6.7-kb Fragment. To analyze the Y-chromosomal 6.7-kb Southern fragment we used BC hybrids carrying a *Dr, Ar Sr-Tu* X/Y translocation chromosome (Fig. 2, lane f) that originated from a *Dr*-deletion X chromosome on which both the *Sd-Tu* complex and the 4.9-kb band were lost and both the *Ar Sr-Tu* and the linked 6.7-kb fragment were gained. Individuals containing this X/Y translocation exhibit phenotypically both the pterinophore patterns "dorsal red" (*Dr*) and "anal red" (*Ar*) and a phenotypically unchanged *Sr-Tu* complex. Following treatment with carcinogens, melanomas develop that are phenotypically the same as those of the BC hybrids carrying the unchanged *Ar Sr-Tu* Y chromosome (lane e). Since the 4.9-kb fragment was not found in these animals, one can assume that the breakpoint of the X chromosome in the X/Y crossover (lane f) was different from that of the *Dr*-deletion X chromosome (lane b) which retained this fragment. On the other hand, the presence of the 6.7-kb fragment indicates its location between the pterinophore locus *Ar* and the *Sr-Tu* complex. A more precise determination of the site of the 6.7-kb fragment comes from a *Sr-Tu* deletion that occurred on the *Dr, Ar Sr-Tu* X/Y translocation chromosome just mentioned (lane f). The resulting *Dr, Ar,* chromosome (lane g) shows neither the 4.9-kb nor the 6.7-kb band, thus indicating that both fragments each must be normally located between to the pterinophore loci and the melanophore loci. The *Dr, Ar* animals lack the capacity to develop melanoma, possibly because of the loss of the both the *Sd-Tu* and *Sr-Tu* chromosome fragments, including their closely linked x-*erb*B Southern fragments.

c) The 11.5-kb Fragment. This fragment is specific to the *Ye Li-Tu* X chromosome of *X. variatus* (Fig. 1, lane c; Fig. 2, lane h). In the course of a *Li-Tu* translocation onto the *Dr*-deletion X chromosome of *X. maculatus* (lane b) this fragment became detached from *Li-Tu* (break point was proximal to *Li-Tu*) resulting in the *Dr, Li-Tu* hybrid chromosome, which exhibits the *X. maculatus*-specific 4.9-kb fragment but lacks the *X. variatus*-specific 11.5-kb fragment (lane i). The pattern of the melanophores of the BC animals carrying this *X. variatus/X. maculatus* hybrid chromosome in their *X. helleri* background genome resemble neither the *Li-Tu* pattern of *X. variatus* nor the *Sd-Tu* pattern of *X. maculatus*. These animals spontaneously develop melanoma of very high malignancy that spreads over and invades the entire body of the fish, indicating that their *Tu* complex is much more out of control than that of any other genotype [14]. The three *Eco*RI restriction fragments comprising 4.9, 6.7, and 11.5 kb are therefore located between the respective pterinophore loci (*Dr, Ar,* or *Ye*) and the remaining parts of the *Tu* complexes (*Sd, Sr,* or *Li*) that are located at the end of the sex chromosomes where the genetic information for melanoma formation (and other types of neoplasia) is encoded.

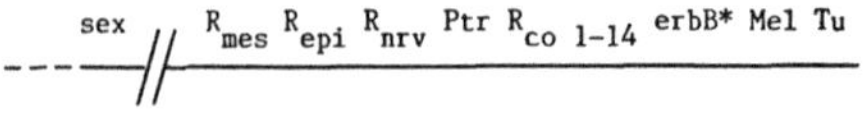

Fig. 3. Preliminary map of the tumor determining region (*Tu* complex) of the sex chromosomes of *X. maculatus* and *X. variatus* based upon 88 deletions, translocations, duplications, and 19 compartment-specific mutations

3. Preliminary Map of the Tumor-Determining Region

A preliminary map of the chromosome region of the *Tu* complex was first proposed 15 years ago [14] and has since that time been subject to several improvements. Based on our earlier observations and the present phenogenetic, cytogenetic, and molecular linkage data we propose one more improvement (Fig. 3):

The tumor-determining region is proximally linked to the sex-determining region. Its constituents are probably located in one Giemsa band which can be lost in total, even homozygously, without creating any detectable disadvantages for the fish [33, 34]. This region is, therefore, considered accessory. Carcinogenesis studies which assigned a large variety of neoplasms other than melanoma to the same *Tu* complex region [17, 35, 36] led to the assumption of a group of tissue-specific regulatory genes (mesenchymal, epidermal, nervous; arbitrary order) adjacent to the sex-determining region. These genes (*R-mes, R-epi, R-nrv*) are followed by at least 14 compartment-specific regulatory genes (*R-co 1–14*) that control the differentiating activity of both the pterinophore (*Ptr*) and melanophore (*Mel*) locus (*Dr Sd, Ar Sr,* etc.). The critical restriction fragments (in Fig. 3 indicated as *erb*B*) are probably very closely linked to both *Ptr* and *Mel*, i.e., they are narrowly intercalated between *Ptr* and *Mel*.

The signal for pigment cell transformation comes certainly from the very end of the chromosome and might possibly be composed of both *erb*B* and *Mel*, which together might represent what was designated arbitrarily as the Mendelian gene *Tu*. In any case, the breaking point data indicate that *erb*B* is not identical with the Mendelian gene *Tu* although it might be involved in its function. Nevertheless we will retain the arbitrary Mendelian symbol *Tu* until we have more information about the biological nature of the entire *Tu* complex.

4. Cloning of the Xiphophorine *erb*B Restriction Fragments

Special information on the nature of these three *Tu* complex-linked fragments comes from studies in which the fragments were molecularly cloned and characterized. As a basis for gene comparison the autosomal *Tu* complex-nonlinked 5.5-kb fragment that is constantly present in all xiphophorine fish irrespective of whether the fish are susceptible to melanoma or insusceptible (see Figs. 1, 2) was also studied. Two λ gt 10 phage libraries were prepared with *Eco*RI digested genomic DNA from *X. maculatus* from Rio Jamapa. By screening the libraries we succeeded in isolation of v-*erb*B homologous clones which contain *Eco*RI inserts representing either the X- and Y-specific 4.9-kb and 6.7-kb fragments (λ x-erb 4.9 gt and λ x-erb 6.7 gt) that are critical for melanoma appearance or the 5.5-kb fragment (λ x-erb 5.5 gt) that appears to be autosomal and, therefore, independent from melanoma appearance.

5. Homology of the Xiphophorine *erb*B Restriction Fragments

Southern blot analysis of restricted DNA from λ x-erb 4.9 gt showed that the v-*erb*B homologous sequences were enclosed in a 0.8-kb *Eco*RI/*Sac*I fragment. Hybridization of this fragment against genomic DNA from *Tu* complex-carrying fish revealed, as shown in Fig. 4 (lanes b–f), that this xiphophorine DNA fragment detects not only the X-chromosomal *Tu* complex-linked 4.9-kb fragment of *X. maculatus* from which it was isolated but also the Y-chromosomal 6.7-kb fragment of the same species and the X-chromosomal 11.5-kb fragment from *X. variatus*. The three fragments are, therefore, highly homologous. No fragment of these lengths could be detected in

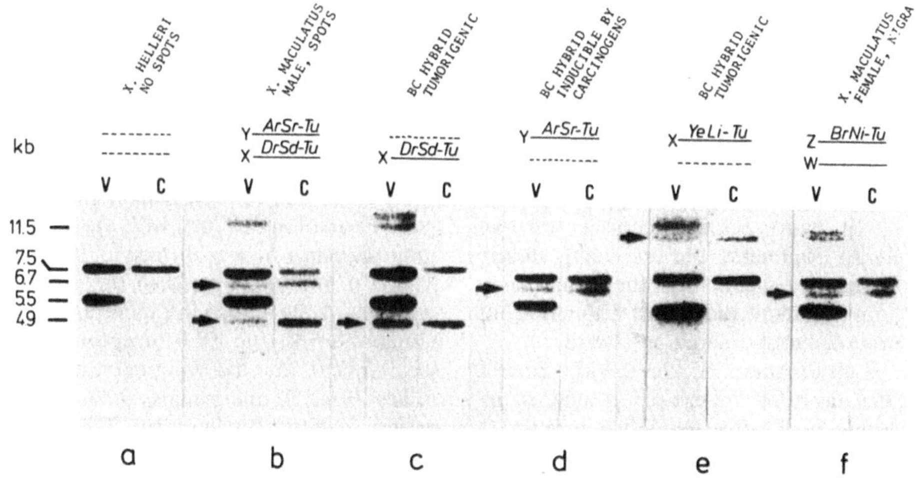

Fig. 4. Hybridization of the cloned 4.9-kb fragment (as indicated in Figs. 1, 2) and hybridization of the v-*erb*B probe against EcoRI digested genomic DNA from purebred and hybrid xiphophorine fish. Note that the 4.9-kb probe detects not only the X-chromosomal 4.9-kb fragment of *X. maculatus* from which it was isolated but also the Y-chromosomal 6.7-kb fragment, the 11.5-kb fragment of the X of *X. variatus*, and an autosomal 7.5-kb fragment. Filters were probed with the v-*erb*B specific fragment under stringent conditions (*V*) and with the 4.9-kb specific fragment under highly stringent conditions (*C*). (For symbols see Table 1)

the genome of *X. helleri* (lane a) or in the genome of the *Tu* complex-lacking purebred and hybrid genotypes (not shown).

As is also shown in Fig. 4, the 0.8-kb *Eco*RI/*Sac*I fragment containing sequences of the 4.9-kb fragment detects not only the other sex-chromosomal *Tu* complex-linked fragments but also an autosomal fragment comprising 7.5 kb which is present in all genotypes irrespective of whether they contain the *Tu* complex (lanes b–f) or not (lane a). This finding is important because earlier and recent carcinogenesis studies suggest that all individuals of *Xiphophorus* contain at least one copy of an autosomal *Tu* complex [3]. Since all deletions of the sex-chromosomal *Tu* complexes are nonlethal in both the heterozygous and the homozygous state, one can conclude that they are accessory for the fish. This is, however, not to say that *Tu* complexes may not be at all essential to the fish. One could, for instance, assume that additional *Tu* complexes present in the autosomes may compensate the loss of the sex-chromosome-linked *Tu* complex according to a gene dosage compensation mechanism, which warrants normal functions. The 7.5-kb fragment could be the indicator of such an indispensable *Tu* complex which is now molecularly approachable. The Southern data obtained with the 4.9-kb fragment and with the v-*erb*B probe under conditions of varied stringency (not shown) suggest that the 7.5-kb band actually consists of two *Eco*RI fragments, the one being closely related to v-*erb*B, and the other being homologous to the 4.9-kb fragment, but more distantly related to v-*erb*B.

Hybridization of the 0.8-kb *Eco*RI/*Sac*I insert of λ x-erb 6.7 gt against genomic *Eco*RI-digested DNA revealed a banding pattern identical to that obtained with the 0.8-kb *Eco*RI/*Sac*I insert of λ x-erb 4.9 gt, indicating more and stronger evidence for a high homology between the X-chromosomal *Tu* complex-linked 4.9-kb fragment and the Y-chromosomal *Tu* complex-linked 6.7-kb fragment.

The restriction map of the λ x-erb 5.5 gt clone showed no similarity to that of the λ x-erb 4.9 gt and λ x-erb 6.7 gt clones which, if compared separately, were very similar. The sequences of λ x-erb 5.5 gt which are homologous to v-*erb*B were enclosed in a 0.8-kb *Xba*I/*Hin*dIII fragment which, when hybridized against *Eco*RI-digested genomic DNA (not shown), detected one single band of 5.5 kb in all individuals of all genotypes, confirming that the insert of this clone represents the always present 5.5-kb fragment, and that the restriction fragment length is always identical.

For further analysis the v-*erb*B homologous regions from λ x-erb 4.9 gt, λ x-erb 6.7 gt and λ x-erb 5.5 gt subcloned in pUC 19 (p x-erb 4.9 gt and p x-erb 6.7 gt, both containing the 0.8-kb *Eco*RI/*Sac*I insert, and p x-erb 5.5 gt, containing the 0.8-kb *Xba*I/*Hin*dIII insert) were further subcloned for sequencing (dideoxy chain termination method).

6. Nucleotide Sequences of Parts of the Xiphophorine *erb*B Restriction Fragments

a) The 4.9-kb Fragment. The nucleotide sequence of the 0.8-kb *Eco*RI/*Sac*I insert of p x-erb 4.9 gt that represents part of the *Tu* complex-linked 4.9-kb fragment of x-*erb*B (see Figs. 1, 2) is shown in Fig. 5. We identified two regions, separated by 88 nucleotides, which share an overall homology of 76% with the nucleotide sequence of v-*erb*B (see [37]). The degree of homology between the partial sequence of the 4.9-kb fragment and the human c-*erb*B1 (the EGF receptor gene; see [38]) is 81% in the first region (nucleotides 70–225) and reaches 76% in the second region (nucleotides 314–391). The homology between the deduced xiphophorine amino acid sequences and that of v-*erb*B was 85% for the first region and 88% for the second region. The degree of homology between the predicted amino acid sequences of the respective regions of human c-*erb*B1 and the xiphophorine 4.9-kb fragment was 81%

```
        10        20        30        40        50        60        70        80        90       100
         .         .         .         .         .         .         .         .         .         .
ATATCTATAGCTCTATCTAGCGGTTAGTTCTGGTTTGTTAAATGCACACACTGTGTCCTGCTGGTTCAGGGGATGAACTACCTGGAAGAGCGCCACCTGG
                                                                     GlyMetAsnTyrLeuGluGluArgHisLeuV
         .         .         .         .         .         .         .         .         .         .
TGCACCGCGACCTGGCAGCCAGGAACGTCCTGCTGAAAAACCCGAACCACGTCAAGATCACAGACTTCGGTCTGTCCAAGCTGCTGACGGCTGACGAGAA
alHisArgAspLeuAlaAlaArgAsnValLeuLeuLysAsnProAsnHisValLysIleThrAspPheGlyLeuSerLysLeuLeuThrAlaAspGluLy
         .         .         .         .         .         .         .         .         .         .
GGAATACCAAGCCGACGGAGGAAAGGTGCCATGGCAATGCCTGACTGGTTTCTGTTTGCTGTTCGGACTGAAAACATGTCAGAGATGAATCACTGCTGCA
sGluTyrGlnAlaAspGlyGlyLys
         .         .         .         .         .         .         .         .         .         .
TCTCTGTGAGCAGGTTCCCATTAAGTGGATGGCTTTGGAGTCGATCCTCCAGTGGACCTACACCCATCAGAGCGACGTGTGGAGCTACGGTGAGGAATCG
             ValProIleLysTrpMetAlaLeuGluSerIleLeuGlnTrpThrTyrThrHisGlnSerAspValTrpSerTyrGly
         .         .         .         .         .
TCCCCACAGCGCCACCTACCTGCCTTCACCCTCTGCTTCCTGTTAGCCGG
```

Fig. 5. Nucleotide sequence and the deduced amino acid sequence of the p x-erb 4.9 gt insert. The sequence contains the exons C and D of the X-chromosomal xiphophorine EGF receptor gene. The exons are separated by an intron comprising 88 nucleotides. Nomenclature of the exons is according to that of the human c-*erb*B-2

for the first region and 88% for the second region. Alignment of the deduced amino acid sequences showed that the first region corresponds to the putative exon C of the human c-*erb*B2 (77% homology), and the second region to exon D (85% homology) [39–41]. Since each region is flanked by AG and GT dinucleotides that border the exons of eukaryotic genes, and since the regions show high homology to v-*erb*B and human c-*erb*B on the amino acid level, we suggest that they represent two exons of a xiphophorine gene related to the human EGF receptor gene. In analogy to the human c-*erb*B2 the exons will be referred to as exons C and D.

b) The 6.7-kb Fragment. Sequencing of the 0.8-kb *Eco*RI/*Sac*I insert of p x-erb 6.7 gt (not shown) revealed two putative coding regions that are identical to those of the p x-erb 4.9 gt insert (exons C and D, according to the human c-*erb*B2 gene). The comparison of the putative exons and introns of both inserts revealed two single-nucleotide substitutions in the region of the introns. We consider the genes corresponding to the X- and Y-chromosomal *Tu* complex-linked 4.9-kb and 6.7-kb fragments as two alleles of a xiphophorine gene related to the human EGF receptor gene. These xiphophorine alleles were designated as *x-egfrB-1* (corresponding to the X-chromosomal 4.9-kb fragment) and *x-egfrB-2* (corresponding to the Y-chromosomal 6.7-kb fragment).

c) The 5.5-kb Fragment. The nucleotide sequence of the 0.8-kb *Xba*I/*Hin*dIII insert of p x-erb 5.5 gt that represents part of the *Tu* complex-independent 5.5-kb fragment (see Figs. 1, 2) contains, as shown in Fig. 6, also two putative coding regions; these are separated by 120 nucleotides, which share an overall 82% nucleotide sequence identity with v-*erb*B and 84% with the human c-*erb*B1. The homology between the deduced amino acid sequences of these two regions and the predicted amino acid sequence of v-*erb*B was 91%. Alignment of the amino acid sequences deduced from human c-*erb*B1 and the *Tu*-nonlinked xiphophorine 5.5-kb fragment nucleotide sequence showed that they share 90% homology. The two putative coding regions are flanked by the splicing consensus sequences AG and GT. In contrast to the sequenced coding regions of the X-chromosomal *x-egfrB-1* and *x-egfrB-2* that correspond to exons C and D of the putative human c-*erb*B2, the coding regions of the 5.5-kb fragment correspond to the putative exons B and C of the human c-*erb*B2.

```
        10        20        30        40        50        60        70        80        90       100
         .         .         .         .         .         .         .         .         .         .
GCTTATGTGATGGCCAGTGTGGAACACCCCCATGTGTGCCGTCTGCTGGGTATCTGCCTCACCTCGACGGTTCAACTCATAACCCAGCTGATGCCGTACG
AlaTyrValMetAlaSerValGluHisProHisValCysArgLeuLeuGlyIleCysLeuThrSerThrValGlnLeuIleThrGlnLeuMetProTyrG
         .         .         .         .         .         .         .         .         .         .
GCTGCCTGCTGGACTACGTCAAAGAAAAAAAGGACAATATTGGCTCCCAGCACCTGCTCAACTGGTGTGTTCAGATAGCCAAGGTGAGGAATCACTTTTA
lyCysLeuLeuAspTyrValLysGluLysLysAspAsnIleGlySerGlnHisLeuLeuAsnTrpCysValGlnIleAlaLys
         .         .         .         .         .         .         .         .         .         .
TTTACTTTTTGCTAGTTATATAAAAACAATGCTTCACCCACCACATTGAACTTTGTTAAAAGATCTGCTCTCATGCCTTAGTTCACTCCTTGTTTGATTA
         .         .         .         .         .         .         .         .         .         .
AAGGGAATGAACTACCTAGAGGAGCGCCACCTAGTGCACCGTGACTTAGCAGCCAGAAACGTCCTGGTCAAGACTCCTCATCATGTCAAGATCACTGACT
   GlyMetAsnTyrLeuGluGluArgHisLeuValHisArgAspLeuAlaAlaArgAsnValLeuValLysThrProHisHisValLysIleThrAspP
         .         .         .         .         .         .         .         .         .         .
TTGGGCTGGCCAAACTCCTCAACGCAGATGAGAAAGAATACCATGCAGATGGAGGAAAGGTCGGTTAGGTCTTAAAGGCGCAGTCTGTTATTTTTGTTGT
heGlyLeuAlaLysLeuLeuAsnAlaAspGluLysGluTyrHisAlaAspGlyGlyLys
         .         .         .
TGTTTTTTATTATGATGGGATTGGGCCATCGAT
```

Fig. 6. Nucleotide sequence and the deduced amino acid sequence of the p x-erb 5.5 gt insert. The sequence contains the exons B and C of the autosomal xiphophorine EGF receptor gene. The exons are separated by an intron comprising 120 nucleotides. Nomenclature of the exons is according to that of the human c-*erb*B-2

These data suggest that the sequence of the xiphophorine *Tu*-nonlinked 5.5-kb fragment contains two exons (defined as exons B and C) that also represent part of a xiphophorine gene related to the human EGF receptor gene. This fish gene was designated as *x-egfr*A. Computer-mediated sequence analysis showed that the putative exon C of *x-egfr*A is homologous to the corresponding sequence of several members of the *src* tyrosine kinase family, whereas the sequence of exon B showed no significant homology. The most striking homology was observed with the tyrosine kinase domain encoding sequence of the human EGF receptor gene (c-*erb*B1; [38]).

The homologies between the *Tu* complex-linked 4.9-kb or 6.7-kb fragment and the *Tu* complex-nonlinked 5.5-kb fragment concern the region of the putative exon C and reach a degree of 89% on the amino acid level. Based on our cytogenetic and molecular data we presume that the 4.9-kb and 6.7-kb fragments (and probably the 11.5-kb fragment) and the 5.5-kb fragment are parts of two different types of xiphophorine genes (*x-egfr*A and *x-egfr*B) encoding two slightly different types of EGF receptors, x-EGFR-A and x-EGFR-B. The existence of two different types of EGF receptor genes in *Xiphophorus* (*x-egfr*A and B), one of which (*x-egfr*B) could be involved in the switch from the normal to the neoplastic state while the other is of minor importance in this context, requires discussion concerning structure and function of the receptor domains encoded: One may ask whether both the *x-egfr*A and B encode a growth factor receptor with an extracellular, transmembrane, and cytoplasmic domain. This question arises since it is known from the human EGF receptor that it consists of three domains, and that lack of them is important for receptor regulation [38, 42–46]. Especially it is of interest to determine whether the xiphophorine EGF receptor genes *x-egfr*A and *x-egfr*B encode an extracellular receptor domain capable of binding EGF or other growth factors, and whether the growth factor binding leads to receptor activation. Differences in kinase activity and activation of the receptor by different growth factors could result in a different type of response of the two types of xiphophorine EGF receptors in question to the humoral signals mediating stimulation or inhibition of cell proliferation. Those cells exhibiting the growth factor receptor x-EGFR-B may respond to internal and external signals inducing cell proliferation and changes in a series of cellular regulatory processes, which together could mediate the switch from the normal to the neoplastically transformed phenotype. In this context we want to mention the positive correlation between the presence of the sex-chromosomal *Tu*-linked *x-erb*B genes and the turnover of phosphoinositides, that was discovered very recently in *Xiphophorus* [47–49].

7. Expression of the Xiphophorine EGF Receptor Genes

RNA dot-blot and Northern blot analysis with a probe specific for exons C and D of x-*egfr*B (*Eco*RI/*Rsa*I fragment excised from p x-erb 4.9 gt) showed expression of the respective genes in testes and embryonic tissue of individuals without accessory *Tu* complexes and enhanced expression of x-*egfr*B in melanoma tissue and in a melanoma cell line (see Figs. 7, 8). These data indicate that the sex-chromosomal x-*egfr*B genes are not only structurally but also functionally related to the melanoma-determining accessory *Tu* complexes. On the other hand, it became obvious that x-*egfr*B genes, namely those genes that are probably linked to the indispensable *Tu* complex (disclosed by the ubiquitous 7.5-kb fragment), fulfill an essential function in normal proliferating tissue. These data suggest that the gene products of the accessory x-*egfr*B and the indispensable x-*egfr*B show differences in structure (e.g., amino acid substitutions) and/or function (e.g., regulation, expression) which in turn might "activate the oncogene potential" of the x-*egfr*B and thereby induce the switch from the normal to the neoplastic transformed state of a cell.

Northern blot analyses with a probe specific for exons B and C of x-*egfr*A (*Hin*dIII/*Cla*I fragment excised from p x-erb 5.5 gt) showed that overexpression of a xiphophorine EGF receptor gene can be specified in melanoma with this probe under stringent (Fig. 7), but not under highly stringent conditions. This indicates that x-*egfr*A genes are neither

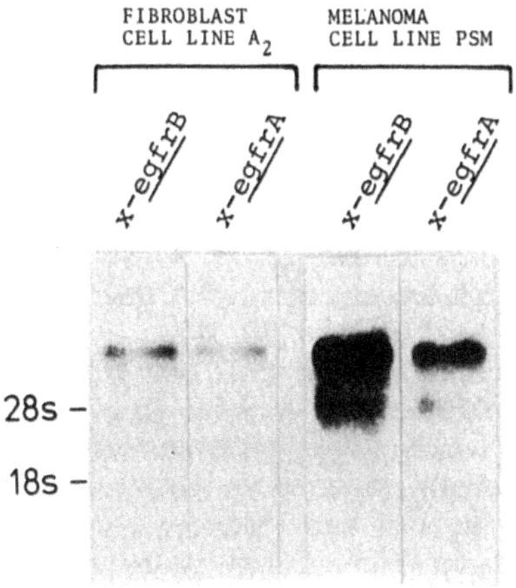

Fig. 7. Expression of xiphophorine EGF receptor genes (Northern blot analysis). Hybridization of probes specific for the x-*egfr*B and the x-*egfr*A against 20 µg total RNA are shown (washing conditions 1 × SSC/1% SDS, 60° C). The hybridization was detected by autoradiography with exposure times of 50 h for the fibroblast cell line RNA and 20 h for the melanoma cell line RNA. Xiphophorine ribosomal RNA of 18 S and 28 S served as internal size markers

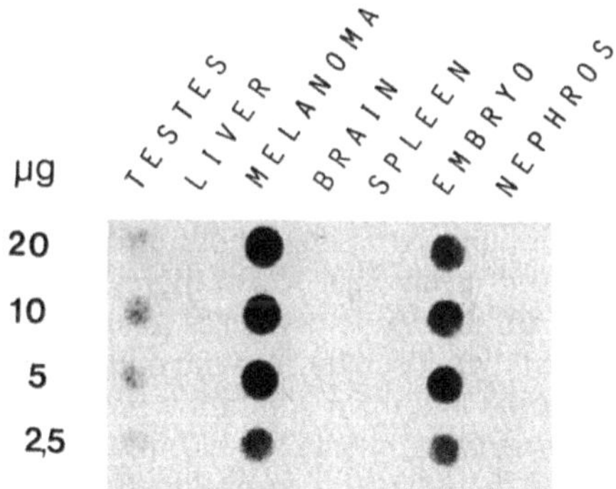

Fig. 8. Expression of the accessory EGF receptor gene x-*egfr*B in normal and transformed tissues (dot-blot analysis). Hybridization of a probe specific for x-*egfr*B against varying amounts of total RNA from different tissues; RNA from whole embryos (stages 17–22, according to [66]) was used. The conditions for the dot-blot analysis were the same as for the Northern blot analysis with the x-*egfr*B specific probe (see Fig. 7).

structurally nor functionally related to the melanoma-determining loci. Expression studies performed with *non*-transformed tissue revealed expression of x-*egfr*A in a fibroblast cell line [22, 28] in eyes, brain and melanoma as well as in a melanoma cell line [22, 28, 30, 31] and a very high amount of x-*egfr*A transcripts in the head nephros [30, 31].

In conclusion, at least two types of xiphophorine EGF receptor genes exist, one of which (x-*egfr*B) is structurally and functionally related to the melanoma-determining loci and therefore could be considered as an oncogene probably critical for the switch from the normal to the neoplastic state of a cell, while the other one (x-*egfr*A) is of minor importance in this context.

II. Oncogenes that Might Be Considered "Critical" for the Late Events in the Manifestation of the Tumor Phenotype

We shall concentrate on genes that might be considered as candidates probably involved in stimulation or repression of proliferation and differentiation of Tr melanocytes.

Not only transforming genes are involved in the causation of spontaneously developing (crossing-conditioned) and induced melanoma. Much more important are the regulatory genes (oncostatic genes) that normally keep the transforming genes and the proliferation genes under negative control [24]. It appears that in the hybridization or in the treatment with the carcinogens some of the *R* genes are lost or impaired, thus permitting an *S* gene-stimulated overexpression of the spotting *Tu* complex that results in the formation of melanoma. It is important to note that a stimulating effect on melanoma formation can also be achieved by tumor promoters such as steroid hormones [15, 36, 50, 51]. These observations led us to the assumption that hormones and hormone receptors, respectively, might be related to the *R* and *S* genes. Since it is known that (a) the members of the steroid/thyroid hormone receptor superfamily act as transcription factors [52, 53], that (b) v-*erb*A is not a direct-acting oncogene but induces the fully transformed phenotype in transformed cells by blocking differentiation [54, 55], and that (c) c-*erb*A encodes a thyroid hormone receptor [56, 57], we started our molecular search for *R* and *S* genes by studies on xiphophorine x-*erb*A genes.

1. Inheritance of Southern Restriction Fragments of the Xiphophorine *erb*A Oncogene

We shall concentrate in particular on certain v-*erb*A homologous DNA fragments which correspond to x-*erb*A genes that probably represent xiphophorine hormone receptor genes (x-*th-r* genes) encoding a receptor which binds thyroid hormone or retinoic acid.

Figure 9 shows a different distribution of several *Eco*RI fragments of x-*erb*A in various purebred and hybrid genotypes: Two fragments comprising 9 and 12 kb are constantly present in all individuals of all populations of *Xiphophorus* tested. Four fragments comprising 2.9, 5.0, 7.5, and 16 kb are restricted to populations of *X. maculatus*. *X. variatus* shows bands of 4.9, 12.0, 9.0 (accessory) and 16 kb. All populations of *X. helleri* tested so far, show species-specific bands of 10 and 14 kb. In addition, it shows species-specific but individually distributed bands comprising 5.2, 5.3, 5.6, and 5.7 kb; at least one, but no more than two of the 5.2-, 5.3-, 5.6-, and 5.7-kb fragments are present in one individual, whereby all combinations of fragments are possible. Southern blot analyses with a probe specific for v-*erb*A revealed a species- and population-specific RFLP for *Hin*dIII-digested genomic DNA (data not shown). Until now, none of the v-*erb*A homologous fragments could be assigned to the *Tu* complex or any *R* or *S* gene. This is not to say that x-*erb*A genes and *R* and *S* genes are not structurally and/or functionally related. Besides the differentiation gene *Diff*, which is molec-

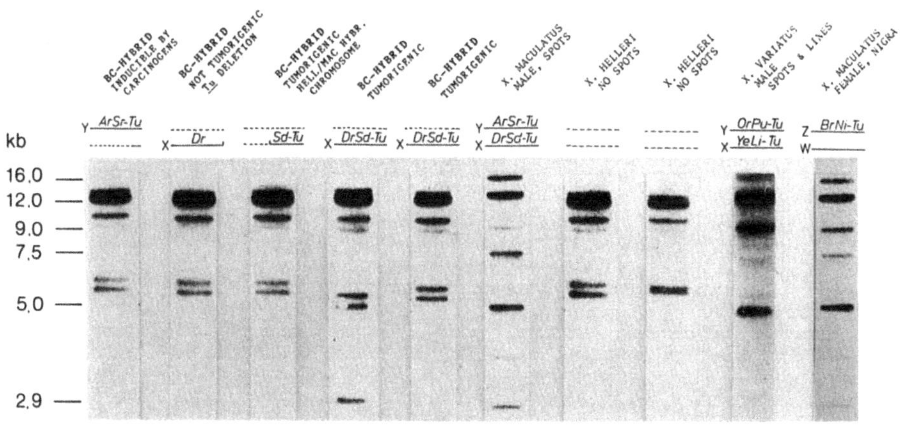

Fig. 9. Detection of an individual- and population-specific RFLP of xiphophorine v-*erb*A homologous sequences (Southern blot analysis). *X,Y,W,Z,* sex chromosomes; *dashes* indicate chromosomes of the recurrent parent *X. helleri* that are homologous to the sex chromosomes of *X. maculatus* and *X. variatus*. For gene symbols and phenotype of the respective animals see Table 1

ularly linked to a locus for esterase 1 [13, 58] and correlated to the appearance of Q base (a highly modified guanine) in certain tRNA species [13, 21], no *Tu*-nonlinked *R* or *S* gene has so far been described to be related to any known molecular or biochemical marker. We wonder whether x-*erb*A-genes themselves might be such markers.

2. Cloning and Sequencing of Xiphophorine v-*erb*A Homologous Restriction Fragments

In order to study x-*erb*A-genes we cloned and characterized four different, distinctly related v-*erb*A homologous restriction fragments, one of which appears to be specific for *X. variatus* [28]. Southern blot and Northern blot analyses confirmed that the cloned sequences are fish specific and represent parts of functional genes.

Two clones representing the v-*erb*A homologous region of the *X. maculatus*-specific 7.5-kb and the ubiquitous 12-kb *Eco*RI fragment [28] were sequenced. Both clones, p x-*erb*A90–3 and p x-*erb* A12–113, contained a stretch of 100 nucleotides exhibiting 75% homology to the v-*erb*A. Alignment of the deduced amino acid sequences of the v-*erb*A homologous regions of the two xiphophorine clones and those deduced from the viral *erb*A [59], the chicken c-*erb*A [56] the human c-*erb*A [57], a human v-*erb*A related sequence representing an open reading frame with hepatitis B virus DNA integration (60) as well as the amino acid sequences predicted for the human retinoic acid receptor h RAR [61, 62], estrogen receptor h ER [63], progesteron receptor h PR [64], and glucocorticoid receptor h GR [65], revealed that both clones contain a sequence probably encoding the first part of the DNA-binding domain (domain C) of two slightly different types of xiphophorine hormone receptors (Fig. 10). The partial sequence of the receptor x-TH-R-1, predicted from the partial sequence of x-*th*-*r*-1 (clone p x-erbA12–113) shows the most striking homology to the h RAR, while that of x-TH-R-2, deduced from the partial sequence of x-*th*-*r*-2 (clone p x-erbA90–3) appears to be most homologous to the h T_3R (thyroid hormone receptor; see Fig. 10).

The homologies between the two xiphophorine sequences concern the re-

	*			*												
hT_3R	Cys	Val	Val	Cys	Gly	Asp	Lys	Ala	Thr	Gly	Tyr	His	Tyr	Arg	Cys	Ile
v-erbA	Cys	Val	Val	Cys	Gly	Asp	Lys	Ala	Thr	Gly	Tyr	His	Tyr	Arg	Cys	Ile
h GR	Cys	Leu	Val	Cys	Ser	Asp	Glu	Ala	Ser	Gly	Cys	His	Tyr	Gly	Val	Leu
h ER	Cys	Ala	Val	Cys	Asn	Asp	Tyr	Ala	Ser	Gly	Tyr	His	Tyr	Gly	Val	Trp
X-TH-R 1	Cys	Val	Val	Cys	Gly	Asp	Lys	Ser	Ser	Gly	Lys	His	Tyr	Gly	Val	Phe
X-TH-R 2	Cys	Val	Val	Cys	Gly	Asp	Lys	Ser	Ser	Gly	Lys	His	Tyr	Gly	Val	Phe
ch PR	Cys	Leu	Ile	Cys	Gly	Asp	Glu	Ala	Ser	Gly	Cys	His	Tyr	Gly	Val	Leu
h RAR	Cys	Phe	Val	Cys	Gln	Asp	Lys	Ser	Ser	Gly	Tyr	His	Tyr	Gly	Val	Ser
ORF[1]	Cys	Phe	Val	Cys	Gln	Asp	Lys	Ser	Ser	Gly	Tyr	His	Tyr	Gly	Val	Ser

		*			*																			
hT_3R	Thr	Cys	Glu	Gly	Cys	Lys	Gly	Phe	Phe	Arg	Arg	Thr	Ile	Gln	Lys	Asn	Leu	His	Pro	Ser	Tyr	Ser	Cys	Lys
v-erbA	Thr	Cys	Glu	Gly	Cys	Lys	Ser	Phe	Phe	Arg	Arg	Thr	Ile	Gln	Lys	Asn	Leu	His	Pro	Thr	Tyr	Ser	Cys	Thr
h GR	Thr	Cys	Gly	Ser	Cys	Lys	Val	Phe	Phe	Lys	Arg	Ala	Val	Glu	Gly	–	–	Gln	His	Asn	Tyr	Leu	Cys	Ala
h ER	Ser	Cys	Glu	Gly	Cys	Lys	Ala	Phe	Phe	Lys	Arg	Ser	Ile	Gln	Gly	–	–	His	Asn	Asp	Tyr	Met	Cys	Pro
x-TH-R 1	Thr	Cys	Glu	Gly	Cys	Lys	Ser	Phe	Phe	Lys	Arg	Ser	Val	Arg	Arg	Asn	Leu	–	–	Ser	Tyr	Thr	Cys	Arg
x-TH-R 2	Thr	Cys	Glu	Gly	Cys	Lys	Ser	Phe	Phe	Lys	Arg	Ser	Ile	Arg	Arg	Asn	Leu	–	–	Asn	Tyr	Ser	Cys	Gln
ch Pr	Thr	Cys	Gly	Ser	Cys	Lys	Val	Phe	Phe	Lys	Arg	Ala	Met	Glu	Gly	–	–	Gln	His	Asn	Tyr	Leu	Cys	Gly
h RAR	Ala	Cys	Glu	Gly	Cys	Lys	Gly	Phe	Phe	Arg	Arg	Ser	Ile	Gln	Lys	Asn	Met	–	–	Val	Tyr	Thr	Cys	His
ORF[1]	Ala	Cys	Glu	Gly	Cys	Lys	–	Val	Ser	His	Thr	Ser	Val	Pro	Asp	–	–	Glu	Leu	Ser	Phe	Ser	Met	Tyr

Fig. 10. Comparison between the partial amino acid sequences deduced from the putative xiphophorine hormone receptor genes encoding x-TH-R-1 and x-TH-R-2 and the corresponding sequences of the human thyroid hormone receptor (*hT_3R*), glucocorticoid receptor (*h GR*), estrogen receptor (*h ER*), retinoic acid receptor (*h RAR*), and the chicken progesterone receptor (*ch PR*), as well as those deduced from the viral *erb*A and a human v-*erb*A homologous open reading frame with hepatitis B virus DNA integration (*ORF*[1]). Homologous amino acids are *boxed; asterisks* indicate the conserved cysteine residues of the first DNA binding finger of the hormone receptors

gion of the putative exon described above (75% homology to v-*erb*A) as well as the region located upstream to the first mentioned sequence (Fig. 11) and reach about 85% homology on the DNA level. The upstream region shows an open reading frame, which probably could represent a sequence encoding a part of the hypervariable domain A/B of a xiphophorine hormone receptor. Since the hypervariable region A/B is not conserved in the receptors of the steroid/thyroid hormone receptor superfamily [52, 53], we cannot determine whether the upstream region identified in the two xiphophorine sequences represents an exon.

Expression studies showed that x-*erb*A-genes are expressed in a fibroblast and in a melanoma cell line (data not shown). The amount of mRNA, as well as the species of mRNA detected was different in the normal and transformed cells [28].

Since it is known that receptors of the steroid/thyroid hormone receptor superfamily specifically stimulate or repress transcription of distinct genes [53], we wonder whether the x-*erb*A genes x-*th-r*-1 and -2 might be involved in the regulation of the proliferation of Tr melanocytes.

The results obtained in Southern blot, Northern blot, and sequencing analyses indicate that x-*erb*A-genes are differentially organized in the genome of different populations of *Xiphophorus* and that these genes probably encode a variety of different hormone receptors related to the steroid/thyroid hormone receptor superfamily. Further experiments will show whether x-*erb*A genes are involved in the manifestation of the tumor phenotype.

```
                     10        20        30        40        50        60
                      .         .         .         .         .         .
x-th-r2
x-th-r1     TAACCAGACGATGGCCATGGTGAGTGGGTCTGGGGAGATCCACACGGGGGCATCAACGGA
x-TH-R1     ***ProSerAspGlyHisGlyGluTrpValTrpGlyAspProHisGlyGlyIleAsnGly
x-TH-R2

                      .         .         .↓        .         .         .
x-th-r2                                            CA   AC  C    G  C  C
x-th-r1     ACTGGGGGACAAGGGCTAACCTATACGGGGGGAGGAGGAGGACGGGTCTCGCAAGCGGGG
x-TH-R1     ThrGlyGlyGlnGlyLeuThrTyrThrGlyGlyGlyGlyGlyArgValSerGlnAlaGly
x-TH-R2                                            Gln   LeuPro

                      .         .         .         .         .         .
x-th-r2     A       TG    C TG  C  C         A
x-th-r1     GGCAGCGACATGGAGGCCGGGGATGAGGACAAGGCCTGCGTGGTGGACTGCGTGGTGTGC
x-TH-R1     GlySerAspMetGluAlaGlyAspGluAspLysAlaCysValValAspCysValValCys
x-TH-R2     Ser      ValAspVal               Thr

                      .         .         .         .         .         .
x-th-r2                   C  G  G              C
x-th-r1     GGGGACAAGTCCAGTGGAAAACACTACGGCGTGTTTACCTGCGAGGGCTGCAAGAGCTTC
x-TH-R1     GlyAspLysSerSerGlyLysHisTyrGlyValPheThrCysGluGlyCysLysSerPhe
x-TH-R2

                      .         .         .         .↓
x-thr-2                 A    GA G     G A    T C
x-th-r1     TTCAAGAGGAGCGTCAGACGTAACCTCAGCTACACATGCAGGTGA
x-TH-R1     PheLysArgSerValArgArgAsnLeuSerTyrThrCysArg***
x-TH-R2                 Ile            Asn   Ser
```

Fig. 11. Partial nucleotide and predicted amino acid sequence of the x-*erb*A clone p x-erb A12-113 (represents part of x-*th-r*-2), and comparison to the x-*erb*A clone p x-erbA90-3 (represents part of x-*th-r*-2). Nucleotides and amino acids of p x-erbA90-3 that are not identical to those of p x-erbA12-113 are shown. *Asterisks* indicate stop codons; *arrows* mark the beginning and the end of the compared region; triangles indicate the beginning of the region homologous to the DNA binding region of steroid and thyroid hormone receptors. The Cys residues corresponding to those conserved in the first DNA binding finger of known hormone receptors are *boxed* (see Fig. 10). The dinucleotide GT that possibly represents a splicing donor site is *underlined*

C. Summary and Conclusions

Southern blot analyses of the xiphophorine genome with probes specific for 15 viral and cellular oncogenes revealed that only three v-*erb*B related *Eco*RI fragments comprising 4.9 kb of a certain X, 11.5 kb of another X, and 6.7 kb of both a Y and a Z chromosome are inherited in parallel with the *Tu* complex and melanoma formation. They are accessory in the genome, and are highly homologous with each other and with an ubiquitous autosomal 7.5-kb fragment. The latter one is probably linked to the indispensable *Tu* complex that is postulated to be present in all individuals of *Xiphophorus* irrespective of whether they possess or lack the capacity to form melanoma in interspecific hybrids. Three restriction fragments, the X-chromosomal 4.9-kb, the Y-chromosomal 6.7-kb and the ubiquitous *Tu*-nonlinked 5.5-kb *Eco*RI fragments were cloned and sequenced. The X- and the Y-chromosomal fragments show perfect identity in the regions of the putative exons C and D of the EGF receptor gene and minor but significant differences to the putative exon C (exon D not identified) of the *Tu*-nonlinked fragment of 5.5 kb, indicating that at least two different types of x-*erb*B genes coding for slightly different EGF-receptors exist in the fish. Northern blot analyses revealed expression of the *Tu*-linked x-*erb*B genes (x-*gfr*B genes) in both transformed and nontransformed tissue, suggesting their essential role in regulation of normal cell proliferation and in carcinogenesis. We conclude that the indispensable x-*egfr*B genes remain unchanged and strictly regulated, while the sex chromosomal accessory x-*egfr*B genes possibly undergo dramatic changes in structure and/or function (e.g., unscheduled expression, ectopic expression, point mutations, truncation) leading to activation of the oncogenic potential of these genes, which in turn could induce several cellular events involved in the switch from the normal to the transformed state of the cell.

In contrast, none of the x-*erb*A restriction fragments could be assigned to the *Tu*-complex or to any regulatory gene (R or S). These results, however, do not exclude the existence of a structural and/or functional relation between x-*erb*A genes and *R* and *S* genes. We therefore analyzed x-*erb*A genes by cloning, sequencing, and expression studies. The data revealed the existence of at least two types of xiphophorine *erb*A genes (x-*th-r* genes) coding for slightly different hormone receptors that are presumably related to the human thyroid hormone and retinoic acid receptor, respectively. It appears that these genes could be involved in the effect of tumor promoters.

We suppose that in analogy to the *erb* A and *erb*B of the avian erythroblastosis virus, xiphophorine *erb*A and *erb*B genes might somehow act in a synergistic way, whereby the x-*erb*B genes are probably involved in the process of cell transformation while the x-*erb*A genes are possibly responsive for regulation of Tr melanophore differentiation.

Acknowledgements. We thank Dr. B. Vennström (Heidelberg, FRG) for providing the plasmid pAE11. Computer analysis of the partial sequence of x-*egfr*A was kindly supported by U. Eisel (Giessen, FRG) and F. Raulf (Munich, FRG). We are grateful to Dr. F. Werner (Heidelberg, FRG) for help with the computer analysis of the partial sequence of x-*th-r*-1, x-*th-r*-2, and x-*egfr*B. We thank Prof. S. Sell (Houston, Texas, USA) for critical reading of the manuscript. We also thank H. Schäfer-Pfeiffer and M. Hündt for excellent technical assistance, K. Krüger for preparation and photographic reproduction of the figures, and S. Lenz for typing of the manuscript.

References

1. Huebner RJ, Todaro GJ (1969) Oncogenes of RNA tumor viruses as determinants of cancer. Proc Natl Acad Sci USA 64:1087–1094
2. Bentvelzen P (1972) In: Emmelot P, Bentvelzen P (eds) RNA viruses and host genome in oncogenesis. Hereditary infections with mammary tumor viruses in mice. North-Holland Publications Amsterdam, pp 309–337
3. Anders A, Anders F (1978) Etiology of cancer as studied in the *platyfish-swordtail* system. Biochim Biophys Acta 516:61–95
4. Duesberg PH (1983) Retroviral transforming genes in normal cells? Nature 304:219–226
5. Duesberg PH (1987) Cancer genes: Rare recombinants instead of activated oncogenes (a review). Proc Natl Acad Sci USA 84:2117–2124
6. Hunter T (1984) Oncogenes and protooncogenes: How do they differ? JNCI 73:773–785
7. Hunter T (1986) Cancer: Cell Growth Control Mechanisms. Nature 322:14–16
8. Knudson AG (1985) Hereditary cancer, oncogenes and antioncogenes. Cancer Res 45:1437–1443
9. Friend SH, Bernards R, Rogelj S, Weinberg RA, Rapaport JM, Albert DM, Dryja TP (1986) A human DNA segment with properties of the gene that predisposes to retinoblastoma and osteosarcoma. Nature 323:643–646
10. Lee WH, Bookstein R, Hong F, Young LJ, Shew JY, Lee EYHP (1987) Human retinoblastoma susceptibility gene: cloning, identification, and sequence. Science 235:1394–1399
11. Gateff E (1978) Malignant neoplasms of genetic origin in *Drosophila melanogaster*. Science 200:1448–1459
12. Lützelschwab R, Müller G, Wälder B, Schmidt O, Fürbass R, Mechler B (1986) Insertion mutation inactivates the expression of the recessive oncogene *lethal* (2)

giant larvae of Drosophila melanogaster. Mol Gen Genet 204:58–63
13. Anders A, Dess G, Nishimura S, Kersten H (1985) In: Bagnara J, Klaus SN, Paul E, Schartl M (eds) Pigment cell 1985 – biological, molecular and clinical aspects of pigmentation. University of Tokyo Press, Tokyo pp 315–324
14. Anders A, Anders F, Klinke K (1973) Regulation of gene expression in the Gordon-Kosswig Melanoma System. I. The distribution of the controlling genes in the genome of xiphophorin fish, *Platypoecilus maculatus* and *Platypoecilus variatus*. In: Schröder HJ (ed) Mutagenesis of fish. Springer, Berlin Heidelberg New York, pp 33–63
15. Anders F, Schwab M, Scholl E (1981) Strategy for breeding test animals of high susceptibility to carcinogens. In: Stich HF, San RHC (eds) Short term tests for chemical carcinogens. Springer, Berlin Heidelberg New York, pp 399–407
16. Anders F, Schartl M, Barnekow A, Anders A (1984) *Xiphophorus* as an in vivo model for studies on normal and defective control of oncogenes. Adv Cancer Res 42:191–275
17. Schwab M, Haas J, Abdo S, Ahuja MR, Kollinger G, Anders A, Anders F (1978) Genetic basis of susceptibility for development of neoplasms following treatment with N-methyl-N-nitrosourea (MNU) or X-rays in the *platyfish-swordtail* system. Experientia 34:780–782
18. Anders F (1967) Tumor formation in platyfish-swordtail hybrids as a problem of gene regulation. Experientia 23:1–10
19. Zander CD (1969) Über die Entstehung und Veränderung von Farbmustern in der Gattung *Xiphophorus* (Pisces). I. Qualitative Veränderungen nach Artkreuzung. Mitt Hamb Zool Mus Inst 66:241–271
20. Vielkind U, Schlage W, Anders F (1977) Melanogenesis in genetically determined pigment cell tumours of *platyfish* and *platyfish-swordtail* hybrids. Z Krebsforsch 90:285–299
21. Kersten H, Schachner E, Dess G, Anders A, Nishimura S, Shindo-Okada N (1983) Quenosine in transfer-RNA in relation to differentiation and pteridine metabolism. In: Curtis HC, Pfleiderer W, Wächter H (eds) Biochemical and clinical aspects of pteridins. de Gruyter, Berlin, pp 367–382
22. Zechel C, Schleenbecker U, Anders A, Anders F (1988) v-*erb*B related sequences in *Xiphophorus* that map to melanoma determining mendelian loci and overexpress in a melanoma cell-line. Oncogene 3:605–617
23. Schartl M, Barnekow A, Bauer H, Anders F (1982) Correlations of inheritance and expression between a tumor gene and the cellular homolog of the Rous Sarcoma Virus-transforming gene in *Xiphophorus*. Cancer Res 42:4222–4227
24. Anders F, Schartl M, Barnekow A, Schmidt CR, Lüke W, Jaenel-Dess G, Anders A (1985) The genes that carcinogens act upon. In: Neth R, Gallo RC, Greaves MF, Janka G (eds) Modern trends in human leukemia VI. Springer, Berlin Heidelberg New York, pp 228–252
25. Anders F, Gronau T, Schartl M, Barnekow A, Jaenel-Dess G, Anders A (1987) Cellular oncogens as ubiquitous genomic constituents in the animal kingdom and as funamentals in melanoma formation. In: Veronesi U, Cascinelli N, Santinami M (eds) Cutaneous melanoma. Academic, New York, pp 351–371
26. Pfütz M (1987) Sequenzierung c-*erb*A-spezifischer Sequenzen aus dem Genom von *Xiphophorus*. Thesis, University of Giessen
27. Schleenbecker U (1988) Moleculare Analyse zellulärer Gene, die für Wachstumsfaktoren und Wachstumsfaktorrezeptoren kodieren. – Untersuchungen an *Xiphophorus* (Pisces; Teleostei). Thesis, University of Giessen
28. Zechel C (1988) Moleculare Analyse von Onkogenen in *Xiphophorus* – *erb*A und *erb*B. Thesis, University of Giessen
29. Barnekow A, Schartl M, Anders F, Bauer H (1982) Identification of a fish protein associated with kinase activity and related to the Rous sarcoma virus transforming protein. Cancer Res 42:2429–2433
30. Mäueler W (1988) Untersuchungen zur tumorspezifischen Genexpression bei *Xiphophorus* (Teleostei; Poeciliidae): 1) Enzyme des Intermediärstoffwechsels. 2) Expression von Proto-Onkogenen. Thesis, University of Giessen
31. Mäueler W, Raulf F, Schartl M (1988) Expression of proto-oncogenes in embryonic, adult, and transformed tissue of *Xiphophorus* (Teleostei: Poeciliidae). Oncogene 2:421–430
32. Martin GS (1986) In: Varmus H, Bishop JM (eds) Cancer surveys 5: advances and prospects in clinical, epidemiological, and laboratory oncology. Oxford University Press, Oxford, pp 199–219

33. Ahuja MR (1979) On the nature of genetic change as an underlying cause for the origin of neoplasms. In: Chandra P (ed) Antiviral mechanisms in the control of neoplasia. Plenum, New York, pp 17–37
34. Ahuja MR, Lepper K, Anders F (1979) Sex chromosome aberrations involving loss and translocation of tumor-inducing loci in *Xiphophorus*. Experientia 35:28–29
35. Schwab M, Abdo S, Ahuja MR, Kollinger G, Anders A, Anders F, Frese K (1978) Genetics of susceptibility in the platyfish-swordtail tumor system to develop fibrosarcoma and rhabdomyosarcoma following treatment with N-methyl-N-nitrosourea (MNU). Z Krebsforsch 91: 301–315
36. Schwab M, Anders F (1981) Carcinogenesis in *Xiphophorus* and the role of the genotype in tumor susceptibility. In: Kaiser HE (ed) Neoplasms – comparative pathology of growth in animals, plants, and man. Williams and Wilkins, Baltimore, pp 451–459
37. Yamamoto T, Nishida T, Miyajima N, Kawai S, Ooi T, Toyoshima K (1983) The *erb*B gene of avian erythroblastosis virus as a member of the *src*-gene family. Cell 35:71–78
38. Ullrich A, Coussens L, Hayflick JS, Dull TJ, Gray A, Tam AW, Lee J, Yarden Y, Libermann TA, Schlessinger J, Downward J, Mayes ELV, Whittle N, Waterfield MD, Seeburg PH (1984) Human epidermal growth factor receptor cDNA sequence and aberrant expression of the amplified gene in A431 epidermal carcinoma cells. Nature 309:418–425
39. Semba K, Kamata N, Toyoshima K, Yamamoto T (1985) A v-*erb*B related protooncogene, c-*erb*B2, is distinct from the c-*erb*B1/epidermal growth factor receptor gene and is amplified in a human salivary gland adenocarcinoma. Proc Natl Acad Sci USA 82:6497–6501
40. King CR, Kraus MH, Aaronson SA (1985) Amplification of a novel v-*erb*B-related gene in a human mammary carcinoma. Science 229:974–976
41. Coussens L, Yang-Feng TL, Liao YC, Chen E, Gray A, McGrath J, Seeburg PH, Libermann TA, Schlessinger J, Francke U, Levinson A, Ullrich A (1985) Tyrosine kinase receptor with extensive homology to EGF receptor shares chromosomal location with *neu* oncogene. Science 230:1132–1139
42. Downward J, Parker P, Waterfield MD (1984) Autophosphorylation sites on the epidermal growth factor receptor. Nature 311:483–485
43. Chen WS, Lazar CS, Poenie M, Tsien RY, Gill GN, Rosenfeld MG (1987) Requirement for intensic protein tyrosine kinase in the immediate and late actions of the EGF receptor. Nature 328:820–823
44. Downward J, Waterfield MD, Parker PJ (1985) Autophosphorylation and protein kinase C phosphorylation of the epidermal growth factor receptor. J Biol Chem 260:14538–14546
45. Hunter T, Ling N, Cooper JA (1984) Protein kinase C phosphorylation of the EGF receptor at a threonine residue close to the cytoplasmic face of the plasma membrane. Nature 311:480–483
46. Soderquist AM, Carpenter G (1984) Glycosylation of the epidermal growth factor receptor in A431 cells: the contribution of carbohydrate to receptor function. J Biol Chem 259:12586–12594
47. Pröfrock A (1988) Untersuchungen zum Phosphatidylinosit-Turnover an ausgewählten *Xiphophorus*-Genotypen. Thesis, University of Giessen
48. Gronau T (1987) Untersuchungen zur Organisation, Aktivität und Wirkung des zellulären Onkogens *c-src* im *Xiphophorus*-Tumorsystem. Thesis, University of Giessen
49. Smith AD, Gronau T, Pröfrock A, Zechel C, Bird JM, Lane PA, Barnekow A, Anders A, Anders F (1989) EGF receptor gene, inositol lipid turnover and *c-src* activity in key processes preceding melanoma in *Xiphophorus*. In: Lynch HT, Fusaro RM (eds) Hereditary Malignant Melanoma. CRC, Boca Raton (in press)
50. Schartl A, Schartl M, Anders F (1982) Promotion and regression of neoplasia by testosterone-promoted cell differentiation in *Xiphophorus* and *Giradinus*. In: Hecker E, Fusenig NE, Kunz W, Marks F, Thielmann HW (eds) Cocarcinogenesis and biological effects of tumor promoters. Raven, New York, pp 427–434
51. Stich HF, Anders F (1988) The involvement of reactive oxygen species in oral cancers of betel quid/tobacco chewers. Mutation Res (in press)
52. Green S, Chambon P (1986) A superfamily of potentially oncogenic hormone receptors. Nature 324:615–617
53. Evans RM (1988) The storoid and thyroid

hormone receptor superfamily. Science 240:889–895
54. Frykberg L, Palmieri S, Beug H, Graf T, Hayman M, Vennström B (1983) Transformation capacity of avian erythroblastosis virus mutants deleted in the v-*erb*A or v-*erb*B oncogene. Cell 32:227–238
55. Kahn P, Frykberg L, Brady C, Stanley J, Beug H, Vennström B, Graf T (1986) V-*erb*A cooperates with sarcoma oncogenes in leukemia cell transformation. Cell 45:349–356
56. Sap J, Munoz A, Damm K, Goldberg Y, Ghysdael J, Leutz A, Beug H, Vennström B (1986) The c-*erb*A protein is a high-affinity receptor for thyroid hormone. Nature 324:635–640
57. Weinberger C, Thompson CC, Ong ES, Lebo R, Gruol DJ, Evans RM (1986) The c-*erb*A gene encodes a thyroid hormone receptor. Nature 324:641–646
58. Ahuja MR, Schwab M, Anders F (1980) Linkage between a regulatory of locus for melanoma cell differentiation and an esterase locus in *Xiphophorus*. J Hered 71:403–407
59. Debuire B, Henry C, Benaissa M, Biserte G, Claverie JM, Saule S, Martin P, Stehelin D (1984) Sequencing the *erb*A gene of avian erythroblastosis virus reveals a new type of oncogene. Science 224:1456–1459
60. Dejean A, Bougueleret L, Grzeschik KH, Tiollais P (1986) Hepatitis B virus DNA integration in a sequence homologous to v-*erb*A and steroid receptor genes in a hepatocellular carcinoma. Nature 322:70–72
61. Giguere V, Ong ES, Segui P, Evans RM (1987) Identification of a receptor for the morphogen retinoic acid. Nature 330:624–629
62. Petkovich M, Brand NJ, Krust A, Chambon P (1987) A human retinoic acid receptor which belongs to the family of nuclear receptors. Nature 330:444–450
63. Green S, Walter P, Kumar V, Krust A, Bornert JM, Argos P, Chambon P (1986) Human oestrogen receptor cDNA: sequence, expression and homology to v-*erb*A. Nature 320:134–139
64. Coneely OM, Sullivan WP, Toft DO, Birnbaumer M, Cook RG, Maxwell BL, Zarucki-Schulz T, Greene GL, Schrader WT, O'Malley BW (1986) Molecular cloning of the chicken progesterone receptor. Science 233:767–769
65. Weinberger C, Hollenberg SM, Rosenfeld MG, Evans RM (1985) Domain structure of human glucocorticoid receptor and its relationship to the v-*erb*A oncogene product. Nature 318:670–672
66. Tavolga WN (1949) Embryonic development of the platyfish (*Platypoecilius*), the swordtail (*Xiphophorus*) and their hybrids. Bull Am Mus Nat Hist 94:167–229

ras Gene Mutations and Clonal Analysis Using RFLPs of X-chromosome Genes in Myelodysplastic Syndromes *

J. W. G. Janssen [1], M. Buschle [1], M. Layton [2], J. Lyons [1], and C. R. Bartram [1]

A. Introduction

Myelodysplastic syndromes (MDS) comprise a heterogeneous group of clonal disorders characterized by quantitative and qualitative abnormalities of hematopoiesis [1, 2].

Up to 30% of cases eventually develop into acute nonlymphocytic leukemia (ANLL). Previous observations by us and others have demonstrated a frequent (30%) activation of the N-*ras* oncogene in ANLL [3–6].

The involvement of the different cell lineages in MDS has been rather controversial. Various approaches such as isoenzyme studies (G-6-PD polymorphisms), cytogenetic analysis (clonal chromosome abnormalities), and autosomal DNA polymorphism as markers of clonality have been employed to study the stem cell origin of MDS. However, these analyses produced conflicting data and have not clearly identified the stem cell origin of MDS [7–12].

Although no accepted effective treatment exists for MDS, some studies have reported response rates of up to 30% to the administration of low-dose cytarabine (LD-AraC).

However, the mechanism of LD-AraC treatment is not clear. Some data suggest that LD-AraC induces differentiation [13, 14], whereas other studies demonstrate a cytotoxic effect of LD-AraC [15, 16].

In this report we address the possible role of *ras* in MDS, the identification of its stem cell origin, and the biological action of LD-AraC treatment in MDS.

B. Materials and Methods

I. Patients

Eighty-one cases of primary MDS were screened for *ras* point mutations. The patients were classified according to FAB criteria: there were 19 cases of refractory anemia (RA), nine cases of refractory anemia with ring sideroblasts (RARS), 22 cases of refractory anemia with excess blasts (RAEB), seven cases of refractory anemia with excess blasts in transformation (RAEB-T), and 24 cases of chronic myelomonocytic leukemia (CMMoL). Peripheral blood or bone marrow samples, as well as a skin biopsy from one patient, were obtained with informed consent.

Moreover, seven of these patients were investigated for clonality, including three patients with RA, two with RARS, one patient with RAEB, and one with CMMoL.

Lymphocyte counts of the samples were between 43% and 52%. Peripheral blood did not contain any blasts except for that of the RAEB patient, who had 1% blasts.

II. Southern Blot Analysis

High-molecular-weight DNA was prepared from bone marrow or peripheral

[1] Section of Molecular Biology, Department of Pediatrics II, University of Ulm, Federal Republic of Germany
[2] Academic Department of Haematology, King's College School of Medicine and Dentistry, London, United Kingdom
* This work was supported by the Deutsche Forschungsgemeinschaft

blood cells by standard techniques. Clonal analyses with the X-linked phosphoglycerate kinase (PGK) probe were performed as described by Vogelstein et al. [17].

III. Detection of *ras* Gene Mutations

About 150 ng of genomic DNA were amplified for *ras* sequences by means of the polymerase chain reaction (PCR) [18].

Amplified DNA was spotted onto nylon filters, fixed by UV illumination, and hybridized with an oligomer panel that is able to detect all possible amino acid substitutions at codons 12, 13, and 61 of N-*ras*, Ki-*ras*, and H-*ras* [19].

C. Results

By means of a rapid dot-blot screening procedure based on a combination of in vitro amplification of *ras*-specific sequences and hybridization to specific oligonucleotide probes, we analyzed 81 cases of MDS for point mutations at codons 12, 13, and 61 in N-*ras*, Ki-*ras*, and H-*ras*. Mutations of Ki-*ras* and N-*ras* were detected in four cases. One RAEB showed a Ki-*ras* mutation, two CMMoLs exhibited N-*ras* mutations, and one other CMMoL scored positive for a mutation of Ki-*ras* at codon 12. In all four *ras*-positive cases, the normal *ras* allele was also present.

As the positive MDS samples contained less than 5% blasts and our dot-blot technique is able to detect a *ras* point mutation only if more than 10% of the cells are positive, we conclude that hematopoietic cells characterized by a *ras* gene mutation have maintained the potential to differentiate in vivo.

In order to determine which cell lineages are involved in MDS we performed cell separation studies of two cases of *ras* mutation-positive CMMoL. Mononuclear cells and granulocytes were separated by standard Ficoll-Hypaque density gradient centrifugation. T- and B-lymphocytes and monocytes were fractionated by positive selection with immunomagnetic beads [20]. Thus we could show that the *ras* mutation was present in all four cell fractions, including granulocyte, monocyte, T- and B-lymphocyte lineages ([6]; unpublished results).

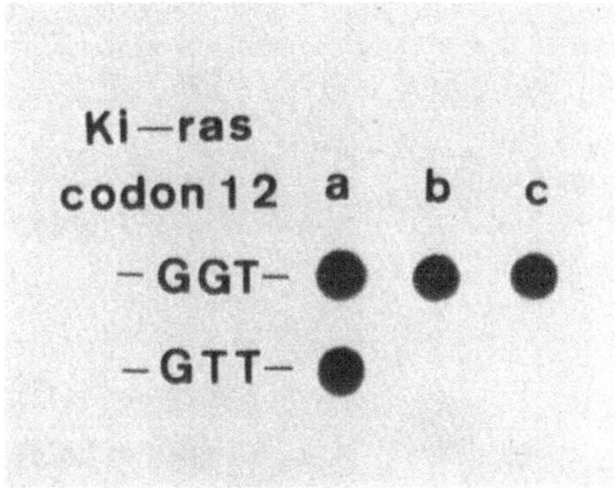

Fig. 1. Disappearance of mutated Ki-*ras* sequence after treatment with low-dose AraC. Five nanograms of amplified DNA from a patient with RAEB was spotted onto a nylon filter and hybridized to oligomers representing the Ki-*ras* wild-type allele (*top row*) and the Ki-*ras* mutated allele (*bottom row*). DNA analysis had shown a mutation of one allele at codon 12, Ki-*ras* substituting valine (*GTT*) for glycine (*GGT*) in peripheral blood cells (*a*) at presentation. Loss of *ras* mutation was observed in peripheral blood (*b*) and bone marrow (*c*) after a second course of LD-AraC

Figure 1 shows the analysis of a *ras*-positive case of MDS, a patient suffering from RAEB who was treated with LD-AraC. The wild-type codon 12 Ki-*ras* allele (glycine) was present before as well as after treatment. However, the mutated *ras* allele had disappeared completely after low-dose AraC treatment.

In seven female MDS patients belonging to different FAB types we investigated clonality by X-chromosome inactivation analyses. For this purpose we employed a recently published technique that utilizes the occurrence of DNA polmorphisms at X-linked loci [17]. The active and inactive alleles can be distinguished from each other by a methylation-sensitive endonuclease, as the 5′ cytosine methylation pattern of

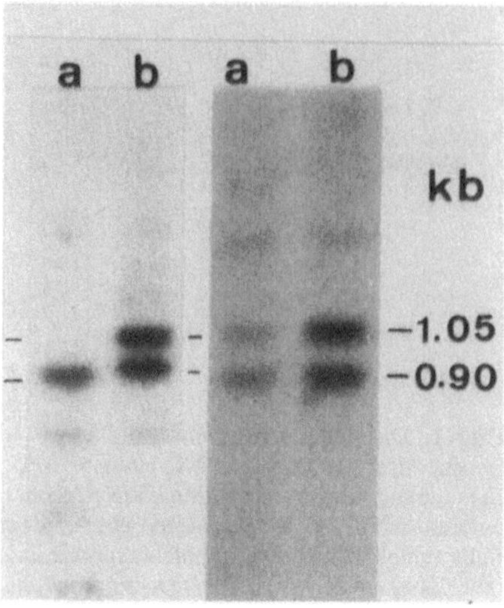

Fig. 2. Clonal analysis using a PGK probe of peripheral blood (*left panel*) and skin fibroblasts (*right panel*) from an RA patient. DNAs were digested with *Bst*XI and *Pst*I to distinguish the maternal and paternal alleles. Subsequently, the DNAs were divided into two equal aliquots; one was not digested further (*lanes b*) and the other was digested with *Hpa*II (*lanes a*) to distinguish between active and inactive X-chromosomes

housekeeping genes is changed after X-chromosome inactivation. All seven MDS patients analyzed showed a monoclonal pattern. Figure 2 shows an example of such an analysis. The right panel represents normal fibroblasts from an MDS patient before and after digestion with the methylation-sensitive enzyme *Hpa*II. Reduction in the intensity of both bands, corresponding to the two different alleles, can be observed after restriction endonuclease *Hpa*II incubation. This is expected in a polyclonal population of fibroblasts in which the X-chromosome is randomly inactivated. In contrast, peripheral blood of an MDS patient analyzed in the same way (left panel), shows the complete disappearance of one band (allele) after digestion with *Hpa*II, characteristic for a clonal population of cells.

D. Discussion

In this study we investigated the frequency of *ras* gene mutations in myelodysplastic syndromes. Point mutations in the Ki-*ras* and N-*ras* gene could be detected in about 5% of the cases tested. Recently, Hirai et al. [21] and Liu et al. [22] identified *ras* mutations in three of eight and two of four patients with MDS, respectively. Both articles suggest that *ras* activation might indicate the imminent conversion of an MDS into a frank leukemia. Analysis of our data has not revealed a correlation between the presence of mutated *ras* genes and the development into AML.

The small numbers of cells required for PCR analysis made it possible to study the cell lineage involvement in two MDS patients. We detected involvement of the granulocytic, monocytic, and B- and T-lymphoid lineages in both CMMoL patients tested. Identical results were obtained in our clonality studies.

This investigation revealed a clear clonal pattern in all seven MDS patients tested, and, as the samples comprised between 40% and 50% lymphocytes, it can be indirectly concluded that in these different MDS cases at least, the lymphocytic lineage was involved. The clinical heterogeneity of different MDS cases may question whether this is a common phenomenon. More cases should be studied in order to clarify this point.

Both methods, the clonality analyses based on X-linked RFLPs and the detection method for activated *ras* sequences, were used by us to investigate the action of LD-AraC in the treatment of MDS. The loss of clonality or mutated *ras* sequence in two cases supports the view that even at low doses cytarabine has a cytotoxic effect.

Acknowledgements. We thank Drs. E. Kleihauer, B. Kubanek, H. Heimpel, H. Drexler, and H. van den Berghe for their support and encouragement. We thank S. Gignac and A. Steenvoorden for expert technical assistance and H. Barro for editing the manuscript.

References

1. Koeffler HP (1986) Myelodysplastic syndromes (preleukemia). Semin Hematol 23:284–299
2. Layton DM, Mufti GJ (1986) Myelodysplastic syndromes: their history evolution and relation to acute myeloid leukemia. Blut 53:429–436
3. Bos JL, Toksoz D, Marshall CJ, Verlaan-de Vries M, Veeneman GH, van der Eb AJ, van Boom JH, Janssen JWG, Steenvoorden ACM (1985) Amino acid substitutions at codon 13 of the N-*ras* oncogene in human acute myeloid leukemia. Nature 315:726–730
4. Bos JL, Verlaan-de Vries M, van der Eb AJ, Janssen JWG, Welwel R, Löwenberg B, Colly LP (1987) Mutations in N-*ras* predominate in acute myeloid leukemia. Blood 69:1237–1241
5. Toksoz D, Farr CJ, Marshall CJ (1987) Ras gene activation in a minor proportion of the blast population in acute myeloid leukemia. Oncogene 1:409–413
6. Janssen JWG, Steenvoorden ACM, Lyons J, Anger B, Böhlke JU, Bos JL, Seliger H, Bartram CR (1987) Ras gene mutations in acute and chronic myelocytic leukemias, chronic myeloproliferative disorders, and myelodysplastic syndromes. Proc Natl Acad Sci USA 84:9228–9232
7. Prchal JT, Throckmorton DN, Carrol AJ, Fuson EW, Gomas RA, Prchal JF (1987) A common progenitor for myeloid and lymphoid cells. Nature 274:590–591
8. Raskind WH, Tirumali N, Jacobson R, Singer J, Fialkow PJ (1984) Evidence for a multistep pathogenesis of myelodysplastic syndrome. Blood 63:1318–1323
9. Shinohara T, Takuwa N, Morishita K, Ieki R, Yokota J, Nakayama E, Asano S, Miwa S (1983) Chronic myelomonocytic leukemia with a chromosome abnormality (46,xy,20q-) in all dividing myeloid cells: evidence for clonal origin in a multi-potent stem cell common to granulocyte, monocyt erythrocyte and thrombocyte. Am J Hematol 15:289–293
10. Amenomori T, Tomonaga M, Yoshida Y, Kuriyama K, Matsuo T, Jinnai I, Ichimaru M, Omiya A, Tsuji Y (1986) Cytogenetic evidence for partially committed myeloid cell origin of chronic myelomonocytic leukaemia and juvenile chronic myeloid leukaemia: evidence to the granulocyte-macrophage precursors and erythroid precursors carry identical marker chromosome. Br J Haematol 64:539–546
11. Lawrence HJ, Broudy VC, Magenis RE, Olson S, Tomar D, Barton S, Fitchen JH, Bagby GC Jr (1987) Cytogenetic evidence for involvement of B lymphocytes in acquired sideroblastic anemia. Blood 70:1003–1005
12. Kere J, Ruutu T, de la Chapelle A (1987) Monosomy 7 in granulocytes and monocytes in myelodysplastic syndromes. N Engl J Med 316:499–503
13. Castaigne S, Daniel MT, Tilly H, Herait P, Degos L (1983) Does treatment with AraC in low dosage cause differentiation of leukemic cells? Blood 62:85–86
14. Sachs L (1978) The differentiation of myeloid cells: new possibilities for therapy. Br J Haematol 40:509
15. Mittermüller J, Kolb HJ, Gerhartz HH, Wilmanns W (1986) In vivo differentiation of leukaemic blasts and effect of low-dose AraC in a marrow grafted patient with leukaemic relapse. Br J Haematol 62:757–762
16. Griffin JD, Spriggs D, Wisch JS, Kufe DW (1985) Treatment of preleukemic syndromes with continuous intravenous infusion of low-dose cytosine arabinoside. J Clin Oncol 3:982–991
17. Vogelstein B, Fearon ER, Hamilton SR, Feinberg AP (1985) Use of restriction fragment length polymorphisms to determine the clonal origin of human tumors. Science 227:642–645
18. Saiki RK, Gelfand DH, Stoffel S, Scharf SJ, Higuchi R, Horn GT, Mullis KB, Erlich HA (1988) Primer-directed enzymatic amplification of DNA with a thermostable DNA polymerase. Science 239:487–491
19. Lyons J, Janssen JWG, Bartram CR, Layton M, Mufti GJ (1988) Mutation of Ki-*ras* and N-*ras* oncogenes in myelodysplastic syndromes. Blood 71:1707–1712
20. Lea T, Vartdal F, Davies C, Ugelstad J (1985) Magnetic monosized polymer particles for fast and specific fractionation of human mononuclear cells. Scand J Immunol 22:207–216
21. Hirai H, Kobayashi Y, Mano H, Hagiwara K, Maru Y, Omine M, Mizoguchi H, Nishida J, Takaku F (1987) A point mutation at codon 13 of the N-*ras* oncogene in myelodysplastic syndrome. Nature 327:430–432
22. Liu E, Hjelle B, Morgan R, Hecht F, Bishop JM (1987) Mutations of the Kirsten-*ras* proto-oncogene in human preleukemia. Nature 300:186–188

Close Localization of the Genes for GM-CSF and IL3 in Human Genome

E. I. Frolova, G. M. Dolganov, M. L. Markelov, and B. Zhumabaeva

Colony-stimulation factors (CSFs), a family of glycoprotein growth factors, have been shown to support clonal proliferation of hematopoietic progenitor cells in vitro [1]. Macrophage-CSF (M-CSF or CSF-1) [2] and granulocyte-CSF (G-CSF) [3] stimulate cells committed to the macrophage and granulocyte lineages respectively, whereas granulocyte-macrophage-CSF (GM-CSF) and interleukin-3 (IL3 or multi-CSF) are capable of stimulating proliferation and differentiation of progenitors along multiple pathways.

Successful cloning of cDNA and genomic copies of mouse and human genes for IL3 [3, 4] and GM-CSF [5, 6], as well as for M-CSF and G-CSF, have had a great impact on the analysis of biological properties of those molecules in vivo and in vitro [7, 8].

The GM-CSF and IL3 genes have been mapped to human chromosome 5 at bands q23-31 [9, 10], a region that is frequently deleted in patients with myeloid disorders [del(5q)] [11]. Several other growth factors and growth-factor receptors – in particular, the CSF-1 gene and proto-oncogene FMS, coding a protein possibly identical to the receptor for CSF-1 – are also located within this region of chromosome 5 [12]. There is a possibility that a family of genes responsible for regulation of cell growth during hematopoiesis is located within the limited segment of chromosome 5 [9]. Precise mapping of this region is essential to the understanding of functional relationships between the genes and could reveal the genes for other growth factors and their receptors that may be located within this region.

For isolation of genomic DNA clones containing genes for human GM-CSF and IL3 we prepared from human leukocyte DNA a genomic library of 1.5×10^6 clones. Using synthetic oligonucleotides, we identified eight individual clones that hybridized with the IL3 probe and five clones that hybridized with the GM-CSF probe; three of these hybridized with both probes. Southern blot hybridization analysis of DNA restriction fragments of individual phages revealed that three independently cloned 20-kb fragments of human genomic DNA contain sequences of both IL3 and GM-CSF genes. Physical maps of the isolated clones are shown in Fig. 1.

Next, we analyzed localization of the two genes in human placental DNA using Southern blot analysis (Fig. 2). The fragments generated by *Xba*I restriction endonuclease hybridized with both IL3 and GM-CSF probes.

The results strongly indicate a close genomic linkage of human IL3 and GM-CSF genes. The distance between the genes is 10 kb, and they are arranged in head-to-tail fashion, the gene for GM-CSF following the gene for IL3.

Close linkage of the two CSF genes may indicate either that they have coordinate regulation during T-lymphocyte gene expression, or that they have diverged from a common ancestral gene by duplication. The latter hypothesis is supported by the fact that both genes have similar exon-intron structures and com-

M. M. Shemyakin Institute of Bioorganic Chemistry, Academy of Sciences of the USSR, Moscow

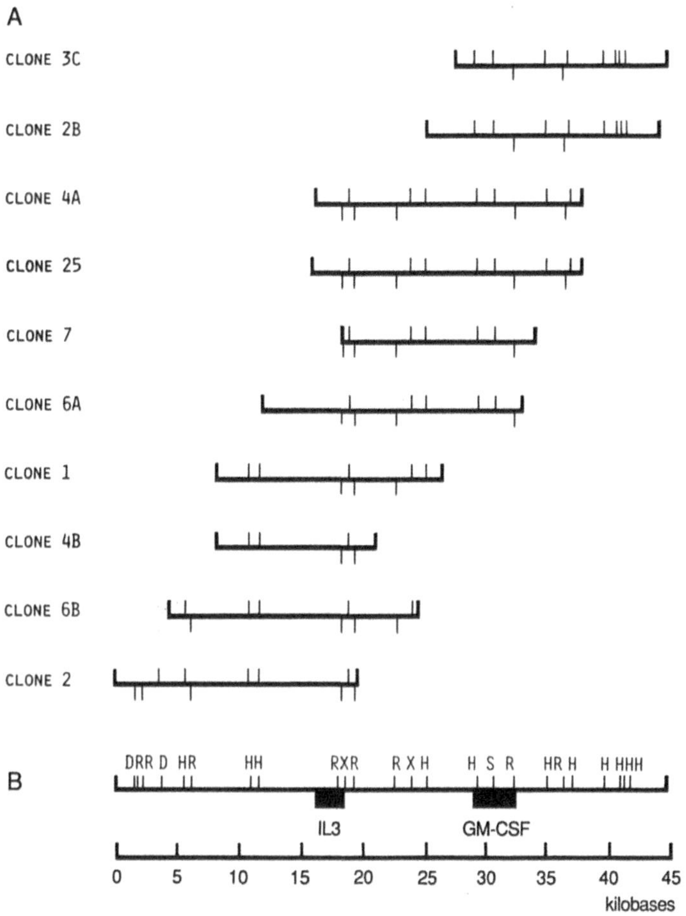

Fig. 1. **A** Physical maps of the individual phage clones containing genes for human GM-CSF and IL3. The positions of the recognition sites for restriction endonucleases *Eco*RI (*R*), *Hind*III (*H*), *Xho*I (*X*), *Sal*I (*S*), and *Xba*I (*D*) are indicated. **B** Physical map of the whole region of human genome containing the genes

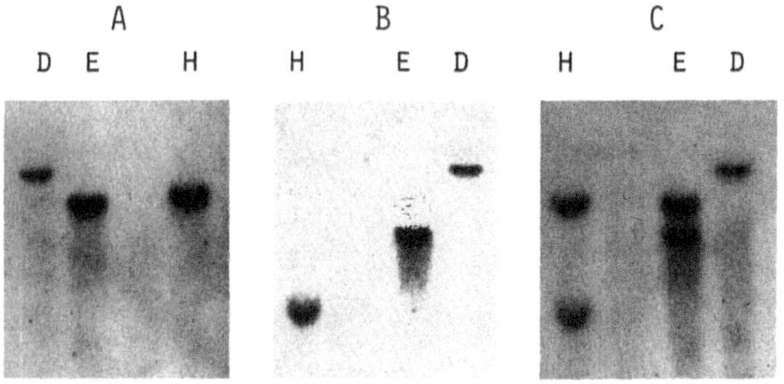

Fig. 2 A–C. Southern blot analysis of human DNA. The DNA from human placenta was cleaved with *Eco*RI (*E*), *Hind*III (*H*), and *Xba*I (*D*) and hybridized with the probes for IL3 **(A)**, GM-CSF **(B)**, and with both probes together **(C)**

mon features in the secondary structure of the two polypeptides displayed in distribution of alfa-helical regions.

References

1. Metcalf D (1984) The hematopoietic colony-stimulating factors. Elsevier, New York, Oxford, Chapter 11. Tissue and cellular sources of the colony stimulating factors, p 309–329
2. Ihle JN, Keller J, Oroszlan S, Henderson LE, Copeland TD, Fitch F, Prystowsky MB, Goldwasser E, Schrader JW, Palaszynski E, Dy M, Lebel B (1983) Biologic properties of homogeneous interleukin 3. I. Demonstration of WEHI-3 growth factor activity, mast cell growth factor activity, P cell-stimulating factor activity, colony-stimulating factor activity, and hystamine-producing cell-stimulating factor activity. J Immunol 131:282–287
3. Cantrell MA, Anderson D, Cerretti DP, Price V, McKeregham K, Tushinski RJ, Mochizuki DY, Larsen A, Grabstein K, Gillis S, Cosman D (1985) Cloning, sequence, and expression of a human granulocyte/macrophage colony-stimulating factor. PNAS 82:6250–6254
4. Kaushansky K, O'Hara PJ, Berkner K, Segal GM, Hagen FS, Adamson JW (1986) Genomic cloning, characterization, and multilineage growth-promoting activity of human granulocyte-macrophage colony-stimulating factor PNAS 83: 3101–3105
5. Fung MC, Hapel AJ, Ymer S, Cohen DR, Johnson RM, Campbell HD, Young IG (1984) Molecular cloning of cDNA for murine interleukin-3. PNAS 307:233–237
6. Yang Yu-Ch, Ciarletta AB, Temple PA, Chung MP, Kovacic S, Witek-Giannotti JS, Leary AC, Kriz R, Donahue RE, Wong GG, Clark S (1986) Human IL-3 (Multi-CSF): identification by expression cloning of a novel hematopoietic growth factor related to murine IL-3. Cell 47: 3–10
7. Donahue RE, Wang EA, Stone DK, Kamen R, Wong GG, Seghal PK, Nathan DG, Clark SC (1986) Stimulation of haematopoiesis in primates by continuous infusion of recombinant human GM-CSF. Nature 321:872–875
8. Kindler V, Thorens B, de Kossodo S, Allet B, Eliason JF, Thatcher D, Farber N, Vassalli P (1986) Stimulation of hematopoiesis in vivo by recombinant bacterial murine interleukin 3. PNAS 83:1001–1005
9. Pettenati MJ, LeBeau MM, Lemons RS, Shima EA, Kawasaki ES, Larson RA, Sherr CJ, Diaz MO, Rowley JD (1987) Assignment of CSF-1 to 5q33.1: Evidence for clustering of genes regulating hematopoiesis and for their involvement in the deletion of the long arm of chromosome 5 in myeloid disorders. PNAS 84:2970–2974
10. LeBeau MM, Epstein ND, O'Brien SJ, Nienhuis AW, Yu-Chung Yang, Clark SC, Rowley JD (1987) The interleukin 3 gene is located on human chromosome 5 and is deleted in myeloid leukemias with a deletion of 5q. PNAS 84:5913–5917
11. Mitelman F (1985) In: Sandberg A (ed) Progress and topics in cytogenetics, vol 5, p 107. Liss, New York
12. LeBeau MM, Westbrook CA, Diaz MO, Larson RA, Rowley JD, Gasson JC, Golde DW, Sherr CJ (1986) Evidence for the involvement of GM-CSF and FMS in the deletion (5q) in myeloid disorders. Science 231:984–987

HIV-1, HTLV-I and the Interleukin-2 Receptor: Insights into Transcriptional Control

E. Böhnlein [1], J. W. Lowenthal [2], Y. Wano [2], B. R. Franza [3], D. W. Ballard [2], and W. C. Greene [2]

A. Introduction

T-cell proliferation is regulated by the induced expression of the cellular genes encoding the T-cell growth factor interleukin 2 (IL-2) and the alpha subunit of the high-affinity IL-2 receptor (IL-2Rα, p55, Tac) [1]. Infection of $CD4^+$ T-lymphocytes by the type I human T cell leukemia virus (HTLV-I), the etiologic agent of adult T cell leukemia (ATL) [2], leads to the deregulation of IL-2Rα gene expression. Virtually all ATL cell lines thus far examined have been found to constitutively display a large number of high- and low-affinity receptors for IL-2. This response appears to involve an action of the 40-kDa transactivator gene product (*tax*, tat-1, $p40^x$) encoded within the pX region of this retrovirus [3, 4]. Transient cotransfection assays have revealed that *tax* markedly augments the activity of the IL-2Rα promoter and partially activates the IL-2 promoter in Jurkat T cells [5–9]. In addition to this effect on these cellular genes, *tax* is also capable of activating the transcriptional enhancer located within the long terminal repeat (LTR) of type 1 human immunodeficiency virus (HIV-1) [10]. In contrast to HTLV-I, infection of $CD4^+$ T cells with HIV-1 leads to cell death and is clinically associated with the acquired immune deficiency syndrome (AIDS) [2]. Like *tax*, various T-cell mitogens including phytohemagglutinin (PHA) and phorbol 12-myristate 13-acetate (PMA) activate both the IL-2Rα promoter and the LTR of HIV-1 [10, 11]. Sequence analysis of the IL-2Rα promoter has revealed the presence of a 12 base pair (bp) element (GGGGAATCTCCC) [11] that shares significant homology with the NF-κB binding site present as a duplicated element in the HIV-1 enhancer [12] and as a single motif in the immunoglobulin κ chain gene enhancer [13]. These findings raised the possibility that a common regulatory intermediate may bind to these conserved sequences and play a role in the induced expression of these cellular and viral genes. We now describe transient transfection studies with 5′ deletion and site-specific mutants of the IL-2Rα promoter, establishing that the κB motif is importantly involved in both *tax* and mitogen inducibility. Furthermore, using gel retardation [14, 15] and DNA footprinting [16] assays, we demonstrate that both *tax* and T-cell mitogens induce the expression of DNA-binding protein(s) that specifically bind to this 12-bp IL-2Rα promoter sequence as well as to the transcriptional enhancer of HIV-1 [17]. Microscale DNA-affinity precipitation assays [18] performed with biotinylated oligonucleotides derived from the IL-2Rα promoter have permitted identification of one of these proteins as an 86-kDa cellular factor termed HIVEN86A [11, 18–20]. Taken together, these results suggest that the inducible cellular protein, HIVEN86A, plays a central role in the transcriptional regulation of both the IL-2Rα gene and the HIV-1 LTR me-

[1] Sandoz Forschungsinstitut, Brunnerstr. 59, A-1235 Vienna, Austria
[2] Howard Hughes Medical Institute, Departments of Medicine, Microbiology, and Immunology, Duke University Medical Center, Durham, NC 27710, USA
[3] Cold Spring Harbor Laboratories, Cold Spring Harbor, NY 11724, USA

diated by either T-cell mitogens or the *tax* protein of HTLV-1.

B. Materials and Methods

I. Cell Culture and Transfection Assays

All cell lines were grown in RPMI-1640 media supplemented with 10% fetal calf serum. Jurkat cell lines stably expressing sense or anti-sense *tax* cDNA were established by electroporation of the respective expression plasmids and subsequent selection for G418 antibiotic resistance [19]. Transient transfection experiments using DEAE dextran were performed as previously described [8]. The nested series of 5′ deletion mutants of the IL-2Rα promoter linked to the chloramphenicol acetyltransferase (CAT) reporter gene was prepared as previously described [9].

II. DNA-Protein Binding Assays

Nuclear extracts were prepared from large-scale cultures of cells grown at 2–5×10^5 cells/ml. The cell pellets were routinely frozen in liquid nitrogen after a single wash in 1×PBS. The cells were thawed on ice in 1 ml buffer I (10 m*M* HEPES of pH 7.9, 1.5 m*M* $MgCl_2$, 10 m*M* KCl, 0.5 m*M* DTT, 0.3 m*M* sucrose, 0.5 m*M* PMSF, 0.1 m*M* EGTA). The cells were then repelleted, homogenized (15 strokes with a loose-fitting Dounce homogenizer), and the nuclei were collected in an Eppendorf centrifuge. The nuclei were then extracted at 4°C with mild agitation in buffer II (20 m*M* HEPES of pH 7.9, 25% glycerol, 0.3 *M* KCl, 1.5 m*M* $MgCl_2$, 0.2 m*M* EDTA, 0.5 m*M* PMSF, 0.5 m*M* DTT, 0.1 m*M* EGTA; 0.3 ml/10^8 cells). Debris was removed by centrifugation, and the supernatant was dialyzed for over 3 h against at least 100 volumes of buffer III (20 m*M* HEPES of pH 7.9, 20% glycerol, 100 m*M* KCl, 0.2 m*M* EDTA, 0.5 m*M* PMSF, 0.5 m*M* DTT). Insoluble material was pelleted (5 min) and 50-μl aliquots were frozen in liquid nitrogen. Protein concentration was determined using the Bio-Rad colorimetric assay. Routinely, 2–10 μg nuclear extract was incubated with radiolabeled DNA at room temperature in the presence of 50 m*M* KCl, 10 m*M* $MgCl_2$, 50 m*M* HEPES of pH 7.9, 1 m*M* EDTA, 1 m*M* DTT, 1 mg/ml BSA, 30% glycerol, and lug poly(d[I-C]) (20 μl final volume). Reaction mixtures were then analyzed by electrophoresis in low ionic strength polyacrylamide gels as previously described [11]. For the in situ DNA footprinting assays, the binding mixture was increased five fold, and the DNA-protein complexes were subsequently separated on 5% nondenaturing polyacrylamide gels. These gels were then exposed to 1,10 phenanthroline copper for 7–12 min, and the samples were purified as described by Ballard et al. [20]. The microscale DNA-affinity precipitation assays were performed as previously reported [11, 18].

C. Results and Discussion

I. Mapping of Functional *cis*-Acting Elements Involved in the IL-2Rα Promoter Activation

Transient transfection of various IL-2Rα promoter deletion mutants revealed that upstream sequences located between nucleotides −281 and −248 were required for mitogen (PMA, 50 ng/ml) induction (Fig. 1). Similarly, an overlapping promoter segment was shown to be essential for activation of the IL-2Rα promoter by the HTLV-I derived *tax* gene product (Fig. 1). Identical results were obtained with transient cotransfection of *tax* cDNA expression plasmids and with transfection of Jurkat cells stably expressing the *tax* protein [19, 20]. In contrast, transfection of the IL-2Rα promoter-CAT plasmids in uninduced Jurkat T-cells or control Jurkat cell lines expressing an anti-sense *tax* cDNA produced only basal levels of CAT activity. As noted above, a κB-like binding site is present between nucleotides −267 and −256 of the IL-2Rα promoter. Site-di-

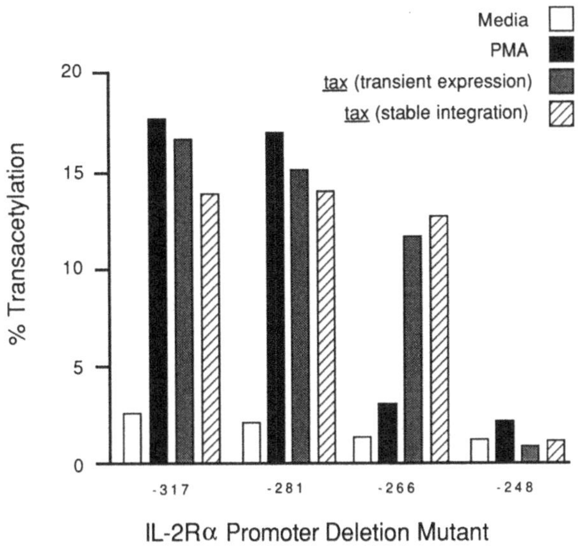

Fig. 1. PMA and *tax* stimulation of 5′ IL-2Rα deletion mutants linked to the CAT indicator gene. IL-2Rα-CAT deletion mutants [9], terminating 5′ at the nucleotide position indicated, were transfected into Jurkat T cells with DEAE dextran as previously described [9]. CAT activity, expressed as percent transacetylation, was measured after 48 h culture [22]. Transcriptional activity (measured by CAT production) obtained with uninduced Jurkat cell extracts (*open bars*) was similar to the level present in anti-*tax* Jurkat cells (data not shown). Stimulation induced by PMA (50 ng/ml) is shown in the *closed bars,* effects of cotransfection of *tax* cDNA in the *dotted bars,* and results in the stable *tax* cell line are shown in the *hatched bars*

rected deletion of this κB element from the active −317 IL-2Rα promoter construct produced a marked decrease in mitogen and *tax* inducibility in Jurkat T cells (data not shown), suggesting that this motif was importantly required for promoter activation. To test whether the κB element was sufficient to support *tax* and mitogen activation of a heterologous promoter, 47-bp or 18-bp oligonucleotides containing this sequence [11] were linked to the thymidine kinase (TK) promoter [21] upstream of the CAT reporter gene. Transfection of these modified TK CAT plasmids into Jurkat cells revealed marked inducibility by either *tax* or T-cell mitogens (Table 1). These oligonucleotides were effective in either orientation, and amplified effects were observed when the binding site was reiterated. Control plasmids containing various mutations in the κB site (IL-2R III M1, IL-2R VII M1, [20]) failed to confer mitogen or *tax* inducibility to the TK promoter (Table 1), further implicating the κB sequence in IL-2Rα promoter induction. Together, these results suggest that the κB-like sequences are both necessary and sufficient for IL-2Rα promoter activation by either *tax* or mitogens. However, these two inducers appear to differ with respect to the requirement for additional sequences flanking the κB element for maximal transcriptional activation (see Fig. 1; [7, 9]).

II. Inducible Nuclear Proteins Bind to the κB Element of the IL-2Rα Promoter

To study the potential interactions of cellular proteins with the κB element of the IL-2Rα promoter, oligonucleotides that conferred *tax* and mitogen inducibility upon the TK promoter were analyzed in gel retardation assays. Nuclear extracts prepared from Jurkat cells induced with

Table 1. IL-2Rα promoter oligonucleotides containing the κB element are sufficient to confer PMA and *tax* inducibility to the TK promoter

Insert	Orientation	J-*tax*-9	J-anti-*tax*-10	Jurkat+PMA
None		1.5	1.3	1.4
IL-2R III	→	10.8	1.1	6.6
	←	16.2	1.4	7.3
	←→	51.8	1.4	29.5
IL-2R III M1	→	2.1	1.7	1.8
	←	1.4	1.1	1.4
	→→	1.8	1.0	1.6
IL-2R VII	→	14.3	1.3	7.2
	←	10.7	1.1	6.3
	→→	28.8	1.5	20.5
	←←	34.2	1.4	22.7
IL-2R VII M1	→	1.7	0.9	1.5
	→→	1.2	1.1	1.5
	←←	1.5	1.0	1.2

Double-stranded oligonucleotides (IL-2R III, −291 to −245; IL-2R III M1, GGG to CTC substitution at −266 to −264 within the κB element; IL-2R VII, −272 to −255; IL-2R VII M1, four substitutions in κB element, see [20]) were inserted upstream of the TK promoter (nucleotide −105) linked to the CAT reporter gene. Arrows denote orientation of the inserts, and copy number is indicated by the number of arrows. Transfections were performed as described in Fig. 1. CAT conversion values for the parental TK CAT construct without an insert are shown in the first line.

PMA (50 ng/ml) and PHA (1 μg/ml) for 5 h yielded two specific DNA-protein complexes with a radiolabeled 47-bp oligonucleotide designated IL-2R III (−291 to −245; Fig. 2A, lane 3). In contrast, nuclear extracts prepared from unstimulated cells failed to mediate detectable complex formation. Analysis of the Jurkat cell lines stably expressing the *tax* gene product also yielded two similarly retarded complexes (Fig. 2B, lanes 3, 4). In constrast, these complexes were not detected with extracts prepared from two independent Jurkat anti-*tax* cell lines (Fig. 2, lanes 1, 2). Competition studies using oligonucleotides mutated in the κB site did not inhibit complex formation while wild-type unlabeled DNA probes competed effectively (data not shown). These findings indicated that the observed DNA-protein complexes were not only sequence specific, but also that the κB element was involved in protein binding. In situ DNA-footprinting experiments confirmed protein contacts over the κB region using nuclear extracts from either *tax*- or mitogen-induced Jurkat cells (Fig. 3). Furthermore, the sequences protected in DNA footprinting experiments with PMA- and *tax*-activated Jurkat nuclear extracts proved virtually identical.

To characterize the protein(s) interacting with this site within the IL-2Rα promoter, microscale DNA-affinity precipitation assays [18] were performed. Consistent with the gel retardation profiles, an 86-kDa cellular protein, HIVEN86A, was detected with the biotinylated IL-2R III probe in extracts obtained from either PHA+PMA induced Jurkat T cells or the *tax*-producing Jurkat cell lines. In contrast, this protein was not detectable in extracts from Jurkat anti-*tax* or unstimulated Jurkat cells (Fig. 4). Binding of this same inducible

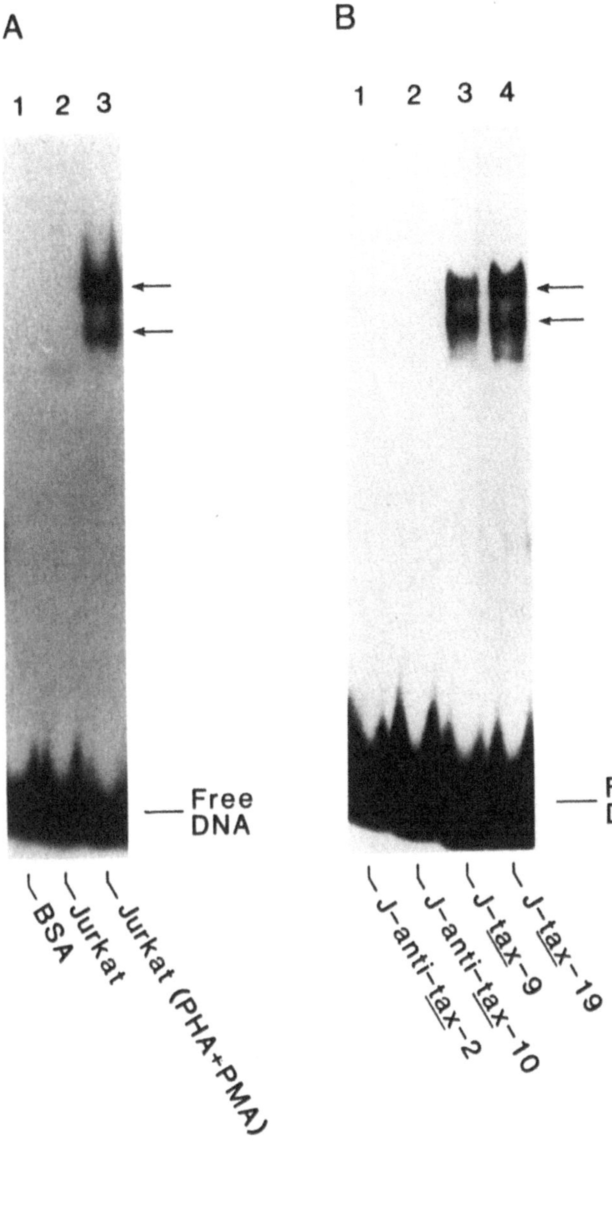

Fig. 2 A, B. Induction of nuclear proteins that bind to the IL-2Rα promoter. **A** Mitogen induction. The 47-bp ^{32}P-radiolabeled IL-2R III [11] oligonucleotide was incubated with bovine serum albumin (*lane 1*), nuclear extracts from uninduced Jurkat T cells (*lane 2*), or nuclear extracts from Jurkat T cells activated with PHA (1 µg/ml) and PMA (50 ng/ml) for 5 h (*lane 3*). **B** *tax* induction. Radiolabeled IL-2R III was incubated with extracts from control anti-sense *tax* cell lines (*lanes 1, 2*) or sense Jurkat *tax* cell lines (*lanes 3, 4*). In both panels, the migration of free DNA and the formation of DNA-protein complexes are indicated by *arrows*

protein to the HIV-1 enhancer has been previously reported [11, 18]. Together, these results indicate that mitogens and *tax* share the capacity to induce the same nuclear protein which in turn interacts with the $\varkappa$B elements present in the IL-2Rα promoter and the HIV-1 enhancer.

D. Summary

In this study, we present direct evidence for the binding of the inducible cellular protein, HIVEN86A, to a 12-bp element present in the IL-2Rα promoter. This element shares significant sequence similarity with the NF-κB binding sites present in the HIV-1 and κ immunoglobulin enhancers. Transient transfection studies indicate that this κB element is both necessary and sufficient to confer *tax* or mitogen inducibility to a heterologous promoter. As summarized schematically in Fig. 5, the findings suggest that the HIVEN86A protein may play a central role in the activation of cellular genes required for T-cell growth, specifically the IL-2Rα gene. In addition, the induced HIVEN86A protein also binds to a similar sequence present in the HIV-1 LTR leading to enhanced viral gene expression and ultimately T-cell death.

Thus, mitogen activation of the HIV-1 LTR appears to involve the same inducible transcription factor(s) that normally regulates IL-2Rα gene expression and T-cell growth. These findings further underscore the importance of the state of T-cell activation in the regulation of HIV-1 replication. Our results also demonstrate that HIVEN86A is induced by the *tax* protein of HTLV-I. Thus, in HTLV-I infected cells, normally the tight control of the transient expression of the IL-2Rα gene is lost. The constitutive high-level display of IL-2 receptors may play a role in leukemic transformation mediated by HTLV-I (ATL). Apparently by the same mechanism, the *tax* protein also activates the HIV-1 LTR through the induction of HIVEN86A. This

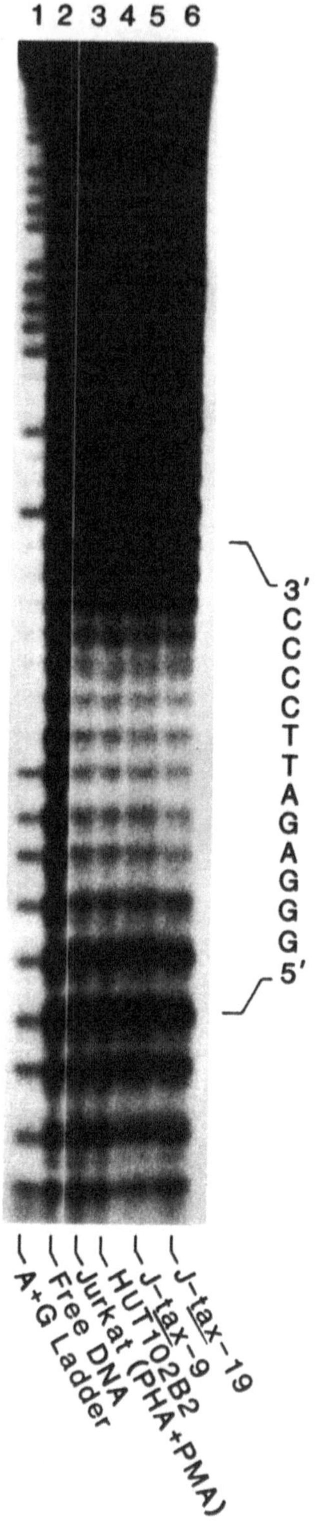

Fig. 3. 1.10 Phenanthroline-copper DNA footprinting of the IL-2Rα promoter. Radiolabeled IL-2R III [11] was incubated with nuclear extracts prepared from induced Jurkat cells (PHA + PMA), two different Jurkat *tax* lines, as well as the HTLV-I infected HUT102B2 T-cell line. The DNA-protein complexes were electrophoretically separated from free DNA on a 5% polyacrylamide gel and subjected to partial chemical digestion for 10 min. Reaction products including free DNA (*lane 2*) were processed as described [20] and subsequently analyzed on a 10% sequencing gel. An A + G ladder (*lane 1*) was prepared and loaded for sequence determination thus permitting identification of the protected area

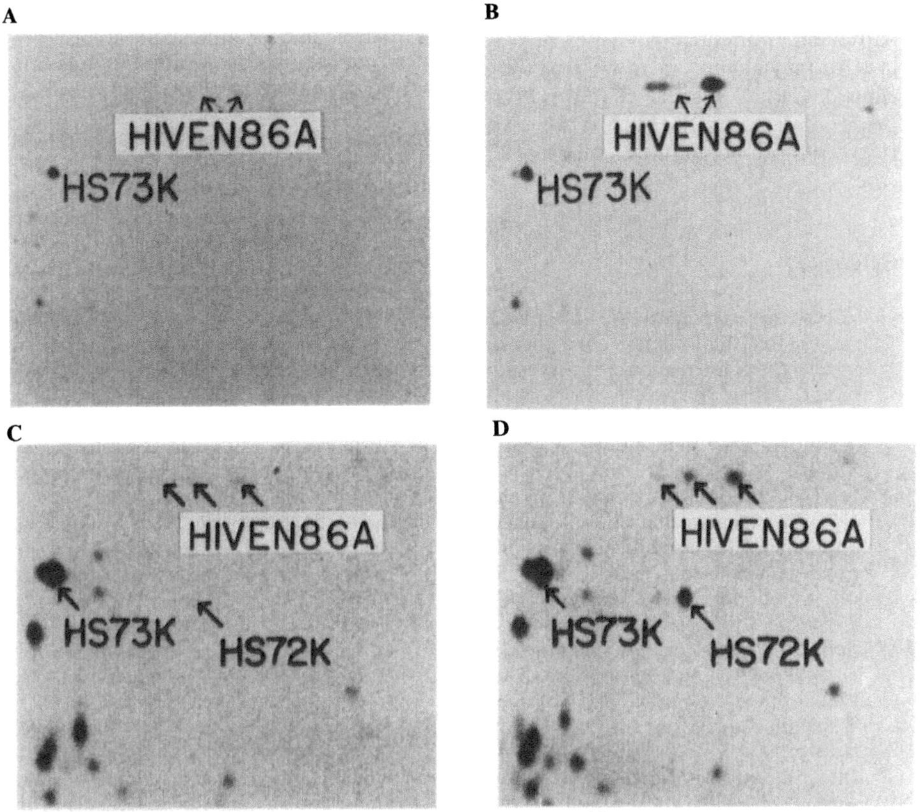

Fig. 4A–D. Microscale DNA-affinity precipitation assays. ^{35}S-Labeled cellular extracts prepared from uninduced Jurkat cells **(A)**, Mitogen-induced Jurkat cells (PMA 50 ng/ml, PHA 1 µg/ml; **B**), an anti-*tax* cell line **(C)**, and Jurkat *tax* line **(D)** were precipitated with a biotinylated IL-2R III DNA probe as previously described [16]. A 73-kDa heat-shock protein (HS73K) is coprecipitated in all assays independent of the biotinylated probe used (data not shown). A second heat-shock protein, HS72K, appears to be induced by *tax* but binds DNA in a non-sequence-specific manner

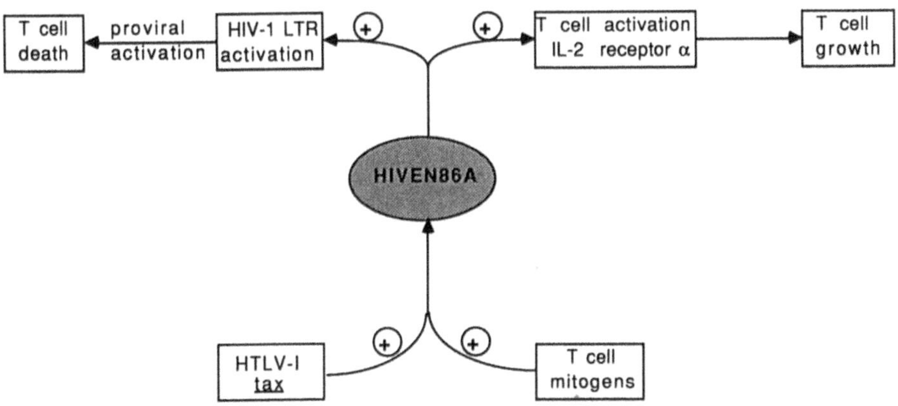

Fig. 5. Working model for the role of HIVEN86A in mitogen and *tax* activation of T-cell growth and HIV-1 replication

molecular interplay between these two pathogenic human retroviruses detected in vitro may also occur in vivo, as recent clinical studies suggest that AIDS patients coinfected with HTLV-I may experience a more accelerated course of disease.

References

1. Greene WC, Leonard WJ, Depper JM (1986) Growth of human T-lymphocytes: an analysis of interleukin-2 and its cellular receptor. Prog Hematol 14:283
2. Wong-Staal F, Gallo RC (1985) Human T-lymphotropic retroviruses. Nature 317:395
3. Seiki M, Hattori S, Hirayama M, Yoshida M (1983) Human adult T-cell leukemia virus: complete nucleotide sequence of the provirus genome integrated in leukemia cell DNA. Proc Natl Acad Sci USA 80:3618
4. Sodroski JG, Rosen CA, Haseltine WA (1984) *Trans*-acting transcriptional activation of the long terminal repeat of human T lymphotropic viruses in infected cells. Science 225:381
5. Inoue J, Seiki M, Taniguchi T, Tsuru S, Yoshida M (1986) Induction of interleukin-2 receptor gene expression by $p40^x$ encoded by human T cell leukemia virus-type I. EMBO J 5:2882
6. Maruyama M, Shibuya H, Harada H, Hatakeyama M, Seiki M, Fujita T, Inoue J, Yoshida M, Taniguchi T (1987) Evidence for aberrant activation of the interleukin-2 autocrine loop by HTLV-I-encoded $p40^x$ and T3/Ti complex triggering. Cell 48:343
7. Cross SL, Feinberg MB, Wolf JB, Holbrook NJ, Wong-Staal F, Leonard WJ (1987) Regulation of the human interleukin-2 receptor α promoter by the *trans*-activator gene of HTLV-I. Cell 49:47
8. Siekevitz M, Feinberg MB, Holbrook N, Wong-Staal F, Greene WC (1987) Activation of interleukin-2 and interleukin-2 receptor (Tac) promoter expression by the *trans*-activator (tat) gene product of human T-cell leukemia virus, type I. Proc Natl Acad Sci USA 84:5389
9. Lowenthal JW, Böhnlein E, Ballard DW, Greene WC (1988) Regulation of interleukin-2 receptor α subunit (Tac or CD25 antigen) gene expression: Binding of inducible nuclear proteins to discrete promoter sequences correlates with transcriptional activation. Proc Natl Acad Sci USA (85:4468)
10. Siekevitz M, Josephs SF, Dukovich M, Peffer N, Wong-Staal F, Greene WC (1987) Activation of the HIV-1 LTR by T-cell mitogens and the *tat*-I protein of HTLV-I. Science 238:1575
11. Böhnlein E, Lowenthal JW, Siekevitz M, Ballard DW, Franza BR, Greene WC (1988) The same inducible nuclear proteins regulate mitogen activation of both the interleukin-2 receptor-alpha gene and type 1 HIV. Cell 53:827
12. Nabel G, Baltimore D (1987) An inducible transcription factor activates expression of human immunodeficiency virus in T cells. Nature 326:711
13. Sen R, Baltimore D (1986) Multiple nuclear factors interact with the immunoglobulin enhancer sequences. Cell 46:705
14. Fried M, Crothers DM (1981) Equilibria and kinetics of *lac* repressor-operator interactions by polyacrylamide gel electrophoresis. Nucleic Acids Res 9:6507
15. Garner MM, Revzin A (1981) A gel electrophoresis method for quantifying the binding of proteins to specific DNA regions: applications to components of the Escherichia coli lactose operon regulatory system. Nucleic Acids Res 9:3047
16. Kuwabara MD, Sigman DS (1987) Footprinting DNA-protein complexes in situ following gel retardation assays using 1,10-phenanthroline-copper ion: *Escherichia coli* RNA polymerase-*lac* promoter complexes. Biochemistry 26:7234
17. Rosen CA, Sodroski JG, Haseltine WA (1985) Location of *cis*-acting regulatory sequences in the human T-cell lymphotropic virus type III (HTLV-III/LAV) long terminal repeat. Cell 41:813
18. Franza RB, Josephs SF, Gilman MZ, Ryan W, Clarkson B (1987) Characterization of cellular proteins recognizing the HIV enhancer using a microscale DNA-affinity precipitation assay. Nature 330:391
19. Wano Y, Feinberg M, Hoskins JB, Bogerd H, Greene WC (1988) Stable expression of the tax gene of type I human T-cell leukemia virus in human T cells activates specific cellular genes involved in growth. Proc Natl Acad Sci USA 85:9733

20. Ballard DW, Böhnlein E, Lowenthal JW, Wano Y, Franza BR, Greene, WC (1988) HTLV-I tax induces cellular proteins that activate the κB element of the IL-2 receptor α gene. Science 241:1652
21. McKnight SL, Gavis ER, Kingsbury R, Axel R (1981) Analysis of transcriptional regulatory signals of the HSV thymidine kinase gene: identification of upstream control region. Cell 25:385
22. Neumann JR, Morency CA, Russian KO (1987) A novel rapid assay for chloramphenicol acetyl transferase gene expression. Biotechniques 5:444

HIV-I Replication Requires an Intact Integrase Reading Frame

H. D. Buchow, E. Tschachler, R. C. Gallo, and M. Reitz Jr.

Human immunodeficiency virus type I (HIV-I), is a human retrovirus that is the causative agent of acquired immune deficiency syndrome (AIDS). Retroviruses replicate through a DNA intermediate. After initial synthesis of unintegrated DNA the insertion of that DNA into the host cell genome is thought to be a required step in the retroviral life cycle [1].

A hallmark of HIV-I infection, however, is the persistent appearance of large amounts of unintegrated DNA. Integrated DNA is often difficult to detect, and is readily detectable only after extensive passage of infected cell lines. Consequently, it is not clear that integration is an obligatory part of the HIV-I life cycle. To test the need for integration in HIV-I replication and expression, we have introduced point mutations including a stop codon into the COOH terminal part of the POL gene containing the integrase coding region and substituted this region for the analogous region of the biologically active molecular clone of HIV-I, pHXB2D [2]. Transfection of cos-I cells with the INT mutant resulted in transient expression of biologically active virus. However, infection of H9 cells with the mutant virus yielded neither detectable persistent media reverse transcriptase activity nor viral GAG antigen. These results suggest that integration of proviral DNA is a necessary part of the productive infection of T-cells.

After infection, retroviral DNA synthesis results in three types of unintegrated DNA: the linear double-stranded DNA in the cytoplasm and two circular species with a single LTR or two tandem LTRs, respectively, in the nucleus (Fig. 1), the former being most likely the immediate precursor of the integrated DNA [3]. The process of integration requires an integrase activity that is encoded by the 3′-terminal region of the POL gene [4]. To evaluate the need for integration in HIV-I replication and expression, we introduced point mutations into this region. Using the M13 oligonucleotide-directed mutagenesis system [5], we changed two bases in a 1.1 kb EcoRI/EcoRI fragment of the biologically active molecular clone pHXB2D (Fig. 2), which had been cloned into the bacteriophage M13 mp18. Synthesized mismatched oligonucleotides containing a premature stop codon and an additional recognition site for the enzyme BamHI were used to prime second-strand mutagenesis with M13 single-stranded DNA. Purified circular double-stranded DNA was isolated using the alkaline lysis method [6]. The replicative forms of M13 mutant clones were grown and the modified EcoRI/EcoRI inserts were cloned back into pHXB2D replacing the wild type EcoRI/EcoRI fragment.

The resultant pHXB2D/INT mutant clones were identified and confirmed by restriction enzyme digested with either KpnI, which yielded identical fragments for both the pHXB2D wild type and the INT mutant, and with BamHI, which gave an additional 3.7 kb fragment only with the INT mutant. Plasmids with the INT mutation were grown in HB101 cells and transfected into cos-I cells. After 62 to 72 h the virus was introduced into H9 cells by either cocultivation with the

National Institutes of Health, National Cancer Institute, Bethesda, USA

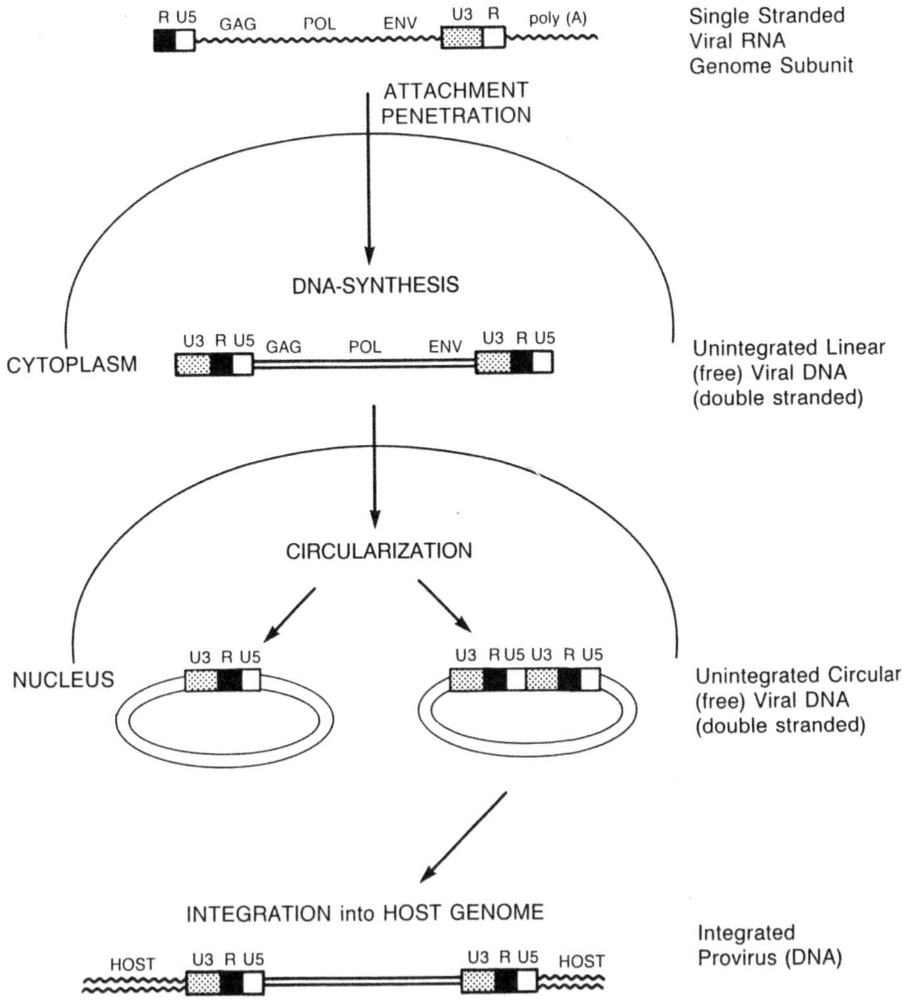

Fig. 1. Integration of retrovirus DNA

transfected cells or by cell-free infection with culture fluids of the same cos-I cell cultures. Transfection of the mutant into cos-I cells yielded media reverse transcriptase (RT) activity, although at a threefold lower level than the wild type HIV-I plasmid.

In radioimmunoprecipitation assays with ^{125}I-protein A, lysates of the transfected cos-I cells were immunoprecipitated with antisera from different HIV-I infected patients. For transfections with both the HIV-I wild type and the mutant virus, the assay resulted in the appearance of at least three bands in the gel at 24, 39 and 55 kDa, that were not found with controls with mock transfected cells or with sera of HIV-I negative patients, most likely representing the GAG protein p24, the GAG precursor protein p39, and the reverse transcriptase p55.

To evaluate the expression of viral GAG antigen by H9 cells that were infected by cocultivation with transfected cos-I cells or cell-free with the appropriate culture fluid as described above, we

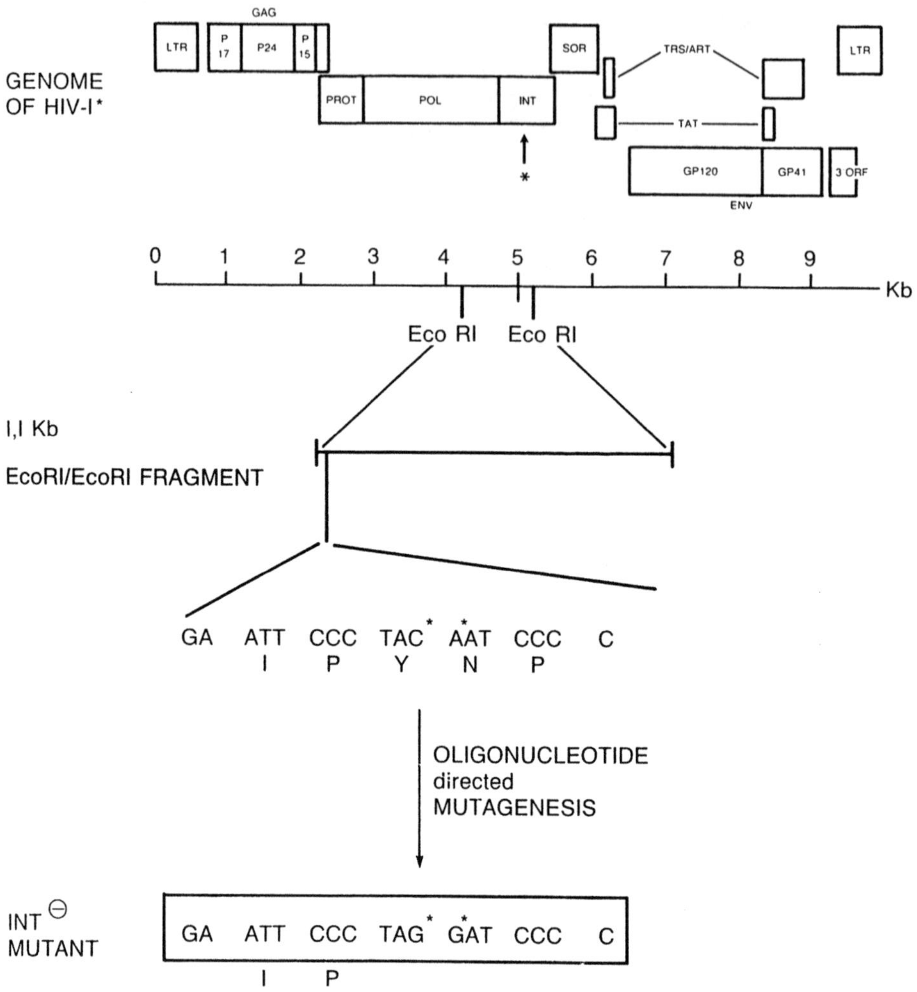

Fig. 2. Construction of INT mutants of HIV-I

used an immunofluorescence assay with anti-p24-antibodies. Both modes of infection resulted in 90% to 100% p24 expression of H9 cells infected with HIV-I wild type, while neither cell-free infection nor infection by cocultivation with the HIV-I/INT mutant showed p24 expression until day 30. Assays for RT activity of the same H9 cultures infected with the HIV-I wild type virus resulted in 78000 and 35000 cpm/ml culture fluid 2 weeks after infection by cocultivation with transfected cos-I cells or with the appropriate cell-free supernatants, respectively. In contrast to this, no RT activity was detected in culture fluids of H9 cells that were cocultivated with or cultured in cell-free supernatant of HIV-I/INT-mutant transfected cos-I cells for various time periods up to 4 weeks.

These results suggest that integration of proviral DNA is a necessary part of the productive infection of T cells. Further studies on expression of viral proteins and mRNA by this mutant are in progress. The requirement for integration for reproduction after infection of non-dividing susceptible cells such as monocyte/macrophages remains to be determined.

Acknowledgement. H. D. Buchow was supported by a grant from the H. and L. Schilling Foundation, F.R.G.

Reference

1. Varmus H, Swanstrom R (1984) Replication of retroviruses. In. Weiss R et al. (eds) RNA tumor viruses, 2nd ed. Cold Spring Harbor Lab, New York
2. Fisher AG et al. (1985) A molecular clone of HTLV-III with biological activity. Nature 316:262–265
3. Panganiban AT (1985) Retroviral DNA integration. Cell 42:5–6
4. Panganiban A, Temin H (1984) The retrovirus POL gene encodes a product required for DNA integration: identification of a retrovirus INT locus. Proc Natl Acad Sci USA 81:7885–7889
5. Zoller MJ, Smith M (1984) Oligonucleotide-directed mutagenesis: a simple method using two oligonucleotide primers and a single-stranded DNA template. DNA 3:479–488
6. Maniatis T, Fritsch EF, Sambrook J (1982) Molecular cloning. A laboratory manual. Cold Spring Harbor Lab, New York

Use of a HIV-1 Retroviral Vector System for Gene Transfer into Human Cells

V. Heisig, G. Jahn, M. Ebeling, and R. Laufs

The transfection of HIV 1 proviral DNA and 3'LTR-CAT plasmids by the DEAE dextran method or protoplast fusion was studied previously in H9 cells, from a T4-lymphocyte cell line. The transient expression of different HIV 1 LTR-CAT DNAs was reproducible for all experiments [1]. However, we failed to establish cell lines of nonreplicating HIV 1 proviruses into H9 cells to study viral genes by this method. Here we investigate the selection for a hygromycin-B gene transfected into H9 cells. Hygromycin B, an aminoglycoside antibiotic produced by *Streptomyces hygroscopicus,* was used to select transfected H9 cells.

Retroviral vectors have been constructed for gene transfer in mammalian and avian cells. However, they are restricted in host range. We describe a retroviral vector system based entirely on a human immunodeficiency virus (HIV 1) with the ability to carry out efficient gene transfer into human cells.

By establishing a helper cell line that produces the *trans*-acting viral gene products, we propagate the *cis*-acting components in them and harvest defective viral particles that contain only the *cis*-acting components. The packaging signal in HIV 1 has not been identified. We postulate that the sequence for the packaging signal is located between the primer binding site (PBS) and the gag genes. Therefore, we constructed deletion mutants in this region and transfected these mutants into H9 cells. The plasmid pHU3d containing the complete proviral genome of HIV 1 and a hygromycin-B gene for selection were used for construction. An additional deletion in the 3'LTR (–138 to –48) has been introduced into all plasmids. We established cell lines with the mutants and studied the reverse transcriptase activity and protein synthesis.

Reference

1. Heisig V, Benter T, Josephs SF, Sadaie MR, Okamoto T, Gallo RC, Wong-Staal F (1987) Interaction of viral and cellular factors with the HTLV-LTR target sequences in vitro. In: Neth R, Gallo RC, Greaves MF, Kabisch H (eds) Modern trends in human leukemia 7. Springer, Berlin Heidelberg New York, p 423

Inst. f. Med. Mikrobiologie und Immunologie, Universität Hamburg, Hamburg, Federal Republic of Germany

Multistage Mastocytoma Model Characterized by Autocrine IL-3 Production

C. Moroni, I. D. Diamantis, A. Wodnar-Filipowicz, and A. P. K. Nair

A. Introduction

Leukaemogenesis, as tumor formation in general, is a multistage process where alterations of proto-oncogenes play some yet unknown role. To understand this process, it is necessary to know how activated proto-oncogenes (i.e., oncogenes) affect and perturb cellular growth control mechanisms from extracellular growth factors down the signal transduction pathway to the nucleus. We and others have previously reported [1, 2] that one of the frequent proto-oncogene alterations in human leukemias and lymphomas are point mutations of the *ras* genes (predominantly N-*ras*), findings which have led us to ask how an activated *ras* gene might affect responsiveness to hemopoietic growth factors.

We have approached this problem by introducing the viral H-*ras* oncogene into an IL-3-dependent, nontumorigenic mouse mastocyte line (PB-3c) [3] and observed the generation of IL-3-secreting, autocrine mastocytomas following a long latency period [4]. In this report, we summarize our data on this multistage tumor system. In addition, we show that IL-3 gene expression can be induced in normal cells in vitro, and autocrine IL-3 production by the mastocytomas can be downregulated by cell fusion.

B. Results and Discussion

Introducing the v-H-*ras* gene via a retroviral *neo*-selectable retroviral vector into IL-3-dependent mouse mastocytes produced short-term and long-term effects. The immediate effect was a reduction of the IL-3 requirement by about 10- to 20-fold in all cell clones tested. The long-range effect was observed in vivo, where infected cells progressed to autocrine, IL-3-secreting mastocytomas. The salient features of this tumor system are summarized in Table 1, and we wish to emphasize the following points. While v-H-*ras*-expressing cells were able to grow at reduced IL-3 levels, this oncogene did not abrogate the requirement for IL-3, in contrast to effects of the *abl* and *myc* genes observed by other workers when analyzing IL-3-dependent cells [5, 6].

Table 1. Features of the v-H-*ras*-induced mastocytoma model

1. *Parental cell:* immortalized, IL-3-dependent and non-tumorigenic mast cell line (PB-3c). Diploid to near diploid [3]
2. *v-H-ras effect:* reduction, but not abrogation, of IL-3 dependence. Observed in all clones tested
3. Two clonable subpopulations within PB-3c: *"transformation-competent"* cells form mastocytomas following v-H-*ras* infection; *"transformation-non-competent"* cells do not
4. Long-latency *mastocytomas* derived in vivo from transformation of competent subclones by the v-H-*ras* gene. Long letancy suggests a progression step taking place in the animal
5. Tumors secrete IL-3, which forms part of an *autocrine loop* as shown by antibody inhibition of growth during cloning in methylcellulose

Institut für Medizinische Mikrobiologie, Petersplatz 10, CH-4003 Basel, Switzerland

In vivo, v-H-*ras*-expressing cells underwent an unknown progression step yielding autocrine, IL-3-secreting mastocytomas. It was of interest to find that only a subpopulation of PB-3 cells (3/20 clones) was transformable by v-H-*ras;* we call them "transformation competent." Non-competent cells produced no tumors following v-H-*ras*-expression, but they still displayed the altered IL-3 requirement described above. While we have no direct evidence, we assume that transformation-competent cells arose from transformation-non-competent ones, as the responsiveness to transforming steps is usually a function of the doubling times undergone by a cell, in other words, transformability increases with time.

The evidence that IL-3 forms part of an autocrine loop and that IL-3 production is a feature relevant to tumor formation is severalfold. All of over 50 mastocytomas produced IL-3, suggesting this trait was selected for. Furthermore, tumor cells were able to grow in culture in the absence of added IL-3; they had acquired growth autonomy. When cloned in methylcellulose in the absence of IL-3, high, but not low, cell numbers seeded were able to produce colonies, a feature characteristic of autocrine tumors. Most importantly, antibody to IL-3 was able to block this colony formation in vitro. Lastly, when IL-3 cDNA was introduced into these cells by a retroviral vector (kindly provided by W. Ostertag), growth autonomy and tumorigenicity occurred together (unpublished data).

An important point to resolve was the temporal relationship between v-H-*ras* and IL-3 expression. Rigorous cloning experiments ruled out that the tumors arose from a preexistent, IL-3-producing cell, as tumors were derived from infected, transformation-competent clones, which were IL-3 dependent prior to infection. Figure 1 shows an analysis of IL-3 expression in transformation-competent cells and tumors derived thereof. IL-3 mRNA is measured by an RNA protection assay. The protected fragment

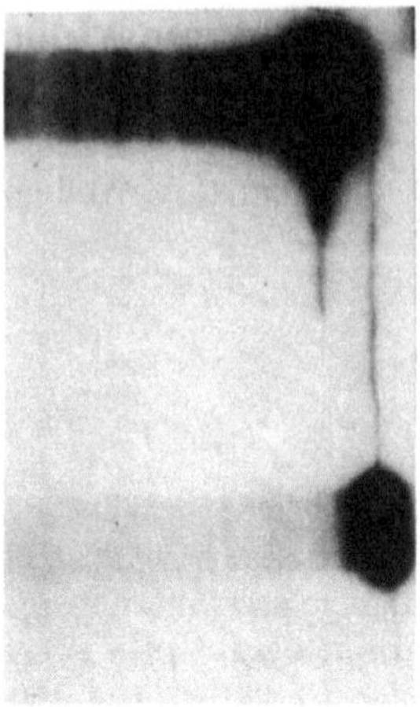

Fig. 1. Analysis of IL-3 mRNA by RNA protection assay. See text for explanation. The probe was a ^{32}P-labeled SP6 transcript of an IL-3 cDNA fragment. (Kindly provided by S. Gough)

(see WEHI-3B, an IL-3-producing line, used for positive control) was not detectable in the transformation-competent clone PB-3c-8, nor in its v-H-*ras*-expressing subclone (8V). IL-3 mRNA was detected in a tumor (8VD3) derived from 8V and in a second tumor analyzed (V4D6). These data show that the expression of IL-3 forms part of a progression step which has taken place in the animal.

Table 2 presents the temporal sequence of our present view by which normal bone marrow cells proceed to the mastocytoma stage. Essential steps are crisis (immortalization), acquisition of transformation competence, v-H-*ras* expression, and the last and important step establishing an autocrine loop involving IL-3. The molecular basis of immortalization, transformation competence and of inducing IL-3 expression are not known at this time.

As IL-3 gene expression becomes activated during mastocytoma formation, we wondered whether IL-3 in PB-3c cells is an inducible gene and tested various potential inducers. A Northern blot analysis (Fig. 2) shows that treatment of these cells with 5 μM calcium ionophore in-

Table 2. Stages of mastocytoma development

Step	Stage	Characteristics
	Bone marrow mast cells	Grow in vitro with IL-3
1. Crisis ⟶	↓	
	Immortalized PB-3c line	IL-3 dependent, not transformable by v-H-*ras*
2. Acquisition of transformation competence ⟶	↓	
	Immortalized transformation competent line	IL-3 dependent transformable by v-H-*ras*
3. v-H-*ras* ⟶	↓	
	Premalignant line	Reduced IL-3 dependency, will progress in vivo
4. Unknown in vivo step ⟶	↓	
	Mastocytoma	Growth autonomy in vitro, secretes IL-3 with autocrine stimulation

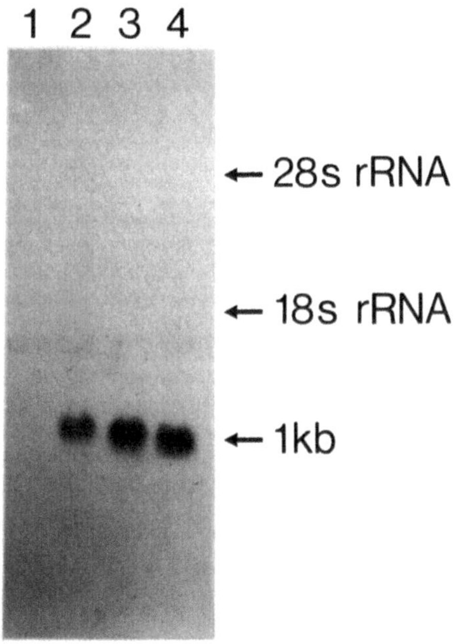

Fig. 2. Induction of IL-3 mRNA by Ca ionophore. *1* control, untreated PB3-c cells; *2* PB3 cells + Ca ionophore 5 μ*M*, 1 h; *3* PB3 cells + Ca ionophore 5 μ*M*, 2 h; *4* PB3 cells + Ca ionophore 5 μ*M*, 4 h. Each lane of the 1.1% agarose-formaldehyde gel contained 20 μg total RNA. The probe was as in Fig. 1

duces IL-3; peak values were seen after 2 h. These data show that the IL-3 locus in PB-3c is subject to regulation, and we are in a position now to analyze the regulating components. It will be interesting to see whether the same induction mechanisms are involved in Ca ionophore induction as in the tumor system.

We turned to cell fusion to study the mechanism by which IL-3 production had become activated in the mastocytomas. The rationale was that analysis of hybrids between IL-3-dependent PB-3c cells and autonomous mastocytomas should indicate whether activation involved a dominant or recessive mechanism. In the case of a dominant mechanism, hybrids should grow autonomously; in the case of a recessive mechanism, hybrids would require IL-3 for growth. A hypoxanthin, aminopterin, thymidin (HAT)-sensitive mastocytoma variant was selected with thioguanine and fused to PB-3c, which are sensitive to G418. (Tumors are resistant to this drug as they carry *neo* from the retroviral vector.) Hybrids were selected in the presence of IL-3 and then tested for growth in the absence of IL-3. Proliferation of sev-

A

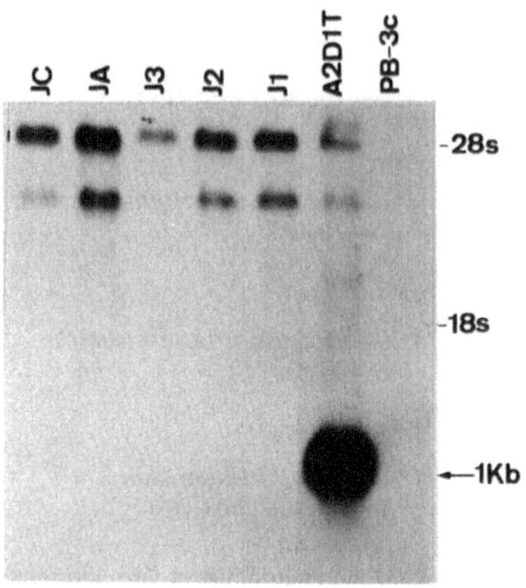

B

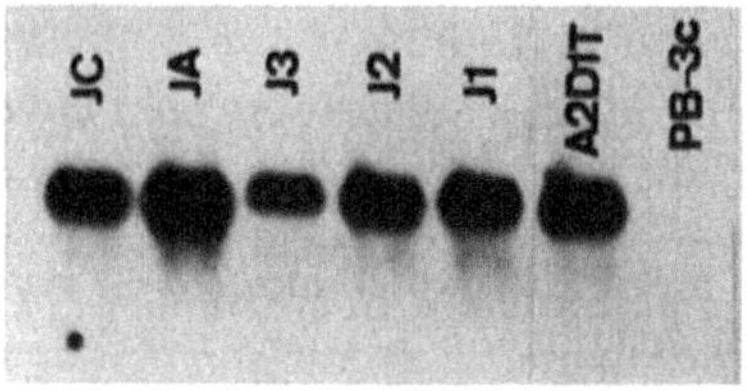

Fig. 3 A, B. Northern blot analysis of somatic cell hybrids. *Lanes* are explained in the text. The IL-3 probe **(A)** was the same as in Figs. 1 and 2. The v-H-*ras* probe **(B)** was a ^{32}P-labeled T7 transcript corresponding to 700 bp v-H-*ras* gene

eral independent hybrid cultures tested was clearly IL-3 dependent, indicating that growth autonomy in the tumors had occurred by a recessive mechanism. We next analyzed the levels of IL-3 mRNA in the hybrids by Northern analysis (Fig. 3 A). The parental tumor A2D1T revealed the expected band of about 1 kb, while no transcript was detectable in PB-3c cells. In five hybrid cultures (JA, JB, J1, J2, J3) IL-3 mRNA was undetectable (Fig. 3 A). When the same samples were analyzed using a v-H-*ras*-specific probe, tumor and hybrids expressed v-H-*ras* (Fig. 3 b). Taken together, these data indicate that cell fusion leads to downmodulation of IL-3 expression by a mechanism not involving downregulation of v-H-*ras* gene expression. This may reflect the activity of a suppressor gene in PB-3c cells, exerting a negative effect on IL-3 gene expression, and lack of this gene function in the tumor. Analysis of this system should not only provide insight into the mechanism by which IL-3 production becomes activated during tumor formation, but also into the mechanism of tumor suppression.

References

1. Senn H-P, Tran-Thang C, Wodnar-Filipowicz A, Jiricny J, Fopp M, Gratwhol A, Signer E, Weber W, Moroni C (1988) Mutation analysis of the N-*ras* proto-oncogene in active and remission phase of human acute leukemias. Int J Cancer 41:59–64
2. Bos JL, Toksoz D, Marshall CJ, Verlaan-de-Vries M, Veeneman GH, van der Eb AJ, van Boom JH, Janssen JWG, Steemvoorden ACM (1985) Amino-acid substitutions at codon 13 of the N-*ras* oncogene in human acute myeloid leukemia. Nature (London) 315:726–730
3. Ball PE, Conroy MC, Heusser CH, Davis JM, Conscience J-F (1983) Spontaneous, in vitro, malignant transformation of a basophil/mast cell line. Differentiation 24:74–78
4. Nair A, Diamantis ID, Conscience JF, Kindler V, Hofer P, Moroni C (1989) A v-H-*ras* dependent hemopoietic tumor model involving progression from a clonal stage of transformation competence to autocrine interleukin-3 production. Mol Cell Biol 9:1183–1190
5. Pierce JH, Di Fiore PP, Aaronson SA, Potter M, Pumphrey J, Scott A, Ihle JN (1985) Neoplastic transformation of mast cell by Abelson-MuLV: abrogation of IL-3 dependence by a nonautocrine mechanism. Cell 41:685–693
6. Rapp UR, Cleveland JL, Brightman K, Scott A, Ihle JN (1985) Abrogation of IL-3 and IL-2 dependence by recombinant murine retroviruses expressing v-*myc* oncogenes. Nature (London) 317:434–438

Purified Bovine NF-κB Recognizes Regulatory Sequences in Multiple Genes Expressed During Activation of T- and B-Lymphocytes

M. J. Lenardo, A. Kuang, A. Gifford, and D. Baltimore

A. Introduction

A crucial event in the differentiation of B-lymphocytes is the transcription of the immunoglobulin light-chain gene which leads to expression of immunoglobulin antigen receptor on the surface of the cell. In an apparently separately regulated arm of the immune response, antigenic stimulation causes proliferation of T-lymphocytes by transcriptional activation of the IL-2 gene and the IL-2 receptor gene. Previous studies have shown that a lymphoid-specific enhancer element plays an important role in achieving high-level transcription of the kappa light-chain gene [1–4]. Molecular genetic dissection of this enhancer has revealed that a DNA sequence which binds to a nuclear factor, NF-κB, is essential for its function [5–8]. NF-κB binding activity is constitutively present only in mature B-lymphocytes and exceptional T-lymphocyte lines [5, 13]. Its binding may be induced in cells early in the B-lymphoid lineage, T-lymphocytes, and in nonlymphoid cells by various treatments such as bacterial lipopolysaccharide, cycloheximide, lectins, and phorbol esters [9]. Previously we have shown that NF-κB binding is critical for the kappa enhancer activity that is constitutively present in mature B-lymphocytes and inducible by lipopolysaccharide or phorbol esters in pre-B cells [7]. Recently, NF-κB has been found to act through an enhancer element in the LTR of human immunodeficiency virus I [10–12]. Evidence has also accumulated that NF-κB is important for the expression of the IL-2 receptor α chain gene (Tac antigen) [16, 19]. We have purified a protein which corresponds to this binding activity and describe its molecular characteristics and binding specificities. Our results suggest that a single protein is sufficient to recognize regulatory sequences in multiple genes that are expressed during the activation of B- and T-lymphocytes.

Whitehead Institute for Biomedical Research, Cambridge, MA 02142 and Department of Biology, Massachusetts Institute of Technology, Cambridge, MA 02139, USA

B. Purification of NF-κB from Bovine Spleen Tissue

To identify a source from which a large quantity of binding activity could be purified, numerous cultured cell lines and various tissues of human, murine, or bovine origin were tested. High levels of NF-κB binding activity were found in spleen tissue of human, murine, or bovine origin which corresponds to its constitutive expression in mature B-lymphocytes. The bovine extracts produced specific complexes with NF-κB binding sites from the kappa enhancer that had a methylation interference pattern identical to that described for the murine binding activity [5, 13]. Because of its ready availability, NF-κB was purified from bovine spleen.

We designed a series of chromatographic steps including S300-Sephacryl, phosphocellulose, hydroxylapatite, *Escherichia coli* DNA Sepharose, and DNA site-specific chromatography to purify the activity over 50000-fold with an approximately 8% recovery. This procedure yielded a 42000-dalton protein spe-

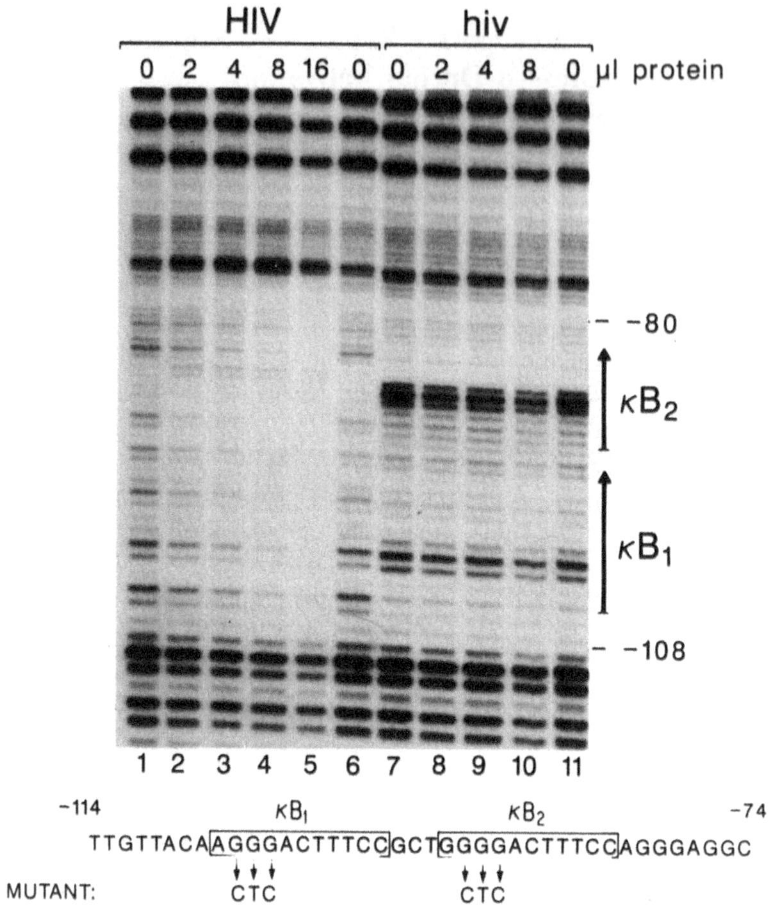

Fig. 1. Binding of purified bovine NF-κB to the HIV-1 enhancer. *Top:* Shown is the DNAse-I cleavage pattern of either the wild type (HIV) or the mutated (hiv) enhancer region following binding with the indicated amounts of affinity-purified bovine protein. The positions of the NF-κB binding motifs and the extent of the footprint (−80 to −108) are indicated. *Bottom:* The DNA sequence of the region containing the enhancer is given. The NF-κB binding motifs are boxed. Base substitutions which abrogate binding in the mutant version are indicated by arrows [10]

cies on an SDS polyacrylamide gel from which binding activity could be recovered by a denaturation-renaturation protocol [13]. Glycerol gradient sedimentation indicated that the protein may exist in solution as a dimer.

I. The Same Polypeptide Recognizes the Kappa Light Chain Gene and Human Immunodeficiency Virus Enhancers

The bovine protein was purified from spleen tissue and is likely to be derived predominantly from B-lymphocytes. In addition, the purified protein was affinity selected using a column containing kappa enhancer sequences and identified by its interaction with the kappa enhancer as NF-κB. Activation of T-lymphocytes greatly increases the levels of a factor that has similar DNA binding specificity to NF-κB [10]. This binding activity has been implicated in HIV-1 transcription through an enhancer comprised of two sequences related to the kappa enhancer binding motif [10–12]

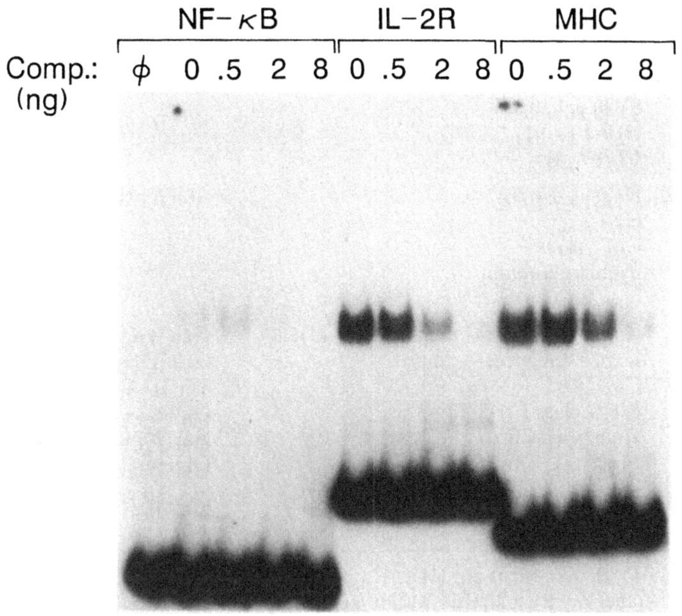

Fig. 2. Competition binding analysis of three NF-κB recognition sites. Shown is a mobility shift electrophoresis assay in which purified bovine protein has been added to all lanes except the first. As indicated at the top, radioactively labeled probes were: lanes 1-5 – kappa enhancer binding fragment; lanes 6-9 – IL-2 receptor α promoter fragment (–298 to –228); and lanes 10-13 – MHC class-I promoter oligonucleotide [15]. Oligonucleotides containing the NF-κB binding motif from the kappa enhancer were added to the binding assays as unlabeled competitor in the nanogram amounts shown. The probe for lanes 1–5 had a specific activity approximately four times lower than the others used

(see Fig. 2). We therefore asked if the same purified polypeptide could interact with the HIV sequence in the absence of any other factors present in T cells. By DNase-I footprint analysis the entire long-terminal repeat enhancer sequence is protected in a pattern virtually identical to that observed with T cell extracts (Fig. 1) [11, 12]. These results support the hypothesis that the same protein may regulate both the HIV and kappa enhancers in T- and B-lymphocytes respectively.

II. The Purified Protein Recognizes a Regulatory Sequence in the IL-2 Receptor α Chain gene

Recent evidence also implicates NF-κB in the expression of the IL-2 receptor α chain (IL-2Rα) in activated T-lymphocytes ([16–19] S. Cross, M. Lenardo, D. Baltimore, and W. Leonard, unpublished observations). Since NF-κB is typically found in activated but not resting T cells, it might have an important role in the program of genes expressed during antigenic stimulation. A sequence closely resembling the NF-κB cognate motif was found in a functionally important region of the IL-2Rα promoter [16]. We found that this sequence was able to bind the purified protein, and that the kappa enhancer site cross-competed equivalently with either itself, the MHC class-I promoter binding site, or the IL-2Rα binding site (Fig. 2).

III. Variation in the DNA Sequences Required for Binding NF-κB

It is now clear that a variety of gene regulatory regions appear to interact pro-

Gene	Sequence
A. Kappa enhancer – mouse SV40 enhancer HIV-1 (−91) CMV[a] (4)[b]	GGGGACTTTCC
B. HIV-1 (−105) HIV-2 CMV (1)[a] β_2-microglobulin	AGGGACTTTCC
C. Variants:	
Kappa enhancer – human CMV (3)[a]	GGGGGATTTCC
Kappa enhancer – rabbit[a]	GCGGGGTTTCC
CMV (2)[a]	TGGGACTTTCC
MHC class I (H2-TF1)	GGGGATTCCCC
Human IL-2Rα	GGGGAATCTCC
Mouse IL-2Rα	GGGGGATTCCT
Consensus:	GGGRNTYCC (T)

Table 1. Consensus sequence for NF-κB recognition

SV40, Simian virus 40; HIV, Human immunodeficiency virus; CMV, cytomegalovirus; MHC, Major histocompatibility complex; IL-2Rα, Interleukin-2 receptor alpha chain.

[a] This sequence has not been tested in a binding assay.

[b] Since there are four putative NF-κB recognition sites in the cytomegalovirus enhancer, these have been numbered 1–4 as they are found from 5′ to 3′ on the coding strand.

ductively with NF-κB. Alignment of the interaction sites reveals significant microheterogeneity in their DNA sequence (see Table 1). This establishes a surprisingly flexible consensus interaction site, derivatives of which can be expected to bind tightly to NF-κB. The essential features of this consensus site are:

1. The motif is composed to two four-nucleotide half-sites separated by a position that can be occupied by any nucleotide. The 5′ half-site is composed of purines and the 3′ half-site is composed of pyrimidines.
2. The first three nucleotides are G residues and the final two nucleotides are C residues. These G:C base pairs at the beginning and end of the motif make important contacts with the protein which can be interrupted by N-7 methylation in all cases examined.
3. The pyrimidine half-site always takes the form of one or two T residues preceding one to three C residues.

These rules can only provide a basis for recognizing putative interaction sites and do not offer insights into how the DNA-protein interactions come about. From this consensus sequence we have predicted putative NF-κB binding sites in the IL-2 gene promoter in both human and murine genes. Preliminary results indicate that these sequences efficiently bind the pruified protein, suggesting a pivotal role for NF-κB in the expansion of T cells during the immune response [13].

C. Summary

To characterize the NF-κB binding factor in molecular terms and to facilitate the cloning of its gene, we have purified this protein from bovine spleen tissue. We have found it is a 42000-dalton protein that exists in solution as a dimer. We were able to use the purified protein to show that the same polypeptide is able to recognize sites important for activation of genes in either B- or T-lymphocytes. Moreover, we were able to define a consensus sequence which allows ascertainment of a wider variety of sequences that are capable of interacting with this protein. The implication of the same protein in gene regulation in two different lineages of lymphoid cells reveals an unexpected unity in the mechanism of gene expression during B- and T-lymphocyte activation. This also suggests that other regulatory events must participate with NF-κB activation in determining B- or T-cell-specific expression.

Acknowledgements. The authors thank Louis Staudt, Kees Murre, Patrick Baeuerle, and Patrick McCaw for their assistance with parts of the purification. We also thank Sharon Cross and Warren Leonard for probes and for making unpublished results available. M.J.L. was supported by a Clinical Investigator Award of the National Cancer Institute and the work was funded by an N.I.H. grant to D.B.

References

1. Queen C, Baltimore D (1983) Immunoglobulin gene transcription is activated by downstream sequence elements. Cell 33:741–748
2. Queen C, Stafford J (1984) Fine mapping of an immunoglobulin gene activator. Mol Cell Biol 4:1042–1049
3. Picard D, Schaffner W (1984) A lymphocyte-specific enhancer in the mouse immunoglobulin kappa gene. Nature 307:80–82
4. Bergman Y, Rice D, Grosschedl R, Baltimore D (1984) Two regulatory elements for immunoglobulin κ light-chain gene expression. Proc Natl Acad Sci USA 81:7041–7045
5. Sen R, Baltimore D (1986) Multiple nuclear factors interact with immunoglobulin enhancer sequences. Cell 46:705–716
6. Atchison M, Perry R (1987) The role of the κ enhancer and the binding factor NF-κB in the developmental regulation of κ gene transcription. Cell 48:121–128
7. Lenardo M, Pierce JW, Baltimore D (1987) Protein-binding sites in Ig gene enhancers determine transcriptional activity and inducibility. Science 236:1573–1577
8. Pierce JW, Lenardo M, Baltimore D (1988) Oligonucleotide that binds nuclear factor NF-κB acts as a lymphoid-specific and inducible enhancer element. Proc Natl Acad Sci USA 85:1482–1486
9. Sen R, Baltimore D (1986) Inducibility of κ immunoglobulin enhancer-binding protein NF-κB by a post-translational mechanism. Cell 47:921–928
10. Nabel G, Baltimore D (1987) An inducible transcription factor activates expression of human immunodeficiency virus in T cells. Nature 326:711–713
11. Wu F, Garcia J, Mitsuyasu R, Gaynor R (1988) Alterations in binding characteristics of the human immunodeficiency virus enhancer factor. J Virol 62:218–225
12. Dinter H, Chiu R, Imagawa M, Karin M, Jones K (1987) In vitro activation of the HIV-1 enhancer in extracts from cells treated with a phorbol ester tumor promoter. EMBO J 6:4067–4071
13. Lenardo M, Kuang A, Gifford A, Baltimore D (1989) NF-κB protein purification from bovine spleen: Nucleotide stimulation and binding site specificity. Proc Natl Acad Sci USA 85:8825–8829
14. Baldwin AS, Sharp PA (1988) Two transcription factors, NF-κB and H2-TF1, interact with a single regulatory sequence in the class-I major histocompatibility complex promoter. Proc Natl Acad Sci USA 85:723–727
15. Singh H, LeBowitz J, Baldwin AS, Sharp PA (1988) Molecular cloning of an enhancer binding protein: isolation by screening of an expression library with a recognition site DNA. Cell 52:415–423
16. Cross SL, Feinberg MB, Wolf JB, Holbrook NJ, Wong-Staal F, Leonard WJ (1988) Regulation of the human interleukin-2 receptor α chain promoter: activation of a nonfunctional promoter by the transactivator gene of HTLV-I. Cell 49:47
17. Böhnlein E, Lowenthal JW, Siekevitz M, Ballard D, Franza BR, Greene WC (1988) The same inducible nuclear proteins regulate mitogen activation of both the interleukin-2 receptor-alpha gene and type-1 HIV. Cell 53:827–836
18. Leung K, Nabel G (1988) HTLV-1 transactivator induces interleukin-2 receptor expression through an NF-κB-like factor. Nature 333:776–778
19. Cross S, Halden N, Lenardo M, Leonardo W (1989) Functionally distinct NF-κB binding sites in the immunoglobulin kappa and IL-2 receptor α chain gene. Science

Hematopoietic Growth Control by the T-Cell CD2 Determinant is Exerted at a Pretranslational Level*

S. Burdach[1,2], N. Zessack[1], and L. Levitt[2]

A. Introduction

Cellular growth control takes a delicate balance between stimulatory and inhibitory signals. Much previous research has focused on the identification of genes which encode for stimulatory signals. Genes that encode for inhibitory signals may be more difficult to analyze [1]. T-cell gene products are capable of inhibition of hematopoiesis in vitro and possibly in vivo [2–6]. The mechanisms regulating the inhibitory hematopoietic T-cell program are not well understood.

We have previously shown that triggering the T-cell antigen receptor associated epitope CD3 induces the p55 chain of the interleukin-2 (IL2) receptor on bone marrow T cells and renders marrow T cells responsive to low concentrations of IL2 [7]. In addition, we have demonstrated that IL2 inhibits the growth of marrow early erythroid progenitor cells (BFU-E) in the presence of IL2 receptor-positive T cells, and that interferon-γ (IF-γ) is an obligatory mediator of IL2-induced inhibition of BFU-E [4, 7]. We have also described a receptor-specific inhibition of myelopoiesis by IL2 which is mediated only in part by IF-γ [8]. Taken together, we have demonstrated a model for molecular regulation of hematopoiesis governed by an array of humoral and cellular signals, termed the lymphokine cascade. We have now examined the role of the early T-cell antigen CD2 in control of hematopoiesis by this lymphokine cascade.

CD2 has been identified for a long time as the receptor mediating sheep erythrocyte binding to T cells [9]. Later studies revealed that CD2 can serve as a receptor for a non-antigen-restricted pathway of T cell activation [10]. Lymphocyte function antigen 3 (LFA-3) has been identified as a natural ligand for CD2 [11, 12] and may induce T-cell activation in conjunction with additional activation signals [13]. LFA-3 is present on various cell types, including T-cells and mature red blood cells [14]. CD2-blocking monoclonal antibodies have been shown to inhibit binding of purified LFA-3 to CD2 [10–16]. Recent data suggest that interactions between CD2 and the antigen receptor may be essential for T-cell activation [17–19]. We utilized the CD2 antibody Leu 5b, which blocks a binding site for LFA-3 to examine the role of CD2 in IL2-induced inhibition of hematopoietic progenitors.

B. Induction of IL2 Receptors by CD2

IL2 receptors (p55) were induced on peripheral blood or marrow T cells via triggering of the antigen receptor associated CD3 epitope, as previously described [4, 8]. In brief, T cells were preincubated with CD2-blocking antibody before activation with CD3 antibody and subsequently cultured for 3–6 days in the presence of IL2. Antibody incubations were performed with T-cell pellets to facilitate interaction between LFA-3 and CD2.

---

[1] Department of Pediatric Hematology/Oncology, Medizinische Einrichtungen der Universität Düsseldorf, Düsseldorf, FRG
[2] and Division of Hematology, Stanford University Medical Center, Stanford, CA, USA
* This work was supported by DFG grant Bu 578/2-1, Elterninitiative Kinderkrebsklinik e.V. and NIH grant 1R01-HL35774

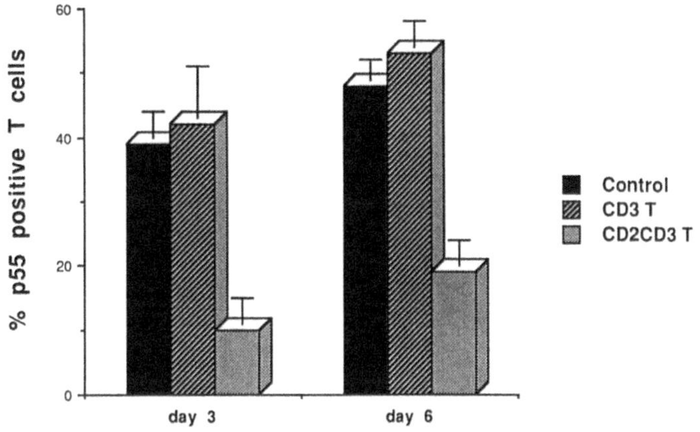

Fig. 1. CD2 blockade inhibits p55 IL2 receptor expression. *CD3 T*, CD3-triggered T cells; *CD2CD3 T*, CD2 blockade

42% of the CD3-triggered T cells expressed p55 at day 3, and 53% expressed p55 at day 6 (Fig. 1). CD2 blockade caused a 75% inhibition of p55 expression at day 3 as compared to preincubation with isotype control CD5 antibody and a 65% inhibition of p55 expression at day 6. Preincubation with isotype control CD5 antibody did not affect CD3-mediated p55 expression. CD2 blockade had no effect on binding of either triggering antibody to CD3 or IL2 receptor antibody to p55.

Next we asked whether regulation of p55 IL2 receptor expression by CD2 is associated with regulation of p55 gene expression. RNA was extracted by phenol extraction in the presence of vanadyl ribonucleotides [21] or by a cesium chloride/guanidium isothiocyanate gradient [22] from immunopurified T cells or monocyte-depleted mononuclear cells (>90% T cells by three-stage indirect immunofluorescence with CD5 antibody). Following gel electrophoresis T-cell RNA was subjected to a modified Northern transfer employing Nylon membranes. Highly sensitive single-stranded probes were constructed utilizing a cDNA which recognizes the 5′ untranslated region and the first exon of p55 (obtained from G. Crabtree, Stanford, California, USA) [23]. Oligonucleotides were utilized as random primers in the presence of DNA polymerase, and the probe was subsequently hybridized to T-cell RNA [45]. The screen was exposed for 48 h.

Figure 2 depicts the results following 16 h of culture. The first lane from the left represents the negative control: RNA extracted from CD3-non-triggered T cells cultured in the absence of IL2. Only minute amounts of p55 3.5-kb mRNA

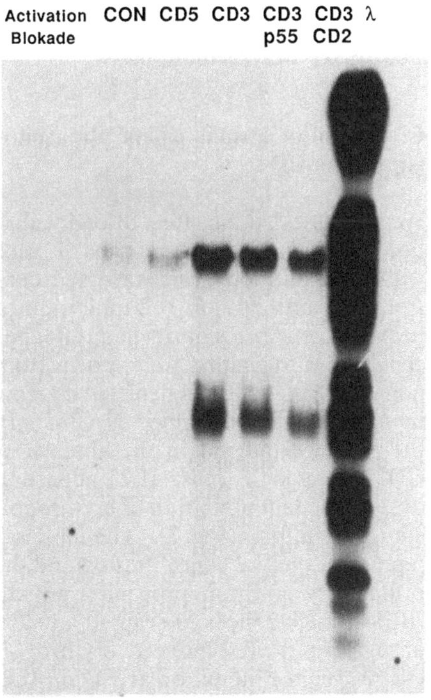

Fig. 2. CD2 blockade down-regulates p55 mRNA

are observed, and a 1.5-kb message is barely detectable. The second lane represents RNA from non-triggered T cells cultured in the presence of 10^2 U/ml: the presence of IL2 did not induce any substantial increase either of 3.5- or of 1.5-kb p55 mRNA. The third lane represents RNA extracted from CD3-triggered T-cells cultured in the presence of IL2: a strong p55 signal is detected. The fourth lane represents RNA from CD3-triggered T cells cultured in the presence of both IL2 (10^2 U/ml) and p55-blocking antibody. The fifth lane represents RNA from CD3-triggered T cells, which were preincubated with CD2-blocking antibody and cultured in the presence of 10^2 U/ml IL2. A definitive decrease of 1.5-kb p55 message is observed. All T cell samples except the CD2-blocked sample were preincubated with an isotype control antibody to rule out possible Fc-mediated effects. The last lane represents the size markers (λ DNA digested with Eco RI/Hind III). Thus, induction of p55 surface expression and accumulation of p55 mRNA are both dependent on the presence of free CD2 determinants.

C. Control of Hematopoietic Progenitor Growth by CD2

We next assessed the effect of preincubation of CD3-triggered marrow T cells with CD2 antibody versus isotype control CD5 antibody on IL2 inhibition of hematopoietic progenitors in autologous coculture. Progenitors were grown from nonadherent and T-cell depleted marrow mononuclear cells (termed NAB-T). In the presence but not in the absence of CD3-triggered T cells, IL2 induced a dose-dependent inhibition of erythropoietic progenitors (Fig. 3). The abscissa represents the IL2 concentration and the ordinate the erythroid progenitor growth (BFU-E). CD2 blockade caused an 87% abrogation of IL2-induced erythropoietic progenitor inhibition at 1 U/ml IL2, a 65% abrogation at 10 U/ml IL2, and a 55% abrogation at 100 U/ml IL2. CD2 blockade had no independent effect on erythropoietic progenitor growth in the absence of IL2 or in the presence of non-CD3-triggered T cells. IL2-induced CD3-triggered T cell mediated inhibition of erythropoietic progenitors was not affected by preincubation of CD3-triggered T-cells with control CD5 antibody.

In contrast to the abrogation of erythropoietic inhibition, CD2 blockade did not abrogate IL2-induced inhibition of monocyte/macrophage progenitors (Fig. 4). Likewise, CD2 blockade did not affect IL2-induced inhibition of total myeloid progenitors (Fig. 4, insert).

Next we assessed whether CD2 blockade modulates IL2-induced release of hematopoietic inhibitors from CD3-triggered T cells. Day 3 supernatants from marrow CD2-non-blocked, CD3-triggered T cells or CD2-blocked, CD3-triggered T cells (Fig. 5) were assessed against nonadherent, T-depleted marrow target cells (NAB-T). All supernatants were established in the presence of IL2. CD3-triggered marrow T-cell supernatants caused a 79% inhibition of erythropoietic progenitors. Blockade of CD2 caused an almost complete abrogation of CD3-triggered T-cell mediated inhibition of erythropoietic progenitors. Preincubation with isotype control antibody had no effect. In contrast, CD2 blockade did not reconstitute growth of monocyte/macrophage progenitors inhibited by IL2.

D. Regulation of Lymphokine Production by CD2

CD2 blockade reduced IF-γ release from CD3-triggered marrow T cells by 81% at day 3 and by 72% at day 6 of culture (Fig. 6). Similar results were obtained with peripheral blood T cells (data not shown).

We then asked whether regulation of IF-γ release by CD2 is associated with regulation of IF-γ gene expression. Total RNA was extracted from immunopurified T cells or monocyte-depleted

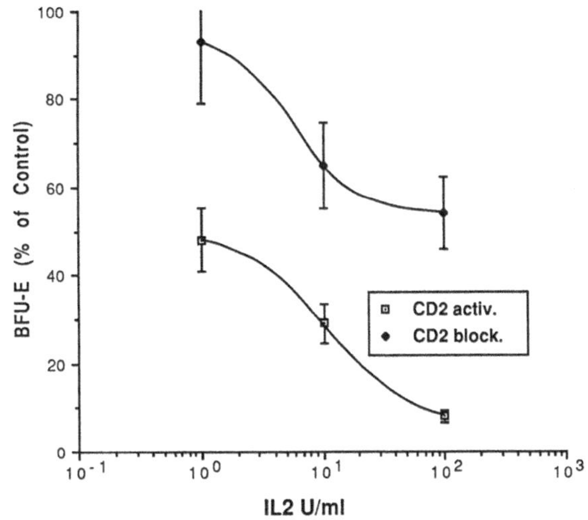

Fig. 3. CD2-mediated control of erythropoiesis

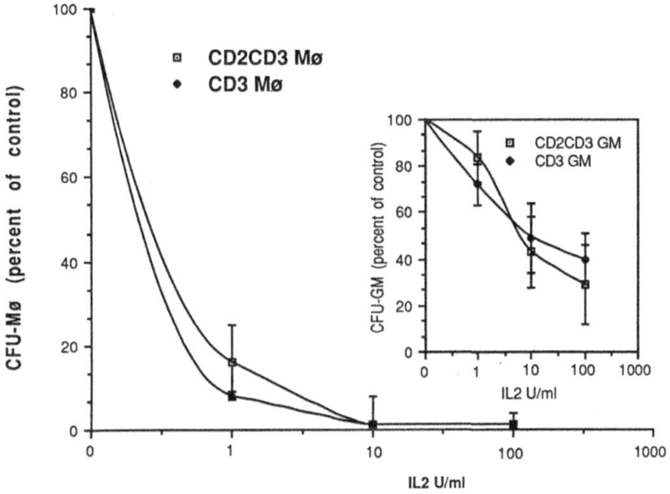

Fig. 4. CD2 blockade does not affect inhibition of colony-forming units – macrophages (*MΦ*). *Insert,* CD2 blockade does not affect inhibition of colony-forming units – granulocyte-macrophages (*GM*) *CD2CD3,* CD2 blockade

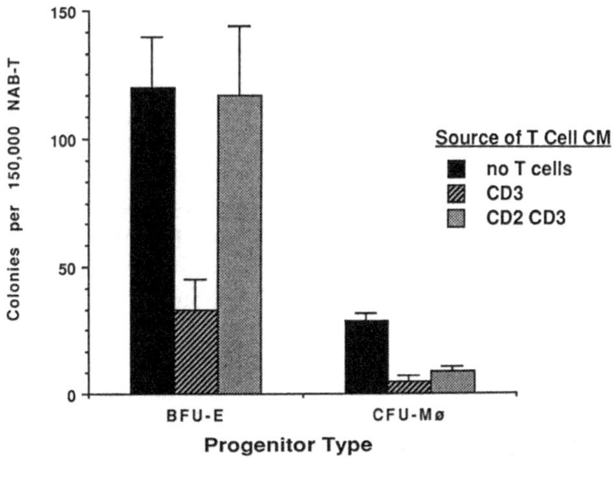

Fig. 5. Differential effect of CD2 blockade on humoral progenitor inhibition

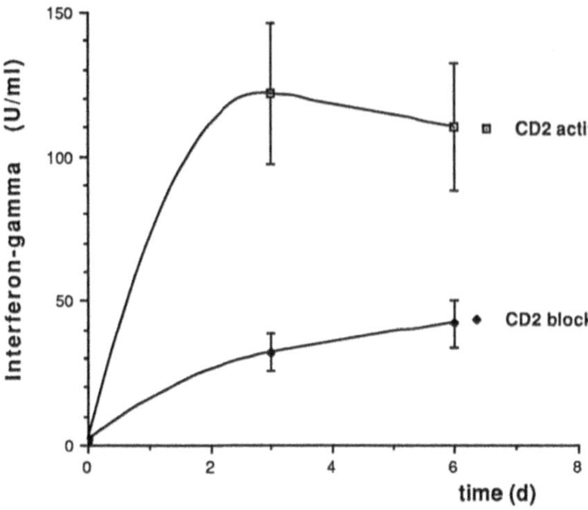

Fig. 6. Abrogation of IF-γ release by blockade of CD2

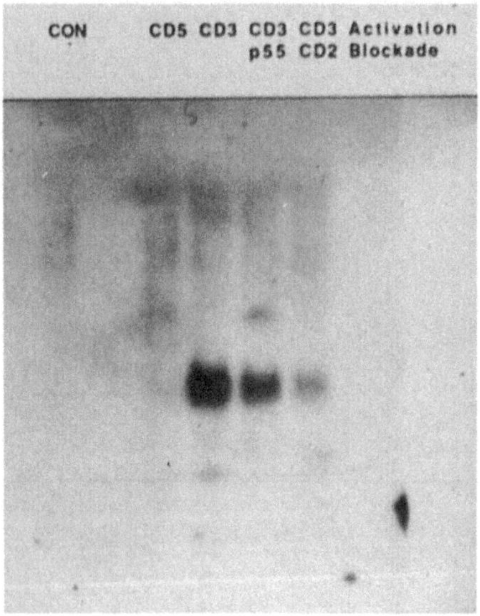

Fig. 7. CD2 blockade down-regulates IF-γ mRNA

mononuclear cells by the cesium chloride/guanidium isothiocyanate method [22]. Following gel electrophoresis of equal amounts, T-cell RNA was subjected to a modified Northern transfer utilizing Nylon membranes. Probes were constructed from a full-length cDNA for IF-γ [25] utilizing oligonucleotides as random primers in the presence of Klenow DNA polymerase. The IF-γ cDNA probes were hybridized to T-cell RNA.

Figure 7 depicts the results of a representative experiment. The first lane on the far left represents the negative control: RNA extracted from T cells before the begin of cultures. The second lane demonstrates that IL2 in the absence of high affinity IL2 receptors does not induce IF-γ mRNA. In contrast, IL2 in the presence of high-affinity IL2 receptors induces a strong signal for IF-γ mRNA (third lane). The fourth lane demonstrates that blocking of the p55 chain of IL2 receptor-positive T cells partially abrogates the IL2-induced increase in IF-γ mRNA. The suboptimal p55-blocking antibody concentration utilized in this study abrogates only about 50% of IL2-induced T-cell proliferation. Of interest, CD2 blockade prior to triggering of CD3 also abrogates the IL2-induced increase of IF-γ mRNA, as indicated in the fifth lane. Thus abrogation of IF-γ protein release by blockade of CD2 is preceded by an abrogation of IF-γ mRNA. This data suggests that CD2-mediated IF-γ production is regulated at a pretranslational level.

E. Conclusion

We conclude that blockade of the T-cell CD2 receptor induces down-modulation of (a) T-cell p55 IL2 receptor mRNA accumulation and membrane receptor expression, (b) IL2-induced inhibition of erythroid but not myeloid progenitors, and (c) IL2-induced marrow and peripheral blood T-cell IF-γ protein release and IF-γ mRNA accumulation.

This study indicates that T-cell erythropoietic immunoregulation is not confined solely to antigen-restricted T-cell activation but also involves an antigen-independent pathway of T-cell activation. These results demonstrate that the alternate receptor not only serves to promote T-cell proliferation and amplification of the immune response against non-self but also participates in the activation of an immunoregulatory T-cell program. The data also indicate that blockade of CD2 down-regulates the whole sequence of inhibitory signals induced by IL2 and provided by the lymphokine cascade for the erythropoietic progenitor cell. On the other hand, failure of CD2 blockade to abrogate inhibition of myelopoiesis indicates that specific regulatory T-cell programs can be triggered via CD3 independently of CD2. CD2 thus participates in hematopoietic differential regulation by the lymphokine cascade.

Acknowledgements. The authors wish to acknowledge the expert technical assistance of Kerstin Bachtenkirch and Marge Shatsky.

References

1. Marx J (1988) Report on the UCLA-Symposium: growth inhibitory and cytotoxic polypeptides. Science 239:975
2. Mamus SW, Beck-Schroeder S, Zanjani ED (1985) Suppression of normal human erythropoiesis by gamma interferon in vitro. J Clin Invest 75:1496
3. Broxmeyer HE, Williams DE, Lu L, Cooper S, Anderson SL, Beyer GS, Hoffman R, Rubin BY (1986) The suppressive influence of human tumor necrosis factors on bone marrow hematopoietic progenitor cells from normal donors and patients with leukemia: synergism of TNF and interferon-gamma. J Immunol 136:4487
4. Burdach S, Levitt LJ (1987) Receptor-specific inhibition of bone marrow erythropoiesis by recombinant DNA-derived interleukin-2. Blood 69:1368–1375
5. Zoumbos NC, Gascon P, Djeu JY, Young NS (1985) Interferon is a mediator of hematopoietic suppression in aplastic anemia in vitro and possibly in vivo. Proc Natl Acad Sci USA 82:188
6. Zoumbos NC, Gascon P, Djeu JY, Trost SR, Young NS (1985) Circulating activated suppressor T lymphocytes in aplastic anemias. N Engl J Med 312:257
7. Burdach S, Levitt LJ (1988) T cell regulated hematopoiesis-molecular interactions in hematopoietic control by CD2 and Interleukin-2. Behring Inst Mitt 83:56–67
8. Burdach S, Shatsky M, Levitt LJ (1987) Receptor-specific modulation of myelopoiesis by recombinant DNA-derived interleukin-2. J Immunol 139:452
9. Romain PL, Schlossman SF (1984) Human T-lymphocyte subsets; functional heterogeneity and surface recognition structures. J Clin Invest 74:1559
10. Meuer SC, Hussey RE, Fabbi M, Fox D, Acuto O, Fitzgerald KA, Hodgdon JC, Protentis JP, Schlossman SF, Reinherz EL (1984) An alternative pathway of T-cell activation: a functional role for the 50 kd T11 sheep erythrocyte receptor protein. Cell 36:897
11. Plunkett ML, Sanders ME, Selvaraj P, Dustin ML, Springer TA (1987) Rosetting of activated human T lymphocytes with autologous erythrocytes. Definition of the receptor and ligand molecules as CD2 and lymphocyte function-associated antigen 3. J Exp Med 165:664
12. Dustin ML, Sanders ME, Shaw S, Springer TA (1987) Purified lymphocyte function associated antigen binds to CD2 and mediates T-lymphocyte adhesion. J Exp Med 165:677
13. Hünig T, Tiefenthaler G, Meyer zum Büschenfelde K-H, Meuer SC (1987) Alternative pathway activation of T cells by binding of CD2 to its surface ligand. Nature 326:298
14. Krensky AM, Sanchez-Madrid F, Robbins E, Nagy JA, Springer TA, Burakoff SJ (1983) The functional significance, distribution, and structure of LFA-1, LFA-2 and LFA-3: cell surface antigens associated with CTL-target interactions. J Immunol 131:611

15. Howard FD, Ledbetter JA, Wong J, Bieber CP, Stinson EB, Herzenberg LA (1981) A human lymphocyte differentiation marker defined by monoclonal antibodies that block E rosette formation. J Immunol 126:2117
16. Peterson A, Seed B (1987) Monoclonal antibody and ligand binding sites of the T-cell erythrocyte receptor (CD2). Nature 329:842
17. Yang SY, Chouaib S, Dupont B (1986) A common pathway for T-lymphocyte activation involving both the CD3-Ti complex and CD2 sheep erythrocyte receptor determinants. J Immunol 137:1079
18. Breitmeyer JB (1987) How T cells communicate. Nature 329:760
19. Breitmeyer JB, Darley JF, Levine AB, Schlossman SF (1987) The T11 (CD2) molecule is functionally linked to the T3/Ti T-cell receptor in the majority of T cells. J Immunol 139:2899
20. Burdach S, Shatsky M, Wagenhorst B, Levitt L (1988) The T-cell CD2 determinant mediates inhibition of erythropoiesis by the lymphokine cascade. Blood (in press)
21. Berger S, Kimmel AR (1987) Guide to molecular cloning techniques. Methods Enzymol 152:33–41
22. Maniatis T, Fritsch EF, Sambrock J (1982) Molecular cloning, a laboratory manual. Laboratory Press, Cold Spring Harbor, p 150
23. Leonard WJ, Depper JM, Crabtree GR, Rudikoff S, Pumphrey J, Rubb RJ, Krönke M, Svetlik PB, Pfeffer NJ, Waldmann TA, Greene WC (1984) Molecular cloning and expression of cDNAs for the human interleukin-2 receptor. Nature 311:626
24. Feinberg AP, Vogelstein B (1983) A technique for radiolabeling DNA restriction endonuclease fragments to high specific activity. Anal Biochem 132:6
25. Gray PW, Leung DW, Pennicia D, Yelverton E, Najarian R, Simonsen CC, Derynck R, Sherwood PJ, Wallace DM, Berger SL, Levinson AD, Goeddel DV (1982) Expression of human immune interferon cDNA in *E. coli* and monkey cells. Nature 295:503

Oncogene Cooperation and B-Lymphoid Tumorigenesis in Eμ-*myc* Transgenic Mice*

W. S. Alexander, O. Bernard, W. Y. Langdon, A. W. Harris, J. M. Adams, and S. Cory

A. Introduction

Alteration of the cellular *myc* oncogene has been strongly implicated in several types of lymphoid tumors. Although its normal role remains largely unknown, c-*myc* seems likely to play a crucial role in the control of cellular proliferation and appears to become oncogenic when genetic alterations deregulate its expression. The gene is activated by retroviral insertion in most avian bursal lymphomas and several mammalian T lymphomas and by chromosomal translocation to the active immunoglobulin heavy-chain (IgH) locus in human Burkitt's lymphomas, mouse plasmacytomas, and rat immunocytomas (reviewed by [3]). Direct evidence that deregulated *myc* expression causes malignancy has been provided by studies with transgenic mice. Mice bearing *myc* coupled to immunoglobulin enhancers [1, 12] or a retroviral long terminal repeat [8, 11] succumb to tumors. Particularly striking are the Eμ-*myc* transgenic mice, which bear c-*myc* driven by the IgH enhancer and invariably develop B-lymphoid tumors. Despite this absolute predisposition to lymphoma, the expression of Eμ-*myc* appears insufficient to tumorigenesis. Although lymphoid tissues in young Eμ-*myc* mice express the transgene at similar levels to the tumor cells [2], unlike the latter, they do not elicit tumors upon transplantation [5]. Furthermore, the time of onset for tumors in individual mice varies widely [5], and the lymphomas are monoclonal, even though many B-lineage cells express Eμ-*myc*. Thus, additional somatic change(s) are presumed necessary to confer a fully tumorigenic potential to Eμ-*myc* cells.

B. Involvement of Additional Oncogenes in Eμ-*myc* Tumorigenesis

An appealing hypothesis is that the somatic change(s) which renders Eμ-*myc* cells malignant involves the activation of other cellular oncogenes. To test this idea directly, we have assayed DNA from several Eμ-*myc* lymphomas for the presence of genes capable of inducing NIH3T3 cells to produce foci in culture and/or fibrosarcomas in nude mice. DNA from one tumor induced focus formation and particularly rapid fibrosarcoma development, and these transfected cells displayed amplified and/or rearranged N-*ras* genes. An N-*ras* cDNA was subsequently cloned from the original Eμ-*myc* lymphoma and shown to contain a GLN→HIS mutation at amino acid 61, a residue commonly implicated in *ras* gene activation. Gene transfer experiments using this clone in a retroviral vector have confirmed that this mutated N-*ras* gene can transform fibroblasts and also cause the tumorigenic conversion of preneoplastic Eμ-*myc* pre-B cells. Thus we have established that at least two oncogenes were involved in the development of an Eμ-*myc* lymphoma (Alexan-

---

Walter and Eliza Hall Institute of Medical Research, P.O. Royal Melbourne Hospital, Victoria 3050, Australia

* This work was supported in part by the Australian National Health and Medical Research Council, the United States National Cancer Institute (Ca-14354) and the American Heart Association

der, Bernard, and Cory, in preparation). Preliminary results suggest that activated c-*raf* and *ras* genes may also be involved in several other Eμ-*myc* tumors.

C. Eμ-*myc* B-Lymphoid Cells Progress Toward Autonomy in Culture

Although Eμ-*myc* appears insufficient for tumorigenesis, its constitutive expression in the B-lineage cells of young transgenic mice causes a dramatic perturbation in B-lymphoid development. Preneoplastic Eμ-*myc* mice display a marked polyclonal expansion of large pre-B cells at the expense or more mature B-lymphocytes and most of these cells appear to be in cycle [5]. To investigate how the B-lineage cells have been disturbed in these mice, we have investigated their growth properties in vitro [6]. Eμ-*myc* B-lineage cells were not autonomous, since they died rapidly when cultured in the absence of feeder cells. Nevertheless, long-term cultures could be established on a layer of bone marrow stromal cells, and these initially behaved similarly to parallel cultures from normal mice. However, after 3 weeks, the Eμ-*myc* cultures consistently grew at slightly higher densities than normal cells. As in vivo, the Eμ-*myc* cells were notably larger and showed considerably greater cell cycle activity.

After about 15 weeks in culture, the Eμ-*myc* populations dramatically shifted to growth at tenfold higher density. Nevertheless, these cells still remained strictly feeder layer-dependent and did not elicit tumors in histocompatible mice. Analysis of immunoglobulin gene rearrangements showed that these changes reflected the emergence of dominant pre-B clones. By 25 weeks, the Eμ-*myc* cultures no longer required stromal layers and now produced tumors upon transplantation. These long-term cultures thus permitted the evolution of Eμ-*myc* cells toward a fully transformed phenotype. Their sequential changes in growth characteristics may indicate that more than one alteration is involved in the development of growth autonomy and tumorigenicity.

D. The v-H-*ras* and v-*raf* Oncogenes Cause Autonomous Growth and Malignant Conversion of Eμ-*myc* Pre-B Cells

Preneoplastic Eμ-*myc* mice provide an ideal system to test directly which oncogenes can synergize with a deregulated *myc* gene for B-lymphoid transformation. Using retroviruses, we have introduced a second activated oncogene into preneoplastic Eμ-*myc* lymphoid cells and compared their growth properties with those of similarly infected normal cells. Uninfected cells from normal or Eμ-*myc* marrow did not proliferate in soft agar or in liquid culture and were not tumorigenic. In contrast, Eμ-*myc* marrow cells infected with helper virus-free stocks of Harvey murine sarcoma virus (H-MSV), which carries the v-H-*ras* gene, or with 3611-MSV, which harbors v-*raf*, produced 10–100 times more lymphoid colonies in soft agar than infected normal marrow cells. Moreover, in liquid culture, nonadherent cell lines emerged from H-MSV or 3611-MSV infected Eμ-*myc* marrow cultures at a higher frequency and with shorter latency than from infected normal marrows. Flow-cytometric analyses for surface markers on cells from the agar clones and the liquid culture populations indicated that they were all pre-B, although some contained subpopulations which had matured to express sIg.

All the Eμ-*myc* cultures established in liquid medium following viral infection were highly polyclonal, and a high proportion of the cells plated as colonies in agar. Furthermore, all 18 clones picked from primary agar cultures and injected into nude mice proved tumorigenic. Thus the activation of an H-*ras* or *raf* gene may be sufficient to convert preneoplastic Eμ-*myc* pre-B cells to malignancy.

Significantly, neither v-*raf* nor v-H-*ras* expression was sufficient to transform normal lymphoid cells. The cells derived by 3611-MSV infection of normal marrow were clearly only partially transformed, since they required feeder cells for growth in vitro, did not clone in soft agar, and were not tumorigenic upon transplantation. Although the H-MSV infected normal marrow cultures did contain tumorigenic cells, these appeared to be rare transformants, since early liquid cultures were only oligoclonal and the cells were density dependent for growth in vitro and cloned with poor efficiency in agar. It seems that somatic changes are required in addition to v-H-*ras* or v-*raf* expression to fully transform lymphoid cells.

Previous studies have established that not all newly derived A-MLV clones from normal marrow are tumorigenic [13, 15]. Our results further indicate that deregulated *myc* expression makes no difference to the emergence of tumorigenic cells, since the proportion of A-MLV infected Eμ-*myc* clones which produced tumors was identical to that of infected normal marrow. Nevertheless, the liquid culture lines derived from Eμ-*myc* marrow with A-MLV plated in soft agar at significantly higher efficiency than those from infected normal marrow. Thus although the deregulated *myc* gene in A-MLV infected transgenic cells may confer some growth advantage in vitro, the combination of *myc* and v-*abl* is clearly insufficient to fully transform early B-lymphoid cells.

E. Lymphoid Transformation by A-MLV Is Reduced in Eμ-*myc* Mice

Abelson murine leukemia virus (A-MLV), which harbors the v-*abl* oncogene, causes the preferential proliferation of pre-B cells from infected bone marrow [10, 14]. Although preneoplastic Eμ-*myc* marrow shows a three- to fourfold elevation in pre-B cells [5], it consistently yielded two- to threefold fewer lymphoid colonies than normal marrow following A-MLV infection. Similarly, while almost every normal marrow culture initiated in liquid medium by A-MLV infection yielded continuously proliferating pre-B cells, five of seven A-MLV infected Eμ-*myc* marrows failed to proliferate.

These results may indicate that a high proportion of cells which coexpress v-*abl* and a deregulated *myc* gene die. Alternatively, the primary target cell for A-MLV transformation may be a primitive B-lymphoid precursor rather than a pre-B cell and merely acquires the pre-B phenotype subsequent to infection. This putative target cell would presumably lack the capacity to activate the IgH enhancer and hence would actually be underrepresented in the Eμ-*myc* marrow in the face of the massive pre-B-cell expansion.

F. C-*myc* Gene Regulation in Infected Cells

As expected, the infected normal and Eμ-*myc* pre-B lines all abundantly expressed the appropriate viral transcripts, as shown for some liquid culture lines in Fig. 1. Furthermore, the Eμ-*myc* lines derived with H-MSV or 3611-MSV exhibited the 3.0-kb transgenic *myc* mRNA but not the smaller transcripts characteristic of the endogenous c-*myc* alleles (Fig. 1). This result confirms the data obtained from spontaneous Eμ-*myc* tumors [1, 2] and is consistent with the proposal that c-*myc* expression is controlled by a negative feedback loop: constitutive production of *myc* protein from a deregulated allele prevents expression of the normal c-*myc* gene [7, 9]. In contrast, the A-MLV infected Eμ-*myc* lines all expressed both the transgenic and endogenous *myc* genes (Fig. 1). It is unlikely that the Eμ-*myc* gene is being deleted (thereby releasing expression of the endogenous c-*myc* alleles), since in all of 16 lines subcloned from these populations, the transgene and the endogenous *myc* genes were still coexpressed in a similar ratio to the parental line. Perhaps v-*abl*

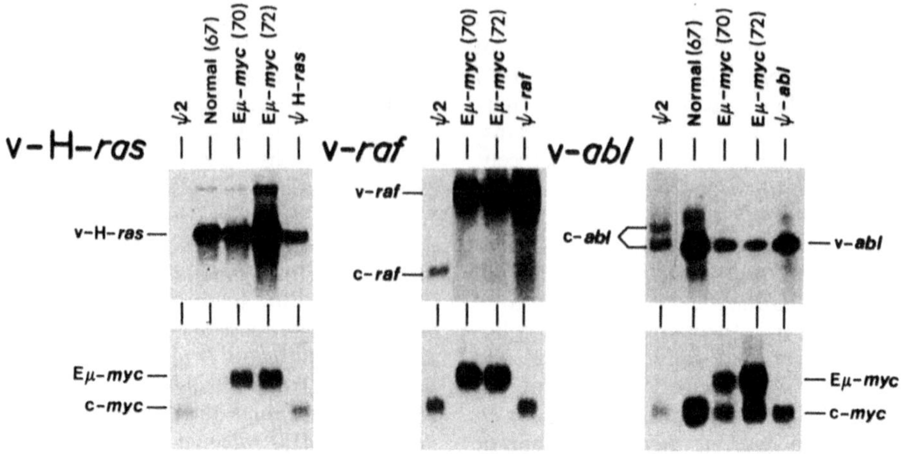

Fig. 1. Oncogene expression in lymphoid cell lines from virus-infected normal and Eμ-*myc* bone marrows. Ψ-H-*ras*, Ψ-*taf*,, and Ψ-*abl* denote the virus-producing fibroblasts and indicate the positions of the 5.4-kb v-H-*ras*, 7.2-kb v-*raf*, and 5.5-kb v-*abl* transcripts, respectively. *Below*, *myc* transcripts are shown; Eμ-*myc* as a 3.0-kb mRNA and the endogenous c-*myc* as the 2.3-kb species

can specifically induce expression of the normal c-*myc* alleles.

In summary, the constitutive expression of *myc* in young Eμ-*myc* mice profoundly disturbs B-lymphoid differentiation, but the onset of B lymphomas requires further somatic change. In vitro studies of the evolution of transgenic pre-B cells toward malignancy suggest that transformation correlates with the loss of growth factor requirements. Analysis of Eμ-*myc* tumor DNA has implicated mutation of the N-*ras* oncogene as a crucial somatic change in at least one tumor. The direct introduction of a second oncogene into preneoplastic lymphoid cells in vitro demonstrated that v-H-*ras* and v-*raf* transformed most (perhaps all) Eμ-*myc* pre-B cells, while v-*abl* appeared unable to cooperate with Eμ-*myc* for B-lymphoid transformation. This model system should help delineate which oncogenes can cooperate with c-*myc* to promote B-lymphoid neoplasia.

References

1. Adams JM, Harris AW, Pinkert CA, Corcoran LM, Alexander WS, Cory S, Palmiter RD, Brinster RL (1985) The c-*myc* oncogene driven by immunoglobulin enhancers induces lymphoid malignancy in transgenic mice. Nature 318:533–538
2. Alexander WS, Schrader JW, Adams JM (1987) Expression of the c-*myc* oncogene under control of an immunoglobulin enhancer in Eμ-*myc* transgenic mice. Mol Cell Biol 7:1436–1444
3. Cory S (1986) Activation of cellular oncogenes in haemopoietic cells by chromosome translocation. Adv Cancer Res 47:189–234
4. Harris AW, Pinkert CA, Crawford M, Langdon WY, Brinster RL, Adams JM (1988) The Eμ-*myc* transgenic mouse: A model for high incidence lymphoma and leukemia of early B cells. J Exp Med 137:353–371
5. Langdon WY, Harris AW, Cory S, Adams JM (1986) The c-*myc* oncogene perturbs B lymphocyte development in Eμ-*myc* transgenic mice. Cell 47:11–18
6. Langdon WY, Harris AW, Cory S (1988) Growth of Eμ-*myc* transgenic B-lymphoid cells in vitro and their evolution towards autonomy. Oncogene Res 3:271–279
7. Leder P, Battey PJ, Lenoir G, Moulding C, Murphy W, Potter H, Stewart T, Taub R (1983) Translocations among antibody genes in human cancer. Science 222:765–770

8. Leder A, Pattengale PK, Kuo A, Stewart TA, Leder P (1986) Consequences of widespread deregulation of the c-*myc* gene in transgenic mice: Multiple neoplasms and normal development. Cell 45:485–495
9. Rabbitts TH, Forster A, Hamlyn P, Baer R (1984) Effect of somatic mutation within translocated c-*myc* genes in Burkitt's lymphoma. Nature 309:592–597
10. Rosenberg N (1982) Abelson leukemia virus. Curr Top Microbiol Immunol 101: 95–126
11. Stewart TA, Pattengale PK, Leder P (1984) Spontaneous mammary adenocarcinomas in transgenic mice that carry and express MTV/*myc* fusion genes. Cell 38: 627–637
12. Suda Y, Aizawa S, Hirai S, Inoue T, Furuta Y, Suzuki M, Hirohashi S, Ikawa Y (1987) Driven by the same Ig enhancer and SV40 T promoter *ras* induced lung adenomatous tumors, *myc* induced pre-B lymphomas and SV40 large T gene a variety of tumors in transgenic mice. EMBO J 6:4055–4065
13. Whitlock CA, Witte ON (1981) Abelson virus-infected cells can exhibit restricted in vitro growth and low oncogenic potential. J Virol 40:577–584
14. Whitlock CA, Witte ON (1985) The complexity of virus-cell interaction in Abelson virus infection of lymphoid and other haematopoietic cells. Adv Immunol 37: 74–98
15. Whitlock CA, Ziegler SF, Witte ON (1983) Progression of the transformed phenotype in clonal lines of Abelson virus-infected lymphocytes. Mol Cell Biol 3:596–604

Even Transcriptionally Competent Proviruses Are Silent in Bovine Leukemia Virus Induced Tumor Cells*

A. van den Broeke[1], Y. Cleuter[2], G. Chen[3], D. Portetelle[3], M. Mammerickx[4], D. Zagury[5], M. Fouchard[5], L. Coulombel[6], R. Kettmann[3], and A. Burny[7]

A. Introduction

Bovine leukemia virus (BLV) is an exogenous retrovirus which induces a chronic disease in cattle, often causing persistent lymphocytosis (PL), with lymphosarcomas developing in a small number of infected animals (for review [1]). The same virus infects sheep, where it induces tumors with very high frequency [2]. In addition to the genes involved in viral replication (*gag, pol, env*), the BLV genome contains an "X" region coding for a 18-kDa protein [3, 4] and a 34-kDa product acting as a transactivator of transcription of the provirus, called Tat protein [5–7]. It is now hypothesized that the BLV Tat protein transactivates some cellular genes, and this is thought to be the key process to initiation of cell transformation. All BLV-induced tumors are clonal and contain at least a portion of a provirus [8, 9] integrated at many sites in the host genome [10, 11]. All deleted proviral copies examined have shown preservation of the X region, stressing again its probable role in the tumoral process. Viral RNA was not detected in fresh lymphocytes isolated from animals with PL or in tumors [9, 12, 13]. When tumor cells were cultivated in vitro, only a very low level of expression, if any, could be detected [14].

In order to investigate the role of proviral integration and expression in cellular transformation induced by BLV, we have isolated and characterized three different tumor proviruses. We show here that these proviruses are silent in the tumor cell. We also sought to determine whether they were able to express viral functions, in particular the Tat protein, once cloned and isolated from their host tumor cells.

B. Results

I. Viral RNA Expression in Tumor Cells Containing Full-Length Proviruses

Two BLV-induced tumors carrying single full-length proviral copies were selected upon restriction and hybridization analysis (T344 and T395). YR2, which is an established cloned lymphoid cell line derived from T395 tumor cells, displayed the same restriction pattern as T395, confirming that the clone proliferating in culture was indeed the clone present in vivo.

[1] Department of Molecular Biology, University of Brussels, 1640 Rhode St-Genèse, Belgium; and Faculty of Veterinary Medicine, University of Liège, 1070 Brussels, Belgium
[2] Department of Molecular Biology, University of Brussels, 1640 Rhode St-Genèse, Belgium
[3] Faculty of Agronomy, 5800 Gembloux, Belgium
[4] National Institute for Veterinary Research, 1180 Brussels, Belgium
[5] Laboratory of Cellular Biology, University Pierre et Marie Curie, Paris, France
[6] Unité Inserm, Hospital Bicêtre, Paris, France
[7] Department of Molecular Biology, University of Brussels, 1640 Rhode St-Genèse, Belgium; and Faculty of Agronomy, 5800 Gembloux, Belgium
* This work was supported by the Fonds Cancérologique de la Caisse Générale d'Epargne et de Retraite, Belgium. R.K. is Maître de recherches FNRS. A.V.D.B. was supported by fellowships from the Lady Tata Memorial trust and the Fondation Rose et Jean Hoguet

To determine whether viral expression took place in tumors 344 and 395 in vivo as well as in the YR2 cell line in vitro, we looked for the presence of viral sequences in their mRNA, using a labeled antisense X RNA as a probe on Northern blots. No hybridization was detected with RNA from tumor 344, tumor 395, or YR2 cells (data not shown). Comparison of these data to results obtained with BLV-producing cell lines allowed us to assume that less than 0.002 copies of viral RNA were present per tumor cell.

Additional evidence for the lack of viral expression in the tumor cells of animal 395 was provided by an experiment in which two sheep and two goats were infected each with 10^7 YR2 cells. After 5 months, no seroconversion occurred, whereas in a previous experiment carried out with PL lymphocytes [15], infection took place after 18 days with as little as 926 inoculated leukocytes.

II. Biological Activity of the Cloned Tumor-Derived Full-Length Proviruses

To establish whether proviruses 344 and 395 would be able to express viral genes once isolated from their natural host cell, the proviral sequences were cloned, generating the recombinant plasmids pV344 and pV395. We looked for viral proteins, in particular the Tat protein, in transient expression assays after transfection of the cloned proviruses in several cell lines. Plasmids pV395 and pV344 carrying the proviral DNA were cotransfected with plasmid pBLVCAT [5] in noninfected mammalian cells: ovine kidney (OVK), Chinese hamster ovary (CHO), and Raji cells. Levels of chloramphenicol acetyltransferase (CAT) enzymatic activity reflect the ability of the proviral DNA to transactivate the LTR sequences located 5′ to the CAT gene in pBLVCAT. As expected, no CAT activity was detected in CHO, OVK, or Raji cells transfected with pBLVCAT alone (Fig. 1, lanes 2–4) or with pBLVCAT and pSP18, a plasmid without any insert (lanes 8, 12, 16). However, in the presence of pV395, a low level of CAT activity was detected in CHO and OVK (lanes 5, 9) but not in Raji cells (lane 13). Upon cotransfection of pV344 and pBLVCAT, much higher levels were observed in CHO and OVK cells. Even in Raji cells an appreciable level was detected (lanes 6, 10, 14).

Production of CAT under control of the BLV long terminal repeats (LTR) in the cotransfected cells showed that the cloned complete tumor proviruses under consideration were able to express a functional protein and transactivate. Moreover, in the culture supernatants and extracts of the cells where the Tat protein was present, Gag p24 and Env gp51 products were also detected by ELISA (data not shown), indicating that structural proteins and viral particles were indeed produced.

III. Analysis of Tumor Cells Containing a Single Heavily Deleted Provirus

Whether unique or multiple in the tumor, proviral copies are complete or harbor deletions. Tumor 1345 was shown to harbor a single provirus with a large 5′ deletion of about 4.4 kb. The proviral sequences were cloned, generating the recombinant plasmid pV1345 and further characterized by sequence analysis. The sequence was compared with that of a complete BLV genome [16], as shown in Fig. 2. The deletion starts within the gene coding for the major internal Gag protein p24 and ends in the middle of the gene coding for the surface glycoprotein gp51. The deletion is 4310 bp long and spans between nucleotides 1022 and 5332.

To determine whether expression took place in tumor 1345, Northern blots were performed as described for tumors 344 and 395. No viral expression could be detected. Transient expression CAT assays with the cloned provirus were negative (Fig. 1, lanes 7, 11, 15) as well as Gag p24 and Env gp51 detection. These results demonstrate that the truncated provirus, even isolated from the tumor, is unable to code for viral proteins including Tat.

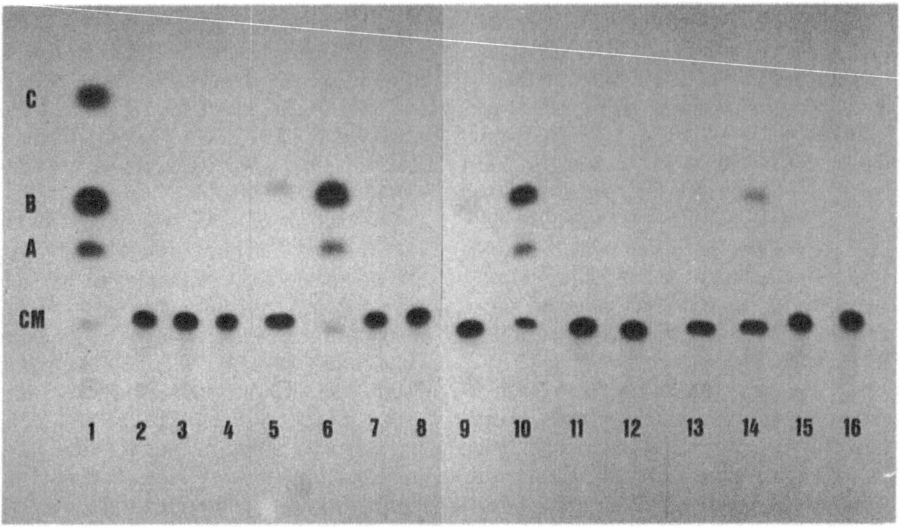

Fig. 1. Assay of CAT activity in fetal lamb kidney (FLK), Chinese hamster ovary (CHO), ovine kidney (OVK), and Raji cells. Transfections and CAT assays were performed as described by Gorman et al. [21]. Chloramphenicol (*CM*) and its acetylated forms (*A* and *B*, monoacetate forms; *C*, diacetate form) were detected by autoradiography. The different lanes show the products of chloramphenicol after incubation with extracts of FLK, a BLV-producing cell line, transfected with pBLVCAT (*lane 1*); CHO, OVK, and Raji transfected with pBLVCAT (*lanes 2–4*), CHO transfected with pBLVCAT and pV395 (*lane 5*), pBLVCAT and pV344 (*lane 6*), pBLVCAT and pV1345 (*lane 7*), pBLVCAT and pSP18 (*lane 8*); OVK transfected with pBLV-CAT and pV395 (*lane 9*), pBLVCAT and pV344 (*lane 10*), pBLVCAT and pV1345 (*lane 11*), pBLVCAT and pSP18 (*lane 12*); and Raji transfected with pBLVCAT and pV395 (*lane 13*), pBLVCAT and pV344 (*lane 14*), pBLVCAT and pV1345 (*lane 15*), pBLVCAT and pSP18 (*lane 16*)

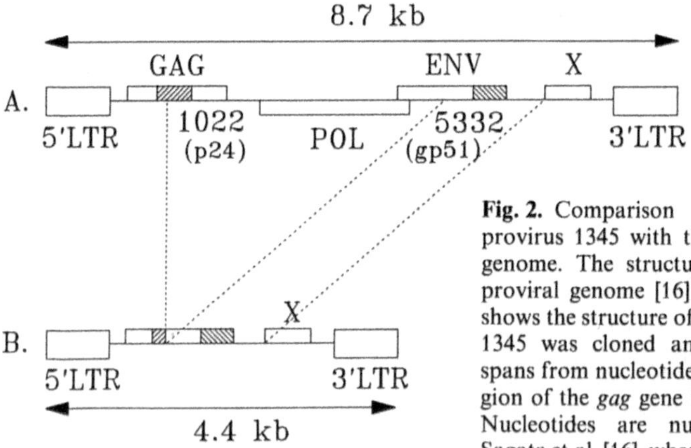

Fig. 2. Comparison of the structure of provirus 1345 with that of a complete BLV genome. The structure of a complete BLV proviral genome [16] is illustrated in (*A*). *B* shows the structure of provirus 1345. Provirus 1345 was cloned and sequenced. Deletion spans from nucleotide 1022 within the p24 region of the *gag* gene to 5332 in the *env* gene. Nucleotides are numbered according to Sagata et al. [16], where nucleotide 1 is the first at the left end of the 5'LTR

C. Discussion

Previous experimental data [9, 12–14] and the results presented here establish that no part of the BLV information is expressed in fresh tumors, in sheep tumor cell lines, or in sheep tumor lines injected into naive recipient sheep. The same conclusion held true whether the unique provirus was apparently intact or carried extensive deletions. Two cases showing integration of a unique apparently intact provirus were studied (tumor 395 and tumor 344). In both cases, the cloned full provirus was transfected and expressed; BLV Gag and Env proteins were detected in culture supernatants, and Tat expression was easily demonstrated in CAT assays. These results demonstrate that the nonexpression of a provirus in a tumor cell does not necessarily imply a structural alteration of the viral information.

Another tumor case (bovine 1345) harbored a single heavily deleted integrated provirus. Cloning and sequencing data showed an extended deletion (4310 nucleotides) expanding from the middle of p24, in the *gag* gene, to the middle of the *env* gene, in the gp51 region. No functional mRNA, even the *tat* gene message, could be transcribed from that unique proviral genome, as the spliced-in segment corresponding to the end of *pol* was missing [17–20]. As expected, the transfected 1345 provirus did not exhibit any Tat activity in the CAT assays. The inescapable conclusion of these experiments is that no viral function is required to maintain the transformed state, even if ample epidemiological and experimental evidence points to BLV as the etiological agent of bovine and ovine leukemia. Whether the provirus is complete or deleted, it is fully repressed in the transformed cell.

Experiments are under way to unravel biochemical mechanisms that lock BLV tumor proviruses in the silent stage. The identification of critical cellular genes activated or repressed by transient expression of Tat should further our understanding of BLV-induced leukemogenesis.

Acknowledgement. We thank Dr. M. Onuma for providing bovine tissues.

References

1. Burny A, Cleuter Y, Kettmann R, Mammerickx R, Marbaix G, Portetelle D, van den Broeke A, Willems L, Thomas R (1987) Bovine leukaemia: facts and hypotheses derived from the study of an infectious cancer. Cancer Surv 6:139–159
2. Mammerickx M, Dekegel D, Burny A, Portetelle D (1976) Study of oral transmission of bovine leukosis to the sheep. Vet Microbiol 1:347–350
3. Rice NR, Simek SL, Dubois GC, Showalter SD, Gilden RV, Stephens RM (1987) Expression of the bovine leukemia X region in virus-infected cells. J Virol 61: 1577–1585
4. Derse D (1988) Trans-acting regulation of bovine leukemia virus mRNA processing. J Virol 62:1115–1119
5. Rosen CA, Sodroski JG, Kettmann R, Burny A, Haseltine WA (1985) Transactivation of the bovine leukemia virus long terminal repeat in BLV-infected cells. Science 227:321–323
6. Rosen CA, Sodroski JG, Willems L, Kettmann R, Campbell K, Zaya R, Burny A, Haseltine WA (1986) The 3′ region of bovine leukemia virus genome encodes a transactivator protein. EMBO J 5:2585–2589
7. Derse D (1987) Bovine leukemia virus transcription is controlled by a virus-encoded trans-acting factor and by cis-acting response elements. J Virol 61: 2462–2471
8. Kettmann R, Cleuter Y, Mammerickx M, Meunier-Rotival M, Bernardi G, Burny A, Chantrenne H (1980) Genomic integration of bovine leukemia provirus: comparison of persistent lymphocytosis with lymph node tumor form of enzootic bovine leukosis. Proc Natl Acad Sci USA 77:2577–2581
9. Kettmann R, Deschamps J, Cleuter Y, Couez D, Burny A, Marbaix G (1982) Leukemogenesis of bovine leukemia virus: proviral DNA integration and lack of RNA expression of viral long terminal repeat and 3′ proximate cellular sequence. Proc Natl Acad Sci USA 79:2465–2469
10. Kettmann R, Deschamps J, Couez D, Claustriaux JJ, Palm R, Burny A (1983) Chromosome integration domain for bo-

vine leukemia virus in tumors. J Virol 47:146–150
11. Gregoire D, Couez D, Deschamps J Heuertz S, Hors-Cayla MC, Szpirer J, Szpirer C, Burny A, Huez G, Kettmann R (1984) Different bovine leukemia tumors harbor the provirus in different chromosomes. J Virol 50:275–279
12. Kettmann R, Marbaix G, Cleuter Y, Portetelle D, Mammerickx M, Burny A (1980) Genomic integration of bovine leukemia provirus and lack of viral RNA expression in the target cells of cattle with different responses to BLV infection. Leuk Res 4:509–519
13. Marbaix G, Kettmann R, Cleuter Y, Burny A (1981) Viral RNA content of bovine leukemia virus-infected cells. Mol Biol Rep 7:135–138
14. Kettmann R, Cleuter Y, Gregoire D, Burny A (1985) Role of the 3′ long open reading frame region of bovine leukemia virus in the maintenance of cell transformation. J Virol 54:899–901
15. Mammerickx M, Palm R, Portetelle D, Burny A (1988) Experimental transmission of enzootic bovine leukosis to sheep: latency period of the tumoral disease. Leukemia 2:103–107
16. Sagata N, Yasunaga T, Tsuzuku-Kawamura J, Ohishi K, Ogawa Y, Ikawa Y (1985) Complete nucleotide sequence of the genome of bovine leukemia virus: its evolutionary relationship to other retroviruses. Proc Natl Acad Sci USA 82:677–681
17. Mamoun RZ, Astier-Gin T, Kettmann R, Deschamps J, Rebeyrotte N, Guillemain BJ (1985) The px region of bovine leukemia virus is transcribed as a 2.1 kilobase mRNA. J Virol 54:625–629
18. Derse D, Caradonna SJ, Casey JW (1985) Bovine leukemia virus long terminal repeat: a cell-specific promoter. Science 227: 317–320
19. Sagata N, Yasunaga T, Ikawa Y (1985) Two distinct polypeptides may be translated from a single spliced mRNA of the X genes of human T-cell leukemia and bovine leukemia viruses. FEBS Lett 192: 37–42
20. Willems L, Bruck C, Portetelle D, Burny A, Kettmann R (1987) Expression of a cDNA clone corresponding to the long open reading frame (XBL-1) of the bovine leukemia virus. Virology 160:55–59
21. Gorman CM, Moffat LF, Howard BH (1982) Recombinant genomes which express chloramphenicol acetyltransferase in mammalian cells. Mol Cell Biol 2: 1044–1051

Restriction Fragment Length Polymorphism of the c-Ha-*ras*-1 Proto-Oncogene as a Marker of Genome Alterations and Susceptibility to the Development of Some Human Carcinomas

P. G. Knyazev, I. F. Nikiforova, O. M. Serova, L. B. Novikov, G. F. Pluzhnikova, A. M. Abramov, and J. F. Seitz

A. Introduction

The c-Ha-*ras*-1 proto-oncogene is highly polymorphic in a human population. A restriction fragment length polymorphism (RFLP) of Ha-*ras*, identified by a set of restriction endonucleases (*Bam*HI, *Msp*I, *Taq*I) was ascribed to change in the number of variable tandem-repeated units (VTR) closely linked to the Ha-*ras* coding sequences from the 3′ end [1, 2]. Restriction analysis established four common and several rare alleles of c-Ha-*ras*-1 [3]. A RFLP of c-Ha-*ras*-1 may be useful in detecting deletions and/or rearrangements of alleles in human DNA. Frequent (20%–60%) nonrandom loss of one of the Ha-*ras* alleles has been shown in Wilms' tumors [4], bladder carcinomas [5], breast carcinomas [6], and rhabdomyosarcomas [7].

The loss of normal cellular sequences is thought to unmask recessive mutations [8]. On the other hand, a deleted locus may represent an "antioncogene" that acts normally to constrain cellular proliferation. The suggestion should not be ruled out that some modifications of the proto-oncogene might turn it into an oncogene, whereas an intact one plays the role of an antioncogene, being involved in the same sequence of molecular events.

This study covered the distribution of c-Ha-*ras*-1 alleles in lung, ovarian, and thyroid cancer patients. Structural alterations of the c-Ha-*ras*-1 proto-oncogene

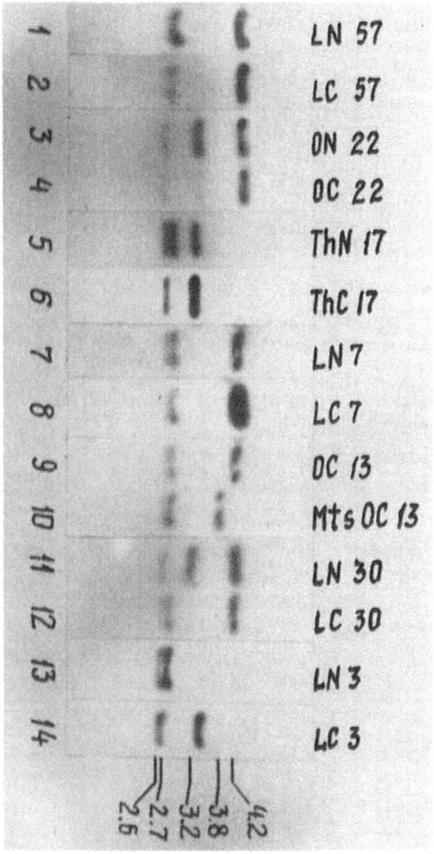

Fig. 1. Alterations of the c-Ha-*ras*-1 locus in tumor DNA of lung, ovarian, and thyroid cancer patients: *1, 3, 5, 7, 9, 11, 13,* constitutional genotypes of cancer patients; *2, 4, 6,* deletion of Ha-*ras* allele with the shorter fragment length; *8,* amplification of a4 allele; *10, 12, 14,* changes of allele fragment length. The size of *Pvu*II restriction fragment (in kb) of each allele is given at the right side of the figure. *OC, LC, ThC,* ovarian, lung and thyroid cancers; *ON, LN, ThN,* normal ovarian, lung and thyroid tissues; *MtsOC,* metastasis of ovarian cancer

The N.N. Petrov Research Institute of Oncology, USSR Ministry of Health, Leningrad, USSR and their role in carcinogenesis in different allele combinations were analyzed.

B. Materials and Methods

Genomic DNAs from tumors of the lung, ovary and thyroid and homologous normal tissues were prepared as previously described [9], digested with restriction endonucleases *Bam*HI, *Pvu*II, *Pst*I, and *Msp*I, electrophoresed in 1% agarose gel, denaturated, and transferred to filters (nylon) [9]. The filters were hybridized with ^{32}P-labeled (nick-translated) hu-c-Ha-*ras*-1 [1] under stringent conditions, then washed, dried, and autoradiographed.

C. Results and Discussion

Restriction analysis of human DNA with the endonucleases *Bam*HI and *Pvu*II identified four common c-Ha-*ras*-1 alleles. The sizes of fragments obtained were as follows: 6.6, 7.1, 7.7, 8.1 and 2.7, 3.2, 3.8, 4.2 kb, respectively. The major a1 allele possessing the shortest VTR region was found in more than 80% of cancer patients, which is consistent with the literature data concerning the distribution of the a1 allele in healthy donors (Table 1). Based on a comparison of the frequency of a2, a3, and a4 alleles in genotypes of lung, ovarian, and thyroid cancer patients and in a normal population, the following conclusions might be drawn: (a) The frequency of the a2 allele was approximately two times higher in thyroid cancer patients coupled with a lower incidence of a3 and a4 alleles; and (b) the a4 allele was more frequently observed in lung and ovarian cancer patients (Table 1).

Deletions of the a1 allele were found in three of seven thyroid carcinomas with an a1/a2 allele combination (Fig. 1, lanes 5, 6). The only two cases of Ha-*ras* alteration (3- to 4-fold and 50- to 80-fold amplifications) were identified in human DNA from 22 thyroid cancer patients lacking the a2 allele.

Table 1. Genotypic distribution of the c-Ha-*ras*-1 gene in lung, ovarian, and thyroid cancer patients and in a normal population

Genotype	Lung cancer patients (41) no. (%)	Ovarian cancer patients (14) no. (%)	Thyroid cancer patients (29) no. (%)	Normal controls (419)[a] (%)
a1/a1	16 (39.0)	5 (36)	12 (41.4)	41.3
a1/a2	6 (14.6)	2 (14)	8 (27.6)	14.9
a1/a3	4 (9.8)	3 (21)	4 (13.8)	14.0
a1/a4	7 (17.1)	3 (21)	3 (10.3)	12.2
a2/a4	2 (4.9)	1 (7)	1 (3.4)	2.2
a2/a2	0 (0)	0 (0)	0 (0)	1.3
a3/a3	0 (0)	0 (0)	0 (0)	1.2
a4/a4	1 (2.4)	0 (0)	0 (0)	0.9
a2/a3	1 (2.4)	0 (0)	0 (0)	2.5
a3/a4	1 (2.4)	0 (0)	0 (0)	2.1
a3/8.5[b]	1 (2.4)	0 (0)	0 (0)	7.1[c]
6.8/a2	1 (2.4)	0 (0)	0 (0)	0.3[d]
6.3/a2	1 (2.4)	0 (0)	0 (0)	
a1/7.5	0 (0)	0 (0)	1 (3.4)	

[a] Summarized data [3, 10–13] calculated according to Hardy-Weinberg test.
[b] *Bam*HI restriction fragment length (kb) of rare alleles.
[c] Common/rare genotypes.
[d] Rare/rare genotypes.

Specific rearrangements (amplification of the a4 allele, deletion of another allele, and change of size of one allele) were established in five of 11 lung tumors (Fig. 1, lanes 3, 4, 7, 8, 11, 12) and in three of four ovarian tumors (Fig. 1, lanes 1, 2, 9, 10) possessing the a4 allele. On the other hand, rearrangements of c-Ha-*ras*-1 were a rare event in tumor DNA obtained from lung and ovarian cancer patients lacking the a4 allele and were detected in two of 40 tumors tested (Fig. 1, lanes 13, 14).

Since the frequency of a2 and a4 alleles were found to be increased in thyroid cancer patients and lung and ovarian cancer patients respectively, and the above changes in Ha-*ras* were observed in a2- and a4-bearing patients, these alleles may perhaps be viewed as genetic markers of predisposition to thyroid, lung, and ovarian cancers in combination with other clinical parameters.

References

1. Goldfarb M, Shimizy K, Perucho M, Wigler M (1982) Isolation and preliminary characterization of a human transforming gene from T24 bladder carcinoma cells. Nature 296:404–407
2. Capon DJ, Chen IW, Levinson AD, Seeberg PH, Goeddel DV (1983) Complete nucleotide sequences of the T24 human bladder carcinoma oncogene and its normal homologue. Nature 302:33–37
3. Krontiris TD, Martino N, Colb M, Parkinson D (1985) Unique allelic restriction fragments of the human Ha-*ras* locus in leukocyte and tumour DNAs of cancer patients. Nature 313:369–374.
4. Fearon IR, Vogelstein B, Feinberg AP (1984) Somatic deletion and duplication of genes on chromosome 11 in Wilms' tumors. Nature 309:176–178
5. Fearon IR, Feinberg AP, Hamilton SH, Vogelstein B (1985) Loss of genes on the short arm of chromosome 11 in bladder cancer. Nature 318:377–380
6. Theillet C, Lidereau R, Escot C, Hutzell P, Brunet M, Gest G, Schlom J, Callahan R (1986) Loss of a c-Ha-*ras*-1 allele and agressive human primary breast carcinomas. Cancer Res 46:4776–4781
7. Scrable HJ, Lampkin BC, Witte DP, Cavenee WK (1987) Loss of heterozygosity of 11p15 in rhabdomyosarcomas. Nature 329:645–647
8. Cavenee WK, Dryia TP, Phillips RQ, Benedict WF, Godbort R, Gallick BI, Murphee QL, Strong LC, White RL (1983) Expression of recessive alleles by chromosomal mechanism in retinoblastoma. Nature 305:779–784
9. Maniatis T, Fritsch IF, Sambrook J (1982) Molecular cloning: a laboratory manual. Cold Spring Harbor Laboratory, New York, p 387
10. Gerhard DS, Dracopoli NC, Bale CJ, Houghoton AN, Watkins P, Payne CI, Green MN, Housman DI (1987) Evidence against Ha-*ras*-1 involvement in sporadic and familial melanomas. Nature 325:73–75
11. Thein SL, Oscier DG, Flint J, Wainscoat JS (1986) Ha-*ras* hypervariable alleles in myelodysplasia. Nature 321:84–85
12. Heighway J, Thatcher N, Cerny T, Hasleton PS (1986) Genetic predisposition to human lung cancer. Br J Cancer 53:453–457
13. Ceccherini-Nelli L, De Re V, Viel A, Molaro G, Zilli L, Clemente C, Boiocchi M (1987) Ha-*ras*-1 restriction fragment length polymorphism and susceptibility to colon adenocarcinoma. Br J Cancer 56:1–5

Analysis of c-*raf* Oncogene Expression in Gastrointestinal Tumor Cells

H. Kalthoff[1], C. Röder[1], K. Schewe[1], W. Schmiegel[1], H.-G. Thiele[1], B. Heimann[2], H. Bading[2], K. Schramm[2], and K. Moelling[2]

A. Introduction

Oncogenes of retroviruses are responsible for the induction of tumors in animals and for the malignant transformation of cells in culture. They are derived from normal cellular proto-oncogenes. Comparison of differences between viral oncogenes and proto-oncogenes has indicated that they are never identical. In human tumors that are not induced by retroviruses, activated proto-oncogenes have been discovered which resemble the viral oncogenes in properties such as mutations, deletions, or amplified expression [1]. Activated oncogenes have been described in many different tumors or tumor cell lines [2]. Often, more than one oncogene seems to be activated, e.g., in HL60 promyelocytic leukemic cells *myc, myb* and *mil/raf* are amplified ([3] and unpublished observation). We are interested in the expression of the activated proto-oncogene c-*mil/raf* in human tumors. It is the homologue to the oncogene v-*mil* of the avian retrovirus MH2 [4]. While v-*mil* is expressed in virus-transformed cells as p100gag-*mil* fusion protein, the cellular homologue exhibits a molecular weight of 74 kD, p74hu-c-*raf* [5]. The v-*mil* protein is located predominantly in the cytoplasm and exhibits a virus-coded protein kinase activity in autophosphorylation reactions specific for the amino acids serine and threonine [6, 7]. It renders the cells independent of growth factors and therefore is involved in alteration of the signal transduction characteristic of many tumor cells.

In order to investigate growth parameters of pancreatic and other gastrointestinal tumors, we analyzed the expression of several proto-oncogenes at the RNA and protein levels. Here we performed immunoperoxidase analysis of cryosections of human pancreatic tissue specimens and cytological preparations of pancreatic, gastric, and colorectal cell lines using specific poly- and monoclonal antibodies against c-*mil/raf*. Additionally, we applied a c-*raf* cDNA probe for Northern blot hybridization of mRNA extracted from pancreatic tumor cell lines.

B. Results and Discussion

We used a DNA construct of v-*mil* and expressed a DNA fragment (Taql/Accl) [8] in a prokaryotic expression vector pPLc24 as a MS2-v-*mil* fusion protein of 43 kD, p43 MS2-*mil* [9]. The protein expressed was used to raise monoclonal antibodies (mab) one of which recognizes a domain conserved among v-*mil* and hu-c-*raf* proteins (Fig. 1). This "mab" allows detection of the p100gag-*mil* protein by indirect immunoprecipitation from metabolically labeled cells (Fig. 2) and the p74hu-c-*raf* protein. A carboxy-terminal synthetic peptide was also used to raise polyvalent antibodies ("C-term ab") against the p100gag-*mil* protein (Fig. 2). It also recognizes the p74hu-c-*raf* protein. Using Western blot techniques (Fig. 3), p74hu-c-*raf* protein was

[1] Abteilung für Immunologie, Universitätskrankenhaus Eppendorf, 2000 Hamburg 20, Federal Republic of Germany [Supported by BMFT (01GA074)]

[2] MPI für Molekulare Genetik, 1000 Berlin-West 33, Germany

detected in the two pancreatic tumor cell lines Capan-1 and QGP-1 by both, "mab" and "C-term ab". The latter recognized a 200-kD protein in addition, which may be due to a cross-reactivity of the polyclonal IgG. In rodent F12C2 cells "p45-50 *raf*" protein(s), representing truncated c-*raf*, were clearly stained

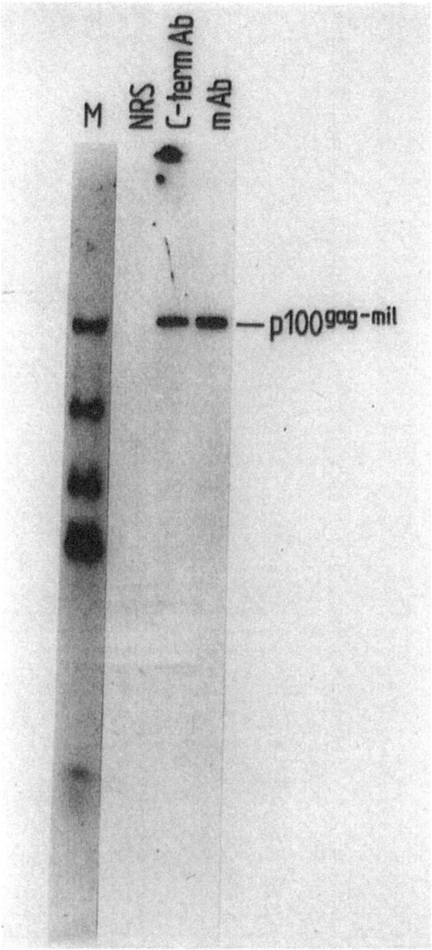

Fig. 1. Antigen specificity of applied antibodies. ^{35}S-radiolabeled MH2-transformed fibroblasts were lysed and immunoprecipitated with normal rabbit serum (NRS), with polyclonal rabbit IgG (C-term ab), or with mouse monoclonal antibody (mab; compare legend of Table 1), run on SDS-polyacrylamide gels, and autoradiographed. M, Molecular weight standards

by both antibodies, "mab" showing two prominent bands in comparison with four bands recognized by "C-term ab". Competition experiments using the two antibodies and their purified respective antigens confirmed the specificity of the two antibodies in immunohistological analysis (data not shown).

With respect to pancreatic tumor cell lines (Table 1) most of them stained positively in the cytoplasm: 8 times clearly positive, 6 times weakly positive, 3 times negative, with both antibodies showing similar staining intensities in most cases ($n=17$). While all colorectal cell lines ($n=5$) were found to be clearly positively stained, the gastric cancer cell lines ($n=8$) reacted less intensely (3 times clearly positive, 2 times weakly positive, 3 times negative) (Table 1). The immunocytological data were in good agreement with the Northern blot results (Table 1, Fig. 4). Only five of 20 samples showed positive hybridization signals in Northern blots, while the corresponding proteins showed only weak immunocytological reactivity if any. Whether this represents different sensitivities of the applied methods or is due to post-transcriptional regulation remains to be shown. In primary ductal adenocarcinomas of the pancreas ($n=18$), fifteen specimens showed positive reactions (mostly of the ductal cells) when tested with the *mil/raf* specific mab in immunohistochemical analyses (Table 2). Interestingly, normal ($n=9$) and chronically inflamed ($n=3$) pancreatic tissues revealed a positive acinar cell reactivity in addition to some ductal cell staining in nearly all samples. Staining was confined to the cytoplasm in most cases, and was particularly prominent in mitotic cells of the QGP-1 pancreatic tumor cell line.

Some cases of pancreatic tumor cell lines, as well as ductal epithelial cells of exocrine pancreatic tumors, revealed a nuclear localization of immunoreactive protein as detected by mab.

These results suggest a possible function of c-*raf* encoded protein expression in proliferating as well as differentiated

Table 1. Analysis of c-*mil*/*raf* oncogene expression in cell lines

Cell line	Immunoperoxidase staining		Northern blot analysis
	C-term Ab	mAb	c-*raf*-1
Pancreatic cancer			
Capan-2	−	−	−
Colo-357	+	+	++
ASPC-I	+	+	++
SW-850	−/±	+/±	++
SW-979	+	+/±	++
QGP-1	+	+/++	+
Panc-89	±	++	n.d.
Panc-Tu I	−/±	±	n.d.
Panc-Tu II	−/+	±	++
HPAF	±/−	±/−	+
BXPC-3	±	±	±
Capan-1	+	+	+++
PT-45-P1	n.d.	−	++
A-818 (−1)	+/±	±	n.d.
A-818(−4)	+	+	±
A-818(−5)	+	+	n.d.
A-818(−7)	+/−	−	−
Gastric cancer			
MKN-7	±/−	±/−	+
MKN-28	n.d.	+	+/±
MKN-45	+	+	±
MKN-74	−	±/−	+
Kawasaki	±/−	−	±/−
Okajima	±	+	n.d.
SCH-1013	n.d.	−	n.d.
MS1-P18	±/−	−	±
Colonic cancer			
WIDR	+	+	+
SW-1116	+/±	+	n.d.
HT-29	+	+	n.d.
E-5583	+/++	+/++	n.d.
Colo-320-HSR	+/++	++	n.d.
Normal fibroblasts			
F2	−	−	n.d.
F14	n.d.	n.d.	n.d.
Transformed fibroblasts			
F12C2	++	++	+

Immunoperoxidase staining was performed using the "mab" or "C-term ab". As a control another monoclonal antibody directed against a v-*mil*-specific determinant, was found to be negative in all samples tested ($n = 32$). Expression of c-*raf* mRNA was studied by Northern blot analysis using a human c-*raf*-specific c-DNA probe (Amersham/Buchler). Relative intensities of reactions are given as "−" (negative), "+/−" (weak), "+ or ++" (moderate or strong), "n.d." (not done), "+++" indicates a strong positivity in Northern blot analysis with an additional 6.2 kb transcript, "/" indicates heterogeneity or variability of results. The pancreatic cancer cell lines tested include five primary cell lines, which have been recently established in our lab. Besides gastric and colonic tumor cell lines normal human fibroblasts as well as transformed rodent fibroblasts were tested.

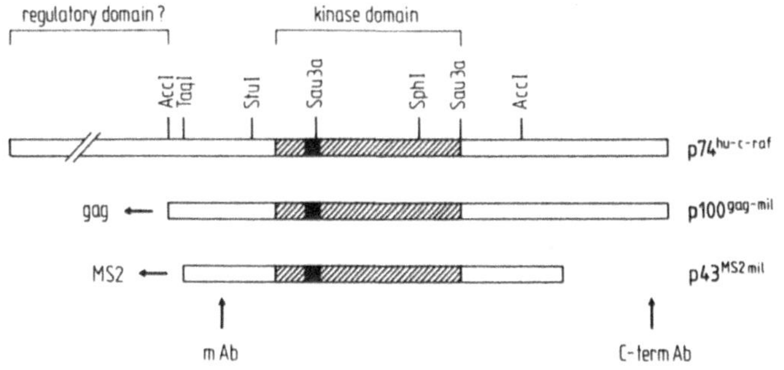

Fig. 2. Schematic epitope localization of applied antibodies. Comparison of p74-human-c-*raf* structure with p100gag-*mil* from MH2 virus and with recombinant bacterial fusion protein p43MS2-*mil*. *Arrows* indicate epitope localization of mouse monoclonal antibody (mAb) and of rabbit polyclonal IgG (C-term Ab). The *black box* represents the ATP-binding site

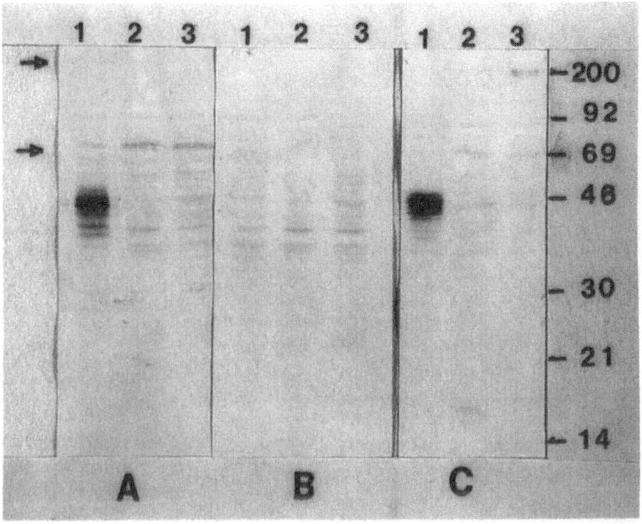

Fig. 3. Western blot analysis of c-*mil*/*raf* encoded protein. Cells were solubilized in 0.1% Triton-X-100 buffer and extracts were clarified by 100000 × g centrifugation. Aliquots of 50 µl were run on 12.5% SDS-polyacrylamide-gels and blotted on nitrocellulose using standard procedures. Lane 1, F12 C2; lane 2, Capan 1; lane 3, QGP-1. Overlay of panel *A* with "mab", *B* with control IgG1, *C* with C-term "ab". Molecular weight markers are given in kD. The *lower arrow* indicates p74hu-c-*raf*, the *upper arrow* points to a 200-kD protein – doublet detected in QGP-1 cells by C-term "ab"

Table 2. Immunoperoxidase studies of c-*mil*/*raf* oncogene expression in pancreatic tissue.

Tissue specimen	Number, score of staining intensity
Ductal adenocarcinoma	15 × +, 3 × – (mainly ductal cells)
Chronic pancreatitis	3 × + (acinar and ductal cells)
Normal pancreas	8 × +, 1 × – (acinar and ductal cells)
Endocrine tumors	3 × +, 1 × – (acinar and tumor cells)

Cryosections were incubated with "mab" (compare legend of Table 1). A negative control antibody showed only some background staining in a few cases (data not shown).

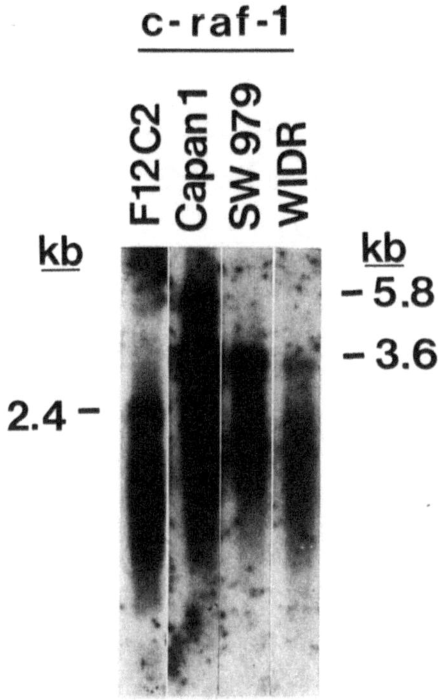

Fig. 4. Northern blot analysis of c-*raf* oncogene expression. The autoradiographs show selected examples of the data summarized in Table 1, comprising different "relative intensities" (+++, ++, +) and sizes of c-*raf*-1 transcripts. Total cellular RNA was extracted by SDS/proteinase K method from *raf*-transformed rodent fibroblasts (F12 C2), human pancreatic tumor cells (Capan 1, SW 979), and colorectal tumor cells (WiDR). Two micrograms polyl $(A)^+$-RNA were applied per lane and a ^{32}P-labeled, c-*raf*-1-specific DNA probe (Amersham/Buchler) was used for hybridization. Molecular weights of the transcripts were determined according to an RNA ladder (BRL)

pancreatic cells; this parallels recent findings reported for the *ras* protein [10]. The prominent immunohistochemical staining of mitotic cells and the positivity of the nuclei in some pancreatic tumor cell lines clearly present further questions on the biological role of c-*mil*/*raf* in the signal transduction pathway and require additional investigations.

References

1. Bishop JM (1985) Viral oncogenes. Cell 42:23–38
2. Moelling K, Heimann B, Bading H, Häder M, Bepler G, Havemann K, Beutler C (1987) Modulation of liver cell expression. MTP/Kluwer, Lancaster, pp 275–285
3. Collins SJ, Ruscetti FW, Galagher RE, Gallo RC (1978) Terminal differentiation of human promyelocytic leukemia cells induced by dimethyl sulfooxide and other polar compounds. Proc Natl Acad Sci USA 75:2458–2462
4. Bunte T, Greiser-Wilke T, Moelling K (1983) The transforming protein of the MC29-related virus CMII is a nuclear DNA-binding protein whereas MH2 codes for a cytoplasmic RNA-DNA binding polyprotein. EMBO J 2:1087–1092
5. Schultz AM, Copeland T, Oroszlan S, Rapp UR (1988) Identification and characterization of c-*raf* phosphoproteins in transformed murine cells. Oncogene 2:187–193
6. Moelling K, Heimann B, Beimling P, Rapp UR, Sander T (1984) Serine- and threonine-specific proteinkinase activities of purified gag-*mil* and gag-*raf* proteins. Nature 312:558–561
7. Denhez F, Heimann B, d'Auriol L, Graf T, Coquillaud M, Coll J, Galibert F, Moelling K, Stehelin D, Ghysdael J (1988) Replacement of Lys 622 in the ATP binding domain of p100gag-*mil* abolishes the in vitro autophosphorylation of the protein and the biological properties of the v-*mil* oncogene. EMBO J 7:541–546
8. Kan NC, Flordellis CS, Mark GE, Duesberg PH, Papas TS (1984) Nucleotide sequence of the avian carcinoma virus MH2: Two potential onc genes, one related to avian virus Mc29 and the other related to murine sarcoma virus 3611. Proc Natl Acad Sci USA 81:3000–3004
9. Moelling K, Pfaff E, Beug H, Beimling P, Bunte T, Schaller HE, Graf T (1985) DNA-binding activity is associated with purified *myb* proteins from AMV and E26 viruses and is temperature-sensitive for E26 ts mutants. Cell 40:983–999
10. Furth ME, Aldrich TH, Cordon-Cardo C (1987) Expression of *ras* proto-oncogene proteins in normal human tissues. Oncogene 1:47–58

c-*ets*-2 and the Mitogenic Signal Pathway

R. J. Fisher, S. Fujiwara, N. K. Bhat, C. W. Schweinfest, and T. S. Papas

A. Introduction

The feature that defines a cancer cell is its ability to propagate under conditions that typically inhibit the growth of the normal cell. In cells where the genetic constitution remains unaltered, restraints on growth are imposed by regulatory activities that take place at the cell surface. These activities, which may be due to either cell contact or a decrease in the production and availability of growth factors, are communicated through the cytoplasm to the nucleus, thus regulating the synthesis of messenger RNAs essential for the unique proteins needed to initiate subsequent DNA synthesis and mitosis. Cells can be relieved of this barrier to mitosis by cell dispersal or addition of growth factors; this molecular environmental change is similarly recognized at the surface of the cell, and is transmitted through the cytoplasm to the nucleus. This signal transduction process activates the transcription and synthesis of specific proteins and other macromolecules that trigger the mitotic process. The process and events leading to cell division are called here the "mitogenic signal pathway".

I. Oncogenes

The RNA tumor viruses (retroviruses) have been particularly useful in identifying and dissecting the metabolic pathways associated with mitosis, as well as for the recognition of macromolecules that are potential determinants of the malignant process [1]. These viruses were found in naturally occurring tumors (although extremely rare) in mice and chickens, it was proposed that such viruses could be components of all cells, and that the part of the virus encoding transforming activity, known as the oncogene, was responsible for the genesis of tumor cells. Since that time, the term "oncogene" has evolved and has been expanded to include all transforming genes in retroviruses that have nucleotide sequences homologous to cellular sequences, as well as all cellular genes that have transforming capability in DNA transfection assays. The complete cellular gene from which the transforming oncogene is derived is called the "proto-oncogene".

Molecular clones of oncogenes have provided an excellent opportunity to examine and compare the structural features of cellular proto-oncogenes in great detail. The general conclusion derived from these studies is that most viral oncogenes are truncated, mutated, or otherwise modified versions of normal cellular proto-oncogenes. The most exciting result of these comparative analyses is the recognition that several oncogene proteins can be identified as altered variants of normal cellular proteins that are involved in signal transduction and growth regulation processes.

II. Oncogenes Are Components of the Mitogenic Signal Pathway

The identification of oncogene to proto-oncogene product and the recognition of their relevance to components of the sig-

Laboratory of Molecular Oncology, National Cancer Institute, Frederick, MD 21701-1013, USA nal transduction pathways have allowed researchers to create a cellular paradigm for the normal mitogenic pathway and helped them to understand how disruption or amplification of the normal pathway can result in tumorigenesis.

The extensive molecular cloning and analysis of the retroviral oncogenes has allowed their grouping according to their biological function, as shown in Fig. 1. Here we see that the oncogenes can be characterized as growth factors (platelet-derived growth factor and *v-sis*), growth factor receptors (epidermal growth factor receptor and *v-erbB* or colony-stimulating factor-1 and *v-fms*), tyrosine kinases (*src* family of kinases), G-proteins (*ras*), and transcriptional activators (*fos, jun/AP-1*). The ligand-responsive transcription factors form a special class, in that these receptors bind their ligand directly and are then able to activate the corresponding genes (thyroid hormone receptor *erbA* and *v-erbA*).

III. Growth Factor Receptor

The prototype receptor, for which the epidermal growth factor (EGF)-receptor is one example, binds growth factor in its external domain. This region is typically located in the amino-terminal half of the protein molecule, a region that is highly glycosylated. Several receptors are known to have an abundance of cysteine residues that are believed to stabilize the conformational structure of the molecule. Although receptors may be specific for individual growth factors, it is likely that certain receptors will respond to more than one ligand. For example, the receptor for EGF can bind with equal affinity to (at least) three different peptides: EGF, transforming growth factor α (TGFα), and vaccinia virus growth factor [2–4]. The ligands appear to have little amino acid sequence homology in common, and presumably have substantial conformational similarities, since they can bind equivalently to the same receptor.

The transmembrane domain of receptors contains a significant proportion of hydrophobic amino acids; such sequences are characteristic of a membrane-associative capability. In addition to being a determinant in membrane localization, the structure of this region may also be important to its receptor activation function; a single amino acid replacement in this hydrophobic region converts a normal receptor molecule into a transforming protein.

The cytoplasmic domain of the receptor appears to have the potential for enzymatic activity; in several cases at least

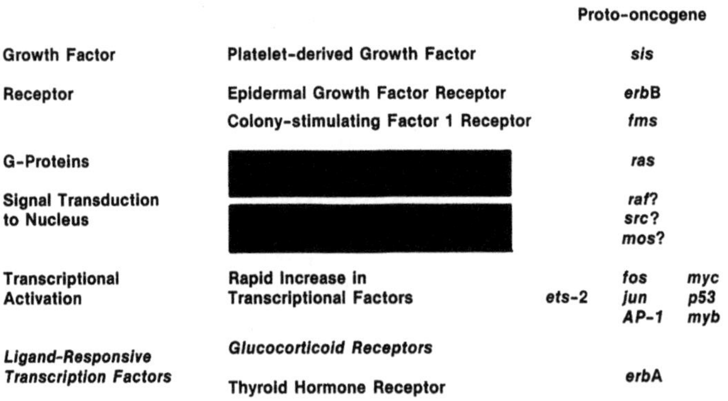

Fig. 1. Signal transduction cascade: The molecular cloning and sequencing of the viral oncogenes has allowed the definition of ■ their corresponding proto-oncogenes. These proto-oncogenes fall into the categories shown. The *black boxes* represent unknown mechanisms but are involved in the mitogenic signal pathway. The ligand-responsive transcription factors are a special class of proteins which are able to bind to sequence-specific elements after ligand binding

this activity is a tyrosine kinase. This internal polypeptide portion of the receptor is phosphorylated at specific serine and threonine residues by protein kinase C, as well as on its tyrosine residue by autophosphorylation. These phosphorylations may, in turn, affect and regulate the kinase activity of the receptor. It appears that ligand binding can activate the kinase activity of receptors. It is known that truncated receptor molecules, as represented by certain oncogene-encoded proteins, are devoid of the binding domain; these defective receptors are constitutively, enzymatically active and independent of growth factors [5].

IV. G-Proteins and *ras*

The activated tyrosine kinases can transmit their effect to nuclear events by either a *ras*-sensitive or a *ras*-insensitive pathway. The *ras*-insensitive pathway is not very well understood but may involve other members of the *scr* family of protein kinases. The *ras*-sensitive pathway involves one of the three *ras* genes (H-*ras*, K-*ras*, and N-*ras*) identified. The products from these genes localize to the cytoplasmic side of the plasma membrane, similar to the subcellular location of G-proteins [6]. Additionally, *ras* and G-proteins have highly conserved amino acid residues at their GTP-binding sites. However, no other amino acid homologies exist between these two classes of proteins. Therefore, while some functional similarities appear to exist for *ras* and the other G-proteins, each class of protein remains clearly distinct. Stimulation of a receptor results in the activation of the G-protein by releasing bound GDP and then binding of cellular GTP. The *ras*-sensitive pathway implicates phospholipid metabolism with generation of second-messenger molecules such as inositol triphosphate (IP3) and diacylglycerol (DG). The IP3 leads to release of calcium from the endoplasmic reticulum. The DG, along with free calcium, will activate protein kinase C [7]. The protein kinase C then could transmit its effect to the nucleus by poorly characterized processes shown by the black boxes in Fig. 1. In fact, the *ras*-sensitive and *ras*-insensitive pathways may converge and use a similar mechanism to activate nuclear regulatory proteins.

Since *ras* proteins bind guanosine nucleotides, it has been strongly suggested that they are related to G-proteins. In the active conformation, the G-protein is able to regulate second messages (cyclic adenosine monophosphate-cAMP, cyclic guanosine monophosphate-cGMP, DG, IP3, and Ca^{++}) by activating or inhibiting the enzymes responsible for their production. Hydrolysis of the bound GTP to GDP returns the activated G-protein to an inactive state, eliminating its regulatory effect. The recently discovered guanosine triphosphatase protein (GAP) regulates the activated *ras* by promoting GTPase activity [8] and may represent a cellular effector molecule for *ras*.

The *ras* oncogene stimulates the mitogenic signal transduction pathway. Rodent cells transformed by v-*ras* exhibit increased levels of both phosphatidylinositol diphosphate (PIP2) and its second messages DG and IP3 [9]. Consistent with this finding is the observation that following microinjection of a transforming $p21^{ras}$ protein into frog oocytes, rapid increases of PIP2, IP, IP2, and particularly DG were seen [10]. Microinjection of a monoclonal antibody to 21^{ras} that effectively neutralizes intercellular $p21^{ras}$ activity prevents the mitogenic activity of a phorbol ester or a calcium ionophore [11], suggesting that *ras* oncogene-encoded proteins can function at more that one critical site in the signal transductions pathway.

V. Nuclear Proto-oncogenes

The nuclear proto-oncogenes share several characteristics, including low abundance, rapid turnover, post-translational modification such as phosphorylation, response to mitogenic stimuli, and DNA binding. The best characterized of these proteins is *fos*, which has recently been shown to be co-induced with the transcription factor *jun/AP-1*. The *jun* binds to a specific DNA sequence (tgactca) found in phorbol ester (TPA)-responsive genes, and the *fos* binds to the *jun*, probably by protein-protein interactions, to begin forming an active transcriptional unit [12].

VI. The Mitogenic Response in Wound Healing

An example of how the mitogenic response pathway works in vivo is given for the platelet-derived growth factor (PDGF), which is found in alpha granules of the circulating platelets and probably functions in the repair of blood vessels in wound healing [13]. Thus, on wounding, platelets adhere to the blood vessel walls and PDGF is released, stimulating the migration of smooth muscle cells from the medial and intimal layers of the artery, where these cells proliferate in response to the injury and subsequent release of PDGF. The presence of the growth factor causes the proliferation of the cells necessary for wound healing, and as the repairs are made, the source of PDGF is removed (the clot disappears) and cell proliferation stops. Thus, the growth factor binds to the PDGF receptor and activates its associated tyrosine kinase activity, resulting in G-protein activation of phospholipase C, which hydrolyses phosphatidylinositol, generating the second messengers IP3 and DG. The IP3 is responsible for the release of Ca^{2+} from the endoplasmic reticulum, and the calcium along with DG can activate the protein C kinase [9]. The DG can be further metabolized by kinases to phosphatidic acid, or by phospholipase A_2 to liberate arachidonic acid – a precursor of prostaglandins. The activation of protein kinase C rapidly induces a nuclear response through transcriptional activation of *fos* and *myc* [14, 15]. The proto-oncogene *ras* is involved in this part of the pathway, since microinjection of monoclonal antibodies to the *ras* protein can block the mitogenic response elicited by phosphatidic acid or by prostaglandin f1α [11], indicating that *ras* may function by coupling a product(s) of phospholipid with cytoplasmic factors, which in turn can activate nuclear regulatory proteins.

B. *ets*-2 and the Mitogenic Signal Pathway

Our interest in the mitogenic signal transduction pathway comes from our investigations on the human *ets*-2 gene [16–22]. This gene has a high degree of homology to the viral oncogene v-*ets*, which was originally identified as a cell-derived sequence transduced into its genome by the avian leukemia virus, E26. Previous observations made in our laboratory have suggested that the *ets*-2 gene has a role in cell proliferation [23, 24]. The product of the *ets*-2 gene is preferentially expressed in a wide variety of proliferative tissue and the level of *ets*-2 expression was generally greater in tissue obtained from a variety of young organs, compared with adult organs of the same type.

We have also studied *ets* expression in quiescent BALB/c 3T3 fibroblasts following serum stimulation; both *ets*-1 and *ets*-2 RNA are increased 3 h after serum addition, while DNA synthesis peaked at 16 h (Fig. 2). The activation of 3T3 cells with serum also increases *fos, myc,* mouse metallothionin (MMT) and heat shock protein (HSP) mRNAs. These changes represent transcriptional activation of these genes. In Fig. 3 we show a post-translational mechanism for the activation of the *ets*-2 involving increase in the half-life ($t½$) of the *ets*-2 protein. These two mechanisms are not mutually exclusive, and the preference for the slower transcriptional or faster post-transcriptional mechanisms for regulation of the *ets*-2 may reflect differences in cell type or tissues examined. Transcriptional activation of the nuclear oncogenes has been described for many systems; the post-translational control of protein concentration by prolongation of $t½$ via

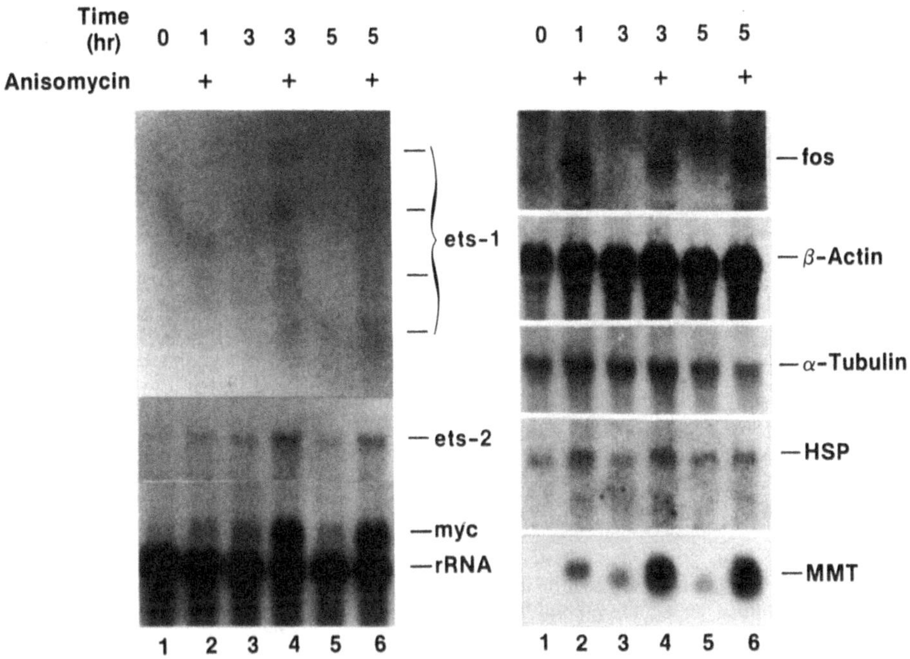

Fig. 2. Gene expression in Balb/c 3T3 fibroblasts after serum stimulation. The *ets*-1 and *ets*-2 mRNA are induced by serum and reach a maximum by 3 h in the presence of the protein synthesis inhibitor, anisomycin. *fos* and heat shock protein (HSP) have reached their maximum at the first time point measured (1 h), while *myc* and mouse metallothionin (MMT) have the same kinetics as do the *ets* genes. The α-tubulin and beta actin housekeeping genes are not induced by this treatment

the protein kinase C pathway is thus far unique for the *ets*-2.

The human *ets*-2 gene product has been identified by means of specific antibodies directed against antigen obtained from the bacterially expressed partial cDNA clone of the *ets*-2 gene [25, 26], as well as an oligopeptide antigen corresponding to a highly conserved hydrophilic region of *ets*. Using both types of sera, a 56-K protein has been identified as the human *ets*-2 gene product; this protein, like the oncogene product $p135^{gag\text{-}myb\text{-}ets}$, is also located in the nucleus. This nuclear localization of the human *ets*-2 protein supports its relationship with other nuclear proto-oncogene products, such as those encoded by c-*fos* and c-*myc*, that have been seen to be expressed in association with cellular proliferation.

Recently, we have found that the *ets*-2 protein is phosphorylated and has a rapid turnover of normally less than half an hour [27, 28]; however, when cells are treated with a tumor promoter such as the phorbol ester 12-O-tetradecanoyl phorbol-13-acetate (TPA), the level of *ets*-2 protein very quickly becomes markedly elevated. This increase in *ets*-2 protein appears to be due to the stabilization of the protein, because the *ets*-2, $p56^{ets\text{-}2}$ product increased its half-life by more than 2 h in the presence of TPA (Fig. 3), while the *ets*-2-specific mRNA did not change. Since an inhibitor of protein kinase C also interferes with the stabilization of $p56^{ets\text{-}2}$, and the effect of TPA could be mimicked by a synthetic diacylglycerol, it appears that the protein kinase C signal pathway is probably involved in the induction of this nuclear

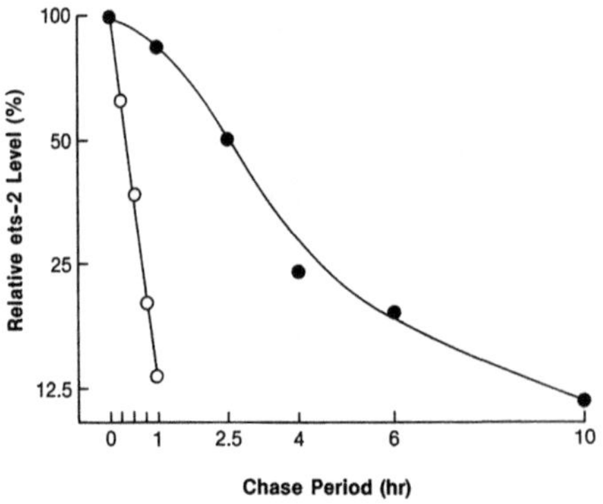

Fig. 3. Effect of 12-O-Tetradecanoylphorbol-13-acetate (TPA) on the *ets*-2 protein turnover. CEM (T-lymphocytic) cells were pulse-labeled with [^{35}S]methionine and chased in excess unlabeled methionine in the presence of TPA (10 n*M*). Cells were chased for 1, 2.5, 4, 6, 10, and 29 h or 15, 30, 45, and 60 min in the absence of TPA. The X-ray film of the immunoprecipitates were analyzed and traced by a densitometer, and areas under peaks were calculated as measures of the amount of the labeled *ets*-2 protein. The labeled *ets*-2 protein levels at various chase periods were normalized to the 100% value obtained at zero time and plotted in semilogarithmic scale. The data were fit to a first-order exponential decay curve with a correlation coefficient of 0.99. *Open circles* are control cells (t½ = 20 min); *closed circles* are the CEM cells chased in the presence of 10 n*M* TPA (t½ = 160 min)

proto-oncogene. In this respect, however, the *ets*-2 protein is unique from the other nuclear oncogene products in its ability to respond to TPA post-translationally. Other nuclear proto-oncogenes respond to TPA at the mRNA level, and subsequently at the protein level, but thus far, only the *ets*-2 protein level is distinct since it increases in the absence of any increase mRNA level. Additionally, consistent with a post-translational mechanism, it should be noted that the protein synthesis inhibitor cycloheximide enhances the effect of TPA on the level of *ets*-2 protein retarding its turnover even further. Taken together, these data suggest that the expression of the proto-*ets*-2 gene and its encoded products are rapidly controlled by signal transduction most probably involving, directly or indirectly, the protein kinase C pathway. This post-translational response implies that such a precisely controlled regulatory mechanism may be an essential feature of the function of the *ets*-2 encoded protein [28]. It can even be speculated that the stabilization of the *ets*-2 protein and the consequent transient elevation of its level may be an intermediary step in the signaling process of the protein kinase C pathway, perhaps interconnecting this activation process with other proliferative gene(s) regulation. The deregulation or any subtle alteration of these controlling mechanisms may then, via the signal transduction pathway, cause a profound change in the intracellular physiology.

Finally, it should be noted that the human *ets*-2 gene is located on the region of chromosome 21 (21q22.3) that is implicated in Down's syndrome [29–31]; trisomy of this small chromosomal domain results in the full manifestation of the Down's syndrome phenotype. Because the *ets*-2 protein level appears to be under precise control, the increase in the *ets*-2 gene dosage resulting from the trisomy may seriously affect the control of the *ets*-2 protein, and it is therefore conceivable that this deregulation is a contributing factor in the development of Down's syndrome.

References

1. Varmus H (1988) Retroviruses. Science 240:1427–1435
2. Marguardt J, Hunkapillar WW, Hood LE, Todaro GT (1984) Rat transforming growth factor type 1: structure and relation to epidermal growth factor. Science 223:1079–1082
3. Stroobant P, Rice AP, Gullick WJ, Cheng DJ, Kerr IM, Waterfield MD (1985) Purification and characterization of vaccinia virus growth factor. Cell 42:383–393
4. Brown JP, Twardzik DR, Marquardt H, Todaro GJ (1985) Vaccinia virus encodes a polypeptide homologous to epidermal

growth factor and transforming growth factor. Nature 313:491–492

5. Yarden Y, Ullrich A (1988) Growth factor receptor tyrosine kinases. Annu Rev Biochem 57:443–478
6. Barbacid M (1987) *ras* genes. Annu Rev Biochem 56:779–827
7. Nishizuka Y (1984) The role of protein kinase C in cell surface signal transduction and tumour promotion. Nature 308:693–698
8. Adari H, Lowy DR, Williamson BM, Der CJ, McCormick F (1988) Guanosine triphosphatase activating protein (GAP) interacts with the p21 *ras* effector binding domain. Science 240:518–521
9. Fleischman LF, Chahwala SB, Cantly L (1986) *ras*-Transformed cells: altered levels of phosphatidylinositol 4,5-biphosphate and catabolites. Science 21:407–410
10. Lacal JC, de la Pena P, Moscat J, Garcia-Barreno P, Anderson PS, Aaronson SA (1987) Rapid stimulation of diacylglycerol production in *Xenopus* oocytes by microinjection of H-*ras* p21. Science 238:533–536
11. Yu C-L, Tsai M-H, Stacey DW (1988) Cellular *ras* activity and phospholipid metabolism. Cell 52:63–71
12. Rauscher FJ, Cohen DR, Curran T, Bos TJ, Vogt PK, Bohmann D, Tjian R, Franza BB (1988) *fos*-Associated protein p39 is the product of the *jun* proto-oncogene. Science 240:1010–1016
13. Ross RJ, Glomset JA, Kariya B, Harker L (1974) A platelet-dependent serum factor that stimulates the proliferation of smooth muscle cells in vitro. Proc Natl Acad Sci USA 71:1207–1210
14. Kelly K, Cochran BH, Stiles CD, Leder P (1983) Cell-specific regulation of the c-*myc* gene by lymphocyte mitogens and platelet-derived growth factor. Cell 35:603–610
15. Kruijer W, Cooper JA, Hunter T, Verma I (1984) Platelet-derived growth factor induces rapid but transient expression of the c-*fos* gene and protein. Nature 312:711–716
16. Watson DK, Smith MJ, Kozak C, Reeves R, Gearhart J, Nunn MF, Nash W, Fowle III jr, Duesberg P, Papas TS, O'Brien SJ (1986) Conserved chromosomal positions of dual domains of the *ets* proto-oncogene in cats, mice and man. Proc Natl Acad Sci USA 83:1792–1796
17. Watson DK, McWilliams-Smith MJ, Nunn MF, Duesberg PH, O'Brien SJ, Papas TS (1985) The *ets* sequence from the transforming gene of avian erythroblastosis virus, E26, has unique domains on human chromosomes 11 and 21: both loci are transcriptionally active. Proc Natl Acad Sci USA 82:7294–7298
18. Watson DK, McWilliams-Smith MJ, Lapis P, Lautenberger JA, Schweinfest CW, Papas TS (1988) Human and mouse *ets*-2 genes: members of a family of *ets* genes. Proc Natl Acad Sci USA 85:7862–7866
19. Rao VN, Papas TS, Reddy ESP (1987) *Erg*, a human *ets*-related gene on chromosome 21: alternative splicing, polyadenylation, and translation. Science 237:635–639
20. Reddy ESP, Rao VN, Papas TS (1987) The *erg* gene: a human gene related to the *ets* oncogene. Proc Natl Acad Sci USA 84:6131–6135
21. Pribyl LJ, Watson DK, McWilliams MJ, Ascione R, Papas TS (1988) The *Drosophila ets*-2 gene: molecular structure, chromosomal localization, and developmental expression. Dev Biol 127:45–53
22. Chen Z-O, Kan NC, Pribyl L, Lautenberger JA, Moudrianakis E, Papas TS (1988) Molecular cloning of the *ets* proto-oncogene of the sea urchin and analysis of its developmental expression. Dev Biol 125:432–440
23. Bhat NK, Fisher RJ, Fujiwara S, Ascione R, Papas TS (1987) Differential regulation of *ets* loci during murine hepatic regeneration. In: Voellmy RW, Ahmad F, Black S, Burgess DR, Rotundo R, Scott WA, Whelan WJ (eds) Advances in gene technology: the molecular biology of development. Cambridge University Press, Cambridge, p 70
24. Bhat NK, Fisher RJ, Fujiwara S, Ascione R, Papas TS (1987) Temporal and tissue-specific expression of mouse *ets* genes. Proc Natl Acad Sci USA 84:3161–3165
25. Fujiwara S, Fisher RJ, Seth A, Bhat NK, Papas TS (1987) Human *ets*-1 and *ets*-2 proteins: identification and intracellular localization. In: Voellmy RW, Ahmad F, Black S, Burgess DR, Rotundo R, Scott WA, Whelan WJ (eds) Advances in gene technology: the molecular biology of development. Cambridge University Press, Cambridge, p 77

26. Fujiwara S, Fisher RJ, Seth A, Bhat NK, Showalter SD, Zweig M, Papas TS (1988) Characterization and localization of the products of the human homologs of the v-*ets* oncogene. Oncogene 2:99–103
27. Fujiwara S, Fisher RJ, Bhat NK, Papas TS (1988) Human *ets*-2 protein: nuclear location, phosphorylation, rapid turnover and induction by TPA. In: Brew K, Ahmad F, Bialy H, Black S, Fenna RE, Puett D, Scott WA, Van Brunt J, Voellmy RW, Whelan WJ, Woessner JF (eds) Advances in gene technology: protein engineering and production. IRL, Oxford, p 107
28. Fujiwara S, Fisher RJ, Bhat NK, Diaz de la Espina SM, Papas TS (1988) A short-lived nuclear phosphoprotein encoded by human *ets*-2 proto-oncogene is stabilized by protein kinase C activation. Mol Cell Biol 8:4700–4706
29. Sacchi N, Gusella JF, Perroni L, Dagna Bricarelli F, Papas TS (1988) Lack of evidence for association of meiotic nondisjunction with particular DNA haplotypes on chromosome 21. Proc Natl Acad Sci USA 85:4788–4794
30. Sacchi N, Nalbantoglu J, Sergovich FR, Papas TS (1988) The *ets*-2 gene in Down syndrome genetic region is not rearranged in Alzheimer's disease. Proc Natl Acad Sci USA 85:7675–7679
31. Patterson D (1987) The causes of Down syndrome. Sci Am 257:52–60

Induction of Lymphokine-Activated Killer (LAK) Cells Against Human Leukemia Cells*

J. V. Teichmann, W.-D. Ludwig, H. Seibt-Jung, G. Sieber, and E. Thiel

A. Introduction

Lymphokine-activated killer (LAK) cells are known to lyse fresh solid tumor cells in vitro [1], and clinical studies suggest that adoptive immunotherapy with LAK cells and interleukin-2 (IL-2) may be a promising approach in the treatment of solid tumors [2, 3]. As only few data are available on the LAK cell system in connection with human leukemia [4, 5], we investigated the induction of LAK cells against human leukemia cells in vitro and studied the augmentation of cytotoxic mechanisms which may be achieved by the combined application of different lymphokines or the coculturing of effector cells with tumor cells.

B. Materials and Methods

I. Lymphokines

Recombinant IL-2 and rIFN-γ were generous gifts from the Glaxo Institute of Molecular Biology, Geneva, and the Ernst-Boehringer Institute, Vienna.

II. Induction of LAK Cells

Allogeneic human LAK cells were generated from peripheral blood mononuclear cells obtained from healthy volunteers and cultured (1×10^6/ml) for 6 days with rIL-2 (1000 U/ml). In some experiments, we induced LAK activity from bone marrow (BM) or peripheral blood (PB) of leukemia patients by long-term culturing (14–24 days) of mononuclear cells in the presence of IL-2.

III. Target Cells

Fresh leukemic cells were obtained from BM or PB of untreated patients by Ficoll-Hypaque gradient centrifugation. Phenotypic analyses were performed by standard indirect immunofluorescence asssays as described elsewhere [6]. K 562 and Daudi cells, maintained in continuous cultures, served as standard target cells. For use in the cytotoxicity assay, fresh leukemic cells and cell lines were labeled with 300 µCi or 50 µCi sodium chromate respectively.

IV. Cytotoxicity Assay

Lymphokine-activated killer cell activity was determined in a standard 4-h ^{51}Cr release assay using 5×10^3 target cells and various effector-to-target (E:T) ratios.

C. Results and Discussion

Leukemic cells from 62 patients were evaluated for their susceptibility to the lytic effect of allogeneic LAK cells. A significant lysis (defined as over 20% specific lysis at an E:T ratio of 100:1) was found in about two-thirds of the leukemias examined (Table 1). No substantial differences could be detected between myeloid and lymphoid leukemias or with regard to the immunological phenotype.

Dept. of Hematology and Oncology, Klinikum Steglitz, Free University of Berlin, D-1000 Berlin, FRG

* Supported by the Deutsche Krebshilfe e.V., Bonn (W 19/87/Te 1)

Target cells	n	Cytotoxicity (% specific lysis) E:T 100:1	
		>20%	<20%
AML	22	15	7
CML-BC	4	3	1
CML chronic phase	3	2	1
ALL	29	17	12
O-ALL	4	2	2
c-ALL	15	13	2
B-ALL	2	1	1
T-ALL	8	1	7
B-CLL	4	2	2
Total	62	39	23

AML, acute myeloblastic leukemia; CML, chronic myelocytic leukemia; ALL, acute lymphoblastic leukemia; CLL, chronic lymphocytic leukemia.

Table 1. Susceptibility of fresh leukemic cells to allogeneic LAK cells

There is growing evidence that leukemia patients have impaired natural killer (NK) cell functions, and this may contribute to leukemogenesis [5, 7]. We therefore studied the possibility of using IL-2 to activate NK cells of leukemic patients. In the presence of IL-2, long-term culturing of mononuclear cells from leukemia patients' BM or PB, containing a high percentage of tumor cells, resulted in induction of highly active cytotoxic cells. During culturing, the number of malignant cells decreased, while residual large granular lymphocytes (LGLs) expanded and developed lytic activity against NK-sensitive (K 562) and NK-resistant (Daudi) target cells (Fig. 1). These cultures of leukemic cells and residual normal mononuclear cells resemble tumor-infiltrating lymphocytes (TILs), which are known to be more active than LAK cells generated from peripheral blood [8]. As demonstrated, the application of IL-2 in vitro can result in an activation and expansion of LGL even in highly leukemic patients, suggesting that the generation of LAK cells by IL-2 in vivo, e.g., after chemotherapy, may be therapeutically useful in preventing the relapse or spread of leukemia.

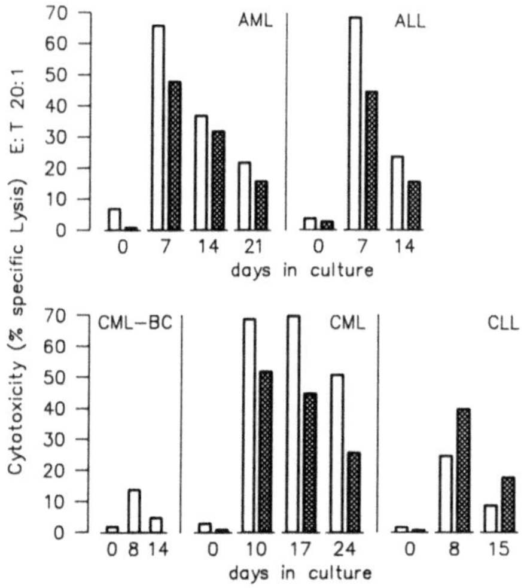

Fig. 1. Induction of LAK cells from BM or PB of leukemia patients by long-term culturing of mononuclear cells in the presence of IL-2 (1000 U/ml) against K 562 (□) and Daudi (▩) target cells

Reports that endogenous IFN-γ is required for IL-2 induction of LAK cells [9] led us to conjecture that it might be possible to augment LAK activity by addition of rIFN-γ during the activation process. The results of our studies show that the combined application of rIFN-γ and rIL-2 can improve the effectivity of cytotoxic mechanisms even at a low E:T ratio of 2.5:1 (Fig. 2). The sequential ad-

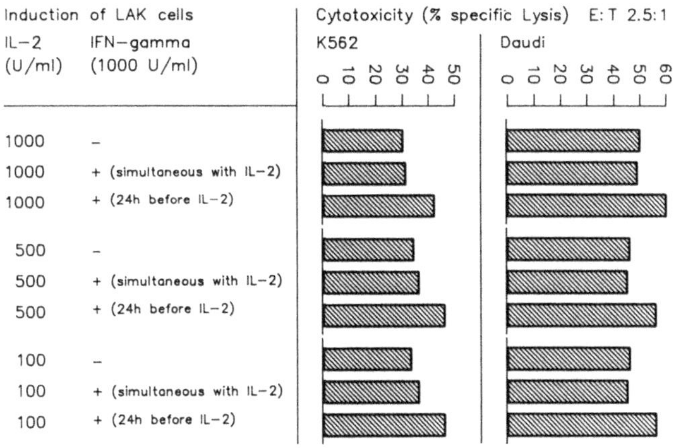

Fig. 2. Improvement of the effectivity of cytotoxic mechanisms: combined sequential application of rIL-2 and rIFN-γ

Table 2. Improvement of the effectivity of cytotoxic mechanisms: coculturing of effector cells (R) with IL-2 and mitomycin-treated tumor cells (S) at an R:S ratio of 5:1

Induction of LAK cells		Cytotoxicity (% specific lysis) E:T 20:1		
IL-2 (1000 U/ml)	Leukemia cells (mitomycin-Tx)	K 562	Daudi	AML
–	–	6	5	0
+	–	50	42	10
–	K 562	1	0	0
+	K 562	69	53	17
–	Daudi	49	25	3
+	Daudi	64	79	22
–	AML	7	6	1
+	AML	79	33	19

AML, acute myeloblastic leukemia.

ministration of rIFN-γ 24 h before IL-2 resulted in a higher augmentation of cytotoxicity than the simultaneous application of both lymphokines. This may be due to induction of IL-2 receptors on the effector cells before they are affected by IL-2, resulting in a more efficient activation or a recruitment of additional cell populations which are not activated by IL-2 alone. Generation of more active LAK cells for clinical application should improve their therapeutic efficacy. This could reduce the necessary dosage and/or treatment time of IL-2 and thus also its toxicity.

The target structures of tumor cells which are recognized by NK cells or LAK cells have not yet been defined. As there are data indicating that in vitro culturing of human peripheral blood mononuclear cells (PBMCs) with B-lymphoblastoid cell lines results in a preferential proliferation of NK cells [10] and that NK cells can be activated by direct stimulation with the NK-sensitive tumor cell line K 562 [11], we hypothesized that LAK cell activity could possibly be augmented by coculturing effector cells with tumor cells in the presence of IL-2. Our studies with mitomycin-treated K 562 or Daudi cell lines as well as fresh leukemia cells in culture with effector cells and IL-2 showed marked augmentation of cytotoxicity compared with cultures of

PBMCs with IL-2 alone (Table 2). This may be due to target cell structures being present during the activation process and giving additional signals for the generation of LAK cells.

The results of these studies suggest that IL-2-induced lymphokine-activated killer (LAK) cells may be of great value in the treatment of leukemia, especially when the tumor burden is low, e.g., during maintenance chemotherapy to eliminate minimal residual disease or in early relapse. The combined application of different lymphokines or the coculturing of effector cells with IL-2 and tumor cells are possible approaches for improving the effectivity of cytotoxic mechanisms.

References

1. Grimm EA, Mazumder A, Zhang HZ, Rosenberg SA (1982) Lymphokine-activated killer cell phenomenon. Lysis of natural killer-resistant fresh solid tumor cells by interleukin 2-activated autologous human peripheral blood lymphocytes. J Exp Med 155:1823–1841
2. Rosenberg SA, Lotze MT, Muul LM, Chang AE, Avis FP, Leitman S, Linehan WM, Robertson CN, Lee RE, Rubin JT, Seipp CA, Simpson CG, White DE (1987) A progress report on the treatment of 157 patients with advanced cancer using lymphokine-activated killer cells and interleukin-2 or high-dose interleukin-2 alone. N Engl J Med 316:889–897
3. West WH, Tauer KW, Yannelli JR, Marshall GD, Orr DW, Thurman GB, Oldham RK (1987) Constant-infusion recombinant interleukin-2 in adoptive immunotherapy of advanced cancer. N Engl J Med 316:898–905
4. Oshimi K, Oshimi Y, Akutsu M, Takei Y, Saito H, Okada M, Mizoguchi H (1986) Cytotoxicity of interleukin 2-activated lymphocytes for leukemia and lymphoma cells. Blood 68:938–948
5. Lotzová E, Savary CA, Herberman RB (1987) Induction of NK cell activity against fresh human leukemia in culture with interleukin 2. J Immunol 138:2718–2727
6. Ludwig W-D, Bartram CR, Ritter J, Raghavachar A, Hiddemann W, Heil G, Harbott J, Seibt-Jung H, Teichmann JV, Riehm H (1988) Abiguous phenotypes and genotypes in 16 children with acute leukemia as characterized by multiparameter analysis. Blood 71:1518–1528
7. Lotzová E, Savary CA, Herberman RB, Dicke KA (1986) Can NK cells play a role in therapy of leukemia? Nat Immun Cell Growth Regul 5:61–63
8. Belldegrun A, Muul LM, Rosenberg SA (1988) Interleukin-2 expanded tumor-infiltrating lymphocytes in human renal cell cancer: isolation, characterization, and antitumor activity. Cancer Res 48:206–214
9. Itoh K, Shiiba K, Shimizu Y, Suzuki R, Kumagai K (1985) Generation of activated killer (AK) cells by recombinant interleukin-2 (rIL-2) in collaboration with interferon-gamma (IFN-γ). J Immunol 134:3124–3129
10. Perussia B, Ramoni C, Anegon I, Cuturi MC, Faust J, Trinchieri G (1987) Preferential proliferation of natural killer cells among peripheral blood mononuclear cells cocultured with B-lymphoblastoid cell lines. Nat Immun Cell Growth Regul 6:171–188
11. Phillips JH, Lanier LL (1985) A model for the differentiation of human natural killer cells. Studies on the in vitro activation of Leu-11$^+$ granular lymphocytes with a natural killer-sensitive tumor cell, K 562. J Exp Med 161:1464–1482

Expression and Chromosomal Assignment of a Novel Protein-Tyrosine Kinase Gene Related to the Insulin Receptor Family

A. R. Bauskin[1], M. Zion[1], J. Szpirer[2], C. Szpirer[2], M. Q. Islam[3], G. Levan[3], G. Klein[4], and Y. Ben-Neriah[1]

A. Introduction

There is growing evidence in recent literature for the involvement of protein-tyrosine phosphorylation in the early events of leukocyte activation. Interleukin-3 (IL-3), a growth factor essential for the growth of hematopoietic stem cells, myeloid cells, and possibly early lymphoid cells induces rapid phosphorylation of several cellular proteins in different IL-3-dependent cells (B. Isfort and J. Ihle, personal communication; [1]). One of the early events in T-cell activation is the tyrosine phosphorylation of a component of the T-cell antigen receptor [2]. Macrophage proliferation in response to M-CSF is mediated through the activation of the tyrosine kinase moiety of its receptor, the *c-fms* proto-oncogene [3].

The protein-tyrosine kinases described to date are either transmembrane proteins having an extracellular ligand binding domain [3–8] or cytoplasmic proteins related to the *v-src* oncogene [9–11]. Most of these proteins are expressed in a wide variety of cells and tissues; few are tissue specific [12]. Previous studies have suggested that lymphokines may mediate hematopoietic cell survival via their action on glucose transport [13, 14] which is regulated in some cells through the protein-tyrosine kinase activity of the insulin receptor [15]. As activation of glucose transport may be essential to hematopoietic cell growth, we investigated the possibility that insulin receptor-like genes are expressed specifically in hematopoietic cells. Using the insulin-receptor related avian sarcoma oncogene, *v-ros* as a probe [16], we isolated and characterized the complementary DNA of a novel gene denoted *ltk* (leukocyte tyrosine kinase), which is expressed mainly in leukocytes and has so far unique structural properties; *ltk* appears to encode a transmembrane protein devoid of an extracellular domain and is related to several tyrosine kinase receptor genes of the insulin receptor family [17].

B. Results and Discussion

The polypeptide predicted from the *ltk* cDNA sequence has a molecular mass of 52212 Daltons, and when plotted according to the hydrophilicity table of Hopp and Woods [18], reveals a predominant stretch of 26 hydrophobic amino acids at the N-terminus of the polypeptide (positions 9–34), followed by three polar amino acids (Fig. 1). This region, albeit having an unusual position, is compatible with the properties of a transmembrane domain and the positively charged peptide (Asn-Gln-Lys) may serve as a membrane transfer stop signal. Following the hydrophobic stretch is a region homologous to the cytoplasmic region of protein kinases having the characteristics of a protein-tyrosine kinase domain [12].

[1] The Lautenberg Center for General and Tumor Immunology, The Hebrew University – Hadassah Medical School, Jerusalem, Israel
[2] Departement de Biologie Moléculaire, Université Libre de Bruxelles, Belgium
[3] Department of Genetics, University of Gothenburg, Göteborg, Sweden
[4] Department of Tumor Biology, Karolinska Institute, Stockholm, Sweden

```
        10                  30                  50
MDLPTTASPLILMGAVVAALALSLLMMCAVLILVNQKCQGLWGTRLPGPELELSKLRSSA

        70                  90                 110
                                          ┌TK
                                           x x  x
IRTAPNPYYCQVGLSPAQPWPLPPGLTEVSPANVTLLRALGHGAFGEVYEGLVTGLPGDS

       130                 150                 170
       x
SPLPVAIKTLPELCSHQDELDFLMEALIISKFSHQNIVRCVGLSFRSAPRLILLELMSGG

       190                 210                 230
DMKSFLRHSRPHPGQLAPLTMQDLLQLAQDIAQGCHYLEENHFIHRDIAARNCLLSCSGA

       250                 270                 290
                   x
SRVAKIGDFGMARDIYQASYYRKGGRTLLPVKWMPPEALLEGLFTSKTDSWSFGVLLWEI

       310                 330                 350
                                                   TK┐
FSLGYMPYPGHTNQEVLDFIATGNRMDPPRNCPGPVYRIMTQCWQHQPELRPDFGSILER

       370                 390                 410
IQYCTQDPDVLNSPLPVEPGPILEEEEASRLGNRSLEGLRSPKPLELSSQNLKSWGGGLL

       430                 450                 470
GSWLPSGLKTLKPRCLQPQNIWNPTYGSWTPRGPQGEDTGIEHCNGSSSSSIPGIQ
```

Fig. 1. Predicted protein sequence of *ltk*. The putative transmembrane region is underlined with a *solid bar*. *Asterisks:* Gly 101, Gly 103, Gly 106, and Lys 128 indicate the potential ATP binding site; and Tyr 260, the putative autophosphorylation site. The tyrosine kinase (TK) domain is indicated by *arrows*

A search through the GenBank protein data-bank reveals that *ltk* shares a high degree of similarity throughout the tyrosine kinase domain with receptor proteins. These include the human *c-ros* protein protein (53%) [19], the *Drosophila* hemeotic protein *sev* (49%) [20], the human receptors for insulin and insulin-like growth factor-1 (IGF-1) (48%) [5, 8] and the receptor portion of the human *trk* protein (45%) [21]. Comparison of *ltk* with other kinase-receptor proteins reveals a lower degree of similarity: 30%–38%. With the former group of receptor proteins *ltk* shares a unique peptide: Tyr-Tyr-Arg-Lys-X-Gly-X-X-Leu-Leu-Pro-Val at the autophosphorylation site. Interestingly, the insulin and IGF-1 receptors are among the few kinases where autophosporylation activates the kinase activity of the protein [22–24]. It therefore seems that *ltk*, together with the former group of kinase-receptor genes, forms a subfamily within the protein-tyrosine kinase family. It remains to be seen whether the members of this subfamily are also functionally related.

In the mouse and human genome *ltk* is transcribed from a single copy gene. Using a series of mouse-human and rat-human cell hybrids we assigned the human *ltk* gene to chromosome 15 (Fig. 2, Table 1). Interestingly, the closely related IGF-1 receptor gene is also located on human chromosome 15.

The expression of *ltk* was analyzed in different murine tissues. Northern blot analysis revealed two transcripts in Rad-LV transformed T-cell lines, Tc38 and Tc23 [25] (Fig. 3a) and in the thymus (not shown) of approximately 2.6 kb and 3.0 kb. The expression of *ltk* in the various mouse tissues and cell lines was quantified by RNA slot-blot hybridization analysis using a 5′ *ltk* probe (excluding the kinase coding region; Fig. 3b), *ltk* was abundant in the thymus and observed in the spleen and kidney. Among the different cell lines, *ltk* was expressed in an interleukin-2 (IL-2) dependent *ltk* line, CTLD4; a macrophage colony stimulating factor (M-CSF) dependent macrophage cell line, Mac26; in the IL-3 dependent cell line, Ba/F3, which is thought to be an early pre-B-lymphocyte line (R. Palacios, personal communication); and in mouse lymphokine activated killer (LAK) cells. Rehybridization of the blot with a *ltk* kinase region probe shows a similar tissue distribution except for an additional faint signal in the brain (not shown). It thus appears that *ltk* is expressed mainly in cells of hematopoietic origin.

Considering the homology of *ltk* to transmembrane protein kinases, it was expected to include an extracellular recognition unit. However, the putative external portion of the protein is only nine amino acids long and unlikely to be sufficient for ligand binding. It is possible that one of the two transcripts detected by the *ltk* probe (Fig. 3a) encodes another protein with a larger extracytoplasmic domain. N-terminal variation has been observed in several other protein-tyrosine kinases as a result of alternative mRNA splicing [11, 26]. The *c-abl* protooncogene generates two mRNA transcripts of 5.5 and 6.5 kb by alternative splicing, each encoding a protein with a

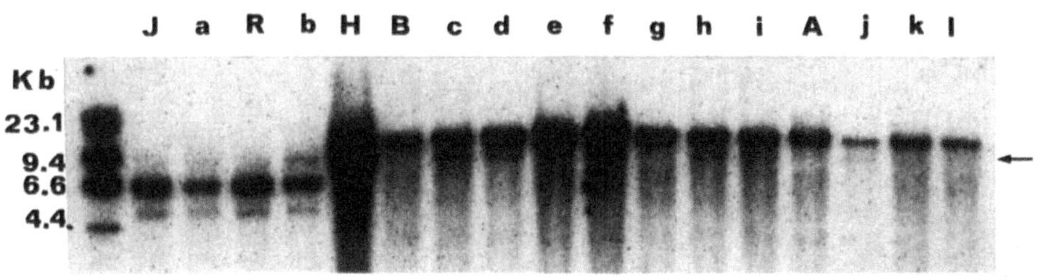

Fig. 2. Mapping of the *ltk* gene to human chromosome 15. Southern blot analysis of DNA (10 μg) from mouse cell lines, A9 (*A*) and B82 (*B*); rat cell lines, JF1 (*J*) and Rat 2 (*R*); human HeLa cell line (*H*); mouse-human and rat-human hybrids, whose chromosomal content is indicated in Table 1. The DNA was digested with an excess of the restriction enzyme HindIII, fractionated in 1% agarose, transferred to a nylon membrane filter (Zetabind), and hybridized to a ^{32}P radiolabeled *ltk* cDNA probe. The position of human *ltk* is indicated by an *arrow*

Table 1. Detection of human *ltk* gene in human × rat and human × mouse hybrids

Hybrids	Human chromosomes																							
	1	2	3	4	5	6	7	8	9	10	11	12	13	14	15	16	17	18	19	20	21	22	X	Y
Positive hybrids																								
j-HA221 [a]	−	−	−	+	−	+	+	−	−	−	+	−	−	−	+	−	+	−	+	−	+	−	+	−
i-HB26 [b]	−	−	+	−	−	+	+	−	−	+	(+)	(−)	+	+	+	+	+	+	(+)	+	+	+	−	−
h-HB29 [b]	−	−	+	+	−	+	+	+	+	+	+	+	+	+	+	+	+	+	+	+	+	+	−	−
g-HB33 [b]	−	−	(+)	(+)	−	+	−	+	−	−	(+)	−	−	+	−	−	(+)	(−)	−	−	−	+	−	−
f-HB43 [b]	+	+	+	+	−	+	+	+	−	−	+	+	−	+	+	+	+	+	+	+	+	+	+	−
d-HB142.2 [b]	−	−	−	+	−	−	−	−	−	−	−	−	−	+	+	−	+	−	−	−	−	−	−	−
c-HB182 [b]	−	−	+	+	+	+	+	+	−	−	+	+	+	+	+	−	+	+	+	+	+	+	−	−
b-HR40C8 [d]	−	−	−	−	−	−	+	−	−	+	−	+	+	+	+	(−)	+	−	+	+	+	(+)	+	−
Negative hybrids																								
1-HA11 [a]	−	−	−	+	+	+	−	+	−	−	+	−	−	−	−	−	+	−	−	+	+	−	−	−
k-HA13 [a]	+	−	−	+	+	+	−	+	−	−	+	+	−	−	−	+	+	−	+	+	+	−	+	−
e-HB112 [b]	−	−	+	(−)	−	−	−	(−)	−	−	+	+	−	+	−	−	+	−	−	−	+	−	−	−
a-JV211 [c]	+	(−)	+	(−)	+	+	+	+	−	+	+	+	−	−	−	+	−	−	−	+	−	+	−	
Percent independent discordant clones	80	60	40	30	70	40	30	60	70	60	20	60	40	20	10	60	20	30	30	60	40	30	50	70

The symbols mean: +, chromosome present in at least 60% of the metaphases; (+), chromosome present in 30% to 50% of the metaphases; (−), chromosome present in 10% to 20% of the metaphases; −, chromosome absent.

[a] Hybrids derived from the mouse A9 parental cell.

[b] Hybrids derived from the mouse B82 parental cell.

[c] Hybrid derived from the rat JF1 parental cell.

[d] Hybrid derived from the Rat2 parental cell.

The hybrids used are described in [28].

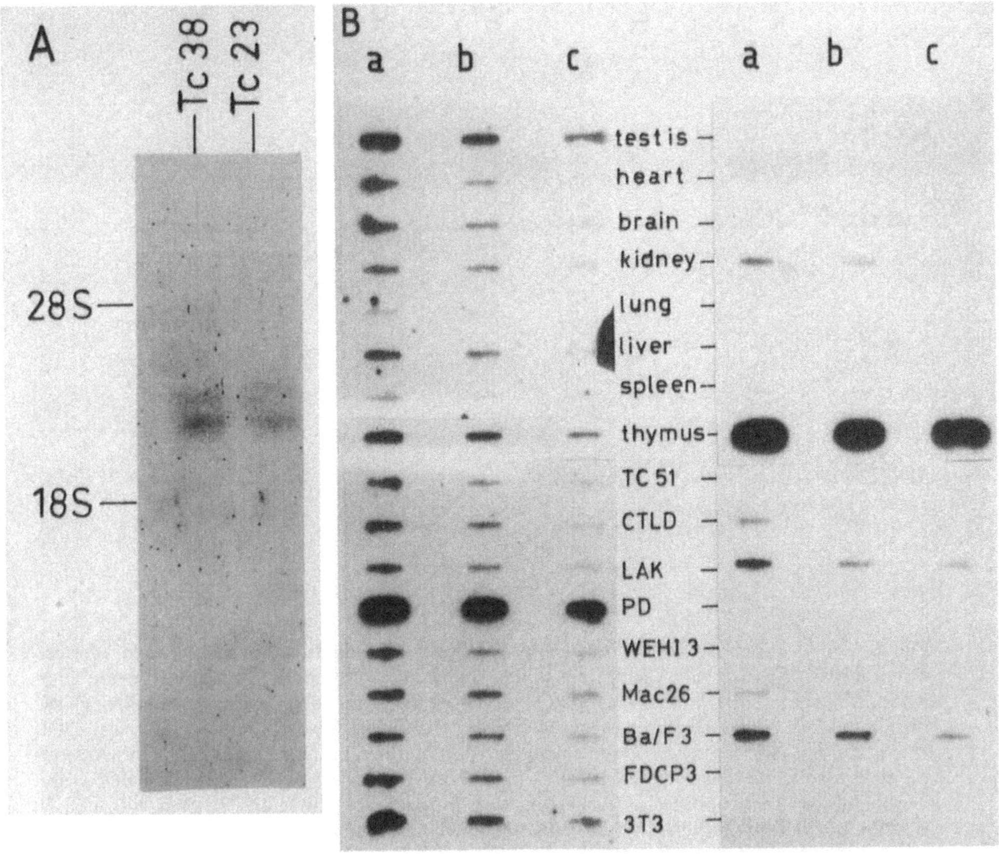

Fig. 3A, B. Expression of *ltk* in different murine tissues and cell lines. **A** Northern blot analysis of poly(A)$^+$ RNA from Rad-LV transformed T-cell lines [22]. Poly(A)$^+$ RNA was extracted by $LiCl_2$-urea lysis and phenol/chloroform extraction, followed by Oligo(dT) selection and separated on a formaldehyde-containing 1% agarose gel (5 µg/lane). The RNA was transferred to Zetabind nylon membrane and hybridized to radiolabeled *ltk* cDNA probe. **B** Slot-blot analysis of total cellular RNA isolated from different murine tissues and cell lines. RNA, 30 µg (*a*), 10 µg (*b*), and 3 µg (*c*) was spotted onto nitrocellulose using the Minifold II slot-blotter (Schleicher and Shuell) and hybridized with a radiolabelled *c-abl* kinase region probe as a control, *left panel;* and a radiolabelled 5′ *ltk* fragment, *right panel.* Cell lines: TC51, Rad-LV transformed T-cell line [22]; CTLD, an IL-2 dependent T-cell line; LAK, lymphokine activated killer cells were prepared by incubating splenocytes with IL-2 (1000 U/ml) for 5 days; PD, Abelson-MuLV transformed pre-B cell line; WEHI 3, macrophage/monocyte line; Mac 26, a M-CSF dependent bone marrow derived macrophage line; Ba/F3, and IL-3 dependent lymphoid line; FDCP 3, a GM-CSF dependent myeloid line; 3T3, a fibroblast line from C3H mouse

different N-terminus [11]. If the origin of the two *ltk* transcripts is similar, we may have characterized one alternative splice form, while the other may encode a different N-terminus. Nevertheless, S1 nuclease analysis showed transcripts fully complementary to the *ltk* cDNA clone in the thymus.

To demonstrate that *ltk* encodes a transmembrane protein, *ltk*-specific antiserum was prepared against an *ltk*/β-galactosidase fusion protein and used to

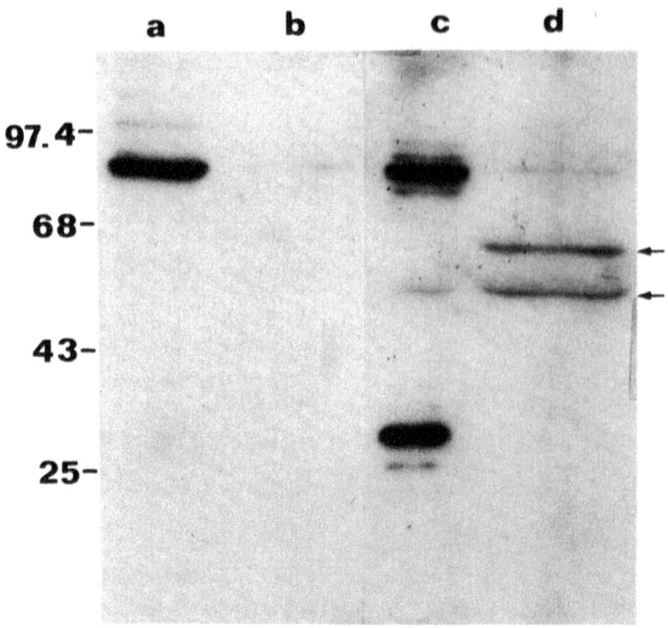

Fig. 4. The *ltk* protein in the thymus has properties of an integral membrane protein. Western blot analysis of the thymus *ltk* protein. Lanes *a* and *b*, reaction with preimmune serum; lanes *c* and *d*, reaction with anti-*ltk* serum; aqueous phase proteins (lanes *a* and *c*); detergent phase proteins (lanes *b* and *d*). Sizes of standards are shown on the *left*. Triton X-114 detergent fractionation[28] was used to separate detergent-soluble membrane proteins from the aqueous phase cytosolic proteins. Protein (50 μg) from both fractions were separated by SDS polyacrylamide gel electrophoresis. transferred to a nitrocellulose filter and reacted with a 1:50 dilution of either preimmune or anti-*ltk* serum which had been preadsorbed on thymus aqueous phase proteins. Blots were developed with ^{125}I goat anti-rabbit antibodies

demonstrate the presence of *ltk* protein in a thymus membrane protein fraction (Fig. 4). The specificity of the serum was shown by immunoprecipitation of *ltk* protein. The largest protein band synthesized in rabbit reticulocyte lysate has a molecular mass of 52 kD in agreement with the predicted size of the protein encoded by the long open reading frame of *ltk* and is immunoprecipitated by *ltk*-specific antisera (not shown).

Western blot analysis of membrane fractionated thymus proteins showed that anti-*ltk* serum reacts specifically with two protein bands of 56 and 64 kD in the thymus membrane fraction (Fig. 4). The 56 kD band is consistent with the expected size of the encoded protein, while the 64 kD band may represent either a second *ltk* gene product due to alternative splicing alternative translation-initiation or a cross-reactive protein.

These results show that authentic *ltk* transcripts exist in thymocytes encoding a protein with features of a transmembrane catalytic subunit. The unusual structure of the putative *ltk* protein implies a role in signal transduction. Thymocytes, B cells, and the various other cells which express the gene, grow and differentiate in response to various extracellular stimuli. Protein-tyrosine phosphorylation appears to be involved in the signal transduction of such stimuli. However, with the exception of the M-CSF receptor, a specific tyrosine-kinase has yet to be identified in the other signal transduction pathways. Therefore, *ltk* is a candidate transducer for early activation events in leukocytes.

Acknowledgement. This work was supported by awards from the Israel Academy of Science – Basic Research Fund and the Israel Cancer Research Fund, the Fund for Medical Scientific Research (Belgium), and the Swedish Cancer Society and Erik Sörensen Foundation. Part of the work was undertaken during the tenure of the International Union Against Cancer (ICRETT) personal exchange award (to Y.B.-N.).

References

1. Morla AD et al. (1988) Haematopoietic growth factors activate the tyrosine phosphorylation of distinct sets of proteins in Interleukin-3-dependent murine cell lines. Mol Cell Biol 8:2214–2218
2. Baniash M et al. (1988) The T-cell antigen receptor zeta chain is tyrosine phosphorylated upon activation. J Biol Chem 263:18225–18230
3. Sherr CJ et al. (1985) The *c-fms* proto-oncogene product is related to the receptor for the mononuclear phagocyte growth factor CSF-1. Cell 41:665–676
4. Ebina Y et al. (1985) The human insulin receptor cDNA: the structural basis for hormone activated transmembrane signalling. Cell 40:747–758
5. Ulrich A et al. (1985) Human insulin receptor and its relationship to the tyrosine kinase family of oncogenes. Nalture 313:756–761
6. Ulrich A et al. (1984) Human epidermal growth factor receptor cDNA sequence and aberrant expression of the amplified gene in A431 epidermoid carcinoma cells.
7. Yarden Y et al. (1986) Structure of the receptor for platelet-derived growth factor helps define a family of closely related growth factor receptors. Nature 323:226–232
8. Ulrich A et al. (1986) Insulin-like growth factor 1 receptor primary structure: comparison with insulin receptor suggests structural determinants that define functional specificity. EMBO J 5:2508–2512
9. Snyder MA, Bishop JM, Colby WW, Levinson AD (1983) Phosphorylation of tyrosine-416 is not required for the transforming properties of pp60[scr]. Cell 32:891–901
10. Cross FR, Hanafuza H (1983) Local mutagenesis of Rous sarcoma virus: The major sites of tyrosine and serine phosphorylation are dispensable for transformation. Cell 34:597–607
11. Ben-Neriah Y, Bernards A, Paskind M, Daley GQ, Baltimore D (1986) Alternative 5′ exons in *c-abl* mRNA. Cell 44:577–586
12. Hunter T (1987) A thousand and one protein kinases. Cell 50:823–829
13. Whetton AD, Bazill GW, Dexter TM TM (1984) Haematopoietic cell growth factor mediates cell survival via its action on glucose transport. EMBO J 3:409–413
14. Dexter MT et al. (1986) The relevance of protein kinase C activation, glucose transport and ATP generation in the response of haematopoietic cells to growth factors. In: Kahn P, Graf T (eds) Oncogenes and growth control. Springer, Berlin Heidelberg New York, pp 163–169
15. Goldfine ID (1987) The insulin receptor: Molecular biology and transmembrane signalling. Endocr Rev 8:235–255
16. Neckameyer WS, Wang LH (1985) Nucleotide sequence of avian sarcoma virus UR2 and comparison of its transforming gene with other members of the tyrosine protein kinase. J Virol 53:879–884
17. Ben-Neriah Y, Bauskin AR (1988) Leukocytes express a novel gene encoding a putative transmembrane protein-kinase devoid of an extracellular domain. 333:672–676
18. Hopp TP, Woods KR (1981) Prediction of protein antigenic determinants from amino acid sequences. Proc Natl Acad Sci USA 78:3821–3824
19. Matsushime H, Wang LH, Shibuya M (1986) Human *c-ros-1* gene homologous to the *v-ros* sequence of UR2 sarcoma virus encodes for a transmembrane receptor-like molecule. Mol Cell Biol 6:3000–3004
20. Hafen E, Basler K, Edstroem J, Rubin GM (1987) *Sevenless,* a cell-specific homeotic gene of *Drosophila,* encodes a putative transmembrane receptor with a tyrosine kinase domain. Science 236:55–63
21. Martin-Zanca D, Hughes SH, Barbacid M (1986) A human oncogene formed by the fusion of truncated tropomyosin and protein tyrosine kinase sequences. Nature 319:743–748
22. Yu KT, Peters MA, Czech MD (1986) Similar control mechanisms regulate the

insulin and type 1 insulin-like growth factor receptor kinases. J Biol Chem 261:11341–11349

23. Rosen OM, Herrara R, Olowe Y, Pelruzelli LM, Cobb MH (1983) Phosphorylation activates the insulin receptor tyrosine kinase. Proc Natl Acad Sci USA 80:3237–3240
24. Sibley DR, Benovic JL, Caron MG, Lefkowitz RJ (1987) Regulation of transmembrane signaling by receptor phosphorylation. Cell 48:913–922
25. Azar Y et al. (1981) Antigen-specific murine T-cell lymphomas: Functional heterogenicity. Cell Immunol 65:194–200
26. Schejter ED et al. (1986) Alternative 5′ exons and tissue specific expression of the *Drosophila* EGF receptor homolog transcripts. Cell 46:1091–1101
27. Bordier C (1981) Phase separation of integral membrane proteins in Triton-X-114 solution. J Biol Chem 256:1604–1607
28. Wathelet MG et al. (1988) Cloning and chromosomal location of human genes inducible by Type 1 Interferon. Somatic Cell Mol Genet 14:415–426

Genetic Characterization of a Human Endogenous Retroviral Element Located on Chromosome 18q21*

C. Leib-Mösch [1], D. E. Barton [2], R. Brack-Werner [3], B. Foellmer [2], T. Werner [4], D. Rohrmeier [1], U. Francke [2], V. Erfle [3], and R. Hehlmann [5]

A. Introduction

A substantial portion of the human genome is thought to have been generated by reverse flow of genetic information from RNA into DNA. Sequence information transposed in this manner has been given the general term "retroposon". Some retroposons show distinct structural and sequence similarities to animal retroviruses. These represent a wealth of retroviral information within the human genome, and inherent regulatory sequences may allow them to effect expression of cellular genes. We have recently discovered a new family of human retroviral sequences by hybridizing human DNA under low-stringency conditions with various DNA probes from the genome of the simian sarcoma-associated virus SSAV [1]. Here we present the structure and chromosomal localization of the human SSAV-related endogenous retroviral element S71.

[1] Medizinische Poliklinik der Universität München, Pettenkoferstr. 8a, D-8000 München, FRG

[2] Department of Human Genetics, Yale University, School of Medicine, New Haven, CT 06510, USA

[3] GSF-Abteilung für Molekulare Zellpathologie, Ingolstädter Landstr. 1, D-8042 Neuherberg, FRG

[4] GSF-Institut für Säugetiergenetik, Ingolstädter Landstr. 1, D-8042 Neuherberg, FRG

[5] III. Medizinische Klinik Mannheim der Universität Heidelberg, Wiesbadenerstr. 7–11, D-6800 Mannheim 31, FRG

* This work was supported by the *Deutsche Forschungsgemeinschaft* (SFB 324), the National Institutes of Health (grant Ror-GM 26105), and by contract B16-156D of the Commission of the European Community

B. Results

The restriction map and genomic organization of molecular clone S71 is shown in Fig. 1. S71 is an incomplete provirus of about 6 kb with sequences related to the SSAV gag and pol genes. Sequence analysis revealed a 535-nucleotide sequence located directly adjacent to the S71 pol region with all salient features of a retroviral long-terminal repeat (LTR). The U3 region of this LTR-like structure contains signal sequences associated with the initiation and termination of transcription.

An S71 probe comprising the pol-LTR region was used for Southern blot analysis of DNA from 13 Chinese hamster × human hybrid cell lines and for in situ hybridization of human metaphase chromosomes. As illustrated in Fig. 2, S71 mapped to chromosome 18 band q21, which also contains the *yes*-1 protooncogene [2] and *bcl*-2, the major breakpoint cluster region of t(14; 18) chromosome translocations [3, 4]. Another human endogenous retroviral sequence, ERV1, has also been mapped to chromosome 18, but more distal, in bands 18q22-qter [5].

Figure 3 shows the hybridization pattern of *Bgl*II and *Eco*RI restricted genomic DNA from nine unrelated European individuals after hybridization with an S71 fragment containing the pol-LTR region. It is evident that the S71 locus shows a restriction fragment length polymorphism (RFLP) for *Bgl*II and *Eco*RI. Both RFLPs comprise two allelic fragments of 13.5 and 6.8 kb (*Eco*RI) and of 11.0 and 6.8 kb (*Bgl*II). The respective frequencies are listed in Table 1.

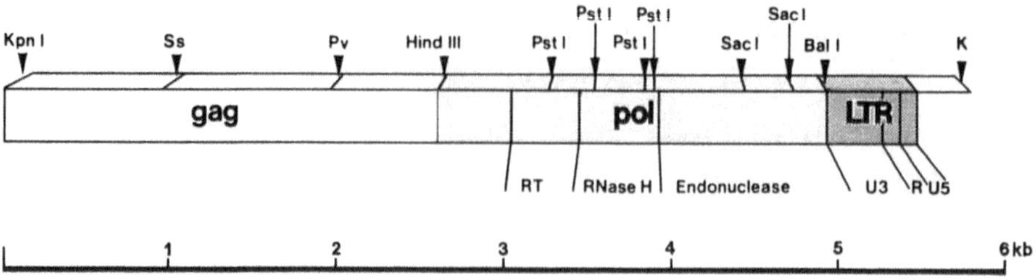

Fig. 1. Genomic organization of human endogenous retroviral element S71

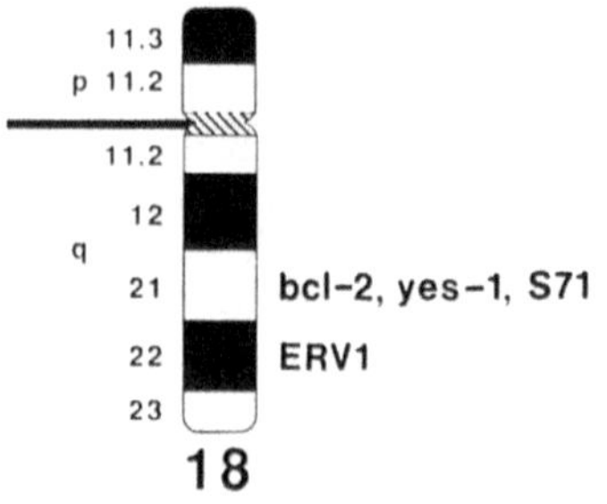

Fig. 2. Chromosomal localization of S71

Table 1. Restriction fragment length polymorphisms detected with S71 in human DNA

Restriction enzyme	Number of DNAs	Allelic fragments (kb)	Frequency
*Eco*RI	14	13.5	0.79
		6.8	0.21
*Bgl*II	9	11.0	0.61
		6.0	0.39

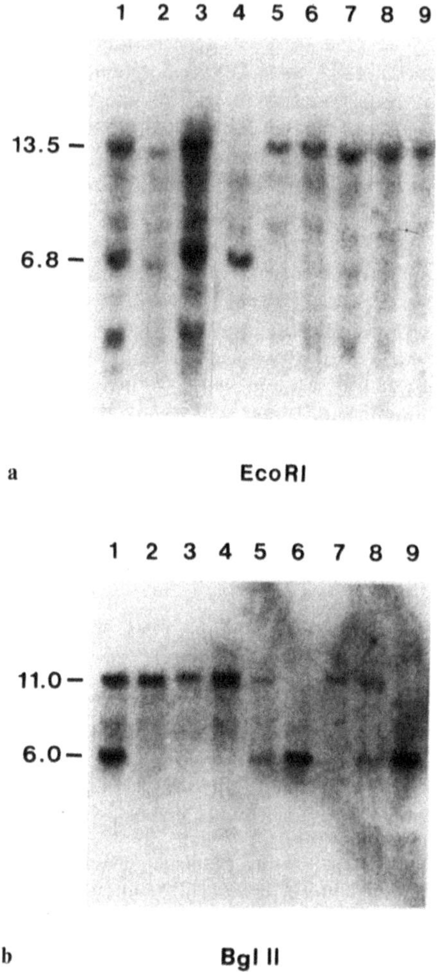

Fig. 3a, b. Restriction fragment length polymorphisms in human DNA detected with S71. EcoRI-(a) and Bgl II-digested DNA (b) from nine unrelated individuals was hybridized with a 3 kb S71 pol-LTR-subfragment

C. Discussion

Endogenous retroviruses and retroviral elements have been detected in the DNA of many vertebrate species, including primates. Although most of them are defective, they represent a reservoir of viral genes which may be activated spontaneously, by recombination events, or by radiation and chemical agents. Once activated, endogenous retroviruses can induce hematopoietic proliferative diseases, e.g., by LTR-directed transcription of adjacent cellular oncogenes [6].

The human endogenous retroviral element S71 is located in a region of high biological significance. T(14; 18) translocations are observed in more than 80% of human follicular B-cell lymphomas and in about 40% of diffuse B-cell lymphomas [7]. This raises the possibility of an involvement of S71 in the genesis of neoplastic diseases. S71 has many closely related sequences dispersed throughout the human genome. These could be sites for recombination events leading to chromosomal rearrangements, as demonstrated recently for *Alu*repeats [8]. Furthermore, the LTR-like sequence in S71 contains signal sequences essential for transcriptional control. Thus, the S71 LTR could play a role in *cis*-acting mechanisms suggested to be involved in the altered transcription of the *bcl*-2 gene as a consequence of t(14; 18) translocation [9]. Since two RFLPs can be detected in human DNA using an S71 fragment as a hybridization probe, this retroviral sequence should prove to be a highly informative marker for this genomic locus.

References

1. Leib-Mösch C, Brack R, Werner T, Erfle V, Hehlmann R (1986) Isolation of an SSAV-related endogenous sequence from human DNA. Virology 155:666–667
2. Yoshida MC, Sasaki M, Semba K, Yamanashi Y, Nishizawa M, Sukegawa J, Yamamoto T, Toyoshima K (1985) Localization of the human cellular oncogene c-*yes*-1 to chromosome 18 band q21. Cytogenet Cell Genet 40:786
3. Tsujimoto Y, Finger LR, Yunis J, Nowell PC, Croce CM (1984) Cloning of the chromosome breakpoint of neoblastic B-cells with the t(14; 18) chromosome translocation. Science 226:1097–1099
4. Bakhshi A, Jensen JP, Goldman P, Wright JJ, McBride OW, Epstein AL, Korsmeyer SJ (1985) Cloning the chromosomal breakpoint of t(14; 18) human lymphomas: clustering around JH on chromosome 14 and near a transcriptional unit on 18. Cell 41:899–906
5. Renan MJ, Reeves BR (1987) Chromosomal localization of human endogenous retroviral element ERV1 to 18q22→q23 by in situ hybridization. Cytogenet Cell Genet 44:167–170
6. Canaani E, Dreazen O, Klar A, Rechavi G, Ram D, Cohen JB, Givol D (1983) Activation of the c-*mos* oncogene in a mouse plasmacytoma by insertion of an endogenous intracisternal A-particle genome. Proc Natl Acad Sci USA 80:7118–7122
7. Ming-Sheng L, Blick MB, Pathak S, Trujillo JM, Butler JJ, Katz RL, McLaughlin P, Hagemeister FB, Velasquez WS, Goodacre A, Cork A, Gutterman JU, Cabanillas F (1987) The gene located at chromosome 18 band q21 is rearranged in uncultured diffuse lymphomas as well as follicular lymphomas. Blood 70:90–95
8. Rouyer F, Simmler M-C, Page DC, Weissenbach J (1987) A sex chromosome rearrangement in a human XX male caused by *Alu-Alu* recombination. Cell 51:417–425
9. Cleary ML, Smith SD, Sklar J (1986) Cloning and structural analysis of cDNAs for *bcl*-2 and a hybrid *bcl*-2/immunoglobulin transcript resulting from the t(14; 18) translocation. Cell 47:19–28

Human Endogenous Retrovirus-like Sequences

R. Brack-Werner [1], C. Leib-Mösch [3], T. Werner [2], V. Erfle [1], and R. Hehlmann [4]

A. Introduction

One of the most salient features of the replication strategy used by retroviruses is the transcription of the retroviral (RNA) genome into DNA followed by integration of this DNA product into the host cell genome. The integrated viral DNA copy, termed "provirus", can then serve as a template for the synthesis of further infectious virus particles. Stably integrated proviruses have been found to also persist in the germ line of animal cells. In this case, they have become an endogenous constituent of their host cell's genome and are passed on as stable Mendelian genes from one generation to the next.

Endogenous retroviruses have been detected in a number of vertebrate species, including primates and birds. As a rule, they persist as silent retroviral copies in their host cell's genome since deletions and mutations in the provirus genome have often led to the loss of their pathogenic potential. There are exceptions, however, and activation of endogenous retroviruses has been found to occur spontaneously, as in the case of the leukemogenic ecotropic provirus of the 101 mouse [31]. Other factors, such as treatment with carcinogens [24] and chemicals such as IUdR (iododeoxyuridine) and BrdU (bromodeoxyuridine) and irradiation can also lead to the production of infectious viral particles from endogenous proviruses [32, 55, 25]. Furthermore, the synthesis of pathogenic retroviruses as a result of recombination events between different endogenous proviral sequences has been shown for the highly leukemogenic murine MCF (mink cell focus-forming) virus [7, 13].

Besides delivering the basis for the induction of potentially pathogenic viral particles, the biological potential of endogenous retroviruses can be found on at least two additional levels. First, even replication-defective proviruses can give rise to products such as the p15E envelope-related proteins, which have been shown to possess immunosuppressive and anti-inflammatory activity [59]. Second, insertion of a proviral sequence can take place within host cell genes, causing changes in expression of the latter (insertion mutagenesis). Furthermore, once the provirus is installed it can influence the expression of adjacent cellular sequences by virtue of its own transcription control signals [reviewed in 40]. Some examples illustrating the mutagenic potential of tumor-associated proviral insertion have been reported for intracisternal A-type particles (IAP) in mouse plasmacytoma [6], MoMuLV-induced tumors [56], and avian leukosis virus (ALV)-induced erythroblastosis [14, 16, 17].

The fact that almost all vertebrate species analyzed to date have been shown to contain endogenous retroviruses makes it highly conceivable that these are also an integral component of the human genome. The evidence pointing to the existence of human endogenous retroviruses runs in three lines. First, particles with

[1] GSF-Abt. für Molekulare Zellpathologie and [2] GSF-Institut für Säugetiergenetik, D-8042 Neuherberg, FRG
[3] Medizinische Poliklinik der Universität München, D-8000 Munich, FRG
[4] III. Medizinische Klinik Mannheim der Universität Heidelberg, Wiesbadenerstr. 7–11, D-6800 Mannheim 31, FRG

Table 1. Copy number and chromosomal localization of human endogenous retroviral sequences

Endogenous retroviral sequence	Length (kb)	Copy no. per haploid genome	Chromosomal localization	Reference
H51 related	4.4	35–50	dispersed to multiple	[61]
4-1 related	8.8	35–50	human chromosomes	
ERV1	8.0	1	18q22–q23	[41, 52]
additional ERV1-related sequences	n.d.	11	n.d.	[2]
ERV3	9.9	1	7	[42]
S71	6	1	18q21–q22	[3]
S71-related	n.d.	35	n.d.	
HuRRS-P	8.1	20–40	n.d.	[29]
RTVL-H	5.8	800–1000	n.d.	[34]
HLM	9.7	50	chromosomes 7, 8, 11, 14, and 17	[23]
HM	6–8	30–40	n.d.	[10]
HERV-K	9.5	50	n.d.	[45]
THE1 repeats	2.3	10000	n.d.	[11]
THE solitary LTRs	0.35	10000	n.d.	

n.d., not determined.

retrovirus-like morphology have been visualized by electron microscopy of various human tissues and cell lines, many of which are of neoplastic origin [1, 28, 33, 38]. The second line of evidence is the detection of proteins related to exogenous animal retroviruses in human tissues or body fluids [18, 21]. We previously reported that antibodies against structural components of the simian sarcoma-associated virus (SSAV) recognize proteins in leukemic sera. Proteins immunologically related to the p30 constituent of the SSAV group-specific antigen were detected only in sera from patients with acute leukemia and CML blast crisis, but not in nonleukemic controls [19]. Furthermore, proteins related to the SSAV envelope gp70 protein seem to be of diagnostic value for the prognosis of patients with acute leukemias or CML blast crisis [20].

The third line of evidence is the existence of numerous retrovirus-like sequences which are indigenous to the human genome. These endogenous retroviral sequences constitute a complex variety of retroviral information in the human genome. A conservative estimate based on the copy number of endogenous retroviral sequences published to date (Table 1) shows that at least 0.6% of the human genome consists of retrovirus-like elements. The actual percentage is probably much higher, since new families of retrovirus-related sequences are being discovered continuously.

B. Identification and Isolation of Human Endogenous Retroviral Sequences

A number of different strategies have been employed to identify retrovirus-related sequences in the human genome (Table 2). Human C-type retrovirus-related sequences were initially discovered by utilizing probes from primate endoge-

Table 2. Identification and isolation of human endogenous retroviral sequences

Source of DNA for human library	Hybridization probe used for screening of human DNA library	Stringency	Identification of	Group of human endogenous retroviral sequences	Reference
Human fetal liver	gag-pol-related fragment from African green monkey endogenous retroviral sequence	low	λH51	C-type-related	[36]
	fragment from λH51-pol-related sequence	high/low	~30 additional retrovirus-related sequences		[53, 58]
	*pol-related fragment from chimpanzee endogenous retroviral sequence	low	ERV1		[2]
	and Baev LTR probe	low	ERV3		[42]
Burkitt's lymphoma	SSAV proviral DNA and various fragments from different regions of the SSAV genome	low	S71		[30]
	DNA fragment containing the retrovirus-related region in S71	high	clones only from S71 genomic locus		
Human male blood cells	Synthetic oligonucleotide complementary to murine $tRNA^{Pro}$	low and medium	Pλ1		[29]
	LTR probe from P1	high	HuRRS-P		
–	–	–	RTVL-H1		[34]
Human embryonic fibroblasts	Various RTVL-H1 fragments	stringent	RTVL-H2		[35]
Human fetal liver	MMTV provirus	low	HLM-2	A-, B-, and D-type-related sequences	[4]
	gag-pol of MMTV provirus	low	HM16		[10]
	pol region of Syrian hamster IAP	low	HERV-K		[45]
Human breast cancer cell line	MMTV provirus as well as gag-pol and LTR region of MMTV provirus	low	NMWV4		[37]
n.s.	Total human genomic DNA and cloned Alu-family member	n.s.	THE1 repeats	retroposons with LTRs	[62]

* ERV3 was isolated by employing the same chimpanzee endogenous retroviral fragment together with the BaEV LTR probe.

nous retroviral sequences for low-stringency hybridization of human genomic libraries. In 1981, Martin and co-workers used a cloned segment of African green monkey DNA which specifically hybridized with C-type murine and primate proviruses to identify related sequences in the human genome [36]. One of these sequences was isolated from a human DNA library (clone λ51-1). High-stringency hybridization of the same library with a retrovirus-related probe from 51-1 yielded over 30 additional type-C retrovirus-related sequences [53]. One of these (4-1) was also shown to contain a full-length provirus [50, 54]. An additional full-length provirus (NP-2) was cloned by low-stringency hybridization using a 51-1 pol probe [58]. Another human C-type retroviral sequence (ERV1) was isolated by Bonner et al. [2] with the help of a fragment from a cloned chimpanzee retrovirus-like sequence homologous to the polymerase genes of the baboon endogenous virus (BaEV) and the Moloney murine leukemia virus (MoMuLV). Low-stringency screening of a human genomic library with the same cloned chimpanzee fragment and a probe containing the BaEV LTR led to the isolation of a full-length human endogenous provirus termed ERV3 [42].

Our initial interest in human endogenous retroviral sequences arose from the observation mentioned above that human sera contain proteins immunologically related to structural components of SSV/SSAV and the closely related gibbon ape leukemia virus (GALV) [19]. Low-stringency Southern blot hybridization of a number of human genomic DNAs with various probes derived from the SSAV genome showed multiple SSAV-related sequences in the human genome [30]. Therefore, we decided to use a direct approach and screen a human DNA library with a probe containing the complete SSAV provirus as well as probes derived from various regions of the SSAV genome under low-stringency conditions. The initial hybridization yielded quite a few positive plaques corresponding to at least 35 copies of SSAV-related sequences per haploid genome. Washing the filters under higher stringency conditions caused a number of the positive signals to grow more or less weaker or to disappear altogether, which indicates that the retrovirus-related sequences detected during initial screening were of varying homologies to SSAV. One clone which gave a particularly strong hybridization signal with an SSAV pol-env probe was termed S71 and chosen for further analysis. The region containing the retrovirus-related sequences in S71 was used for renewed screening of the human DNA library, this time under high-stringency conditions. All positive clones obtained overlapped with clone S71 to some extent, comprising about 36 kb of the S71 genomic locus. Contrary to Repaske et al. [53], we had not been able to isolate any additional retrovirus related sequences by high-stringency screening of a human DNA library with S71 probes. This suggests that the SSAV-related human endogenous retroviral sequences are less similar to each other than the members of the 51-1/4-1 family.

A further family of C-type retrovirus-related sequences was isolated by virtue of the fact that retroviruses contain short sequences complementary to tRNA molecules, which are used as primers for reverse transcription. Screening of a human DNA library with an oligonucleotide complementary to $tRNA^{Pro}$ (murine) yielded a human LTR-like sequence which could be utilized for renewed screening and isolation of a retrovirus-like sequence termed HuRRS-P [29]. Finally, one multicopy endogenous retrovirus-like element termed "RTVL-H" was discovered fortuitously during attempts to clone a region of the human β-globin gene cluster region [34]. Additional RTVL-H elements were isolated by screening a human DNA library with RTVL-H1 probes [35].

The strategy of direct screening of human DNA libraries with probes derived from recombinant rodent proviruses was used to initially identify a second large group of human endogenous retroviral elements (Table 2). This group consists of sequences related to the B-type mouse mammary tumor virus (MMTV) as well as to the Syrian hamster IAP and to the D-type squirrel monkey retrovirus (SMRV). Members of this group were isolated by low-stringency hybridization with DNA probes encompassing various regions of the MMTV genome [5, 10, 37] or by employing a probe from the polymerase gene of the Syrian hamster IAP [45].

The final group of human endogenous retroviral sequences consists of elements flanked by two sequences with the hallmarks of retroviral long-terminal repeats (LTRs); [48]. This group of elements, designated THE 1 repeats by Sun et al. [62], was isolated from a human DNA library as clones hybridizing to human genomic DNA but not to an Alu family member. Like the other endogenous retroviral sequences discussed here, these elements possess features indicative of having been generated by the reverse flow of genetic information from RNA to DNA. Such elements are known collectively as retroposons [67].

C. Chromosomal Localization

Some human retroviral elements occur singly or in a few copies in the human genome enabling their assignment to distinct chromosomes (Table 1). Hybridization of DNA from rodent x human somatic cell hybrids revealed that the full-length retroviral sequence ERV3 resides at a single locus on human chromosome 7 [42]. The long arm of chromosome 18 carries two incomplete proviral sequences: S71 at band q21 [3] and ERV1 at bands q22–q23 [41, 52]. The chromosomal location of these retroviral elements was determined by Southern blot analysis of DNA from hybrid cell lines as well as by in situ hybridization. The members of the closely related 4-1 and 51-1 families were found to be widely dispersed over the human genome, indicating that the 50–100 copies of these sequences may have been generated by amplification processes [61]. Clone λNP-2, a full-length proviral sequence related to 4-1 and 51-1, was localized in two copies on the Y chromosome. Conservation of cellular flanking sequences suggests that the second copy results from gene duplication, rather than from provirus insertion [58]. Some members of the B-type-related multicopy HLM-family were mapped to chromosomes 1, 5, 7, 8, 11, 14, and 17 [23]. The RTLV-H elements and the THE 1 repeats occur in much higher copy numbers in the human genome than the other retroviral elements (Table 1).

D. Organization of Human Endogenous Retrovirus-like Sequences

Hybridization studies and nucleotide sequence analysis showed that each group of human endogenous retroviral sequences has one or more members resembling full-length proviruses; i.e., their retroviral sequences are arranged 5′LTR-gag-pol-env-LTR3′ as in proviruses resulting from infection with exogenous viruses (Fig. 1, MoMuLV). In the group of C-type-related retroviral sequences, 4-1 and ERV3 show a proviral organization [54, 42], and in the group of B-type-related sequences this holds true for the HERV-K family [46] (Fig. 2). However, 4-1 and ERV3 both contain stop codons and frame shifts in their nucleotide sequence, precluding the synthesis of infectious virus particles. In 4-1, complete nucleotide sequence analysis revealed these mutations to be dispersed over the whole genome inactivating all three retroviral genes [54; see also 22]. It seems that these sequences are of sufficient danger to the human cell to warrant an efficient blockade of their expression.

A great proportion of human endogenous retroviral elements consist of retrovirus-related sequences organized in a manner suggestive of truncated provirus-

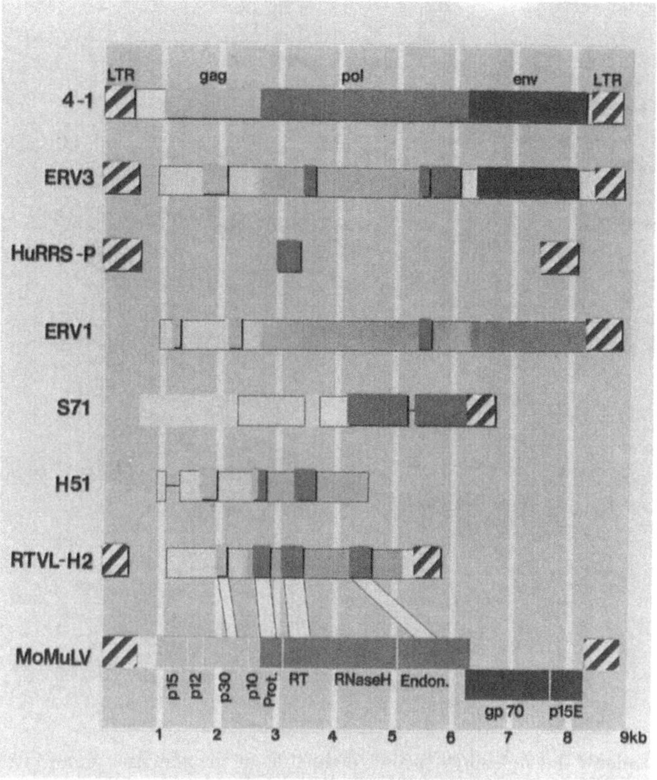

Fig. 1. Genomic organization of C-type-related human endogenous retroviral sequences. The genomic organization of C-type-related human retroviral elements was deduced primarily from sequence comparison with the MoMuLV genome (depicted at *bottom*). The gene assignment of *lightly shaded regions* was inferred on the basis of their location between sequenced regions or from hybridization data. LTR-like sequences are *hatched. Horizontal lines* in H51 and S71 delineate deleted sequences. The *left (gray) box* in the S71 element marks the gag region. The *right (white) box* immediately adjacent to the S71 pol sequence shows the minimal extent of nonretroviral sequences in S71. References: 4-1 [54, 60], ERV3 [42, 43], HuRRS-P [29], ERV1 [2], S71 [3, 30], H51 [53], RTVL-H2 [35], MoMuLV [57]

es. These elements may lack only a small part of the retroviral genome, such as one of the two LTRs at either end (ERV1; Fig. 1), or they may be completely devoid of sequences corresponding to one or more proviral genes. We have found the SSAV-related human retroviral element S71 to provide a good example for such a truncated endogenous provirus. By hybridization of molecular clone S71 with probes derived from various SSAV genes, the S71 retroviral element was delineated to a region of approximately 6 kb. Since a full-length C-type provirus ranges from 8.5 to 9.5 kb in length, the S71 retroviral element is obviously lacking part of the retroviral genome. Interestingly, the retroviral region in S71 is surrounded by Alu repeats, which, although nonviral, are also retroposons. Other human C-type-related retroviral elements have also been reported to be associated with retroposons, such as the Alu or the Kpn I family of reiterated sequences [53, 60].

The retroviral region in S71 contains sequences related to the gag and pol genes of SSAV. In addition, hybridization with an SSAV LTR probe suggested the presence of an LTR-like sequence. To obtain a better idea of the organization of the pol-related sequences in S71 we compared the sequence of the 3′ half of the S71 retroviral element with the pol-gene sequence of the Moloney murine leukemia virus [57]. The pol-gene sequence of retroviruses codes for three ac-

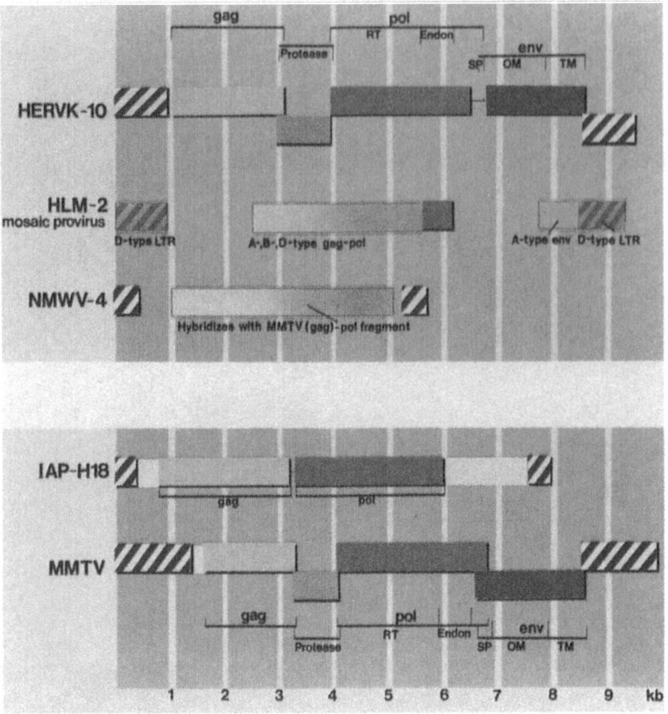

Fig. 2. Genomic organization of A-, B-, and D-type-related human endogenous retroviral sequences. The genomic organization of the human retroviral elements was deduced from sequence comparison with the genome of the Syrian hamster IAP H18 and/or the MMTV provirus (both shown at the *bottom*) or from hybridization data (*lightly shaded regions* of HLM-2 and NMWV-4). The B-type-related endogenous element HM16 [10] was omitted since, aside from the presence of a 2.1-kb pol sequence and restriction fragments containing repeated sequences, the data available did not allow further deduction of the genomic organization of HM16. References: HERVK-10 [46], HLM-2 [5], NMWV-4 [37], IAP-H18 [44], MMTV [39]

tivities: the RNA-directed DNA polymerase (reverse transcriptase), a ribonuclease H, responsible for degradation of viral RNA in RNA·DNA hybrids, and an endonuclease which is essential for integration of the viral information into the host cell genome. In the polymerase genes of C-type retroviruses these activities are arranged 5′ reverse transcriptase – RNAse H – endonuclease 3′ [26]. We found the polymerase-related sequences in S71 to correspond to a region of the MoMuLV pol gene beginning in the 3′ half of the reverse transcriptase domain and extending through the RNase H and most of the endonuclease domain (Fig. 1). With the exception of a small deletion at the 5′ terminus of the endonuclease domain (indicated by a horizontal line in Fig. 1), the S71 pol sequence aligns to the corresponding region of the MoMuLV pol gene in a colinear manner. Thus, the S71 pol sequences show the same structural organization as the corresponding sequences of infectious C-type retroviruses. Translation of the S71 pol nucleotide sequence yields an amino sequence which is 40%–60% identical with the MoMuLV pol sequence, depending on the region of the polymerase gene used for comparison. The S71 pol amino acid sequence contains three stop codons, one each in the deduced reverse transcriptase and RNAse H domains and one in the endonuclease sequence. Therefore, the situation in S71 is similar to the pol region in the C-type-related 4-1 [54] and the B-type-related HM16 element [10], in that numerous stop codons seem to serve the purpose of preventing synthesis of functional polymerase proteins from these endogenous retroviral sequences. Indeed, the polymerase sequence of only one human endogenous retroviral element, the B-type-related HERV-K [46], has yet been reported to constitute an open reading frame long enough to allow synthesis of full-length polymerase proteins.

In a biological sense, expression of sequences enabling random reverse flow of genetic information from RNA to DNA would pose a great threat for the evolutionary stability of the human genome. A prerequisite for the maintenance of such sequences in the human genome is therefore a very rigid control mechanism precluding their random expression. The numerous stop codons and frame shifts observed in the pol sequences of C-type-related human endogenous retroviral elements may be a significant factor contributing to this stringent control.

Replication of viral RNA in the host cell leads to the duplication of sequences specific for the 5′ and 3′ ends of the viral RNA. Therefore, the integrated provirus is flanked by long terminal repeats (LTRs) which in the case of mammalian C-type retroviruses are typically 500–600 bp in length [8]. It is still not clear whether endogenous retroviral sequences were generated via the same replication mechanism essential for the spread of infectious retroviruses. However, it is remarkable that quite a few human endogenous retroviral elements are also flanked by LTRs (Figs. 1 and 2). Like the LTRs of infectious proviruses, the endogenous LTRs contain signal sequences implicated in transcriptional control. Indeed, the LTR of the C-type-related endogenous retroviral sequence ERV3 was recently demonstrated to drive transcription of the retroviral and adjacent cellular sequences in a tissue-specific manner [43, 27]. In some human retroviral elements, e.g., ERV1 (Fig. 1), LTR-like sequences were not discovered as duplicated sequences at both ends of the retroviral element. Rather, they were identified as possessing the same sequence features and structural organization as the LTRs of infectious proviruses.

Figure 3 shows the LTR structure of a typical mammalian C-type provirus. The boundaries of retroviral LTRs are formed by inverted repeats, beginning with TG and ending with CA. The LTRs consist of three entities: the U3, R, and the U5 region. The U3 region contains signal sequences necessary for transcription initiation, including the CCAAT and TATAA boxes, and an enhancer region, which often contains directly repeated sequences. However, it should be pointed out that at least three human endogenous LTRs lack a CCAAT box (hsRTVL-H [34]; O-LTR [48]; 4-1 [60].) The beginning of the R region is marked by the cap site, a G nucleotide. As a rule, the R region also contains a poly A signal, although this signal seems to be dispensable for LTR function in some cases [64], and a poly A addition site (CA) which marks the end of the R region. The remaining sequence, including the 3′ inverted repeat counterpart, makes up the U5 region.

We determined the nucleotide sequence of a 535 bp region located at the 3′ terminus of the S71 retroviral element directly adjacent to the pol-related sequences. By comparison of the S71 sequence with the aligned nucleotide sequences of 11 LTRs, six of which were derived from human endogenous retroviral elements and four from infectious proviruses [3], we were able to identify all salient features characteristic for mammalian C-type proviral LTRs. In addition, alignment of the human endogenous LTR sequences demonstrated a common sequence motif, all or part of which is reflected in five of the six human endogenous LTRs analyzed. In the S71 LTR-like sequence this motif contains a 9-bp region with eight matches to the enhancer core consensus sequence present in a number of viral enhancers [66]. Sequences with potential enhancer function, such as this common motif or direct repeats, two of which are also contained in the S71 LTR-like sequence, may enable human endogenous LTRs to influence the expression of adjacent cellular genes in *cis*.

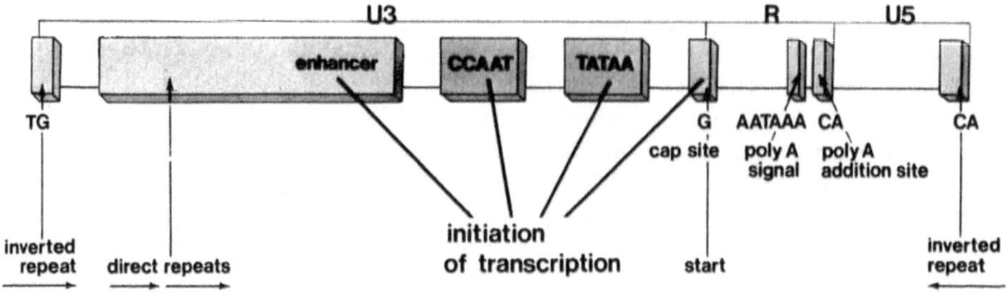

Fig. 3. Structure of mammalian C-type proviral LTRs

Hybridization analysis of a 3-kb restriction fragment directly bordering the 5′ terminus of the S71 pol sequences shows this region to contain sequences related to the gene coding for the group-specific antigen (gag) of SSAV. Preliminary sequence analysis disclosed the S71 gag-related region to encompass about 1 kb (H. Backhaus, personal communication; Fig. 1). Furthermore, the gag- and pol-related sequences in S71 are separated from each other by a sequence about 0.5–1 kb in length. This part of S71 discloses no similarity to any known retroviral genes or control elements (Fig. 1), and therefore is most likely of cellular origin.

Overall, the structure of the S71 retroviral element shows remarkable parallels to the genomic organization of the *sis* oncogene-transducing retrovirus SSV (simian sarcoma virus [12]). This acutely transforming virus is thought to have arisen by recombination of SSAV with the cellular homologue of the v-*sis* sequence which codes for a component of the platelet-derived growth factor. The SSV genome lacks a substantial portion of the pol gene, and most of the envelope gene has been replaced by cellular sequences *(sis)*. In analogy, the S71 retroviral element is likewise missing part of the pol gene – although the pol deletion is not as extensive as in SSV – as well as envelope sequences. In addition, the S71 element also contains nonretroviral sequences embedded in retroviral sequences. These analogies imply that the generation of the oncogene-transducing retrovirus SSV and the human endogenous SSV/SSAV-related sequence in S71 may have involved similar mechanisms.

E. Expression of Human Endogenous Retroviral Sequences

Although all human endogenous retroviral elements examined so far are replication defective, some of them have been shown to be transcriptionally active in human tissues and cell lines. Several discrete mRNA species hybridizing to LTR and env DNA probes derived from the 4-1 element were detected in human placenta, spleen, normal colon mucosa, and primary colon cancers, as well as in colon cancer cell lines (SW1116, HCT, Caco2), in a breast carcinoma cell line (T47D), and in a T-cell acute lymphocytic cell line (8402) [15, 50, 51]. In colon tumors an increase of env-LTR-related 1.7- and 3.0-kb transcripts was observed compared with normal colon tissue, whereas a 3.6-kb transcript abundant in normal colon mucosa was decreased in tumor cells [15]. Partial cDNA clones of 4-1 env-related mRNA transcripts were isolated from human placenta. Sequence analysis of two placental cDNA clones, however, revealed in-frame termination codons, so that neither of them could encode full-length env proteins [51].

The env region of another C-type-related full-length retroviral element, ERV3, contains a long open reading frame corresponding to approximately

650 amino acids. This potential polypeptide was found to exhibit features characteristic of retroviral glycoproteins, including several potential glycosylation sites and sequences indicative of transmembrane proteins [9]. An ERV3 env-specific c-DNA of 2.85 kb was isolated from a human fetal cDNA library and found to be identical to ERV3 by DNA sequence analysis. Three polyadenylated RNAs of 9, 7.3, and 3.5 kb were identified in human placental chorion and characterized by Northern blotting and S1 nuclease mapping [27]. The RNAs were found to be spliced mRNAs lacking the gag and most of the pol gene. The two larger mRNAs extended through the polyadenylation site in the 3′ LTR and contained adjacent cellular sequences.

We have also identified S71-related cDNA clones in a human osteosarcoma and a placenta cDNA library, indicating that S71-related sequences are expressed in these tissues. Sequence analysis of the cDNA inserts in these clones is currently in progress (Leib-Mösch et al. manuscript in preparation).

The MMTV-related human proviral sequence HERV-K was found to be expressed as an 8.8-kb full-length mRNA transcript in cell lines from breast carcinoma (T47D), gastric carcinoma (Kato-III), malignant melanoma (HMT-2), and epidermoid carcinoma (HEp-2, Hela). Stimulation of HERV-K expression was observed in steroid-treated T47D cells [47].

In spite of abundant transcription of endogenous retroviral sequences in various cells, the corresponding proteins have not yet been identified. The only case in which there is at least some indirect evidence for expression at the protein level is the truncated retroviral element ERV1. Antibodies were raised against a synthetic undecapeptide, the sequence of which was derived from the gag-related region of ERV1. These antibodies identified a 75 kD protein in renal adenocarcinoma, placenta, and trophoblastic tumors [63, 65].

However, it should be pointed out that mammalian C-type retroviral gag proteins were recently shown to share an antigenic determinant with the snRNP-associated 70 kD protein [49]. This most likely represents an example of molecular mimicry resulting from convergent evolution of otherwise unrelated proteins. Although the sequence of the antigenic determinant common to retroviral $p30^{gag}$ and the 70 kD protein is not contained in the ERV1 gag synthetic peptide, involvement of a similar phenomenon cannot be ruled out at this stage.

F. Concluding Remarks

To summarize briefly, endogenous retroviral elements are a substantial component of the human genome. In structure, they resemble either full-length or truncated proviruses. Retrovirus-related sequences seem to be dispersed to all human chromosomes; however, single-copy retroviral elements could be assigned to distinct chromosomal loci. Although their function is still unknown, RNA expression has been detected in various human materials, including tumor-derived tissues and cell lines as well as placenta. Human retroviral elements exhibit a number of features giving them a potential for involvement in carcinogenesis. One of them is their likelihood of being transposed, thereby enabling them to act as insertional mutagens. Other intrinsic properties of retroviral elements relevant for their tumorigenic potential reside in their sequence information. These include the potential immunosuppressive activity of p15E envelope-related proteins and the ability of retroviral LTRs to influence transcription of adjacent cellular genes. Besides the enlightenment of a possible contribution of retroviral elements to the evolutionary versatility of the human genome, the possible role of human endogenous retroviral sequences in pathogenesis is currently a subject of great interest.

References

1. Boller K, Frank H, Löwer J, Löwer R, Kurth R (1983) Structural organization of unique retrovirus-like particles budding from human teratocarcinoma cell lines. J Gen Virol 64:2549–2559
2. Bonner TI, O'Connell C, Cohen M (1982) Cloned endogenous retroviral sequences from human DNA. Proc Natl Acad Sci USA 79:4709–4713
3. Brack-Werner R, Barton DE, Werner T, Foellmer BE, Leib-Mösch C, Francke U, Erfle V, Hehlmann R (1989) Human SSAV-related endogenous retroviral element: LTR-like sequence and chromosomal localization to 18q21. Genomics 4, 68–75
4. Callahan R, Drohan W, Tronick S, Schlom J (1982) Detection and cloning of human DNA sequences related to the mouse mammary tumor virus genome. Proc Natl Acad Sci USA 79:5503–5507
5. Callahan R, Chiu IM, Wong JF, Tronick SR, Roe BA, Aaronson SA (1985) A new class of endogenous human retroviral genomes. Science 228:1208–1211
6. Canaani E, Dreazen O, Klar A, Rechavi G, Ram D, Cohen JB, Givol D (1983) Activation of the c-mos oncogene in a mouse plasmacytoma by insertion of an endogenous intracisternal A-particle genome. Proc Natl Acad Sci USA 80:7118–7122
7. Chattopadhyay SK, Cloyd MW, Linemeyer DL, Lander MR, Rands E, Lowy DR (1982) Cellular origin and role of mink cell focus-forming viruses in murine thymic lymphomas. Nature 295:25–31
8. Chen HR, Barker WC (1984) Nucleotide sequences of the retroviral long terminal repeats and their adjacent regions. Nucleic Acids Res 12:1767–1779
9. Cohen M, Powers M, O'Connell C, Kato N (1985) The nucleotide sequence of the env gene from the human provirus ERV3 and isolation and characterization of an ERV3-specific cDNA. Virology 147:449–458
10. Deen KC, Sweet RW (1986) Murine mammary tumor virus pol-related sequences in human DNA: characterization and sequence comparison with the complete murine mammary tumor virus pol gene. J Virol 57:422–432
11. Deka N, Willard CR, Wong E, Schmid CW (1988) Human transposon-like element insert at a preferred target site. Nucleic Acids Res 16:1143–1151
12. Devare SG, Reddy EP, Law JD, Robbins KC, Aaronson SA (1983) Nucleotide sequence of the simian sarcoma virus genome: demonstration that its acquired cellular sequences encode the transforming gene product $p28^{sis}$. Proc Natl Acad Sci USA 80:731–735
13. Evans LH, Cloyd MW (1985) Friend and Moloney murine leukemia viruses specifically recombine with different endogenous retroviral sequences to generate mink cell focus-forming viruses. Proc Natl Acad Sci 82:459–463
14. Fung YK, Lewis WG, Crittenden LB, Kung HJ (1983) Activation of the cellular oncogene c-erbB by LTR insertion: molecular basis for induction of erythroblastosis by avian leukosis virus. Cell 33:357–368
15. Gattoni-Celli S, Kirsch K, Kalled S, Isselbacher KJ (1986) Expression of type C-related endogenous retroviral sequences in human colon tumors and colon cancer cell lines. Proc Natl Acad Sci USA 83:6127–6131
16. Goodwin RG, Rottman FM, Callaghan T, Kung H-J, Maroney PA, Nilsen TW (1986) c-erbB activation in avian leukosis virus-induced erythroblastosis: multiple epidermal growth factor receptor mRNAs are generated by alternative RNA processing. Mol Cell Biol 9:3128–3133
17. Hayward WS, Neel BG, Astrin SM (1981) Activation of a cellular onc gene by promoter insertion in ALV-induced lymphoid leukosis. Nature 290:475–480
18. Hehlmann R (1976) RNA-tumorviruses and human cancer. Current Topics in Microbiology and Immunology Vol 73, Springer Verlag, Berlin-Heidelberg-New York:141–215
19. Hehlmann R, Schetters H, Erfle V, Leib-Mösch C (1983) Detection and biochemical characterization of antigens in human leukemic sera that crossreact with primate C-type viral proteins (Mr 30000). Cancer Res 43:392–399
20. Hehlmann R, Erfle V, Schetters H, Luz A, Rohmer H, Schreiber MA, Pralle H, Essers U, Weber W (1984) Antigens and circulating immune complexes related to the primate retroviral glycoprotein SiSVgp70. Cancer 54:2927–2935
21. Hehlmann R, Schetters H, Leib-Mösch C, Erfle V (1984) Current understanding of

viral etiology in leukemia. Recent Results in Cancer Res Vol 93, Springer Verlag, Berlin-Heidelberg-New York 93:1–28
22. Hehlmann R, Brack-Werner R, Leib-Mösch C (1988) Human endogenous retroviruses. Leukemia 2 (12S):167S–177S, 1988
23. Horn TM, Huebner K, Croce C, Callahan R (1986) Chromosomal locations of members of a family of novel endogenous human retroviral genomes. J Virol 58:955–959
24. Irons RD, Stillman WS, Cloyd MW (1987) Selective activation of endogenous ecotropic retrovirus in hematopoietic tissues of B6C3F1 mice during the preleukemic phase of 1,3-butadiene exposure. Virology 161:457–462
25. Janowski M, Merregaert J, Boniver J, Maisin JR (1985) Proviral genome of radiation leukemia virus: molecular cloning of biologically active proviral DNA and nucleotide sequence of its long terminal repeat. J Virol 55:251–255
26. Johnson MS, McClure MA, Feng D-F, Gray J, Doolittle RF (1986) Computer analysis of retroviral pol genes: assignment of enzymatic functions to specific sequences and homologies with nonviral enzymes. Proc Natl Acad Sci USA 83:7648–7652
27. Kato N, Pfeifer-Ohlsson S, Kato M, Larson E, Rydnert J, Ohlsson R, Cohen M (1987) Tissue-specific expression of human provirus ERV3 mRNA in human placenta: two of the three ERV3 mRNAs contain human cellular sequences. J Virol 61:2182–2191
28. Keydar I, Ohno T, Nayak R, Sweet R, Simoni F, Weiss F, Karby S, Mesa-Tejada R, Spiegelman S (1984) Properties of retrovirus-like particles produced by a human breast carcinoma cell line: immunological relationship with mouse mammary tumor virus proteins. Proc Natl Acad Sci USA 81:4188–4192
29. Kröger B, Horak I (1987) Isolation of novel human retrovirus-related sequences by hybridization to synthetic oligonucleotides complementary to the $tRNA^{Pro}$ primer-binding site. J Virol 61:2071–2075
30. Leib-Mösch C, Brack R, Werner T, Erfle V, Hehlmann R (1986) Isolation of an SSAV-related endogenous sequence from human DNA. Virology 155:666–667
31. Leib-Mösch C, Schmidt J, Etzerodt M, Pedersen FS, Hehlmann R, Erfle V (1986) Oncogenic retrovirus from spontaneous murine osteomas. Virology 150:96–105
32. Lesser J, Lasneret J, Canivet M, Emanoil-Ravier R, Peries J (1986) Simultaneous activation by 5-azacytidine of intracisternal R particles and murine intracisternal-A particle related sequences in Syrian hamster cells. Virology 155:249–256
33. Löwer R, Löwer J, Frank H, Harzmann R, Kurth R (1984) Human teratocarcinomas cultured in vitro produce unique retrovirus-like viruses. J Gen Virol 65: 887–898
34. Mager DL, Henthorn PS (1984) Identification of a retrovirus-like repetitive element in human DNA. Proc Natl Acad Sci USA 81:7510–7514
35. Mager DL, Freeman JD (1987) Human endogenous retrovirus-like genome with type C pol sequences and gag sequences related to human T-cell lymphotropic viruses. J Virol 61:4060–4066
36. Martin MA, Bryan T, Rasheed S, Khan AS (1981) Identification and cloning of endogenous retroviral sequences present in human DNA. Proc Natl Acad Sci USA 78:4892–4896
37. May FEB, Westley BR (1986) Structure of human retroviral sequence related to mouse mammary tumor virus. J Virol 60:743–749
38. Mondal H, Hofschneider PH (1982) Isolation and characterization of retrovirus-like elements from normal human fetuses. Int J Cancer 30:281–287
39. Moore R, Dixon M, Smith R, Peters G, Dickson C (1987) Complete nucleotide sequence of a milk-transmitted mouse mammary tumor virus: two frameshift suppression events are required for translation of gag and pol. J Virol 61:480–490
40. Nusse R (1986) The activation of cellular oncogenes by retroviral insertion. TIG 2, 244–247
41. O'Brien SJ, Bonner TI, Cohen M, O'Connell C, Nash WG (1983) Mapping of an endogenous retroviral sequence to human chromosome 18. Nature 303:74–77
42. O'Connell C, O'Brien S, Nash WG, Cohen M (1984) ERV3, a full-length human endogenous provirus: chromosomal localization and evolutionary relationships. Virology 138:225–235
43. O'Connell CD, Cohen D (1984) The long terminal repeat sequences of a novel human endogenous retrovirus. Science 226:1204–1206

44. Ono M, Toh H, Miyata T, Awaya T (1985) Nucleotide sequence of the Syrian hamster intracisternal A-particle gene: close evolutionary relationship of type A particle gene to types B and D oncovirus genes. J Virol 55: 387–394
45. Ono M (1986) Molecular cloning and long terminal repeat sequences of human endogenous retrovirus genes related to types A and B retrovirus genes. J Virol 58: 937–944
46. Ono M, Yasunaga T, Miyata T, Ushikubo H (1986) Nucleotide sequence of human endogenous retrovirus genome related to the mouse mammary tumor virus genome. J Virol 60: 589–598
47. Ono M, Kawakami M, Ushikubo H (1987) Stimulation of expression of the human endogenous retrovirus genome by female steroid hormones in human breast cancer cell line T47D. J Virol 61: 2059–2062
48. Paulson KE, Deka N, Schmid CW, Misra R, Schindler CW, Rush MG, Kadyk L, Leinwand L (1985) A transposon-like element in human DNA. Nature 316: 359–361
49. Query CC, Keene JD (1987) A human autoimmune protein associated with U1 RNA contains a region of homology that is cross-reactive with retroviral $p30^{gag}$ antigen. Cell 51: 211–220
50. Rabson AB, Steele PE, Garon CF, Martin MA (1983) mRNA transcripts related to full-length endogenous retroviral DNA in human cells. Nature 306: 604–607
51. Rabson AB, Hamagishi Y, Steele P, Tykocinske M, Martin MA (1985) Characterization of human endogenous retroviral envelope RNA transcripts. J Virol 56: 176–182
52. Renan MJ, Reeves BR (1987) Chromosomal localization of human endogenous retroviral element ERV1 to 18q22→q23 by in situ hybridization. Cytogenet Cell Genet 44: 167–170
53. Repaske R, O'Neill RR, Steele PE, Martin MA (1983) Characterization and partial nucleotide sequence of endogenous type C retrovirus segments in human chromosomal DNA. Proc Natl Acad Sci USA 80: 678–682
54. Repaske R, Steele PE, O'Neill RR, Rabson AB, Martin MA (1985) Nucleotide sequence of a full-length human endogenous retroviral segment. J Virol 54: 764–772
55. Schmidt J, Luz A, Erfle V (1988) Endogenous murine leukemia viruses: frequency of radiation-activation and novel pathogenic effects of viral isolates. Leukemia Res 12: 393–403
56. Shen-Ong GLC, Morse III HC, Potter M, Mushinski JF (1986) Two modes of c-myb activation in virus-induced mouse myeloid tumors. Mol Cell Biol 6: 380–392
57. Shinnick TM, Lerner RA, Sutcliffe JG (1981) Nucleotide sequence of Moloney murine leukaemia virus. Nature 293: 543–548
58. Silver J, Rabson A, Bryan T, Willey R, Martin MA (1987) Human retroviral sequences on the Y chromosome. Mol Cell Biol 7: 1559–1562
59. Snyderman R, Cianciolo GJ (1984) Immunosuppressive activity of the retroviral envelope protein P15E and its possible relationship to neoplasia. Immunol Today 5: 240–244
60. Steele PE, Rabson AB, Bryan T, Martin MA (1984) Distinctive termini characterize two families of human endogenous retroviral sequences. Science 225: 943–947
61. Steele PE, Martin MA, Rabson AB, Bryan T, O'Brien SJ (1986) Amplification and chromosomal dispersion of human endogenous retroviral sequences. J Virol 59: 545–550
62. Sun L, Paulson KE, Schmid CW, Kadyk L, Leinwand L (1984) Non-Alu family interspersed repeats in human DNA and their transcriptional activity. Nucleic Acids Res 12: 2669–2690
63. Suni J, Närvänen A, Wahlström T, Aho M, Pakkanen R, Vaheri A, Copeland T, Cohen M, Oroszlan S (1984) Human placental syncytiotrophoblastic M_r 75000 polypeptide defined by antibodies to a synthetic peptide based on a cloned human endogenous retroviral DNA sequence. Proc Natl Acad Sci USA 81: 6197–6201
64. Trainor CD, Scott ML, Josephs SF, Fry KE, Reitz MS, Jr (1984) Nucleotide sequence of the large terminal repeat of two different strains of gibbon ape leukemia virus. Virology 137: 201–205
65. Wahlström T, Närvänen A, Suni J, Pakkanen R, Lehtonen T, Saksela E, Vaheri A, Copeland T, Cohen M, Oroszlan S (1985) M_r 75000 protein, a tumor marker in renal adenocarcinoma, reacting with antibodies to a synthetic peptide based on

a cloned human endogenous retroviral nucleotide sequence. Int J Cancer 36: 379–382
66. Weiher H, König M, Gruss P (1983) Multiple point mutations affecting the simian virus 40 enhancer. Science 219: 626–631
67. Weiner AM, Deininger PL, Efstratiadis A (1986) Nonviral retroposons: genes, pseudogenes, and transposable elements generated by the reverse flow of genetic information. Ann Rev Biochem 55: 631–661

Early Life and Evolution

Fantasy of a "Virus" from the Inorganic World: Pathogenesis of Cerebral Amyloidoses by Polymer Nucleating Agents and/or "Viruses"

D. C. Gajdusek

A. Introduction

Kuru and the transmissible virus dementias are in a group of virus-induced slow infections that we have described as subacute spongiform virus encephalopathies (SSVEs) because of the strikingly similar histopathological lesions they induce. Scrapie, mink encephalopathy, and the chronic wasting disease with spongiform encephalopathy of captive mule deer and of captive elk all appear, from their histopathology, pathogenesis, and the similarities of their infectious agents, to belong to the same group (Gajdusek and Gibbs 1975; Gajdusek et al. 1965, 1966; Masters et al. 1981 a, b; Williams and Young 1980, 1982; Williams et al. 1982). The basic neurocytological lesions in all these diseases are a progressive vacuolation in the dendritic and axonal processes and cell bodies of neurons and, to a lesser extent, in astrocytes and oligodendrocytes; an extensive astroglial hypertrophy and proliferation; and, spongiform change or status spongiosis of gray matter and extensive neuronal loss (Beck et al. 1975, 1982; Klatzo et al. 1959).

These atypical infections differ from other diseases of the human brain, that have been subsequently demonstrated to be slow virus infections, in that they do not evoke a virus-associated inflammatory response in the brain (i.e., no perivascular cuffing or invasion of the brain parenchyma with leukocytes); they usually show no pleocytosis nor marked rise in protein in the cerebrospinal fluid throughout the course of infection (Gajdusek 1985 b; Gajdusek and Zigas 1957, 1959; Traub et al. 1977). Furthermore, they show no evidence of an immune response to the causative virus and there are no recognizable virions in sections of the brain visualized by electron microscopy, whereas in other virus encephalopathies virions have been readily observed. Instead, they show ultrastructural alteration in the plasma membrane that lines the vacuoles (Beck et al. 1982), piled up neurofilament in some swollen nerve cells (Beck et al. 1975, 1982; Klatzo et al. 1959; Lampert et al. 1971) and strange arrays of regularly arrayed tubules that look like particles in cross-section in postsynaptic processes (Baringer et al. 1979, 1981; David-Ferreira et al. 1968; Field and Narang 1972; Field et al. 1969; Lamar et al. 1974; Narang 1973; 1974 a, b; Narang et al. 1972, 1980; Vernon et al. 1970; ZuRhein and Varakis 1976).

The pursuit of the transmissibility and virus etiology of kuru (Gajdusek and Zigas 1957, 1959; Gajdusek et al. 1966; Klatzo et al. 1959) and the presenile dementia of the Creutzfeldt-Jakob disease (CJD) type (Gajdusek 1977; Gajdusek and Gibbs 1975; Gibbs et al. 1968) has led to the definition of the unconventional viruses as a new group of microbes, which, because of their very atypical physical, chemical, and biological properties, has stimulated a worldwide quest to elucidate their structures and resolve the many paradoxes they present to the basic tenets of microbiology and to solve the enormous clinical and epidemiological problems these viruses pose. The

Laboratory of Central Nervous System Studies, National Institute of Neurological Disorders and Stroke, National Institutes of Health, Bethesda, MD 20892, USA unanticipated ramifications of the discovery of these slow infections and the peculiar properties of the unconventional viruses, which have even challenged the central dogma of modern molecular biology, have led to a series of discoveries each of which have wide implications to microbiological and neurobiological research (Braig and Diringer 1985; Diringer et al. 1983; Gajdusek 1977, 1984, 1985a, b, c; Gajdusek and Gibbs 1975; Goldgaber et al. 1987a, b; Masters et al. 1981a, b, 1985a, b; Multhaup et al. 1985; Oesch et al. 1985; Prusiner 1982, 1984; Prusiner et al. 1983, 1984; Robertson et al. 1985; Rohwer 1984a, b, c; 1985, unpublished work; Rohwer and Gajdusek 1980; Rohwer et al. 1979). These are summarized below.

B. Interference with Axonal Transport. Amyloid Formation from Precursor Protein in Alzheimer's Disease and Normal Aging and in Slow Virus Infections

The cytoskeleton of all cells contains three ultrastructurally distinct elements made of fibrous macromolecules; microtubules 24 nm in diameter, intermediate filaments 10 nm in diameter, and microfilaments about 5 nm in diameter and composed of polymerized actin.

Neurofilaments, also called neuronal intermediate filaments, are antigenically distinct from the intermediate filaments of other cells. They extend from the cell body down the whole length of the axon; they are composed of three proteins of 200, 150, and 68 kilodaltons (kDa), respectively. Our work on the etiology of kuru (Gajdusek 1977, 1984, 1985a, b, c; Gajdusek and Gibbs 1975; Gajdusek and Zigas 1957, 1959; Gajdusek et al. 1965, 1966; Klatzo et al. 1959) and on the cause of amyotrophic lateral sclerosis (ALS) and parkinsonism with dementia (PD) with the early appearance of neurofibrillary tangles (NFT; Anderson et al. 1979; Chen 1981) in the populations in high incidence foci in the Western Pacific (Gajdusek 1988a; Gajdusek and Salazar 1982; Garruto et al. 1982, 1984, 1986; Perl et al. 1982) has pointedly emphasized that this molecular complex is not a static cytoskeletal structure, but moving fibers, responsible for the slow component of axonal transport of lysosomes, enzymes, and transmitter molecules to the presynaptic terminals (Gajdusek 1984, 1985a, b, c).

Interference with axonal transport may be responsible for a stagnation or pooling of cytoskeletal elements, and subsequent degradation of the sequestered cytoskeletal molecule(s) or matrix protein(s) in which they rest to form amyloid fibrils of paired helical filaments (PHFs) in the neurofibrillary tangles (NFTs) and the neuritic plaques that characterize Alzheimer's disease (Autilio-Gambetti et al. 1983; Gambetti et al. 1983a; Bizzi et al. 1984; Dahl and Bignami 1985; Gambetti et al. 1983a, b; Hirano and Inoue 1980; Hirano et al. 1984a, b; Inoue and Hirano 1979; Klatzo et al. 1965; Rasool and Selkoe 1985; Selkoe et al. 1986; Sotelo et al. 1980b; Terry and Pena 1965). Furthermore, it now appears that amyloid deposits in the nervous system, particularly the amyloid plaques of Alzheimer's disease and those of Down's syndrome, Pick's disease, and normal aging, and the perivascular accumulations of amyloid in the CNS, and in the vascular walls extending out into the meninges, are derived from a precursor matrix protein (amyloid B-protein or A4) trapped in these cytoskeletal accumulations, while the paired helical filaments of NFTs may represent yet further intracellular degradation of the same precursor matrix protein in the stagnated cytoskeletal elements (Glenner and Wong 1984a, b; Gredert et al. 1988; Guiroy et al. 1987; Kidd et al. 1985; Ksiezak-Reding et al. 1988a, b; Masters et al. 1985a, b; Schubert et al. 1988; Wischik et al. 1988a, b). We earlier presumed that the precursor for these brain amyloid fibrils is the 200-kDa component of the protein triad from which 10-nm neurofilaments are formed or microtubule-associated proteins tau (MAP-tau; Anderton et al. 1982; Gajdusek 1984, 1985a). However Goldgaber et al. (1987a, b) have isolated, sequenced, characterized, and localized on chromosome 21 cDNA clones coding for this precursor protein from the adult human brain cDNA library. Kang et al. (1987) have also isolated and sequenced the cDNA coding for the same protein from the fetal human brain cDNA library, and Tanzi et al. (1987) and Robakis et al. (1987) have isolated the same gene. Down's syndrome patients carry three copies of this gene and overexpress this precursor and thus express NFTs, amyloid plaque cores (APCs), and congophilic angiopathy (CA) 50 years earlier than normal subjects (Delabar et al. 1987), but we cannot reconfirm that AD patients carry three copies of the gene as we first reported (Delabar et al. 1987). However, since virtually everyone shows NFTs, APCs, or CA by 100 years of age, no abnormal gene is needed for the production of this amyloid.

The normal amyloid precursor B-protein gene is overexpressed in brains of Alzheimer's disease patients (Cohen et al. 1988; Higgins et al. 1988; Schmechle et al. 1988). It also encodes several different mRNAs that are identical except for the expression of a 168-bp long exon, or an added 57-bp long exon immediately following, at position 289 (between the G and TT) of a GTT valine codon) of the full mRNA without either of these inserts. The 168-bp long exon shares 50% homology to Kunitz-type protease inhibitors, with all six cysteine residues conserved (Goldgaber et al. 1988; Kitaguchi et al. 1988; Ponte et al. 1988; Tanzi et al. 1988).

The 4100 Dalton subunit polypeptide (42 amino acids) of vascular amyloid (Glenner and Wong 1984a, b), APCs (Masters et al. 1985b), and also the PHF from NFTs of Alzheimer's disease patients (Masters et al. 1985a) all have the same amino acid sequence with progressively more N-terminal heterogeneity (Masters et al. 1985a, b). This indicates that vascular amyloid deposits are least degraded from the parent host-B-protein, core amyloid of amyloid plaques next, and the amyloid polypeptide of PHF is most degraded from this same precursor B-protein. Although protein components of microtubules (and tubulin or MAP proteins) might well be the precursor or parent protein we seek, we now find that in all conditions where these masses of amyloid appear (perivascular or in neuritic or amyloid plaques and NFTs) there is a pooling or piling up of neurofilament in perikaryon and axonal swellings. It appears that the B-protein precursor of amyloid, the gene for which we have identified on chromosome 21 in man, is a membrane anchored or excreted matrix element caught in this mass of collapsed, pooled cytoskeletal elements (Schubert et al. 1988).

In fact, Hirano has demonstrated ultrastructurally minute masses of amyloid fibers and of regular paracrystalline arrays of particles or tubules within packed masses of piled up NF in spheroids that have formed from such swollen perikarya or axonal swellings in motor neurons of the spinal cord in ALS (Inoue and Hirano 1979; Hirano and Inoue 1980; Hirano et al. 1984a, b). Kirschner et al. (1986) have pointed out that the helical structure of the 200-kDa component of neurofilaments does not lend itself to degeneration to the B-pleated sheet structure common to all brain amyloids, and that perhaps MAP-tau is the more likely precursor. It too is accumulated in pooled masses of neurofilaments (Grundke-Iqbal et al. 1986; Kosik et al. 1986; Wood et al. 1986). This now seems not to be the case, although copolymerization or firm associations of the amyloid B-protein fibrils with a B-pleated peptide derived from MAP-tau appears to be likely (Gajdusek 1988b; Guiroy and Gajdusek 1988). However, we do not yet know the normal function of the precursor protein for amyloid polypeptide formation.

Thus, interference with axonal transport of neurofilaments may be a basic mechanism of pathogenesis that leads to

(a) pooling of the cytoskeletal elements associated proteins and matirx proteins in the perikaryon or axonal cylinders and lysis of the neuron as in ALS and other motor neuron diseases; (b) amyloid and neuritic plaque formation, from degradation in Alzheimer's disease and many other CNS degenerations of a precursor B-protein, which is anchored to the external membrane surface or excreated from the cell; and, finally, (c) neurofibrillary tangle formation with the same precursor protein further modified to form paired helical filaments probably copolymerized or otherwise associated with numerous proteins of the cytoskeletan. We know that the precursor B-protein is synthesized in neurons and probably also in microglial and oligodendroglial and some cerebrovascular endothethial cells (Bahmanyar et al. 1987; Fukatsu et al. 1984a, b; Goldgaber et al. 1987a, b; Schmechle et al. 1988). It could be released into the extracellular space by all of these cells.

C. Two Forms of Amyloid in Cerebral Plaques

The larger, more regular amyloid plaques of kuru, of Creutzfeldt-Jakob disease (CJD) and its Gerstmann-Sträussler variant, and of scrapie are composed of a scrapie-specific amyloid protein (PrP_{27-30}), a degradation product of a host-specified larger protein (PrP_{35-37}). The cDNA for this precursor protein has been sequenced (Oesch et al. 1985) and much of the oligonucleotide sequence confirmed by amino acid sequencing of parts of the isolated precursor (Multhaup et al. 1985). The gene for this precursor protein of the amyloid of the transmissible dementias is located on chromosome 20 in man, 2 in mouse.

Prusiner calls this scrapie-specified host-specified protein his "prion" protein (PrP_{27-30}; Bendheim et al. 1984; Bolton et al. 1985; Multhaup et al. 1985; Oesch et al. 1985; Prusiner 1982, 1984; Prusiner et al. 1983, 1984; Rohwer 1984c). In CJD the amyloid in plaque cores carries the same immunologic specificity as those of kuru and scrapie (Bendheim et al. 1984; Bockman et al. 1985; Braig and Diringer 1985; Brown et al. 1986a; Manuelidis 1985; Manuelidis et al. 1985) and amino acid sequences as does the purified 27- to 30-kDa protein of scrapie-associated fibrils (SAF; or "prion protein" PrP_{27-30}). The microheterogeneity of the CJD plaque polypeptide is the result of cleavage from different regions of the same host precursor protein (PrP_{35-37}; Multhaup et al. 1988).

The amyloid of Alzheimer's and Down's syndrome is composed of a self-aggregating 4.1-kDa amyloid polypeptide subunit of 42 amino acids (B-protein or A4 protein; Glenner and Wong 1984a; Masters et al. 1985a, b; Wong et al. 1985). The cDNA clones coding for this amyloid subunit have been isolated and characterized by Goldgaber and Lerman and their coworkers (1987a, b) and by Kang et al. (1987), Robakis et al. (1987), and Tanzi et al. (1987). This is a precursor protein for a different amyloid from that of kuru-CJD-scrapie. It is specified by a gene on chromosome 21 in man, 16 in mouse. On the other hand the amyloid of CJD is made by polymerization of a heavily glycosylated 27- to 30-kDa glycoprotein closely related to the scrapie-specific protein from scrapie-associated fibrils (SAFs; Bendheim et al. 1985; Bolton et al. 1985; Multhaup et al. 1985).

Thus, there are two forms of brain amyloid: that of the transmissible dementias and that of Alzheimer's disease, aged Down's syndrome, Guamanian ALS/PD, and normal aging (Merz et al. 1986). The respective precursors are specified by different genes located in man in chromosome 20 and 21, respectively (in mouse on chromosome 2 and 16, respectively). To determine whether these amyloids are formed from a neuronal, microglial, or serum-borne precursor has been the problem. It now appears that the amyloid in NFTs is formed from neuronal synthesized precursor; extracellular amyloid of plaques and congophilic angiopathy may be of microglial and vascular endothelial origin (Fukatsu 1984a, b; Higgins et al. 1988).

The mechanism of processing that produces the regularly oriented birefringent configuration of B-pleated sheets of amyloid proteins is not known. The known sequences of the amyloid in perivascular deposits (Glenner and Wong 1984a, b; Wong et al. 1985), plaque cores (Masters et al. 1985b), and PHFs of neurofibrillary tangles (Masters et al. 1985a; Guiroy et al. 1987), which are all alike, do not correspond with the amino acid sequence of the SAF protein (PrP; Multhaup et al. 1985; Oesch et al. 1985; Prusiner et al. 1984). Furthermore, neither precursor shows any homology with the sequences for the major components of the cytoskeleton: the three protein component neurofilaments, α- or β-tubulin or MAP II or MAP-tau or actin (Geisler et al. 1985; Lewis and Cowan 1985). Thus, we are dealing with two different precursor proteins and two different amyloid polypeptides, or small proteins, derived from them in the transmissible and the nontransmissible dementias. The two host-specified precursor proteins are not yet identified with a known function or structure in normal cells.

In the course of scrapie infection the tertiary and quarterinary structure of the precursor protein of 35–37 kDa is altered to render it insoluble and protease resistant. Once cleaved to the 27–30 kDa fragment, it is easily polymerized into fibrils. Both forms are apparently infectious.

D. Scrapie-Associated Fibrils

In preparations of scrapie-affected brain suspensions in a density gradient, Merz and Somerville have demonstrated amyloid-like two-stranded fibers – each fiber composed of two protofibrils – that increase in quantity with virus titer (Merz – 1981; 1984a, b). We have found these structures in brains of CJD patients and in brains of primates with experimental CJD and kuru but not in normal control brains or brains of patients with other neurodegenerative diseases (Brown et al. 1985; Gajdusek 1985c; Merz et al. 1984a, b). It has been postulated that these structures may represent the scrapie or CJD or kuru infectious agent (Gajdusek 1985b; Merz et al. 1981, 1984a, b; Prusiner 1984; Prusiner et al. 1983). Such structures bring to mind the filamentous plant viruses and filamentous phage fd which are of about the same diameters. However, no nucleic acid has been demonstrated in purified preparations of SAF proteins (PrPs). SAFs are yet the more intriguing since they are the central core of "cigar-like" tubulofilamentous structures in scrapie and CJD brains (Narang et al. 1987), but obscured by an outer coat of proteinaceus material and an inner coat of single-stranded host-derived DNA (Narang et al. 1988).

These scrapie-associated fibrils (SAFs) which may be the infectious agents are distinguishable ultrastructurally from the paired helical filaments (PHFs) of neurofibrillary tangles and the fibrils of brain amyloid (Merz et al. 1981; 1984a, b). However, their similarity is misleading since these do not share antigenicity with the PHFs of NFTs or with the amyloid fibrils in amyloid plaques of aging, Alzheimer's disease, and Down's syndrome. Thus, some antisera, both polyclonal and monoclonal, to the PHFs of Alzheimer's disease NFTs cross-react with the purified subunit protein of amyloid from plaque cores of senile plaques (Autilio-Gambetti et al. 1983; Gambetti et al. 1983a, b; Rasool and Selkoe 1985; Selkoe et al. 1986). However, most antisera to amyloid plaque cores do not react with NFTs. Neither of these antisera react, however, with SAFs of scrapie (Kingsbury et al. 1985; Manuelidis 1985).

Antibodies to the 27- to 30-kDa subunit protein of SAFs (or Prusiner's "prion proteins", PrP_{27-30}) cross-react strongly on Western blots with the subunit protein of SAFs from CJD- (Bendheim et al. 1985; Brown et al. 1985) and kuru- (Brown et al. 1986a) affected brains. However, such SAF-specific sera do not react with neurofilaments or with PHfs or plaque core amyloid from Alzheimer's disease (Bendheim et al. 1985; C. J. Gibbs and D. C. Gajdusek unpublished data).

E. Viruses Provoking No Immune Response and Evidencing No Nonhost Antigen

The CJD-kuru-scrapie-like slow viruses first invade the reticuloendothelial cells and particularly low-density lymphocytes in the spleen. Yet, they provoke no antibody response which can be demonstrated using as antigen live virus preparation of highly infectious titers (Gajdusek 1985a, b; Gajdusek and Gibbs 1975; Kasper et al. 1981; McFarlin et al. 1971). With the inability to demonstrate any antiviral antibody response or any immune response directed against nonhost viral components or capable of neutralizing the virus activity, these unconventional viruses are unique in their immunologic behavior. Natural and experimental infections with these viruses elicit no antibody response in the host nor does immunosuppression with whole-body radiation, cortisone, antileukocytic serum, or cytotoxic drugs alter the incubation period, progress, or pattern of disease, or duration of illness to death. Finally, in vivo and in vitro study of both B-cell and T-cell function revealed no abnormality early or late in the course of illness and no in vitro sensitization of the cells taken from diseased animals to high-titer preparations of these viruses (Gajdusek 1977, 1985b, c; Gajdusek and Gibbs 1975). Since high-titer infective material in both crude suspension and highly purified also fails to elicit an immunologic response against nonhost components, even when used with adjuvants, this becomes the first group of microbes in which such immunologic inertness has been demonstrated, which has evoked the speculation that the replication of these viruses does not involve production of a virus-specified nonhost antigen (Gajdusek 1977; Prusiner 1982). Instead, their protein component must be specified by host genes and thus be recognized as self.

The amyloid 27- to 30-kDa protein obtained from highly purified preparations of SAFs (prion protein, PrP_{27-30}) has now been shown to be infectious (Ceroni et al. 1989; Piccardo et al. 1989; Safar et al. 1989a, b, c) and is a subunit of the SAFs which are a fibrillary aggregation of such subunits. It aggregates into dimer, tetramer, octomer, and hexadecamer polymers, as does the different subunit polypeptide (4.1 kDa) of amyloid of Alzheimer's disease and aging brain (Braig and Diringer 1985; Masters et al. 1985b; Multhaup et al. 1985). Antibody to this same scrapie amyloid protein has been made in rabbits and such polyclonal antibody reacts well with SAFs by an enzyme-linked immunosorbent assay (ELISA; Brown et al. 1985), Western blotting technique (Brown et al. 1985, 1986a; Manuelidis et al. 1985), and gold-bead decoration immunoelectron-microscopy (Manuelidis et al. 1985). Such antibodies to the scrapie SAFs cross-react well with the SAFs of kuru and of CJD and the Gerstmann-Sträussler form of CJD (Bendheim et al. 1985; Brown et al. 1985; Manuelidis et al. 1985) and already provide a quick means of diagnosis of these diseases (Brown et al. 1986a). These antisera to SAFs cross-react with the amyloid plaques of kuru, Creutzfeldt-Jakob disease, and scrapie, but they do not cross-react with the amyloid plaques of Alzheimer's disease or the aging brain (Brown et al. 1985; C. J. Gibbs, D. C. Gajdusek, unpublished data; Kitramoto et al. 1986).

F. Enormous Resistance to Physical and Chemical Inactivation

The demonstration of the resistance of the unconventional viruses to high concentrations of formaldehyde or glutaraldehyde, psoralens, and most other antiviral and antiseptic substances (Brown et al. 1982a, 1986b), and to ultraviolet

(UV) and ionizing radiation, ultrasonication, and heat, and the further demonstration of iatrogenic transmission through implanted surgical electrodes, contaminated surgical instruments, and corneal transplantation, injections of human growth hormone derived from pituitary glands obtained from cadavers (Brown et al. 1985), and dura mater obtained from cadavers and "sterilized" by ionizing radiation, and possibly through dentistry, has led to the necessity of changing autopsy room and operating theater techniques throughout the world as well as the precautions used in handling older and demented patients. Many of the gentle organic disinfectants, including detergents and the quarterinary ammonium salts, often used for disinfection and even hydrogen peroxide, formaldehyde, ether, chloroform, iodine, phenol and acetone, are inadequate for sterilization of the unconventional viruses, as is the use of the ethylene oxide sterilizer. More recently, it has been shown that formaldehyde-fixed brain tissue is much more resistant to inactivation by autoclaving than is unfixed fresh scrapie infected brain (Taylor and McConnell 1988). This demands revision of previously acceptable procedures for decontamination and disinfection (Brown et al. 1982a, b, 1984, 1986b).

These unconventional viruses are also resistant, even when partially purified, to all nucleases, to β-propiolactone, ethylenediaminetetraacetic acid (EDTA), and sodium deoxycholate. They are moderately sensitive to most membrane-disrupting agents in high concentration such as phenol (60%), chloroform, ether, urea (6 *M*), periodate (0.01 *M*), 2-chloroethanol, alcoholic iodine, acetone, chloroform-butanol, hypochlorite, and alkalai, to chaotropic ions such as thiocyanate and guanadinium and trichloroacetate, and to proteinase K and trypsin when partially purified (Prusiner 1982), but these only inactivated 99% to 99.9% of the infectious particles leaving behind highly resistant infectivity (Rohwer 1984b). Sodium hydroxide (1.0 *N*) and hypochlorite (5%), however, quickly inactivate over 10^5 ID_{50} of the virus (Brown et al. 1984). They have a UV inactivation action spectrum with a six fold increased sensitivity at 237 nm over that at 254 nm or 280 nm, and 50-fold increased sensitivity at 220 nm (Gibbs et al. 1977; Haig et al. 1969; Latarjet 1979; Latarjet et al. 1970). Moreover, they show remarkable resistance to ionizing radiation that would indicate a target size, if such a naive calculation is applicable to a highly aggregated "semisolid" array of associated proteins, of under 100000 kDa (Gibbs et al. 1977; Latarjet 1979; Latarjet et al. 1970; Rohwer and Gajdusek 1980).

However, many investigators have seen regular arrays of particles that appear to be tubular structures seen in cross-section, in postsynaptic terminals of neurons in experimental animals infected with CJD, kuru, and scrapie (Baringer et al. 1979; 1981; David-Ferreira et al. 1968; Field and Narang 1972; Field et al. 1969; Lamar et al. 1974; Narang 1973; 1974a, b; Narang et al. 1972, 1980, 1987, 1988; Vernon et al. 1970; ZuRhein and Varakis 1976). Structures more typical of virions are not recognized on electron microscopic study of infected cells in vivo or in vitro, nor are they recognized in highly infectious preparations of virus concentrated by density-gradient banding in the zonal rotor.

These atypical properties have led to the speculation that the infectious agents lack a nucleic acid, and that they may be a self-replicating protein (perhaps by derepressing or causing misreading of cellular DNA bearing information for their own synthesis), even a self-replicating membrane fragment which serves as a template for laying down abnormal plasma membrane, including itself (Bendheim et al. 1985; Bolton et al. 1982, 1984, 1985; Gajdusek 1984, 1985a, b, c; Oesch et al. 1985; Prusiner 1982, 1984; Prusiner et al. 1983, 1984). I have often suggested that they are catalyzing and organizing the specific degradation of a host-specified precursor protein, autocatalytically producing themselves in the process (Gajdusek 1977, 1984, 1985a, b, c). More recently I have suggested that the fibril amyloid enhancing factors offer a good model for scrapie replication (Gajdusek 1988b; Guiroy and Gajdusek 1988).

Analogies with defective of "contaminated" seed crystals of simple nucleating molecules specifying the crystallization of their own distinct crystal structure come to mind as to mineral nucleation of protein crystallization (McPherson and Shlichta 1988). The presence of mineral deposits in neurons in the form of hydroxyapatites often containing aluminum (Bizzi et al. 1984; Nikaido et al. 1972; Perl and Brody 1980), silicon (Austin 1978; Austin et al. 1973; Garruto et al. 1986; Iler 1985; Nikaido et al. 1972), and other atoms as the antecedents to NFT formation with the aymloid protein of PHFs has been shown. Such deposits are exceptionally intense in the high incidence foci of ALS, parkinsonism-dementia, and associated early appearance of NFTs in the Western Pacific (Gajdusek and Salazar 1982; Garruto et al. 1982, 1984, 1986; Perl et al. 1982). More recently, Masters et al. (1985a) and Candy et al. (1986) have found silicon and aluminum deposits in the center of amyloid plaque cores in Alzheimer's disease. The aluminum silicate, perhaps in the form of montmorillonites, are in the center of amyloid plaque cores. Candy et al. have thus suggested because of this location, that they are the initiating elements of the amyloid deposition (Candy et al. 1986). Thus, we wonder whether a nucleus of a cation-binding mineral lattice may initiate the change to amyloid configuration of the normal host scrapie precursor protein (Iler 1985; Rees and Cragg 1983; Weiss 1981).

G. Mendelian Single Gene Autosomal Dominant Inheritance Determines Expression in Familial CJD

CJD became the first human infectious disease in which a single gene was demonstrated to control susceptibility and occurrence of the disease. The CJD virus is isolated from the brain of such familial cases. The autosomal dominant behavior of the disease in such families, including the appearance of the disease in 50% of siblings who survive to the age at which the disease usually appears, has evoked the possibility of virus etiology in other familial dementias. The presence of CJD patients in the families of well-known familial Alzheimer's disease, and the familial occurrence of the spinocerebellar ataxic form of Creutzfeldt-Jakob disease, the Gerstmann-Sträussler syndrome, which is also transmissible, have led to renewed interest in familial dementias of all types (Masters et al. 1981a, b; Traub et al. 1977). Thus, we are trying to determine the chromosomal location of the gene determining familial focus of CJD, in order to discover the effect this gene has on the processing of the precursor protein (PrP_{35-37}) of scrapie amyloid (PrP_{27-30}).

H. Autoantibody to 10-nm Neurofilament in SSVE Patients

The demonstration by Sotelo et al. of a very specific autoantibody directed against 10-nm neurofilaments and no other component of the CNS in over 60% of the patients with kuru and CJD as a phenomenon appearing late in the disease, was the first demonstration of an immune phenomenon in the SSVEs and an exciting new avenue of approach for the study of the transmissible dementias (Aoki et al. 1982; Bahmanyar et al. 1983, 1984; Sotelo et al. 1980a, b). This autoantibody behaves like many other autoantibodies such as the rheumatoid factor and the anti-DNA antibody in lupus and the antithyroglobulin antibody in Hashimoto's thyroiditis in that it is often present in normal subjects, and more often present in subjects closely related to the patients. Although found in more than one-half of patients with transmissible virus dementia, it was not detected in

40% of patients with classical CJD. It does develop in other gray matter diseases, including Alzheimer's and Parkinson's diseases, but at far lower incidence than in CJD (Bahmanyar et al. 1983; Sotelo et al. 1980a). Furthermore, it was not detected in patients with other immune diseases such as disseminated lupus erythematosis and chronic rheumatoid arthritis (Bahmanyar et al. 1983). We have demonstrated that on Western blots separating the three proteins comprising the 10-nm neurofilament triad of 200 kDa, 150 kDa, and 68 kDa, most positive sera have antibodies directed against the 200-kDa protein with some cross-reaction with the 150-kDa protein, some sera react better with the 150-kDa protein, and rare sera only with the 68-kDa protein, thought to be an internal component of the neurofilament (Bahmanyar et al. 1984; Toh et al. 1985a, b). Sheep with scrapie, however, often react best with a 62-kDa neurofilament-associated protein (Toh et al. 1985b). Some authors found a higher incidence than we have of these specific antibodies in normal subjects (Stefansson et al. 1985). Nonetheless, the same problem is posed. Why are there antibodies to the neurofilament proteins and not to other CNS antigens?

I. Unconventional Viruses: Subviral Pathogens, Perhaps Devoid of a Nucleic Acid or a Non-host Protein

The scrapie virus has been partially purified by density-gradient sedimentation in the presence of specific detergents. Scrapie virus has been over 1000-fold purified relative to other quantifiable proteins in the original brain suspension (Bolton et al. 1982, 1984; Diringer et al. 1983; Manuelidis and Manuelidis 1983; Multhaup et al. 1985; Prusiner et al. 1984; Rohwer and Gajdusek 1980; Rohwer et al. 1979). In such preparations the virus is susceptible to high concentrations of proteinase K and trypsin digestion, but it is not inactivated by any nuclease (Prusiner 1982). Sedimented, washed, and resuspended virus has been banded into peaks of high infectivity with the use of cesium chloride, sucrose, and metrizamide density gradients in the ultracentrifuge. Attempts to demonstrate a nonhost nucleic acid in scrapie-virus preparations using DNA homology and transfection and nuclease inactivation have been unsuccessful (Borras and Gibbs 1986; Borras et al. 1982, 1986; Hunter et al. 1976). No significant quantities of nucleic acid are present in purified preparations of 27- to 30-kDa SAF associated-protein (PrP_{27-30}), and such preparations were first found to be noninfectious (Diringer et al. 1983; Manuelidis 1985; Multhaup et al. 1985; Oesch et al. 1985), but have been shown to be highly infectious (Ceroni et al. 1989; Piccardo et al. 1989; Safar et al. 1989a, b, c).

The atypical action spectrum for inactivation of scrapie virus by UV should not be taken as proof that no genetic information exists in the scrapie virus as nucleic acid molecules, since Latarjet has demonstrated similar resistance to UV and a similar UV action spectrum for microsomes (Gibbs et al. 1977; Haig et al. 1969; Latarjet 1979; Latarjet et al. 1970). Ultraviolet resistance also depends greatly on small RNA size, as has been shown by the high resistance of the purified, very small, tobacco ring spot satellite virus RNA (about 80 kDa). However, we may read this UV-resistance at face value as the first clear evidence that we were dealing with an infectious polypeptide.

Moreover, the unconventional viruses possess numerous properties in which they resemble classical viruses (Gajdusek 1977, 1985b; Rohwer 1984a, b; 1985, unpublished work; Rohwer and Gajdusek 1980), and some of these properties suggest far more complex genetic interaction between virus and host than one might expect for genomes with a molecular mass of only 10^5 kDa. Rohwer has shown that the scrapie virus replicates in hamster brain at a constant rate, with no eclipse phase, and with a doubling time of 5.2 days (Rohwer 1985, unpublished work). Examination of the kinetics of its inactivation and the demonstrated association or aggregation of scrapie virus particles into polymers or clusters that can be disrupted by ultrasonication have cast doubt on the calculation of its small size from ionizing radiation inactivation data and inferences about its structure from resistance to chemical inactivating agents. Thus, aggregates make necessary "multiple hits" for inactivation, whereas free virus is killed by a single event (Rohwer 1985, unpublished work).

In plant virology we have been forced to modify our concepts of a virus to include subviral pathogens such as the newly described viroids causing 11 natural plant diseases – potato spindle tuber disease, chrysanthemum stunt disease, citrus exocortis disease, Cadang-Cadang disease of coconut palms, cherry chloratic mottle, cucumber pale fruit disease, hop stunt disease, avacado sunblotch disease, tomato bunchy top disease, tomato "planta macho" disease, and burdock stunt disease – and the virusoids of four natural plant diseases (velvet tobacco mottle virus, solanum nodiflorum mottle virus, lucerne transient streak virus, subterranean clover mottle virus) to which we may turn for analogy (Diener and Hadidi 1977; Sänger 1982). All of the viroids are small circular RNAs containing no structural protein or membrane and they have all been fully sequenced and their fine structures determined. They have only partial base pairing as the circle collapses on itself. They contain only 246 to 574 ribonucleotides and replicate by a "rolling circle" copying of their RNA sequences in many sequential rotations to produce an oligomeric copy, which is then cut into monomers or sometimes dimers. No protein is synthesized from their genetic information, and only the replication machinery of the cell is used. These subviral pathogens have caused us to give much thought to possible similarities to the unconventional viruses. However, we and others have shown that the unconventional viruses differ markedly from the plant viroids on many counts (Diener and Hadidi 1977; Gajdusek 1985b, c; Prusiner 1982; Sänger 1982); in fact, many of their properties are diametrically opposite to those of the viroids. Thus, the intellectually stimulating analogies of the unconventional viruses to viroids and virusoids prove to be spurious, yet these subviral pathogens of plants have served to alert us to the possibility of extreme departure from conventional virus structures.

Recent work on amyloid enhancing factors, particularly fibril amyloid enhancing factor (Niewold et al. 1987) strongly suggests that an autocatalytic nucleating process directing fibril growth according to its own specified fibril structure appears to give us the most challenging model for scrapie replication (Gajdusek 1988b; Guiroy and Gajdusek 1988). The newer work of Safar and his coworkers (Ceroni et al. 1989; Piccardo et al. 1989; Safar et al. 1989a, b, c) clearly shows that the normal host-specified scrapie precursor protein (PrP_{35-37}) is converted to an infectious form by configurational change in secondary and tertiary structure of the noninfectious precursor. It will require the work of crystallographers to define molecularly how this de novo configurational change to an infectious polypeptide is autoinduced and autopatterned.

J. Concluding Hypothesis – Fantasy of a "Virus" from the Inorganic World

We are at an exciting moment in the study of the unconventional viruses. Either the SAF-associated protein (PrP_{27-30}) and its infectious progenitor, the infectious form of the scrapie precursor protein (PrP_{35-37}) are the infectious agent directing its own synthesis by nucleation and autopatterned crystallization or by augmentation (and alternative splicing) of its host gene, or this protein is simply an elegant molecular biological "high-tech" demonstration of what we have known for a long while, namely, that amyloid is found in the CNS in all of these diseases and is a distant byproduct of the cell damage caused by the virus. In that case we are still in search for the atypical virus.

If the alteration of a host precursor protein to the self-polymerizing, insoluble, protease-resistant amyloid-like infectious scrapie-specific 35- to 37-kDa protein from a host protein by autoinduced configurational change in secondary structures or by posttranslational processing, a glycosylation (Bolton et al. 1985; Manuelidis et al. 1985; Multhaup et al. 1985), phosphorylation (Sternberger et al. 1985), peptide bond hydrolysis, cleavage, with proteolytic truncation at both termini, cross-linkage, altered splicing and repacking (Connors 1985; Masters et al. 1985a) is the basic growth process of scrapie replication, then the hydroxyapatites-aluminum silicate inorganic nidi in NFTs and in the center of amyloid plaque cores in Alzheimer's disease may signal that *this mineral-protein complex is the nucleating agent that has proved so elusive.* We must allow for the possibility that such a mineral-amyloid complex might in the proper milieu of the interior of a cell replicate slowly and regularly as it degrades a 35- to 37-kDa host precursor matrix protein (Oesch et al. 1985) to the amyloid we see in SAFs (PrP) and the amyloid plaques of these infections. In Alzheimer's and Pick's diseases and Down's syndrome a 4.1-kDa polylpeptide or its polymers complexes as an amyloid protein to a calcium-aluminum silicate apparently can self-replicate and self-aggregate as it autocatalytically degrades a precursor protein, presumably the amyloid B-protein precursor now identified, to the mineral-amyloid aggregates or paracystalline arrays we see in neurofibrillary tangles (Garruto et al. 1984; Perl et al. 1982;) and the amyloid plaque cores (Austin 1978; Candy et al. 1986; Masters et al. 1985a; Nikaido et al. 1972). Only in the nondividing neuron does this slow degenerative process eventually kill the cell. Thus, our atypical slow "virus" may simply be similar to a crystal template directing its own crystallization or "crystal lattice" from a source of presynthesized host protein precursors and an inorganic cation receptor nucleant. This remains a still-tenable hypothesis. If so, we wonder whether inorganic polymer chemistry and crystallography may provide better insights than the normal paradigms of modern molecular biology (Connors 1985; Iler 1985; Weiss 1981).

The calcium, aluminum, and silicon deposits have only been found in the center of cores of amyloid plaques and in neurofibrillary tangles. Thus, they remain candidates for the initiation or nucleation phase of amyloidogenesis in these degenerative amyloidoses of brain. In the slow virus infections the microfibril or oligomer of the scrapie-associated protein (PrP_{27-30}) may be its own nucleating agent and crystallization template. In these infections no mineral deposits have been found in the center of cores of the amyloid plaques (unpublished data).

I would prefer to call the infectious agent of scrapie a *virus,* even if it proves to be as romantically exotic as a polypeptide directing an auto-catalytically patterned degradation of a stagnated, pooled host-specified matrix protein to a glycosylated amyloid. The potent abstract concept of a virus as a self-specifying transmissible entity requiring the machinery of the host for its replication does not specify any specific structure. Mathematicians playing with computers have not hesitated to use the term "virus" for the "virus infections" of computer memories they have produced (Dewdney 1985a, b). Dewdney (1984), with his Core Wars program, initiated computer virology. The facts that software viruses contain no nucleic acid nor are nucleic acids in any way involved in the pathology that these viral diseases produce has not prevented computer scientists from appropriately calling them viruses (Denning 1988).

References

Anderson FH, Richardson EP, Okazaki H, Brody JA (1979) Neurofibrillary degeneration on Guam. Frequency in Chamorros and non-Chamorros with no known neurological disease. Brain 102 (1): 65–77

Anderton BH, Breinburg D, Downes MJ, Green PJ, Tomlinson BE, Ulrich J, Wood JN, Kahn J (1982) Monoclonal antibodies show that neurofibrillary tangles and neurofilaments share antigenic determinants. Nature 298: 84–86

Aoki T, Gibbs, CJ, Sotelo J, Gajdusek, DC (1982) Heterogenic autoantibody against neurofilament protein in the sera of the animals with experimental kuru and Creutzfeldt-Jakob disease and natural scrapie infection. Infect Immun 38: 316–324

Austin JH (1978) Silicon levels in human tissues. In: Bendz G, Lindqvist I (eds) Biochemistry of silicon and related problems. Plenum, New York, pp 255–268

Austin JH, Rinehart R, Williamson J, Burcar P, Russ K, Nikaido T, Lafrance M (1973) Studies in ageing of the brain. III. Silicon levels in postmortem tissues and body fluids. Prog Brain Res 40: 485–495

Autilio-Gambetti L, Gambetti P, Crane RC (1983) Paired helical filaments: relatedness to neurofilaments shown by silver staining and reactivity with monoclonal antibodies. In: Katzman R (ed) Biological aspects of Alzheimer's disease. Banbury report no 15. Cold Spring Harbor, New York, pp 117–124

Bahmanyar S, Moreau-Dubois MC, Brown P, Gajdusek DC (1983) Serum antineurofilament antibodies in patients with neurological and non-neurological disorders and healthy controls using rat spinal cord. J Neuroimmunol 5: 191–196

Bahmanyar S, Liem RKH, Griffin JW, Gajdusek DC (1984) Characterization of antineurofilament autoantibodies in Creutzfeldt-Jakob disease. J Neuropathol Exp Neurol 43(4): 369–375

Bahmanyar S, Higgins GA, Goldgaber D, Lewis DA, Morrison JH, Wilson MC, Shankar SK, Gajdusek DC (1987) Localization of amyloid B-protein messenger RNA in brains from Alzheimer's disease patients. Science 237(4810): 77–80

Baringer JR, Wong J, Klassen T, Prusiner SB (1979) Further observations of the neuropathology of experimental scrapie in mouse and hamster. In: Prusiner SB, Hadlow WJ (eds) Slow transmissible diseases of the nervous system, vol 2. Academic, New York, pp 111–121

Baringer JR, Prusiner SB, Wong JS (1981) Scrapie-associated particles in postsynaptic processes. Further ultrastructural studies. J Neuropathol Exp Neurol 40: 281–288

Beck E, Bak IJ, Christ JF, Gajdusek DC, Gibbs CJ, Hassler R (1975) Experimental kuru in the spider monkey. Histopathological and ultrastructural studies of the brain during early stages of incubation. Brain 98: 595–612

Beck E, Daniel PM, Davey AJ, Gajdusek DC, Gibbs CJ (1982) The pathogenesis of spongiform encephalopathies: an ultrastructural study. Brain 105: 755–786

Bendheim PE, Barry RA, DeArmond SJ, Stites DP, Prusiner SB (1984) Antibodies to a scrapie-prion protein. Nature 310: 418–421

Bendheim PE, Bockman JO, McKinley MP, Kingsbury DT, Prusiner SB (1985) Scrapie and Creutzfeldt-Jakob disease prion proteins share physical properties and antigenic determinants. Proc Natl Acad Sci USA 82: 997–1001

Bizzi A, Crane RC, Autilio-Gambetti L, Gambetti P (1984) Aluminum effect on slow axonal transport: a novel impairment of neurofilament transport. J Neurosci 4(3): 722–731

Bockman JM, Kingsbury DT, McKinley MP, Bendheim PE, Prusiner SB (1985) Creutzfeldt-Jakob disease prion proteins in human brains. N Engl J Med 312: 73–78

Bolton DC, McKinley MP, Prusiner SB (1982) Identification of a protein that purifies with the scrapie prion. Science 218: 1309–1311

Bolton DC, McKinley MP, Prusiner SB (1984) Molecular characteristics of the major scrapie prion protein. Biochemistry 23: 5898–5905

Bolton DC, Meyer RK, Prusiner SB (1985) Scrapie PrP 27–30 is a sialoglycoprotein. J Virol 53: 596–606

Borras MT, Gibbs CJ (1986) Molecular hybridization studies with scrapie brain nucleic acids. I. Search for specific DNA sequences. Arch Virol 88: 67–78

Borras MT, Kingsbury DT, Gajdusek DC, Gibbs CJ (1982) Inability to transmit scrapie by transfection of mouse embryo cells in vitro. J Gen Virol 58: 263–271

Borras MT Merendino JJ, Gibbs CJ (1986) Molecular hybridization studies with scrapie brain nucleic acids. II. Differential expression in scrapie hamster brain. Arch Virol 88: 79–90

Braig HR, Diringer H (1985) Scrapie: concept of a virus induced amyloidosis of the brain. Eur J Mol Biol 4: 2309–2311

Brown P, Gibbs CJ, Amyx HL, Kingsbury DT, Rohwer RG, Sulima MP, Gajdusek DC (1982a) Chemical desinfection of Creutzfeldt-Jakob disease virus. N Engl J Med 306(21): 1279–1282

Brown P, Rohwer RG, Green EM, Gajdusek DC (1982b) Effect of chemicals, heat and histopathological processing on high infectivity hamster-adapted scrapie virus. J Infect Dis 145: 683–687

Brown P, Rohwer RG, Gajdusek DC (1984) Sodium hydroxide decontamination of Creutzfeldt-Jakob disease virus. N Engl J Med 310: 727

Brown P, Gajdusek DC, Gibbs CJ, Asher DM (1985) Potential epidemic of Creutzfeldt-Jakob disease from human growth hormone therapy. N Engl J Med 313: 728–731

Brown P, Coker-Vann M, Pomeroy BS, Asher DM, Gibbs CJ, Gajdusek DC (1986a) Diagnosis of Creutzfeldt-Jakob disease by Western blot identification of marker protein in human brain tissue. N Engl J Med 314: 547–551

Brown P, Rohwer RG, Gajdusek DC (1986b) Newer data on the inactivation of scrapie virus or Creutzfeldt-Jakob disease virus in brain tissue. J Infect Dis 153(6): 1145–1148

Candy JM, Oakley AE, Klinowski J, Carpenter TA, Perry RH, Atack JR, Perry EK, Blessed G, Fairbairn A, Edwardson JA (1986) Aluminosilicates contribute to senile plaque formation in Alzheimer's disease. Lancet i: 354–356

Ceroni M, Piccardo P, Safar J, Gajdusek DC, Gibbs CJ Jr. (1989) Scrapie infectivity and scrapie amyloid protein are distributed in the same pH range in agarose isoelectric focusing. Neurology (in press)

Chen Leung (1981) Neurofibrillary change on Guam. Arch Neurol 38(1): 16–18

Cohen ML, Golde TE, Usiak MF, Younkin LH, Younkin SG (1988) In situ hybridization of nucleus basalis neurons shows increased B-amyloid mRNA in Alzheimer's disease. Proc Nat Acad Sci USA 85: 1227–1231

Connors LH (1985) In vitro formation of amyloid fibrils. Biochem Biophys Res Commun 131(3): 1063–1068

Dahl D, Bignami A (1985) Two different populations of neurofibrillary tangles in Alzheimer's dementias revealed by neurofilament immunoreactivity and Congo Red staining. In: Bignami A, Bolis L, Gajdusek DC (eds) Molecular mechanisms of pathogenesis of central nervous system disorders, Disc Neurosci 3(1): 80–82

David-Ferreira JF, David-Ferreira KL, Gibbs CJ, Morris JA (1968) Scrapie in mice: ultrastructural observations in the cerebral cortex. Proc Soc Exp Biol Med 127: 313–320

Delabar J-M, Goldgaber D, Lamour Y, Nicole A, Huret J-L, DeGrouchy J, Brown P, Gajdusek DC, Sinet P-M (1987) B-amyloid gene duplication in Alzheimer's disease and karyotypically normal Down syndrome. Science 235: 1390–1392

Denning PJ (1988) The science of computing. Computer viruses. Am Sci 76: 236–238

Dewdney AK (1984) Computer recreations: in the game called Core Wars hostile program's engage in a battle of bits. Sci Am 250, 5; 14, 18–20, 22.

Dewdney AK (1985a) Computer recreations: a core war bestiary of viruses, worms and other threats to computer memories. Sci Am 252, 3; 14, 19–23

Dewdney AK (1985b) Analog gadgets that solve a diversity of problems and array of questions. Sci Am 252, 6; 18, 22, 24, 25, 28, 29

Diener T, Hadidi A (1977) Viroids. In: Fraenkel-Conrat II, Wagner RR (eds) Comprehensive virology, vol 11. Plenum, New York, pp 285–337

Diringer H, Gelderblom H, Hilmert H, Özel M, Edelbluth C, Kimberlin RH (1983) Scrapie infectivity, fibrils, and low molecular weight protein. Nature 306: 476–478

Field EJ, Narang HK (1972) An electron microscopic study of natural scrapie in the rat: further observations on "inclusion bodies" and virus-like particles. J Neurol Sci 17(3): 347–364

Fukatsu R, Gibbs CJ, Amyx HL, Gajdusek DC (1984a) Amyloid plaque formation along the needle track in experimental murine scrapie. J Neuropathol Exp Neurol 43(3): 313

Fukatsu R, Gibbs CJ, Gajdusek DC (1984b) Cerebral amyloid plaques in experimental murine scrapie. In: Tateishi J (ed) Proceedings of the workshop on slow transmissible diseases. Research Committee on Slow Virus Infections. Japanese Ministry of Health, Aug 31, Tokyo, pp 71–84

Gajdusek DC (1977) Unconventional viruses and the origin and disappearance of kuru. Science 197:943–960

Gajdusek DC (1984) Interference with axonal transport of neurofilament: the underlying mechanism of pathogenesis in Alzheimer's disease, amyotrophic lateral sclerosis and many other degenerations of the CNS. The Merrimon Lecture, School of Medicine, University of North Carolina, Chapel Hill

Gajdusek DC (1985a) Hypothesis: interference with axonal transport of neurofilament as a common pathogenetic mechanism in certain diseases of the central nervous system. N Engl J Med 312 (11):711–719

Gajdusek DC (1985b) Unconventional viruses causing subacute spongiform encephalopathies. In: Fields BN (ed) Virology. Raven, New York, chap 63, pp 1519–1557

Gajdusek DC (1985c) Subacute spongiform virus encephalopathies caused by unconventional viruses. In: Maramorosch K (ed) Subviral pathogens of plants and animals: viroids and prions. Academic, New York, chap 20, pp 483–544

Gajdusek DC (1988a) Cycad toxicity not the cause of high incidence amyotrophic lateral sclerosis/Parkinsonism-dementia on Guam, Kii Peninsula of Japan or in West New Guinea. In: Hudson AJ (ed) Amyotrophic lateral sclerosis: concepts in pathogenesis and etiology. University of Toronto Press, Toronto

Gajdusek DC (1988b) Etiology versus pathogenesis: the causes of posttranslational modifications of host specified brain proteins to amyloid configuration. In: Genetics and Alzheimer's Disease, Sinet PM, Lamour Y and Christen Y, editors. Research and Perspectives in Alzheimer's Disease, Fondation IPSEN pour la Recherche Therapeutique, Springer-Verlag, Berlin pp 174–176.

Gajdusek DC, Gibbs CJ (1975) Slow virus infections of the nervous system and the laboratories of slow, latent and temperate virus infections. In: Tower DB (ed) The nervous system, vol 2: the clinical neurosciences. Raven, New York, pp 113–135

Gajdusek DC, Salazar A (1982) Amyotrophic lateral sclerosis and parkinsonian syndromes in high incidence among Auyu and Jakai peoples of West New Guinea. Neurology 32 (2):107–126

Gajdusek DC, Zigas V (1957) Degenerative disease of the central nervous system in New Guinea. The endemic occurrence of "kuru" in the native population. N Engl J Med 257:974–978

Gajdusek DC, Zigas V (1959) Kuru: clinical, pathological and epidemiological study of an acute progressive degenerative disease of the central nervous system among natives of the Eastern Highlands of New Guinea. Am J Med 26:442–469

Gajdusek DC, Gibbs CJ, Alpers M (eds) (1965) Slow, latent and temperate virus infections. National Institutes of Health, NINCDB Monogr 2 (PHS Publ no 1378) US Government Printing Office, Washington

Gajdusek DC, Gibbs CJ, Alpers M (1966) Experimental transmission of a kuru-like syndrome in chimpanzees. Nature 209:794–796

Gambetti P, Autilio-Gambetti L, Perry G, Shecket G, Crane RC (1983a, b) Antibodies to neurofibrillary tangles of Alzheimer's disease raised from human and animal neurofilament fractions. Lab Invest 49(4):430–435

Gambetti P, Shecket G, Ghetti B, Hirano A, Dahl D (1983b) Neurofibrillary changes in human brain. An immunocytochemical study with a neurofilament antiserum. J Neuropathol Exp Neurol 42(1):69–79

Garruto RM, Yanagihara R, Arion DM, Daum CA, Gajdusek DC (1982) Bibliography of amyotrophic lateral sclerosis and Parkinsonism-dementia of Guan. National Institutes of Health, Bethesda, US Government Printing Office, Washington

Garruto RM, Fukatsu R, Yanagihara R, Gajdusek DC, Hook G, Fiori CE (1984) Imaging of calcium and aluminum in neurofibrillary tangle-bearing neurons in parkinsonism-dementia of Guam. Proc Nat Acad Sci USA 81(6):875–879

Garruto RM, Swyt C, Yanagihara R, Fiori CE, Gajdusek DC (1986) Intraneuronal colocalization of silicon with calcium and aluminum in amyotrophic lateral sclerosis and parkinsonism with dementia of Guam. N Engl J Med 315:711–712

Geisler N, Plassmann Um Weber K (1985) The complete amino acid sequence of the major mammalian neurofilament protein (NF-L). Fed Eur Biol Soc 182(2):475–478

Gibbs CJ, Gajdusek DC, Asher DM, Alpers MP, Beck E, Daniel PM, Matthews WB (1968) Creutzfeldt-Jakob disease (subacute spongiform encephalopathy): transmission to the chimpanzee. Science 161:388–389

Gibbs CJ, Gajdusek DC Latarjet R (1977) Unusual resistance to UV and ionizing radiation of the viruses of kuru, Creutzfeldt-Jakob disease, and scrapie (unconventional viruses). Proc Natl Acad Sci USA 75:6268–6270

Glenner GG, Wong CW (1984a) Alzheimer's disease: report of the purification and characterization of a novel cerebrovascular amyloid protein. Biochem Biophys Res Commun 120:885–890

Glenner GG, Wong CW (1984b) Alzheimer's disease and Down's syndrome: sharing of a unique cerebrovascular fibril protein. Biochem Biophys Res Commun 122(3):1131–1135

Goedert M, Wishcik CM, Crowther RA, Walker JE, Klug A (1988) Cloning and sequencing of the cDNA encoding a core protein of the paired helical filament of Alzheimer disease: identification as the microtubule-associated protein tau. Proc Natl Acad Sci USA 85:4051–4055

Goldgaber D, Lerman M, McBride W, Saffiotti U, Gajdusek DC (1987a) Isolation, characterization and chromosomal localization of cDNA clones coding for the precursor protein of amyloid of brain in Alzheimer disease, Down's syndrome, and aging. In: Corkin SH, Growden JH (eds) therapies. Proceedings of the fourth meeting of the International Study Group on the pharmacology of memory disorders associated with aging, Zurich, 16–18 Jan 1987. Center for Brain Science and Metabolism Charitable Trust, Cambridge, MA, pp 209–217

Goldgaber D, Lerman M, McBride W, Saffiotti U, Gajdusek DC (1987b) Characterization and chromosomal localization of cDNA encoding brain amyloid of Alzheimer's disease. Science 235:877–880

Goldgaber D, Teener JW, Gajdusek DC (1988) Alternative cDNA clones of amyloid B-protein precursor gene encode proteinase inhibitor. Disc Neurosci (5:3; 40–46)

Gorevic P, Goni F, Pons-Estel B, Alvarez F, Peress R, Frangione B (in press) Isolation and partial characterization of neurofibrillary tangles and amyloid plaque cores in Alzheimer's disease. Immunological studies. J Neuropathol Exp Neurol (in press)

Grundke-Iqbal I, Iqbal K, Tung Y-C, Quinlan M, Wiesniewski H, Binder LI (1986) Abnormal phosphorylation of the microtubule-associated protein t (tau) in Alzheimer cytoskeletal pathology. Proc Natl Acad Sci USA 83:4913–4917

Guiroy DC, Gajdusek DC (1988) Fibril-derived amyloid enhancing factors as nucleating agents in Alzheimer's disease and transmissible virus dementias. Disc Neurosci 5:3; 69–73

Guiroy DC, Miyazaki M, Multhaup G, Fischer P, Garruto RM, Beyreuther K, Masters C, Simms G, Gibbs CJ, Gajdusek DC (1987) Amyloid of neurofibrillary tangles of Guamanian parkinsonism-dementia and Alzheimer's disease share identical amino acid sequence. Proc Natl Acad Sci 84:2073–2077

Haig DC, Clarke MC, Blum E, Alper T (1969) Further studies on the inactivation of the scrapie agent by ultraviolet light. J. Gen. Virol. 5:455–457

Higgins GA, Lewis DA, Bahmanyar S, Goldgaber D, Gajdusek DC Young WG, Morrison JH, Wilson MC (1988) Differential regulation of amyloid B-protein mRNA expression with hippocampal neuronal subpopulations in Alzheimer disease. Proc Natl Acad Sci USA 85(4):1297–1301

Hirano A, Inoue K (1980) Early pathological changes in amyotrophic lateral sclerosis. Electron microscopic study of chromatolysis, spheroids, and Bunina bodies. Neurol Med (Tokyo) 13:148–160

Hirano A, Donnenfeld H, Sasaki S, Nakano I (1984a) Fine structural observations of neurofilamentous changes in amyotrophic lateral sclerosis. J Neuropathol Exp Neurol 43(5):461–470

Hirano A, Nakano I, Kurland LT, Mulder DW, Holley PW, Saccomanno G (1984b) Fine structural study of neurofibrillary changes in a family with amyotrophic lateral sclerosis. J Neuropathol Exp Neurol 43(5):471–480

Hunter GD, Collis SC, Millson GC, Kimberlin RH (1976) Search for scrapie-specific RNA and attempts to detect an infectious DNA or RNA. J Gen Virol 32:157–162

Iler RK (1985) Hydrogen-bond complexes of silicon with organic compounds. In: Bendz G, Lindqvist I (eds) Biochemistry of silicon and related problems. Plenum, New York, pp 53–76

Inoue K, Hirano A (1979) Early pathological changes in amyotrophic lateral sclerosis. Autopsy findings of a case of ten months duration. Neurol Med (Tokyo) 11:448–455

Kang J, Lemaire H-G, Unterbeck A, Slabaum JM, Masters CL, Grzeschik K-H, Multhaup G, Beyreuther K, Muller-Hill B (1987) The precursor of Alzheimer's disease amyloid A4 protein resembles a cell-surface receptor. Nature I (325):733–736

Kasper KC, Bowman K, Stites DP, Prusiner SB (1981) Toward development of assays for scrapie-specific antibodies. In: Hamster immune responses in infectious and oncogenic diseases. Streilein JW, Hart DA, Stein-Sterilein J, Duncan WR, Billingham RE (eds) Plenum, New York, pp 401–413

Kidd M, Allsop D, Landon M (1985) Senile plaque amyloid, paired helical filaments and cerebrovascular amyloid in Alzheimer's disease are all deposits of the same protein. Lancet i:278

Kingsbury DT, Prusiner SB, Bockman JM, McKinley MP, Barry RA (1985) Reply to the Editor: Laura Manuelidis: Creutzfeldt-Jakob disease prion proteins in human brains. N Engl J Med 312(25):1644–1645

Kirschner DA, Abraham C, Selkoe DJ (1986) X-ray diffraction from intraneuronal paired helical filaments and extraneuronal amyloid fibers in Alzheimer disease indicates cross B-conformation. Proc Nat Acad Sci USA 83:503–507

Kitaguchi N, Takahashi Y, Tokushima Y, Shiojiri S, Ito H (1988) Novel precursor of Alzheimer's disease amyloid protein shows protease inhibitory activity. Nature 331:530–532

Kitramoto T, Tateishi J, Tashima T, Takeshita I, Barry RA, DeArmond SJ, Prusiner SB (1986) Amyloid plaques in Creutzfeldt-Jakob disease stain with prion protein antibodies. Ann Neurol 20:204–208

Klatzo I, Gajdusek DC, Zigas V (1959) Pathology of kuru. Lab Invest 8:799–847

Klatzo I, Wisniewski H, Streicher EJ (1965) Experimental production of neurofibrillary degeneration. 1. Light microscopic observations. J Neuropathol Exp Neurol 24:187–199

Kosik KS, Joachim CL, Selkoe DJ (1986) Microtubule-associated protein tau (t) is a major antigenic component of paired helical filaments in Alzheimer disease. Proc Natl Acad Sci USA 83:4044–4048

Ksiezak-Reding H, Binder LI, Yen S-H (1988 a) Immunochemical and biochemical characterization of proteins in normal and Alzheimer's disease brains with Alz 50 and Tau-1. J Biol Chem 263(17):7948–7953

Ksiezak-Reding H, Davies P, Yen S-H (1988 b) Alz 50, a monoclonal antibody to Alzheimer's disease antigen, cross-reacts with proteins from bovine and normal human brain. J Biol Chem 263(17):7943–7947

Lamar CH, Gustafson DP, Krashnovich M, Hinsman EJ (1974) Ultrastructural studies of spleens, brains and brain cell cultures of mice with scrapie. Vet Pathol 11:13–19

Lampert PW, Gajdusek DC, Gibbs CJ (1971) Experimental spongiform encephalopathy (Creutzfeldt-Jakob disease) in chimpanzees. Electron microscopic studies. Presented at a meeting of the American Association of Pathologists and Bacteriologists, St Louis, March 7, 1970. J Neuropathol Exp Neurol 30:20–32

Latarjet R (1979) Inactivation of the agents of scrapie, Creutzfeldt-Jakob disease and kuru by radiation. In: Slow transmissible diseases of the nervous system, vol 2. Prusiner SB, Hadlow WJ (eds) Academic, New York, pp 387–407

Latarjet R, Muel B, Haig DA, Clarke MC, Alper T (1970) Inactiviation of the scrapie agent by near-monochromatic ultraviolet light. Nature 227:1341–1343

Lewis SA, Cowan NJ (1985) Genetics, evolution, and expression of the 68,000-molecular-weight neurofilament protein: isolation of a cloned cDNA probe. J Cell Biol 100:843–850

Manuelidis L (1985) Letter to the Editor. Creutzfeldt-Jakob disease prion proteins in human brains. N Engl J Med 312(25):1643–1644

Manuelidis L, Manuelidis EE (1983) Fractionation and infectivity studies in Creutzfeldt-Jakob disease. In Katzman R (ed) Biological aspects of Alzheimer's disease. Banbury report no 15. Cold Spring Harbor, New York, pp 399–412

Manuelidis L, Valley S, Manuelidis EE (1985) Specific proteins associated with Creutzfeldt-Jakob disease and scrapie share antigenic and carbohydrate determinants. Proc Natl Acad Sci USA 82:4263–4267

Masters CL, Gajdusek DC Gibbs CJ (1981 a) The familial occurrence of Creutzfeldt-Jakob disease and Alzheimer's disease. Brain 104: 535–558

Masters CL, Gajdusek DC, Gibbs CJ (1981 b) Creutzfeldt-Jakob disease virus isolations from the Gerstmann-Sträussler syndrome, with an analysis of the various forms of amyloid plaque deposition in the virus-induced spongiform encephalopathies. Brain 104: 559–588

Masters CL, Multhaup G, Simms G, Pottgiesser J, Martins RN, Beyreuther K (1985 a) Neuronal origin of a cerebral amyloid: neurofibrillary tangles of Alzheimer's disease contain the same protein as the amyloid of plaque cores and blood vessels. Eur J Mol Biol 4: 2757–2763

Masters CL, Simms G, Weinman NA, Multhaup G, McDonald BL, Beyreuther K (1985 b) Amyloid plaque core protein in Alzheimer's disease and Down's syndrome. Proc Natl Acad Sci USA 82(12): 4245–4249

McFarlin DE, Rott MC, Simpson L, Nehlson S (1971) Scrapie in immunologically deficient mice. Nature 233: 336

McPherson A, Schlichta P (1988) Heterogeneous and epitaxial nucleation of protein crystals on mineral surfaces. Science 239: 385–387

Merz PA, Somerville RA, Wisniewski HM Iqbal K (1981) Abnormal fibrils from scrapie-infected brain. Acta Neuropathol (Berl) 54: 63–74

Merz PA, Rohwer RG Kascsak R, Wisniewski HM, Somerville RA, Gibbs CJ, Gajdusek DC (1984 a) Identification of a disease-specific particle in scrapie-like slow virus diseases. Science 225: 437–440

Merz PA, Somerville RA, Wisniewski HM, Manuelidis L, Manuelidis EE (1984 b) Scrapie associated fibrils in Creutzfeldt-Jakob disease. Nature 306: 474–476

Merz PA, Wisniewski HM, Rubenstein R, Kascsak RJ (1986) Immunological studies on paired helical filaments and amyloid of Alzheimer's disease. Disc Neurosc 3(1): 58–68

Multhaup G, Diringer H, Hilmert H, Prinz H, Heukeshoven J, Beyreuther K (1985) The protein component of scrapie-associated fibrils is a glycosylated low-molecular-weight protein. Eur J Mol Biol 4: 1495–1501

Multhaup G, Burke D, Beyreuther K, Brown P (1988) Protein composition and sequence analysis of cerebral amyloid fibrils of unconventional slow virus diseases. In: Court LA, Cathala F (eds) Slow transmissible diseases of the nervous system. Abbaye de la Mellerye, Paris (in press)

Narang HK (1973) Virus-like particles in natural scrapie of the sheep. Res Vet Sci 14: 108–110

Narang HK (1974 a) An electron microscopic study of natural scrapie sheep brain: further observations on virus-like particles and paramyxovirus-like tubules. Acta Neuropathol (Berl) 28(4): 317–329

Narang HK (1974 b) An electron microscopic study of the scrapie mouse and rat: further observations on virus-like particles with ruthenium red and lanthanum nitrate as a possible trace and negative stain. Neurobiology 4(4): 349–363

Narang HK, Shenton BK, Giorgi PP, Field EJ (1972) Scrapie agent and neuron. Nature 240(5376): 105–107

Narang HK, Chandler RL, Anger HS (1980) Further observations on particulate structures in scrapie affected brain. Neuropathol Appl Neurobiol 6: 23–28

Narang HK, Asher DM, Gajdusek DC (1988) Evidence that DNA is present in abnormal tubulofilamentous structures found in scrapie. Proc Natl Acad Sci USA 85: 3575–3579

Narang HK, Asher DM, Gajdusek DC (1987) Tubulofilaments in negatively stained scrapie-infected brains: relationship to scrapie-associated fibrils. Proc Natl Acad Sci USA 84(2): 7730–7734

Niewold TA, Hol PR, van Andel ACJ, Lutz ETG, Gruys E (1987) Enhancement of amyloid induction by amyloid fibril fragments in hamsters. Lab Invest 56: 544–549

Nikaido T, Austin J, Truch L, Reinhart R (1972) Studies in ageing of the brain. II. Microchemical analyses of the nervous system in Alzheimer patients. Arch Neurol 27: 549–554

Oesch B, Westaway D, Wälchli M, McKinley MP, Kent SBH, Aebersold R, Barry RA, Tempst P, Teplow DB, Hood LE, Prusiner SB, Weissmann C (1985) A cellular gene encodes scrapie PrP 27–30 protein. Cell 40: 735–746

Perl DP, Brody AR (1980) Alzheimer's disease: X-ray spectrometric evidence of aluminum accumulation in neurofibrillary tangle-bearing neurons. Science 208: 297–299

Perl DP, Gajdusek DC, Garruto RM, Yanagihara RT, Gibbs CJ (1982) Intraneuronal

aluminum accumulation in amyotrophic lateral sclerosis and parkinsonism-dementia of Guam. Science 217(4564): 1053–1055
Piccardo P, Safar J, Ceroni M, Gajdusek DC, Gibbs CJ Jr. (1989) Immunohistochemical localization of scrapie amyloid protein in spongiform encephalopathies and normal brain tissue. Neurology (in press)
Ponte P, Gonzalez-DeWhitt P, Shilling J, Miller J, Hsu D, Greenberg B, Davis K, Wallace W, Lieberburg I, Fuller F, Cordell B (1988) A new A4 amyloid mRNA contains a domain homologous to serine proteinase inhibitors. Nature 331: 525–527
Prusiner SB (1982) Novel proteinaceous infectious particles cause scrapie. Science 216: 136–144
Prusiner SB (1984) Some speculations about prions, amyloid and Alzheimer's disease. N Engl J Med 310: 661–663
Prusiner SB, McKinley MP, Bowman KA, Bolton DC, Bendheim PD, Groth DF, Glenner GG (1983) Scrapie prions aggregate to form amyloid-like birefrigent rods. cell 35: 349–358
Prusiner SB, Groth DF, Bolton DC, Kent SB, Hood LE (1984) Purification and structural studies of a major scrapie prion protein. Cell 38: 127–134
Rasool CG, Selkoe DJ (1985) Sharing of specific antigens by degenerating neurons in Pick's disease and Alzheimer's disease. N Engl J Med 312(11): 700–705
Rees S, Cragg B (1983) Is silica involved in neuritic (senile) plaque formation? Acta Neuropathol (Berl) 59: 31–40
Robakis NK, Wisniewski HM, Jenkins EC, Devine-Gage EA, Housck GE, Xiu-Lan Yao, Ramakrishna N, Wolfe G, Silverman WP, Brown WT (1987) Chromosome 21q21 sublocalization of gene encoding B-amyloid peptide in cerebral vessels and neuritic (senile) plaques of people with Alzheimer disease and Down syndrome. Lancet i(8520): 384
Robertson HD, Branch AD, Dahlberg JE (1985) Focusing on the nature of the scrapie agent. Cell 40: 725–727
Rohwer RG (1984a) Scrapie shows a virus-like sensitivity to heat inactivation. Science 223: 600–602
Rohwer RG (1984b) Scrapie: virus-like size and virus-like susceptibility to inactivation of the infectious agent. Nature 308: 658–662
Rohwer RG (1984c) Letter to the Editor. Scrapie-associated fibrils. Lancet ii(8393): 36
Rohwer RG, Gajdusck DC (1980) Scrapie-virus or viroid: the case for a virus. In: Boese A (ed) Search for the cause of multiple sclerosis and other chronic diseases of the central nervous system. Proceedings of the first international symposium of the Hertie Foundation, Frankfurt am Main, Sept 1979. Verlag Chemie, Weinheim, pp 333–355
Rohwer RG, Brown PW, Gajdusek DC (1979) The case of sedimentation to equilibrium as a step in the purification of the scrapie agent. In: Prusiner SB, Hadlow WJ (eds) Slow transmissible diseases of the nervous system. Academic, New York, pp 465–478
Safar J, Ceroni M, Piccardo P, Gajdusek DC, Gibbs CJ Jr. (1989a) Scrapie precursor proteins: antigenic relationship between species and immunocytochemical localization in normal, scrapie and Creutzfeldt-Jakob disease brains. Neurology (in press)
Safar J, Ceroni M, Piccardo P, Liberski PP, Miyazaki M, Gajdusek DC, Gibbs CJ Jr. (1989b) Subcellular distribution and physicochemical properties of scrapie precursor protein and relationship with scrapie infectivity. Neurology (in press)
Safar J, Wang W, Paogett MP, Ceroni M, Piccardo P, Zopf D, Gajdusek DC, Gibbs CJ Jr. (1989c) Molecular mass, biochemical composition and physicochemical behavior of the infectious form of the scrapie precursor protein monomer. J Virol (in press)
Sänger HL (1982) Biology, structure, functions, and possible origins of plant viroids. In: Parthier B, Bolter D (eds) Nucleic acids and proteins in plants. II. Springer, Berlin Heidelberg New York, pp 368–454 (Encyclopaedia of plant physiology, NS 14B)
Schmechle DE, Goldgaber D, Burkhart DS, Gilbert JR, Gajdusek DC, Roses AD (1988) Cellular localization of amyloid-beta-protein messenger RNA in postmortem brain in Alzheimer's disease patients. Int Alzheimer's Dis Rel Dis 2:2; 96–111
Schubert D, Schroeder R, LaCorbiere M, Saiton T, Cole G (1988) Amyloid B-protein precursor is possibly a heparin sulfate proteoglycan core protein. Science 241: 223–226
Selkoe DJ, Abraham CR, Podlinsky MD, Duffy LK (1986) Isolation of low molecular weight proteins from amyloid plaque fibers in Alzheimer's disease. J Neurochem 146: 1820–1834
Sotelo J, Gibbs CJ, Gajdusek DC (1980a) Autoantibodies against axonal neurofila-

ments in patients with kuru and Creutzfeldt-Jakob disease. Science 210:190–193

Sotelo J, Gibbs CJ, Gajdusek DC, Toh BH, Wurth M (1980b) Method for preparing cultures of central neurons: cytochemical and immunochemical studies. Proc Natl Acad Sci USA 77:653–657

Stefansson K, Marton LS, Dieperink ME, Molnar GK, Schlaepfer WW, Helgason CM (1985) Circulating autoantibodies to the 200000-dalton protein of neurofilaments in the serum of healthy individuals. Science 228:1117–1119

Sternberger NH, Sternberger LA, Ulrich J (1985) Aberrant neurofilament phosphorylation in Alzheimer's disease. Proc Natl Acad Sci USA 82(12):4274–4276

Tanzi RE, Gusella JF, Watkins PC, Bruas GAP, St. George-Hyslop, P, VanKeuren ML, Patterson D, Pagan S, Kuruit DM, Neve RL (1987) Amyloid B-protein gene: cDNA, mRNA distribution, and genetic linkage near the Alzheimer locus. Science 235:880–884

Tanzi RE, McClatchey AI, Lamperti ED, Villa-Komaroff L, Gusella JF, Neve RL (1988) Protease inhibitor domain encoded by an amyloid protein precursor mRNA associated with Alzheimer's disease. Nature 331:528–530

Taylor DM, McConnell I (1988) Autoclaving does not decontaminate formal-fixed scrapie tissues. Lancet i(8600):1463–1464

Terry RD, Pena CJ (1965) Experimental production of neurofibrillary degeneration 2. Electron microscopy, phosphatase histochemistry and electron probe analysis. Neuropathol Exp Neurol 24:200–210

Toh BH, Gibbs CJ, Gajdusek DC, Goudsmit J, Dahl D (1985a) The 200- and 150-kDa neurofilament proteins react with IgG autoantibodies from patients with kuru, Creutzfeldt-Jakob disease and other neurologic diseases. Proc Nat Acad Sci USA 82:3485–3489

Toh BH, Gibbs CJ, Gajdusek DC, Tuthill DD, Dahl D (1985b) The 200- and 150-kDa neurofilament proteins react with IgG autoantiboides from chimpanzees with kuru, Creutzfeldt-Jakob disease and 62-kDa neurofilament-associated protein reacts with sera from sheep with natural scrapie. Proc Natl Acad Sci USA 82:3894–3896

Traub R, Gajdusek DC, Gibbs CJ (1977) Transmissible virus dementias. The relation of transmissible spongiform encephalopathy to Creutzfeldt-Jakob disease. In: Kinsbourne M, Smith L (eds) Aging and dementia. Spectrum, Flushing, New York, pp 91–146

Troncoso JC, Hoffman PN, Griffin JW, Hess-Kozlow KM, Price DL (1985) Aluminum intoxication: a disorder of neurofilament transport in motor neurons. Brain Res 342:172–175

Vernon ML, Horta-Barbosa, L, Fuccillo DA, Sever JL, Barringer JR, Burnbaum G (1970) Virus-like particles and nuclear protein type filaments in brain tissue from two patients with Creutzfeldt-Jakob disease. Lancet i(7564):964–967

Weiss A (1981) Replication and evolution of inorganic systems. Angew Chem Int Ed Engl 20:850–860

Williams ES, Young S (1980) Chronic wasting disease of captive mule deer: a spongiform encephalopathy. J Wildl Dis 16:89–98

Williams ES, Young S (1982) Songiform encephalopathy of Rocky Mountain elk. J Wildl Dis 18:465–471

Williams ES, Young S, Marsh RF (1982) Preliminary evidence of the transmissibility of chronic wasting disease of mule deer. In: Proceedings of the Wildlife Disease Association Annual conference, Aug 19; 1982, Madison, WI (Abstract no 22)

Wischik CM, Novak M, Edwards PC, Klug A, Tichelaar W, Crowther RA (1988) Structural characterization of the core of the paired helical filament of Alzheimer disease. Proc Natl Acad Sci USA 85:4884–4888

Wischik CM, Novak M, Thofweawn H-C, Edwards PC, Runswick MJ, Jakes R, Walker JE, Milstein C, Roth M, Klug A (1988b) Isolation of a fragment of tau derived from the core of the paired helical filament of Alzheimer disease. Proc Natl Acad Sci USA 85:4506–4510

Wong CW, Quaranta V, Glenner GG (1985) Neuritic plaques and cerebral vascular amyloid in Alzheimer disease are antigenically related. Proc Natl Acad Sci USA 82:8729–9732

Wood JG, Mirra SS, Binder LI (1986) Neurofibrillary tangles of Alzheimer disease share antigenic determinants with the axonal microtubule-associates protein tau (t). Proc Natl Acad Sci USA 83:4040–4043

ZuRhein GM, Varakis J (1976) Subacute spongiform encephalopathy In: Kimberlin RH (ed) Slow virus diseases of animal and man. North-Holland, Amsterdam, pp 359–380

Clay Minerals – Blueprints of Early Life

E. T. Degens

A. Background

Aside from helium, living and cosmic matter are principally composed of the same chemical elements, hydrogen – oxygen – carbon – nitrogen, reading in order of abundance, as: C – O – H – N vs H – O – C – N. In contrast, the Earth has an entirely different bulk composition iron – oxygen – silicon – magnesium.

In comparing the various patterns one might suggest that living matter sprouts directly out of the ancestral universal matrix, whereas terrestrial matter must be a late derivation product of that matrix. Only oxygen ranks high in all three compartments: cosmos, life, and earth. Thus, it is tempting to look for the roots of life in outer space. Actually, such an attempt is in strong contrast to textbook dogma where the origin of life presupposes a synthesis of vital monomers in a reducing terrestrial atmosphere by means of high energy radiation. A subsequent "raining out" is assumed to have led to a "primordial organic soup" from which physiologically interesting polymers and eventually the first living cell arose (Oparin 1953).

To prime the pump of my antithesis, a few details on the distribution of matter in interstellar space and its fractionation in the course of solar and planetary evolution are needed. Its further assemblage in the direction of a workable cellular system by way of mineral templates and globular aggregates will represent the essence of my subsequent presentation.

B. On Matter in Dark Molecular Clouds

Our galaxy, the Milky Way, would look if seen from the outside somewhat like a galaxy in Ursa Major, M81, just 9 million light years away from us (Fig. 1). About 100 billion stars are contained in this magnificent spiral about 100000 light years across, and quite a few may have planets like the Sun. Our solar system would occupy a place close to the medium plain of the spiral about one-third the distance of the total diameter removed from the galactic center. We need roughly 250 million Earth years to circumnavigate the galactic center. Accordingly, this period of time has been defined as 1 cosmic year.

Of the mass we can account for in our galaxy, about 95% resides in stars and the rest is interstellar gas and dust in a ratio of about 99 to 1. However, this interstellar matter is not uniformly distributed throughout the galaxy but is concentrated in the form of clouds in regions close to the galactic plane of symmetry, where new stars are born. To most of us, clouds are seen more in conjunction with weather and climate. Namely, water vapor will condense and form clouds of different sizes and shapes, which are freely moving, can become dispersed or aggregate, and form a cloud cover. Some regions are more cloudy than others. In due course, they will rain out and the meteorological cycle will commence again with the evaporation of water bodies. In analogy, interstellar clouds, a few

---

Institute of Biogeochemistry and Marine Chemistry at the Center of Marine Research and Climatology, University of Hamburg, Hamburg, F.R.G.

Fig. 1. Galaxy in Ursa Major, M81, about 9 million light years away from us

light years or more in diameter, are in constant slow motion around the galactic center. In their perpetual wandering through galactic space, they may grow in size or collapse. During 1 cosmic year, an average cloud could conceivably double its mass in sweeping up dispersed interstellar gas and dust. Once a critical size is reached or a nearby supernova explosion occurs a cloud can rain out, that is collapse, yielding stars and satellites of various sorts.

Now a word has to be said on the chemistry of cloud complexes. Research on giant interstellar clouds by means of astrospectrographs and radiotelescopes was initiated in 1963 by a team of the Massachusetts Institute of Technology and the Lincoln Laboratory. The most crucial finding was the observation that cloud complexes contained in addition to H_2 a variety of molecules. By including all isotopic species, close to 100 different types of molecules were recognized. A molecule of particular relevance is carbon monoxide because its radiation properties at low temperatures make the molecules an excellent signpost for the mapping of cloud complexes. However, the most dominant molecule is hydrogen, in a concentration of about 10^4 H_2 per cm^3. This value represents ca. 99% of the mass of a giant molecular cloud. The remainder of the molecules are just "impurities", but considering the size of a cloud, they can pile up to a rather substantial stack of matter. For instance, trace constituents such as carbon monoxide, water, methane, formaldehyde, methyl alcohol, ethyl alcohol, hydrogen cyanide, ammonia, or the hydroxyl radical far exceed all the mass contained in our solar system. Ethyl alcohol by itself could readily fill up the whole Earth with 100-proof "whiskey."

In a cloud complex all carbon is molecularly bound, that is, no atomic carbon remains. For oxygen, 30% has entered a complex organic molecule. For nitrogen and sulfur compounds present in the interstellar medium, the data bank is not sufficient to make a tentative assignment as yet.

As far as the "dust" goes which represents about 1% of the total matter, we are essentially dealing with ordinary minerals, <0.5 micron in size, such as Fe-Mg silicates, native iron, and graphite – the same stuff earth is principally made of. The presence of solid particles has numerous consequences for the synthesis and protection of organic molecules in the environment of space. For example, the probability of collision between atoms and molecules is enhanced, and three-body reactions become feasible. Moreover, mineral surfaces may provide not only a convenient "resting place" for certain atoms and molecules, but by virtue of their crystalline order, catalysis and epitaxis may ensue. The generation of more complex molecules such as sugars, amino acids, or the bases of the purines and pyrimidines is conceivable, but their detection requires more sophisticated technologies.

In brief, the chemistry of giant molecular cloud complexes is basically one of hydrogen, oxygen, carbon and nitrogen, judged by the prevalence of molecules containing C-O-H-N. It is noteworthy that in the presence of a mineral catalyst, simple organic molecules such as formaldehyde or hydrogen cyanide are expected to yield biochemically interesting monomers, for instance, sugars, amino acids, purines, and pyrimidines. Laboratory experiments done under low-temperature conditions have indeed shown the feasibility of a rapid synthesis of these compounds. In Fig. 2, the steps involved in the synthesis of common sugars in a formaldehyde-clay system are schematically shown. All of this can be used as

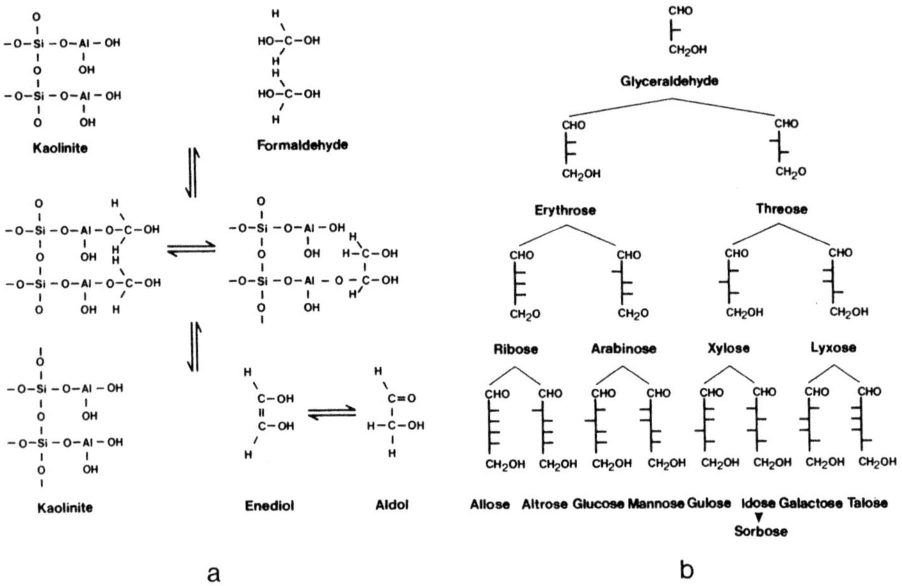

Fig. 2. **a** Possible sequences for kaolinite-catalyzed reactions. **b** Addition of one unit of formaldehyde at a time to D-glyceraldehyde in the presence of kaolinite would result in the distribution of the sugars illustrated. Thermodynamic factors, such as steric repulsions of hydroxyl groups, play an important role in the distribution of sugars

Fig. 3a–l. Diagram showing major steps in the evolution of the solar system (from upper left ► to lower right): **a** Whirlpool galaxy, similar to Milky Way, about 100000 light years in diameter; frontal and edge-on views. **b** Contours of a giant molecular cloud complex, a few light years in

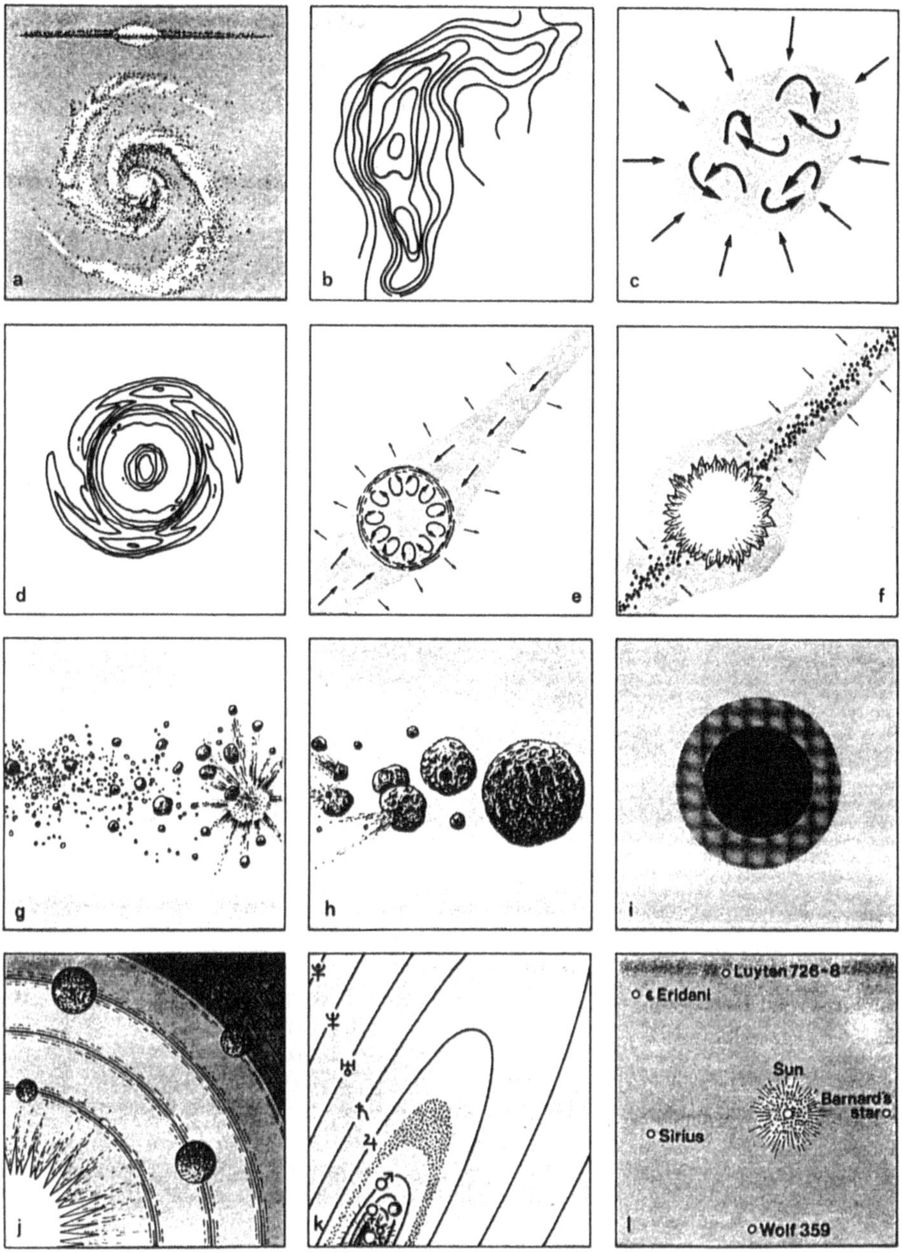

diameter revealed by the radiation of carbon monoxide at 2.7 mm ^{13}CO radio line. **c** Collapse of molecular cloud complex possibly triggered by supernova event. **d** Rotating solar nebula in statu nascendi with the proto-Sun evolving in its center. **e** Formation of accretion disk; *arrows* indicate motion of gas. **f** Aggregation of particles along midplane of accretion disk. **g** and **h** Accretion from dust, to planetesimals, to planets. **i** Retention of primordial atmosphere by a large planet (e.g., Jupiter) **j** T Tauri phase, sweeping off "excess" primordial gases from the solar system leaving atmosphere-free terrestrial planets behind, **k** Spacing of the orbits of the planets *(astrological symbols)* and the asteroids *(dotted area)*. **l** Oort's cloud (comet reservoir) surrounding the Sun with a radius of about 1 light year in relation to the nearest stars

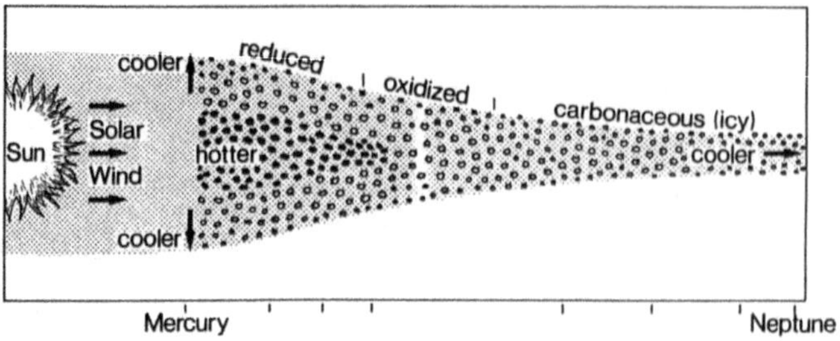

Fig. 4. Accretion disk across solar system from Mercury to Neptune. Variations in temperature and oxidation states in the assumed growth regions of meteorites and cometary bodies

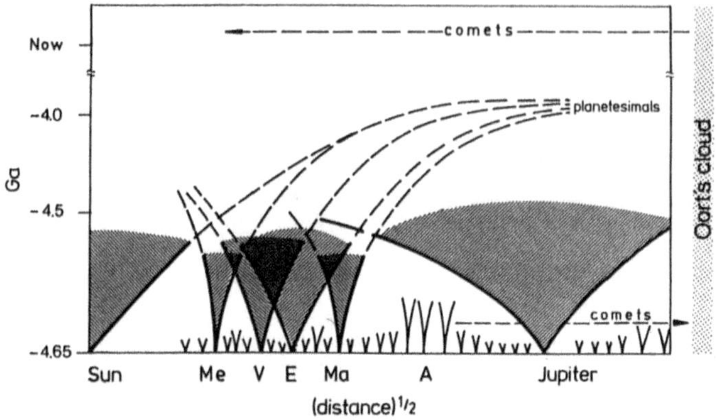

Fig. 5. Diagram showing planetesimals to planets. Feeding zones widen with time in accordance with the growth rate of the protoplanet, eventually yielding overlap of zones. Major surviving feeding zones give rise to planets and asteroids which are displayed at the square root of distance from the Sun. Planetesimals ejected from giant planets yield cometary bodies, which accumulate in the Oort cloud, from where they become episodically ejected and reenter the solar system. Bombardment of the terrestrial planets by comets and asteroids throughout the ages has been omitted for graphical reasons

indication that a dark molecular cloud is a gigantic ice box for all sorts of organic molecules just waiting to become defrosted, processed, and utilized for the construction of a cell, once the environment turns hospitable to the creation of life.

The first step to achieve this goal is the collapse of a molecular cloud and to illustrate this I have drawn three diagrams. The first (Fig. 3) shows critical events – one by one – that have led to the formation of our solar system. The second (Fig. 4) illustrates the wide range in temperature and oxidation states that prevailed during the formative years of the planets. The third (Fig. 5) elucidates the logic behind the enormous variations in the size of the planets. The situation revealed in Fig. 5 somehow reminds me of a little boy throwing small rocks horizontally across a water surface, from where the flat pebbles bounce off numerous times – high and low – till their momentum runs out.

Earth has received a tiny share of all types of incoming debris: first the hot and highly reduced irons, then the stones, and finally the icy and oxidized material containing gases, water, organic molecules, and clay minerals. The last dowry may be viewed as frosting on top of a cake which in due course led to air, sea, and life.

The tripartition of Earth into core, mantle, and crust is a reflection of the stepwise accretion of the proto-Earth.

Since the last incoming material – our "frosting" – is assumed to have formed a layer about 700 km thick around the globe, there certainly were organic molecules galore available to trigger life.

C. Crystalline Blueprints

Rock-forming minerals are principally composed of oxygen ions which have as their main coordination partners silicon, aluminum, iron, calcium, magnesium, sodium, and potassium ions. Crust and upper mantle may thus be viewed as an ionic oxysphere. Crystals may contain charge deficiencies, structural irregularities, lattice defects, and, in hydrated varieties such as clays, may even develop hydrogen bonds.

To structure our discussion, I will begin with a process commonly described under the heading "epitaxis," a term derived from the Greek *tassein,* meaning to arrange or to organize.

The growth of crystalline material on other crystal surfaces is a well-studied subject in the field of crystallography. Epitaxis can also proceed on organic templates with the resultant formation of biominerals in teeth, bones, or shells. Furthermore, organic polymers can promote the synthesis of other organic polymers, and the living cell is vivid proof of that. Finally, mineral surfaces may provide sites for activation and protection of functional groups displayed by organic molecules, and may accordingly serve as polymerization matrix. Thus, one can distinguish between four systems in which one partner represents the template and the other partner the epitaxial product (Table 1).

Table 1. Four systems of templates and epitaxial products

Template	Product
Mineral	Mineral
Biopolymer	Mineral
Biopolymer	Biopolymer
Mineral	Biopolymer

Epitaxis on solid-state surfaces should be viewed in relation to catalysis because both processes follow a similar reaction path. Catalysis represents a process in which a solid-state surface "tries" to establish a thermodynamically favorable phase transition structure with the adsorbent. Phase-transition structures emerge which, when chemically stable, lead to oriented intergrowth. In contrast, should transition structures introduce a chemical change of the adsorbent such as polymerization, hydrogenation, dehydrogenation, etc., we are dealing with catalysis. Principles of cellular catalysis, as, for example, executed by enzymes, are identical to those observed in mineral systems. Catalysis constitutes a flow of epitaxial associations, whereas epitaxis involves a "frozen in" transition structure provided by a morphological catalyst.

With the help of clay minerals, chemical synthesis of a number of physiologically interesting polymers has been successful; this particularly concerns the formation of peptides. The mechanism involves carboxyl activation and the inactivation of functional groups not participating in the formation of the amide bond by so-called protective groups displayed along mineral surfaces. The relationships for a kaolinite-amino acid system are schematically illustrated in Fig. 6. In the presence of kaolinite, amino acids will be picked up from an aqueous solvent and brought into solid solution. Amino groups become hydrogen bonded to structural oxygen, or in the case of basic amino acids, occur as positive ions. They are tightly fixed to the silicate surface, and thus rendered inactive. Carboxyl groups associated with charged Al-oxy-hydroxy groups by means of ionic bridges become directly attached to the aluminum.

In water, amino acids cannot polymerize because of dipole-dipole interactions. In solid solution, however, amino acids will polymerize, because the solvent medium does not interfere, and because

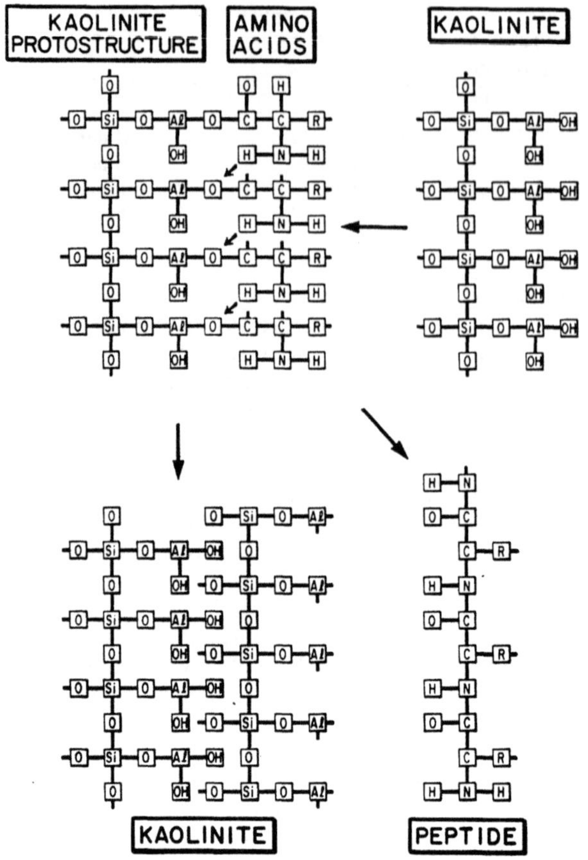

Fig. 6. Polymerization of amino acids along clay templates

this reaction step is favored energetically. In the kaolinite experiment, about 1000 times more amino acids were polymerized to peptides than could conceivably become adsorbed to the clay surface. In consequence, a flow of freshly polymerized molecules across the catalytically effective mineral surface has to be postulated. Kaolinite is also instrumental in preferentially synthesizing pentoses and hexoses from formaldehyde and transforming them into polysaccharides. Kaolinite can also generate fatty acids and entertain esterification reactions leading to glycerides. Furthermore, addition of calcium phosphate to an aqueous mixture of kaolinite, glycerol, and palmitic acids may yield phospholipid monolayers which are deposited in epitaxial order with a 40 Å periodicity on the crystal surface of kaolinite as ascertained by transmission electron microscopy. In essence, all principal building blocks of life have been synthesized by employing a variety of crystalline blueprints.

A remarkable characteristic of life is that all peptides and proteins are exclusively composed of the L-optical isomers of amino acids. Preferential polymerization of L-amino acids on kaolinite can be attributed to the inherent enantiomorphism of the edges of the octahedral layer of kaolinite (Fig. 7), and to the fact that kaolinite crystals are either entirely right-handed or entirely left-handed. Quite a number of ordinary minerals exhibit chirality. The fact that common carbohydrates such as cellulose, starch, or cane and beet sugars (sacrose) are composed

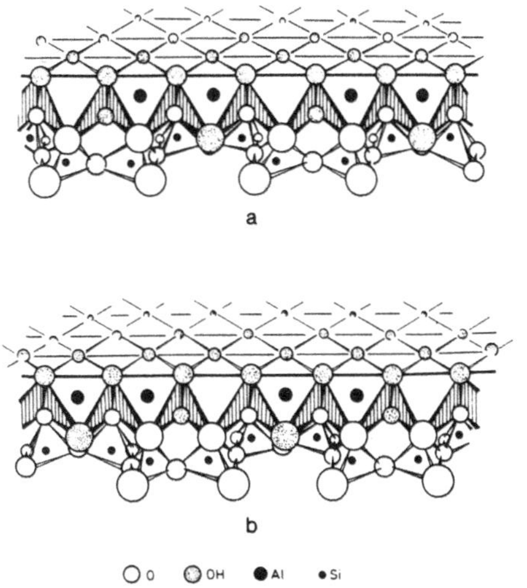

Fig. 7a, b. Schematic representation of the edge of a kaolinite crystal of ideal composition **(a)**, and its mirror image **(b)** viewed along the *a*-axis

of D-configurated monomers can also be related to the optical activity of the archaic mineral matrix.

The structural shape of biomolecules, which is a key element of cellular function, is asymmetric. It is conceivable that on the prebiotic Earth left- and right-handed polymers were generated by mineral printing machines having either D or L block letters. Once the first organism had chosen the L-configurated amino acid polymer, or the D-configurated sugar polymer, their mirror images had no chance to evolve further.

Summing up, crystalline blueprints are effective devices for generating leading biomolecules and for promoting chirality. Clays are outstanding in this respect, but they only deliver semifinished products, not life itself.

D. Towards the First Living Cell

Considering the complexity of life it may come as a surprise that only about 300 monomeric building blocks, lots of water, and some salt are needed to generate all the vital stuff in the genetic and metabolic apparatus. However, this stuff has to become neatly packaged into a cellular envelope in a manner able to fire the engine of life. So it seems that the most critical question has to do with the way membranes arise and of how that system is energized and autocatalytically maintained (evolution is hereby of no relevance because thermodynamically it constitutes just a disorder phenomenon). The answer is simple, life has adopted biophosphates for various structural and functional assignments.

In the biological sciences a quiet revolution is presently going on, namely the recognition that the architectural principles observed in biophosphates are identical to those established in the inorganic phosphates. The early adoption of the name phosphorus, the carrier of light, has not lost its true meaning over the centuries. Phosphorus, in the form of phosphate bonds, is the carrier of energy in the living system. It is the "energy currency" as George Wald so nicely put it, which becomes printed, exchanged, and converted at various rates in the perpetuating cycle of life. It is phosphorus which controls the structure and shape of cellular material and which thus selects the energy transfer sites. Thus, in the element phosphorus lies the answer to the question of what distinguishes life from an ordinary mineral. Life is based on PO_4 units, and rock-forming minerals on SiO_4 units:

$$\begin{array}{c} O \\ | \\ O - P = O \\ | \\ O \end{array} \qquad \text{versus} \qquad \begin{array}{c} O \\ | \\ O - Si - O \\ | \\ O \end{array}$$

It is essentially the π electron – the high energy bond – in PO_4 which maintains the animated world. Since the π bond can lie "parallel" to any of the four sigma bonds, giving rise to a variety of differently shaped tetrahedra, a flexible and dynamic tetrahedra network can be created. It is principally the type of metal ion adopted by PO_4 that shapes the tetrahedron. In contrast, the SiO_4 unit has "just" four sigma bonds which only per-

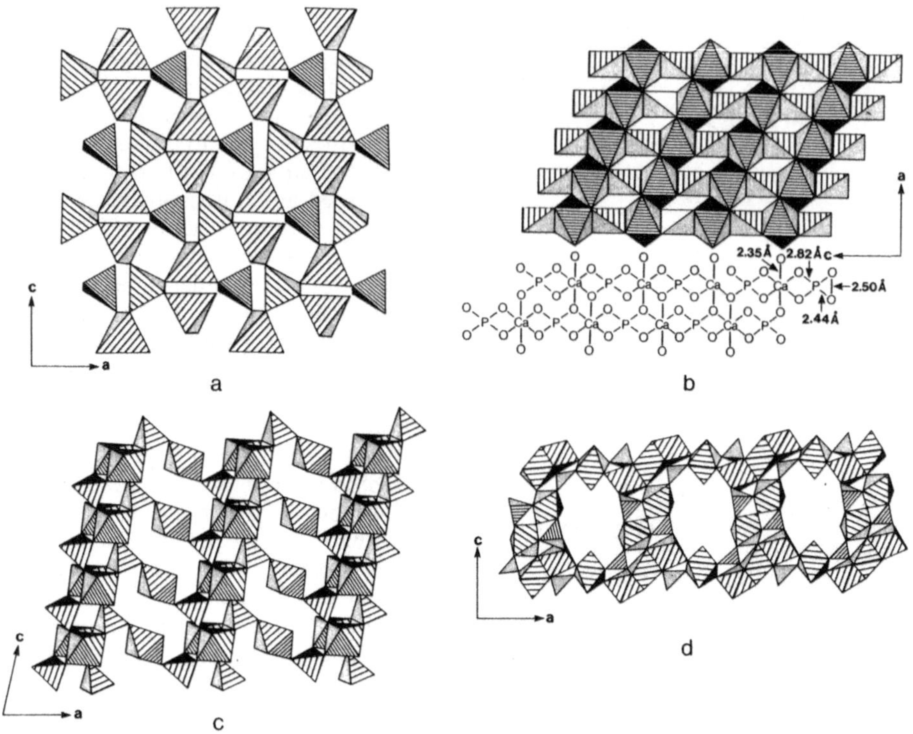

Fig. 8 a. Gallium phosphate, $GaPO_4$. Each PO_4 tetrahedron is joined to four GaO_4 tetrahedra. **b** Molecular structure of a sheet composed of phosphate-calcium chains developed in dicalcium phosphates: $CaHPO_4 \times 2\ H_2O$, $CaHPO_4 \times H_2O$; water is not shown for graphical reasons. **c** Iron phosphate: $Fe_3(PO_4)_2 \times 8\ H_2O$. The PO_4 groups are shown as tetrahedra. **d** Molecular structure of beraunite: $Fe_{0.5}{}^{II}Fe_{2.5}{}^{III}(OH)_{2.5}(PO_4)_2 \times 3\ H_2O$

mit the establishment of rigid networks. How can we possibly imagine such networks to look alike? To overcome difficulties in visualizing order phenomena, for instance, in biological membranes, examples of known layered phosphate structures in ordinary minerals are presented for illustration (Fig. 8). Phosphate tetrahedra and metal ion oxygen polyhedra can combine to a variety of geometries including undulating surfaces or concave/convex perforated surfaces. These loosely arranged space fabrics exhibit selective molecular sieve properties and ion exchange characteristics.

Phospholipid membranes are expected to exhibit identical properties and structures as they exist in inorganic phosphate crystals, even including holes and surface granularities. It is proposed that the interchangeable nature of metal ions causes membranes to act as *dynamic molecular sieves*. Their pore size and shape must be quite variable as a function of type and availability of metal ions which in the last instances are enzyme controlled. Assuming adenosine triphosphate (ATP) is capable of trapping metal ions but adenosine diphosphate (ADP) is not, a periodic pulsation of the membrane lattice is the consequence.

What we learned in this brief discussion on the structural and functional relationship between inorganic and biotic phosphate "membranes" may now permit us to understand better the mechanisms behind the origin of cellular structures at the dawn of time. It all has to do with the ability of phospholipids to jointly with metal ions construct stable fabrics and become separated from the aqueous medium. Experimental data on emulsions and foams show that micelles equipped with anisotropic membranes are able to grow at the expense of other micelles by consuming them through surface attachment (lowering in surface energy) in a process called emulsification. These globules, soaps, or emulsions as they are termed exhibit an optimal critical diameter in the order of 10^3 to 10^4 Å. The main feature of this water-organic system is an anisotropic and charged phase boundary layer. The newly generated macromicelles created in a process termed *coacervation,* will envelop water droplets, whereby the original micelle content is exchanged but according to laws different from those established in aqueous systems. That is, condensation of lipid membranes towards a rigid membrane is achieved by the uptake, for instance, of cholesterol or metal ions. The expulsion of water proceeds during intercellular attachment by means of oxygen-coordinated metal bridges. Due to the ionic fabric closely attached to the coacervate, dissolved organic molecules such as peptides or carbohydrates are bonded and precipitated on the membrane surface. In the course of coacervate development structures will arise which are enclosed by a double phospholipid skin (bilayer membrane). Phosphate groups become oriented towards the aqueous phase and double layers may combine to multilayered stacks. Membrane pouches come into existence, resulting in the formation of multichambered coacervates bearing striking resemblance to mitochondria. Judged by the conservative nature of mitochondria, it appears that, as a system, it still carries relics of its abiotic origin. The development of such a self-controlled reaction agrees with the thermodynamics of system behavior. A stable cyclic process can exist in the vicinity of a stationary phase and may operate repeatedly an infinite number of times without ever passing through the stationary phase itself.

In conclusion, the primordial metabolism of the coacervate was in all probability maintained by means of a reversible phosphorylation cycle. In consequence, the origin of metabolism is in no way linked to the development of the genetic transcription apparatus. It must be considered an independent formation process. For this reason the abiotic origin of phosphorylation must be regarded as an equally important step towards the creation of the primordial cell. The problem, therefore, centers around the question of how to polymerize the common monophosphates into di-, tri-, or tetraphosphates, since polyphosphates are unstable in natural environments. The only reasonable choice left is to place the polymerization event within the coacervates. It is conceivable that phospholipid solid-state surfaces served in this capacity, because inorganic mineral surfaces too act as templates and furthermore may catalyze phosphorylation as has been demonstrated for apatite crystals.

The establishment of an interconnected and chemical reaction pattern for the coacervate system as a whole exists when phosphorylation can be maintained. This requires a constant supply of organic molecules and metal ions that are consumed, or utilized during this development. In this manner, a certain modus vivendi is established. Sources of energy were oxidizable organic compounds in the ambience. Molecules such as amino acids or sugars must have been present in huge quantities in the environment in view of their mineral fabricated origin.

So far, however, no vital power was involved. It is postulated that a primitive heterotrophic metabolism improved progressively. Its development took shape independently of the evolution of the nucleic acids and the genetic code. By superimposing the two separately developed entities, (a) the genetic apparatus, and (b) the heterotrophic metabolism, the primordial cell came into existence.

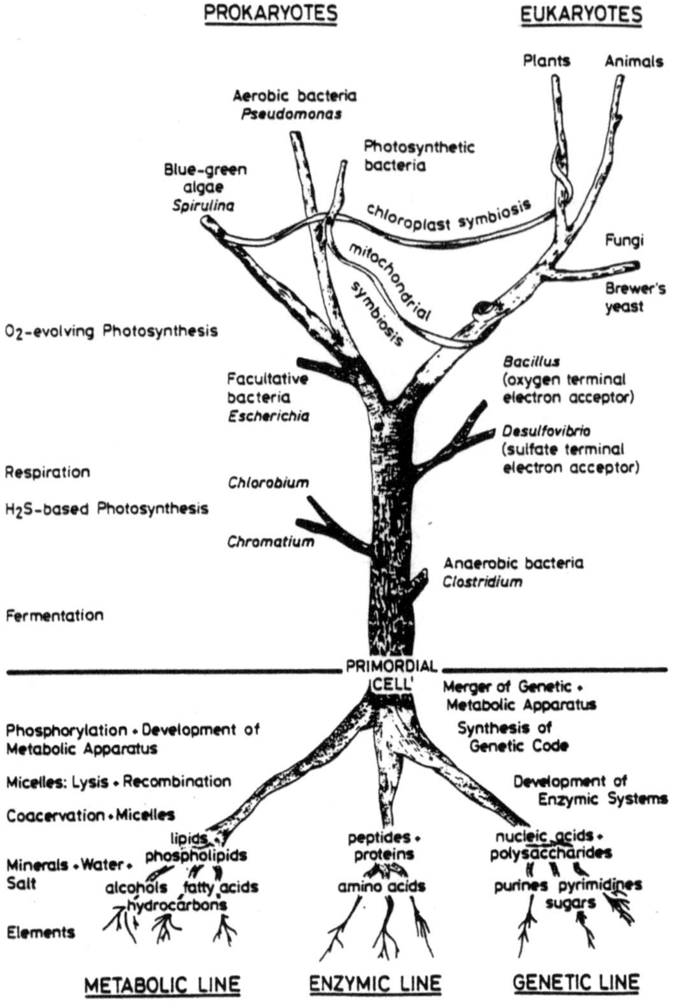

Fig. 9. Composite evolutionary tree (schematic) summarizing the principal steps in chemical and biological evolution. The sequence of events depicted for chemical evolution follows from the discussion in the text. The upward progression from anaerobic to facultative to aerobic forms is indicated in the *shading pattern.* Mitochondrial and chloroplast invasions are roughly drawn between points of suggested origin and uptake, respectively

The primitive metabolism of coacervates was kept "alive" via phosphorylation processes and became embodied by the self-reproducing cycles. It is likely that the nucleic acids were able to encode polypeptides utilizing certain metal ions. Alternately, nucleic acids succeeded in adopting the available metalloproteins in their environs – among which must have been enzymes in the billions – for their own reproduction. In any event, the link between the two independently developed events, (a) the primitive metabolism, and (b) the genetic reproduction apparatus is represented by the peptides. They are the essential tool by which coacervate metabolism – for the purpose of nucleic acid replication – was utilized.

The structure of phospholipid membranes and the genetic code are archaic elements – biochemists generally use the term "universal elements" – which remained steadfast in the course of evolution. In Fig. 9, the three modes of life leading to the first living cell, the progenote, are schematically depicted. The compost of life started to form in dark molecular clouds. Once on Earth, the metabolic, enzymic, and genetic lines took shape, and progressed independent-

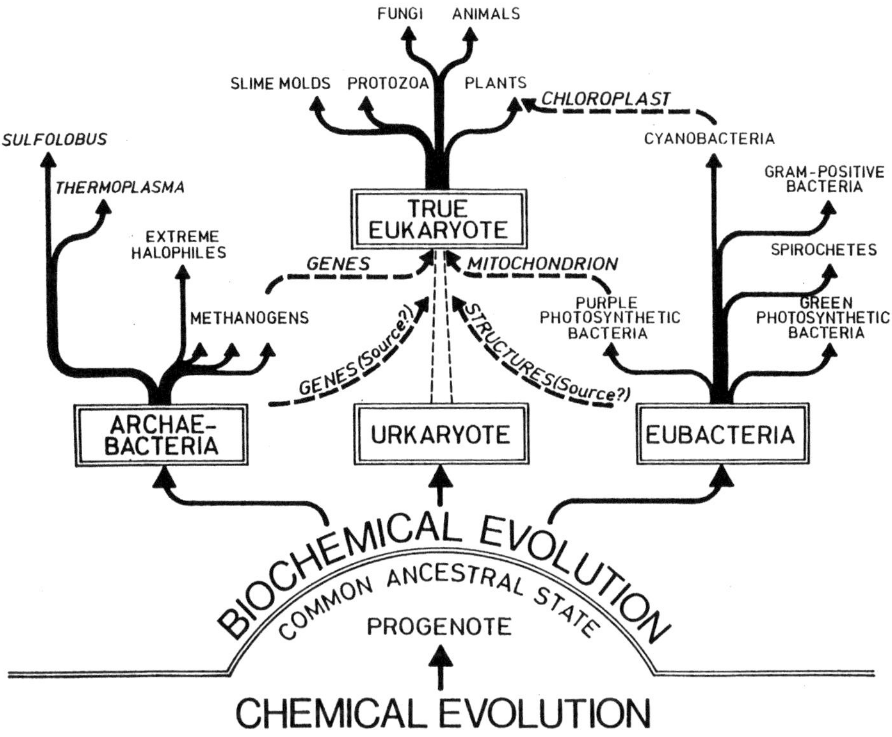

Fig. 10. Biochemical evolution starting from a common ancestral state: the progenote (highly schematic)

ly to eventually merge into the *progenote*. The subsequent evolution based on protein and nucleic acid sequence data, has been constructed in the form of a composite "evolutionary tree" thus linking the eukaryotes directly to one kingdom of the prokaryotes, the true bacteria or eubacteria. Although it is tempting to use such a tree in order to draw conclusions with respect to the sequence of events such as the start of (a) photosystem I, (b) photosystem II, (e) respiration, (d) sulfate reduction, etc., recent work using the 16 S ribosomal RNA sequence suggests a different scenario. Data indicate that three lines of descent diverged before the level of complexity usually associated with the prokaryotic cell was reached, that is: archaebacteria, eubacteria, and 'ur'eukaryotes. All three lineages were independently derived from a common progenote (Fig. 10).

E. Final Comment

We have come a long way during this presentation. I have "crudely" abstracted from the wealth of data available on the origin of the first living cell, but still hope that my "nutshell" approach has provided at least some idea of the work being done in a field of science involved with unraveling the mysteries of life. A more comprehensive treatise may be consulted (Degens 1989) for special references or to obtain further details on the roots and evolution of the biological cell in the course of more than 4 billion years of Earth's history.

References

Degens ET (1989) Perspectives on biogeochemistry. Springer, Berlin Heidelberg New York, pp 423

Oparin AI (1953) The origin of life. Dover, New York, pp 270

Modern Coding Sequences Are in the Periodic-to-Chaotic Transition

S. Ohno

A. Introduction

It seems as though biologists are extraordinarily fond of randomness. A population is defined as one, randomly mating, interbreeding unit, although truly random mating would hardly be practiceable in a reasonably large population. Similarly, spontaneous mutations are viewed as randomly sustained base substitutions, in spite of our knowledge of mutational hot spots. I suspect that this extraordinarily strong belief in randomness stems from our too strong faith in the power of natural selection. The unpredictable world of randomness is the world of chaos. Yet in recent times, there has been increasing realization that there is order in chaos as well. This realization started with three equations by Laurenz to describe meteorological phenomena. No one would dispute the unpredictability of weather. Yet, these three equations describing heat reflected by the earth and frictions caused by rotation of the earth revealed the presence of the strange attractor. The presence of the attractor, no matter how strange, is a sure indication of order. Thus, Feigenbaum's conjecture on chaos came about [1].

There are many different ways of viewing these developments. Nevertheless, I will present one version pertinent to the present discussion: the chaotic state is the degenerate form of the ordered (periodical) state, and this degeneration is due primarily to progressive step-wise increase of the original periodicity. Keeping the above in mind, now let us examine the 173-codon-long chicken lens αA-crystallin which is primarily made of β-sheet structures [2].

Beckman Research Institute of the City of Hope, Duarte, California, U.S.A.

B. CCTG Tetramer as the Primordial Repeating Unit of the Crystallin Coding Sequence

As shown at the top of Fig. 1, this GC rich coding sequence contained more pyrimidines than purines because of the abundance of C (32.4%). After this realization, the frequency distribution of C-containing dimers (C X and X C) were obtained. The procedure forced C C dimers to be overrepresented, for the C C C trimer was counted as 2 C C dimers. This was because th C C X trimer C C A, e.g., had to be counted as a 1 C C and a 1 C A dimer. If C C C was regarded as 1 C C dimer, the recurrence rate of the C C dimer is reduced to 47 X. Since all sequences, no matter how short, are translatable by three reading frames, ⅓ of them should serve as Pro codons C C X. This predicts the presence of 16 Pro (9%) in this protein. Indeed, there were 14 Pro residues. Next to C C, the more frequently recurring C-containing dimers were T C (41 X), C T (39 X), and C A (39 X). The above suggested relative abundance of Ser and Leu but not of Gln and His, for ⅓ of 39 C A X are to be split evenly between Gln codons C A G, C A A, and the His codons C A C and C A T. Indeed, there were 5 Gln and 6 His residues. The very fact that the amino acid composition of the protein is fairly predictable by recurrent rates of base dimers in its coding sequence immediately places in grave doubt the conventional wisdom of genes

CHICKEN ALPHA-A-CRYSTALLIN (BETA-SHEET PROTEIN)
173-CODON-LONG

A T/G C = 0.78

A: 109, G: 123, T: 119, C: 168 (32.4%)

DISTRIBUTION OF C X-DIMERS

C C: 51 X	C T: 39 X	C A: 39 X	C G: 23 X
14 PRO	15 LEU	5 GLN & 6 HIS	13 ARG

DISTRIBUTION OF X C-DIMERS

C C: 51 X	T C: 41 X	A C: 27 X	G C: 28 X
	25 SER	6 THR	4 ALA

USAGE OF SYNONYMOUS CODONS

17 + 8 SER			12 + 3 LEU			10 + 3 ARG		
	41 T C X			39 C T X			23 C G X	
TRIMER		CODON	TRIMER		CODON	TRIMER		CODON
18	T C C	10	15	C T G	9	9	C G G	4
14	T C T	4	12	C T C	3	7	C G C	4
5	T C A	1	8	C T T	0	5	C G T	2
4	T C G	2	4	C T A	0	2	C G A	0
	29 A G X			33 T T X			29 A G X	
10	A G C	7	8	T T G	3	8	A G G	2
6	A G T	1	4	T T A	0	5	A G A	1

Fig. 1. At the *top*, the AT/GC ratio and base composition of the 519-base-long chicken αA-crystallin coding sequence [2] are given, followed by the recurrent rates of C X and X C dimers. The rate for the C C dimer is an overestimate for the reason given in the text. In the case of Leu, Gln, His, and Thr, the recurrence of a relevant dimer divided by 3 gives a resonable estimation of the number of amino acids. At the *bottom*, 6 codons each for Ser, Leu, and Arg are shown in three vertical columns. The recurrent rates of each as a trimer and as a codon are shown. In each instance, the most preponderant among the synonymous codons also recurred most frequently as the base trimer

evolving by natural selection operating upon individual codons. Indeed, three columns at the bottom of Fig. 1 show that with regard to Ser, Leu, and Arg, encodable by 6 codons each, preponderant among the synonymous codons sharing the first two bases invariably is the one that recurred most frequently as base trimer. Thus, codon usages too are determined merely by recurrent rates of pertinent base trimers. Figure 1 and data not shown also suggested that the most frequently recurring base tetramer should be C C T G. This was due to the fact that the 21 X recurring trimer C C T and the 15 X recurring C T G overlap with each other.

Indeed, C C T G was the most frequently recurring base tetramer; 9 X recurrence (Fig. 2). This tetramer was translated in all three different reading frames to encode two Pro, five Leu, and one each of Trp and Cys. As might be deduced from Fig. 1, the next most frequently recurring base tetramer was 7 X recurring T C T C as shown boxed in Fig. 2. T C T C, however, can be regarded as two successive T C dimers. Nevertheless, this tetramer would soon be mentioned again. How significant was the 9 X recurrence of C C T G? The expected recurrence rate of this tetramer can be computed in two different ways. If based upon the 15 X recurrent rate of C T G,

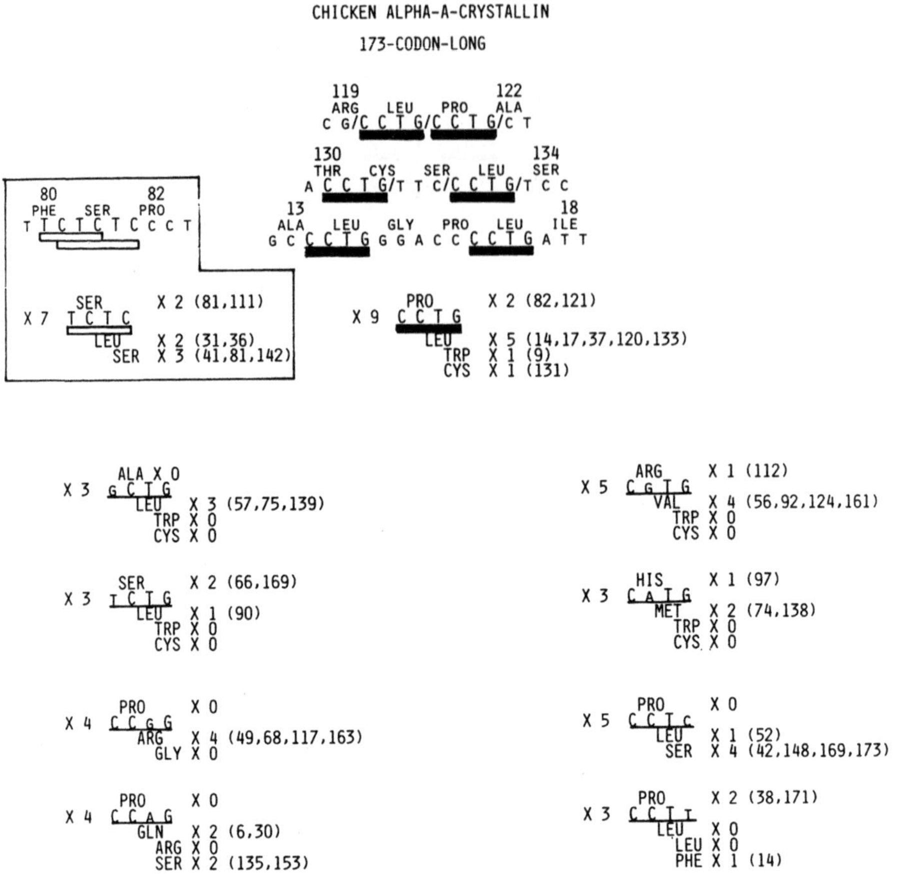

Fig. 2. C C T G as the primordial heptamer is underlined by the *thick solid bar*. It recurred 9 X and was translated in three different reading frames as shown in the *upper center stage*. In its 1st reading frame, it encoded 2 Pro (the positions of these Pro in the amino acid sequence are shown in parentheses), 5 Leu in its 2nd, and 1 each of Trp and Cys in its 3rd reading frames. Shown at the *top* are three pairs of C C T G that recurred in succession. Placed inside the *box* at the left is the 2nd most frequently recurring base tetramer T C T C that recurred 7 X. This, however, is a T C dimer X 2 and in one place a T C dimer recurred three times in succession. Shown in two columns *near the bottom* are 8 of the 12 single-base-substituted derivatives of G C T G that recurred 3 X or more

the expected recurrent rate for C C T G becomes $0.324 \times 15 = 4.86$. As shown at the top of Fig. 1, 0.324 of the 519 bases were Cs. If based upon the 21 X recurrence of the C C T trimer, the expected recurrence rate of C C T G now becomes $21 \times 0.237 = 4.9$. Clearly the 9 X recurrence of the C C T G tetramer was not by chance. Due to single base substitutions affecting one or the other of four positions, C C T G was expected to yield 12 different kinds of single-base-substituted derivatives. As shown in the bottom half of Fig. 2, 8 of them recurred 3 to 5 times, while the remainder recurred twice each. It follows then that not counting several overlapped bases twice, C C T G and its single-base-substituted derivatives occupied 35% of the entire coding sequence.

It would thus appear that the αA-crystallin coding sequence was ultimately derived from C C T G tetrameric repeats

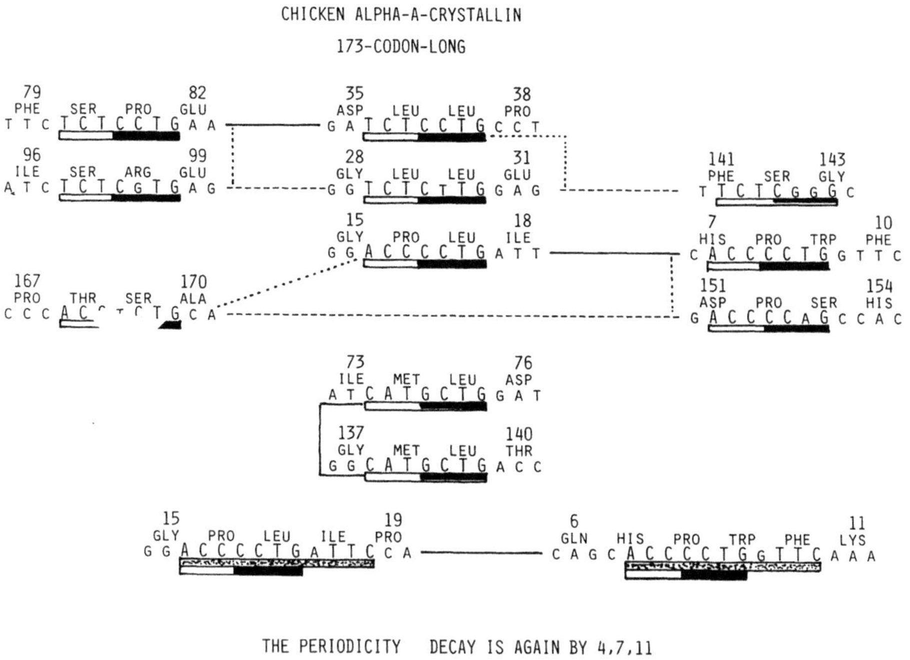

Fig. 3. Shown in the 1st and 3rd rows are two pairs of identical C C T G containing heptamers, while shown in the 3rd and 4th rows are each one's respective single-base-substituted copies. Identical heptamers are connected by the *solid line* and single-base-substituted derivatives by *broken lines*. Those translated in the 1st reading frame are shown in the *left column*, while the *center column* contains those translated in 2nd reading frame, and the *right column* those in 3rd reading frame. Two identical G C T G-containing heptamers, both translated in the 2nd reading frame, are shown in the 5th and 6th rows. Two identical heptamers shown in the 3rd row were actually parts of 11-base-long repeating units as shown at the *bottom*

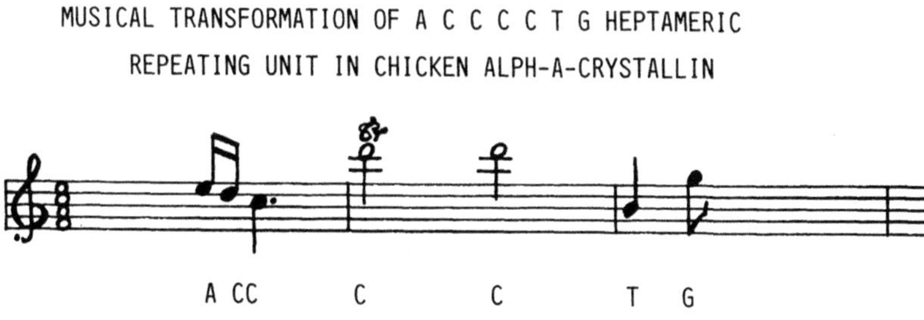

Fig. 4. The particular musical transformation given to the recurring heptamer A C C C C T G according to the set rule previously put forward [5]

that existed in the prebiotic world of eons ago [3]. Three consecutive copies of C C T G should have given the tetrapeptidic periodicity Pro-Ala-Cys-Leu to the original peptide chain. Indeed, the 120th and 121st Leu-Pro of the chicken αA-crystallin were still encoded by two consecutive copies of C C T G as shown at the very top of Fig. 2. As the periodicity decayed in the periodic-to-chaotic transition, the original exact tetrameric periodicity should have yielded to longer and

CHICKEN LENS ALPHA-A CRYSTALLIN

Fig. 5a

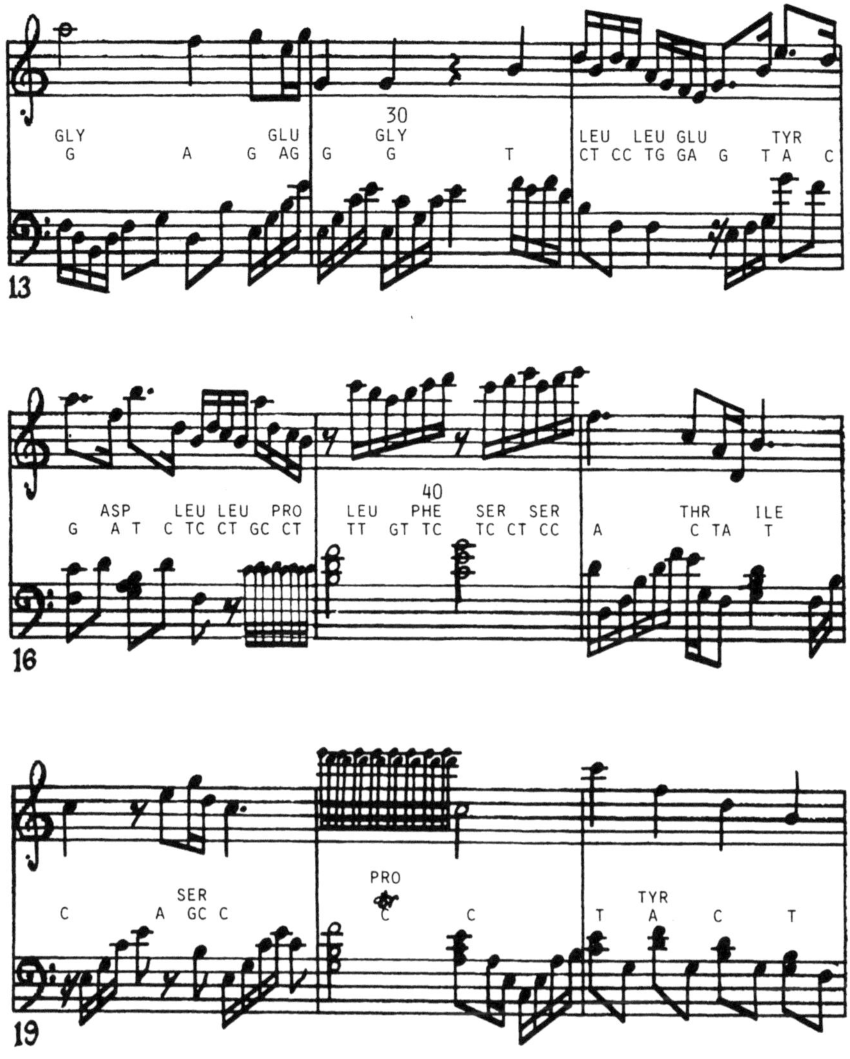

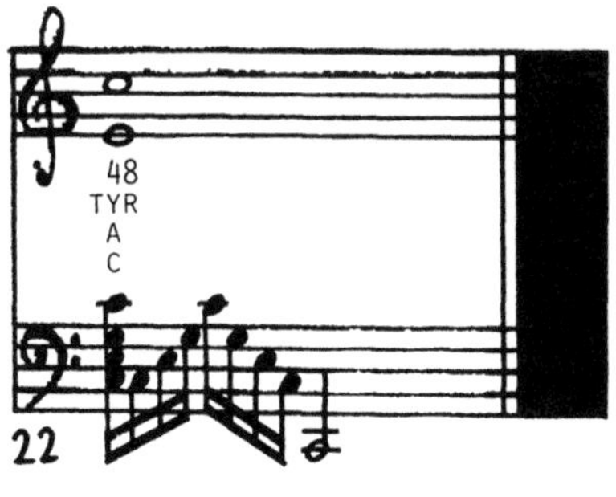

b

Fig. 5a, b. The musical transformation based on the melodic heptamer (Fig. 4) of 7th to 48th codons of the chicken αA-crystallin coding sequence [2] longer less exact periodicities. Indeed, two pairs of C C T G shown near the top of Fig. 2 were now separated by 3 and 5 bases.

C. Periodicity Decay by the Golden Mean: the 3, 4, 7, 11, 18 Rule

Of the consecutive numbers the first four are 1, 2, 3, 4. At this point, we begin to add previous two numbers to obtain the next number; i.e., $3+4=7$. If we keep doing this, the series of numbers form: 3, 4, 7, 11, 18, 29, 47, 76, 123, etc. Now we divide 7 by 4, 11 by 7, 18 by 11, and so on. Then we see that results begin to approach 1.618 and reach that goal at 123 divided by 76, and remains 1.618 forever thereafter. Now, 1.618 is the well-known golden ratio expressed as

$$\frac{a+b}{a}=\frac{a}{b} \quad \text{which is} \quad \frac{1+\sqrt{5}}{2}.$$

In a previous paper, we have shown that the periodicity decay in coding sequences is according to the above-noted golden mean [4]. Of the nine C C T G tetramers, two recurred in immediate succession of each other as shown in Fig. 2. The remaining seven, on the other hand, recurred as parts of recurring heptamers. Two such pairs are shown in Fig. 3, because members of each pair are translated in different reading frames. Shown at the top row of Fig. 3 are two identical copies of the heptamer T Z C T C C T G, yielding the 80th and 81st Ser-Pro when translated in the first reading frame, while encoding Leu-Leu dipeptide in the second reading frame. A pair of single-base-substituted copy T C T C G G G; the translation of this heptamer in its third reading frame encoding the 141st to 143rd Phe-Ser-Gly. It is pointed out that each of these five heptamers (one identical pair and a triplet derived from that pair) contained the second most frequently recurring base tetramer T C T C already noted. Thus, five of the 7 X recurring T C T C combined with the most frequently recurring tetramer C C T G and its derivatives to become parts of recurring heptamers. Shown in the third row of Fig. 3 are another identical pair of C C T G-containing heptamers A C C C T G encoding the 16th and 17th Pro-Leu in its second reading frame, while encoding the seventh to tenth His-Pro-Trp in its third reading frame. This identical pair of heptamers on the one hand yielded its single-base-substituted derivatives (Fig. 3, fourth row) while, on the other hand, becoming parts of the pair of 11-base-long repeating units that differed from each other by a single-base substitution (Fig. 3, bottom row). Thus, the periodicity decay by the chicken lens αA-crystallin coding sequence is indeed according to the golden mean: 4, 7, 11 rule. Needless to say, single-base-substituted derivatives of the primordial heptamer C C T G have often become parts of the identical pair of heptamers. One such G C T G-containing pair of identical heptamers encoding a pair of Met-Leu dipeptides of the 74th, 75th and 138th and 139th positions is shown in the fifth and sixth rows of Fig. 3.

When modern coding sequences are analyzed in the above manner, one can not help but realize that natural selection operating upon individual codons has mainly contributed to the conservation of a fait accompli by eliminating function-depriving, therefore, deleterious mutations. But this had very little to do with the initial acquisition of functions by proteins encoded by ancestral coding sequences of eons ago. For this, I contend that the universal principle of periodic-to-chaotic transition is responsible.

D. Musical Transformation of the 7th to 48th Codons of the Chicken αA-Crystallin Coding Sequence

Some time ago, I came to the realization that the periodic-to-chaotic transitional state of modern coding sequences can best be appreciated by their musical transformation under the set rule [5]. The 5' region of the 519-base-long chicken αA-crystallin coding sequence [2] is the domain ruled by the heptamers A C C C C T G and T C T C C T G as shown in Fig. 3. By giving the melody shown in Fig. 4 to the former, the 7th to 48th codons of this coding sequence have been transformed to the musical composition for piano shown in Fig. 5a and b). By listening to it, one can readily realize the periodicity decay by the 4, 7, 11, 18 rule.

E. Summary and Conclusions

Modern coding sequences are in the periodic-to-chaotic transition. In the case of αA-crystallin coding sequence of the chicken, the initial tetrameric periodicity of the primordial tetramer C C T G has been decaying by the golden mean: the 4, 7, 11 rule. Thus, the tetramer has become parts of recurring heptamers, and some heptamers have become parts of the 11-base-long repeating units.

References

1. Feigenbaum MJ (1985) The universal metric properties of nonlinear transformations. J Stat Physics 21:669–706
2. Okazaki KM, Yasuda K, Kondoh H, Okada TS (1985) DNA sequences responsible for tissue-specific expression of a chicken alpha-crystalling gene in mouse lens cells. EMBO J 4:2589–2595
3. Ohno S (1987) Evolution from primordial oligomeric repeats to modern coding sequences. J Mol Evol 25:325–329
4. Ohno S (1988) Codon preference is but an illusion created by the construction principle of coding sequences. Proc Natl Acad Sci USA 85ff.
5. Ohno S, Ohno M (1986) The all pervasive principle of repetitious recurrence governs not only coding sequence construction but also human endeavor in musical composition. Immunogenetics 24:71–78

Translation to Human Temperaments of the Tyrosin-Kinase Active Site of the Human Insulin Receptor β-Chain

By
Dennis Busby, Papenhuderstraße 25, 2000 Hamburg 76, FRG
Dennis Swing Club

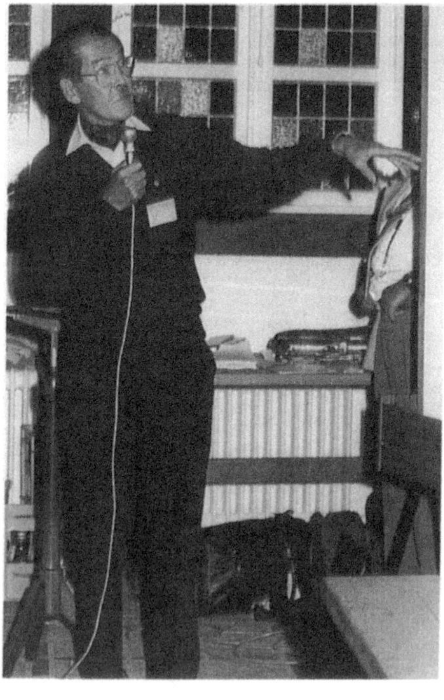

Music and Genes, Translation to Human Temperaments in the Emmenhof

TYROSINE-KINASE ACTIVE SITE
OF
HUMAN INSULIN RECEPTOR β-CHAIN

TENOR
B
C
FREE TEMPO
Free
Tempo

Trumpet
Funk
A
B
C
FREE
TEMPO
FREE
TEMPO

Piano
Funk
A
B
Solo 8 bars (whole tone)
C
Free
Tempo
Free
Tempo

CONTRA BASS
FUNK
A
B
C
FREE
TEMPO
Free
TEMPO
Star Nr. 6, 12 Systeme®

DRUMS
FUNK
A
B
PIANO SOLO (PP)
SFZ
C
FUNK
Free
Tempo
Free
Tempo

Subject Index